McGRAW-HILL YEARBOOK OF
Science &
Technology

2005

McGRAW-HILL YEARBOOK OF
Science &
Technology

2005

Comprehensive coverage of recent events and research as compiled by
the staff of the McGraw-Hill Encyclopedia of Science & Technology

McGraw-Hill
New York   Chicago   San Francisco   Lisbon   London   Madrid   Mexico City   Milan
New Delhi   San Juan   Seoul   Singapore   Sydney   Toronto

Library of Congress Cataloging in Publication data

McGraw-Hill yearbook of science and technology.
1962–   . New York, McGraw-Hill.

    v. illus. 26 cm.
    Vols. for 1962–   compiled by the staff of the
McGraw-Hill encyclopedia of science and technology.
    1. Science—Yearbooks.   2. Technology—
Yearbooks.   1. McGraw-Hill encyclopedia of
science and technology.
Q1.M13   505.8   62-12028

ISBN 0-07-144504-8
ISSN 0076-2016

1 2 3 4 5 6 7 8 9 0   DOW/DOW   0 10 9 8 7 6 5 4

This book was printed on acid-free paper.

*It was set in Garamond Book and Neue Helvetica Black Condensed by
TechBooks, Fairfax, Virginia. The art was prepared by TechBooks.
The book was printed and bound by RR Donnelley, The Lakeside Press.*

Contents

Editing, Design, and Production Staff

Consulting Editors

Dr. Sally E. Walker. *Associate Professor of Geology and Marine Science, University of Georgia, Athens.* INVERTEBRATE PALEONTOLOGY.

Prof. Pao K. Wang. *Department of Atmospheric and Oceanic Sciences, University of Wisconsin–Madison.* METEOROLOGY AND CLIMATOLOGY.

Dr. Nicole Y. Weekes. *Pomona College, Claremont, California.* NEUROPSYCHOLOGY.

Prof. Mary Anne White. *Killam Research Professor in Materials Science, Department of Chemistry, Dalhousie University, Halifax, Nova Scotia, Canada.* MATERIALS SCIENCE AND METALLURGIC ENGINEERING.

Prof. Thomas A. Wikle. *Head, Department of Geography, Oklahoma State University, Stillwater.* PHYSICAL GEOGRAPHY.

Prof. Jerry Woodall. *School of Electrical and Computer Engineering, Purdue University, West Lafayette, Indiana.* PHYSICAL ELECTRONICS.

Dr. James C. Wyant. *University of Arizona Optical Sciences Center, Tucson.* ELECTROMAGNETIC RADIATION AND OPTICS.

Article Titles and Authors

The 2005 *McGraw-Hill Yearbook of Science & Technology* provides a broad overview of important recent developments in science, technology, and engineering as selected by our distinguished board of consulting editors. At the same time, it satisfies the nonspecialist reader's need to stay informed about important trends in research and development that will advance our knowledge in the future in fields ranging from agriculture to zoology and lead to important new practical applications. Readers of the *McGraw-Hill Encyclopedia of Science & Technology*, 9th edition (2002), also will find the *Yearbook* to be a valuable companion publication, supplementing and updating the basic information.

In the 2005 edition we continue to chronicle the rapid advances in cell and molecular biology with articles on topics such as cloning, helper and regulatory T cells, lipidomics, olfactory system coding, and protein networks. Reviews in topical areas of biomedicine, such as HIV vaccines, hearing disorders, severe acute respiratory syndrome (SARS), and West Nile fever, are presented. Key scientific and technical issues relating to security are covered in articles on aviation security, bioterrorism, and transportation system security. Advances in computing and communication are documented in articles on advanced wireless technology, computational intelligence, XML databases, extreme programming and agile methods, mobile satellite services, serial data transmission, and Voice over IP, among others. Noteworthy developments in materials science, chemistry, and physics are reported in reviews of atomic-scale surface imaging, combinatorial materials science, macromolecular engineering, polymer recycling and degradation, pentaquarks, and tetraquarks. Articles on change blindness, experience and the developing brain, language and the brain, and visual illusions and perception survey recent advances in psychology. And reviews in atmospheric modeling, environmental sensors, glaciology, hurricane-related pollution, and satellite climatology are among the articles in the earth and environmental sciences.

Each contribution to the *Yearbook* is a concise yet authoritative article prepared by one or more experts in the field. We are pleased that noted researchers have been supporting the *Yearbook* since its first edition in 1962 by taking time to share their knowledge with our readers. The topics are selected by our consulting editors in conjunction with our editorial staff based on present significance and potential applications. McGraw-Hill strives to make each article as readily understandable as possible for the nonspecialist reader through careful editing and the extensive use of graphics, much of which is prepared specially for the *Yearbook*.

Librarians, students, teachers, the scientific community, journalists and writers, and the general reader continue to find in the *McGraw-Hill Yearbook of Science & Technology* the information they need in order to follow the rapid pace of advances in science and technology and to understand the developments in these fields that will shape the world of the twenty-first century.

Mark D. Licker
PUBLISHER

Acoustic viscometer

The Greenspan acoustic viscometer, first investigated by M. Greenspan and F. N. Wimenitz in 1953, is a relatively compact and robust device for measuring the viscosity of gases using the principles of acoustic resonance. The viscometer was further developed both experimentally and theoretically during the past decade.

Viscosity is a physical parameter which characterizes frictional processes in moving fluids. It has units of pressure times time, and is used to calculate the flow of gases in pipes and the frictional drag of gases on moving objects. Viscosity can be determined by several methods, including measurements of gas flow in capillaries and the damping of oscillating bodies in the gas. The acoustic viscometer is an alternative to these classical methods.

Current versions of the viscometer are fabricated entirely of metal. The only moving parts are the gas sample and acoustic transducers mounted external to the resonator. The viscometer is normally mounted in a sample chamber with provision for temperature and pressure control, evacuation, and change of gas composition. Because the gas sample contacts only metal parts, the viscometer is particularly suitable for use on corrosive gases when a corrosion-resistant metal is used for fabrication. An individual measurement of gas viscosity at a fixed temperature and pressure takes several minutes. The experiment can be fully automated so that a series of measurements can be made under computer control without a human attendant.

In principle, the Greenspan viscometer is an absolute instrument from which the viscosity can be determined using apparatus dimensions and other well-known thermophysical properties of the gas. In practice, the accuracy can be improved through calibrating with a substance for which the viscosity is well known, like helium or argon. An apparatus calibrated with helium was recently used to measure the viscosity of four polyatomic gases (methane, nitrogen, propane, and sulfur hexafluoride). The results agreed with well-accepted reference values for these gases within ±0.5%. Measurements of similar accuracy should be possible for a wide range of gases.

The viscometer has been tested in the temperature range 273–373 K (32–212°F) and the pressure range 0.2–3 MPa (2–30 atm).

Design. The viscometer is an acoustic resonator designed to make the resonance response sensitive to the viscosity of the gas filling the resonator. A simplified resonator (**Fig. 1**) can be machined in three pieces. The center piece forms a circular duct and a plate that separates two chambers. This system,

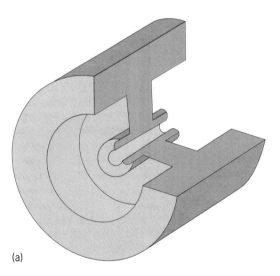

(a)

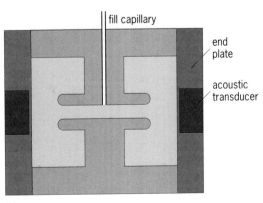

(b)

Fig. 1. Simplified Greenspan viscometer. (a) Perspective view. (End plates, acoustic transducers, and fill capillary are omitted.) (b) Cross-sectional view. The size of the fill capillary is exaggerated for clarity.

known as a double-Helmholtz resonator, has a low-frequency mode of oscillation in which the gas oscillates transversely in the duct. As the gas moves to the right, the gas in the right chamber is compressed and the gas pressure increases. The pressure in the left chamber decreases as the gas moves to the right. Both the higher-pressure gas in the right chamber and the lower-pressure gas in the left chamber exert a leftward force on the gas in the duct, which decelerates the gas, eventually reversing its motion and completing a half cycle of oscillation. During the second half cycle, the gas initially moves to the left, is decelerated by the increasing pressure in the left chamber and the decreasing pressure in the right chamber, and eventually reverses direction. Oscillating systems of this type are referred to as Helmholtz resonators, named after Hermann Helmholtz, who studied similar (but single-chambered) resonators in the nineteenth century.

Resonance frequency. The resonance frequency f_0 of the Greenspan viscometer is given approximately by Eq. (1), where c is the speed of sound, A is the

$$f_0 = \frac{c}{2\pi} \sqrt{\frac{2A}{LV}} \qquad (1)$$

cross-sectional area of the duct, L is the effective length of the duct, and V is the chamber volume. As an example, with a duct diameter of 5 mm, length 35 mm, and chambers of radii and length 20 mm, in air at ambient conditions, the resonance frequency is 367 Hz. The wavelength of acoustic waves of this frequency, 940 mm, greatly exceeds the dimensions of the viscometer. The wavelength determines the scale of spatial variations of the gas velocity and pressure. Accordingly, the gas motion is very uniform within the duct, and the acoustic pressure is very uniform within the chambers.

Frequency response. The oscillations can be excited and detected by acoustic transducers mounted outside the resonator and coupling to the gas sample through thin regions machined into the chamber walls. The frequency response of the resonator can be measured by sweeping the frequency of the excitation transducer through f_0 while measuring the signal with the detector transducer. The resonance profile (**Fig. 2**) has a width f_0/Q, where Q, known

as the quality factor of the resonance, can be determined experimentally by fitting a theoretical resonance curve to the measured response. The quality factor of an oscillating system is approximately equal to the average energy stored in the system divided by the energy dissipated in one oscillation cycle. The acoustic viscometer is designed so that the major source of energy dissipation is viscous friction at the walls of the duct. If only viscous dissipation contributed to the resonance width, the viscosity η could be determined from the measured Q using Eq. (2), where ρ is the gas density. In practice, this

$$\eta = \rho f_0 A / Q^2 \qquad (2)$$

formula overestimates the viscosity because of another important dissipation process.

Effect of thermal dissipation. When a gas is compressed rapidly, the gas temperature rises. The normal response to an elevated temperature is heat flow, which tends to restore temperature uniformity. However, the compressions and rarefactions of gas in an acoustic wave are sufficiently rapid that there is insufficient time for significant heat flow to keep the temperature uniform. Instead, the gas temperature oscillates in phase with the pressure. Although the temperature oscillations cause negligible heat flow within the bulk of the gas, heat currents are generated near solid boundaries, because solids are much better conductors of heat than gases.

A full theoretical analysis shows that the oscillating heat currents are confined to a thin region near the resonator walls. This region, known as the thermal boundary layer, has an effective thickness δ_t dependent upon the sound frequency f, the gas thermal conductivity K, the gas density ρ, and the specific heat c_p, as given by Eq. (3).

$$\delta_t = \sqrt{K/(\pi \rho f c_p)} \qquad (3)$$

Within the duct, the gas velocity varies from zero at the duct walls to a spatially uniform value well within the duct. This increase occurs within the viscous boundary layer, a region of effective thickness given by Eq. (4). The lengths δ_v and δ_t have similar

$$\delta_v = \sqrt{\eta/(\pi \rho f)} \qquad (4)$$

magnitudes; they also depend on frequency and density in the same way.

An improved formula for the viscometer quality factor is given by Eq. (5), where r_d is the duct

$$\frac{1}{Q} = \frac{\delta_v}{r_d} + (\gamma - 1)\frac{S\delta_t}{2V} \qquad (5)$$

radius, S the internal surface area of one chamber, and γ the ratio of the constant-pressure specific heat to the constant-volume specific heat. This quantity (γ) is greater than unity for all gases, equal to 5/3 for ideal monatomic gases, and trends toward unity as the complexity of the gas molecules increases. The first term in Eq. (5) is the ratio of the viscous boundary layer to the duct radius. This ratio is the approximate fraction of the duct where viscous dissipation

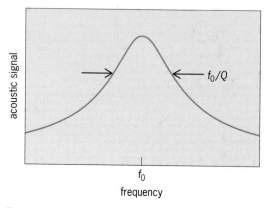

acoustic signal

f_0/Q

f_0

frequency

Fig. 2. Resonance profile.

takes place. The second term is the ratio of the effective volume $S\delta_t$ where thermal dissipation takes place to the total chamber volume V, multiplied by $(\gamma - 1)/2$.

Thermal dissipation can be reduced by making the chambers large compared with the duct so that V/S is much larger than r_d. Also, for many gases of practical interest, γ is close to unity so the thermal dissipation is further reduced by the factor $(\gamma - 1)$. Other criteria enter into a practical design. The compromises chosen for a recent design led to a thermal contribution to $1/Q$ under typical measurement conditions that was 36% of the viscous term for argon ($\gamma \approx 1.67$) and 7% of the viscous term for propane ($\gamma \approx 1.14$). Although the ratio of the thermal conductivity to the specific heat must be known in order to determine the viscosity, these quantities need not be known to high accuracy, owing to the low sensitivity of $1/Q$ to thermal dissipation for gases with γ near unity.

Use of theoretical model. The preceding discussion of the contributions to $1/Q$ illustrates the physical principles of the viscometer operation. In actual practice, a full theoretical model of the viscometer is used. This model accurately predicts a complex resonance response function that is directly fitted to the measured resonance response. The gas viscosity is determined as one of the fit parameters.

Experimental techniques. In laboratory environments, gas viscometers are generally placed in a regulated thermal environment so that the temperature of the measurement can be well controlled. Gas-handling plumbing is provided for filling the apparatus with test gases, for controlling the pressure of the test gas, and for evacuation of the apparatus prior to changing test gases. Materials and techniques suitable for maintaining gas purity are used. Actual viscosity measurements are made under computer control. Appropriate electronic devices are used to generate sound within the viscometer and to store the resonance profiles in computer memory. Further computer analysis of the resonance profiles using the full experimental model is applied to yield values of viscosity.

For background information *see* ACOUSTIC RESONATOR; CONDUCTION (HEAT); KINETIC THEORY OF MATTER; Q (ELECTRICITY); RESONANCE (ACOUSTICS AND MECHANICS); SPECIFIC HEAT; VISCOSITY in the McGraw-Hill Encyclopedia of Science & Technology.
James B. Mehl

Bibliography. K. A. Gillis, J. B. Mehl, and M. R. Moldover, Theory of the Greenspan viscometer, *J. Acous. Soc. Amer.*, 114:166–173, 2003; J. J. Hury et al., The viscosity of seven gases measured with a Greenspan viscometer, *Int. J. Thermophys.*, 24:1441–1474, 2003.

Active Thermochemical Tables

Active Thermochemical Tables (ATcT; a software suite) are based on a distinctively different paradigm of how to derive accurate, reliable, and internally consistent thermochemical values, and are rapidly becoming the archetypal approach to thermochemistry.

Knowledge of the thermochemical stability of various chemical species (encompassing both stable substances and ephemeral moieties such as free radicals) is central to most aspects of chemistry and critical in many industries. This is because the outcome of all chemical reactions ultimately hinges on the underlying thermodynamics. In particular, the availability of accurate, reliable, and internally consistent thermochemical values for a broad range of chemical species is essential in various branches of physical chemistry, such as chemical kinetics, construction of reaction mechanisms, or formulation of descriptive chemical models that have predictive abilities. Similarly, the availability of properly quantified uncertainties for thermochemical properties is essential for correctly assessing hypotheses, models, and simulations.

Traditional thermochemical tables. Traditional thermochemical tables are tabulations of thermochemical properties conveniently sorted by chemical species. Their quality ranges from fully documented critical data evaluations, through compilations containing references but not explaining the reasons for favoring a particular value, to lists providing fully referenced multiple values without attempting an evaluation, to tabulations that select values from other compilations. The selection of listed properties varies considerably. For example, the enthalpy of formation is usually provided, and frequently the Gibbs energy of formation, heat capacity, entropy, enthalpy increment (also known as the integrated heat capacity), and so on. The temperatures for which these properties are listed typically include room temperature (298.15 K or 77°F) and sometimes a selection of other temperatures. Some compilations present the thermochemistry in the form of polynomials rather than tables. The most reliable thermochemical values come from critical data evaluations, which include an independent review of the original determinations (subject to random and systematic errors) of the physical quantities involved.

Customary sequential approach. The thermochemical properties for any chemical species are derived from basic determinations that fall into two categories: species-specific and species-interrelating. Certain properties (such as heat capacity, entropy, and enthalpy increment) can be derived directly from species-specific information such as spectroscopic measurements, electronic structure computations of the electronic states for gas-phase species, or direct measurements of selected properties (for example, heat capacity) for condensed-phase species. However, the enthalpy and Gibbs energy of formation are obtained from determinations that express these quantities relative to other chemical species. Examples of such species-interrelating determinations are bond dissociation energies, enthalpies of chemical reactions, kinetic equilibria, electrode potentials, and solubility data.

Species-interrelating determinations are traditionally solved via a stepwise process usually following the "standard order of elements" (oxygen, hydrogen, noble gases, halogens, and so on) pioneered by the National Bureau of Standards (NBS) compilation. During each step, a new chemical species is adopted and available scientific information is scrutinized. The "best" species-interrelating measurements are selected through a critical evaluation process and then are used to obtain, at one temperature, either the enthalpy or Gibbs energy of formation for that species. One clear limitation is that only species-interconnecting determinations that link the current species exclusively to species that have been compiled during previous steps can be used. Once the enthalpy or Gibbs energy of formation is determined, the temperature dependence and the other thermochemical properties can be computed from the available species-specific information. The thermochemical information for the chemical species under consideration is then adopted and used in subsequent steps as a constant.

While the "standard order of elements" strategy helps alleviate some of the problems of the sequential process, the resulting tabulations have a number of difficulties. The biggest problem is the hidden progenitor/progeny dependencies across the tabulation. Consequently, traditional tabulations are nearly impossible to update with new knowledge. At best, one can use new species-interrelating data to update the properties of one species, which is tantamount to revising one of the steps in the middle of the original sequence that produced the tabulation. While this improves things locally (for the species in question), it introduces new inconsistencies across the tabulation. Generally, there will be other species in the table that are linked directly or indirectly to the old value of the revised species. These species also need to be updated, but it is not explicit which species these may be. Other difficulties, such as cumulative errors, are caused by lack of corrective feedback to the thermochemistry of species that have been determined in previous steps and frozen. The sequential approach also produces uncertainties that are not necessarily properly quantified and do not reflect the information content that is being used in other parts of the tabulation. Even under the best circumstances, the available information is exploited only partially in this approach.

Thermochemical network approach. The recognition that species-interconnecting information forms a thermochemical network, together with general pointers to proper statistical treatment, was formulated a long time ago but never fully developed or properly utilized.

Global thermochemical network. The thermochemical network represented in **Fig. 1** corresponds to a global thermochemical network, where all primary vertices except for reference elements in standard states (H_2 and O_2 at the bottom of the figure) are treated as unknowns. Figure 1 provides a graphical representation of a small thermochemical network

related to the recent discovery that the generally accepted value of the bond dissociation energy in water is in error. The primary vertices of the graph (boxes) represent the enthalpies of formation which need to be determined, while the secondary vertices (ovals) represent the chemical reactions. Each of the numbers contained in the ovals reference one thermochemically relevant species-interrelating determination at a particular temperature. The directed edges (arrows) and their weights define participation in the chemical reactions.

The graph conforms to a number of additional rules. First-neighbor vertices are always of a different kind, and second neighbors are always of the same kind. The topology of the graph is driven by the reactions it describes, which fixes the weights for the edges and defines which vertices can be first and second neighbors.

In general, secondary vertices may have multiple degeneracies (reflecting competing species-interrelating determinations of the same chemical reaction). The graph can be mapped onto one adjacency matrix and two column vectors (**Fig. 2**). The columns and rows of the adjacency matrix correspond to primary and secondary vertices (with each degenerate component forming a different row), and the matrix elements reflect the weight and direction of the edges. The two column vectors have the same number of rows as the adjacency matrix, and their elements contain the values and adjunct uncertainties of the species-interrelating determinations. Removal of primary vertices that are considered fixed (such as reference elements in their standard states), coupled to modifications of the column vectors, as necessary, is a legitimate procedure in handling the graph. The adjacency matrix and the two vectors correspond algebraically to a system of linear equations. The matrix is sparse, and the system is usually heavily overdetermined (the number of secondary vertices including all degeneracies normally exceeds the number of primary vertices).

Local thermochemical network. Another variant is a local thermochemical network, obtained by removing the primary vertices for which solutions are considered to be firmly known from prior considerations. While treatment of the global thermochemical network is the preferred approach, the use of local networks can be helpful in some cases.

A visual examination of the graph quickly reveals that there are many allowed paths between the two arbitrarily selected primary vertices. The traditional sequential approach uses only a small subset of possible paths. Starting at a reference element in a standard state (which can be removed from the graph), each of the sequential steps generally corresponds to selecting a trivial subgraph consisting of one secondary vertex and one primary vertex, solving for the primary vertex, fixing its value, and then removing it from the network (together with any secondary vertices that become disjointed during the procedure). From the statistical viewpoint, the best solution for the thermochemical network is obtained not

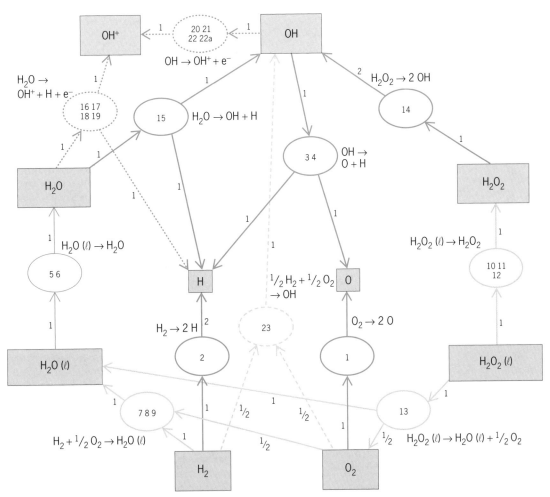

Fig. 1. Graphical representation of a small thermochemical network. The thermochemical network contains two types of vertices and has directed and weighted edges. The primary vertices (denoted by boxes) represent the enthalpies of formation of species which need to be determined (where the same chemical compound in a different aggregate state is properly distinguished as a thermochemically different species). The two liquids bear the label (ℓ), while the customary label indicating gaseous state, (g), pertinent to all other species, has been omitted. The secondary vertices (ovals) represent the chemical reactions. Each of the numbers contained in the ovals references one thermochemically relevant species- interrelating determination at a particular temperature. The directed edges (arrows) define participation in the chemical reactions: arrows pointing away from a species indicate reactants, and arrows pointing toward a species indicate products. The weights of the edges (indicated next to the arrows) are determined by the stoichiometry of the reaction. The electrons balancing the chemical reactions are not shown explicitly since the thermochemical network conforms to the "stationary electron convention."

by selecting a particular path but by considering all paths. This is accomplished by finding the simultaneous solution of the whole system via minimization of a suitable statistical measure, such as the chi-square distribution χ^2, if the adjunct uncertainties are an honest representation of the underlying confidence in the species-interrelating determination present in the thermochemical network. However, if the latter is not fulfilled, "optimistic" uncertainties create outliers that will skew the results. Hence, simultaneous solution needs to be preceded by a statistical analysis that will detect and correct possible optimistic uncertainties. Among various statistical approaches that are possible, the "worst offender" analysis has proven to be very successful. In this iterative approach, trial solutions are repeatedly used to uncover optimistic uncertainties in the thermochemical network. The isolation of offenders is made possible by

the presence of alternative paths in the graph. During each iteration the uncertainty of one secondary vertex (the current worst offender) is slightly expanded (usually by a few percent), and the procedure is repeated until the thermochemical network is self-consistent, at which point the final solutions for all primary vertices can be found. These are then coupled to species-specific information to develop the full complement of thermochemical information.

Advantages of thermochemical network approach. The solutions and their adjunct uncertainties are guaranteed to be consistent both internally and with all the knowledge stored in the thermochemical network, thus producing thermochemical values that are superior to those obtained from a sequential approach. In addition, the resulting thermochemistry is easily updated with new knowledge. New information is simply added to the network and the system is solved again, propagating properly the consequences throughout the resulting thermochemical tables.

Another exciting feature is the possibility of running "what if" scenarios (hypothesis testing). New or tentative data are easily tested for consistency against the existing body of knowledge. If the tested hypothesis is inconsistent with the existing knowledge, the statistical analysis will identify the conflicting information, allowing its reanalysis. Similarly, a statistical analysis of the thermochemical network is capable of isolating the "weakest links" in the existing knowledge, thus identifying new experiments or computations that will enhance the resulting thermochemistry. This alone is a new paradigm for the efficient use of limited laboratory and computational resources.

The **table** lists the enthalpies of formation at 298.15 K (77°F) obtained by solving the small thermochemical network shown in Fig. 1, and provides an illustration of what can be expected from this approach. The species involved in this network are of fundamental importance and are generally considered to be already well-established via conventional approaches. Not surprisingly, the literature values for water (which is historically one of the most studied substances) are not significantly affected. However, the improvements for the other species are quite significant.

Web and grid servers. Active Thermochemical Tables are based on the thermochemical network approach. While it is possible to run ATcT as a self-standing application on a personal computer, ATcT is currently available through a framework of Web and grid services using servers at Argonne National Laboratory near Chicago. These services are an integral part of the Collaboratory for Multi-Scale Chemical Science (CMCS) and are accessible to users via its Web portal (**Fig. 3**).

CMCS is one of the Scientific Discovery through Advanced Computing (SciDAC) National Collaboratory projects funded by the Office of Science of the U.S. Department of Energy (DOE). CMCS brings together leaders in scientific research and technological development across multiple DOE

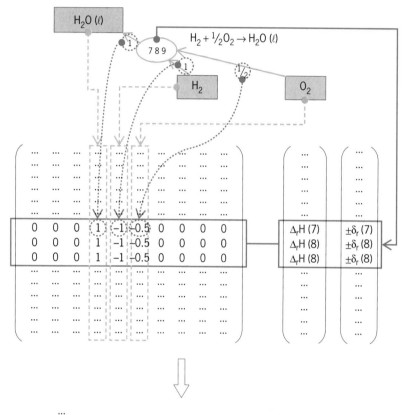

Eq. (7): $\Delta_f H(H_2O <\ell>) - \Delta_f H(H_2) - 0.5\,\Delta_f H(O_2) = \Delta_r H\,(7) \pm \delta_r\,(7)$

Eq. (8): $\Delta_f H(H_2O <\ell>) - \Delta_f H(H_2) - 0.5\,\Delta_f H(O_2) = \Delta_r H\,(8) \pm \delta_r\,(8)$

Eq. (9): $\Delta_f H(H_2O <\ell>) - \Delta_f H(H_2) - 0.5\,\Delta_f H(O_2) = \Delta_r H\,(9) \pm \delta_r\,(9)$

Fig. 2. Mapping of a selected portion (a subgraph) of the small thermochemical network depicted in Fig. 1 onto the adjacency matrix and column vectors. This selected subgraph corresponds to the chemical reaction $H_2 + {}^1/_2\,O_2 \rightarrow H_2O\,(\ell)$ and hence contains three primary vertices (gray boxes in the graph) denoting the chemical species and one secondary vertex denoting the reaction (the upper oval in the graph). The secondary vertex has a triple degeneracy, reflecting the fact that there are three competitive experimental measurements for the enthalpy of this reaction, denoted by equations (7), (8), and (9). The primary vertices denoting the enthalpies of formation of H_2, O_2, and H_2O (ℓ) correspond to the marked columns of the adjacency matrix. The triple degeneracy of the secondary vertex produces the three marked rows of the adjacency matrix and the two column vectors. The styles and directions of the edges (arrows) that connect the primary and secondary vertices of the subgraph determine the algebraic signs and the numeric values of the nonzero elements of the adjacency matrix. The appropriate elements of the two column vectors are populated by the three similar (but not quite identical) measured values and the adjunct uncertainties for the enthalpy of this reaction. The rows of the adjacency matrix and the column vectors map onto the linear equations at the bottom of the figure. It should be noted that the vertices that correspond to enthalpies of formation of the species that are fixed (in the present case, H_2 and O_2, which are elements in their reference states) can be removed from the graph, modifying, when necessary, the values of the secondary vertex. The action is equivalent to moving the appropriate enthalpies of formation from the left side of the equations to the right side, and modifying accordingly the elements of the column vectors (leading in this particular case to no numerical changes to the right side of the equations, since the enthalpies of formation of H_2 and O_2 are fixed to zero by thermochemical conventions).

Enthalpies of formation at 298.15 K, $\Delta_f H°$ (298 K), obtained by solving the small thermochemical network shown in Fig. 1, together with literature values

Chemical species[a]	$\Delta_f H°$ (298 K) from ATcT, kJ/mol	$\Delta_f H°$ (298 K) from literature, kJ/mol
O (g)	249.2291 ± 0.0021	249.18 ± 0.10[b,c,d]
H (g)	217.99781 ± 0.00010	217.998 ± 0.006[b,c,d]
H$_2$O (ℓ)	−285.826 ± 0.039	−285.830 ± 0.040[b,c,d]
H$_2$O (g)	−241.822 ± 0.039	−241.826 ± 0.040[b,c,d]
OH (g)	37.325 ± 0.047	38.99 ± 1.2[c]
		39.349 ± 0.21[d]
		37.3 ± 0.3[e]
OH$^+$ (g)	1293.060 ± 0.052[f]	1310.9 ± 0.5[f,c]
		1284.1 ± 5.0[f,d]
		1293.0 ± 0.3[f,e]
H$_2$O$_2$ (ℓ)	−187.695 ± 0.065	−187.78 ± 0.08[d]
H$_2$O$_2$ (g)	−135.805 ± 0.086	−136.106[c]
		−135.88 ± 0.22[d]

[a] The suffix (g) denotes a species gaseous state, and (ℓ) denotes a liquid.
[b] J. D. Cox, D. D. Wagman, and V. A. Medvedev, *CODATA Key Values for Thermodynamics*, Hemisphere, New York, 1989.
[c] M. W. Chase, Jr. (ed.), *NIST-JANAF Thermochemical Tables*, 4th ed., J. Phys. Chem. Ref. Data, Monog. No. 9, 1998.
[d] L. V. Gurvich, I. V. Veyts, and C. B. Alcock, *Thermodynamic Properties of Individual Substances*, vol. 1, pts. 1 and 2, Hemisphere, New York, 1989.
[e] B. Ruscic et al., *J. Phys. Chem. A*, 106:2727, 2002.
[f] The value is given in the "stationary electron convention"; the corresponding 298.15 K value in the "thermal electron convention" can be obtained by adding 6.197 kJ/mol.

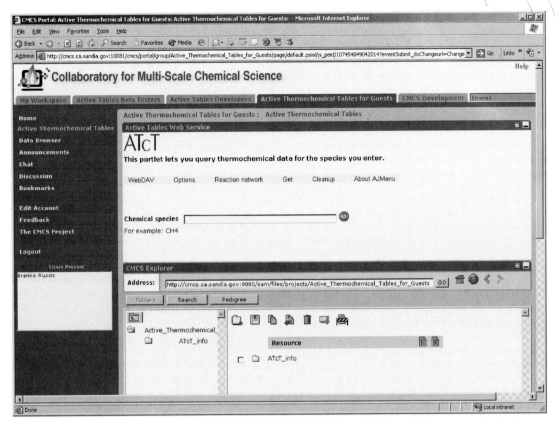

Fig. 3. Active Thermochemical Tables front-end portlet within the community portal of the Collaboratory for Multi-Scale Chemical Science (http://www.cmcs.org). Users can access the full functionality of ATcT and associated services run at a remote server. The configurable sidebar shows the various software tools available in the workspace, while the lower half of the environment allows remote handling of data in the CMCS database.

laboratories, other government laboratories, and academic institutions to develop an open "knowledge grid" for multiscale informatics–based chemistry research.

ATcT is the central application on the thermochemical scale of CMCS. ATcT consists of several parts: the software kernel, the user interface with the framework of services, and the underlying databases. Both the software and the databases are under continuous development at Argonne National Laboratory. The software suite allows queries of existing data, as well as visualization and manipulation of thermochemical networks. Users can set a variety of options that direct the mode in which the databases are queried, in which the results presented, or in the handling of thermochemical networks.

The ATcT databases are organized as a series of libraries. The main library contains a thermochemical network that currently has over 200 thermochemical species and is growing daily. The auxiliary libraries provide historical reference data from standard compilations. Users have the opportunity to establish their own "mini libraries" in which they can store data that supplements or modifies the information contained in the central and auxiliary libraries. As new scientific results are obtained from the thermochemical network's main library, they will be published in the scientific literature.

The development of Active Thermochemical Tables is supported by the U.S. Department of Energy, Division of Chemical Sciences, Geosciences and Biosciences of the Office of Basic Energy Sciences, and by the Mathematical, Information, and Computational Science Division of the Office of Advanced Scientific Computing Research, under Contract No. W-31-109-ENG-38.

For background information *see* CHEMICAL KINETICS; CHEMICAL THERMODYNAMICS; ENTHALPY; ENTROPY; GIBBS FUNCTION; HEAT CAPACITY; PHYSICAL CHEMISTRY; PHYSICAL ORGANIC CHEMISTRY; THERMOCHEMISTRY in the McGraw-Hill Encyclopedia of Science & Technology.

Branko Ruscic

Bibliography. B. Ruscic et al., Introduction to Active Thermochemical Tables: Several "key" enthalpies of formation revisited, *J. Phys. Chem. A*, 108(45):9979–9997, 2004; B. Ruscic et al., Ionization energy of methylene revisited: Improved values for the enthalpy of formation of CH_2 and the bond dissociation energy of CH_3 via simultaneous solution of the local thermochemical network, *J. Phys. Chem. A*, 103(43):8625–8633, 1999; B. Ruscic et al., On the enthalpy of formation of hydroxy radical and gas-phase bond dissociation energies of water and hydroxyl, *J. Phys. Chem. A*, 106(11):2727–2747, 2002; B. Ruscic et al., Simultaneous adjustment of experimentally based enthalpies of formation of CF_3X, X = NIL, H, Cl, Br, I, CF_3, CN, and a probe of G3 theory, *J. Phys. Chem. A*, 102(52):10889–10899, 1998.

Adaptive wings

A wing is the primary lift-generating surface of an aircraft. Whether it is rigidly attached to the fuselage for fixed-wing airplanes or rotating for helicopters, a primary design objective of such a surface is to maximize the lift-to-drag ratio, which is achieved by controlling the airflow around the wing. Other design objectives for the wing include improving maneuverability and minimizing vibrations and flow-induced noise. The wing can have a set design optimized for specific flight conditions, or it can change shape to conform to a variety of conditions. Chosen judiciously, minute dynamic changes in the wing's shape can, under the right circumstances, greatly affect the airflow and thus the aircraft's performance.

Adaptive wings—also known as smart, compliant, intelligent, morphing, controllable, and reactive wings—are lifting surfaces that can change their shape in flight to achieve optimal performance at different speeds, altitudes, and ambient conditions. There are different levels of sophistication, or intelligence, that can be imbued in a particular design.

Flying insects and birds, through millions of years of evolution, can change the shape of their wings, subtly or dramatically, to adapt to various flight conditions. The resulting performance and agility are unmatched by any human-made airplane. For example, the dragonfly can fly forward and backward, turn abruptly and perform other supermaneuvers, hover, feed, and even mate while aloft (**Fig. 1**). Undoubtedly, its prodigious wings contributed to the species' survival for over 250 million years.

Among human-made flyers, the Wright brothers changed the camber of the outboard tip of their aircraft's wings to generate lateral or roll control

Fig. 1. Male and female Cardinal Meadowhawk dragonflies following airborne mating. The male has towed the just-inseminated female to a pond and is dipping her tail in the water so she can deposit her eggs. (*Reprinted with permission, The Press Democrat, Santa Rosa, CA*)

(combined with simultaneous rudder action for banked turn), thus achieving in 1903 the first heavier-than-air, controlled flight. The R. B. Racer built by the Dayton Wright Airplane Company in 1920 allowed the pilot to change the wing camber in flight using a hand crank. The wings of today's commercial aircraft contain trailing-edge flaps and leading-edge slats to enhance the lift during the relatively low speeds of takeoff and landing, and ailerons for roll control, all engaged by the pilot via clumsy, heavy, and slow servomechanisms. To equalize the lift and prevent rolling in forward flight, the rotary wings of most helicopters are cyclically feathered to increase the pitch on the advancing blade and decrease it on the retreating blade.

While bird wings are quite smart, human-made ones are not very intelligent. With few exceptions, the level of autonomous adaptability sought in research laboratories is some years away from routine field employment. Intelligent control of the wing shape involves embedded sensors and actuators with integrated control logic; in other words, the wing's skin is made out of smart materials. The sensors detect the state of the controlled variable, for example the wall shear stress, and the actuators respond to the sensors' output based on a control law to effect the desired in-flight metamorphosing of the wing. For certain control goals, for example skin-friction drag reduction, required changes in the wing shape can be microscopic. For others, for example morphing the wing for takeoff and landing, dramatic increases in camber may be needed. *See* FLAPPING-WING PROPULSORS.

Smart materials. Adaptive wing design involves adding smart materials to the wing structure and using these materials to effect flow changes. Smart materials are those that undergo transformations through physical interactions. Such materials sense changes in their environment and adapt according to a feedforward or feedback control law. Smart materials include piezoelectrics, electrostrictors, magnetostrictors, shape-memory alloys, electrorheological and magnetorheological fluids, and optical fibers. For no rational reason, several other types of sensors and actuators that fall outside those categories are not usually classified as constituting elements of smart structures.

The piezoelectric effect is displayed by many non-centrosymmetric ceramics, polymers, and biological systems. The direct effect denotes the generation of electrical polarization in the material in response to mechanical stress; the poled material is then acting as a stress or strain sensor. The converse effect denotes the generation of mechanical deformation upon the application of an electrical charge; in this case the poled material is acting as an actuator. The most widely used piezoceramic and piezopolymer are, respectively, lead zirconate titanate (PZT) and polyvinylidene fluoride (PVDF). Piezoelectrics are the most commonly used type of smart materials and the only ones that can be used readily as both sensors and actuators.

Electrostrictive materials are dielectrics that act similarly to piezoelectric actuators, but the relation between the electric charge and the resulting deformation in this case is nonlinear. Examples of such materials are lead magnesium niobate (PMN) compounds, which are relaxor ferroelectrics. Magnetostrictive materials, such as Terfenol-D, are actuators that respond to a magnetic field instead of an electric field.

Shape memory alloys, such as a nickel–titanium alloy known as Nitinol, are metal actuators that can sustain large deformation and then return to their original shape by heating without undergoing plastic deformation. Electrorheological and magnetorheological fluids rapidly increase in viscosity—by several orders of magnitude—when placed in, respectively, electric or magnetic fields. Both kinds of fluids can provide significant actuation power and are therefore considered for heavy-duty tasks such as shock absorbing for large-scale structures. Finally, optical fibers are sensors that exploit the refractive properties of light to sense acoustical, thermal, and mechanical-strain perturbations.

Outstanding issues to be resolved before the use of smart materials for aircraft wings becomes routine include cost; complexity; the required computer's memory, speed, and software; weight penalty; maintenance; reliability; robustness; and the integrity of the structure on which the sensors and actuators are mounted. Sensors and actuators that have length scales between 1 and 1000 micronmeters constitute a special domain of smart materials that in turn is a cornerstone of micro-electro-mechanical systems (MEMS).

Flow control. An external wall-bounded flow, such as that developing on the exterior surfaces of a wing, can be manipulated to achieve transition delay, separation postponement, lift increase, skin-friction and pressure drag reduction, turbulence augmentation, mixing enhancement, and noise suppression. These objectives are multiply interrelated (**Fig. 2**). If the boundary layer around the wing becomes turbulent, its resistance to separation is enhanced and more lift can be obtained at increased incidence. On the other hand, the skin-friction drag for a laminar boundary layer can be as much as an order of magnitude less than that for a turbulent one. If transition is delayed, lower skin friction and lower flow-induced noise are achieved. However, a laminar boundary layer can support only very small adverse pressure gradients without separation and, at the slightest increase in angle of attack or some other provocation, the boundary layer detaches from the wing's surface and subsequent loss of lift and increase in form drag occur. Once the laminar boundary layer separates, a free-shear layer forms, and for moderate Reynolds numbers transition to turbulence takes place. Increased entrainment of high-speed fluid due to the turbulent mixing may result in reattachment of the separated region and formation of a laminar separation bubble. At higher incidence, the bubble breaks down, either separating completely or forming a

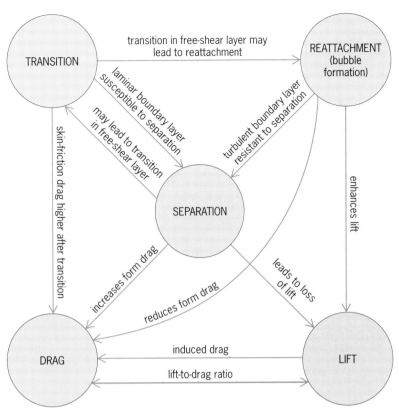

Fig. 2. Partial representation of the interrelation between flow-control goals. (*After M. Gad-el-Hak, Flow Control: Passive, Active and Reactive Flow Management, Cambridge University Press, London, 2000*)

longer bubble. In either case, the form drag increases and the lift curve's slope decreases. The ultimate goal of all this is to improve the airfoil's performance by increasing the lift-to-drag ratio. However, induced drag is caused by the lift generated on a wing with a finite span. Moreover, more lift is generated at higher incidence, but form drag also increases at these angles.

Flow control is most effective when applied near the transition or separation points, where conditions are near those of the critical flow regimes where flow instabilities magnify quickly. Therefore, delaying or advancing the laminar-to-turbulence transition and preventing or provoking separation are easier tasks to accomplish. Reducing the skin-friction drag in a nonseparating turbulent boundary layer, where the mean flow is quite stable, is a more challenging problem. Yet, even a modest reduction in the fluid resistance to the motion of, for example, the worldwide commercial airplane fleet is translated into annual fuel savings estimated to be in the billions of dollars. Newer ideas for turbulent flow control focus on targeting coherent structures, which are quasiperiodic, organized, large-scale vortex motions embedded in a random, or noncoherent, flow field (**Fig. 3**).

Future systems. Future systems for control of turbulent flows in general and turbulent boundary layers in particular could greatly benefit from the merging of the science of chaos control, the technology of microfabrication, and the newest computational tools collectively termed soft computing. Control of chaotic, nonlinear dynamical systems has been

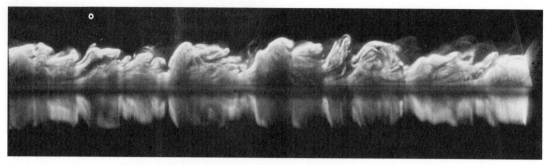

Fig. 3. Large-eddy structures (one type of coherent structure) in a turbulent boundary layer. The side view is visualized using a sheet of laser light and a fluorescent dye. (*From M. Gad-el-Hak, R. F. Blackwelder, and J. J. Riley, On the interaction of compliant coatings with boundary layer flows, J. Fluid Mech., 140:257–280, 1984*)

demonstrated theoretically as well as experimentally, even for multi-degree-of-freedom systems. Microfabrication is an emerging technology which has the potential for mass-producing inexpensive, programmable sensor–actuator chips that have dimensions of the order of a few micrometers. Soft computing tools include neural networks, fuzzy logic, and genetic algorithms. They have advanced and become more widely used in the last few years, and could be very useful in constructing effective adaptive controllers. Such futuristic systems are envisaged as consisting of a colossal number of intelligent, interactive, microfabricated wall sensors and actuators arranged in a checkerboard pattern and targeted toward specific organized structures that occur quasirandomly (or quasiperiodically) within a turbulent flow. Sensors would detect oncoming coherent structures, and adaptive controllers would process the sensors' information and provide control signals to the actuators that in turn would attempt to favorably modulate the quasiperiodic events. A finite number of wall sensors perceives only partial information about the flow field. However, a low-dimensional dynamical model of the near-wall region used in a Kalman filter can make the most of this partial information. Conceptually all of that is not too difficult, but in practice the complexity of such control systems is daunting and much research and development work remain. *See* COMPUTATIONAL INTELLIGENCE.

Control strategies. Different levels of intelligence can be imbued in a particular control system (**Fig. 4**). The control can be passive, requiring no auxiliary power and no control loop, or active, requiring energy expenditure. Manufacturing a wing with a fixed streamlined shape is an example of passive control. Active control requires a control loop and is further divided into predetermined or reactive. Predetermined control includes the application of steady or unsteady energy input without regard to the particular state of the system—for example, a pilot engaging the wing's flaps for takeoff. The control loop in this case is open, and no sensors are required. Because no sensed information is being fed forward, this open control loop is not a feedforward one. This subtle point is often confused, blurring predetermined control with reactive, feedforward control.

Reactive, or smart, control is a special class of active control where the control input is continuously adjusted based on measurements of some kind. The control loop in this case can either be an open, feedforward one or a closed, feedback loop. Achieving that level of autonomous control (that is, without human interference) is the ultimate goal of smart-wing designers. In feedforward control, the measured variable and the controlled variable are not necessarily the same. For example, the pressure can be sensed at an upstream location, and the resulting signal is used together with an appropriate control law to actuate a shape change that in turn influences the shear stress (that is, skin friction) at a downstream position. Feedback control, on the other hand, necessitates that the controlled variable be measured, fed back, and compared with a

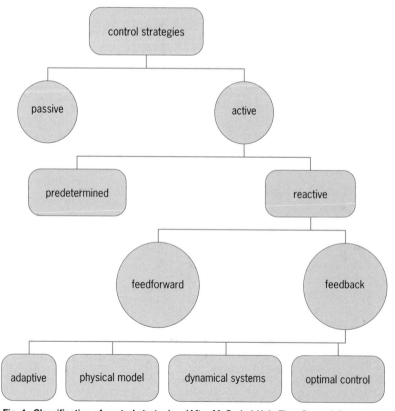

Fig. 4. Classification of control strategies. (*After M. Gad-el-Hak, Flow Control: Passive, Active and Reactive Flow Management, Cambridge University Press, London, 2000*)

reference input. Reactive feedback control is further classified into four categories: adaptive, physical model–based, dynamical systems–based, and optimal control. An example of reactive control is the use of distributed sensors and actuators on the wing surface to detect certain coherent flow structures and, based on a sophisticated control law, subtly morph the wing to suppress those structures in order to dramatically reduce the skin-friction drag.

Prospects. In the next few years, the first field applications of adaptive wings will probably take place in micro-air-vehicles (MAV) and unmanned aerial vehicles (UAV). The next generations of cruise missiles and supermaneuverable fighter aircraft will probably have adaptive wings. In a decade or two, Commercial aircraft may have smart wings, fuselage, and vertical and horizontal stabilizers. *See* LOW-SPEED AIRCRAFT.

For background information *see* AERODYNAMIC FORCE; AILERON; AIRFOIL; BOUNDARY-LAYER FLOW; CHAOS; CONTROL SYSTEMS; ELECTROSTRICTION; ELEVATOR (AIRCRAFT); ESTIMATION THEORY; FIBEROPTIC SENSOR; FLIGHT; FLIGHT CONTROLS; FUZZY SETS AND SYSTEMS; GENETIC ALGORITHMS; HELICOPTER; MAGNETOSTRICTION; MICRO-ELECTROMECHANICAL SYSTEMS (MEMS); NEURAL NETWORK; PIEZOELECTRICITY; REYNOLDS NUMBER; SHAPE MEMORY ALLOYS; TURBULENT FLOW; WING in the McGraw-Hill Encyclopedia of Science & Technology.   Mohamed Gad-el-Hak

Bibliography. H. T. Banks, R. C. Smith, and Y. Wang, *Smart Material Structures: Modeling, Estimation and Control*, Wiley, New York, 1996; M. Gad-el-Hak, *Flow Control: Passive, Active and Reactive Flow Management*, Cambridge University Press, London, 2000; M. Gad-el-Hak (ed.), *The MEMS Handbook*, CRC Press, Boca Raton, FL, 2002; M. Gad-el-Hak, R. F. Blackwelder, and J. J. Riley, On the interaction of compliant coatings with boundary layer flows, *J. Fluid Mech.*, 140:257–280, 1984; M. Schwartz (ed.), *Encyclopedia of Smart Materials*, vols. 1 and 2, Wiley-Interscience, New York, 2002.

Adoptive tumor immunotherapy

Adoptive tumor immunotherapy is a new technology being developed for the treatment of cancer. This technology utilizes cells from the patient's immune system to destroy the cancerous cells that cannot be surgically removed. This unique therapy employs three steps. First, cells from the immune system that have potential to fight the tumor are isolated from the patient's blood, tumor, or lymph nodes. Second, these specialized cells are expanded in number and possibly functionally enhanced in culture. Third, these cells are reintroduced into the patient to kill the remaining tumor cells. The reintroduction of the cells into the patient has been termed adoptive cell transfer (ACT) therapy.

Immune Response to Tumors

Typically, the immune system is inadequate at clearing tumors because tumors are constructed from the same components as the surrounding normal tissues. However, many types of cells in the immune system have been shown to contribute to an antitumor response. Genetic changes in the tumor cells and changes within the local environment induced by the tumor can be detected by the immune system. For example, T cells can recognize tumor-associated antigens that are expressed both on the tumor and on antigen-presenting cells, which have taken up part of the tumor. Thus, two common cell types are being exploited for ACT therapy. First, T cells that recognize tumor-associated antigens are used to directly kill the tumor. Second, antigen-presenting cells (for example, dendritic cells) are used to activate T cells that recognize tumor-associated antigens.

T-cell activation. T cells receive the signal to differentiate into "killer T cells" (that is, cytotoxic T lymphocytes) via interactions with antigen-presenting cells in the lymph nodes (**illus.** *a*). Antigen-presenting cells are activated through the binding of toll-like receptors. Positive signals on activated antigen-presenting cells stimulate T cells to differentiate into cytotoxic T lymphocytes: (1) T-cell receptors bind tumor-associated antigens (hypothetically, an infinite number of different T-cell receptors can be generated to bind to every imaginable antigen) and (2) T cells bind molecules that provide co-stimulation (for example, CD28 on the T cell binds B7 family members on the antigen-presenting cell). These interactions lead to multiple rounds of T-cell division, and the T cells acquire the ability to kill target tumor cells.

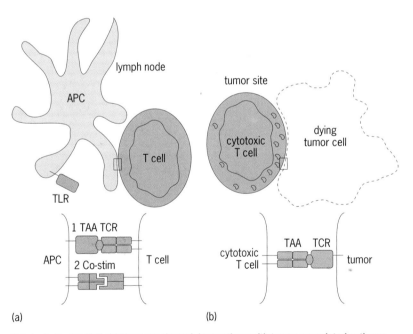

Cytotoxic T cells (CTLs) kill tumors through interactions with tumor-associated antigens (TAAs). (*a*) In the lymph node, antigen-presenting cells (APCs) are activated through the binding of toll-like receptors (TLRs). (The TLR shown is on the surface of the APC; however, some TLRs function inside the APC.) Activated APCs stimulate T cells to differentiate into CTLs: (1) T-cell receptors (TCRs) bind TAAs on APCs and (2) T cells bind co-stimulatory molecules on APCs. (A closeup of the interaction between the APC and T cell is shown below.) These interactions cause T cells to undergo expansion in numbers, and the activated TAA-specific T cells migrate to the tumor. (*b*) Then, tumor cells expressing antigens that bind to the TCRs on CTLs can be killed. (A closeup of the interaction between the CTL and tumor cell is shown below.)

The cytotoxic T lymphocytes then respond to signals from the tumor environment and migrate to the tumor site (illus. *b*).

T-cell destruction of tumors. When cytotoxic T lymphocytes encounter tumor cells expressing tumor-associated antigens, they kill via three possible mechanisms. First, they secrete pore-forming and protein-degrading molecules (perforin and granzyme molecules, respectively). The pore-forming molecules become lodged in the membrane of the target cells, and the protein-degrading molecules enter via the pores to initiate programmed cell death (apoptosis). Second, programmed cell death is initiated through direct binding of ligands that deliver a death signal (for example, the Fas ligand on the T cell binds to its receptor on the tumor cell). Third, cytotoxic T lymphocytes kill tumor cells indirectly by producing cytokines (such as interferon-γ or tumor necrosis factor-α) that stimulate other cells to phagocytose ("eat") tumor cells.

Improving Therapy Efficiency

Although clinical trials testing ACT therapy have not shown complete remission in cancer patients, promising results have been obtained. In the most notable trial to date, ACT therapy combined with other treatment modalities resulted in a significant decrease in metastatic disease volume in 6 of 13 melanoma patients. Although we currently have not reached the ultimate goal—cure—there are many potential methods that may vastly improve the efficiency of the therapy.

Selecting better T cells. Researchers speculate that the perfect T cells to use in ACT therapy will reflect a combination of several improvements. For example, these T cells may be genetically engineered to have T-cell receptors with the most affinity for the tumor-associated antigen, to have a lifespan adequate to kill the tumor, and to provide the most effective killer function to destroy all of the tumor cells. The therapy would also include treatment with T-cell growth factors to support the transferred T cells, as determined to be helpful in initial trials testing ACT therapy.

More effective antigen binding. Generally, the strength of the T-cell receptor–antigen interactions correlates with the strength of the killer function. During T-cell development, T cells expressing T-cell receptors that efficiently bind antigens from normal tissues are physically or functionally deleted to prevent autoimmunity. Since tumor-associated antigens are typically normal proteins expressed at the wrong time or place, interactions between the tumor-associated antigens and T cells are usually weak. Thus, T cells that recognize tumor-associated antigens are characteristically ineffective in controlling tumor growth.

Initial clinical trials testing ACT therapy isolated these weak T cells from the tumor, expanded their numbers in culture, and then introduced them back into patients. Patients were also treated with a T-cell growth factor (interleukin-2). Although these trials were not significantly effective, they provided promise for future trials. One hypothesis being tested

in the next generation of trials is that the stronger the tumor-associated antigen binds to T cells, the better the antitumor response. Research is currently being conducted to determine the most effective binding properties of tumor-associated antigens that may activate the T cells. It has been noted in mouse and human model systems that the highly avid recognition of antigen is not the only necessary consideration for the most effective immune response.

Longer lifespan. Results of clinical trials have also indicated that the T cells were not surviving once they were reintroduced into the patient. After the expanded T cells were transferred back into the patient, they were not detectable, indicating that they had died. Options to improve the lifespan of T cells have become plentiful. For example, the T cells could be genetically altered to express antideath genes (that is, antiapoptotic genes such as *Bcl-2*).

Enhanced killing capacity. In addition to antigen binding and T-cell survival, the functions of the transferred T cells could also be improved. Markers have been identified on T cells with the most effective killing capacity. Sorting these cells from other less functional cells may improve therapy.

Engineering better antigen-presenting cells. The capacity of T cells to kill tumor cells is initiated by interactions with antigen-presenting cells. One might suppose that if an antigen-presenting cell were fused with a tumor cell, the resulting cell fusion would be efficient at activating tumor-associated antigen–specific T cells. However, this type of antigen-presenting cell is ineffective in generating a consistent antitumor response using ACT therapy. Again, the tumor is mainly composed of material that is in the normal tissue. In addition, the antigens on these antigen-presenting cells are not being presented to the T cells with the proper activation signals. The perfect antigen-presenting cell for ACT therapy requires high levels of antigen, costimulatory molecules, and activated toll-like receptors.

Increased expression of antigen. The antigen-presenting cells must express high levels of tumor-associated antigens from the tumor. Antigen-presenting cells obtain the tumor-associated antigens by ingesting dying tumor cells. Alternatively, the tumor-associated antigen can be added to the antigen-presenting cells before they are reintroduced.

Increased expression of co-stimulatory molecules. Antigen-presenting cells also express co-stimulatory molecules that signal the T cells to kill cells that express the antigen (illus. *a*). In the absence of co-stimulatory molecules, T cells do not become functional killers. Since tumors do not typically express co-stimulatory molecules, the tumor cannot activate T cells. Animal experiments in which these co-stimulatory molecules were introduced into a tumor showed increased immunity to the tumor.

Activated toll-like receptors. Toll-like receptors on the antigen-presenting cells must be bound for the antigen-presenting cell to activate T cells. Toll-like receptors bind pathogen-associated molecule patterns on bacterial or viral products. These molecules are foreign to the immune system and vigorously activate antigen-presenting cells. Recent research has shown that stimulating antigen-presenting cells with such molecules increases the activity of the responding T cells toward a tumor.

Greater cell expansion. Once cells have been generated for therapy, they require expansion to an effective level (greater than 10^{10} cells—more than 10 grams of cells—for T cell ACT therapy). In the past decade, the ability to expand these immune system cells has been a priority for tumor immunologists. The combinations of growth factor cocktails, antibodies, and conditions for expansion have greatly improved.

Increasing effectiveness of cell injection. The best method for injecting the cells into the patient has also been closely examined. For example, injection of the T cells into the arteries of the tumor is more effective than injection of these T cells into the veins. The other route of delivery that may be effective is direct injection of the cells into the tumor. The number of antigen-presenting cells required and the most effective injection site is less clear, although antigen-presenting cells function best in the lymph node closest to the tumor.

Removing immune response inhibitors. Additional protocols to augment ACT therapy are being developed. For example, improved clinical outcomes are observed when ACT therapy is given after partial or complete depletion of the patient's immune system. In this paradigm, cells from the patient are isolated, expanded, and selected in culture while the patient's immune system is depleted via therapeutic drug treatments or irradiation. This therapy removes any cells that may inhibit the immune response to tumors. In addition, the cells transferred into the depleted host will expand to regenerate a more effective immune system.

Other issues. The probability of inducing an immune response to normal tissues is greatly increased using immune therapies. For example, the side effect of autoimmunity has been noted in many clinical trials for melanoma treatment. [The T-cell response is predominantly directed to proteins that produce pigment in the skin; autoimmunity is evident by pigment-free spots (vitiligo)]. If the tumor is killed, such side effects may be acceptable if the target molecules are not required for life. Another issue is that expanding the cells of each patient individually may be prohibitively expensive. However, once the methods have been optimized, the price of the treatments may drop.

Understanding how to produce a specific T-cell response for conditions in which the immune system is normally ineffective will provide more specific, less invasive treatment options than those currently available. Lessons learned in successfully treating cancer with this therapy will apply to the development of therapies for other diseases that contain an immunological component such as human immunodeficiency virus (HIV)/acquired immune deficiency syndrome (AIDS).

For background information *see* CANCER (MEDICINE); CELLULAR IMMUNOLOGY; IMMUNITY; IMMUNOLOGIC CYTOTOXICITY; IMMUNOTHERAPY in the McGraw-Hill Encyclopedia of Science & Technology.

Jill E. Slansky

Bibliography. V. Cerundolo et al., Dendritic cells: A journey from laboratory to clinic, *Nature Immunol.*, 5(1):7–10, January 2004; M. E. Dudley et al., Adoptive-cell-transfer therapy for the treatment of patients with cancer, *Nat. Rev. Cancer*, 3(9):666–675, September 2003; M. E. Dudley et al., Cancer regression and autoimmunity in patients after clonal repopulation with antitumor lymphocytes, *Science*, 298(5594):850–854, Oct. 25, 2002; W. Y. Ho et al., Adoptive immunotherapy: Engineering T cell responses as biologic weapons for tumor mass destruction, *Cancer Cell*, 3(5):431–437, May 2003; L. W. Kwak et al., Adoptive immunotherapy with antigen-specific T cells in myeloma: A model of tumor-specific donor lymphocyte infusion, *Semin. Oncol.*, 31(1):37–46, February 2004; R. M. Steinman et al., Exploiting dendritic cells to improve vaccine efficacy, *J. Clin. Invest.*, 109(12):1519–1526, June 2002; Y. Yang et al., Persistent toll-like receptor signals are required for reversal of regulatory T cell-mediated CD8 tolerance, *Nature Immunol.*, 5(5):508–515, May 2004; C. Yee et al., Adoptive T cell therapy using antigen-specific CD8+ T cell clones for the treatment of patients with metastatic melanoma: In vivo persistence, migration, and antitumor effect of transferred T cells, *Proc. Nat. Acad. Sci. USA*, 99(25):16168–16173, December 10, 2002.

Advanced wireless technology

Ultrawideband (UWB) and Bell Labs Layered Space Time (BLAST) technology (also known as multiple-input, multiple-output, or MIMO technology) are two of the wireless technologies that promise to have significant impact on the communications world over the next 10 years. These wireless techniques are in the very early stages of commercial deployment. BLAST/MIMO approaches first appeared in the literature in the mid-1990s. UWB techniques, which emerged in the early 1960s, have been used for some time in military and underground object location applications but are just starting to appear in commercial applications.

Ultrawideband. The acronym UWB is used to designate wireless technology that utilizes a transmission bandwidth that is a significant fraction of the carrier frequency. In other words, the fractional bandwidth, defined as $(F_u - F_l)/F_c$, where F_c is the center frequency of the utilized band, F_u is the upper frequency, and F_l is the lower frequency, is equal to or greater than 25%. By contrast, code-division multiple access (CDMA) cellular in the United States (ITU Standard IS-95) has a fractional bandwidth of 1.5%; and GSM (Global System for Mobile Communications), the cellular technology in Europe based on time-division multiple access (TDMA), uses 0.3%.

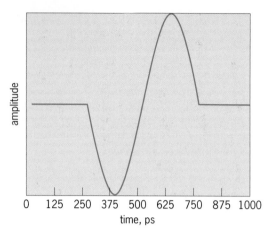

Fig. 1. Individual ultrawideband (UWB) pulse. 1 ps = 10^{-12} s.

Although UWB has four major application areas—radar, positioning, imaging, and communications—the following discussion will be limited to communications. UWB technology originated in the early 1960s, with the use of impulse response analysis to characterize the transient behavior of microwave networks. Up until the mid-1990s, work in UWB in the United States was done primarily in classified government programs. Commercial applications were enabled by a Federal Communications Commission (FCC) UWB Report and Order, adopted in February 2002, which allows the unlicensed deployment of commercial UWB systems subject to certain limitations on power, frequency, and other application-specific factors. Commercial UWB communications systems can operate (up to the limit for emissions of

electronic equipment specified by the FCC) in the 3.1–10.6-GHz range; outside this frequency range, they must operate at least 10 decibels below the FCC limit. UWB is not currently approved for commercial use in other countries.

Unlike conventional wireless communication systems, which use continuous waveforms for communication of information, UWB uses narrow pulses, less than 1 ns in duration (**Fig. 1**), with uniform or nonuniform pulse spacing (**Fig. 2**). Nonuniform spacing flattens the spectrum by eliminating regularly spaced frequency components and makes the signal more noiselike. This is desirable since there is then less interference to other wireless systems using the same spectrum. Nonuniform spacing requires synchronization to properly decode the signal. A concern in any wireless system is multipath reflections, signals that are reflected from objects between the transmitter and the receiver. These signals travel a longer path than the signal that propagates directly from the transmitter to the receiver (the line-of-sight signal), and their arrival at the receiver is therefore delayed. Multipath reflections and the line-of-sight signal may interfere with each other, causing decreases in the signal level, a phenomenon known as multipath fading. However, because the duty cycle of the UWB waveform (the percentage of the time during which waveform is nonzero) is so low, it is highly unlikely that this will happen in a UWB system.

UWB systems may use differential phase-shift keying (DPSK), a variant of phase-shift keying in which only the differences of phases (not the actual phases) are encoded, as a signal modulation format. When this is done, the received pulses are correlated with the expected pulse shape to detect the received symbol. In time-modulated UWB, pulse position is varied by a small time difference (less than the pulse width) to indicate a binary or *m*-ary value. Multiple pulses are used to form a symbol and are coherently integrated (meaning that information, including phase information, from the multiple pulses is combined) to achieve an improvement in detection performance, which is known as processing gain. Processing gain can be measured in decibels and is determined by the number of pulses used to form a usable symbol. (Coherent integration yields greater processing gain than noncoherent integration, where the phase information is not used.) Binary phase-shift keying (BPSK), pulse-amplitude modulation (PAM), on-off keying (OOK), and other modulations are also used. So-called rake receivers can be used to improve performance by combining the received multipath signals, but require accurate estimation of the signal paths and increase implementation complexity.

Advantages. UWB has several important advantages over other radio communication schemes. UWB energy is spread over a very wide frequency range, making the power spectral density very low. Unlike other radio-frequency technologies, UWB systems are designed to coexist in the same frequency bands

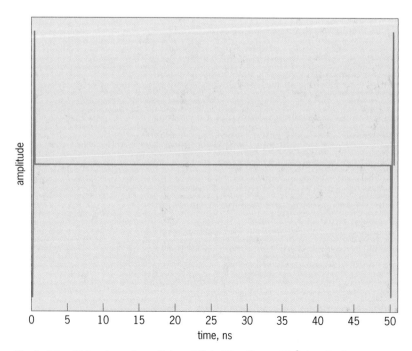

Fig. 2. Ultrawideband waveform. Pulse width is 0.5 ns (1 ns = 10^{-9} s), and pulse rate is 2×10^{7} pulses per second.

as other radio-frequency applications. The FCC Report and Order limits UWB communication devices to the same level as FCC Part 15, which specifies the maximum limit of any unintentional radio emissions from commercial electronic equipment, making the likelihood of interfering with other devices extremely low. UWB systems are advertised to have better material penetration properties, but this is more an artifact of the frequency range used than UWB itself. Low susceptibility to multipath fading improves performance and quality of service compared to the wireless local-area network (LAN) standard IEEE-802.11 (also known as WiFi). IEEE-802.11 is widely deployed and was designed for data. As such it has trouble efficiently carrying streaming video (sequences of moving images that are sent in compressed form and displayed by the viewer as they arrive) and voice over Internet Protocol (VoIP) traffic. Because UWB has not yet been deployed in standards-based products, UWB protocols can be designed with efficient quality-of-service implementation in mind.

Another benefit of UWB is its ability to transmit at much higher bit rates at a given power level than other radio-frequency techniques. This ability is a consequence of the Shannon equation (1), which

$$C = B \log_2 \left(1 + \frac{P}{BN_0} \right) \qquad (1)$$

lies at the foundation of communication theory. Here C is the channel capacity in bits per second, B is the channel bandwidth in hertz, P is the received signal power in watts, and N_0 is the noise power spectral density in watts per hertz.

It can be seen that the channel capacity increases as a function of bandwidth much faster than as a function of signal power. Alternatively, UWB systems can transmit at lower power levels and dissipate less power at a given data rate than narrowband systems. Lower power enables increased battery life and the use of less expensive integrated electronics for radio receivers.

A major cost of conventional radios is the radio-frequency power amplifier, its dc power supply, and associated filters needed to suppress unwanted out-of-band radio emissions often produced by the power amplifier. By trading off power for bandwidth, UWB devices can potentially be very inexpensive. UWB technology has the unique potential of enabling the output of a complementary metal-oxide semiconductor (CMOS) chip to directly drive the antenna itself. No analog components are required except a strip line and antenna. Thus the dream of a true single-chip radio could be realized with as little as a one-chip (CMOS) circuit, a battery, and a simple antenna.

WPAN. A current commercial focus for UWB communication is in the wireless personal area networking (WPAN) application. Due to FCC power constraints, ranges of less than 10 m (33 ft) are targeted, with bit rates of up to 500 Mbit/s. Potential applications include home distribution of multiple streams of high-quality video and audio to and from video displays, recorders, and players; wireless connectivity to replace the wired IEEE-1394 (Firewire, iLink) connections in consumer electronic devices such as camcorders, digital cameras, and audio devices; wireless connectivity of personal computer peripherals, such as such as mice, keyboards, and printers; replacement of existing Bluetooth and IEEE-802.11 products; and home networking of all of these applications.

Transmitter constraints. A major limitation of UWB applications is transmitter power constraints. UWB systems generally operate at the noise floor specified by the FCC for other services utilizing the frequency band of the UWB device. This limits the range of UWB products. Concern about interference of UWB devices with other wireless products such as Global Positioning System (GPS) receivers has prompted the FCC to be particularly conservative in UWB emissions specifications. Concern has also been raised over radio interference of UWB devices with television relay receivers in the 3–4-GHz band and with satellite receivers in the 1.4-, 2.3-, and 2.6-GHz bands. *See* GPS MODERNIZATION.

BLAST/MIMO. The terms BLAST and MIMO refer to new wireless technologies that take advantage of multipath scattering through the use of multiple transmitting and receiving antennas and sophisticated signal processing to achieve higher bandwidth efficiency for wireless communications. MIMO is the most general case, including schemes in which there is one transmitting antenna and one receiving antenna, multiple transmitting antennas and one receiving antenna, and one transmitting antenna and multiple receiving antennas, as well as multiple transmitting antennas and multiple receiving antennas. There are several ways of using these multiple antennas, including beam steering, for example. BLAST is a particular family of algorithms for receiver processing, including vertical BLAST (VBLAST) and horizontal BLAST (HBLAST). (The latter two terms refer to the coding of information across the two dimensions of space and time, with time depicted on the horizontal axis and space across the vertical. The original version of BLAST is known as diagonal BLAST or DBLAST. Vertical BLAST is a simplified version where no coding is performed. In horizontal BLAST, symbols from each transmitting antenna are coded across time only.)

Conventional wireless systems, with a single transmitting antenna and a single receiving antenna (known as SISO, for single-input, single-output), have the Shannon capacity, given by Eq. (1). With multiple receiving antennas, the receiver can capture ever greater amounts of signal power in direct proportion to the number of antennas. Therefore, capacity grows logarithmically with the number of receiving antennas, M, according to Eq. (2). With a MIMO

$$C = B \log_2 \left(1 + M \frac{P}{BN_0} \right) \qquad (2)$$

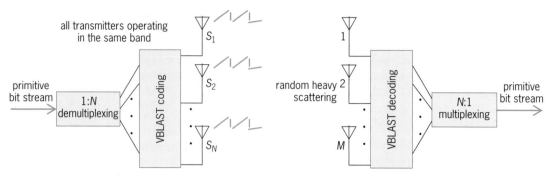

Fig. 3. BLAST/MIMO system. Total power P is constant (P/N per channel, and bit rate, R_b, is constant on each channel. (The N-fold self-interference is a minor degradation.) The total transmission rate is proportional to NR_b.

system consisting of L transmitting and M receiving antennas, the channel capacity approaches the value given in Eq. (3), or a value that is proportional to

$$C = \min(M, L)\log_2\left(1 + M\frac{P}{LBN_0}\right) \qquad (3)$$

the number of antennas in the transmit-receive array. These gains are achievable assuming sufficient multipath transmission and low correlation between the signals received at the different antennas. Hence, in BLAST/MIMO systems, unlike conventional communication systems, multipath enhances the data rate. Thus, BLAST/MIMO systems are capable of bandwidth efficiency many times that of other wireless techniques.

In a BLAST/MIMO system (**Fig. 3**), bit streams are demultiplexed into a number of substreams to the different transmitters, encoded, and transmitted using conventional techniques (such as DPSK). All transmitters operate in the same frequency band and at the same time. The transmitted signals are received, along with their multipath reflections, at each of the receiving antennas. Assuming flat fading conditions (that is, assuming that the delay of the multipath reflections is not a significant portion of the signal wavelength, so that fading is frequency independent), a single scalar can be used to represent the gain coefficient (the number used to multiply the received signal to compensate for attenuation in the channel) for a transmit-receive antenna pair. The received signal streams can be recovered using an $L \times M$ matrix of these scalar coefficients. Where flat fading cannot be assumed, multicarrier and equalization approaches can be applied. Alternative approaches trade data rate for diversity maximization (that is, using the multiplicity of transmission paths to optimize the signal). In the most extreme approach, the data rate of a single antenna system can be targeted, with transmitting antennas receiving a redundant version of the same signal to minimize the outage probability.

BLAST/MIMO techniques have limitations. There is increased complexity in applying these techniques to rapidly fading channels (as in the case of mobile receivers). BLAST/MIMO systems require antenna arrays, but antenna elements can be closely spaced [as close as 0.1 wavelength, which is 1 cm (0.4-in.)

at a frequency of 3 GHz] and still provide excellent performance, making implementation in small portable devices practical. With the use of polarization, BLAST/MIMO performance gains can be achieved with antennas in essentially the same physical location. Unfortunately, multiple antennas and radio-frequency sections for transmitters and receivers increase cost. Computational complexity is also a cost concern, but Moore's law will address this in the future. There currently are no standards for BLAST/MIMO, making it difficult to develop systems for wide deployment. However, several standards bodies are looking into the use of MIMO for next-generation wireless systems, including 3GPP and 3GPP2 (cellular), IEEE-802.11 (WiFi), and IEEE-802.16 (WiMAX). BLAST/MIMO also appears to have near-term commercial potential in fixed wireless applications. The proliferation of wireless hotspots is an application area well suited to MIMO, both for communication from laptop computers to access points and for the communication link that connects the access point to the Internet service provider (known as access point backhaul). Notebook computers are excellent physical platforms for MIMO antenna arrays.

Most importantly, BLAST/MIMO techniques allow multiple users on a cellular network to share the same spectrum, as well as the same spreading codes, thus preserving the precious code and spectrum resources available to a wireless service provider.

For background information *see* BANDWIDTH REQUIREMENTS (COMMUNICATIONS); GROUND-PROBING RADAR; INFORMATION THEORY; MOBILE RADIO; MODULATION; MULTIPLEXING AND MULTIPLE ACCESS; PHASE MODULATION; SPREAD SPECTRUM COMMUNICATION in the McGraw-Hill Encyclopedia of Science & Technology.   Michael D. Rauchwerk

Bibliography. R. J. Fontana, *Recent Applications of Ultra Wideband Radar and Communication Systems*, Kluwer Academic/Plenum Publishers, May 2000; G. J. Foschini and M. J. Gans, On limits of wireless communications in a fading environment when using multiple antennas, *Wireless Pers. Commun.*, 6:311–335, March 1998; C. E. Shannon and W. Weaver, *The Mathematical Theory of Communication*, University of Illinois Press, 1949.

Arctic mining technology

Mineral extraction in the Arctic, or other frozen regions, presents tremendous technical challenges to natural resource industries. There is good potential for finding economically attractive mineral deposits within the largely unexplored arctic-shield-type geology. Shields, also known as cratons, are large exposed areas of the oldest rocks on Earth and occur on all continents. Because of their age and the changes due to faulting, volcanism, metamorphism, and erosion, which have taken place through hundreds of millions of years, shields are favorable areas for mineral deposits. Similarly, offshore sedimentary basins offer the prospect of significant petroleum reserves. The challenges to the resource industry are sociological, environmental, economic, and technical.

Sociological and environmental challenges. Exploitation of minerals has driven the development of transportation infrastructure, communities, and commerce. In the northern circumpolar nations, recent development has been northward, making previously inaccessible regions available for exploration and subsequent exploitation. This development often can encroach on areas which have traditionally been the domain of aboriginal peoples, whose very existence has depended upon a carefully managed system of exploitation of the resources available to them.

Arctic ecosystems are delicate and easily damaged. Considering that it may take generations, or even centuries, for damage to be repaired (if it ever can be), development must proceed with a culturally sensitive, well-managed scientific and technical approach. Mining practices that inflict damage upon the tundra by heavy-wheeled vehicles—leaving refuse and garbage at former exploration and mining camps, and abandoning fuel depots and other facilities—must fade away. Today, we must consider the possible impacts of plans for exploration and exploitation before proceeding.

In Arctic regions, the choice of a transportation route is not simply a matter of physical geography and economics. One must also consider the migration routes of land and sea animals. There are certain periods of the year during which the routes must be closed to access. On land, most of the annual transportation will likely take place during the winter months, when it becomes possible to drive trailer trains—a series of trucks or other mobile equipment proceeding as a group—over the frozen tundra and lakes. The design of such roads must consider the possible effects of vehicular movement on the stability of the ice. During the summer, depending on the location, it may be possible to have access to marine navigation for a limited period. This requires incorporation of icebreaker technology into the design of suitable ships.

During the twentieth century, it was common that a town or mining camp would accompany the development of a mine or a number of contiguous mines. Many of these communities became ghost towns once the economic potential of the mineral deposits was exhausted. Others have attained new life through diversification of the economy by attracting other industries or as educational, government, retirement, or recreation centers. In all instances, there may be geotechnical risk potential due to the construction of structures and transportation corridors above mine openings. There also have been economic hardships as mine workers and local commerce have had to adjust to the economic consequences of mine closure.

Because of this, and because the development of a mine in a remote region can seriously affect the way of life of local peoples, the trend today is toward "no trace" mining camps, where the facilities are present only during the life of the mining operation and are removed on mine abandonment. If these camps are in very remote locations, with no communities close by, "fly-in-fly-out" operations are used in which parts of the work force arrive and depart on a rotational basis. If, on the other hand, there are communities in the area, the local peoples must be informed of the limited production lifespan and the eventual abandonment of the mine. A formal abandonment plan is an essential part of the planning process.

Economic and technical challenges. At mines in the north, and especially in the Arctic, the provision of energy is one of the costliest items in the budget. In the absence of a road or rail infrastructure, all hydrocarbon fuels and most supplies must be transported either by trailer train in the winter or by sea in the summer. Transportation by air is possible but can be prohibitively expensive.

Energy and water considerations. Accurate estimation of the quantities of supplies and energy required to support the annual operation therefore is critical, and energy conservation is essential.

In the design of the buildings, there must be adequate insulation in order to prevent heat loss. Modular construction is common, as it is advantageous to orientate the facilities into the prevailing wind so that wind resistance and heat loss are reduced. Buildings are constructed on insulated pilings so that heat will not be transferred into the frozen ground below. The high cost of obtaining water for processing the minerals or to supply the camp generally requires as much water as possible be recycled. Any water that is discharged into the environment must be treated in order to avoid chemical pollution.

Thermal considerations. The same practices apply during mining. All of the High Arctic and the Arctic are in the continuous permafrost zone. Thus, the ground is perennially frozen, often to depths of hundreds of meters. In the Subarctic, much of the terrain is in the discontinuous permafrost zone, in which frozen ground occurs intermittently. In either case, depending on the nature of the geological material being mined (rock or soil), the effects of heat being transferred from the mining process into the surrounding natural material must be assessed.

In instances when the ore-bearing rock is a hard rock with relatively few weaknesses, discontinuities,

Refrigeration plant at the portal of Teck Cominco's Polaris mine on Little Cornwallis Island in the Canadian High Arctic. Closed in August, 2002, it was the northernmost metal (lead-zinc) mine in the world.

and contained ice, it is likely that thawing will have few effects on its stability. The opposite is true for highly friable or fractured or "soillike" rock. Thawing is likely to result in substantial reductions of in-situ strength, and subsequent instabilities and "falls of ground" (collapse of loosened blocks of rock or soil). To prevent this, a refrigeration plant to cool the incoming ambient air was installed at the portal of Canada's northernmost mine, the Polaris lead-zinc mine, on Little Cornwallis Island in the High Arctic (see **illustration**). Likewise, when backfill was placed in the completed primary mine workings from which one was extracted (called stopes), it was placed according to a schedule so that the thermal damage to the surrounding rock mass could be minimized. The filled opening was then allowed to freeze before mining proceeded into the adjacent workings. In this case, it was essential that the ore and the surrounding rock should remain in the frozen state during mining.

The same considerations also apply to both open pit and placer mining. In both cases, the slopes exposed by mining may be weakened substantially through thawing. A frozen soillike material with a substantial content of ice can become supersaturated on thawing, and can flow, threatening the mining operations. Very significant losses of the strength of the geological material can result from thawing. Finally, cyclical freezing and thawing can result in a geomorphological process known as frost-jacking, through which blocks and slabs are exfoliated from the rock mass due to the expansion of ice in cracks.

Stability considerations. If refrigeration, cooling, or other measures necessary to ensure stability are not an option, the mining plans must be adjusted to the stability conditions which are foreseen. Slope angles must be reduced to within safe limits. Monitoring

becomes necessary to provide warnings of slope movements or potential falls of ground. Blasting practices are also affected by the presence of significant quantities of frozen water in the ore. This can necessitate secondary blasting in order to cope with oversized chunks. In some instances, the transportation of the ore from the mine to the processing plant can be problematic due to thawing and settling of the broken material in conveyances.

In many mining operations, the ore-bearing rock is processed to produce a concentrate of the valuable minerals. During the milling process the ore is crushed and ground to the sizes at which the valued minerals can be efficiently separated from the other mineral components of the rock (known as waste or gangue). Milling involves a substantial expenditure of energy. In Arctic mining, every effort must be made to ensure that valued heat is not released to the surrounding environment. Modern engineering practice is such that, through sound planning and design, almost all of the heat can be captured for further use.

Waste disposal. Following the separation of the waste from the ore, the waste must be disposed of in the vicinity of the mine and processing plant. Some may be used to fill abandoned openings in the mine, or to fill abandoned open pits, or may be placed in tailings ponds and waste piles. The options available to the designer will depend upon the features of the operation, the environmental requirements, and the local terrain.

Historical mining practices have left a legacy of abandoned mining sites which present both physical and chemical hazards to the environment. In much of the world, these past practices are no longer acceptable, and mine operators must ensure that a close-out and site remediation plan is in place before mining begins.

In particular, every effort must be made to ensure that harmful chemicals are not discharged into the environment. All ores contain waste minerals, and some of these, particularly sulfides, can combine with air and water to produce acidic leachates. Acid mine drainage can be devastating to the environment, and especially so in the delicate Arctic ecosystems.

Fortunately, the freezing Arctic climate conditions can slow down, or even arrest, the rate of chemical reactions. From the design point of view, the current practice is to ensure that piles of waste rock and tailings ponds are frozen as quickly as possible, and will remain in the frozen state for the foreseeable future. Known as cap and cover, the technique involves the placement of impermeable materials immediately above the wastes, thereby preventing contact with air and water, and covering this with a thick layer of inert soil or rock. The intent is to ensure that the surface of the permanently frozen ground beneath (the permafrost) will rise and adapt to the new topography, thereby ensuring that the mine wastes becomes, and will remain, frozen. In instances where earth dams have been constructed to contain the mine wastes, thermosiphons can be used to draw heat from the dam, thereby ensuring that it too remains frozen.

During the past few years, a number of mining operations in the Canadian Arctic and Subarctic have been concluded. At Schefferville, in the Labrador Trough, Quebec, the town site and mine plants, which once supported a major iron mining operation, were removed. At Deception Bay on the Ungava Peninsula, the port facilities, which once served the Asbestos Hill mining operation, are now being used to transport the mineral concentrates from New Quebec Raglan's Kattiniq mines.

In the High Arctic, both the Polaris and Nanisivik lead-zinc mines on Little Cornwallis Island and Baffin Island, respectively, have now closed. Both sites are being decommissioned and will be monitored for the foreseeable future.

For background information *see* ARCTIC AND SUBARCTIC ISLANDS; MINING; SURFACE MINING; TUNDRA; UNDERGROUND MINING in the McGraw-Hill Encyclopedia of Science & Technology.   John E. Udd

Bibliography. O. B. Andersland and B. Ladanyi, *Frozen Ground Engineering*, 2d ed., Wiley, New York, 2004; Arctic Mining (theme issue), *CIM Bull.*, Canadian Institute of Mining, Metallurgy and Petroleum, Montréal, Québec, vol. 89, no. 1005, November/December 1996.

Atmospheric modeling, isentropic

In order to study atmospheric motions, we need to establish spatial coordinates so that the movement of air can be described as a function of space and time. There are many ways to set up such coordinate systems to represent air motions in the atmosphere. The most common one is the classical cartesian coordinate system, the (x,y,z) system, where x, y, and z usually (in meteorology) represent the eastward, northward, and upward (vertical) coordinates.

Another common practice in atmospheric science is to replace the vertical coordinate z by pressure p. This is because the atmosphere as a whole, either at rest or in motion, is generally stably stratified and obeys the hydrostatic condition (that is, the pressure gradient force is exactly balanced by the gravitational force) to a high degree. In such an atmosphere, the pressure always decreases with height (although the decrease rate may be different in different places) and there is always one p value corresponding to a z value for a given (x,y) point. Theoretical meteorological models using pressure as the vertical coordinate are called isobaric models.

There are other possible choices for the vertical coordinate, and one of those is the isentropic (unchanging entropy) coordinate. This coordinate system appears to be gaining popularity and deserves a closer look at its benefits and limitations. Before we describe such a system in some detail, let us examine the isentropic process.

Adiabatic and nonadiabatic processes. An adiabatic process in thermodynamics is a physical process in which the total energy of a system remains constant. This implies that no exchange of energy occurs between a system and its environment, and no additional heat source or sink is present (although there may be one within the system). Since we are talking about the atmosphere here, let us use an air parcel as a concrete example of such a system.

Imagine an air parcel moving in the atmosphere. The environment of this air parcel consists of the rest of the atmosphere, so there is no rigid boundary between the air parcel and its environment. If this air parcel does not mix with its environmental air and exchange any heat, the motion is adiabatic.

In reality, there are many ways that the motion of this air parcel can become nonadiabatic. For example, if mixing occurs between the air parcel and its environmental air (for example, due to different velocities or turbulence), the energy of the resulting mixture will differ from the original parcel and the motion is not adiabatic. Another way is that the air parcel loses or gains heat by radiation or latent heat during its motion.

There are many examples of nonadiabatic processes in the atmosphere. The formation of radiation fog in a calm summer morning is one such example because the moist air parcel near the surface is cooled by contact with the cold surface that, in turn, cools by radiating heat away overnight. The air parcel thus reaches saturation, and fog drops form. This obviously is a nonadiabatic process. In addition, the formation of fog drops is a phase-change process during which latent heat of condensation is released. This is again a nonadiabatic process even though the source is within the air parcel itself.

Another example is the formation of convective clouds such as cumulus. The most common way cumulus clouds form is that moist air parcels, rising from lower levels, cool due to expansion and become saturated. During the ascent of such an air parcel (usually warmer and more humid than its environment), turbulent mixing (called entrainment in this case) usually occurs due to the different velocities between the parcel and its environmental air, and the whole process is indeed nonadiabatic.

Whereas real atmospheric motions are rarely truly adiabatic, we often consider adiabatic motions because the nonadiabatic contributions are either negligible or unessential to the heart of the problem. In the case of cumulus formation, the essential part of the process is the adiabatic expansion cooling due to the ascent of the air parcel in a stratified atmosphere. Consequently, we often approximate the cumulus formation as an adiabatic process.

Equivalence of adiabatic and isentropic processes. During an adiabatic process, the total energy of the air parcel is conserved. Thus the net change of the total energy, measured by the quantity TdS, where T is the temperature of the air parcel and dS is the change of its entropy, should be zero. Since the temperature cannot be zero here, dS must vanish.

This is to say that the entropy S must remain unchanged. Hence, an adiabatic process must also be an isentropic process. Thus when an air parcel rises adiabatically, its entropy remains constant no matter how high it goes. In meteorology, the potential temperature is usually used as a surrogate for entropy. Therefore, if the potential temperature of an air parcel stays the same after the air parcel has moved from one place to the other, we know it has performed an adiabatic motion.

Representing atmospheric motions in isentropic coordinates. Since the large-scale stratification of the atmosphere is stable, it is possible to use potential temperature as the vertical coordinate, whereas the horizontal coordinates can be the regular (x,y) coordinates as mentioned before. In a stably stratified atmosphere, the potential temperature varies monotonically with height z. At any given z, there is only one corresponding potential temperature value θ (theta), and hence there is no ambiguity. We shall see later that this is not so when the stratification is unstable.

In a coordinate system using θ as the vertical axis, the interpretation of atmospheric motions is, of course, different from that in cartesian coordinates. If an air parcel moves along a constant θ surface (an isentropic surface), it is performing an adiabatic motion. On the other hand, if an air parcel moves from an isentropic surface θ_1 to another isentropic surface θ_2 this parcel must have gone through some nonadiabatic, or diabatic, processes. This implies that the air parcel has either gained or lost energy. Thus by merely examining the changes in the vertical coordinate, we can discern the energy state of the air parcel and possibly even identify the process that is responsible for such changes. This is a great benefit in terms of analysis in atmospheric dynamics, a benefit not easily realized in other coordinate systems.

Example of atmospheric convection in isentropic coordinates. **Figure 1** represents a convective storm, using the results of a cloud model simulation of an actual storm. Figure 1*a* shows a snapshot of the distribution of relative humidity in the central east-west cross section of the simulated storm in regular cartesian (x,z) coordinates. Figure 1*b* shows the same cross section but in isentropic coordinates. Compared to the cartesian coordinates, the representation in isentropic coordinates exaggerates the upper portion of the system and hence is better in resolving upper-atmospheric processes.

There are also disadvantages in an isentropic coordinate system. In an unstably stratified atmosphere, the isentropic surfaces may intersect each other, resulting in two or more values of θ corresponding to a certain z (and vice versa) and in the vertical coordinate becoming ill-defined. The other difficulty is that the lower boundary (that is, the earth surface in most cases) usually does not conform to an isentropic surface, making the analysis complicated there.

Hybrid isentropic coordinate models. In order to alleviate the complication at the lower boundary (which is a problem not just to the isentropic representation but to isobaric representation as well), hybrid models, which are based on some kind of mixture vertical coordinates, have been developed. The most common of these is probably the hybrid sigma-isentropic coordinate model, in which the lower part of the vertical coordinate is the terrain-following σ coordinates while the upper part is the isentropic coordinate. Naturally, the two vertical coordinates have to

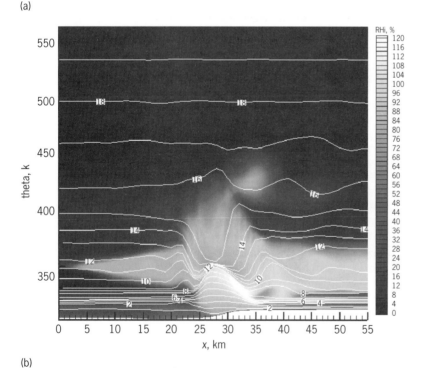

(a)

(b)

Fig. 1. Profile of relative humidity with respect to ice (RHi) in the central east-west cross section of a simulated storm using a 3D cloud model (see P. K. Wang, 2003). (*a*) Profile in cartesian *x-z* coordinates. White contour lines indicate the potential temperatures. (*b*) Same profile but in *x-θ* coordinates. White contour lines indicate the height in kilometers.

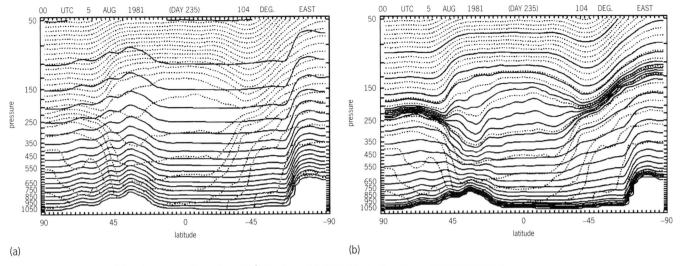

Fig. 2. Schematic of meridional cross sections along 104°E for August 5, 1981. The broken lines represent potential temperature in each panel. (*a*) Solid lines represent scaled sigma model surfaces. (*b*) Solid lines represent θ-σ model surfaces. (*Courtesy of D. R. Johnson*)

merge somewhere in the middle. One such hybrid model is the hybrid θ-σ model developed by D. R. Johnson and coworkers. **Figure 2** shows an example of the cross sections of hybrid models.

Hybrid models have been increasingly popular, especially for studying global climate and transport processes. This is mainly because of the quasiadiabatic nature of large-scale motions in the atmosphere, especially when long-range material transport is concerned. Such transports tend to move along isentropic rather than quasihorizontal surfaces, and three-dimensional advection becomes quasi-two-dimensional, resulting in greatly reduced computational errors. The σ coordinate used near the surface helps resolve the problem of a pure isentropic coordinate. There are also some other advantages.

But there are also limitations of hybrid models. For one, a hybrid coordinate does not preserve adiabatic flow in the boundary layer because it is not isentropic. In addition, the blending of the σ coordinate and θ coordinate, if not done properly, may introduce spurious mass, momentum, or energy sources. Modelers will have to weight advantages versus the limitations to suit their purposes.

For background information *see* ADIABATIC PROCESS; COORDINATE SYSTEMS; ENTROPY; ISENTROPIC PROCESS; ISENTROPIC SURFACES; ISOBARIC PROCESS; METEOROLOGY; THERMODYNAMIC PROCESSES in the McGraw-Hill Encyclopedia of Science & Technology.

Pao K. Wang

Bibliography. D. R. Johnson et al., A comparison of simulated precipitation by hybrid isentropic-sigma and sigma models, *Mon. Weath. Rev.*, 121:2088–2114, 1993; L. W. Uccellini, D. R. Johnson, and R. E. Schlesinger, An isentropic and sigma coordinate hybrid numerical model: Model development and some initial tests, *J. Atm. Sci.*, 36:390–414, 1979; P. K. Wang, Moisture plumes above thunderstorm anvils and their contributions to cross tropopause transport of water vapor in midlatitudes, *J. Geophys. Res.*, 108(D6):4194, 2003 (doi: 10.1029/2003JD002581).

Atom economy

A broad imperative for synthetic chemists today is the need to develop innovative and sophisticated synthetic methods to keep pace with the continually changing variety of molecular targets. In 1991, Barry M. Trost presented a set of coherent guiding principles to evaluate the efficiency of a specific chemical process, which has subsequently altered the way many chemists design and plan their syntheses. He proposed that within synthetic efficiency there are two broad categories: selectivity and atom economy. Selectivity is defined as chemo-, regio-, and stereoselectivity, while atom economy seeks to maximize the incorporation of the starting materials into the final product for any given reaction. An additional corollary is that if maximum incorporation cannot be achieved, then ideally the side-product quantities should be minute and environmentally innocuous.

In the past, attaining the highest yield and product selectivity were the governing factors in chemical synthesis. As a result, multiple reagents often were used in stoichiometric quantities that were not incorporated into the target molecule, resulting in significant side products and waste. Today, the underlying standard is synthetic efficiency—the ability to maximize the incorporation of the starting materials into the final product and to minimize by-products.

The reaction yield and the atom economy yield are calculated by different means. The reaction yield [Eq. (1)] is concerned only with the quantity of desired product isolated, relative to the theoretical quantity of the product, and does not provide information in terms of the efficiency of the transfer of molecular weight from the reactants into the desired product, which is the concern of atom economy. Atom economy takes into account all the reagents used and the unwanted side products, along with the desired product [Eq. (2)].

Reaction yield

$$= \frac{\text{quantity of product isolated}}{\text{theoretical quantity of product}} \times 100\% \quad (1)$$

Atom economy

$$= \frac{\text{molecular weight of products}}{\text{molecular weight of all products}} \times 100\% \quad (2)$$

Atom-uneconomic reactions. Examples of atom-uneconomic reactions are the Wittig reaction (formation of triphenylphosphine oxide), Grignard reaction, and substitution and elimination reactions (3).

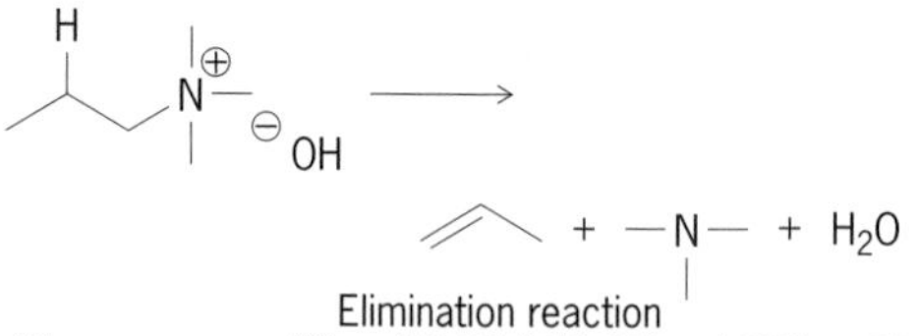

Elimination reaction
[Atom economy (%) = (42.08/119.21) × 100% = 35.3%]

Atomic-economic reactions. Some atomic-economic reactions are the Diels-Alder reaction, rearrangements [reaction (4)], and concerted reactions. Not

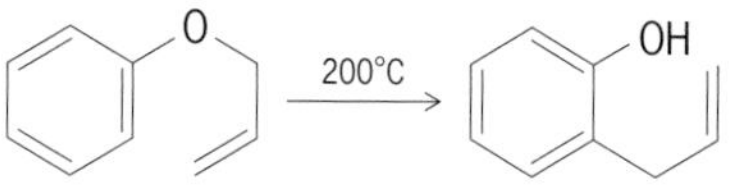

Claisen rearrangement
[Atom economy (%) = (134.18/134.18) × 100% = 100%]

all rearrangements are equal in that some require large amounts of acids to push the reaction forward (for example, the Beckmann rearrangement), and so those acids are considered to be ancillary reagents and factor into the economy of the reaction.

Currently, many chemists are working on increasing the number of reactions in the "toolbox" by developing ideal and innovative methodologies that combine both selectivity and atom economy. Transition metals have proven to be excellent catalysts for the stereoselective transformation of many organic molecules, and since they are used catalytically, they are ideal for the development of atomic-economic

methodologies. Reaction scheme (5) is an example of an transition-metal-catalyzed, atom-economic reaction.

The ruthenium catalyst complex is easily accessible, and its generality makes it the catalyst of choice. With this catalyst, the reactions are usually complete in 30 minutes at ambient temperature, and based upon the substituent there is regioselective control. In addition, this reaction is 100% atom-economic.

Synthesis of ibuprofen. A practical application of atom economy in industry is the synthesis of ibuprofen. The traditional synthesis of ibuprofen was developed in the 1960s. It was a six-step synthesis that used stoichiometric amounts of reagents and generated large quantities of waste by-products that needed further treatment [reaction scheme (6)]. There are numerous waste by-products, and the overall atom economy of the synthesis is 40%, or conversely, there is an overall waste of 60%.

Reaction scheme (7) shows a more atom-economic synthesis of ibuprofen, having an atom economy percentage of 77% (or 99% if the acetic acid generated in the first step is recovered).

For both reaction schemes (6) and (7), the starting materials and products are the same for the first step, with the only difference being the catalyst. In reaction scheme (6), AlCl$_3$ is not truly a catalyst but an auxiliary reagent that is needed in stoichiometric amounts and produces aluminum trichloride hydrate as a waste by-product. Hydrogen fluoride used in reaction scheme (7) is a catalyst, which is recovered and repeatedly reused. In addition, the Raney nickel and palladium catalysts are also recovered and reused. As a result, almost no waste is generated.

Since reaction scheme (7) requires three steps [relative to six for reaction scheme (6)], there is the additional elimination of waste by not having three extra steps. This results in the ability to produce larger quantities of ibuprofen in less time and with less capital expenditure. Therefore, besides having an atom-economic and environmentally friendly green synthesis, the product has a larger profit capability.

Outlook. In industry and academia, chemists have seen the value of atom economy and have integrated this idea into the development of new methodologies, which is a component of designing a more environmentally friendly synthesis within the framework of green chemistry. Therefore, with all the innovation and ingenuity of synthetic chemists today, the

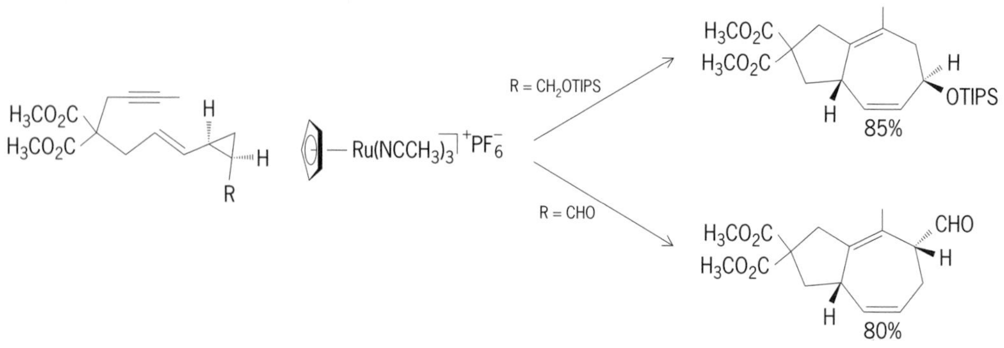

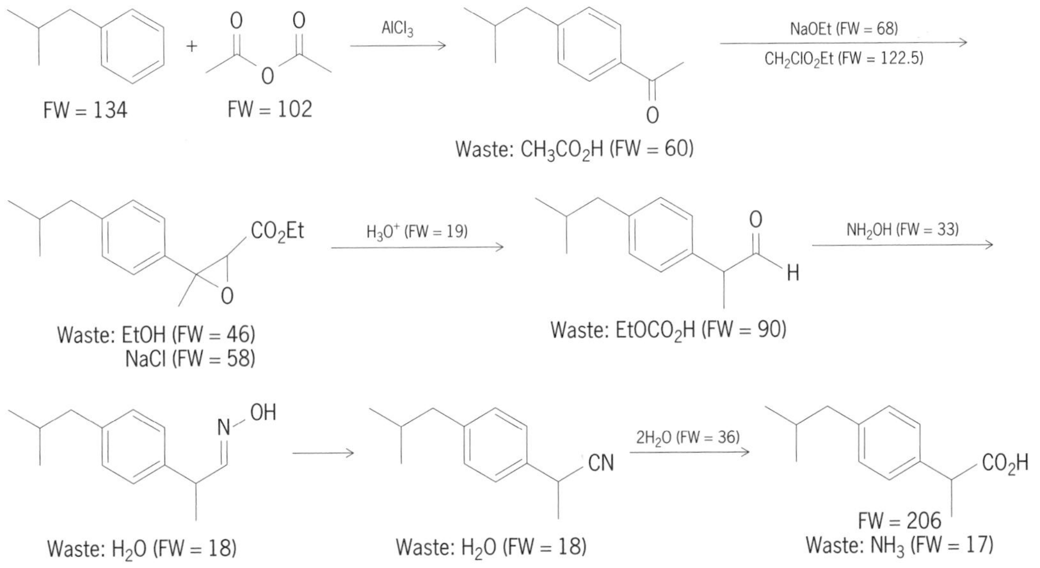

FW = 134   +   FW = 102 $\xrightarrow{AlCl_3}$

Waste: CH_3CO_2H (FW = 60)

$\xrightarrow[CH_2ClO_2Et\ (FW = 122.5)]{NaOEt\ (FW = 68)}$

Waste: EtOH (FW = 46)
NaCl (FW = 58)

$\xrightarrow{H_3O^+\ (FW = 19)}$

Waste: $EtOCO_2H$ (FW = 90)

$\xrightarrow{NH_2OH\ (FW = 33)}$

Waste: H_2O (FW = 18)

Waste: H_2O (FW = 18)

$\xrightarrow{2H_2O\ (FW = 36)}$

FW = 206
Waste: NH_3 (FW = 17)   (6)

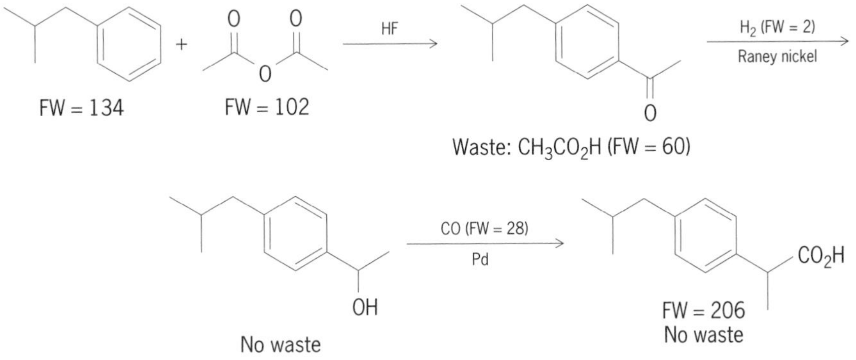

FW = 134   +   FW = 102 $\xrightarrow{HF}$

Waste: CH_3CO_2H (FW = 60)

$\xrightarrow[Raney\ nickel]{H_2\ (FW = 2)}$

No waste

$\xrightarrow[Pd]{CO\ (FW = 28)}$

FW = 206
No waste   (7)

ideal situation of 100% atom efficiency for each synthetic step is confidently within our grasp in the near future.

For background information *see* CATALYSIS; ORGANIC SYNTHESIS; STEREOCHEMISTRY; STEREOSPECIFIC CATALYST in the McGraw-Hill Encyclopedia of Science & Technology.   Charlene C. K. Keh; Chao-Jun Li

Bibliography. P. T. Anastas and J. C. Warner, *Green Chemistry: Theory and Practice*; Oxford, New York, 1998; B. M. Trost, The atom economy—a search for synthetic efficiency, *Science*, 254:1471–1477, 1991; B. M. Trost, On inventing reactions for atom economy, *Acc. Chem. Res.*, 35:695–705, 2002.

Atomic-scale imaging of dynamic surfaces

The advent of high-spatial-resolution scanning probe microscopes, such as the scanning tunneling microscope, has afforded a remarkable view of atomic- and molecular-scale surface dynamics. In particular, use of scanning tunneling microscopy (STM) for the detailed investigation of surface behavior has extended the boundaries associated with the measurement, comprehension, and ultimately application of nanoscale phenomena. The fundamental understanding of surface dynamics at the atomic and molecular

scale influences catalyst customization and use, structural alloy compositions, microelectronic device fabrication, and molecule-based nanoelectronics.

Methods. Scanning tunneling microscopy is capable of generating topographical surface maps and resolving the local (as opposed to surface-averaged) electronic structure of materials with subangstrom (less than 10^{-10} m) resolution. When used in combination with controlled environments, such as ultrahigh vacuum (10^{-10} torr or 10^{-8} Pa range) and cryogenic and elevated temperature regimes (<10 to 1200 K; −263 to 927°C; −442 to 1700°F), a wealth of surface dynamics can be described with incredible precision. By exploring a surface process at variable temperatures, it becomes possible to track the response of the system to various levels of thermal energy (kinetics) or to quench the system so that its dynamics are slowed to the point where individual reaction transition states and products can be identified.

Surface reconstructions and phase transitions. Surface reconstructions include changes in surface morphology, topography, and crystallinity. A variety of experiments using local imaging probes, such as the scanning tunneling microscope, have uncovered many of the crucial electronic and morphological features that influence the interactions and organization of atoms and molecules attached to surfaces. The intercommunication between surface-bound molecules and the underlying surface can give rise to and drive surface reconstructions and molecule ordering. Surface morphology and structure, especially defects such as surface steps, kinks, and other crystalline deviations, have been found to play large roles in determining surface reaction rates and pathways, including chemical passivation, catalysis, nucleation, and corrosion, to name some. Chemical passivation includes oxide layers grown on metallic and semiconducting surfaces, as well as chemically inert (nonreactive) monolayer (single-molecule-thick) films. Nucleation is the growth of a surface thin film by aggregation of molecules to a localized point on the surface (nucleation center).

In order to address many of these issues at the atomic level, STM has been used to observe how surfaces and molecules bound to those surfaces behave as functions of crystallinity and defect density, as well as adsorption and absorption of molecules. Aspects of real or practical applications that involve polycrystalline and alloy systems with uncontrolled conditions on an atomic scale can be studied on a rigorous fundamental level using accurately defined and characterized atomic crystallographies under ultrahigh-vacuum conditions.

The other impetus for investigating the influence of microscopic features or local morphology is to understand how atomic-level properties and interactions can be harnessed and manipulated in novel ways on a microscopic scale. Atoms and molecules are capable of self-organization and self-assembly in the proper chemical environments and underlying substrate structure. More routes to nanoscale engineering and design become possible as researchers understand how local properties influence surface and interfacial behavior.

Metallic oxidation has been probed as a function of surface properties, such as the orientation of surface atoms and defects, in order to more fully understand the fundamental mechanisms involved in oxide formation. The effect of a defect such as a monatomic step (a change in height on a surface by one atomic layer) on surface behavior, including the creation of a metal oxide, can be assessed by employing an intentionally stepped surface—a staircaselike surface with steps and risers (a periodic array of terraces with an average width). These systems have been the subject of an assortment of experimental and theoretical studies. The properties of the clean surface and the nature of the interaction of adsorbates (such as oxygen) with the step array are significant departures from the behavior of a nominally flat surface from which the stepped surface is derived. These surfaces are prone to reconstruct where steps on the surface begin to coalesce to form taller and wider steps due to the adsorption of molecules and thermal energy.

Figure 1 is a snapshot of the initiation of a surface reconstruction event that involves the joining of two metallic steps, that is, faceting on a stepped nickel (Ni) surface as a result of the adsorption of a small amount of oxygen (<2% of surface cover-

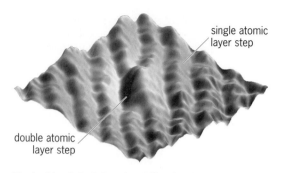

Fig. 1. Adsorbate-induced metallic microfaceting on a vicinal, or stepped surface, Ni(977) captured using time-lapse scanning tunneling microscopy.

age) and elevated temperature (375–500 K; 102–227°C; 215–440°F). The details of the surface dynamics and mechanism associated with this morphological transformation were made available only by time-resolved STM. It was possible to track the overall transformation of a surface region, as well as the evolution of isolated step coalescence events, as a function of oxygen exposure and temperature. The thermal regime and oxygen surface coverage associated with this surface reconstruction allowed for its dynamics to be captured with a relatively slow recording tool such as the scanning tunneling microscope. Imaging rates for this type of microscope are about 1 min per frame, which is a very long time in reaction dynamics. But if the system is probed at the proper conditions, where the activated processes are essentially slowed down, the scanning tunneling microscope has a chance to uncover real-time information with exquisite resolution.

Single molecules. It is possible to study the dynamics of individual and small groups of molecules interacting with the surface to which they are attached. Thin films on surfaces, such as an ordered single layer of molecules arranged on a surface, have been shown to be tailorable and definable such that they act as host matrices for the isolation and selective attachment of single molecules and nanostructures. One example is the formation of a self-assembled monolayer of organic molecules on a flat metal surface. **Figure 2** shows an STM image collected at high gap impedance (~100 GΩ) and at low temperatures (4 K; −269°C; −452°F) which resolves a single molecule, 4-(2-nitro-4-phenylethynyl-phenylethynyl)-benzenethiol, that was intentionally inserted in a defect present in a decanethiolate self-assembled monolayer on gold [Au(111)]. The inset shows the relationship between the inserted molecule and the monolayer matrix. Currently, this type of system is under intense study in molecular electronic device design, where the surface is used to isolate and position other molecules that have been synthesized with distinct electronic characteristics. In the example (Fig. 2), the molecule inserted in the film conducts electrons more effectively than the film and is structurally longer. This molecule is chemically tethered on one end, but its other end is free to move, depending on the nature (for example, volume and chemical identity) of the film surrounding the molecule. The dynamics of conformational motion of this larger molecule in a film of other molecules is crucial to understanding how this system can potentially be integrated and used in a device. This system has been studied using a variable-temperature scanning tunneling microscope that has been able to resolve the real-time dynamics of single-molecule motion as a function of surface thin-film composition, order, and temperature.

Bridging the pressure gap. Many experiments in heterogeneous catalysis are done under ultrahigh vacuum, but the conventional environments for common catalytic systems involve pressures that are above >1000 torr (>133 kPa). The challenge for catalysis experiments is to bridge this pressure gap, while closely mimicking real reaction conditions in order to observe how the composition and function of a catalyst is influenced by optimized reaction conditions.

To further improve understanding of known catalytic systems, especially with regard to what features are the most active in directing a chemical reaction, high-resolution STM has been employed in environments that more closely mimic true reaction conditions with respect to ambient pressure and temperature. Relatively simple catalytic reactions, such as the oxidation of carbon monoxide on metal surfaces, have been studied at hyperbaric pressures [as high as 4000 torr (533 kPa)] and temperatures as high as 425 K (152°C; 305°F) on well-defined metallic surfaces. What makes operating a scanning tunneling microscope under these environments remarkably challenging is that the performance of the instrument is strongly dependent on thermal stability and is typically more capable of probing relatively static chemical interfaces.

Technological importance. Atomic- and molecule-scale understanding is required if nanoscale properties are to be integrated with technologies such as chemical sensors, alternative electronic architectures, and selective catalytic arrays and surfaces. The synergistic approach of using combinatorial or surface-averaged approaches with local probe microscopes to address surface-bound behavior will ultimately propel much of nanotechnology.

Future challenges. In order to further understand and utilize the properties of single molecules and the nanoscale behavior of surfaces, new high-resolution local probes working in a complementary fashion are necessary. Scanning tunneling microscopy has been used to uncover a broad spectrum of surface behavior and dynamics and has spurred the development of related probe microscopes that are capable of measuring chemical bonding forces, magnetic spin, single-molecule vibrational states, as well as photon interactions. These efforts are aimed at resolving nanoscale behavior with multifaceted approaches for understanding the linkage between macroscopic and nanoscopic phenomena, as well as finding ways in which intrinsically nanoscale surface dynamics can be harnessed and manipulated.

For background information *see* CATALYSIS; NANOSTRUCTURE; SCANNING TUNNELING MICRO-

Fig. 2. Scanning tunneling microscopy image resolving a single, long-chained organic molecule supported in a self-assembled monolayer on Au(111). The inset shows the relationship between the molecule and the monolayer.

SCOPE; SURFACE AND INTERFACIAL CHEMISTRY; SURFACE PHYSICS in the McGraw-Hill Encyclopedia of Science & Technology.   Thomas P. Pearl

Bibliography. Z. J. Donhauser et al., Conductance switching in single molecules through conformational changes, *Science*, 292:2303–2307, 2001; H.-J. Freund, Clusters and islands on oxides: From catalysis via electronics and magnetism to optics, *Surf. Sci.*, 500:271–299, 2002; W. Ho, Single molecule chemistry, *J. Chem. Phys.*, 117:11033–11061, 2002; T. P. Pearl and S. J. Sibener, Spatial and temporal dynamics of individual step merging events on Ni(977) measured by scanning tunneling microscopy, *J. Phys. Chem. B*, 105:6300–6306, 2001; G. A. Somorjai, *Introduction to Surface Chemistry and Catalysis*, Wiley, New York, 1994; P. Thostrup et al., CO-induced restructuring of Pt(110)-(1 × 2): Bridging the pressure gap with high-pressure scanning tunneling microscopy, *J. Chem. Phys.*, 118:3724–3730, 2003.

Autoimmunity to platelets

Idiopathic thrombocytopenic purpura (ITP) is a hematologic disorder of unknown origin (that is, idiopathic) in which the blood does not clot properly due to a decreased number of platelets (thrombocytopenia). It is characterized by the presence of purpura, small hemorrhages of capillaries that appear as purple bruises on the skin or mucous membranes. There are two forms of the disorder: adult (or chronic) ITP and childhood (or acute) ITP. Adult ITP generally lasts 6 months or longer and usually results from the development of an autoantibody (an antibody formed by an individual against the individual's own tissues) directed against a structural platelet antigen. Childhood ITP generally lasts less than 6 months and is thought to be caused by a viral infection that triggers the synthesis of an antibody that may react with a viral antigen associated with the platelet surface.

Autoimmunity. Autoimmunity results from the body being unable to distinguish self from nonself. The job of the immune system is to defend the body from foreign invaders such as viruses and bacteria. Autoimmunity occurs when the immune cells mistake the body's own cells for invaders and attack them. It is not known why the immune system treats some body parts like germs and, thus, precisely why autoimmunity occurs; however, it is thought that both genes and the environment are involved. Autoimmunity can affect almost any organ or body system. An individual's exact problem with autoimmunity (or its related diseases) depends on which tissues are targeted.

Role of platelets. Platelets (also known as thrombocytes) are the smallest cell type in mammalian blood, averaging 2–3 micrometers in diameter. They have a circulating life span of 9–12 days, and the normal count in the blood varies between 150,000 and 300,000/mm^3. Platelets have three major structural and functional components: (1) the membrane, where adhesion and receptor function occurs; (2) secretory granules; and (3) an intrinsic contractile system. Platelets exhibit a variety of functions, including the production of cytokines (intercellular messenger proteins) such as platelet activation factor (PAF) as well as the release of several key factors that stimulate the coagulation cascade (for example, platelet factor 3).

Platelets demonstrate a high propensity to adhere to damaged blood vessels as well as to artificial surfaces, such as the plastics used in extracorporeal circuits during dialysis and cardiac surgery. Platelet adherence to damaged surfaces often leads to vessel narrowing and formation of a thrombus, an aggregation of blood factors (primarily platelets and fibrin) which may dislodge, leading to myocardial infarction, stroke, or deep vein thrombosis (the formation of blood clots in deep veins within the body, typically in the lower legs and thighs). Platelets are initially activated by contact with collagen-containing subendothelial basement membranes, which occurs after blood vessels have been damaged. The platelet changes shape from a disc to a spiny sphere with multiple pseudopods (cellular extensions). Pseudopod formation enables the platelet to aggregate with other platelets as well as to adhere to either endothelium (inner lining of blood vessels) or leukocytes (white blood cells).

Diagnosis. Physical examination results from ITP patients are normal except for the presence of petechiae (small red spots on the skin that usually indicate a low platelet count), purpura, and bleeding, which may be minimal or extensive. Circulating blood is normal except for a reduced platelet count. The bone marrow is also normal, although the number of megakaryocytes (cells found in the bone marrow, each of which gives rise to 3000–4000 platelets) may be elevated.

Treatment. Childhood ITP is mostly short lived and will resolve on its own without treatment. Treatment of adults usually begins with an oral corticosteroid (such as prednisone). Because corticosteroids help the body maintain the integrity of the walls of veins and arteries, they are helpful in stopping or preventing unwanted bleeding, allowing for the platelet count to rise to normal within 2–6 weeks. The corticosteroid dosage is then tapered. However, most patients fail to respond adequately or relapse as the corticosteroid dose is tapered. Corticosteroids also have strong side effects such as hypertension, osteoporosis, and psychiatric syndromes.

For an ITP patient with life-threatening bleeding, a physician may prescribe intravenous immunoglobulin (IVIg) in an attempt to suppress clearance of antibody-coated platelets by phagocytic cells of the reticuloendothelial system (RES). IVIg is prepared from large pools of plasma from healthy donors (1000–50,000 donors per batch). It is composed primarily of immunoglobulin G (IgG), the predominant immunoglobulin (antibody) in plasma and the main source of humoral (antibody-mediated) immunity. IVIg is indicated for use in primary immunodeficiencies, genetic disorders in which the body is unable to produce adequate amounts of IgG. Another labeled use is for chronic lymphocytic leukemia, particularly in situations in which the leukemia has suppressed bone marrow function and has impaired the body's ability to produce IgG. Other labeled uses for IVIg include the treatment of children with active acquired immune deficiency syndrome (AIDS), and in bone marrow transplantation. IVIg is also approved by the U.S. Food and Drug Administration for the treatment of two autoimmune conditions: ITP and Kawasaki syndrome (a rare inflammatory illness that occurs in children).

Mechanism of action of IVIg. Despite extensive clinical use, the mechanism of action of IVIg remains a mystery. However, several theories have been proposed to explain how administration of IVIg to individuals with ITP increases the platelet count.

RES blockade. It has been postulated that the success of IVIg in treating ITP is due to competitive inhibition of the RES by sensitized red blood cells (see **illustration**). The RES is part of the immune system and consists of a variety of phagocytic cells, primarily macrophages and monocytes, that degrade foreign antigenic substances. In normal blood, there is no contact between platelets and macrophages. However, in ITP the platelet is coated (opsonized) with autoantibodies. The Fc (fragment crystallizable) region of the autoantibody is recognized by the Fc receptor on both monocytes and macrophages, and the platelet is engulfed, decreasing platelet counts. IVIg may work by opsonizing red blood cells, which phagocytic cells (monocytes and macrophages) then work to remove from the blood. Since there are many more red blood cells (4.6–6.2 million/mm³ blood) than platelets in the blood, opsonized red cells can block the ability of the RES to clear platelets.

Phagocytic cells of the spleen and of the RES generally possess three classes of Fcγ receptors, which recognize immunoglobulins of the IgG class. FcγRI binds noncomplexed IgG, and FcγRII and FcγRIII bind IgG complexed with antigen. Blocking FcγRI seems to have no effect in ITP, but an antibody that blocks both FcγRIII and FcγRII has been found to increase platelet counts. A study in animals, using erythrocytes (red blood cells) instead of platelets, has shown that administration of a monoclonal antibody which specifically blocks FcγRII and FcγIII within the RES can prevent clearance of IgG-sensitized erythrocytes.

An example of an antigen-specific intravenous IgG preparation (and therefore an IVIg) is anti-D. Anti-D appears to be effective in patients whose red cells are Rh D antigen-positive. Anti-D can react with and opsonize Rh D antigen-positive red blood cells, which then saturate the RES, competing with sensitized autoantibody-coated platelets for RES clearance, thereby increasing the platelet count in ITP patients.

There are studies in which IVIg attenuates ITP independent of the RES blockade, leading to other possible mechanisms being involved such as anti-idiotypic antibodies, cytokine production, or apoptosis (programmed cell death).

Anti-idiotypic antibodies. One alternative mechanism involves the regulatory properties of a subset of antibodies called anti-idiotypic antibodies (antibodies which bind to the antigen-combining region of other antibodies). One of the major targets of the autoantibodies in ITP is the platelet membrane glycoprotein (gp) IIb/IIIa. Therapeutic preparations of IVIg contain antibodies which can interact with, and neutralize the effects of, anti-gpIIb/IIIa autoantibodies, and it has been observed ex vivo that platelet-reactive autoantibodies from patients with ITP can bind IVIg. This neutralizing effect would prevent new platelets from encountering anti-gpIIb/IIIa autoantibodies, which would result in a decrease in the amount of antibody attached to the platelets, a reduction in platelet RES sequestration and destruction, and reversal of the thrombocytopenia.

Cytokine regulation. IVIg has also been demonstrated to have effects on the cellular immune response itself. Specifically, long-term responses following IVIg administration are associated with enhanced regulatory T-lymphocyte function and decreased autoantibody production. In one study, IVIg was shown to reduce the number of CD4+ T-helper cells in vivo in some patients. IVIg has been shown to be capable of inducing immune tolerance (a state of unresponsiveness to the autoantibody) in both B cells

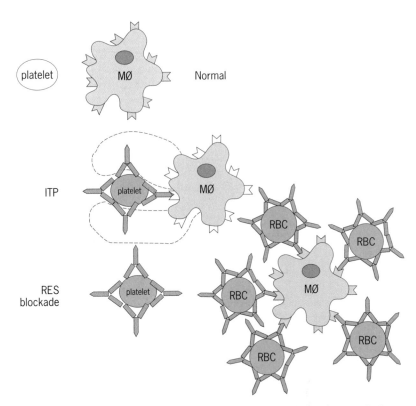

Blockade of the reticuloendothelial system (RES). In normal blood, there is no contact between platelets and macrophages (MØ), which are part of the RES. In ITP, the platelet is coated (opsonized) with autoantibodies, which are recognized by Fc receptors on the macrophages. The coated platelets are engulfed (broken line) by the macrophages, decreasing platelet counts. IVIg may work by coating red blood cells (RBC), which greatly outnumber platelets in the blood, blocking the ability of the RES to clear the platelets.

and T cells. The mechanism(s) of how IVIg exerts its regulatory functions on T cells has not yet been definitively established, but IVIg has been shown to affect both cytokine and cytokine receptor levels in vitro and in vivo by suppressing monocyte functions. IVIg infusions have been associated with favorable changes in cytokine network patterns, leading to the proposal that some of the beneficial effect of IVIg may be indirect, through pathways involving cytokine regulation. One of the in-vitro effects of IVIg is the growth arrest of many cytokine-secreting cells, such as fibroblasts (connective tissue cells), hematopoietic cells (which give rise to blood cells), lymphocytes, and endothelial cells.

Apoptosis. IVIg has also been thought to work by causing apoptosis via a Fas (CD95)–dependent effect. Recent studies suggest tight control of apoptosis by various gene products, and have reported abnormalities of a cell-surface protein receptor known as Fas (which is able to trigger apoptosis when bound to its ligand, FasL) and apoptosis in some autoimmune diseases, specifically in systemic lupus erythematosus, an inflammatory connective tissue disease that may affect many organ systems, including the skin, joints, and internal organs. Defects in the Fas gene resulted in autoimmunity in lupus-prone mice, and Fas-related apoptosis occurred in peripheral blood lymphocytes from patients with systemic lupus erythematosus. Furthermore, recent studies have demonstrated that sFas (soluble Fas) is increased in sera of lupus patients, and suggest that this increase leads to altered lymphocyte activation or proliferation. Since the immunological abnormalities in ITP are similar to those found in systemic lupus erythmatosus, it is possible that IVIg interacts with Fas on the surface of monocytes and macrophages, causing them to undergo apoptosis, which would allow the platelet concentration to increase.

Conclusion. An increasing body of evidence implicates inflammation in many if not most diseases, especially chronic ones. It may be that IVIg—with its ability to block the phagocytic and inflammatory properties of the RES (IVIg, in attaching to red cells, will keep the monocytes and macrophages from engulfing the platelets without causing these cells to release proinflammatory cytokines)—functions as a universal anti-inflammatory agent. Unfortunately, it is much more expensive than familiar anti-inflammatory drugs, such as aspirin, but it might be safer to use. As more is learned about the mode of action of IVIg, it is likely that drugs mimicking its actions will be developed.

For background information *see* ANTIBODY; ANTIGEN; AUTOIMMUNITY; BLOOD; CELLULAR IMMUNOLOGY; HEMATOLOGIC DISORDERS; IMMUNO-GLOBULIN in the McGraw-Hill Encyclopedia of Science & Technology.   Zoë Cohen; Alison Starkey

Bibliography. A. H. Lazarus, Mechanism of action of IVIG in ITP, *Vox Sanguinis*, 83(Suppl. 1):53–55, 2002; A. H. Lazarus and A. R. Crow, Mechanisms of action of IVIG and anti-D in ITP, *Transfusion Apheresis Sci.*, 28:249–255, 2003; S. Song et al., Monoclonal IgG can ameliorate immune thrombocytopenia in a murine model of ITP: An alternative to IVIG, *Blood*, 101:3708–3713, 2003.

Aviation security

Since the terrorist attacks of September 11, 2001, the United States government has sought to impose measures intended to protect the traveling public.

Air marshals. The Federal Air Marshall Program is an updated version of a program that has been in existence since the 1970s, when hijacking of aircraft to Cuba was a relatively common event. Then it was referred to as the Sky Marshall program. The agents became some of the most highly trained federal officers and were specifically schooled in the firing of weapons in a pressurized airborne environment. The program proved to be ineffective and had been significantly reduced in size prior to September 2001, but was reinvigorated by the Bush administration. The first new officers were already trained federal officers transferring from other government agencies.

Cockpit doors. As mandated by the Aviation and Transportation Security Act, the government also required United States–based passenger airplanes to install reinforced cockpit doors to prevent intruders from gaining access to the flight deck. The entire

United States fleet has been so equipped, costing hundreds of thousands of dollars. New, reinforced and ballistic-resistant doors, composed of a variety of metals and composites, meet even more stringent security standards than those mandated by the act. The mandate covered some 5800 domestic aircraft and the Transportation Security Administration (TSA) reimbursed 58 domestic air carriers for the cost. However, the pilots must exit the cockpit to use the restrooms, and the door is not closed during boarding. It is also questionable whether a pilot in command would permit a passenger or flight attendant to be killed for failure to open the door upon a terrorist demand. Hence, the United States fleet is still vulnerable to unauthorized access. Even so, there are softer targets which terrorists can exploit.

Screening of passengers and carry-on luggage. One such target is carry-on luggage. Congress had legislated that a federal screening force be in place by November 2002. However, efforts to upgrade the quality of the screening force proved more difficult than expected. Since federal assumption of the screening function, accusations of failure to complete required background inspections, cheating on competency tests and theft, and dereliction of duty have plagued the endeavor.

Congress had also legislated the requirement to have all carry-on baggage screened by explosive-detection machines by the end of 2003, but later reworded the law to mandate explosive detection equipment or "other appropriate means of detection," which could include trace-detection equipment, manual searches, and bomb-sniffing dogs. The cost of a government-approved explosive-detection machine exceeds a million dollars per unit, which may account for the approval by the TSA of the use of alternative detection systems.

Explosive detection machines. An explosive-detection machine operates and looks like a medical computerized axial tomography (CAT) scanner. The system first produces an x-ray scan similar to the conventional airport x-ray scanner. An automated inspection algorithm determines the locations within the baggage where the absorption indicates a suspicious area; cross-sectional computerized tomography (CT) slices then determine the size, shape, mass, density, and texture of any suspect object. Dual-energy CT, a theoretically possible although not yet implemented option, would also provide information on the nature of the explosive. If no high-density areas are detected, a single slice through the bag is made to look for any explosives that may not have been seen in the projection scan. Since the CT scan produces true cross-sectional slices, it is able to identify objects that are surrounded by other materials or hidden by innocuous objects. Three-dimensional rendering may also be applied. The machine is programmed to recognize various chemical compositions.

Trace-detection equipment. Trace-detection equipment can analyze a swipe or air sample, detecting and identifying minute traces of substances. Some equipment can access the human convection plume, a natural airflow radiating from the human body, to collect any threatening particles. If a person has explosives strapped to his or her body or has even handled explosives, trace particles will contaminate clothing and register.

The process takes 4 seconds to collect the trace particles and another 8 seconds to analyze them. As the person being inspected stands in the center of an archway, gradually stronger puffs of air come from four surrounding columns, positioned to direct the air from the lower to the upper parts of the body, accelerating the plume at a faster rate than it would naturally rise. The plume is collected in an overhead detector hood, and the collected particles are vaporized. The molecules are either positively or negatively charged, and the resulting ions are pulsed down a drift tube. The equipment measures how fast the ions travel from point to point. This acts as a thumb print of the substance, since each specific type of ion has its own particular travel time. This enables the machine to identify a broad range of organic matter, including explosives and materials associated with chemical and biological weapons.

Neutron-based detection. There is also research on the use of neutron technology to pinpoint chemicals. Pulsed fast neutron analysis (PFNA) technology is based on the detection of signature radiation (gamma rays) induced in material scanned by a beam of neutrons.

Biometric systems. In 2004, the U.S. House of Representatives introduced a bill to establish biometric standards for use at United States airports. Biometric systems are designed to recognize biological features to facilitate identity verification. The individual whose identity is being verified places his or her finger, hand, retina, or face onto or near a scanner and provides data which are compared with a database. The systems can be based on fingerprints (optical scanning of a finger), signature recognition (including measurement of the motion and pressure used in writing the signature), hand geometry (the physical attributes of the hand, such as the length of the figures), speaker verification (utilizing the uniqueness of voice patterns), or the blood vessel pattern of the retina. The only drawback is that medical procedures can now be used to alter physical characteristics.

Cargo screening. The cargo hold remains the most vulnerable part of the aircraft and presents the greatest threat to the passenger. Most of it is loaded unscreened. In addition, many problems revolve around the shipment of known and unknown cargo. Known cargo is generically defined as cargo presented for carriage by a known freight forwarder who has been a customer for over a year. Otherwise, the definition is quite open to interpretation. The United States must soon seek to implement air cargo security programs already in use in Europe, where European Union regulations and ICAO (International Civil Aviation Organization) Annex 17 now mandate a complete air cargo security program. Air cargo agents are investigated and approved and regulated agents conduct security investigations not only of the delivery component (usually truck) but also of the manufacturer. The entire supply chain has a security element authorized by law. In the United States, by contrast, over 4000 unidentified freight forwarders operate virtually on trust. A freight forwarder is a company that is in business for the purpose of accepting and shipping items on commercial airlines. Freight forwarders typically execute all aspects of the shipping process, from handling goods and paperwork to clearing shipments through Customs and making delivery to the end consignee. The airlines are more than willing to accept the business, and the cargo is freely loaded onboard all cargo aircraft as well as passenger aircraft.

Arming pilots. The U.S. Congress has also addressed the concepts of arming pilots and equipping aircraft with missile defense systems. Some experts have indicated that arming pilots will serve as deterrence; others feel that it is more likely to provide a ready-made armory for terrorists. Nonetheless, in the week of September 8, 2002, the TSA trained its first class of Federal Flight Deck Officers (FFDO) at its training facility in Artesia, New Mexico. Following the prototype class, the TSA launched full-scale training for commercial pilots.

Antimissile defenses. The concept of equipping the commercial carrier fleet with antimissile defenses presents an entirely different set of challenges. Officials are addressing the possibility that terrorists could use shoulder-fired missiles to bring down an airliner or attack other forms of transportation. Patrols of perimeters have increased, but facilities remain vulnerable.

An aircraft antimissile defense would be based on a Doppler radar system made up of four antennas at the front of the aircraft, two on the sides, and four at the back. The antennas would be capable of giving 360° of radar coverage around the aircraft. Within seconds of a missile being detected, an onboard computer would release flares, firing at different angles to act as a diversion. The system would be completely automated; that is, there would be no involvement of the pilot or copilot.

The SA-7, or a variant of the shoulder-fired missile, is currently being produced under license in several countries, including China, Egypt, North Korea, and the former Yugoslavia. It can be bought for approximately $5000, and according to *Jane's Defense Weekly* several thousand are in circulation all over the world. Israeli authorities confiscated four SA-7's in 2003 being smuggled on the Lebanese-flagged ship *Santorini*. Hezbollah, a fundamentalist Islamic terrorist group, is also thought to have acquired Stingers from the Afghani Muhajideen, as well as some Chinese-made QW-1's. The threat is real, but the feasibility of equipping aircraft with antimissile systems has a prohibitive cost in conjunction with significant technical challenges relating to the speed required to react (the pilot does not have it) and the fact that commercial planes are not as maneuverable as fighter military aircraft.

The United States economy is dependent on the aviation industry, which clearly constitutes a critical national strategic asset. More focused and time-tested threat and risk assessments should be conducted and changes made to security based on those evaluations.

For background information *see* AIR TRANSPORTATION; AVIATION SECURITY; DOPPLER RADAR in the McGraw-Hill Encyclopedia of Science & Technology.

Kathleen M. Sweet

Bibliography. C. Combs, *Terrorism in the Twenty First Century*, 2d ed., Prentice Hall, Little Saddle River, NJ, 2000; C. E. Simonson and J. R. Spindlove, *Terrorism Today, The Past, The Players, The Future*, Prentice Hall, Upper Saddle River, NJ, 2000; K. M. Sweet, *Terrorism and Airport Security*, Edwin Mellen Press, March 2002; K. M. Sweet, *Aviation and Airport Security: Safety and Terrorism Concerns*, Prentice Hall, Upper Saddle River, NJ, November 2003; J. White, *Terrorism: An Introduction*, 3d ed., Wadsworth, Belmont, CA, 2002.

Biological pest control

Biological control (biocontrol) is the use of living organisms to reduce pest abundance and damage. Biocontrol agents are often consumers of pests; thus herbivores are used to reduce weeds, while predators or parasites are used to reduce insect and other animal pests. Pathogens and competitors that interact with pests are also used in biocontrol. Biologically based pest management includes biocontrol and other techniques, such as the use of pheromones to disrupt mating by pests. Agricultural and forest pests are the main targets of biocontrol, although biocontrol is also used against threats to human health (such as mosquitoes) and the environment (such as certain weeds). Biocontrol may operate as an alternative to pesticides and other control techniques or may be included in an integrated pest management system. The successful practice of biocontrol typically requires knowledge of the ecology of pests and their biocontrol agents and may also involve techniques based on other scientific disciplines, including physiology, animal behavior, genetics, and biochemistry. The three major methods of biocontrol are importation, augmentation, and conservation.

Importation biocontrol. Many pest species are not indigenous to the locations where they are problematic. Some of the earliest scientifically documented biocontrol successes involved controlling such pests by importing species that consume the pest in its indigenous range. Thus, scientists often use the term "classical biocontrol" when referring to the importation of a nonindigenous biocontrol agent to suppress a nonindigenous pest. A notable example of classical biocontrol is the introduction of beneficial insects to control the cottony cushion scale, *Icerya purchasi*. The cottony cushion scale is an insect indigenous to Australia, where it was not considered a pest. However, in 1868 the scale was found in California on citrus trees and was considered a serious pest by 1886.

Importation of predators and parasites, such as the vedalia beetle (*Rodolia cardinalis*), from Australia into California to control the scale began in 1887. Within 2 years of the introduction of the vedalia beetle, the cottony cushion scale was under satisfactory control.

Research on aspects of importation biocontrol continues for many of today's notorious pest insects. For example, between 1999 and 2001, scientists working for the Agricultural Research Service of the U.S. Department of Agriculture studied potential importation biocontrol against the gypsy moth, *Lymantria dispar*; the Russian wheat aphid, *Diuraphis noxia*; the red imported fire ant, *Solenopsis invicta*; the Mediterranean fruit fly, *Ceratitis capitata*; and other pests. Their work includes accurately identifying the target pests and potential biocontrol agents; exploring foreign countries for potential biocontrol agents; quarantining, rearing, and predicting the success of the imported agents; predicting the effects that biocontrol agents will have on nontarget species; and evaluating biocontrol agents after their release. Similar procedures are applied for importation biocontrol of weeds, plant pathogens, and vertebrate pests.

Augmentation biocontrol. Repeated releases of biocontrol agents are referred to as augmentation biocontrol. Both the augmented agents and the target pests may be indigenous or nonindigenous. The released biocontrol agents may be reared in a laboratory or greenhouse setting, or may be relocated from another site where they occur naturally (see **illustration**). Many types of augmentative biocontrol agents are commercially available, including bacteria, nematodes, mites, insects and fish for arthropod control, herbivorous insects for weed control, and fungi and bacteria for control of plant pathogens. A widely used augmentative biocontrol agent is the bacterium *Bacillus thuringiensis* (Bt). Different varieties, or isolates, of Bt cause disease in different types of insects. Bt produces a toxin that aids in killing its host. The Bt toxin can be extracted and used as an insecticide without living bacteria. In addition, the gene that produces the Bt toxin has been inserted into some crop varieties, which causes the plant to produce the toxin and resist pests.

Augmentation biocontrol is increasingly used to control greenhouse pests. Greenhouses offer some practical advantages for augmentation biocontrol relative to open-field habitats. The greenhouse enclosure somewhat restricts the activities of the biocontrol agent to the crop, so higher densities of biocontrol agents can be maintained and the agents are less likely to wander into areas where they will not be useful. It is also easier to make precise targeted releases in greenhouses, and environmental settings can be manipulated to favor biocontrol agents. In Europe, biocontrol is an integral method for managing insect pests on greenhouse tomatoes. For example, aphid pests are controlled with predatory midges in the genus *Aphidoletes* and with *Aphidius* and *Aphelinus* wasps; caterpillar pests are controlled with *Trichogramma* wasps and Bt; and whitefly pests are controlled with *Macrolophus* bugs, *Encarsia* and *Eretmocerus* wasps, and several fungal pathogens.

Conservation biocontrol. The third major method of biocontrol involves modifying environments to preserve, protect, and enhance biocontrol agents that are indigenous or were originally released in the area through importation or augmentation efforts. Conservation biocontrol has been used to control arthropod pests, weeds, and plant pathogens. Methods include improving the crop habitat; providing or preserving refuges or food adjacent to crops; and facilitating the movement of biocontrol agents within fields, between fields, and between fields and refuges.

One of the most effective ways of improving crop habitats for biocontrol agents is to reduce or modify the use of pesticides, as exemplified by insect pest management in rice. Misuse of pesticides in rice kills predators as well as pests, causing resurgent pest problems and further need for pesticides. Limiting pesticide use early in the growing season increases predator populations, which then control pests and greatly reduce the need for subsequent pesticide use. A similar example is found in the increasingly popular practice of organic agriculture, which abstains from using synthetic chemicals, including pesticides. A recent comparison of organic versus conventional tomatoes found similar levels of pest damage and greater abundance and diversity of biocontrol agents on organic farms. Further work is needed to determine whether such results will be obtained in other crops.

Other habitat modifications that can contribute to conservation biocontrol include growing alternative crops or noncrop plants within or adjacent to fields, reducing tillage, and spraying sugar water to supplement the diets of biocontrol agents. Similar to pesticide reduction, these methods often increase the abundance of several potential biocontrol agents rather than a single agent. Potential biocontrol agents are not always compatible with each other. For example, potential biocontrol agents that consume other biocontrol agents can weaken pest control. However, in other cases, enhancing a group of beneficial species still improves pest control even when some interference occurs. Ongoing research is identifying ecological interactions that improve pest control while enhancing multiple species of potential biocontrol agents, as well as mechanisms of interference that can weaken pest control by multiple species.

Advantages and limitations. Depending on the target pest, successful biocontrol reduces pest damage below an economic threshold, improves public health, or benefits the environment. The specific merits and drawbacks of biocontrol compared with other pest control methods, or with taking no action, must be determined on a case-by-case basis. However, certain advantages and limitations of biocontrol occur repeatedly. For example, biocontrol may substitute for using pesticides, thus providing

The convergent ladybird beetle, *Hippodamia convergens*, is a predatory insect indigenous to North America. In northern Idaho, *H. convergens* aggregates in forests to overwinter and then disperses into other habitats during the spring. During the growing season, it can be found consuming pestiferous aphids in crops such as wheat and peas. Dealers of augmentative biocontrol agents sometimes raid overwintering aggregations to collect large numbers of the beetle. However, without the proper environmental cues, *H. convergens* will disperse from areas where it is released. An attempt at conservation biocontrol might include preserving or improving overwintering habitats and making crop habitats more hospitable to ladybird beetles. (*Courtesy of Gary Chang*)

environmental benefits and offering a control option when pests develop resistance to chemical controls.

High economic benefit-to-cost ratios can be obtained from importation biocontrol, in part because successful importation biocontrol agents are self-sustaining. Once a successful importation biocontrol agent is established, few or no additional measures are needed for control. Although most imported biocontrol agents are not successful in controlling pests, the economic benefits from successful programs outweigh the costs of failed programs.

Augmentative biocontrol agents are not intended to be self-sustaining, so releases and their expenses need to be repeated. Expensive greenhouse-reared biocontrol agents may be too costly for certain large-scale, open-field applications. Scientists continue to seek cost-cutting methods for rearing and releasing biocontrol agents.

Nonindigenous species can create environmental problems even when they have been deliberately introduced to an area. Imported biocontrol agents can have negative environmental impacts when they attack nonpest species. Conservation biocontrol methods avoid many of the environmental risks inherent to importation because the biocontrol agents are species that are already in an area. However, relatively few practices have been deliberately adopted with conservation biocontrol as a primary goal. Instead, practices such as organic agriculture have gained popularity for other reasons, and conservation biocontrol may be a secondary benefit. Another limitation of some conservation biocontrol strategies is that growing alternative plants for biocontrol agents may require too much land to be taken away from crop production.

Compatibility between control measures. Researchers continue to work toward integrating the use of different biocontrol agents with each other and with other control options. Pesticides that kill biocontrol agents are particularly difficult to integrate with biocontrol. Recently, selective pesticides have been developed, including some that are biologically based. Although the ecological effects of many selective pesticides are not well documented, selective pesticides are intended to inflict less harm upon nontarget organisms. Selective pesticides may increase biocontrol usage while retaining chemical control as an option.

For background information *see* AGRICULTURE; AGROECOSYSTEM; FOREST PEST CONTROL; INSECT CONTROL, BIOLOGICAL; PLANT PATHOLOGY; PREDATOR-PREY INTERACTIONS in the McGraw-Hill Encyclopedia of Science & Technology.   Gary Chang

Bibliography. P. Barbosa (ed.), *Conservation Biological Control*, Academic Press, San Diego, 1998; G. Gurr and S. Wratten (eds.), *Biological Control: Measures of Success*, Kluwer Academic Publishers, Norwell, MA, 2000; R. L. Metcalf and R. A. Metcalf, *Destructive and Useful Insects: Their Habits and Control*, McGraw-Hill, New York, 1993; R. G. Van Driesche and T. S. Bellows, Jr., *Biological Control*, Chapman & Hall, New York, 1996.

Bioterrorism

In the United States, the Centers for Disease Control and Prevention recently defined bioterrorism as "the intentional or threatened use of viruses, bacteria, fungi, or toxins from living organisms or agents to produce death or disease in humans, animals, and plants."

Historical background. The Black Death epidemic, which swept through Europe, the Near East, and North Africa in the midfourteenth century, was probably the greatest public health disaster in recorded history. It is widely believed that the Black Death, which was probably caused by bubonic plague, reached Europe from the Crimea (a region of the Ukraine) in 1346 as a result of a biological warfare attack. During this attack, the Mongol army hurled plague-infected cadavers into the besieged Crimean city of Caffa (now Feodosija, Ukraine), thereby transmitting the disease to the inhabitants. Fleeing survivors of the siege spread the plague from Caffa to the Mediterranean Basin. Such transmission would have been especially likely at Caffa, where cadavers would have been badly mangled by being hurled, and many

of the defenders probably had cut or abraded hands from coping with the bombardment. The large numbers of cadavers that were involved greatly increased the opportunity for disease transmission. In addition, the Mongol army used their hurling machines to dispose of the bodies of victims, increasing the number of cadavers to the thousands. Caffa was thus the site of the most spectacular incident of biological warfare in history to that point, with the Black Death epidemic as its disastrous consequence—a powerful illustration of the horrific consequences when disease is successfully used as a biological weapon.

More recently, in the New World at the beginning of the nineteenth century, British and American colonists attempted to use smallpox as a biological weapon to weaken their Native American opponents. They gave the Native Americans bed linens used by smallpox victims. However, it is not known whether the contaminated linens were in fact the source of the horrendous toll that smallpox eventually took on the Native Americans or whether frequent exposure of the Native Americans to the British and American colonists infected with smallpox caused the outbreak of smallpox.

At the end of the twentieth century, Iraq used chemical weapons on Iran and on its own citizens, and attempted to develop a biological weapons program. During this same period, the hitherto-unknown Japanese cult Aum Shinrikyo used nerve gas in a Tokyo subway. Currently, terrorist incidents and hoaxes involving toxic or infectious agents have been on the rise. Although the arsenal of the potential terrorist is large, this discussion of bioterrorism focuses only on the biological threats.

Biological weapons defense. Among weapons of mass destruction, biological weapons are more destructive than chemical weapons, including nerve gas. In certain circumstances, biological weapons can be as devastating as a nuclear explosion—a few kilograms of anthrax can kill as many people as a Hiroshima-size nuclear bomb. In 1998, the U.S. government launched the first national effort to create a biological weapons defense. The initiatives include (1) the first-ever production of specialized vaccines and medicines for a national civilian protection stockpile; (2) invigoration of research and development in the science of biodefense; (3) investment of more time and money in genome sequencing, new vaccine research, and new therapeutic research; (4) development of improved detection and diagnostic systems; and (5) preparation of clinical microbiologists and the clinical microbiology laboratory as members of the "first responder" team, which is to respond in a timely manner to acts of bioterrorism.

Biological agents posing greatest threat. The list of biological agents that could pose the greatest public health risk in the event of a bioterrorist attack includes viruses, bacteria, parasites, and toxins (see **table**). The list includes agents that, if acquired and properly disseminated, could become a public health challenge, making it difficult to limit the number of casualties and control damage.

Biological agents associated with bioterrorism or biocrimes		
Type	Traditional biological warfare agents	Agents of biocrimes and bioterrorism
Pathogens	Smallpox virus (variola) Viral encephalitides Viral hemorrhagic fevers *Bacillus anthracis* *Brucella suis* *Coxiella burnetii* *Francisella tularensis* *Yersinia pestis*	*Ascaris suum* *Bacillus anthracis* *Coxiella burnetii* *Giardia lamblia* HIV *Rickettsia prowazekii* (typhus) *Salmonella* spp. *Schistosoma* spp. *Vibrio cholerae* Viral hemorrhagic fevers (Ebola) Yellow fever virus *Yersinia enterocolitica* *Yersinia pestis* (plague)
Toxins	Botulinum Ricin Staphylococcal enterotoxin B	Botulinum Cholera endotoxin Diphtheria toxin Nicotine Ricin Snake toxin Terodotoxin
Anticrop agents	Rice blast Rice stem rust Wheat stem rust	

SOURCE: Adapted from the 2000 *NATO Handbook on the Medical Aspects of NBC Defense Operations*.

Anthrax. The disease anthrax is caused by the bacterium *Bacillus anthracis*. The most severe outcome results from the inhalation of the airborne endospores (the form of anthrax used in the bioterrorism events that occurred in the U.S. Senate Chambers in 2001). Once the spores are inhaled, the fatal systemic illness may result. Anthrax can be prevented by vaccination, but this option is not widely available. Prophylaxis with antimicrobial medications is possible for those who might have been exposed, but this requires prompt recognition of the exposure. Fortunately, person-to-person transmission does not occur.

Botulism. Botulism is caused by ingestion of the toxin produced by the bacterium *Clostridium botulinum*. Botulism is not contagious; however, since mucous membranes in the lungs readily absorb the toxin, it could be aerosolized and used as a weapon. Botulism can probably be prevented by vaccination, but this possibility remains investigational. An antitoxin is also available.

Pneumonic plague. The disease pneumonic plague results from inhalation of the bacterium *Yersinia pestis*. Although no effective vaccine is available, postexposure prophylaxis with antibiotics is possible. Very strict isolation precautions must be used for patients who have pneumonic plague, since the disease is easily transmitted by respiratory droplets from person to person.

Smallpox. Smallpox is caused by the variola virus. Although a vaccine is available, routine vaccinations were stopped over 20 years ago because the natural disease was eradicated from our planet. As is the case with most viral diseases, effective drugs are not available. Special isolation precautions must be used for smallpox, since the virus can be acquired through droplet, airborne, or contact transmission from person to person.

Current trends in counter-bioterrorism. Overall, vaccination has been the single most cost-effective public health intervention that has shielded populations from a dozen serious and sometimes fatal naturally transmitted illnesses. In evaluating the role of vaccines for protecting civilian populations from bioterrorism, new problems arise. Despite the protective efficacy of vaccines against individual microorganisms, the very high costs and the great difficulties involved in vaccinating large populations, along with the broad spectrum of potential agents, make it practically impossible to use vaccines to protect a population against bioterrorism.

In the United States, the Food and Drug Administration has been charged to foster the development of vaccines, drugs, and diagnostic products, food supply safeguards, and other measures needed to respond to bioterrorist threats. Activities of the Food and Drug Administration to counter bioterrorism are as follows:

1. Enhancing the expeditious development and licensure of new vaccines and biological therapeutics through research and review activities—anthrax vaccine and antisera to botulinum, for example.

2. Enhancing the timeliness of application reviews of new drugs and biological products and new and existing products.

3. Participation in the planning and coordination of public health and medical responses to a terrorist attack involving biological or chemical agents.

4. Participation in the development of rapid detection and decontamination of agents of bioterrorism such as *Clostridium botulinum* toxins, *Yersinia pestis*, and *Bacillus anthracis*.

5. Ensuring the safety of regulated foods, drugs, medical devices, and biological products; arranging for seizure and disposal of affected products.

6. Developing techniques for detection of genetic modifications of microorganisms that make them more toxic or more antibiotic- or vaccine-resistant.

7. Rapidly determining a microorganisms's sensitivity to drug therapies.

8. Determining the mechanism of replication and pathogenicity or virulence of identified microorganisms, including elements that can be transferred to other organisms to circumvent detection, prevention, or treatment.

9. Enhancing bioterrorism agent reporting and surveillance capabilities.

The expected outcomes of these activities include safe and effective products to treat or prevent toxicity of biological and chemical agents; methods for the rapid detection, identification, and decontamination of hazardous microorganisms; greater ability to ensure the safety of the food supply; and greater capacity to provide appropriate medical care and a public health response.

Misuse of genomics. Despite the positive social impact of deoxyribonucleic acid (DNA) technology, dangers may be associated with DNA sequencing and analysis and gene cloning, that is, genomics. The genome is the full set of genes present in a cell or a virus. Genomics is the study of the molecular organization, informational content, and gene products encoded by the genome.

The potential to alter an organism genetically raises serious scientific and philosophical questions, many of which have not been adequately addressed. The use of genetic engineering in biological warfare and bioterrorism is a case in point. Although there are international agreements that limit research in this area to biological weapons defense, the knowledge obtained in such research can easily be used in offensive biological warfare. Effective vaccines constructed using DNA technology can protect the attacker's troops and civilian populations. Because it is easy and inexpensive to prepare bacteria capable of producing massive quantities of toxins or to develop particular virulent strains of viral and bacterial pathogens, even small countries and terrorist organizations might acquire biological weapons. Many scientists and nonscientists are also concerned about increases in the military DNA research carried out by the major powers.

Genomics has greatly enhanced our knowledge of genes and how they function, and it promises to improve our lives in many ways. Yet, problems and concerns remain to be resolved. Past scientific advances have sometimes led to unanticipated and unfortunate consequences, such as environmental pollution, nuclear weapons, and biological weapons used in bioterrorism. With prudence and fore-thought we may be able to avoid past mistakes in the use of this new genomic technology.

For background information *see* ANTHRAX; BIO-TECHNOLOGY; BOTULISM; GENETIC ENGINEERING; PLAGUE; SMALLPOX in the McGraw-Hill Encyclopedia of Science & Technology.   John P. Harley

Bibliography. R. M. Atlas, Bioterrorism: From threat to reality, *Annu. Rev. Microbiol.*, 56:167–185, 2002; P. Berche, The threat of smallpox and bioterrorism, *Trends Microbiol.*, 9(1):15–18, 2001; S. M. Block, The growing threat of biological weapons, *Amer. Scientist*, 89:28–37, 2001; R. Casagrande, Technology against terror, *Sci. Amer.*, 287(4):83–87, 2002; D. A. Henderson et al., *Bioterrorism: Guidelines for Medical and Health Management*, AMA Press, Chicago, 2002; J. Milleer, *Germs: Biological Weapons and America's Secret War*, Simon and Schuster, New York, 2001; R. Preston, *The Cobra Event*, Ballantine Books, New York, 1998.

Brain size (genetics)

The size of the brain varies substantially among different species of animals. Across the kingdom of mammals, brain weight (and therefore size) is positively correlated with body weight: in general, for every doubling of the body weight, the brain weight increases by 68%. Primates, porpoises, and dolphins, however, have larger brains than expected for their body sizes, and the human species is one of the most prominent outliers. The human body is only about 20% heavier than that of our closest relative, the chimpanzee, but the human brain is 250% larger. The expansion of the human brain began 2–2.5 million years ago (m.y.a.) when the average weight was 400–450 grams (14–16 oz) and ended about 0.2–0.4 m.y.a. with an average weight of 1350–1450 g (48–51 oz). This represents one of the fastest morphological changes in evolutionary history. It is widely believed that brain expansion in humans set the stage for the emergence of language and other high-order cognitive functions and that it was an adaptation to support the complex social structure that emerged in hominid evolution. Indeed, since large brains are costly to maintain because of their high energy requirements, they are unlikely to have emerged without some adaptive value. Although paleoanthropologists and comparative anatomists have learned a lot about the morphological changes that accompanied human brain expansion, the genetic basis of the expansion has remained elusive. However, in recent years human geneticists, molecular biologists, molecular evolutionists, and anthropologists have joined forces to help this fundamentally human phenomenon reveal its history.

ASPM gene. Genetic studies usually start from analyses of mutants, that is, individuals with abnormal but heritable phenotypes. Humans with exceptionally small heads, a condition known as primary microcephaly, are a well-studied population with a heritable abnormal phenotype. Primary microcephaly is an autosomal recessive genetic disease that occurs in 4–40 children out of every million live births in western countries. It is defined as a head circumference more than three standard deviations smaller than the population-age-corrected mean, but with no associated malfunctions other than mild-to-moderate mental retardation. The reduction in head circumference correlates with a markedly reduced brain size. In fact, because the brain size (weight) of a typical microcephaly patient (430 g) is comparable with that of early hominids, such as the 2.3–3.0-m.y.a. *Australopithecus africanus* (420 g) of which "Lucy" is the best-known specimen, it has been hypothesized that the genes mutated in microcephaly patients might have played a role in human brain expansion.

Microcephaly is associated with mutations in at least six genetic loci (*MCPH1-6*), two of which have been molecularly identified. That is, the genes have been identified on the chromosomes and have been sequenced. *MCPH5* is located in chromosome 1 and the identified gene is named *ASPM* (abnormal spindlelike microcephaly associated). About 20 different homozygous mutations that introduce premature stop codons in *ASPM* (and thus shortened proteins) have been found in microcephaly patients, but none are seen in normal individuals. The molecular function of *ASPM* has been studied in several model organisms. In mouse, the *ASPM* gene is highly expressed in the embryonic brain, particularly during growth of the cerebral cortex. The corresponding gene in the fruit fly, *asp*, is involved in organizing and binding together microtubules at the spindle poles and in forming the central mitotic spindle during cell division. Mutations in *asp* cause dividing neuroblasts to arrest in metaphase, resulting in reduced central nervous system development. Thus, the basic physiological role of *ASPM* appears to be conserved across animals. It affects brain size by regulating somatic cell division (mitosis) during brain development.

The hypothesis that *ASPM* may have contributed to human brain expansion can be tested using a molecular evolutionary approach. Indications of positive Darwinian selection on *ASPM* during human evolution would strongly support this hypothesis, because the action of positive selection suggests a modification in gene function that results in elevated organismal fitness. Analysis has been primarily based on a comparison of the rates of synonymous (d_S) and nonsynonymous (d_N) nucleotide changes in gene evolution. Synonymous nucleotide changes do not alter the coding for amino acids in a protein sequence, do not affect protein function, and are therefore selectively neutral. Nonsynonymous changes alter the protein sequence and may be subject to positive or negative selection. Positive selection for nonsynonymous changes may lead to an increase in the proportion of nonsynonymous relative to synonymous changes ($d_N/d_S > 1$), while negative selection against nonsynonymous changes results in $d_N/d_S < 1$. Analysis of the *ASPM* gene sequences from the human, chimpanzee, orangutan, and mouse showed that d_N/d_S is slightly higher than 1 in the human lineage after the human-chimpanzee split but lower than 1 in all other lineages. This result could be interpreted in two drastically different ways. It might indicate that *ASPM* has become unimportant in humans and that there is no longer selection against nonsynonymous changes, leading to a d_N/d_S ratio of ~1. Alternatively, it could indicate the presence of negative selection at some sites of the gene and of positive selection at other sites, resulting in an average d_N/d_S ratio of ~1 for the entire gene.

Four lines of evidence argue against the former hypothesis and support the latter. First, there is tremendous variation in d_N/d_S across different regions of the *ASPM* gene, supporting the action of both negative and positive selection and contradicting the hypothesis of no selection. Second, *ASPM* mutations produce a severe microcephalic phenotype, indicating that the gene is important and under selective constraints. Third, human population genetic surveys have shown that the d_N/d_S ratio is significantly lower for common polymorphisms than for rare polymorphisms. This is because many nonsynonymous mutations are slightly deleterious and negative selection prevents them from reaching high frequencies in a population. This observation provides convincing evidence of the presence of negative selection on *ASPM* in current human populations. Finally, the d_N/d_S ratio for the *ASPM* nucleotide changes that have been fixed in hominid evolution (that is, are shared by all individuals) significantly exceeds the ratio for common polymorphisms. This comparison suggests that, relative to a background of largely neutral common polymorphisms, positive selection has led to the fixation of relatively more nonsynonymous than synonymous changes. It is estimated that about 12 adaptive amino acid substitutions in *ASPM* have occurred in humans since they evolved away from chimpanzees 6–7 m.y.a. We can estimate that the positive selection on *ASPM* occurred at least 100,000–200,000 years ago; however, when highly favorable mutations sweep through a population, genetic traces of the event do not persist forever, so a more accurate date is difficult to estimate.

Subsequent studies of *ASPM* in more species of primates have supported the above findings and revealed additional evolutionary lineages in which *ASPM* may have been subject to positive selection. In addition, it is interesting to note that *FOXP2*, the first gene shown to be associated with human speech and language development, was also found to undergo adaptive changes within the last 100,000–200,000 years. It is tempting to suggest that the brain expansion brought about by adaptive evolution of *ASPM* and other genes provided the necessary hardware for the origin of human language.

Microcephalin. *Microcephalin* is the second gene (*MCPH1*) shown to be responsible for some cases of primary microcephaly. This gene, located on chromosome 8, is expressed in the developing cerebral cortex of the fetal brain, but its molecular function is unknown. A homozygous mutation that generates a premature stop codon was found to cosegregate with microcephaly in two families who had previously been determined via linkage analysis to have a mutant gene in the *MCPH1* region. Applying a molecular evolutionary approach similar to that used in the studies of *ASPM*, two groups studied the evolution of *microcephalin* in humans and other primates. The investigators found that the d_N/d_S ratio was ~0.5 in the human lineage and was not higher than the average among several other primates examined. Interestingly, the highest d_N/d_S (>1) was found in the ancestral lineages to apes (great apes and gibbons) and great apes (humans, chimpanzees, gorillas, and orangutans). These evolutionary patterns are more complex and difficult to interpret than those for *ASPM*. One hypothesis is that *microcephalin* played a more important role in brain-size variation in the evolution of nonhuman apes than in that of hominids.

Other genes. In addition to *ASPM* and *microcephalin*, several other genes have been implicated in the control of brain size and brain expansion. One of these genes, *ARFGEF2* (ADP-ribosylation factor guanine nucleotide-exchange factor 2), is involved in vesicle trafficking during neural cell progenitor proliferation and migration, and mutations in this gene have been implicated in a condition known as autosomal recessive periventricular heterotopia with microcephaly. This condition, also marked by the small head size seen in primary microcephaly, causes seizures and is thought to be due to the failure of neurons to migrate to the surface of the brain during fetal development. The evolution of *ARFGEF2* has not been studied. Another gene hypothesized to play an important role in human brain expansion is *MYH16*, a myosin gene expressed in masticatory (chewing) muscles. Approximately 2.4 million years ago, a two-nucleotide deletion in this gene led to its inactivation in the hominid lineage. It has been proposed that the loss of MYH16 function led to a substantial reduction in jaw size, removing the constraint on brain size and allowing the brain to expand. Although the estimated date of this mutation appears to coincide with that of human brain expansion, the association between jaw size and brain size still needs to be firmly established. After all, Neanderthals had big brains as well as big jaws.

Outlook. Brain size is apparently a complex trait controlled by many genes, but this does not preclude the presence of some genes that have major effects. The detection of positive selection in the *ASPM* and *microcephalin* genes associated with primary microcephaly implies that the fitness gains due to modifications of the two genes were quite large. This strongly suggests that the human brain expansion may have been initiated by functional changes in a small number of genes. The impending publication of the chimpanzee genome sequence and the continuing effort to delineate the molecular genetic architecture of microcephaly will provide new clues to the genetic basis of human brain expansion.

For background information *see* HUMAN GENETICS; MOLECULAR ANTHROPOLOGY; PHYSICAL ANTHROPOLOGY; VERTEBRATE BRAIN (EVOLUTION) in the McGraw-Hill Encyclopedia of Science & Technology.
Jianzhi Zhang

Bibliography. J. Allman, *Evolving Brains*, Freeman, New York, 2000; J. E. Bond et al., ASPM is a major determinant of cerebral cortical size, *Nat. Genet.*, 32:316–320, 2002; P. D. Evans et al., Reconstructing the evolutionary history of microcephalin, a gene controlling human brain size, *Hum. Mol. Genet.*, 13: 1139–1145, 2004; C. G. Woods, Human microcephaly, *Curr. Opin. Neurobiol.*, 14:112–117, 2004; J. Zhang, Evolution of the human ASPM gene, a major determinant of brain size, *Genetics*, 165:2063–2070, 2003.

Catalytic hydroamination

Nitrogen-containing compounds are widespread in nature, are pharmacologically interesting compounds, and are used in the synthesis of fine chemicals. Because of their pervasiveness in the chemical industry, the selective generation of amines remains of fundamental importance to the synthetic organic chemist. Among the standard methods for their synthesis, *N*-alkylation of ammonia, reductive amination of aldehydes or ketones, reduction of amides, and aminations of olefins using stoichiometric promoters are the most common. Although these routes are well developed, they are often not implemented when large amounts of amines are needed. These reactions suffer from the production of vast amounts of by-products created during the process, the elimination of which is costly and environmentally unfriendly.

Industrially, substituted amines are generally produced from alcohols with a solid acid catalyst by the elimination of water. The alcohols used in these reactions are often produced from alkene hydrocarbons. Ideally, a synthetic route directly from the amine and the alkene would eliminate the need for the alcohol intermediate, thereby avoiding the cost and energy consumption required for separation and purification of the alcohol.

The hydroamination of unactivated alkenes and alkynes offers a particularly attractive alternative to the limitations described above. Hydroamination represents a fundamentally simple transformation that involves the formal addition of ammonia or an amine across a carbon-carbon multiple bond. These processes convert inexpensive, readily available starting materials into the desired products in a single reaction without the formation of side products or the need to isolate intermediates. Depending on the reaction conditions, the Markovnikov (branched) or the anti-Markovnikov (linear) product is obtained from the process for alkenes [reaction (1)] and alkynes [reaction (2)]. Although this reaction is

$$R' \equiv\!\!\!\!\!\!\!\!\xrightarrow{\;HNR_2\;}\;\; \underset{\text{Anti-Markovnikov}}{R'\!\!\diagup\!\!\diagdown\!NR_2}\;\; H \;+\;\; \underset{\text{Markovnikov}}{R'\!\!\diagup\!\!\diagdown\!H}\;\; NR_2 \qquad (2)$$

fundamentally simple, progress in this field has met great successes within the past decade.

The addition of amines to alkenes is thermodynamically favorable because the reactions are slightly exothermic or approximately thermoneutral. Kinetically, the energy barrier to achieve the direct addition of amines to carbon-carbon multiple bonds is quite high. To overcome the generally unfavorable characteristics of the reaction, it is necessary to find a catalytic process to overcome the high energy of activation and realize a successful hydroamination process.

Since 2000, over 200 papers on catalytic hydroaminations have been published. Research in this field has provided the organic chemist not only with methods to synthesize simple acyclic amines but also with a means to assemble relatively complex nitrogen-containing molecules.

Modes of activation. Two modes of activation are available to achieve catalytic hydroamination: alkene/alkyne activation and amine activation. Alkene/alkyne activation is generally accomplished with late-transition-metal catalysts, for example, those derived from palladium, platinum, or copper [reaction scheme (3)]. Attack of amines on an unsatu-

$$= \xrightarrow[-L]{+ML_n} L_{n-1}M\!-\!\| \xrightarrow[-H^+]{+HNR_2} \left[L_{n-1}M \diagup\!\!\diagdown NR_2 \right]^-$$

$$\xrightarrow{+H^+,\,L} \diagup\!\!\diagdown NR_2 \;+\; ML_n \qquad (3)$$

rated carbon-carbon bond is facilitated by interaction of the alkene or alkyne with the metal center. The resultant (2-aminoethyl) metal complex is protonated to regenerate the catalytically active transition-metal complex and liberate the free amine.

Activation of the amine moiety can be accomplished in several ways. Typically, late-transition metals can oxidatively add an amine to afford a metal amide hydride, which can undergo reaction of an alkene with the metal-nitrogen or metal-hydrogen bond [reaction scheme (4)].

Actinide and early-transition-metal complexes convert amines to a coordinated metal imide M=NR and enable the reaction of carbon-carbon multiple bonds with the metal-nitrogen bond [reaction scheme (5)]. Additionally, strong bases or strongly electropositive metals such as alkali, alkaline earth, or the lanthanide group elements undergo protonolysis with amines [reaction scheme (6)]. The metal amides thus generated are more nucleophilic and can undergo addition to unsaturated moieties.

$$H\!-\!NR_2 \xrightarrow{+ML_n} L_nM\!\!\diagdown\!\!\overset{H}{\underset{NR_2}{}} \begin{array}{c} +\,= \\ \\ +\,= \end{array} \begin{array}{c} L_nM\!\!\diagup\!\!\overset{H}{\diagdown}\!NR_2 \\ L_nM\!\!\diagdown\!\!\overset{H}{\diagup}\!NR_2 \end{array} \longrightarrow \diagup\!\!\diagdown NR_2 + ML_n \qquad (4)$$

$$H_2NR \xrightarrow[-2R'H]{+L_nMR'_2} L_nM\!=\!NR \xrightarrow{+\equiv} L_nM\overset{R}{\underset{}{\triangle}}N \xrightarrow{+H_2NR} \diagup\!\!\diagdown NR_2 + L_nM\!=\!NR \qquad (5)$$

$$H\!-\!NR_2 \xrightarrow[-R'H]{+MR'} M\!-\!NR_2 \xrightarrow{+\,=} M\diagup\!\!\diagdown NR_2 \xrightarrow{+HNR_2} \diagup\!\!\diagdown NR_2 + MNR_2 \qquad (6)$$

Reactions of alkenes. The origin of catalytic hydroamination chemistry dates back to 1971, when D. Coulson and coworkers disclosed that rhodium trichloride catalyzes the addition of secondary amines to ethylene [reaction (7)]. Since then, progress on the intermolecular hydroamination of olefins has been limited to activated alkenes (for example, vinylarenes or α, β-unsaturated carbonyl compounds). In 1988, researchers at DuPont reported that the addition of aniline to norbornene can be accomplished using an iridium-based catalyst [reaction (8)]. Almost a decade later, A. Togni and coworkers discovered that fluoride ions accelerate the rate of the reaction. In addition, M. Beller reported that the hydroamination of vinyl pyridine with morpholine can be catalyzed by [Rh(COD)$_2$]BF$_4$/2PPh$_3$, where COD = cyclooctadiene [reaction (9)].

$$= \;+\; HN\!\!\bigcirc \xrightarrow[200°C,\,3\,h,\,70\%]{RhCl_3 \cdot 3H_2O \,(cat.)} \diagup\!N\!\!\bigcirc \qquad (7)$$

$$\text{(norbornene)} \;+\; H_2N\!\!\bigcirc \xrightarrow[2\%\,ZnCl_2,\,THF,\,reflux,\,3d]{10\,mol\,\%\,Ir(PEt_3)_3Cl} \text{(norbornyl)}\!-\!NHPh \qquad (8)$$

$$\text{(vinylpyridine)} \;+\; HN\!\!\bigcirc\!O \xrightarrow[THF,\,reflux,\,98\%]{[Rh(COD)_2]BF_4/2PPh_3} \text{(pyridyl)}\!\diagup\!\diagdown\!N\!\!\bigcirc\!O \qquad (9)$$

J. F. Hartwig and coworkers demonstrated that various vinylarenes can undergo palladium-catalyzed hydroamination with aryl amines to afford the Markovnikov amine adducts as the major regioisomer. Reactions of electron rich anilines or electron-poor vinylarenes occurred with the fastest rates and highest yields [reaction (10)].

The anti-Markovnikov addition of dialkylamines to vinylarenes was also accomplished in the Hartwig laboratories [reaction (11)]. The rhodium-catalyzed process constitutes a mild method by which terminal amines can be obtained as the major product in good yields.

The scope of the organolanthanide-catalyzed intermolecular hydroamination of olefins was investigated by T. Marks and coworkers. Organolanthanide metallocenes of the type Cp$_2^*$LnCH(SiMe$_3$)$_2$ (Cp* = Me$_5$C$_5$; Ln = La, Nd, Sm, Lu) and Me$_2$SiCp$_2'$-LnCH(SiMe$_3$)$_2$ (Cp$'$ = Me$_4$C$_5$; Ln = Nd, Sm, Lu) serve as efficient precatalysts for the regioselective intermolecular hydroamination of various alkenes

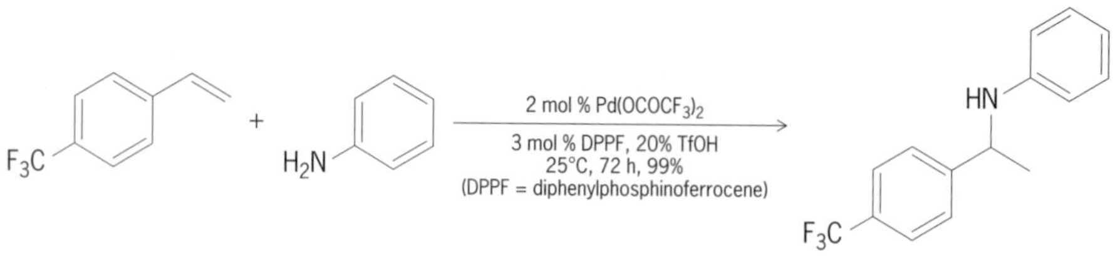

(10)

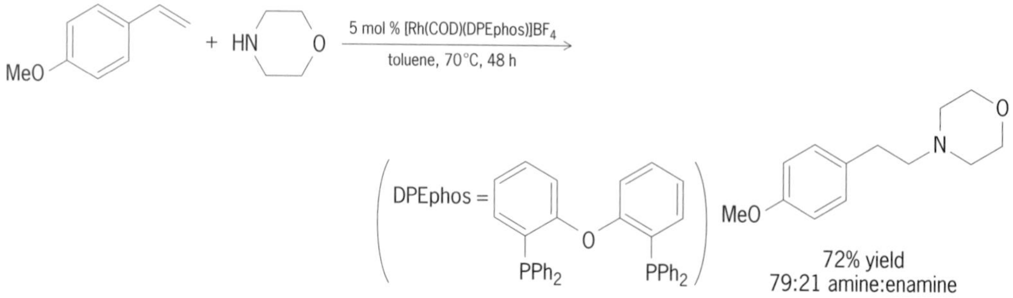

(11)

$$\left(DPEphos = \right)$$

72% yield
79:21 amine:enamine

(Table 1). The hydroamination of 1-pentene, a non-activated olefin, proceeded to give the branched, Markovnikov product (entry 1 in table). In contrast to organolanthanide-catalyzed processes of unfunctionalized alkenes, the hydroamination of polarized olefins, such as vinylsilane, vinylarenes, and 1,3-butadiene, with propylamine provided the terminal amine adducts.

Reactions of alkynes. Although the hydroamination of alkenes is primarily limited to activated alkenes, great progress has been achieved in the case of unactivated alkynes. The hydroamination of alkynes provides access to relatively reactive enamines and imines. As a result, these compounds are often reduced in a subsequent step to provide amine products or are hydrolyzed to afford carbonyl compounds, thus offering high synthetic flexibility.

The catalytic hydroamination of alkynes was first reported by J. Barluenga and coworkers. In a span of 5 years, Tl(OAc)$_3$ and HgCl$_2$ were revealed as ef-

ficient catalysts for the hydroamination of phenylacetylene with various primary and secondary amines. Since then, various catalysts have been developed to avoid the high toxicity associated with the thallium- and mercury-based catalysts. R. G. Bergman and coworkers reported that a zirconium-based catalyst promotes the intermolecular addition of 2,6-dimethylaniline to alkynes and allenes. Unsymmetrically disubstituted alkynes react with 2,6-dimethylaniline with good to moderate regioselectivities [reaction (12)]. In each case, the favored hydroami-

(12)

nation product bears the smaller alkyne substituent α to the nitrogen atom.

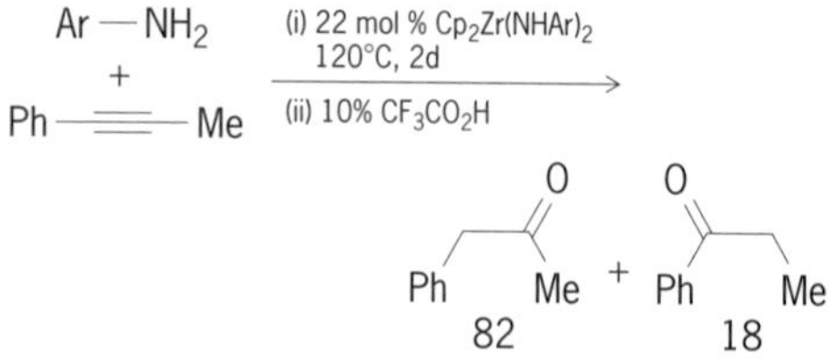

TABLE 1. Organoneodymium-catalyzed intermolecular hydroamination of alkenes*

Substrate $\xrightarrow[\text{propylamine, benzene, 60°C, 3 d}]{\text{1.3 mol \% Me}_2\text{SiCp}'_2\text{NdCH(SiMe}_3)_2}$ product

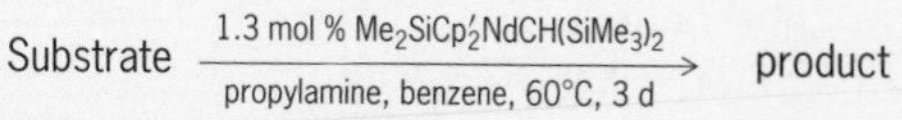

Entry	Substrates	Product	% yield
1		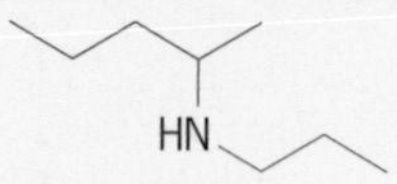	90
2	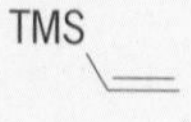	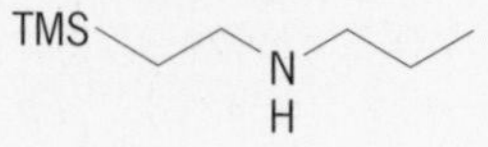	93
3†	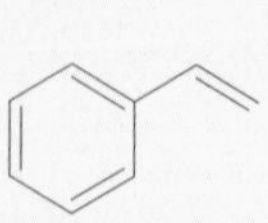	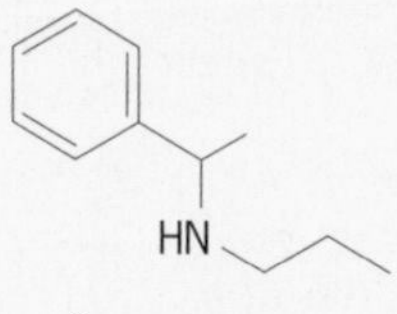	90
4		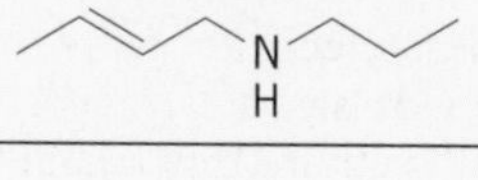	90

*Cp′ = C$_5$Me$_4$.
†Reaction catalyzed by 10 mol % Cp$^*_2$LaCH(SiMe$_3$)$_2$ at 90°C in toluene.

In 1996, M. Eisen and coworkers published the catalyzed hydroamination of terminal alkynes with aliphatic amines using actinide complexes of the type $Cp_2^*AcMe_2$ (Ac = U, Th). Reactions employing the uranium-based catalyst afforded the corresponding aldimines, resulting from anti-Markovnikov addition, in excellent yields from alkyl terminal alkynes [reaction (13)]. Ketimines derived from Markovnikov addition are obtained in low yields (10%) when the analogous thorium based catalyst is used.

Great advances in group 4 chemistry were achieved when S. Doye and coworkers found that Cp_2TiMe_2 (Cp = C_5H_5) is a widely applicable, inexpensive, low-toxicity catalyst for the intermolecular alkyne hydroamination reaction. At elevated temperatures, the addition of hindered alkyl and aryl amines to symmetrical and unsymmetrical alkynes was effected [reaction (14)]. In the case of unsymmetrical alkynes, the reaction occurs with high regioselectivity ($\geq$98:2) to afford the product bearing the smaller substituent α to the nitrogen atom. Optimization efforts revealed that microwave heating dramatically reduced the long reaction times of the Cp_2TiMe_2-catalyzed process. Futher developments in catalyst structure revealed that minor changes in the catalyst structure provided improved yields, higher regioselectivities, and decreased reaction times.

A major limitation to the chemistry developed by Doye is the inability to react with terminal alkynes. A. L. Odom and coworkers developed a complementary process employing $Ti(NMe_2)_4$ to effect the regioselective hydroamination of 1-hexyne [reaction (15)]. Reactions of internal alkynes (for example, diphenylacetylene) are not catalyzed by these complexes under comparable conditions. Despite the numerous titanium-based complexes known to catalyze the hydroamination of alkynes, the highly oxophilic (affinity for oxygen) character of titanium is often incompatible with many oxygen-containing functional groups.

Late-transition-metal complexes offer an alternative to the highly oxophilic titanium complexes. The first ruthenium-catalyzed hydroamination of alkynes was developed by Y. Uchimaru. In a 1999 publication, the hydroamination of phenylacetylene with excess N-methylaniline was realized using 2 mol % $Ru_3(CO)_{12}$. Concurrently, Y. Wakatsuki introduced a $Ru_3(CO)_{12}$/acid catalyst system that allowed the reaction of anilines with terminal phenylacetylenes to give the corresponding imines [reaction (16)]. Small amounts of acid (for example, HBF_4) or ammonium salts (for example, NH_4PF_6) greatly increase the activity of the ruthenium catalyst.

M. Beller and coworkers reported the efficient rhodium-catalyzed reaction of alkyl terminal alkynes with anilines [reaction (17)]. The desired imines were obtained with good regioselectivity and good to excellent yields. Advantageously, the reactions proceed smoothly at room temperature, thus limiting oligomerization and polymerization often observed with other catalysts.

A 2003 report by Y. Yamamoto and coworkers revealed the first palladium-catalyzed intermolecular hydroamination of alkynes with o-aminophenol. Using 15 mol % $Pd(NO_3)_2$, the successful hydroamination of a wide range of alkynes was accomplished, and upon hydrolysis of the reaction mixture, the corresponding carbonyl compounds were obtained in moderate to excellent yields [reaction (18)]. Although poor regioselectivities (1:1 to 3:1) were observed for the hydroamination of unsymmetrical alkynes, only ketone products were obtained for the reaction of terminal alkynes, albeit in moderate yields. Interestingly, the use of o-aminophenol as the amine partner dramatically enhances the reaction rate, affording a complete reaction after 10 hours in comparison to less than 50% yield when variations upon the aniline are made.

The addition of aniline derivatives to aromatic and aliphatic alkynes can also be promoted by (Ph_3P)-$AuCH_3$ in conjunction with acidic promoters to afford the corresponding imines in good to excellent yields. Both internal and terminal alkynes were reactive under the reaction conditions. The reaction proceeds more smoothly when the amine substituent

TABLE 2. Organoneodymium-catalyzed intermolecular hydroamination of alkynes

$$R \!\!-\!\!\!\equiv\!\!\!-\!\! CH_3 \;+\; amine \xrightarrow[\text{benzene, 60°C, 3 d}]{\text{1.3 mol \% Me}_2\text{SiCp}_2'\text{NdCH(SiMe}_3)_2{}^*} product$$

Entry	R	Amine	Product	% yield
1	CH₃	 H₂N⌒⌒	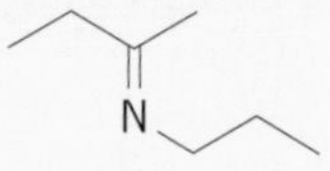	91
2	TMS	 H₂N⌒⌒⌒	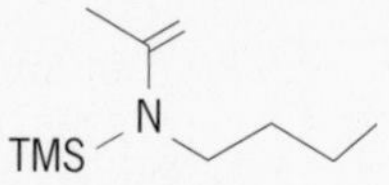	62
3	Ph	 H₂N⌒⌒	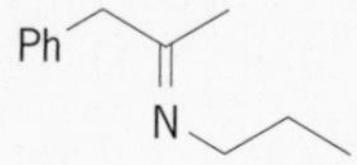	85

*Cp′ = C₅Me₄.

is more electron-withdrawing and the alkyne substituent is more electron-donating [reaction (19)].

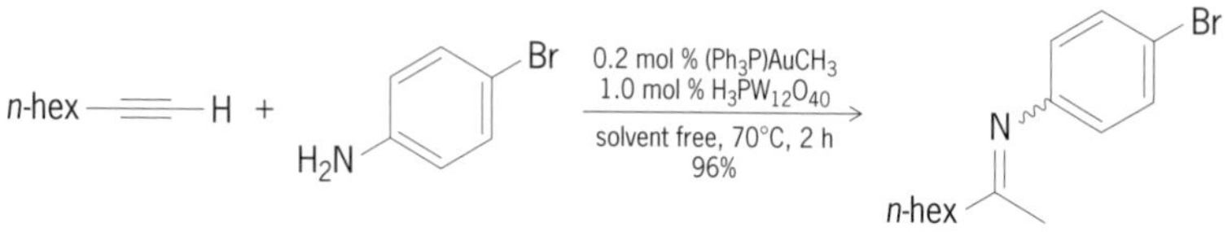

$$(19)$$

Several examples of the regioselective intermolecular hydroamination of aliphatic and aromatic alkynes have also been accomplished utilizing group 3 metallocenes (**Table 2**). As with the hydroamination of α-functionalized olefins, silyl- and aryl-directive effects play an important role in determining the regioselectivity of the reaction.

Intramolecular hydroamination reactions. Of great synthetic utility is the intramolecular cyclization of amine-tethered unsaturated substrates. Intramolecular hydroamination reactions provide a powerful means to construct nitrogen heterocycles with high selectivity. Although various catalyst systems have been developed for hydroaminations, the lanthanide-based catalysts are particularly remarkable because of the extensive body of research in this area.

Early examples of the group 3 metallocene–catalyzed intramolecular hydroamination demonstrated that five-, six-, and seven-membered rings were efficiently created, and a variety of substitution patterns about the amine were tolerated [reaction (20)].

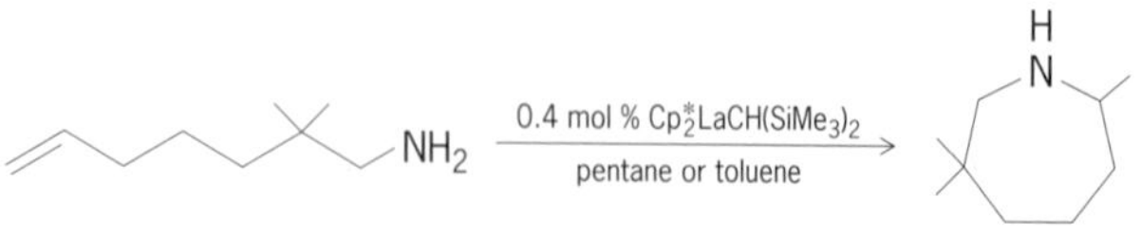

$$(20)$$

More complex heterocycles are possible by making minor changes to the ligand system of the catalyst. Monocyclic amines [reaction (21)], fused

bicyclic systems [reaction (22)], and bridged heterobicycles [reaction (23)] created by hydroamination of 1,1-disubstituted alkenes could all be assembled in excellent yields. The potent anticonvulsant and neuroprotective agent MK-801 was synthesized using this strategy [reaction (23)].

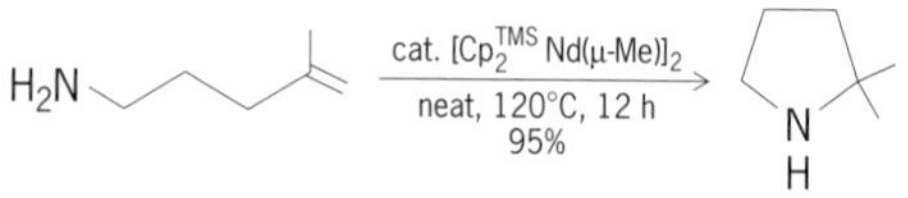

$$(21)$$

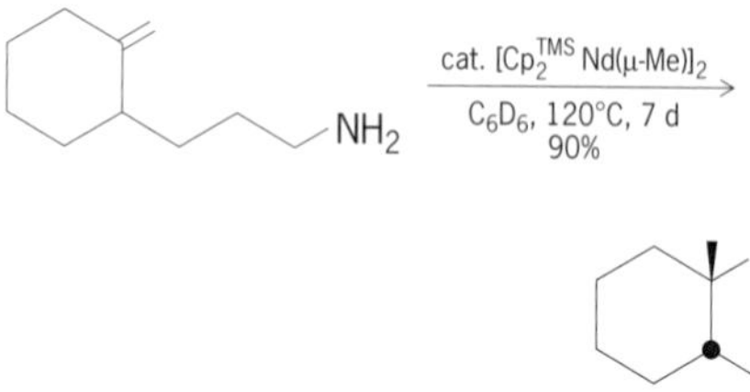

$$(22)$$

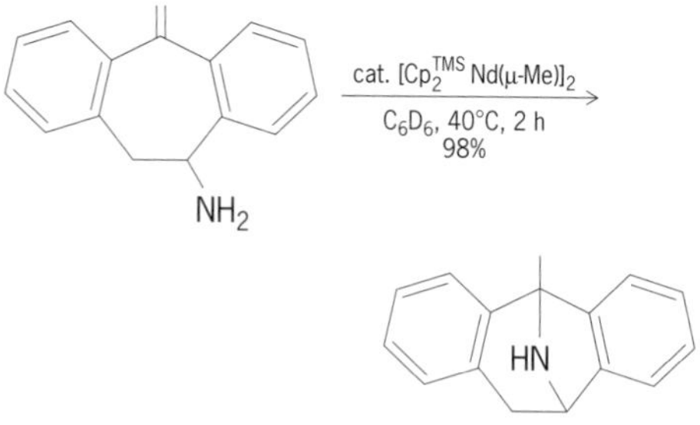

$$(23)$$

MK-801

Recently, unactivated 1,2-disubstituted alkenes have succumbed to cyclohydroamination by organolanthanide catalysts. Although high reaction temperatures and prolonged reaction times were required, pyrrolidines and piperidines with various substitution patterns were constructed in good to excellent yields and high diastereoselectivity [reaction (24)].

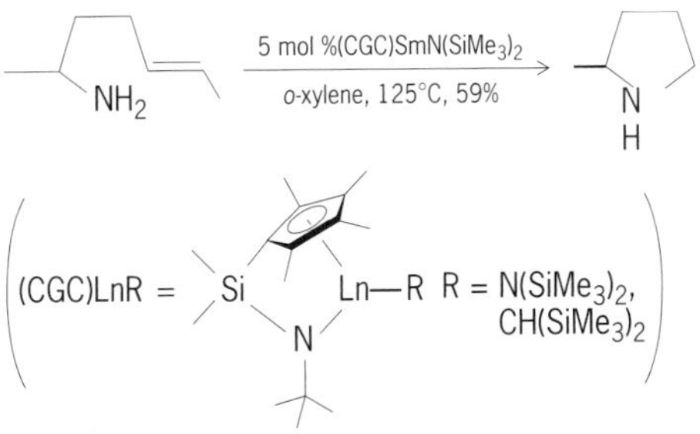

(24)

(25)

(26)

(27)

(28)

R = H, CH₃, TMS, Ph
n = 1, 2, 3

yield = 47–95%

(29)

(30)

(+)-197B

(31)

(+)-Xenovenine

The high diastereoselectivity of the intramolecular hydroamination was demonstrated by G. A. Molander and coworkers. To exemplify the value of the transformation, a total synthesis of the alkaloid (−)-pinidinol was carried out [reaction scheme (25)].

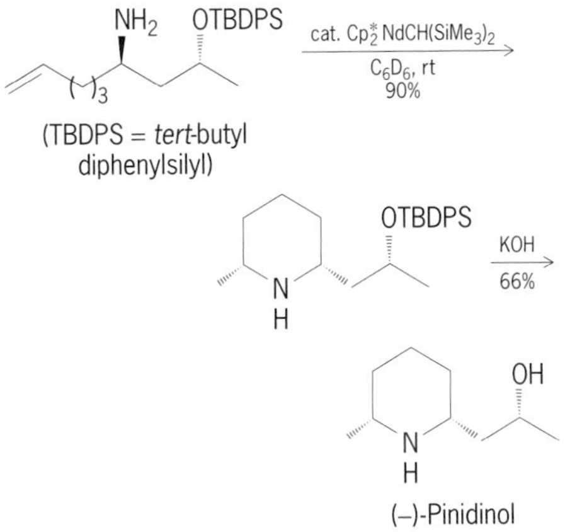

(TBDPS = *tert*-butyl diphenylsilyl)

(−)-Pinidinol

Double hydroamination reactions on dienyl primary amines have been carried out successfully. Using catalytic [Cp₂^{TMS}Nd(μ-Me)]₂, pyrrolizidine scaffolds were realized in good to excellent yields and high diastereoselectivities [reaction (26)]. More reactive constrained-geometry catalysts allowed for the stereoselective construction of the indolizidine skeleton [reaction (27)].

Alkynes also undergo smooth regiospecific intramolecular hydroamination [reaction (28)]. Enamines generated in the cyclization tautomerize to the thermodynamically favored imines at room temperature. Similar to the intramolecular cyclization onto alkenes, five-, six-, and seven-membered rings can be formed. Unlike intermolecular reactions, terminal alkynes readily undergo organolanthanide-catalyzed intramolecular hydroamination.

In addition to isolated alkenes and alkynes, allenes have proven to be excellent substrates for the intramolecular hydroamination process. A variety of interesting and useful pyrrolidines and piperidines can be constructed using this method. High diastereoselectivities at the ring stereocenters, affording *cis*-piperidines [reaction (29)] or *trans*-pyrrolidines [reaction (30)], were observed in all cases, but *Z:E* diastereoselectivity was highly variable. Utilizing more reactive constrained-geometry catalysts, the double hydroamination occurs with facility, providing the nitrogen heterobicyclic system in excellent yield [reaction (31)]. The lanthanocene-catalyzed

hydroamination reaction of aminoallenes has been utilized as the key cyclization step in the construction of the pyrrolidine alkaloid (+)-197B [reaction (30)] and the pyrrolizidine alkaloid (+)-xenovenine [reaction (31)].

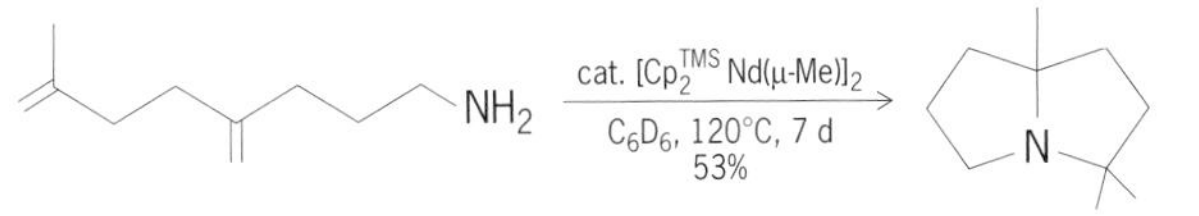

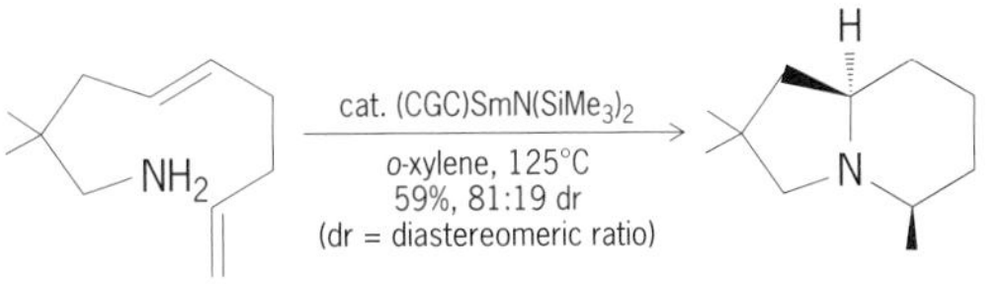

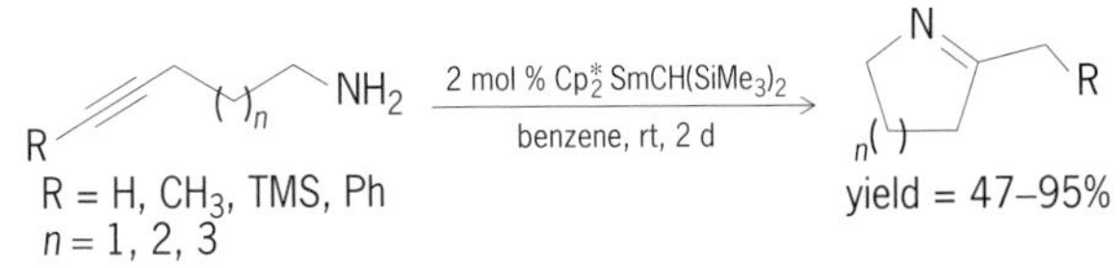

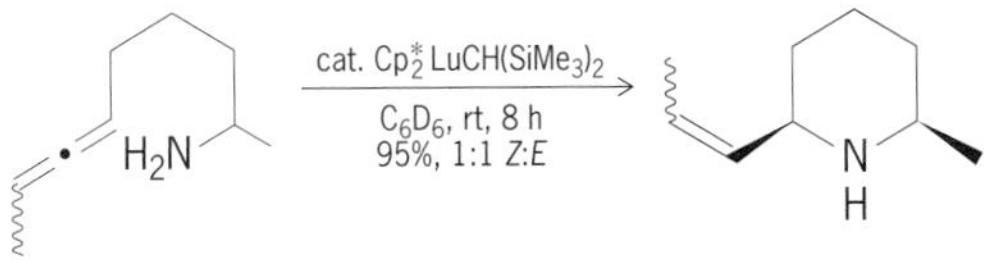

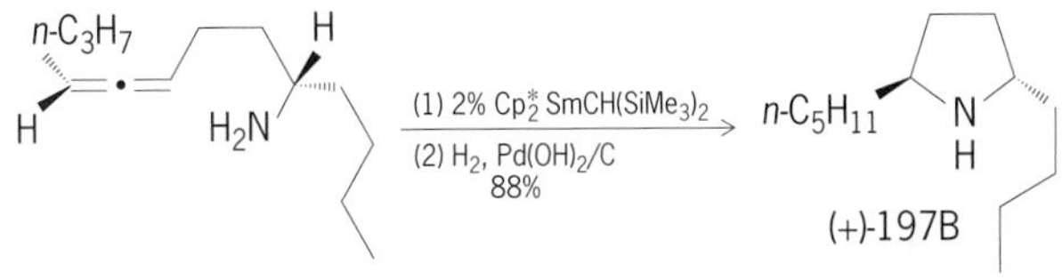

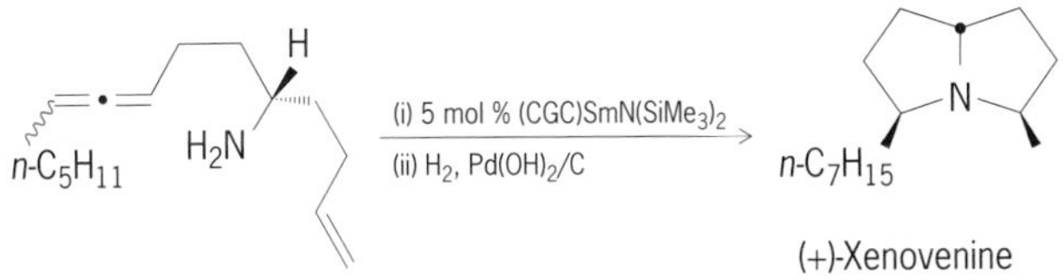

For background information *see* ALKENE; ALKYNE; AMINE; CATALYSIS; ORGANIC SYNTHESIS in the McGraw-Hill Encyclopedia of Science & Technology.

Jan Antoinette C. Romero; Gary A. Molander

Bibliography. I. Bytschkov and S. Doye, Group-IV metal complexes as hydroamination catalysts, *Eur. J. Org. Chem.*, no. 935, 2003; G. A. Molander and J. A. C. Romero, Lanthanocene catalysts in selective organic synthesis, *Chem. Rev.*, 102:2161, 2002; T. E. Müller and M. Beller, Metal-initiated amination of alkenes and alkynes, *Chem. Rev.*, 98:675, 1998; F. Pohlki and S. Doye, The catalytic hydroamination of alkynes, *Chem. Soc. Rev.*, 32:104, 2003; P. W. Roesky and T. E. Müller, Enantioselective catalytic hydroamination of alkenes, *Angew. Chem., Int. Ed. Engl.*, 42:2708, 2003.

Cell membrane sealing

The cell membrane, made up of roughly 70% lipids and 30% protein, serves as an effective barrier to maintain the differences in composition between the intracellular and the extracellular space. The majority of the energy required to sustain cellular function is expended in maintaining large differences in electrolyte ion concentrations across the cell membrane. The lipid bilayer of the membrane serves this barrier role remarkably well by establishing a nonpolar region through which an ion must pass to cross the membrane. The energy required to move a hydrated ion from the aqueous phase into the nonpolar lipid phase is extremely high, thus providing a strong impediment to passive ion diffusion across the lipid bilayer. On the other hand, many of the membrane proteins facilitate and regulate membrane ion transport. The presence of these proteins thus enhances the ability of the mammalian cell membrane to actively transport certain ions.

The mammalian cell membrane lipid bilayer is in essence a two-dimensional structured fluid, held intact only by differential hydration forces. Occasionally, small separations in the lipid packing order occur, producing transient structural defects with lifetimes on the order of nanoseconds. This lifetime is sufficient to permit passage of small solutes, including water. The lifetime and size of these transient pores are influenced by external parameters such as temperature and electric field strength.

Cell membrane damage. Despite its critical role in supporting life, the lipid bilayer is quite fragile compared to other biological macromolecular structures. Many forms of trauma can disrupt the transport barrier function of the cell membrane. Loss of structural integrity occurs in tissues at supraphysiologic temperatures in case of thermal burns, with intense ionizing radiation exposure, in frostbite, in barometric trauma, and with exposure to strong electrical forces in electrical shock or lightning injury. Perhaps exposure to reactive oxygen intermediates (that is, free radicals) is the most common etiology. Resuscitated victims of major trauma often experience ischemia-reperfusion injury (tissue damage due to fluctuating blood supply) that involves the effects of superoxide free radical ($\cdot O_2^-$)-mediated lipid peroxidation on cell membrane integrity. Reactive oxygen-mediated membrane breakdown is the mechanism of high-dose radiation–induced acute necrosis (localized cell death). Under freezing conditions, ice nucleation in the cytoplasm can lead to very destructive mechanical disruption of the membrane. Sudden changes in very strong barometric pressures can lead to acoustic disruption of the cell membrane. Electrical shock is the paradigm for necrosis primarily mediated by membrane permeabilization. Skeletal muscle and nerve tissue exposed to strong electrical fields (greater than 50 V/cm) can experience membrane damage by at least three distinct physiochemical processes: thermal burns secondary to Joule heating, permeabilization of cell membranes, and denaturation of macromolecules such as proteins.

When the bilayer structure is damaged, ion pumps cannot keep pace with the increased diffusion of ions across the membrane. Under these circumstances, the metabolic energy of the cell is quickly exhausted, leading to biochemical arrest and necrosis. Defects formed in the membrane can be stabilized by membrane proteins anchored in the intra- or extracellular space. It has been demonstrated that stable structural defects—"pores" in the range of 0.1 μm—occur in electroporated cell membranes (membranes subjected to pulses of electricity). In other cases, the translateral motion of the lipids, normally restricted by anchored proteins, may cause the membrane to form bubbles as a result of the expansion of electroporated cell membranes.

Surfactant sealing. Sealing of permeabilized cell membranes is an important, naturally occurring process. Fusogenic proteins induce the sealing of porated membranes following exocytosis (release of material from the cell) by creating a low-energy pathway for the flow of phospholipids across the defect or by inducing fusion of transport vesicles to plasma membranes. However, in the case of trauma such as electrical shock and lightning injury, the natural sealing processes can be impeded or too slow to cope with the sudden loss of membrane integrity. An artificial means of membrane sealing is therefore of vital importance for trauma victims.

It turns out that membrane sealing can be accomplished using surfactants. The amphiphilic nature of poloxamer surfactants, a family of surface-active block copolymers of hydrophobic poly(propylene oxide) moieties and capped with hydrophilic poly-(ethylene oxide) moieties on the two ends (**Fig. 1**), enable the surfactants to interact with lipid bilayers and restore the structural integrity of damaged membranes. Poloxamer 188 (P188) has been used widely in medical applications since 1957, mainly as an emulsifier and antisludge agent in blood. Thus, most investigations on the sealing capabilities of synthetic surfactants have focused on P188 due to its already established medical safety record. In 1992, R. C. Lee and coworkers first demonstrated that P188 could seal cells against loss of carboxyfluorescein dye after electroporation. It has since been demonstrated that P188 can also seal membrane pores in skeletal muscle cells after heat shock and enhance the functional recovery of lethally heat-shocked fibroblasts (connective tissue cells). More recently, P188 has been shown to protect embryonic hippocampal neurons against death due to neurotoxic-induced loss of membrane integrity and to reduce the leakage of normally membrane-impermeant calcein dye from

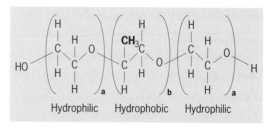

Fig. 1. Chemical structure of poloxamers. The series of different poloxamers is constituted through varying numbers and ratios for a and b.

high-dose irradiated primary isolated skeletal muscle cells. Other surfactants, such as Poloxamine 1107, have been shown to reduce testicular ischemia-reperfusion injury, hemoglobin leakage from erythrocytes after ionizing radiation, and propidium iodine uptake of lymphocytes after high-dose ionizing irradiation. Although the observed sealing actions have been attributed to the interaction of the surfactant with the permeabilized cell membrane, the mechanisms of interaction are only beginning to emerge.

Lipid-poloxamer interaction. Recent research efforts have focused on the interactions between poloxamers and cell membranes at the molecular level. To gain insight into the mechanism of interaction between poloxamer and damaged membrane, model lipid systems such as lipid monolayers, planar supported bilayers, as well as unilamellar vesicles have been used. The advantage of these model systems is that they provide excellent platforms for systematic investigation of the mode of lipid-poloxamer interaction responsible for the sealing phenomena, and help provide answers to the questions: Does P188 seal by selectively inserting into damaged membranes? What is its effect on lipid packing? What is P188's fate when the membrane regains its integrity? Each type of model system has its distinct advantages and shortcomings.

A Langmuir lipid monolayer, a two-dimensional layer of lipid molecules of single-molecular thickness adsorbed at the air-fluid interface, serves as a good model for the outer leaflet of the cell membrane. By controlling the surface area available for the lipid monolayer, the disrupted portion of membrane (with loosely packed lipid molecules), as well as intact membrane (with tightly packed lipid molecules), can be effectively mimicked. Monolayer results show that P188 is a highly surface-active polymer; this high surface activity no doubt aids in its adsorption and facilitates its insertion into lipid films. Experimental results clearly indicate that the poloxamer readily inserts into lipid monolayers when the lipid packing density falls below that of an intact cell membrane, but fails to do so when lipid films have lipid density equivalent to that of a normal membrane. The poloxamer thus selectively inserts into damaged portions of the membrane where the local lipid packing density is reduced, thereby localizing its effect. This observation also suggests that P188 does not nonspecifically interact with membranes that are not damaged. The once-inserted poloxamer, however, is squeezed out of the lipid film when the lipid packing density is increased (**Fig. 2**).

The advent of intense and well-collimated x-ray beams from synchrotron sources has made it possible to use surface x-ray scattering as a molecular probe for direct structural information on the organization of lipid molecules at the air-water interface. X-ray data show that when lipid molecules are loosely packed, P188 inserts into the lipid matrix but remains phase-separated from the lipid molecules.

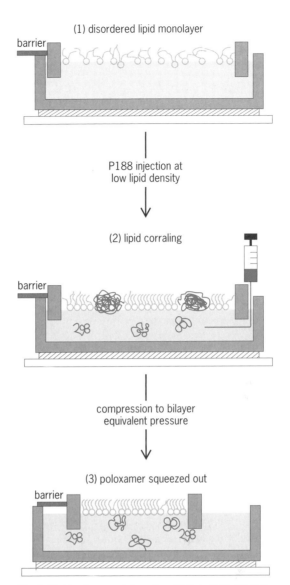

Fig. 2. Lipid-poloxamer interaction. (1) For a lipid with low packing density, the poloxamer readily inserts into it. (2) Upon insertion, the poloxamer is phase-separated from the lipid and thus corrals the lipid molecules to pack more tightly. (3) When the lipid packing density is increased, the poloxamer is squeezed out.

By physically occupying part of the surface area, the poloxamer reduces the amount of space available for the lipid molecules to span. In effect, the poloxamer "corrals" the lipid molecules to pack tightly, a mechanism by which the membrane reestablishes its barrier function. When normal membrane lipid packing density is resumed after the cell has successfully undergone a self-healing process, x-ray data confirms that P188 is squeezed out of the lipid film, indicating that the poloxamer only associates with membranes whose structural integrity has been compromised.

The incapability of P188 to remain in the lipid film at normal membrane lipid packing density can be beneficial in terms of its application as a membrane sealant. For a cell membrane that has been structurally damaged, the cell responds by activating a self-healing process aimed at restoring the integrity of the cell membrane. As the cell heals and the lipid packing of the membrane is restored, these recent data suggest a graceful exit mechanism for the poloxamer from the membrane.

For background information *see* CELL MEMBRANES; CELL PERMEABILITY; COPOLYMER; LIPID; SURFACTANT in the McGraw-Hill Encyclopedia of Science & Technology. Ka Yee C. Lee

Bibliography. R. C. Lee et al., Biophysical injury mechanisms in electrical shock trauma, *Annu. Rev. Biomed. Eng.*, 2:477–509, 2000; R. C. Lee et al., Surfactant-induced sealing of electropermeabilized skeletal muscle membranes in vivo, *Proc. Nat. Acad. Sci. USA*, 89:4524–4528, 1992; J. D. Marks et al., Amphiphilic, tri-block copolymers provide potent, membrane-targeted neuroprotection, *FASEB J.*, DOI: 10.1096/fj.00-0547fje, 2001; S. A. Maskarinec et al., Comparative study of poloxamer insertion into lipid monolayers, *Langmuir*, 19:1809–1815, 2003; S. A. Maskarinec et al., Direct observation of poloxamer 188 insertion into lipid monolayers, *Biophys. J.*, 82: 1453–1459, 2002; G. Wu et al., Lipid corralling and poloxamer squeeze-out in membranes, *Phys. Rev. Lett.*, 93(2):028101, 2004.

Change blindness (psychology)

Most people have a strong belief in the accuracy and completeness of their visual experience. Indeed, it is often said that "seeing is believing," indicating that visual perception is considered to be a trustworthy means of obtaining information about the world, distorting little and missing less. However, research has shown that visual perception does not capture as much of the world as we think. For example, drivers might believe that simply by looking around they would always be able to see an oncoming car, always notice the sudden veering of a nearby child on a bicycle, or always see an animal that suddenly rushes in front of their car. But they would be wrong. Even if viewing conditions were excellent, they could still miss such events—for example, if they were talking on a cell phone.

Much of this shortcoming is based on the existence of change blindness, the inability to notice large changes that occur in clear view of an observer. This effect is extremely robust and can be produced under many different conditions, including instances when the changes are repeatedly made and the observer knows that they will occur. Consequently, change blindness is believed to reflect limitations on the way we see—limitations with important consequences for how much of the world we actually perceive in everyday life.

Basic effect. Change blindness can be produced in a variety of ways. One way is shown in **Fig. 1**. In this example, a picture of a real-world scene is presented to observers for a half second or so, followed by the same picture changed in some way—for example,

an object changed in color or size, or removed altogether. These two pictures then alternate, with a blank field appearing for a fraction of a second before each appearance of each picture. Observers are asked to watch this flickering display and find the change.

Although observers generally believe that they can easily see any change that is large enough, experimental results paint a very different picture. Under these flicker conditions, even large changes can remain unseen for long stretches of time, with observers sometimes requiring 10 or 20 seconds (20 or 40 alternations) before they notice the change, even when that change is extremely easy to see once noticed.

Why do these conditions create change blindness? The key physical factor is believed to be the appearance of the blank fields. Normally, a change produces a motion signal at its location. This motion signal is a change in the intensities or colors anywhere in the retinal image, that is, a change in the pixels in the image. This could be restricted to a small part of the image (for example, if it is a small object moving) or quite a large part (for example, if there is a sudden change in lighting). When a blank field appears, it produces motion signals throughout the display. If these global signals are strong enough, they will overwhelm the local signals and make changes more difficult to notice.

Extent of the effect. Although the flicker conditions of Fig. 1 seem somewhat artificial, counterparts can be found in everyday life. For example, in motion pictures changes are often accidentally introduced during cuts between scenes—such as a sudden closeup or switch of viewpoint—owing to changes that may occur while the lighting and camera are being set up for the new view. (For example, cigarettes keep burning.) A cut in a film acts like a flicker, in that it swamps the local motion signals arising from any such change, making it difficult to notice. Consequently, viewers can miss large changes in the position of actors, changes to their clothing, and sometimes even changes of the actors themselves if these are made during a cut.

A change can also go unnoticed if it is made during an eye movement or an eye blink. Under these conditions, the local motion signals accompanying the change are again overwhelmed by global signals due to the movement of the eye or due to the blink. Such techniques were used by film directors decades ago to deliberately make unnoticed changes in the middle of a scene. For example, a change might be made at the moment a major character appeared at the side of the screen, and thus at the moment the audience moved their eyes. A change might also be made during a sudden loud noise (such as a gunshot), which would cause the audience to blink.

Change blindness can also occur if the change is made while the changing item is occluded by another object. Again, this condition is such that the local motion signals accompanying the change are lost. Change blindness created this way can be found in the real world. For example, observers often fail to notice changes to other people in real-life situations (including changes of conversation partner!) if the change is made at the moment a panel passes in front of the observed person or when that person briefly ducks behind a tall counter.

Visual attention. The key factor in causing change blindness appears to be the effective removal of the local changes that accompany a change, this being accomplished either by overwhelming the local changes with global motion signals or by hiding the local changes altogether. But why should this be? The prevailing explanation is that visual attention is needed to see change. Under normal conditions, the local motion signal created by a change automatically draws attention, allowing it to be seen. But if this signal is swamped by global signals, attention will no longer be automatically drawn to the change. Instead, the viewer must send his or her attention around the scene on an item-by-item basis, until it reaches the item that is changing. Only then will the viewer be able to see the change, regardless of how much time has passed.

This proposed explanation has major implications for our understanding of visual attention. Previously, attention was believed to integrate information only across space—for example, creating the percept of an object from a set of disconnected lines in an image. But it now appears that attention also integrates information across time, allowing that object to be seen to move or change in a dynamic way. The large size of these effects also allows change blindness (using carefully controlled images) to be used as a tool to explore the nature of attention. Among other things, it has been discovered that no more than four or five items—and only a few properties (for example, color or shape) of each—can be attended at a time.

This account can also explain many of the effects created by magicians, in which objects suddenly appear, disappear, or are transformed in ways that seem to defy the laws of physics. But a careful examination shows that many of these effects rely critically on a manipulation of attention. Although many things are in play during a performance, only a few can be attended at any one time. The only events seen to occur are, therefore, those to which the magician has directed attention; all others pass by unnoticed.

Scene perception. If the perception of change requires attention, and if attention is limited to just four or five items, our representation of events in the world cannot be very complete. Why, then, do we have such a strong impression of seeing everything that happens in a scene?

One possibility is that scene perception is based on a "just-in-time" visual information processing system in the brain in which detailed representations of objects and events are created only when requested **(Fig. 2)**. If the allocation of attention were well managed, the appropriate representation would always

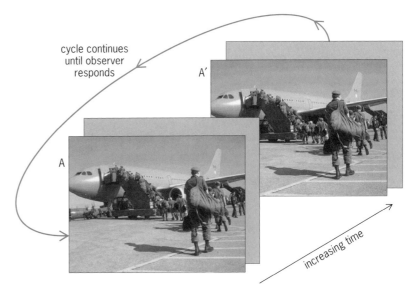

Fig. 1. Flicker technique. The observer views a continuous cycling of images between original picture A and altered picture A′, with blank fields briefly appearing between them. Each image is typically on for about a half second, while the blanks are on for a quarter second or so. (The strength of the effect does not depend greatly upon the exact durations used.) The images alternate until the observer responds that he or she sees the change that is occurring. In this example, the change is the appearance and disappearance of the airplane engine.

be ready, appearing to higher levels of processing (for example, object recognition, thoughts about what to do with the object, and acting on the object) as if all representations were present simultaneously. This scheme is somewhat like that used in lighting a refrigerator: if the light is always on when needed

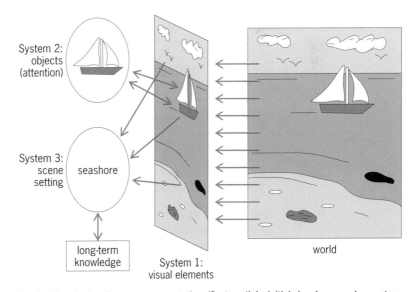

Fig. 2. "Just-in-time" scene representation. (System 1) An initial visual processing system analyzes visual input into a set of simple elements. These elements are very short lived and are continually regenerated as long as light continues to enter the eyes. (System 2) Visual attention "grabs" a subset of these elements and enters them into a more coherent object representation that allows changes to be tracked. At most, only a few such representations exist at any time. After attention releases them, the "held" items return to their original short-lived state. (System 3) At the same time as objects are being formed, other aspects of the scene are rapidly computed without attention, based directly on the visual elements. Together with knowledge in long-term memory, this forms a "setting" capable of guiding attention to those items relevant for the task at hand.

(that is, when the door is open), it will appear as if the light is on all the time.

In this account, the representation of a scene may be much sparser and more dynamic than previously believed, with only a few objects and events represented in detail at any time. Since the choice of which objects and events to represent in detail depends on the goal and the knowledge of the observer, different people could literally see the same scene in different ways. Indeed, studies have shown that experts can perceive a situation differently than novices and that differences can exist between observers from different cultures.

Future research. Although change blindness has already provided a great deal of information about how people see, considerably more can still be done. For example, the relative ease in detecting a change in an object indicates how quickly attention is sent to that object, and thus how important it is to the viewer. Mapping out the relative ease in detecting changes in the objects in a scene can, therefore, provide information about how any individual perceives the world and how this depends on factors such as emotional state or task.

Change blindness can also be used to explore those aspects of perception that are not accessible to conscious awareness. For example, the eyes of an observer can track an object and respond to changes in its position, even when these changes are undetected at a conscious level. Recent studies suggest that observers may do better than chance allows at guessing the existence of changes they do not consciously see. In addition, some observers are able to sense changes before they can see them, which may partly explain the common belief in a sixth sense. Further research of this kind is likely to provide new insights into the nature of perception, and perhaps even the nature of consciousness.

For background information *see* BRAIN; COGNITION; INFORMATION PROCESSING (PSYCHOLOGY); PERCEPTION; VISION in the McGraw-Hill Encyclopedia of Science & Technology.   Ronald A. Rensink

Bibliography. V. Coltheart, *Fleeting Memories: Cognition of Brief Visual Stimuli*, MIT Press, Cambridge, MA, 1999; R. A. Rensink, Change detection, *Annu. Rev. Psychol.*, 53:245–277, 2002; R. A. Rensink, Seeing, sensing, and scrutinizing, *Vision Res.*, 40:1469–1487, 2000; D. J. Simons (ed.), *Change Blindness and Visual Memory: A Special Issue of the Journal of Visual Cognition*, Psychology Press, Hove, U.K., 2000; D. J. Simons and D. T. Levin, Change blindness, *Trends Cognitive Sci.*, 1:261–267, 1997.

Chimpanzee behavior

The behavior of wild chimpanzees (*Pan troglodytes*) is fascinating because of its similarity to human behavior. These behavioral similarities are due in part to our close evolutionary relationship.

Taxonomy. Along with the bonobo, or pygmy chimpanzee (*P. paniscus*), chimpanzees are human-kind's closest living relatives (**Fig. 1**). Current evidence suggests that chimpanzees and bonobos shared a common ancestor with humans about 5–8 million years ago. A third African ape, the gorilla, diverged from the human family tree earlier in time, while the orangutan, an Asian ape, is more distantly related to us. Chimpanzees are distributed across the African continent, ranging from Tanzania in the east to Senegal in the west. Three types or subspecies of chimpanzees in the eastern, central, and western regions of Africa are generally recognized. Recent genetic studies suggest the existence of a fourth subspecies of chimpanzee in West Africa, near the Cross River that forms the boundary between Nigeria and Cameroon.

Ecology. Chimpanzees are highly adaptable and live in several different kinds of habitats, including forests, woodlands, and savannas. Chimpanzees typically rely on a diet composed largely of ripe fruit, although they also eat animals as diverse as insects and monkeys. In their search for seasonally scarce fruit, chimpanzees move over relatively large territories of 5–30 km^2. Because of the large size of these territories, chimpanzee population densities are generally low, rarely exceeding more than five individuals per square kilometer.

Demography. Some unusual demographic characteristics contribute to these low population densities. Female chimpanzees give birth for the first time relatively late in life at an average age of 14 years. They reproduce slowly, having offspring once every 5 to 6 years. During the first 2 years of life, infant mortality is high, ranging up to 30%. Disease, predation, infanticide, and maternal death are factors that contribute to this high rate of infant mortality. Slow reproductive rates correlate with long life spans, however. Chimpanzees in the wild occasionally reach ages of 50 years.

Social organization. Like most other primates, humans included, chimpanzees are social creatures. Chimpanzees live in groups, or communities, that vary in size from 20 to 150 individuals. Members of these communities are never found in a single spot at a single time, but form temporary subgroups, or parties, that change in size and in composition. Male chimpanzees are more social than females. Male chimpanzees frequently associate in parties and cooperate by grooming each other (**Fig. 2**) and by forming coalitions during which two individuals direct aggression toward others. In contrast, female chimpanzees often move alone in the company of only their dependent offspring. The asocial nature of female chimpanzees makes it difficult to study them in the wild, consequently, we possess much more information about male social behavior.

Human behavioral similarities. Behavioral similarities between humans and chimpanzees include tool use, hunting, food sharing, male competition, and territoriality.

Tool use. Tool manufacture and use was one of the earliest and most startling observations of chimpanzee behavior in the wild. Chimpanzees use tools to get difficult-to-obtain foods, to direct aggression toward others, to communicate, to inspect the environment, and to clean their own bodies. Tool use appears to be a universal feature of chimpanzee behavior, but tool kits differ between populations of chimpanzees. For example, in some populations in West Africa, chimpanzees use stones to crack open nuts that are difficult to access. In contrast, East African chimpanzees do not display this behavior, even though the same species of nuts are available to them. In East Africa, chimpanzees modify stems and branches to extract termites from mounds; however, chimpanzees in West Africa do not show this "fishing" behavior in the presence of the same kinds of termites.

These regional variations in tool use are not easily explained by differences in the physical and biotic environments. Some observations suggest that local variations in tool use are learned through social transmission, leading to the proposition that the variations reflect incipient forms of culture as practiced by humans. The best evidence for social learning comes from a recent study that documents the development of termite fishing and points to some fascinating sex differences in behavior. Female chimpanzees fish for termites at an earlier age, and significantly more than males. After acquiring the habit, females are more proficient fishers than males. The significance of these sex differences is currently unclear, but they mirror the proclivity of females to fish for termites more than males in adulthood.

Hunting. Though primarily fruit eaters, chimpanzees are also adept hunters. An additional sex difference exists in hunting behavior, with male chimpanzees engaging in this activity more than females. Other primates, principally red colobus monkeys, are the favored prey of chimpanzees (**Fig. 3**). Chimpanzees pursue these monkeys during seasonal binges, during which hunting takes place virtually every day.

Food sharing. Since meat is a scarce and valuable resource that is prized by chimpanzees, why do they share it so readily with others? Recent field observations indicate that male chimpanzees share meat nonrandomly, doling out pieces to friends and allies. Meat is shared reciprocally, such that if one male gives meat to another, then the latter will likely share with the former. In addition, a male chimpanzee will not only swap meat for itself but also exchange meat for support in coalitions. These observations are consistent with the idea that male chimpanzees use meat as a political tool to curry the favor of other individuals in their communities.

Male competition. There may be good reasons for male chimpanzees to be nice to others. Males compete vigorously for status. They form linear dominance hierarchies, with males often reaching the top of the hierarchy with the coalitionary support of other individuals. Behavioral observations indicate that high-ranking male chimpanzees mate sexually fertile females more often than do lower-ranking individuals,

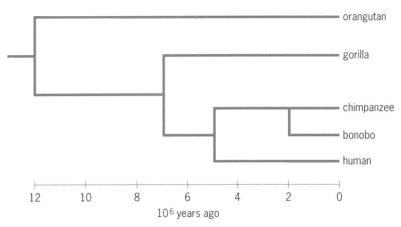

Fig. 1. Evolutionary tree of apes and humans.

and recent genetic analyses reveal that these same high-ranking males produce a disproportionately high number of infants born into any given community.

Despite the reproductive benefits of life at the top, additional physiological studies show that high-ranking male chimpanzees pay some costs. These individuals possess high levels of corticosteroids, the so-called stress hormones. Further research is needed to determine whether this relationship is due to increased psychosocial stress incurred by dominant males, the higher energetic demands placed on them, or both.

Male territoriality. The meat sharing and coalitionary ties that bind male chimpanzees together within communities are put on display in their territorial behavior. Chimpanzees are highly territorial. Male chimpanzees launch gang attacks on members of other communities. These aggressive interactions

Fig. 2. Two male chimpanzees grooming. (*Courtesy of John C. Mitani*)

Fig. 3. Adult male chimpanzee feeding on the body of a red colobus monkey. (*Courtesy of John C. Mitani*)

sometimes escalate to the point at which a chimpanzee falls victim. Lethal intercommunity aggression has been reported from several study sites across the African continent. Boundary patrolling behavior is an integral part of chimpanzee territoriality. In these patrols, male chimpanzees typically gather and move in direct and single-file fashion toward the periphery of their territory. Once there, they begin to seek signs, if not contact, with members of other communities. During these encounters, male chimpanzees communally attack members of other groups, with such episodes resulting in fatalities.

Male chimpanzees occasionally meet a female from another community during patrols. In these instances, infants are sometimes taken from their mothers and killed by the invading males. For reasons not currently understood, male chimpanzees typically cannibalize infants after killing them. Why male chimpanzees kill individuals from other communities is also not well understood. One hypothesis suggests that lethal intercommunity aggression is used to reduce the strength of neighbors. Aggressors are able to expand their territories and improve their feeding conditions, thereby increasing the ability of females in their communities to reproduce. More long-term observations of chimpanzees in the wild will be necessary to validate this intriguing hypothesis.

Conservation. Recent fieldwork continues to reveal astonishing aspects of chimpanzee behavior,

but is threatened by the dwindling populations of these animals in the wild. Low population densities, slow reproduction rates, and encroaching human populations conspire against the preservation of chimpanzees in their natural habitats. Habitat destruction due to human activity is the single largest cause for concern. Commercial activities, such as logging and subsistence farming, contribute to the vast majority of habitat loss. Hunting constitutes a growing problem, especially in West and Central Africa where a "bush meat" trade severely threatens some populations of chimpanzees. Poaching of apes for food continues to accelerate despite laws that protect chimpanzees in host countries.

There are other compelling reasons to stop humans from eating chimpanzees. Chimpanzees harbor a retrovirus similar to the human acquired immune deficiency syndrome (AIDS) virus, and it is now generally agreed that the origin of the AIDS pandemic began with a cross-species transmission event when a human hunter butchered an infected chimpanzee. It is likely that chimpanzees possess other diseases that may jump the species barrier with their continued consumption by humans. A well-documented outbreak of another epidemic involving the Ebola virus has recently decimated large populations of chimpanzees in West and Central Africa.

Threats to chimpanzees continue to grow, and they face a very uncertain future. Unless we intervene to save our closest living relatives and their habitats in the wild, it is likely that they will soon be driven to extinction.

For background information *see* APES; BEHAVIORAL ECOLOGY; PRIMATES; SOCIAL MAMMALS; SOCIOBIOLOGY; TERRITORIALITY in the McGraw-Hill Encyclopedia of Science & Technology.   John C. Mitani

Bibliography. C. Boesch, G. Hohmann, and L. Marchant (eds.), *Behavioral Diversity in Chimpanzees and Bonobos*, Cambridge University Press, 2002; W. McGrew, L. Marchant, and T. Nishida (eds.), *Great Ape Societies*, Cambridge University Press, 1996; J. Mitani, D. Watts, and M. Muller, Recent developments in the study of wild chimpanzee behavior, *Evol. Anthropol.*, 11:9–25, 2002; R. Wrangham et al., *Chimpanzee Cultures*, Harvard University, Cambridge, 1994.

Chinese space program

The flight of China's first astronaut, Yang Liwei, on October 15–16, 2003, answered many questions about China's ability to explore space, and raised new questions about the future of space activities in China and in other countries.

The spaceship *Shenzhou 5* was launched from the Jiuquan Space Center near the edge of the Gobi Desert in northern China (**Fig. 1**). The spacecraft, whose name means "divine vessel," was 8.55 m (28 ft) long and weighed 7.8 metric tons (8.5 short tons), substantially larger than the Russian *Soyuz* space vehicle still in use, and similar in size to the

Fig. 1. *Shenzhou 5* launch vehicle, a Long March (Chang Zheng or CZ) 2F rocket. (*Paolo Ulivi, Zhang Nan*)

National Aeronautics and Space Administration's planned *Constellation* spacecraft (whose final design has not yet been selected).

The first crewed flight of the *Shenzhou* has already had profound political, social, and diplomatic echoes. In addition to gaining international prestige, China hopes that its human space flight program will stimulate advances in the country's aerospace, computer, and electronics industries. Space successes will raise the attractiveness of exports and enhance the credibility of military power.

China's near-term space plans focus on establishing its own space station in Earth orbit. Within a decade, China's space activities may well surpass those of Russia and the European Space Agency, and if China becomes the most important space power after the United States, an entirely new "space race" may begin.

Relation to other space programs. A significant factor in China's success, and a major influence on its future space achievements, is the degree to which its program depends on foreign information. The crewed Chinese spaceship used the same general architecture as both the Russian *Soyuz* and the American *Apollo* vehicles from the 1960s. The cabin for the astronauts, called the command module on Apollo, and the reentry module or descent module by the Chinese (**Fig. 2** and **3**) lies between the section containing rockets, electrical power, and other supporting equipment (the service module) and a second inhabitable module, in front, to support the spacecraft's main function (for the Russians, the orbital module, and for Apollo, the lunar module; **Fig. 4**). However, despite superficial resemblances, the Shen-

zhou is not merely a copy of the Russian *Soyuz*—nor is it entirely independent of the experiences of the Russian and American space programs.

Its service module, for example, has four main engines, whereas *Apollo*'s service module had only one, and *Soyuz* has one main and one backup engine. Also, *Shenzhou*'s large solar arrays generate several times more electrical power than the Russian system.

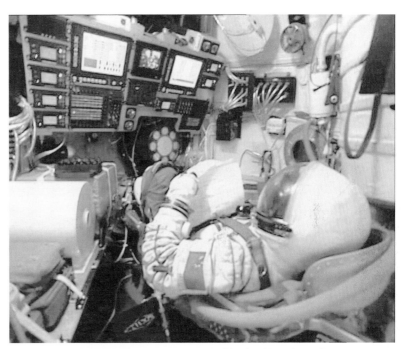

Fig. 2. Interior of the *Shenzhou 5* reentry module. (*Hsinhua News Agency*)

Fig. 3. *Shenzhou 5* capsule (reentry module) after its return to Earth. (*Hsinhua News Agency*)

orbital
module

reentry
module

service
module

Fig. 4. Assembled *Shenzhou 5* spacecraft. (*Brian Harvey*)

And unlike *Soyuz*, the Chinese orbital module carries its own solar panels and independent flight control system, allowing it to continue as a free-flying crewless minilaboratory long after the reentry module has brought the crew back to Earth.

On the other hand, in the case of cabin pressure suits, used to protect the astronauts in case of an air leak during flight (a much more sophisticated suit is used for spacewalks), the Chinese obtained samples of Russia's Sokol design and duplicated it exactly, down to the stitching and color scheme. Other hardware systems that are derived from foreign designs include the ship-to-ship docking mechanism and the escape system that can pull a spacecraft away from a malfunctioning booster during launching.

However, a Chinese team worked out the design of the aerodynamic stabilization flaps of the escape system on their own after Russian experts set too high a price on this unit. This pattern of studying previous work but then designing the actual flight

hardware independently was followed on most other *Shenzhou* systems as well.

Development plans. China's long-range strategy was laid out in a white paper (a guiding policy description) issued in 2000 by the Information Office of the State Council. It stated that the space industry was an integral part of the state's comprehensive development strategy. Instead of developing a wide variety of aerospace technologies, China would focus on specific areas where it could match and then outdo the accomplishments of other nations. Further, China would develop all the different classes of applications satellites that have proven highly profitable and useful in other countries: weather satellites, communications satellites, navigation satellites, recoverable research satellites, and Earth resources observation satellites. It also will launch small scientific research satellites.

Earth orbit plans. In 2002, Chinese press reports stated that after the first crewed space flight China would launch a "cosmic experimental capsule" capable of supporting astronauts' short stays in orbit. This would be followed later by a space station designed for long-term stays.

Several months before the flight of astronaut Liwei, another Chinese space official stated that the second crewed flight would carry two astronauts for a week-long mission. After Liwei's mission, these predictions were repeated, but surprisingly it would be as much as 2 years before the second mission, not the 4 to 6 months that Western experts had expected.

When officials devised a plan for China's space rendezvous and docking experiments, they remembered that United States and Russian tests in the 1960s had required launching two different spaceships—a target and a crewed chaser. The Chinese plans involve detaching the forward section of the *Shenzhou*, pulling away from it, and then returning to it for a docking.

This would be followed by flights by *Shenzhou 7* and *Shenzhou 8*, where China would launch a space station of larger scale with greater experimental capacity. A photograph of what appears to be a mockup of this module resembles the Soviet *Salyut 6* space station (1977–1980), but with a more modern ship-to-ship docking mechanism.

Lunar plans. In the enthusiasm surrounding the Shenzhou program, many Chinese scientists made bold promises to domestic journalists about ambitious future projects, especially concerning the Moon. Western observers also expected major new Chinese space missions. One aerospace "think tank" predicted that China intended to conduct a mission to circumnavigate the Moon in a manner similar to that of the American *Apollo 8* in 1968. And at a trade fair in Germany, spectacular dioramas showed Chinese astronauts driving lunar rovers on the Moon. However, those exhibits seem to be only copies of *Apollo* hardware with flags added. There is little if any credible evidence that such hardware is being designed in China for actual human missions to the Moon.

Real plans are much more modest. China's first lunar mission will be a small orbiting probe called *Chang'e* (the name of a moon fairy in an ancient Chinese fable). Pictures of the probe suggest that it is to be based on the design of the *Dong Feng Hong 3* communications satellite, which has already been launched into a 24-h orbit facing China. This probe is expected to reach the Moon in 2007.

Chinese press reports also describe widespread university research on lunar roving robots, and especially on the robot manipulators (the arm and hand) to be installed on them. These probes, and a long-range plan for an automated sample return mission by 2020, will not be direct copies of previous missions by Russian and American spacecraft. Chinese Moon probes will aim at questions not addressed by previous missions.

More powerful rocket. The key to more ambitious Chinese Moon plans—to the rover mission, for example, or even a flyby of the Moon by a crewed spacecraft—is the development of a new and more powerful booster called the CZ (Chang Zheng or Long March) 5. Comparable to the European Ariane 5 booster, it will not be a simple upgrade of previous vehicles in this series, where more power was obtained by adding side-mounted boosters, stretching the fuel tanks, and installing high-energy upper stages. Those incremental advances have reached their limits, and an entirely new design of large rocket sections and bigger engines must be developed over the next 5 years.

China has stated that it intends to develop this mighty rocket for launching larger applications satellites into 24-h orbits and for launching its small space station. The components are too large to move by rail to the existing inland launching sites, so they will be shipped by sea to an entirely new launch facility on Hainan Island, on China's southern flank.

This new launch vehicle is a quantum jump in the Long March family and presents very formidable engineering challenges. It will take tremendous efforts, significant funding, and some luck to develop the schedule that has been announced.

For background information *see* ROCKET PROPULSION; SPACE FLIGHT; SPACE STATION in the McGraw-Hill Encyclopedia of Science & Technology.

James Oberg

Bibliography. I. Chang, *Thread of the Silkworm*, Basic Books, New York, 1995; B. Harvey, *China's Space Program—From Conception to Manned Spaceflight*, Springer Praxis Books, 2004; J. Johnson-Freese, *The Chinese Space Program—A Mystery Within A Maze*, Krieger Publishing, Malabar, FL, 1998; J. Oberg, China's great leap upward, *Sci. Amer.*, 289(4):76–83, October 2003.

Cirrus clouds and climate

Cirrus are thin, wispy clouds that appear at high altitude and consist of ice crystals. At midlatitudes, clouds with base heights above about 6 km (20,000 ft) are designated as high clouds, a category that includes cirrus (Ci), cirrostratus (Cs), and cirrocumulus (Cc). Cirrus clouds are globally distributed at all latitudes over land or sea at any season of the year. They undergo continuous changes in area coverage, thickness, texture, and position. The most striking cirriform cloud features are produced by weather disturbances in midlatitudes. In the tropics, cirrus clouds are related to outflows from tower cumulus associated with the convective activity over the oceans. The global cirrus cover has been estimated to be about 20–25%, but recent analysis using the satellite infrared channels at the 15-micrometer carbon dioxide (CO_2) band has shown that their occurrence is more than 70% over the tropics.

Cirrus composition. Cirrus clouds usually reside in the upper troposphere, where temperatures are generally colder than -20 to $-30°C$ (-4 to $-22°F$). Because of their high location, direct observation of the composition and structure of cirrus clouds is difficult and requires a high-flying aircraft platform. In the 1980s, comprehensive information about cirrus composition became available because of the development of several airborne instruments to sample their particle-size distribution with optical imaging probes using a laser beam, high-resolution microphotography, and replicators, which preserve cloud particles in chemical solutions.

Ice crystal growth has been the subject of continuous laboratory, field, and theoretical research in the atmospheric sciences discipline over the past 50 years. It is the general understanding now that the shape and size of an ice crystal in cirrus is primarily controlled by the temperature and relative humidity inside the cloud. If ice crystals undergo collision and coalescence due to gravitational pulling and turbulence, more complicated shapes can result. In midlatitudes, where most of the observations have been made, cirrus clouds have been found to be composed of primarily nonspherical ice crystals with shapes ranging from solid and hollow columns to plates, bullet rosettes, and aggregates, with sizes spanning from about ten to thousands of micrometers. Observations in midlatitudes also revealed that at cloud tops pristine small columns and plates are predominant, whereas at the lower part of the cloud bullet rosettes and aggregates are most common (see **illusration**).

Limited measurements from high-flying aircraft in tropical cirrus clouds, which extend as high as 15–18 km (8–11mi), show that their ice crystal sizes range from about 10–2000 μm with four predominant shapes—bullet rosettes, aggregates, hollow columns, and plates—similar to those occurring in midlatitudes. In the tropics, observations reveal that large ice crystal sizes are associated with warmer temperatures or the development stage of clouds related to convection. Ice crystal data in arctic cirrus have also been collected which show their shapes to be a combination of pristine and irregular types with sizes appearing to be larger than about 40 μm. In the Antarctic, the extensive collection of ice

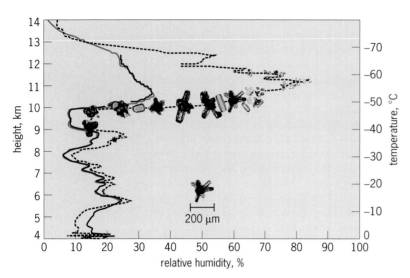

Ice crystal size and shape as a function of height, temperature, and relative humidity captured by a replicator balloon sounding system in Marshall, Colorado, on November 10, 1994. The broken and solid lines denote the relative humidity measured by cryogenic hygrometers and Vaisala RS80 instruments, respectively. (*Graphic by Andrew Heymsfield, National Center for Atmospheric Research. data from K. N. Liou, An Introduction to Atmospheric Radiation, 2d ed., Academic Press, 2002*)

particles at a surface station reveals the prevalence of long, needle-shaped ice crystals.

Ice crystals vary substantially in size and shape from the tropics, to midlatitudes, to the polar regions. In addition to the variety of intricate shapes and a large range of crystal sizes, the horizontal orientation of some cirrus columnar and plate crystals has been observed from a number of lidar backscattering depolarization measurements, as well as limited polarization observations from satellites. Also, the fact that numerous halos (sundogs) and bright arcs surrounding the Sun have been observed demonstrates that a specific orientation of the ice particles must exist in some cirrus. An understanding of the climatic effect of cirrus clouds must begin with a comprehensive understanding of their microscopic composition and associated radiative properties.

Cirrus radiative forcing. The amount of sunlight that cirrus clouds reflect, absorb, and transmit depends on their coverage, position, thickness, and ice crystal size and shape distributions. Cirrus clouds can also reflect and transmit the thermal infrared emitted from the surface and the atmosphere and, at the same time, emit infrared radiation according to the temperature structure within them. The ice crystal size and shape distributions and cloud thickness are fundamental cirrus parameters that determine the relative strength of the solar-albedo (reflecting of sunlight) and infrared-greenhouse (trapping of thermal radiation) effects, which are essential components of the discussion of cirrus clouds and climate. These radiative effects are determined by the basic scattering and absorption properties of the ice crystals. Unlike the scattering of light by spherical water droplets (which can be solved by Lorenz-Mie theory), an exact solution for the scattering of light by nonspherical ice crystals, covering all sizes and shapes that occur in the Earth's atmosphere, does

not exist in practical terms. Recent advances in this area have demonstrated that the scattering and absorption properties of ice crystals of all sizes and shapes, which commonly occur in the atmosphere, can be calculated with high precision by a unified theory for light scattering. This theory combines the geometric optics approach for large particles and the finite-difference time-domain numerical method for small particles. Results of this theory have been used to assist in the remote-sensing and climate-modeling programs involving cirrus clouds.

To comprehend the impact of cirrus clouds on the radiation field of the Earth and the atmosphere and thus climate, the term "cloud radiative forcing" is used to quantify the relative significance of the solar-albedo and infrared-greenhouse effects. Cloud radiative forcing is the difference between the radiative fluxes at the top of the atmosphere in clear and cloudy conditions. The addition of a cloud layer in a clear sky would lead to more sunlight reflected back to space, reducing the amount of solar energy available to the atmosphere and the surface. In contrast, the trapping of atmospheric thermal emission by nonblack (-body) cirrus clouds enhances the radiative energy, or heat, available in the atmosphere and the surface. Based on theoretical calculations, it has been shown that the infrared greenhouse effect for cirrus clouds generally outweighs their solar albedo counterpart, except when the clouds contain very small ice crystals on the order of a few micrometers, which exert a strong solar-albedo effect. The relative significance of the solar-albedo versus infrared-greenhouse effects is clearly dependent on the ice crystal size and the amount of ice in the cloud. Because of the complexity of sorting cirrus signatures from satellite observations, actual data to calculate the global cirrus cloud radiative forcing is not yet available.

Cirrus and greenhouse warming. An issue of cirrus clouds and greenhouse warming produced by the increase in greenhouse gases, such as carbon dioxide (CO_2), methane (CH_4), nitrous oxide (NO_2), chlorofluorocarbons (CFC), and ozone (O_3), is the possible variation in their position and cover. Based on the principles of thermodynamics, the formation of cirrus clouds would move higher in a warmer atmosphere and produce a positive feedback in temperature increase because of the enhanced downward infrared flux from higher clouds. A positive feedback would also be evident if the high cloud cover increased because of greenhouse perturbations. Climate models have illustrated that high clouds that move higher in the atmosphere could exert a positive feedback, amplifying the temperature increase. However, the extent and degree of this feedback and temperature amplification have not been reliably quantified. The prediction of cirrus cloud cover and position based on physical principals is a difficult task, and successful prediction using climate models has been limited. This difficulty is also associated with the uncertainties and limitations of inferring cirrus cloud cover and position from current satellite radiometers. In fact, there is not sufficient cirrus cloud data to correlate with the greenhouse warming that has occurred so far.

Another issue that determines the role that cirrus play in climate and greenhouse warming is related to the variation of ice water content and crystal size in these clouds. Based on aircraft observations, some evidence suggests that there is a distinct correlation between temperature and ice water content and crystal size. An increase in temperature leads to an increase in ice water content. Ice crystals are smaller at colder temperatures and larger at warmer temperatures. The implication of these microphysical relationships for climate is significant. For high cirrus containing primarily nonspherical ice crystals, climate model results suggest that the balance of solar-albedo versus infrared-greenhouse effects, that is, positive or negative feedback, depends not only on ice water content but also on ice crystal size. This competing effect differs from low clouds containing purely water droplets in which a temperature increase in the region of these clouds would result in greater liquid water content and reflect more sunlight, leading to a negative feedback.

Contrail cirrus. In addition to naturally occurring cirrus, the upper-level ice crystal clouds produced by high-flying aircraft, known as contrails, or condensation trails, have also been frequently observed. Contrails are generated behind aircraft flying in sufficiently cold air, where water droplets can form on the soot and sulfuric acid particles emitted from aircraft or on background particles and then freeze to become ice particles. Based on a number of recent field experiments, contrails were found to predominantly consist of bullet rosettes, columns, and plates, with sizes ranging from about 1 to 100 μm. Persistent contrails often develop into more extensive contrails in which the ice supersaturation is generally too low to allow cirrus clouds to form naturally. Consequently, contrails may enhance the extension of the natural cirrus cover in the adjacent areas where the relative humidity is too low for the spontaneous nucleation of ice crystals to occur, although this is an indirect effect that has not yet been quantified.

The climatic effect of contrail cirrus also includes their impact on the water vapor budget in the upper troposphere, which is important in controlling the thermal infrared radiation exchange. It has been estimated that aircraft line-shaped contrails cover about 0.1% of the Earth's surface on an annually averaged basis, but with much higher values in local regions.

An analysis of cirrus cloud cover in Salt Lake City based on surface observations revealed that in the mid-1960s a substantial increase in cirrus clouds coincided with a sharp increase in domestic jet fuel consumption. A similar increase has also been detected at stations in the midwestern and northwestern United States that are located beneath the major upper tropospheric flight paths. Satellite infrared imagery has recently been used to detect contrail cirrus, but long-term observations are needed for assessment purposes. Analysis of contrail cirrus

and radiative forcing indicates that the degree and extent of net warming or cooling would depend on the cloud optical depth (a nondimensional term denoting the attenuation power of a light beam) and the ice crystal sizes and shapes that occur within them. Projections of air traffic show that the direct climatic effects of contrails could be on the same order as some tropospheric aerosol types. It appears that the most significant contrail effect on climate would be through their indirect effect on cirrus cloud formation, a subject requiring further observational and theoretical modeling studies.

Indirect effects. The indirect aerosol-cloud radiative forcing has usually been connected to low clouds containing water droplets via modification of the droplet size and cloud cover/precipitation. Recent analyses of ice cloud data, however, suggest that mineral dust particles transported from Saharan Africa and Asia are effective ice nuclei capable of glaciating supercooled middle clouds. Thus, it appears that major dust storms and perhaps minor eolian (wind) emissions could play an important role in modulating regional and global climatic processes on the formation of cirrus clouds through an indirect effect.

Currently, it is certain that, through greenhouse warming and indirect effects via high-flying aircraft and aerosols, cirrus clouds play a pivotal role in shaping climate and climate change of the Earth and the atmosphere system in connection with the solar-albedo and infrared-greenhouse effects. However, there is not sufficient global data from satellite observations to ascertain the long-term variability of cloud cover, cloud height, and cloud composition to enable the construction of a climate model to assess the impact of their changes in terms of temperature and precipitation perturbations. Moreover, many thin and subvisual cirrus clouds, with an optical depth less than about 0.1, have not been detected by the present satellite radiometers and retrieval techniques. The subject of cirrus clouds and climate is a challenging problem and requires substantial observational and theoretical research and development.

For background information *see* AEROSOL; ALBEDO; ATMOSPHERE; CLIMATOLOGY; CLOUD; CLOUD PHYSICS; GREENHOUSE EFFECT; HALO; HEAT BALANCE, TERRESTRIAL ATMOSPHERIC; INFRARED RADIATION; METEOROLOGICAL OPTICS; RADAR METEOROLOGY; SATELLITE METEOROLOGY; TERRESTRIAL RADIATION in the McGraw-Hill Encyclopedia of Science & Technology.   K. N. Liou

Bibliography. K. K. Liou, Influence of cirrus clouds on weather and climate processes: A global perspective, *Mon. Weather Rev.*, 114:1167–1198, 1986; K. N. Liou, *An Introduction to Atmospheric Radiation*, 2d ed., Chap. 8, Academic Press, 2002; K. Sassen, Saharan dust storms and indirect aerosol effects on clouds: CRYSTAL-FACE results, *Geophys. Res. Lett.*, 30, 10.1029/2003GL017371, 2003; U. Schumann, Contrail cirrus, in D. K. Lynch et al. (eds.), *Cirrus*, Oxford University Press, pp. 231–255, 2002.

Clinical yeast identification

The classification of clinically important yeasts and diagnosis of the role they play in the infectious process are dependent on culture of the organism from clinical material and/or on visualizing structures consistent with their morphology in diseased tissues. However, these methods may provide ambiguous results. Cultures may be positive or negative (that is, the organisms may or may not grow in the culture from a sample). Especially when yeasts are isolated from specimens such as sputum and skin, several problems arise in determining the clinical relevance of these findings. Healthy humans frequently harbor yeasts in the absence of disease as part of the indigenous flora of the oral mucosa; and women, under certain metabolic and physiological conditions, may be culture-positive for yeasts in vaginal swab samples. In addition, the lower gastrointestinal tract serves as a reservoir for yeasts and these organisms may transiently colonize the skin surrounding the tract exit.

During the past two decades, life-threatening opportunistic fungal infections caused by a variety of yeasts have increased in importance, particularly in patients who are immunocompromised for various reasons such as acquired immunodeficiency syndrome (AIDS), those who are undergoing organ transplantation, and those who are being treated for various malignancies. In this clinical setting, the early and accurate diagnosis of invasive disease is of paramount importance. Medical practices that contribute to increased susceptibility to yeasts are the widespread use of broad-spectrum antibiotics and treatments that employ corticosteroids, anticancer drugs, radiation, and indwelling catheters. In addition, chronically bedridden patients who develop ulcers on the skin, those who are obese, and individuals with poorly fitting dentures or poor oral hygiene are subject to infections with yeasts. As a result, the physician, in determining the significance of yeast isolated from clinical material, must decide whether the yeast occurs as a result of infection or simply represents transient colonization.

Classification of yeast. Various laboratory schemes are employed in the classification of yeasts. Historically, the criteria used are the physiological ability of yeasts to assimilate and ferment various carbon compounds. For any yeast isolate that is tested, a pattern of assimilation and/or fermentation will result after a suitable period of incubation. Depending upon the number of different carbon substrates that are employed in the test panel, the resulting pattern of assimilation determines the probability that the tested yeast is identified as a certain genus and species. Additional characteristics—such as production of specific morphologic structures during growth in various media such as serum, formation of distinct pigments utilizing chemically defined substrates, and detection of various enzymes—further aid in identification and classification. Unfortunately the classical methods are cumbersome, must be standardized with controls, are time-consuming to perform,

require careful monitoring, may yield variable results, and often require a high degree of mycological expertise to interpret. To circumvent some of these problems and simplify the method, several yeast identification systems have been fabricated commercially in kit form. These identification systems are, for the most part, based solely on the ability of the test organism to assimilate a variety of different carbon sources along with defined media containing specific substrates. They provide results relatively rapidly and are widely used by clinical microbiology laboratories. However, in comparative studies many of these products lack sensitivity, address only a portion of the identification procedure, and, since they are based on the principle of probability, require supplementary tests in order to provide proper identification. Another problem that limits the utility of these commercial identification kits is that some of them are semiautomated and require specialized equipment to yield the results. Within specified limits they adequately provide the purposes for which they were designed; however, there is a critical need for improvements.

Clinically important species. *Candida albicans* is the species of medically important yeasts most frequently isolated from patients, but *C. parapsilosis*, *C. tropicalis*, *C. krusei*, *C. lusitaniae*, *C. glabrata*, and *C. guilliermondii* have also been implicated as etiologic (disease-causing) agents. Yeasts belonging to the *Candida* genus cause a spectrum of diseases from superficial skin infections to systemic life-threatening disease. On various occasions, less known species of *Candida* have been isolated from patients. These include *C. kefyr*, *C. viswanathii*, *C. utilis*, *C. rugosa*, *C. famata*, *C. holmii*, *C. inconspicua*, *C. haemulonii*, *C. norvensis*, *C. zeylanoides*, and *C. dubliniensis*. The emergence of these species as significant pathogens pose clinical problems in management because some of them appear to be less susceptible to the antifungal medications, such as amphotericin B and the azole derivatives, that are commonly used in treatment of infections. Proper clinical management of patients infected with these yeasts requires that suitable specimens be submitted to the clinical laboratory for culture. All colonies that appear must be confirmed to be yeast by microscopic examination.

In contrast to the various species of *Candida*, the important yeast *Cryptococcus neoformans* is rarely isolated in the absence of disease and is the predominant cause of fungal infections of the brain and meninges. Several distinguishing morphological and physiological features of *C. neoformans* set it apart from other medically important yeasts and are useful characteristics for clinical identification. This yeast possesses an acidic mucopolysaccharide capsule that is readily visualized by microscopic examination using a contrast background such as India Ink. Importantly, it does not ferment carbohydrates. There are four serotypes that can be detected by specific antisera. The yeast elaborates various enzymes, such as urease and phenoloxidase. Most importantly the yeast has a teleomorphic (sexual) phase when mated with a compatible strain and is taxonomically placed in the genus *Filobasidiella*, class Basidiomycetes. There are two varieties, *C. neoformans* var. *neoformans* and var. *gattii*.

Nonculture diagnostic procedures. In many clinical situations the early diagnosis of invasive yeast infection is difficult because cultures are negative and the patient's clinical situation requires a presumptive diagnosis, supported by sound microbiological evidence, so that therapy can be instituted promptly. A number of experimental approaches that do not require cultures have been undertaken to develop novel methods for diagnosing infections caused by these organisms. They include the use of monoclonal antibodies to detect cellular breakdown products of the yeast in body fluids of patients. Such products include mannans, $1,3-\beta$-glucans, chitin, unique cytoplasm enzymes such as enolase, and heat shock proteins. While progress has been made in the use of such methods, for the most part they have limitations of sensitivity and specificity.

Clinically relevant tests based on deoxyribonucleic acid (DNA) technology have the potential of reducing the time needed to perform laboratory identification of yeasts. Detection of *Candida* spp. by in-situ hybridization has received little attention to date, but this approach has the potential of providing rapid specific diagnosis. Much effort has been expended in designing oligonucleotide primers optimal for the detection of the fungus. However, when applied to clinical material, detection of DNA of *Candida* has been less sensitive than culture in identifying the yeast. At the present time, detection of *Candida* spp. by polymerase chain reaction (PCR) is a research tool. While these approaches appear logical, there is an inherent problem in interpreting the results due to the ubiquity of *Candida* spp.

For background information *see* CULTURE; FUNGAL INFECTIONS; MEDICAL MYCOLOGY; MOLECULAR PATHOLOGY; YEAST in the McGraw-Hill Encyclopedia of Science & Technology.   G. S. Kobayashi

Bibliography. E. J. Anaissie, M. R. McGinnis, and M. A. Pfaller, *Clinical Mycology*, pp. 212–218, Churchill Livingstone, New York, 2003; N. J. W. Kreger-van Rij, Taxonomy and systematics of yeast, in A. H. Rose and J. S. Harrison (eds.), *The Yeasts*, vol. 1, pp. 5–78, 1969; L. J. Wickerham, *Taxonomy of Yeasts*, Tech. Bull. 1029, U.S. Department of Agriculture, Washington, DC, 1951; D. M. Wolk and G. D. Roberts, Commercial methods for identification and susceptibility testing of fungi, in A. L. Truant (ed.), *Manual of Commercial Methods in Clinical Microbiology*, pp. 225–255, ASM Press, Washington, DC, 2002.

Cloning

Cloning, the asexual creation of a genetic copy, is a capability possessed by plants but not by most animals. Thus, plants generate genetic copies spontaneously, and rooting "cuttings" is widely used by horticulturists to propagate millions of clones annually. In animals, only some lower invertebrates can be cloned by "cutting"; for example, earthworms when bisected will regenerate the missing half, resulting in two whole, genetically identical individuals. However, asexual reproduction and cloning do not normally occur in vertebrates except for the special case of identical twinning. This is despite the fact that individual cells, called blastomeres, within the very early embryo are totipotent; that is, each is capable, if evaluated on its own, of developing into a viable term pregnancy and infant.

A major scientific interest in cloning revolves around the question of whether the hereditary material in the nucleus of each cell remains intact throughout development, regardless of the cell's fate. On a more practical level, the production of genetic copies of mammals could support the rapid improvement of livestock herds by propagation of valuable founder animals, the creation and production of disease models or transgenic animals for biomedical research, and the preservation of the genetic contribution of a particularly valuable animal, even after death. Therapeutic cloning, a variation that involves the isolation of embryonic stem cells, may provide new cell-based medical approaches to the treatment of human diseases or degenerative conditions.

Reproductive cloning. Scientific inquiry into reproductive cloning in animals began with a "fantastical experiment" suggested by Hans Spemann in 1938 that involved the insertion of a nucleus into an ovum bereft of its own genetic material. This experiment was eventually conducted in 1952 by Robert Briggs and Thomas King in an amphibian, the northern leopard frog, and the technology was quickly extended to a number of other lower vertebrates and invertebrates, and eventually to mammals. The first step in mammalian reproductive cloning is removal of the genetic material from an egg by micromanipulation to create an enucleated egg called a cytoplast. Then genetic material from a donor cell is added, in the form of an intact cell or an isolated nucleus, to produce a diploid, reconstructed embryo. The cell cycle of the nuclear donor cell may be temporarily slowed or stopped in advance of nuclear transfer. Development of the nuclear transfer embryo is triggered chemically, and the cloned embryo is subsequently transferred into a host mother in order to establish a pregnancy (see **illustration**).

When the nuclear donor cell originates from an embryo, the process is called embryonic cloning, first reported in cattle and sheep in 1986 and in primates in 1997. When the donor nucleus is derived from a fetus or a juvenile or adult animal and is not a germ cell, the process is called somatic cell cloning. This form of cloning was thought to be impossible in mammals even as recently as 1995, when scientists in Scotland cloned sheep from differentiating embryonic cells. In 1996, the birth of Dolly, a cloned sheep, was the first successful generation of a viable

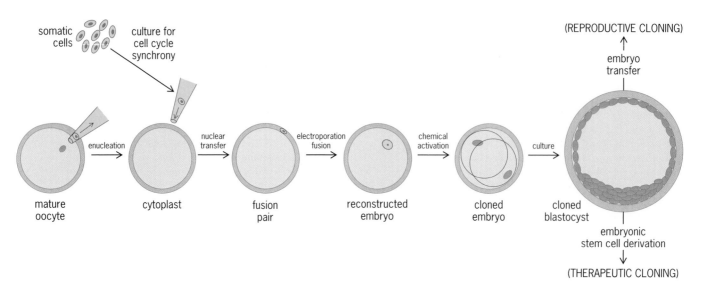

Mammalian somatic cell cloning. The starting requirements include a cohort of mature fertile eggs and a nuclear donor source. Reproductive cloning in mammals involves the transfer of a nucleus or an entire donor cell into an enucleated egg or cytoplast, followed by the subsequent reprogramming of that nucleus by the egg, to the degree that the events of development are recapitulated in the cloned embryo. Therapeutic cloning is accomplished by conducting nuclear transfer with a nuclear donor cell derived from the patient, a skin fibroblast perhaps, transferred into a cytoplast. The resulting nuclear transfer embryo, after culture to the blastocyst stage, would then provide a source of inner-cell-mass cells from which an embryonic stem cell line could be derived.

mammal derived from the transfer of an adult cell (a mammary gland cell).

Successful somatic cell cloning in other species (including the cow, mouse, goat, pig, rabbit, cat, horse, mule, and rat) quickly followed. A wide variety of somatic cells—even those that are unequivocally, terminally differentiated—have been used as nuclear donors. It now appears logical that cloning will be possible in all mammals, given an adequate opportunity to define the unique conditions required by each species.

While dramatic progress has been realized in somatic cell cloning, virtually nothing is known about how the process works, and significant limitations continue to undermine efforts to produce viable offspring. Thus, somatic cell cloning is characterized by high fetal and neonatal wastage, along with age-onset health problems (including obesity, hepatic and immune impairment, and premature death) in surviving clones. It is in fact amazing that somatic cell cloning works at all, given that over the course of normal development as a cell specializes, some genes are silenced temporarily or in some cases permanently. In order for the events of development to be accurately recapitulated following nuclear transfer of a differentiated cell, the donor nucleus must be reprogrammed so that it ceases to express genes of importance to a differentiated or somatic cell in favor of those important to a young embryonic cell; reprogramming is a responsibility of the cytoplast that must occur in a timely manner following nuclear transfer, perhaps along with dramatic remodeling of chromosomal architecture. In addition, nuclear reprogramming must ensure retention of the pattern of expression of imprinted genes, that is, those genes that are differentially expressed depending on their maternal or paternal origin. The inappropriate or abnormal expression of imprinted genes, as regulated by epigenetic changes (factors that alter the expression of genes rather than the genes themselves) such as DNA methylation, can impact placental development and fetal wastage.

Other possible explanations for the losses associated with somatic cell cloning are that clones age prematurely because the nuclear donor cells have shortened telomeres (chromosome tips) and because of heteroplasmy, the presence of more than one type of mitochondrial DNA in a cell. Since mitochondrial DNA is contributed by both the cytoplast and by the injection of a donor cell or nucleus preparation, clones are not exact genetic copies; indeed, they can be phenotypically quite different from one another. Fetal exposure to unique environments is another factor that results in phenotypic differences.

Therapeutic cloning. In therapeutic cloning, embryos are produced by nuclear transfer, as in reproductive cloning, but with the intent of isolating embryonic stem cells rather than transferring cloned embryos into a host to produce a pregnancy. Embryonic stem cells are undifferentiated cells present in the embryo that are pluripotent; that is, they are capable of giving rise to all of the major cell lineages in the adult body but unable to generate a functioning organism. Adult stem cells exist and can be recovered in limited numbers from a host of different tissues or organs. They can, in theory, be used in cell-based therapies to treat human disease, but their low numbers and limited developmental potential are drawbacks. It is now clear that the pre-implantation-stage embryo contains a cohort of stem cells confined to the inner cell mass at the blastocyst stage, and these embryonic stem cells can be harvested, propagated, and established as permanent cell lines. These embryonic stem cells could be stored frozen until needed and then thawed and used directly for transplantation into the patient, or after their fate was directed into the desirable cell type in vitro by exposure to the appropriate chemical signals (see illustration).

Embryonic stem cells are the most primitive, most mitotically capable stem cells, and they hold virtually unlimited promise in the treatment of diseases attributable to the loss or dysfunction of known cell types. This cell-based therapy, as envisioned, would involve the transplantation of embryonic stem-cell-derived progeny that had been induced to differentiate into the missing or dysfunctional cells. Transplantation of allogeneic stem cells (stem cells from another individual of the same species) or their progeny, however, is complicated by the likelihood of host-graft rejection. As an alternative to overcome this problem, the patient could generate his or her own embryonic stem cells by contributing the donor nucleus or cell in somatic cell cloning, thereby creating self-cells transparent to the immune system such that their transplantation would not elicit an immune response. Proof-of-principle experimentation with adult cells in therapeutic cloning has now been reported in both mice and humans.

Future research. An immediate focus of cloning research is to overcome the low success rates with somatic cell cloning in order to extend this technology to other nonhuman species. Undoubtedly, improvements will be realized with additional experience and increased surveillance at birth, with cloned pregnancies treated as high risk. This is already happening with cattle as substantial financial incentive exists in the cattle industry to produce cloned animals, and with mice due to their amenability to laboratory experimentation. Scientists are asking what kinds of cells make the best donors, what environments are most conducive to early development of the cloned embryo, what cell cycle stages are optimal for reprogramming, and how reprogramming can be initiated in the test tube. It seems reckless to extend reproductive cloning to humans at present because of the unacceptable risks imposed on the embryo, fetus, and neonate. Additional ethical considerations have been addressed extensively, for instance, in a United States National Bioethics Advisory Commission report issued in 1997. Therapeutic cloning in humans is less controversial, since it is not associated with efforts to produce cloned children and may represent a viable approach to treat human disease. Clearly, though, ethical concerns do persist with therapeutic cloning, based on the requirement to recruit egg donors in order to create a cloned embryo and then the need to destroy the embryo in the process of deriving embryonic stem cell lines. An alternative to creating individual embryonic stem cell lines by therapeutic cloning may ultimately be the development of stem cell banks that can be used to match the needs of all patients.

While many challenges lie ahead for somatic cell cloning, the mere demonstration that this process is possible in mammals is a highly significant beginning. Ongoing research efforts will lead to a better understanding of the molecular correlates to reprogramming, information that in turn will allow the development of in vitro conditions for reprogramming (for instance, the use of readily available cytoplasm as might be harvested from amphibian eggs). Ultimately, reprogramming will be accomplished with chemically defined reagents and, in the process, we will have developed profound new insights into the events underlying development in mammals, humans included.

For background information *see* CELL DIFFERENTI-ATION; CELL DIVISION; GENE; GENETIC ENGINEERING; SOMATIC CELL GENETICS; STEM CELLS in the McGraw-Hill Encyclopedia of Science & Technology.

Don P. Wolf

Bibliography. M. A. DiBerardino, R. G. McKinnell, and D. P. Wolf, The golden anniversary of cloning: A celebratory essay, *Differentiation*, 71:398–401, 2003; K. Hechedlinger and R. Jaenisch, Monoclonal mice generated by nuclear transfer from mature B and T donor cells, *Nature*, 415:1035–1038, 2002; W. S. Hwang et al., Evidence of a pluripotent human embryonic stem cell line derived from a cloned blastocyst, *Science*, 303:1669–1674, 2004; A. McLaren, Cloning: Pathways to a pluripotent future, *Science*, 288:1775–1780, 2000; National Bioethics Advisory Commission, *Cloning Human Beings: Report and Recommendations of the National Advisory Commission*, National Bioethics Advisory Commission, Rockville, MD, 1997; E. Pennisi and G. Vogel, Clones: A hard act to follow, *Science*, 288:1722–1727, 2000; I. Wilmut et al., Somatic cell nuclear transfer, *Nature*, 419:583–586, 2002.

Collective flux pinning

Type II superconductors exhibit two superconducting phases in applied magnetic fields. The first is a state of perfect diamagnetism (Meissner phase), the same state as is found in type I superconductors, with complete exclusion of the magnetic flux. However, if the value of the applied magnetic field is increased, a transition to a second state develops, in which the magnetic field threads inside the superconductor and is associated with a regular array of supercurrent vortices, a vortex lattice, with each vortex surrounding one quantum of magnetic flux ($\Phi_0 = 2.07 \times 10^{-7}$ gauss cm^2 = 2.07×10^{-15} weber). This state is called the mixed, vortex, or Abrikosov state. In the core of these vortices the superconducting state vanishes and the current carriers are electrons; outside the vortices the sample remains superconducting and the carriers are Cooper pairs. Therefore, if the vortices are pushed and they move, electrical resistance and dissipation will develop in the sample, and superconductivity will be lost. This will, indeed, occur in the presence of a very small current if the vortices are not anchored by some mechanism. Thus, one of the most relevant topics in applied superconductivity is to look for pinning mechanisms that will anchor or pin the vortices and preserve the superconducting state, although the sample is in the mixed state. Ordinarily, pinning is provided through randomly distributed defects in the crystal lattice of the superconductor. New fabrication techniques in the realm of nanotechnology allow the fabrication of nanostructured superconductors with periodic arrays of pinning centers. In the presence of such arrays, the phenomenon of collective flux pinning can occur: The whole vortex lattice can be pinned collectively. This phenomenon offers a promising approach to tailor pinning mechanisms at will.

Vortices and vortex lattices. Vortices are present in nature in a wide variety of forms, from the rotating water in a bathtub to tornadoes and hurricanes. They range in size from nanometers to meters and are encountered in research areas from nanoscience (Bose-Einstein condensates) to aviation (airplane wings). Vortices appear in fluids, in superconductors, in the atmosphere, in magnetism, in crystal growth, and so on. The vortices in superconductors and superfluids are quantized, and they are generated, move, and vanish in the nanoscale range.

In 1957, Alexei Abrikosov worked out the main properties of type II superconductors. He found the criteria to classify superconductors as belonging to type I or type II. He showed that a triangular vortex lattice appears in the type II superconducting mixed state, with a lattice constant a_0 given by the equation below, where Φ_0 and B are the magnetic flux quan-

$$a_0 = \{(2/\sqrt{3})\Phi_0/B\}^{1/2}$$

tum and the magnetic field, respectively. According to Abrikosov, each point (vortex) of this lattice carries a magnetic flux quantum. This structure is a real lattice that can be experimentally imaged by appropriate techniques, such as making the vortices visible with ferromagnetic nanoparticles, and various types of electronic and tunneling microscopy.

The vortex lattice has the same general behavior as any other lattice. In the mixed state, the system expends energy breaking the superconducting state and generating vortices, but this energy is lowered if the vortex is pinned to a defect. The ideal triangular vortex lattice (Abrikosov lattice) deforms in the presence of pinning centers. Deformations of the vortex lattice are controlled by changes in its elastic energy. The vortex lattice has compression, tilt, and shear moduli, exactly as a solid lattice. (This scenario is very different in the case of high-temperature superconductors, where the interplay between the high temperature and the vortex lattice energy leads to a very complex vortex phase diagram, with vortex liquid and vortex glass phases, and so on. This article focuses on low-temperature superconductors, which have a very well defined vortex solid phase.)

Collective flux pinning theory is a first approach to the behavior of the vortex lattice under the influence of random distributions of pinning defects. A. I. Larkin and Yu. N. Ovchinnikov (1974) calculated, at $T = 0$, the region (Larkin domain) in which the vortex pinning is still correlated. These lengths could be estimated taking into account the competition between pinning and elastic deformation energies. The comparison of these lengths with the sample dimensions governs the volume where the vortex lattice remains correlated.

When a magnetic field (along the z axis) and an electric current (along the x axis) are applied to a type II superconductor in the mixed state, a Lorentz force (along the y axis) develops, pushing the vortices, but a vortex moves only when the pinning force is overcome by this Lorentz force. The vortex core electrons are in the normal state. Motion of normal electrons gives rise to resistance and dissipation, and if the vortex lattice moves, the superconducting state vanishes in the sample. The electric current needed to unpin the vortex lattice is called critical current. The competition between the pinning force and the Lorentz force governs the value of the critical current and therefore the region where the sample is in the superconducting state. The transition of a superconductor to the normal state at the critical current is a major factor limiting the application of superconductivity. Hence, it is desirable to enhance collective pinning in order to increase the superconducting critical current and thereby facilitate the many useful applications of superconducting materials.

Artificial nanostructures as pinning centers. Many different techniques could be used to fabricate submicrometric patterns in superconducting materials. Among them, electron beam lithography is a very powerful tool that allows the fabrication of arrays of periodic nanometric defects in a superconducting sample. The sizes of these defects are ideal for pinning the vortices. These artificially fabricated periodic pinning centers could add to the intrinsic random defects, and they could help to increase the critical current, consequently decreasing the dissipation. This effect is shown in **Fig. 1**. Figure 1*a* shows the increasing dissipation in a superconducting niobium film with the usual randomly distributed defects when the applied magnetic field is increased; the graph is a smooth curve. Figure 1*b* shows the same effect, but now the niobium film is grown on top of a triangular array (the side of the unit cell is 600 nanometers) of nickel dots (with diameters of 200 nm). A series of dissipation minima appears, which is all the more striking when one considers that the vertical axis of the graph is a logarithmic scale, meaning that the resistivity at one of the minima can be a tenth or less of its value at nearby values of the magnetic field. Minima in dissipation mean enhancement in the critical current values. These sharp minima are equal-spaced in the value of the magnetic field, providing a strong hint of what is taking place. The triangular array of nickel dots mimics

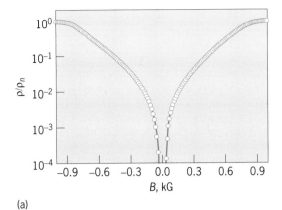

(a)

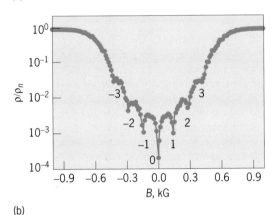

(b)

Fig. 1. Plot of mixed-state normalized resistivity (ρ/ρ_n, where ρ is the resistivity and ρ_n is the resistivity of the normal metal) versus applied magnetic field (B) in (a) a niobium film, and (b) a niobium film with a triangular array of nickel dots. The numbers below the minima of the curve indicate the number of vortices per unit cell of the array. $T/T_c = 0.99$, where T is the absolute temperature and T_c is the critical temperature. 1 kG = 0.1 tesla.

the triangular Abrikosov vortex lattice. The value of the applied magnetic field governs both the number of vortices inside the sample and the lattice parameter of the vortex lattice. When both lattices—the triangular lattice of dots and the Abrikosov (triangle) vortex lattice—match, the array of artificially grown pinning centers induces a synchronized pinning of the vortex lattice. At this specific magnetic field, an enhancement of the critical current (a minimum in dissipation) occurs. The geometric matching occurs when the vortex density is an integer multiple of the pinning center density. The size of the vortices and the size of the nickel dots are similar. The first minimum is at the magnetic field at which there is one vortex per unit cell of the array, the second minimum when there are two vortices per unit cell, and so on. At these values of the magnetic field, the whole vortex lattice is pinned collectively and the pinning force is effectively much stronger.

In **Fig. 2**, the pinning centers are the same nickel dots as before, but now the array is rectangular. The experimental data show again equally spaced minima, but there are two different spacings, and there are very sharp minima for one spacing, which occurs at the lower values of the magnetic field, and

very shallow minima for the second spacing, which occurs at the higher magnetic-field values. Two new results arise. First, at low magnetic fields, the periodic pinning due to the array is strong enough to distort the regular Abrikosov (triangular) lattice to a rectangular one. Second, increasing the number of vortices per unit cell of the array (increasing the applied magnetic field) causes the vortex lattice to relax. The sharp minima correspond to a geometrical matching with the rectangular array, and the shallow minima correspond to a square array matching the short side of the rectangular nickel dot array. Increasing the applied magnetic field results in the sample holding too many vortices per rectangular unit cell, and the competition between the pinning (lock-in) energy and the vortex lattice elastic energy changes the vortex lattice from a rectangular to a square lattice. In the square lattice configuration, there are many interstitial vortices, and therefore the matching conditions are not so well defined and the minima are blurred. In summary, it is possible to tailor, almost at will, the vortex lattice symmetry using a nanostructured array of pinning centers.

Motion on asymmetric pinning potentials. The collective pinning of the vortices is a very interesting topic, since it offers the possibility of enhancing the critical current, but the dynamics of the vortex lattice on a periodic array of pinning potentials can lead to interesting effects as well. These dynamics are related to the electrical properties of the superconductor. The vortex lattice moves when the applied current is higher than the critical current. An applied current is equivalent to an applied force on the vortex. If the voltage in the sample is measured, the vortex velocity can be easily calculated.

A striking vortex dynamics effect is based on the ratchet effect. The ratchet effect is encountered in very different fields, from biology to physics. In general, the ratchet effect occurs when the system in question has particles moving on an asymmetric potential. The particles exhibit a net flow when they are subjected to external fluctuations. Raising and lowering the barriers and wells of the potential, via an external time-dependent modulation, provide the energy necessary for net motion.

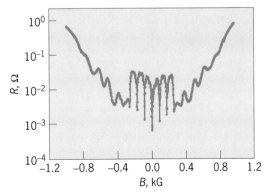

Fig. 2. Plot of mixed-state resistance (R) versus applied magnetic field (B) in a niobium film with a rectangular array of nickel dots. 1 kG = 0.1 tesla.

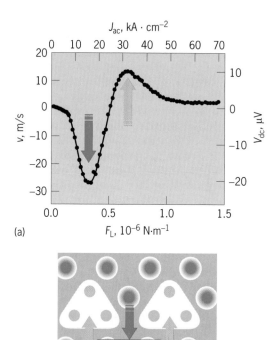

J_{ac}, kA · cm^{-2}

(a)

F_L, 10^{-6} N·m^{-1}

(b)

Fig. 3. Superconducting rectifier. (*a*) Plot of dc voltage (V_{dc}) and vortex velocity (*v*) versus alternating current (J_{ac}) and force on the vortices (F_L). (*b*) Niobium film with array of nickel triangles, showing the situation and motion of pin vortices (nickel triangles) and interstitial vortices (ghost triangles).

In the case of a mixed-state type II superconductor, the particles are vortices, the asymmetric potentials are arrays of nickel triangles (**Fig. 3***a*), and an injected alternating current provides the external time-dependent modulation. The most relevant and complete experimental results are shown in Fig. 3. If the applied magnetic field corresponds to six vortices per unit cell, three of them are pin vortices in the nickel triangles and three more are cage or interstitial vortices. The alternating current (equivalently, the force on the vortices) produces, through the ratchet effect, a net dc voltage (equivalently, a vortex velocity). The sample could be seen as a sample with two arrays, one of them consisting of real nickel triangles, and the other with reversed ghost triangles (Fig. 3*b*). The interstitial vortices on these ghost triangles move first, and when the ac current amplitude is enough the pin vortices begin to move but in the opposite direction because the ratchet potential is reversed.

The niobium film on an array of nickel triangles acts as a superconducting rectifier since injecting alternating current to the device produces a dc voltage. This rectifier has the unique property that its polarity can be tuned with the applied magnetic field.

For background information *see* CRYSTAL DEFECTS; NANOTECHNOLOGY; SUPERCONDUCTIVITY; VORTEX

in the McGraw-Hill Encyclopedia of Science & Technology.   José Luis Vicent

Bibliography. G. Blatter et al., Vortices in high-temperature superconductors, *Rev. Mod. Phys.*, 66: 1125–1388, 1994; J. E. Evetts, *Concise Encyclopedia of Magnetic and Superconducting Materials*, Pergamon Press, Oxford, 1992; J. I. Martin et al., Artificially induced reconfiguration of the vortex lattice by arrays of magnetic dots, *Phys. Rev. Lett.*, 83:1022–1025, 1999; J. E. Villegas et al., A superconducting reversible rectifier that controls the magnetic flux quanta, *Science*, 302:1188–1191, Nov. 14, 2003.

Combinatorial materials science

Traditionally in materials science, researchers performed an experiment, analyzed the results, and then used these results to decide on the next experiment. This model of experimentation is a clear application of the scientific method and has been the foundation of many significant discoveries. However, the use of computers for both data acquisition and analysis has enabled high-throughput approaches to materials discovery experiments, that is, parallel rather than serial sample preparation, characterization, and analysis. Such experimentation is known as the combinatorial method, whereby many nearly simultaneous experiments are performed and analyzed at each step. This approach can greatly accelerate the rate at which science can be done and knowledge acquired. Combinatorial experimental methods are becoming common in a number of areas, including catalyst discovery, drug discovery, polymer optimization, phosphor development, and chemical synthesis. For example, complex catalysts have been discovered 10 to 30 times faster using the combinatorial method rather than conventional approaches. Perhaps the biggest challenge of combinatorial science is in analyzing rapidly the vast amount of data (hundreds, thousands, or millions of experiments) for important trends and results. This is typically accomplished by using data mining software.

Combinatorial approach. The experiments discussed here illustrate the effectiveness and challenges of using combinatorial techniques to explore the physical parameters, such as elemental composition (stoichiometry), substrate temperature during deposition, substrate choice, and oxygen partial pressure during processing, which effect the growth of transparent conducting oxide (TCO) thin films. Transparent conducting oxide thin films are a key technology for photovoltaic cells, flat panel displays, low-e (thermal reflecting) windows, and organic light-emitting diodes. For comparison, the work reported here would have taken over a year using traditional methods, while combinatorial techniques enabled it to be completed in 2 months. This large improvement in throughput enables the screening of compositionally complex materials that would be impractical to study with traditional techniques.

The basic methodology of combinatorial science is a four-step process of sample array (library) creation, library measurement, data analysis, and data mining. For screening TCOs, a composition-spread (a gradient of the different elements) library is created using a materials deposition technique such as sputtering, pulsed laser deposition, or inkjet printing. Relevant properties of the library are then measured using high-throughput characterization techniques, with an emphasis on identifying how material properties vary with composition. In the research presented here, the variation in the optical properties and electrical conductivity was particularly important. For other applications of TCOs, properties such as etchability, hardness, corrosion resistance, or compatibility with other materials might be measured. The point of these initial, high-throughput measurements was to identify general trends across a large range of compositions so that promising material compositions could be identified for further study. Analyzing data from high-throughput measurements is often a challenge, as the amount of data makes sample-by-sample analysis too time-consuming. Instead, software tools must be developed to extract selected performance or property parameters from the data. The transformation of large data sets, which are too big to visualize, into smaller, more useful nuggets of information is a key component and major challenge of combinatorial work.

In most cases, it is useful to measure more fundamental material properties, which may be "one step removed" from the properties directly relevant to the applications. For example, the crystal structure of the TCO films was measured because it is well-known that crystal structure affects optical and electrical properties. Comparing a fundamental film property (crystal structure) with important application properties (such as transparency and conductivity) provides insight into why certain material compositions are better. The use of combinatorial methods does not replace the science-based approach. Instead, it expands the scale of what can be experimentally tested, making complex materials and processes feasible to study.

The full compositional range of $In_2O_3 \leftrightarrow ZnO$ was examined in a search for improved transparent conducting oxides that can be deposited at low temperatures (100°C or 212°F), such as is necessary for depositing TCOs onto flexible polymer substrates. This compositional range is of great interest since both In_2O_3 and ZnO are important TCOs that are used in many commercial applications. This particular choice is motivated by recent reports that an intermediate composition (nominally reported as 10% by weight ZnO in In_2O_3) is becoming important in commercial flat-panel displays because of its low deposition temperature and the ease of patterning, which reduce manufacturing costs. Combinatorial methods allowed a ready exploration of compositions between the two endpoint constituents, In_2O_3 and ZnO.

Library deposition. Compositionally graded indium-zinc-oxide (IZO) libraries were deposited via direct-current (dc) magnetron sputtering. Sputtering is similar to atomic-scale spray painting, and by using two different targets at the same time, mixed composition libraries can be made.

As shown in **Fig. 1**, the 2-in.-diameter (5-cm) In_2O_3 (material 1) and ZnO (material 2) metal oxide targets were positioned on either side of the substrate at an angle of $\sim 30°$. The orientation of the targets, combined with control of the deposition powers, determines the composition range and gradient for each IZO library. In practice, four or five libraries are needed to cover the full compositional range. All the libraries were deposited with a substrate temperature of 100°C (212°F) at a pressure of 4.5 millitorr (0.6 pascal) in argon gas. Measurements of the

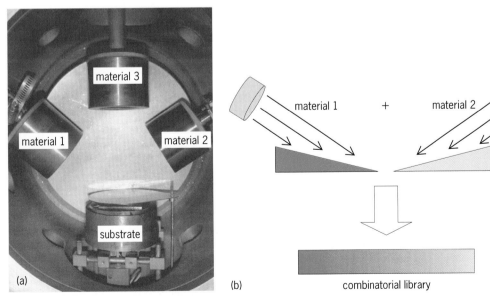

Fig. 1. Library deposition. (*a*) Inside of the three-gun combinatorial sputtering system. (*b*) Codeposition method.

combinatorial samples (using both targets) confirmed that the total film thickness was accurately predicted using a linear combination of single-target deposition rates. After deposition, the actual In/Zn ratio for each library was confirmed as a function of position by automated electron probe microanalysis mapping.

Combinatorial analysis. Conductivity, which is just the inverse of the resistivity, was determined from resistance measurements made using an automated four-point current-voltage probe. The film thickness was determined from the optical measurements. Optical reflection and transmission spectra maps were measured using fiber optically–coupled spectrometers equipped with either a Si CCD-array detector for wavelengths between 0.2 and 1.0 μm or an InGaAs diode array detector for wavelengths between 1 and 2 μm. In addition, an automated Fourier-transform infrared (FTIR) spectrometer was used for measurements at wavelengths longer than 2 μm. (The visible spectrum covers ~0.4–0.7 μm.) The crystalline structure of the libraries was measured using x-ray diffraction with a two-dimensional x-ray detector array. For the data presented here, each of these measurements was done on the same set of 11 physical locations on each library with the 11 points chosen to span the full compositional gradient of each 2 × 2 in. library.

Figure 2a shows the conductivity of the libraries as a function of materials composition. For the four initial library depositions, the general dependence of the conductivity on indium content was determined and showed a peak in the conductivity near 70% In for Zn, comparable to the best commercial materials. However, because the 70% In composition happened to occur near the edge of two libraries, a fifth library centered on this critical composition was deposited. The conductivity data for this followup library confirmed the conductivity maximum near 70% In. While the data overlap between libraries in Fig. 2a generally is good, some disagreement between similar compositions from different libraries is observed. This is quite clear near the maximum at 70% In where three libraries overlap. This data mismatch, found to be reproducible and systematic, depended on the proximity to the sputtering gun running at high power, which affected crystallinity. This type of unintended but practically important observation was made only because the high-throughput combinatorial approach was used.

Figure 3 shows some of the x-ray diffraction data measured on the followup library centered near 70% In. The inset shows the raw megapixel detector image, and for each analysis spot six such images are taken. Using a two-dimensional x-ray detector array, this image represents only 60 seconds of data acquisition. The horizontal (2θ) axis is related to the crystalline lattice spacing, and the vertical axis (χ) relates to the orientation of the crystalline axis. In this example, the shift of the peak position upward from $\chi = 0$ (horizontal broken line) shows that the crystalline axes are preferentially tilted toward the In sputtering gun. As a first step toward simplifying the data, the measured data are integrated in χ to yield 2θ spectra such as the two shown in the top panel of Fig. 3. The top spectra is for a 87% In composition and corresponds to the detector image shown in the inset. This spectra has a sharp peak at $2\theta \approx 30°$ indicating that the sample is crystalline (ordered atomic structure). The bottom spectra, for a 46% In sample, shows no such peak indicating that the sample is amorphous (random atomic structure). Using a color intensity scale, the bottom panel shows the eleven full 2θ spectra measured for this library. There is a clear transition from amorphous to crystalline material at ~75% In. Still, a scalar parameter is needed for further data analysis. Because the absolute x-ray diffraction intensity is very sensitive to the distribution of directions of crystalline orientation, the width W of the crystalline x-ray diffraction peak is a better choice (Fig. 2b). For the amorphous regions $W \approx 2$, whereas for the crystalline regions $W < 1$. Based on this, the approximate amorphous region from 46 to 80% In is shaded in Fig. 2. In addition, this allows W to be used as a simple scalar parameter, which can be queried in subsequent data mining to characterize whether the sample is amorphous or crystalline.

Figure 4 shows the ultraviolet, visible, and near-infrared (IR) reflectance R for the library centered near the 70% In composition. In the top panel, a representative reflectance spectra is shown for an

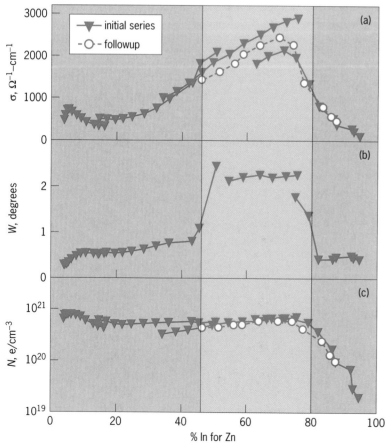

Fig. 2. Materials properties of IZO versus composition. (*a*) **Electrical conductivity** (σ). (*b*) **X-ray diffraction peak width** (*W*). (*c*) **Optically determined conducting electron concentration** (*N*).

indium composition of 64%. The corresponding transmission T spectra, which is normalized to the substrate and shown in black, is cut off at 5 μm because the glass substrate is completely opaque at longer wavelengths. These spectra are typical of a transparent conducting oxide. In the visible region (0.4–0.7 μm), oscillations around $R \sim 25\%$ occur due to interference effects in the film. In addition, the film thickness can be determined from the spectral position of the maxima and minima in these oscillations. For example, in the bottom panel the general shift of these oscillatory features to shorter wavelengths from the top (87% In) to the bottom (46% In) of the figure indicates a continuous decrease in the sample thickness as the indium content decreases for this particular library.

At longer wavelengths, the reflectivity decreases to a local minimum before monotonically increasing again in the infrared along with a corresponding decrease in the transmission. The high infrared reflectivity is due to the excitation of the nearly free electrons in the film. These electrons, which are also responsible for the electrical conductivity of the film, can be excited when the light frequency is lower than the characteristic electron gas oscillation frequency, known as the plasma frequency. The corresponding plasma wavelength λ_P, which can be estimated as the reflectivity minimum just prior to the large long-wavelength increase in the reflectance (see R_{min} in Fig. 4), is related to the density of free conducting electrons in the material n, as $n : 1/\lambda_P^2$. Thus, a spectral shift in the minimum of the reflectivity indicates a change in the carrier concentration. The black line overlaid on the spectra in the bottom panel of Fig. 4 shows λ_P for the measured spectra. The corresponding carrier concentration for this library, shown in Fig. 2c with open symbols, shows a maximum at an indium concentration of 74%, correlating to the peak in the conductivity. This shows how, for transparent conducting oxides, the infrared optical spectra mapping can be used as the basis for a noncontact combinatorial electron concentration probe.

Figure 2 summarizes the correlation between the electrical and structural properties of the In-Zn-O TCO libraries presented here. The composition range from $In_{0.05}Zn_{0.95}O_x$ to $In_{0.95}Zn_{0.05}O_x$ was covered with four initial libraries and one additional focused followup library. Figure 2a shows that the conductivity has a broad peak with the maximum at $\sim$70% In. The x-ray diffraction peak width shown in Fig. 2b indicates an amorphous region (shaded) from % In $\approx$ 46–80, which is well correlated with the range of the broad conductivity peak. Figure 2c shows the optically determined carrier concentration on a logarithmic scale. The carrier concentration is nearly constant up to 80% In, beyond which it abruptly drops. This abrupt drop, the start of which is correlated with the amorphous to crystalline transition at 80% In, results in the decrease in conductivity at high In % shown in Fig. 2a.

Summary. The data yielded the surprising results that the best electrical properties occur between 46

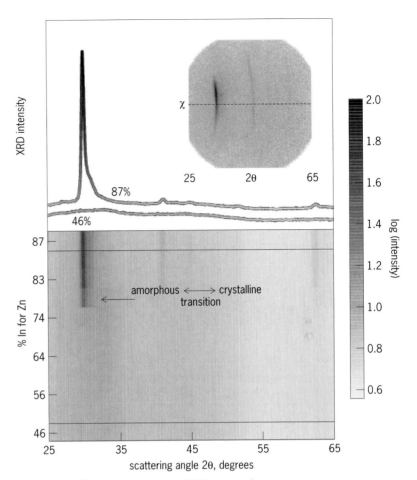

Fig. 3. X-ray diffraction analysis of an IZO library.

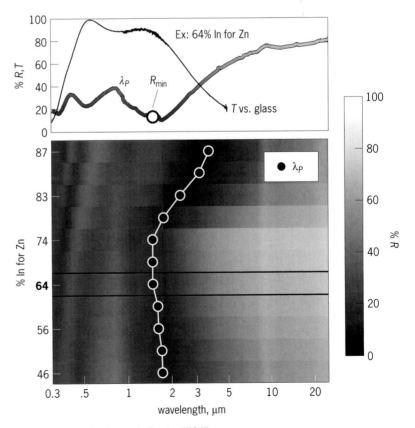

Fig. 4. Optical reflection analysis of an IZO library.

and 80% In where the thin films are amorphous, as deposited. A maximum in the desired properties occurs at a composition of about 70% In. This unexpected observation clearly illustrates the power of the combinatorial method for materials discovery. In this case, not only was a large range of compositions covered rapidly, but also a promising region was spotted that would have been missed if only the typical perturbative doping regimes within 10% of each endpoint were examined. This study illustrates the increasing potential for amorphous materials in systems that have classically been dominated by crystalline materials as transparent conducting oxides.

[This work was supported by the U.S. Department of Energy under Contract no. DE-AC36-99-G010337 through the National Center for Photovoltaics and by the U.S. Air Force, Wright Patterson.]

For background information *see* COMBINATORIAL CHEMISTRY; CRYSTAL GROWTH; CRYSTAL STRUCTURE; MATERIALS SCIENCE AND ENGINEERING; OXIDE; SPUTTERING in the McGraw-Hill Encyclopedia of Science & Technology.   David Ginley; Charles Teplin; Matthew Taylor; Maikel van Hest; John Perkins

Bibliography. T. J. Coutts, D. L. Young, and X. Li, *Characterization of Transparent Conducting Oxides*, MRS Bull. 25, p. 58, 2000; D. S. Ginley and C. Bright, *Transparent Conducting Oxides*, MRS Bull. 25, p. 15, 2000; T. Moriga et al., Structures and physical properties of films deposited by simultaneous DC sputtering of ZnO and In_2O_3 or ITO targets, *J. Solid State Chem.*, 155:312, 2000; J. D. Perkins et al., Optical analysis of thin film combinatorial libraries, *Appl. Surf. Sci.*, 223:124, 2004; T. Sasabayashi et al., Comparative study on structure and internal stress in tin-doped indium oxide and indium-zinc oxide films deposited by r.f. magnetron sputtering, *Thin Solid Films*, 445:219, 2003; M. P. Taylor et al., Combinatorial growth and analysis of the transparent conducting oxide ZnO/In (IZO), *Macromol. Rapid Commun.*, 25:344, 2004.

Comparative bacterial genome sequencing

The first complete genome sequence of a bacterial pathogen, *Haemophilus influenzae*, was published in 1995. Since then the sequencing of well over 100 bacterial genomes has been completed. In the early days of this enterprise, the primary approach was to sequence genomes from very different pathogens that were responsible for major disease problems. More recently there has been a recognition of the value of sequencing whole genomes from closely related bacteria. The enormous potential for understanding pathogenesis by comparison of genomes becomes apparent when bacterial species that appear to be very similar cause different diseases in different hosts. At the simplest level, if two bacterial types are more or less identical yet one causes disease and the other does not, then it is reasonable to suggest that the disease-causing bacterium may have extra genes that enable it to bring about the partic-

ular pathology. However, the situation in reality is a good deal more complex. Not only must the presence or absence of particular genes be considered, but also factors such as how different levels of gene expression and subtle changes in the amino acid sequence of the proteins they encode can contribute to the overall disease phenotype of the pathogen.

To illustrate the potential of comparative genome sequencing of pathogens, this article concentrates on comparisons between bacteria that mainly infect via the intestinal tract (enterobacteria) and comparisons between species of bordetellae (respiratory pathogens).

Escherichia coli strains. *Escherichia coli* is found mainly in the intestinal tract of many different hosts, where it can live without causing any pathology at all or where it can cause a range of disease syndromes. Some *E. coli* strains can also live in extraintestinal sites, where they are almost always associated with disease (for example, the bloodstream or the urinary tract). *Escherichia coli* is also the bacterium that has been used most extensively as a model in laboratory investigations of basic bacterial physiology and genetics. The genome sequence of a standard harmless laboratory strain (called K-12) was recently completed, followed by the sequence of an enterohemorrhagic strain (EHEC), which can cause intestinal bleeding, and then a uropathogenic strain (UPEC), which can cause urinary tract infections. EHEC generally live in the guts of ruminants such as cattle and sheep without causing disease; however, they can cause severe intestinal disease in humans, which may develop into the life-threatening hemolytic uremic syndrome. The completion of these three genome sequences has allowed extensive comparisons to be drawn and some basic principles to emerge.

Comparing the genomes of the EHEC and K-12 strains showed that they share a backbone of 4.1 megabases (Mb) of deoxyribonucleic acid (DNA) but that they both also contain scattered islands of unique sequence. These islands comprise 1.34 Mb and 0.53 Mb in the EHEC and K-12 strains, respectively. Some of these islands contain genes known to be required for virulence, for example the locus of enterocyte effacement (LEE) in the EHEC strain, which has been shown experimentally to be essential for the interaction of the pathogen with host cells. However, the situation is not simply that the pathogen contains virulence genes in addition to the backbone sequence that it shares with K-12. Some of the islands in the K-12 strain contain genes that may also be involved in virulence, although the pathogenic potential of K-12 for different hosts and for different disease states is unknown. Analysis of DNA base composition suggests that the islands in both species have been acquired via horizontal genetic exchange between different bacteria (for example, via bacterial viruses) during evolution, a common theme in the genomics of several bacterial genera. A similar picture can be drawn from comparison of the UPEC strain with the K-12 laboratory strain. Again there is a shared backbone sequence, this time of 3.92 Mb, with islands specific to each of the strains. The UPEC-specific islands total about 1.3 Mb, whereas the K-12 specific sequences account for about 0.7 Mb. Interestingly, the differences between the UPEC and EHEC strains were of a similar magnitude to the differences between each of the pathogens and the K-12 strain. Also, many of the islands found in the pathogen genomes were inserted at the same relative positions in the backbone but contained different genes, indicating that these positions have been more receptive than others to insertion of horizontally acquired DNA.

Overall, the proportion of proteins shared by all three *E. coli* strains was only 39%. Each of the *E. coli* strains has a complement of genetic islands that has allowed it to exploit the particular ecological niche in which it has evolved, and thus analysis of the genes in the islands should give insight into the pathogenesis of the different bacteria. However, there may also be common capabilities coded by the backbone genes, which the pathogens need in order to be successful in their different lifestyles. In other words, the genes encoded in the islands may be advantageous to the pathogen only in the context of its complete genome.

Shigella flexneri. The genome sequence of *Shigella flexneri* provides an interesting example with which to compare the *E. coli* genomes. A great deal of evidence suggests that *Shigella* and *E. coli* are closely related. In fact it has been proposed that *Shigella*, instead of having its own genus, should be regarded as a species within the genus *Escherichia*. *Shigella* causes bacillary dysentery and is pathogenic only for humans. On analysis of the *S. flexneri* genome, once again the "backbone and island" structure seen in the different *E. coli* is observed, but the amount of horizontally acquired DNA is much less than in the pathogenic *E. coli* species. *Shigella flexneri* has 37 more islands than K-12, whereas the EHEC and UPEC strains have over 100 extra islands. A distinct feature of the *S. flexneri* genome is the presence of a large number of specialized mobile DNA elements called insertion sequences (IS) as well as a large number of pseudogenes (genes that were once functional but are no longer due to having acquired mutations). The presence of these components is a common theme among several pathogens that have evolved, or are evolving, to cause disease in a single host. While a great deal is known about the pathogenic mechanisms that *S. flexneri* uses to invade and grow inside cells, much less is known about its host specificity and tissue preferences. Candidate genes for these processes emerge from genome comparisons. For example, a number of outer membrane proteins, putative adhesons (adhesive proteins), and fimbrial (hairlike surface fiber) genes have been identified in several of the unique genomic islands analyzed, and these, along with other mechanisms, may well underlie the nature of cellular attraction in this pathogen.

Bordetella. The genus *Bordetella* contains eight species, three of which are of greatest interest in terms of human and animal disease. *Bordetella pertussis* and *B. parapertussis* cause whooping cough in children and are restricted to the human host. *Bordetella bronchiseptica* infects a wide range of mammalian hosts, with a spectrum of consequences from no obvious effect to clinical respiratory diseases such as kennel cough in dogs, snuffles in rabbits, and atrophic rhinitis in pigs. The genome sequences of these three bordetellae have recently been completed and compared. All three bordetellae share a large number of genes, and the shared genes are very similar to each other at the nucleotide level. However, the overall genetic complement varies considerably between the three species. The *B. bronchiseptica* genome consists of 5.3 Mb, *B. parapertussis* has 4.8 Mb, and *B. pertussis* is shortest with 4.1 Mb. The intial assumption given these data might be that there has been considerable gene acquisition by the bacteria with the larger genomes. This is, however, incorrect. Genetic analysis suggests that the two human-adapted species arose from a common ancestor, most similar to *B. bronchiseptica*, via independent events at different times. After independent acquisition and expansion of ISs, subsequent intragenome recombination between identical copies of the ISs resulted in gene deletions and chromosomal inversions. This is particularly clear when the gene order in *B. pertussis* is examined. It appears to be completely scrambled in comparison with the other two species.

Given the major differences in host specificity between the different bordetellae, it was hoped that whole genome comparisons would suggest some possible mechanisms underlying the variation. Unfortunately the situation did not prove to be that simple. Adaptation to a single host occurred with wholesale gene loss, not gene gain, and there are no obvious candidates for adhensins or toxins that explain the varying host and tissue preferences. What did become obvious from all three sequences is that there are many more membrane proteins, adhesins, fimbriae, and the like than was thought to be the case in the bordetellae. Thus much more sequence-based research into basic pathogenetic mechanisms is required before the understanding of *Bordetella* virulence will be complete.

Conclusions. Comparative genome sequence analysis is fertile ground for exploration, but it has not yet produced a simple explanation for differences in bacterial pathogenesis. There is much more information to be mined in the genomes that have been published, and the continuing efforts to sequence, for example, several different serovars of *Salmonella enterica* may produce a dataset that will clarify some of the mysteries underlying the different host adaptations seen within this set of organisms. Rather than providing immediate answers, comparative genome analysis is a tool for generation of hypotheses and questions about pathogenicity and virulence that need to be further investigated experimentally. Without the comparative analyses, however, many of these questions would have never been asked.

For background information *see* BACTERIAL GENET-ICS; BACTERIAL PHYSIOLOGY AND METABOLISM; BAC-TERIAL TAXONOMY; ESCHERICHIA; RECOMBINATION (GENETICS); TRANSDUCTION (BACTERIA); VIRULENCE; WHOOPING COUGH in the McGraw-Hill Encyclopedia of Science & Technology.   Duncan J. Maskell

Bibliography. J. Parkhill et al., Comparative analysis of the genome sequences of *Bordetella pertussis, Bordetella parapertussis* and *Bordetella bronchiseptica, Nat. Genet.*, 35:32–39, 2003; N. T. Perna et al., The genomes of *Escherichia* K-12 and pathogenic *E. coli*, in M. S. Donnenberg, *Escherichia coli Virulence mechanisms of a versatile pathogen*, Academic Press, Boston, 2002; A. Preston, J. Parkhill and D. J. Maskell, The Bordetellae: Lessons from genomics, *Nat. Rev. Microbiol.*, 2:379–390, 2004; C. M. Fraser, T. Read, and K. E. Nelson (eds.), *Microbial Genomes*, Humana Press, Totowa, NJ, 2004.

Computational intelligence

Computational intelligence is a term that was coined in the early 1990s by James C. Bezdek to provide a perspective on technologies different from classical artificial intelligence (AI) approaches. Computational intelligence refers to a set of techniques that produce systems which when given input data react in ways that might be considered intelligent. There is significant continuing research in the area of computational intelligence, as well as a number of practical, everyday systems that make use of computational intelligence approaches.

Computational intelligence has been used to enable a car to autonomously navigate from Washington, DC, to San Diego with a human driver intervening only a few times. It has been used in automatic transmissions and antilock braking systems. It has been used to determine whether one might grant credit card purchase requests. One of the best checkers-playing computer programs in the world was developed using a computational intelligence approach.

Originally, the term "computational intelligence" was used to differentiate neural network approaches from symbolic, artificial intelligence approaches. Over time the term has come to embrace neural networks, fuzzy systems, and evolutionary computation. Each of these approaches relies on mathematical computation to produce intelligent systems.

In the classical approach to intelligent systems, known as artificial intelligence, all reasoning may be done using high-level symbols. For example, one can use rules such as: if it is raining then bring an umbrella. These types of rules (called heuristic rules) may also be used in fuzzy systems with the difference that approximate matches are allowed. So, one might still bring bring an umbrella on a cloudy overcast day where it was not raining at the moment.

Systems that are described by the term "computational intelligence" make use of lower-level computational features even when they may have a high-level interpretation, for example in the case of fuzzy systems which may be expressed in heuristic rules. These rules have been successfully used in fuzzy control of antilock brakes, for example.

Fuzzy systems. A fuzzy system is one that makes use of approximate reasoning using the mathematical basis of fuzzy logic. The concept of fuzzy logic is concerned with membership in the classes of true or false. For example, a person with some bald spots might be said to be balding. However, he would completely belong neither to the class of bald people nor to the class of people with hair.

To see how this differs from probability, consider that you have the opportunity to acquire one of two lottery tickets after the winning tickets from the lottery have been chosen. The first ticket is described as having a 90% chance of having won a lot of money, and the second one is described as having a 0.9 membership in the class of tickets which have won a lot of money. In the first case you have a 90% chance of winning a lot of money and a 10% chance of winning nothing. In the second case, you will have approximately won a lot of money. This is likely to translate into something between a lot of money and a fair amount of money. If you had to put a probability distribution on the likely winnings in the second case, you might have a probability of 0 of winning no money and a probability of 0 of winning the entire lottery, but a probability of 80% of being runner-up, 10% of being second runner-up, and a 10% probability of being third runner-up. Most people would choose the ticket with the 0.9 membership in the class of tickets which have won a lot of money because they would be guaranteed a good sum of money. Of course, some gamblers would choose the all-or-nothing route.

Neural networks. Perhaps the most widely used technology within computational intelligence approaches is that of neural networks. Neural networks are inspired by the underlying processing done in the human brain. There are nodes which correspond (loosely) to neurons in the brain and have interconnections which correspond (loosely) to the synapses that connect neurons in the brain. A simple three-layer neural network is shown in the **illustration**.

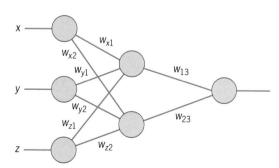

Feedforward neural network with three inputs, one output, and one hidden layer of two nodes. The connection weights are denoted by w_{ij}.

A neural network learns from a set of labeled examples to produce intelligent responses. Neural networks are also used to intelligently group data into clusters. Thus, a neural network is one attempt to mimic the underlying learning processes of the human brain. Clearly, a system that can learn to improve its performance will be quite valuable. There are many approaches to training a neural network. Many of them rely on the feedforward–backpropagation approach which was popularized in the mid-1980s. In this approach, examples are presented to the neural network, and the connection weights between nodes are updated to make the outputs of the network closer to those of the examples. The connection weights loosely correspond to the synaptic strengths of connections between neurons in the human brain.

Neural networks have a powerful property of universal approximation. This means that they can arbitrarily closely approximate any function, in theory. This has been proven mathematically. In practice, the theorems do not show how to configure the network or what type of functions should exist within a node to determine its output or, perhaps most importantly, do not guarantee that the appropriate weights can be learned. So, it is known that there is a neural network that arbitrarily closely approximates a particular function, but it is not known if this network can ever be discovered.

Consider the function that emulates a person who decides whether a credit card transaction should be allowed. Given what is known about neural networks, it is possible to develop one that is as good as any human expert. It is unknown if this can actually be done, but in practice such systems have been built. The function to be approximated is the binary human decision of whether to grant credit or not grant credit. A highly accurate system will save money for the credit card company, and it will have the side effect of preventing people from getting in too much debt.

Fuzzy systems and neural networks have been combined to produce neuro-fuzzy systems. Under certain conditions they have been shown to be universal approximators also. Typically, a set of fuzzy rules can be extracted from a neuro-fuzzy system which may provide an explanation capability that is lacking in a typical neural network where all information is stored in a set of weights. The extent of the explanation capability will depend on the complexity of the fuzzy rules. There is, typically, a trade-off between complexity and accuracy. So, more understandable but less complex rules will be less accurate than the more complex rules in the usual case.

Evolutionary computation. Evolutionary computation embodies a set of methods to evolve solutions to problems. Evolutionary computation has typically been applied to optimization problems. For example, excellent solutions to the traveling salesman problem have been evolved. The traveling salesman problem is to minimize the distance traveled by a salesman who has to make calls in a set of cities.

Evolutionary computation considers potential solutions to a problem as members of a population. Mimicking how natural evolution is considered to work, members of the population can be chosen to mate. They then produce offspring, typically with the use of a crossover operator in which parts of each solution become a part of the offspring. After crossover, a mutation operator is often applied to the child which may randomly change some of its characteristics. The offspring compete for survival to the next generation based upon a fitness function. Those that are more fit survive.

There are at least three major paradigms within evolutionary computation. Genetic algorithms, which merge parts of solutions to form new solutions and then apply random mutation to them, can be applied to both binary bit strings and real-valued solutions. Evolutionary strategies make heavy use of the mutation operator. There is also genetic programming in which computer programs are evolved, typically in the Lisp language. The latter approach provides a computer-executable solution to a problem.

Evolutionary computation has been used to evolve the weights, and sometimes the structure, of neural networks. It has also been used to evolve fuzzy systems and fuzzy neural networks. Thus, these approaches are all complementary. An interesting application of evolutionary computation was its use to evolve an expert checkers-playing program that did not require any human knowledge. The system evolved weights for a neural network which served as a board evaluation function. After evolution, it was able to beat an expert-level checkers player.

In summary, all the computational intelligent approaches discussed here can be used alone or together to produce systems that exhibit human expertise or intelligence.

For background information *see* ARTIFICIAL INTELLIGENCE; EVOLUTIONARY COMPUTATION; FUZZY SETS AND SYSTEMS; GENETIC ALGORITHMS; NEURAL NETWORK in the McGraw-Hill Encyclopedia of Science & Technology.   Lawrence O. Hall

Bibliography. J. C. Bezdek, On the relationship between neural networks, pattern recognition and intelligence, *Int. J. Approx. Reasoning*, 6(2):85–107, February 1992; D. B. Fogel, *Evolutionary Computation: Toward a New Philosophy of Machine Intelligence*, 2d ed., IEEE Press, Piscataway, NJ, 1999; D. E. Goldberg, *Genetic Algorithms in Search, Optimization, and Machine Learning*, Addison-Wesley, Reading, MA, 1989; W. Pedrycz and F. Gomide, *An Introduction to Fuzzy Sets: Analysis and Design (Complex Adaptive Systems)*, MIT Press, Cambridge, MA, 1998.

CT scanning (vertebrate paleontology)

CT (computed, computer, or computerized tomography) is a technology that enables highly detailed and nondestructive three-dimensional examination of solid objects using x-rays. It is perfectly suited for paleontology, as fossil material is often precious and irreplaceable, and information on internal structures is crucial for placing ancient species into their proper anatomical, and ultimately evolutionary, context. In the past, comparable data could be obtained only though serial grinding, in which a specimen was slowly ground away and illustrations made at small intervals, a process that is slow, laborious, and destructive.

CT is also an ideal technology for the digital age, as the imagery can be quickly and easily distributed via the Internet. CT data can also be used for a wide variety of computational analyses, from three dimensional measurements to finite element modeling (an engineering tool that can be used in paleontology to study the mechanical behavior and function of bones).

Operating principles. Tomography is a technique for digitally cutting a specimen open to reveal its cross section. A CT image is typically called a slice, in analogy to a slice from a loaf of bread. A CT slice has a thickness corresponding to a certain thickness of the object being scanned. Therefore, whereas a typical digital image is composed of pixels (picture elements), a CT slice image is composed of voxels (volume elements).

The gray levels in a CT slice correspond to x-ray attenuation, or the proportion of x-rays scattered or absorbed as they pass through each voxel. X-ray attenuation is primarily a function of density, atomic number, and x-ray energy. A CT image is created by directing x-rays through the slice plane from multiple orientations and measuring their resultant decrease in intensity. A specialized algorithm is then used to reconstruct the distribution of x-ray attenuation in the slice plane.

By acquiring a stacked, contiguous series of CT images, data describing an entire volume can be obtained, in much the same way as a loaf of bread can be reconstructed by stacking all of its slices. The resulting "data brick" can be visualized in a number of ways, including reslicing along an arbitrary plane and volume rendering, in which each voxel is assigned a color and an opacity, allowing some of the object to be made transparent so that inner structures can be studied.

Development. CT was originally developed for medical diagnosis in the late 1960s–early 1970s. Typical medical CAT (computed axial tomography) scanners are not well suited for imaging most fossil material, however, as they are optimized for living subjects, which have low density, low tolerance for radiation and are prone to move. A typical medical CAT scanner uses a low-energy (<140 keV), high-intensity, large-focal-spot (1-2 mm; 0.04-0.08 in.) x-ray source and large (1-2 mm) detectors to acquire images quickly and with minimal dosage to the patient.

Industrial CT scanners were developed for non-destructive examination of inanimate objects. They typically utilize x-ray sources with higher energies and smaller focal spots, combined with smaller detectors and more precise positioning, to allow imaging of dense objects at much greater resolution than is possible with medical devices. The field is still advancing rapidly, with innovations in x-ray sources, detectors, computers, and algorithms serving to increase imaging capabilities.

Early applications in paleontology. In the mid-1980s, a number of paleontologists began to explore the use of medical CAT scanners for imaging fossil material. Although success was generally limited due to the low x-ray energy and coarse resolution of these systems, a number of studies foreshadowed the potential of tomography. In 1984, in the first published paleontology study using CT, a matrix-filled skull of the Middle Oligocene ungulate (hoofed mammal) *Stenopsochoerus* was scanned to generate fifty 256 × 256 slices at 2-mm (0.08-in.) intervals. The contrast between the fossilized bone and the sandstone matrix was sufficient to allow digital removal of the cavity-filling material, permitting observation and measurement of internal structures such as the hard palate and the basicranial axis (the line connecting the forward-most point of the upper jaw to the front of the opening for the spinal cord).

In 1988, another noteworthy early study featured CT imagery of the skull region of the Eichstätt specimen (which is housed at the Jura-Museum in Eichstätt, Germany) of the earliest known bird, *Archaeopteryx*. Image generation was challenging due to the slab geometry of the fossil and the similarity in density between the bone material and enclosing Solnhofen Limestone (a famous fossil-bearing lithology in Bavaria, Germany). The authors studied, in particular, the form and articulations of the quadrate (a small bone at the back of the skull that forms part of the upper jaw joint), and interpreted it to be double-headed (birdlike) rather than single-headed (reptilelike); however, the data were not clear enough to be definitive.

Two milestones were reached in 1993 in a study of a skull of *Thrinaxodon*, an early synapsid (mammal-like reptile). It was the first specimen imaged using an industrial CT scanner, which provided data at a resolution of 0.2 mm (0.008 in.). The resulting series of 153 horizontal slices, plus associated products such as orthogonal reslicings (slice images generated along the other perpendicular axes) and 3D models, were too extensive for standard journals. Instead, the image data and accompanying analysis were published on compact disk, making it the first digital paleontology publication.

Recent applications. From these beginnings, the use of CT in paleontology has proliferated rapidly. The examples given here are only a small sampling of recent work and are intended to illustrate the range of circumstances in which CT has been applied.

Large specimens. The skull of the *Tyrannosaurus rex* specimen known as Sue was scanned in 1998 with a CT system built around a linear accelerator, which generates x-rays at 6 MeV, a requirement when imaging such large specimens. A series of 748 slices was collected at 2-mm (0.08-in.) intervals in the coronal plane (which divides a specimen into front and rear sections). These data revealed numerous previously undocumented features and served as the basis for the creation of a digital endocast showing the internal structure of the central nervous system.

Small specimens. A 4-mm-diameter (0.16-in.) specimen of the Carboniferous mitrate *Jaekelocarpus* was scanned using a microfocal CT system, which provided a slice spacing of 16 μm (0.0006 in.) and inplane resolution of 8 μm (0.0003 in.). (A mitrate is an early echinoderm; modern echinoderms include sea urchins and starfish.) Analysis of these data published in 2002 demonstrated the presence of paired gill slits internally, whose form and function provide evidence for this taxon belonging to the crown group (that is, being related to the last common ancestor) of chordates.

Specimens in amber. In most cases of specimens encased in amber, particularly insects, microfocal CT reveals that no internal details are preserved, probably because microbes inside the organism consumed the tissues after it was trapped. However, two scans of small amber-encased lizards have revealed skeletal details. A *Sphaerodactylus* gecko trapped in Miocene amber from the Dominican Republic was scanned at 60 μm (0.0024 in.) resolution, revealing numerous fine details of its skeletal structure. A second study examined the structure of an *Anolis* lizard skull also in Dominican amber.

Internal morphology. Pneumatic paranasal cavities (airfilled cavities in the skull adjacent to the nasal cavity) are distinctive features of modern crocodylians, but have been difficult to observe in the fossil record. CT scans of the single known specimen of the Early Jurassic crocodylian *Calsoyasuchus* revealed an extensive network of sinus cavities, placing the evolution of these features at a time much earlier than previously documented (**Figs. 1** and **2**).

Forensic paleontology. An unexpected application of CT occurred in the scanning of the *Archaeoraptor* fossil which, with a curious combination of a birdlike wrist and dinosaurlike tail, was purported to be an important "missing link" between the two groups. Scanning revealed a layer of grout beneath the upper surface of the specimen, indicating that it had been reconstructed from many pieces. Further analysis of the misfit between adjacent fragments was used to establish which ones were indeed related, and ultimately revealed that the specimen was reconstructed with elements from two to five different individuals, likely of different species.

Neuroanatomy. Digital endocasts of the brain and vestibular (inner ear) apparatus of two pterosaurs (winged reptiles), *Rhamphorynchus* and *Anhanguera*, which were generated from ~0.2–0.4 mm (0.008–0.016 in.)-resolution CT data, were used to reconstruct their neurological bases for flight control. Although their brains share a number of

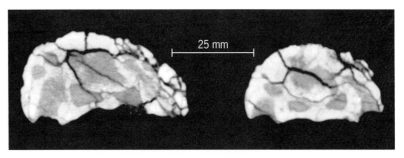

Fig. 1. Two CT images through the snout of the Early Jurassic crocolylian *Calsoyasuchus valliceps*. Brighter regions are fossilized bone and darker regions are matrix-filled cavities. Cavities on each side are tooth roots. The three central regions are the principal nasal cavity (center) and accessory pneumatic paranasal cavities (lower left and right).

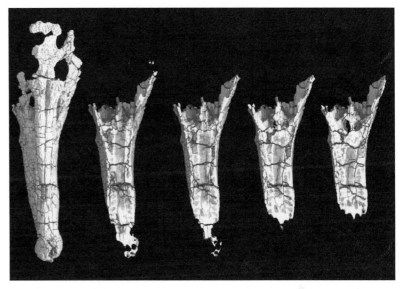

Fig. 2. 3D reconstructions derived from *Calsoyasuchus* CT data, showing the top of the specimen in its entirety (left) and selected cutaways showing tooth roots on each side and nasal cavities in the center. Entire skull is 380 (15 in.) long; cutaways are spaced at 2.5-mm (0.1-in.) intervals.

structural characteristics with modern birds, possibly linked with the demand of acquiring sensory information during flight, this study also revealed that pterosaur brains are proportionally smaller than those of birds but larger than those of living reptiles.

Trabecular structure. Trabecular bone, the spongy tissue found in many joints, dissolves and reprecipitates throughout the life of an organism, and in so doing, reorients itself to better accommodate the stresses that it experiences. This principle is being used to study the locomotive patterns of fossil primates. In one example, high-resolution (>40 μm; >0.0016 in.) scans of the femor (thigh bones) of the Eocene primates *Shoshonius* (**Fig. 3**) and *Omomys* were analyzed to determine variations in trabecular density, orientation, and degree of anisotropy (variation in strength with respect to direction). By comparing the observed patterns to those of similarly sized extant primates whose behavior patterns are known, it was possible to evaluate the extent to which they were specialized leapers as opposed to more generalized quadrupeds.

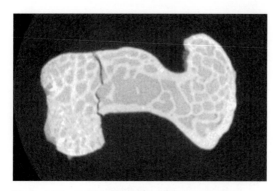

Fig. 3. A 9.1-μm-thick (0.00036-in.) CT slice showing trabeculae inside the femoral head of *Shoshonius cooperi*, an Early Eocene primate. Specimen width is 5.9 mm (0.23 in.). (*Image courtesy of Tim Ryan*)

Challenges. The primary challenge of CT scanning of fossils lies in their variability, which precludes the development of standard imaging procedures or protocols. Scanning success depends on a number of factors, including composition, size, and shape. Because CT detects compositional differences, successive episodes of remineralization will often diminish the contrast between bone and matrix; as a result, older (Paleozoic) specimens are often more problematic than younger ones. Likewise, scanning is frequently unsuccessful for invertebrates, as specimens usually consist of carbonate structures in a carbonate matrix. Specimen size is important because it is necessary to balance the need for sufficient x-ray penetration, which requires more and higher-energy x-rays for larger samples, with the desire for detail and material contrast, which benefits from the use of more tightly focused, lower-energy x-rays. Specimen shape is influential, as x-ray paths can have very divergent lengths in flat or irregular specimens, potentially causing scanning artifacts.

Another challenge inherent in the use of CT data is the size of the data sets. Typical CT image stacks have grown from a few megabytes in the 1980s to gigabytes today. The computational resources required to manipulate these data are at the leading edge of contemporary technology. This trend is likely to continue, as imaging capabilities roughly keep pace with computational ones.

Future. The use of CT in paleontology is continuing to grow rapidly, as appropriate scanners and expertise in both scanning and utilization of imagery become more widespread. One of the great potentials being realized is for dissemination of CT data via the Internet. Whereas in the past paleontologists would need to travel to museums around the world to study and take measurements of fossils species they are researching, the digital archive provided by CT data allows frequently superior observations to be made from one's own desk.

For background information *see* ARCHAEORNITHES; COMPUTERIZED TOMOGRAPHY; DINOSAUR; FOSSIL; PALEONTOLOGY; PTEROSAURIA; SKELETAL SYSTEM in the McGraw-Hill Encyclopedia of Science & Technology.   Richard A. Ketcham

Bibliography. F. Mees, *Applications of X-ray Computed Tomography in the Geosciences*, Geological Society, London, 2003; T. Newton and D. Potts, *Technical Aspects of Computed Tomography*, Mosby, St. Louis, 1981; T. Rowe et al., *Thrinaxodon Digital Atlas of the Skull*, University of Texas Press, Austin, 1993.

Denali earthquake

Strike-slip faults are vertical fractures of the Earth's crust parallel to the relative motion (usually horizontal) between two crustal blocks. Rupture of such faults occurs when accumulated shear stresses due to frictional resistance between the sliding blocks is relieved by sudden lateral shifting (slipping) of the blocks. The rupture of the Denali fault in Alaska during the moment magnitude (M_W) 7.9 earthquake of November 3, 2002, ranks among the largest strike-slip ruptures of the past two centuries. Its rupture length and slip magnitude are comparable with those of the great California earthquakes of 1906 and 1857. Because surface ruptures accompanying large-magnitude earthquakes on strike-slip faults are rare, detailed observations of these phenomena have been scarce. The Denali rupture may provide the best modern analog for large events on the San Andreas Fault in California.

Earthquake and rupture pattern. The earthquake occurred along the Denali Fault, an active intracontinental fault that accommodates some of the oblique collision between the Pacific and North American plates, in a sparsely populated region, about 130 km (80 mi) south of Fairbanks, Alaska (**Fig. 1**). Due to the remote location, few were injured and no major buildings were damaged. The earthquake was preceded by a M_W 6.7 foreshock, about 22 km (14 mi) west of the earthquake epicenter. Surface rupture occurred along three faults, the Susitna Glacier, Denali, and Totschunda faults (Fig. 1). The rupture pattern is complex, with about 320 km (200 mi) of surface rupture beginning in the west along the Denali and Susitna Glacier faults and extending southeast on the Totschunda Fault (**Fig. 2**).

The westernmost surface rupture occurred along 40 km (25 mi) of a previously unknown fault named the Susitna Glacier Fault, located about 15 km (9 mi) south of the Denali Fault. Surface rupture along this reverse fault (a fault in which one block moves up relative to the other) lifted the northern side of the Alaska Range relative to the south side. Vertical displacements average about 1.5 m (5 ft) and peak at about 4 m (13 ft) along the 40-km (25-mi) surface rupture (Fig. 2 and **Fig. 3**). Rupture then propagated to the east along the Denali Fault, producing large right-lateral surface rupture—that is,

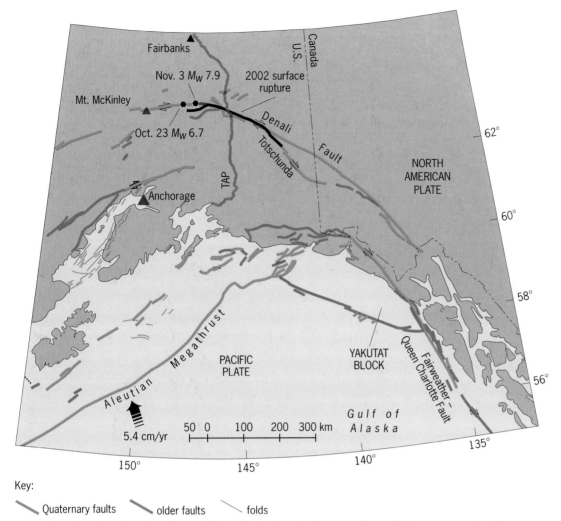

Fig. 1. Tectonic setting of southern Alaska. Plate motion is shown by the arrow: 54 mm (2 in.)/yr convergent rate (*from C. DeMets et al., Effect of recent revisions to the geomagnetic reversal time scale on estimates of current plate motions, Geophys. Res. Lett., 21:2191–2194, 1994*). Surface rupture of the 2002 Denali earthquake is in black. TAP, trans-Alaska pipeline. (*Modified from D. Eberhart-Phillips et al., The 2002 Denali fault earthquake, Alaska: A large magnitude, slip-partitioned event, Science, 300:1113–1118, 2003*)

the crustal block on the far side (relative to the Susitna Fault) moved right relative to the near side.

Right-lateral slip along the Denali Fault from the 2002 earthquake diminished from approximately 9 m (29.5 ft) in the east to about 3 m (10 ft) in the west (Fig. 2). Along the 210-km-long (5-mi) stretch of the Denali Fault, right-lateral offsets average about 4 m (13 ft) and reach values as high as about 9 m (29.5 ft), 170 km (106 mi) from the epicenter (**Fig.** 4). Vertical displacements along the Denali Fault are commonly about 20% of the strike-slip offset. At the eastern limit of surface rupture, right-lateral slip drops dramatically from about 5 m (16 ft) to about 2 m (6.5 ft) across a broad extensional valley between the Denali and Totschunda faults. Surface rupture continued to the southeast along the Totschunda Fault for about 76 km (47 mi); right-lateral offset along the Totschunda fault averages a modest 1.5 m (5 ft). The region between the Denali and Totschunda faults consists of a series of nor-

mal faults (one block moves down relative to the other), with displacements as high as 2.7 m (9 ft). The Denali Fault did not rupture east of the Totschunda Fault intersection.

The locations of large vertical and lateral surface offsets correlate with locations of high seismic moment release (essentially, energy radiated by seismic waves), as determined from geodetic and strong-motion data. Geodetic measurements use Global Positioning System (GPS) observations to determine changes in ground positions. Strong-motion data include data from strong-motion seismographs. Seismic moment calculations from these data agree with geologic moment estimates of M_W 7.8. Geodetic modeling suggests shallow slip to a depth of about 15 km (9 mi) and a deep slip patch about 110 km (68 mi) east of the hypocenter (the point inside the Earth, as opposed to its surface, where an earthquake originates).

Effects. The earthquake triggered thousands of landslides or rock avalanches. The largest rock

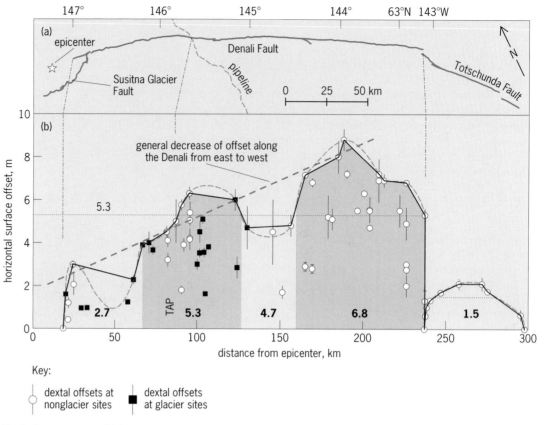

Fig. 2. Rupture pattern. (*a*) Surface rupture of the Denali Fault in November 2002. Rupture initiated on the Susitna Glacier thrust fault and propagated to the east along the Denali and Totschunda faults. (*b*) Right-lateral surface offset along the Denali and Totschunda faults. Bold numbers indicate average slip of maximum values in each fault segment of the Denali and Totschunda faults. Broken line shows general westward decrease in offset along the Denali. TAP, trans-Alaska pipeline. (*Modified from D. Eberhart-Phillips et al., The 2002 Denali fault earthquake, Alaska: A large magnitude, slip-partitioned event, Science, 300:1113–1118, 2003*)

Fig. 3. View to the northeast of 2002 surface rupture along the Susitna Glacier reverse fault.

avalanche consisted of about 30 million cubic meters of rock and ice that collapsed from McGinnis Peak and traveled about 10 km (6 mi) along the McGinnis Glacier. In general, landslides were concentrated along a narrow, 30-km (19-mi) zone along the surface ruptures.

The earthquake also triggered bursts of local seismicity to distances as great as thousands of kilometers. Most regions with triggered seismicity are located in active volcanic and geothermal areas. For example, strong surface waves from the earthquake had a profound impact on the hydrothermal systems of Yellowstone National Park, over 3100 km (1926 mi) away. In addition, seismic waves from the earthquake caused remotely triggered earthquakes in northern and central California (for example, Mammoth Lakes and Coso geothermal field) and along the Wasatch Fault Zone in Utah.

Implications. The trans-Alaska pipeline, supplying about 17% of the United States' domestic petroleum supply, withstood about 5.5 m (18 ft) of right-lateral crustal offset without substantial damage. This is perhaps the first case where a large-diameter crude oil pipeline, specifically designed for major earthquake hazards, substantially withstood near-fault ground shaking approaching or exceeding design criteria.

Although the 2002 earthquake ruptured mainly along the Denali Fault, only the central part of the fault ruptured during the earthquake. Why did slip end to the west, before reaching Denali National Park? It is possible that the rupture was impeded by a sharp 20° bend in the fault near the epicenter. Alternatively, perhaps the western stretch of the Denali Fault did not fail because it sustained a large rupture in the Recent, yet prehistoric past.

The 2002 rupture did not continue eastward on the Denali Fault but transferred slip onto the nearby Totschunda Fault. Preexisting fault scarps (landforms showing evidence of vertical movement) along the complex ruptures between the Denali and Totschunda surface ruptures attest to the recurring activity through the stepover (transitional region) between these two faults. This stepover may be analogous to a similar stepover between the San Andreas and San Jacinto faults of California and suggests that rupture of the northern part of the San Jacinto Fault could be part of a future rupture of the San Andreas.

For background information *see* EARTHQUAKE; FAULT AND FAULT STRUCTURES; SEISMOLOGY; TRANSFORM FAULT in the McGraw-Hill Encyclopedia of Science & Technology.   Charles M. Rubin

Bibliography. G. Anderson and C. Ji., Static stress transfer during the 2002 Nenana Mountain-Denali Fault, Alaska, earthquake sequence, *Geophys. Res., Lett.*, 30(6):1310, 2003; D. Eberhart-Phillips et al., The 2002 Denali fault earthquake, Alaska: A large magnitude, slip-partitioned event, *Science*, 300:1113–1118, 2003; S. Hreinsdóttir et al., Coseismic slip distribution of the 2002 MW 7.9 Denali fault earthquake, Alaska, determined from GPS measurements, *Geophys. Res. Lett.*, 30:1670, 2003; R. Kayen, D. Keefer, and B. Sherrod, Landslides and liquefaction triggered by the M 7.9 Denali Fault earthquake of 3 November 2002, *GSA Today*, 13(8):4–10, 2003.

Fig. 4. Right-lateral offest of about 10 m (33 ft) along the Denali Fault. The photograph was taken in June 2003. The geologist is shown for scale. View is upstream, to northeast.

Dinosaur growth

An amazing feature of many dinosaurs was their size, and the group included some of the largest terrestrial animals to inhabit Earth. But how did the largest dinosaurs get so big? Did they grow slowly, like living reptiles, reaching adulthood over periods perhaps exceeding a century? Or did they grow very rapidly, like living birds and mammals, attaining adult size in just a few decades? How did birds, now known to be direct descendants of carnivorous dinosaurs, acquire their exceptionally fast growth rates? Was it through their dinosaurian ancestry?

For the past 150 years scientists have pondered questions of dinosaur growth. However, only in the last few years have they been able to accurately assess dinosaur growth rates. This achievement was made possible through the merging of several techniques: (1) determining the age of dinosaur specimens throughout development using growth line counts, (2) making of whole-animal body mass estimations using leg bone measurements, and (3) analyzing growth data with respect to evolutionary history. The result has been the first quantified growth curves for dinosaurs from which basic dinosaur life history parameters (such as growth rates, longevity, and age at maturity) can be discerned. These methodological advancements are a gateway to a much more comprehensive understanding of dinosaur biology in the future.

History of investigation. Throughout the history of dinosaur research, opinions about biological aspects of these animals have differed tremendously.

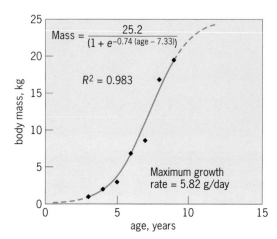

$$\text{Mass} = \frac{25.2}{(1 + e^{-0.74\,(\text{age} - 7.33)})}$$

$R^2 = 0.983$

Maximum growth
rate = 5.82 g/day

Fig. 1. S-shaped growth curve for the ceratopsian dinosaur, *Psittacosaurus mongoliensis*. This curve was generated from longevity and body-mass estimates for a growth series (juvenile through adult stage specimens) of this species. The steepest portion of the curve is known as the exponential stage of development. During this period animals are growing with their maximum potential and acquire the majority of their adult size. When contrasting growth rates between different animals, comparative animal physiologists standardize to size by using adult body mass and make the comparison using exponential stage rates.

Resemblance to living reptiles. The first scientists to study the enormous remains of dinosaurs believed that they resembled the slow-growing reptiles of today, just scaled up to giant size. Others pointed to their erect posture as evidence of their physiological kinship with rapid-growing birds and mammals. By the middle of the twentieth century the

former hypothesis won favor in professional circles. The reason was twofold. First, because of the great weight and fragility of fossilized bones, mounting very large centerpiece dinosaur specimens in the major museums of the world posed a considerable engineering challenge. Thus, museums were forced to mount their specimens in mechanically stable, tail-dragging poses, which fostered the perception of these animals as giant, lethargic versions of today's slow-growing reptiles. Second, early in the twentieth century, evolutionary biologists professed that dinosaurs were evolutionary failures, outcompeted by metabolically superior, fast-growing birds and mammals that consequently replaced them as the dominant denizens on earth.

Resemblance to living birds and mammals. During the late 1960s, in a period nick-named the "dinosaur renaissance," a reanalysis of dinosaur biology occurred, and the notion of these animals as physiological equivalents of living birds and mammals was rekindled. This change of heart was fueled by, among other things, (1) the discovery of *Deinonychus*, a dinosaur with an extremely athletic build (some scientists considered this proof for its being physiologically similar to birds); (2) evidence garnered from anatomical studies of *Deinonychus* and other theropods (carnivorous dinosaurs) confirming that birds are in fact the direct descendants of dinosaurs and hence are dinosaurs themselves; and (3) analogies drawn between the community structure of dinosaurs and ecosystems from present-day Africa dominated by large, fast-growing, endothermic mammals.

At this time there was also considerable interest in the metabolic status of dinosaurs. Researchers seized on the fact that animals today that grow exceptionally fast, such as birds and mammals, have bones that are highly vascularized. On the other hand, bones of exceptionally slow-growing animals, such as living amphibians and reptiles, are nearly avascular and solid. By showing that dinosaur bones were highly vascularized, these investigators provided strong evidence that dinosaurs grew more rapidly than living reptiles. Some took this as documentation that dinosaurs were physiologically similar to extant birds. It was proposed that the growth rate of avians was inherited from their dinosaurian ancestry, not via the evolution of birds proper, as was once believed.

Subsequent research has shown that although the vascularization of dinosaur bones was similar to that of birds and mammals, there were also growth lines resembling those typically found in reptiles. Such lines indicate that growth completely stopped for part of the year. (This is a common occurrence in ectothermal reptiles whose activities come to a standstill during adverse times, such as annual droughts, or at the onset of winter.) This discovery forced a reanalysis of how dinosaurs and early birds really grew. Some interpreted the blend of reptilian and avian/mammalian features to mean that dinosaurs grew at rates intermediate between the rapid rates seen in living birds and mammals and the slow rates

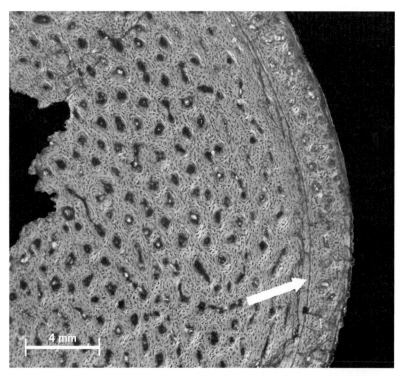

Fig. 2. Section of the femur from *Shuvuuia deserti*, a tiny carnivorous dinosaur, showing highly vascularized bone and a growth line. Vascular canals appear black. An annual growth line used in assessing age is denoted by the arrow.

4 mm

of extant reptiles. Implicit in this argument is that birds acquired some of their rapid growth rates through dinosaurian ancestry, and the rest presumably with the evolution of birds proper.

Establishing age/body-mass growth curves. Clearly, solving the mystery surrounding dinosaur growth rates and the genesis of rapid avian growth required the generation of data comparable to that used by animal physiologists when contrasting growth rates among living animals. This is done by establishing age/body-mass growth curves for animals throughout development. From the S-shaped curves (**Fig. 1**), the maximal growth rates can be determined at the point of maximal slope during the exponential stage of development. Such values are standardized to body mass (that is, same-sized animals are compared) to negate the effects of shape differences between animals. Comparisons are made only between animals of equivalent size. Using this methodology, it is possible to quantitatively show, for instance, that a typical 20-kg (44-lb) reptile grows at rates 15 times slower than a living eutherian (placental) mammal or precocial (active and mobile from birth, for example chickens and ducks) bird of comparable size.

In the early 1990s it was shown that dinosaur longevity could be determined through growth line counts. Fossilized bones were sectioned with diamond saws, affixed to microscope slides, and polished to a thickness of less than 1 mm so that light could be passed through them using polarized light microscopy. Annual growth lines were revealed by using this technique (**Fig. 2**), total growth line counts were made, and the age of each specimen was determined. Some of the giant sauropods were shown to have lived 70 years, whereas the tiniest theropods were not found to have reached more than 5 years of age. Application of the same technique to a growth series of dinosaur specimens (juveniles, subadults, and adults) allowed for the first aging of animals throughout development.

The final missing piece in making growth curve reconstruction feasible was assessing the body mass of the various animals representing each growth stage. The mass of the largest specimen could readily be estimated using regression equations for predicting adult mass of living animals from leg bone circumference measurements. However, such equations are invalid for subadult animals. (Changes in bone size within a species differ from changes that occur between species.) A means around this problem was found by applying a scaling principle called developmental mass extrapolation (DME). This principle is based on the observation that in some bones, such as the femur (thigh bone) in crocodilians (the closest living relatives to dinosaurs) and birds (the closest living relatives to carnivorous dinosaurs), increases in element length (a linear measure, x) correspond proportionally to increases in body mass (a cubic measure, x^3)—that is, as an animal's length doubles, so does its width and height in a very predictable manner throughout development. Therefore, it is possible to infer animal sizes for all individuals in a growth

series by using femur measurements as a proxy for body mass.

The DME principle was first applied to the ceratopsian (horned-dinosaur) *Psittacosaurus*. First, a measurement of a femur length for each specimen was made. Next, each value was cubed (x^3) and the proportion of each value compared with that of the adult (x^3) value for which body mass was known from the bone circumference equation. The subadult proportions were then multiplied by the adult mass value, revealing the corresponding value for each subadult animal. By coupling these data with the longevity estimates from the growth line counts, the first age/mass-growth curve for a dinosaur was revealed (Fig. 1). For the first time, the whole-animal maximal growth rate of a dinosaur could be compared with the rates of living vertebrates (**Fig. 3**).

Testing the competing hypotheses. Testing the various hypotheses of dinosaur growth rates and the evolution of avian rapid growth required a two-step approach: (1) data from a diversity of dinosaurs was obtained to establish rates typical for the group, and (2) these rates were then compared with those typical of living reptiles, birds, and mammals. To do this, researchers utilized a growth series for a diversity of dinosaurs, including some of the smallest and largest

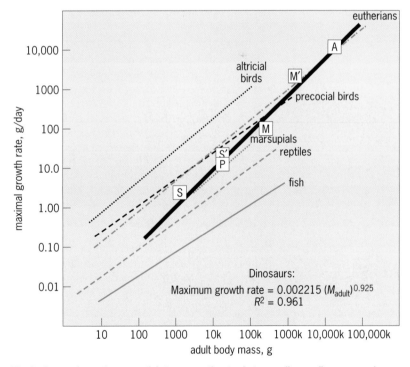

Dinosaurs:
Maximum growth rate = $0.002215 (M_{adult})^{0.925}$
$R^2 = 0.961$

Fig. 3. Comparison of exponential stage growth rates between diverse dinosaurs and living animal groups. The letters represent dinosaurs for which exponential stage rates were discerned from growth curve reconstructions: S = *Shuvuuia*, P = *Psittacosaurus*, S' = *Syntarsus*, M = *Massospondylus*, M' = *Maiasaura*, A = *Apatosaurus*). The bounds of the dinosaur regression line represent predicted rates for the smallest known dinosaur, *Microraptor*, and the largest, *Argentinosaurus*. All dinosaurs grew more rapidly than living reptiles. Dinosaurs appear to have had a unique growth trajectory that cannot be analogized with the major living animal groups. Notably, the rates for small dinosaurs from which birds evolved are considerably lower than the precocial rates seen in the most primitive birds today. This, coupled with evidence that early bird bone was slow-growing, suggests that precocial birds acquired their rapid growth rates in a two-step fashion—partially through their dinosaurian ancestry and then later after they diverged from other dinosaurs.

known dinosaurs, some of the earliest and latest, and representatives from each of the major groups, as well as those closely related to birds. Growth curves for each were obtained, and from these curves the exponential stage rates could be discerned and plotted with respect to adult body mass. These data were then compared with those for living groups (Fig. 3).

Dinosaur growth rates. The results revealed that dinosaur growth rates ranged from 0.33 g/day for the tiny carnivorous dinosaur *Microraptor* to as high as 56 kg/day for the enormous long-necked herbivore *Argentinosaurus*. All dinosaurs grew more rapidly than living reptiles of similar size. Dinosaurs as a whole had a characteristic growth pattern that is atypical of living reptiles, birds, and mammals. Notably, the trajectory of their pattern crosses the regression lines of birds and mammals. Clearly, the new data on dinosaur growth rates do not substantiate any of the current hypotheses for how dinosaurs grew.

Evolution of avian rapid growth. The data show some interesting findings regarding the evolution of avian rapid growth. DME results from the smallest dinosaurs sampled (which represent carnivorous forms most closely related to birds) show growth rates two to seven times slower than the condition seen in precocial birds. This suggests that birds attained a small portion of their rapid growth rates through their dinosaurian ancestry, but the jump to modern avian levels came either with or after the evolution of the first birds.

Current work on fossil bird bone histology indicates that the earliest birds had bone histology similar to their smaller dinosaur ancestors, that is, moderately vascularized but with prevalent growth lines. This finding directly contrasts with the condition in living birds, in which highly vascularized bones without growth lines are encountered. This suggests that the earliest birds, such as *Archaeopteryx*, were physiologically little more than feathered dinosaurs with typical dinosaurian growth rates and that the rapid growth rates of living birds evolved millions of years later.

For background information *see* ANIMAL GROWTH; AVES; DINOSAUR; FOSSIL; ORGANIC EVOLUTION; REPTILIA in the McGraw-Hill Encyclopedia of Science & Technology.   Gregory M. Erickson

Bibliography. A. Chinsamy et al., Growth rings in Mesozoic birds, *Nature*, 368:196–197, 1994; G. M. Erickson et al., Dinosaurian growth patterns and rapid avian growth rates, *Nature*, 412:361–462, 2001; G. M. Erickson and T. A. Tumanova, Growth curve of *Psittacosaurus mongoliensis* Osborn (Ceratopsia: Psittacosauridae) inferred from long bone histology, *Zool. J. Linn. Soc.*, 130:551–566, 2000; D. K. Thomas and E. C. Olson (eds.), *A Cold Look at Warm-Blooded Dinosaurs*, Westview, Boulder, 1983.

Directed evolution

Directed evolution is an attempt to replicate in the laboratory nature's method of developing protein catalysts (enzymes) with desirable properties or for particular reactions. In this technology, the amino acid sequence of a known enzyme is randomly mutated to create millions of different versions; the variants are then screened to reveal which has particular attributes or is best suited to perform a specific task. Over the last decade, directed evolution has become an important tool for enzyme discovery in both academia and industry.

Enzymes. Enzymes are nature's catalysts, carrying out in cells the chemistry that keeps organisms alive. Critically, they accelerate the rate of biological reactions (some by as much as 10^{17}-fold over the uncatalyzed reactions) which otherwise would be too slow to sustain life. Despite their high efficiency, they can reliably select one or a few molecules from among many they encounter, and carry out very specific reactions on these substrates. Unlike traditional chemical synthesis, these reactions are accomplished under very mild conditions of temperature and pressure, while generating little waste.

Given their useful properties, there is enormous interest in using enzymes for reactions outside cells. Enzymes are already used with great success in industry, in chemical synthesis, as reagents for biomedical research, in bioremediation, and more recently as therapeutics. However, a continuing barrier to exploiting these catalysts is that they have been specialized for particular functions within living organisms and so are generally unsuited for other activities.

Enzymes are selective about which molecules they will accept (that is, they have substrate specificity) and which reactions they will catalyze. They are designed to operate in the watery environment of the cell and so may not tolerate nonbiological conditions. In addition, because of participating in tightly controlled metabolic networks within cells (in which feedback inhibition prevents overproduction), many stop working when they encounter even small amounts of their own products. To be useful for biotechnology, enzymes should accept just about any substrate, be functional and stable under a wide range of unnatural conditions including organic solvents, and be active for long enough to make lots of product. In fact, the most desirable enzymes would be able to carry out reactions never performed in nature.

Engineering enzymes. For the past few decades, researchers have addressed these limitations by attempting to engineer enzymes—to change their structures at the amino acid level through appropriate modification of their encoding deoxyribonucleic acid (DNA). Typically, they aim to increase robustness and activity under a desirable set of conditions; to introduce selectivity toward a new substrate; or most ambitiously, to elicit an entirely new function from an existing catalyst.

In principle, it would seem that there is considerable scope for modifying or improving enzymes, because evolution can have sampled only a tiny fraction of the possible protein sequences. However, most completely random sequences are devoid of function. Therefore, the most useful starting point for engineering is an enzyme that is already very similar to the desired catalyst in terms of its properties, reaction mechanism, or substrate preference. Nonetheless, it is still difficult to decide where, in the vast sequence space of the existing enzyme, to look for a new activity or an enhanced feature. Any laboratory-based exploration of this space can reach only a small subset of the possible sequences and so must be carefully designed.

Directed evolution. For the last 10 years, researchers have used a methodology called directed evolution. The selling point of directed evolution is that it does not require any understanding of the complex relationship between protein structure and function. Instead, it lets the experiment reveal the best solution to the engineering problem by implementing a form of laboratory-based natural selection. The rationale here is that natural selection has a proven record in enzyme development, having successfully adapted natural catalysts to a wide range of environments, including volcanic acids, alkaline lakes, arctic tundra, and ocean depths. In the lab, however, the selection takes weeks instead of millions of years.

As in the natural process, directed evolution requires a large pool of enzymes with different sequences. In nature these variants result from random mutation of the coding DNA, while in the lab genetic diversity is deliberately created (the larger the number of enzymes, the better the chance of finding one molecule with the desired behavior). In nature, "the winning" enzyme (the one that makes it into the next generation) is the one which happens to confer the greatest reproductive benefit on its host; there is no particular direction in mind. In contrast, the lab version is strictly goal-oriented. The challenge is to select for a particular new trait or ability, many examples of which would not be essential for an organism's survival in the wild.

Experimental design. The first step in any directed-evolution experiment is to create large numbers (typically tens of thousands) of gene sequences on which the selection regime can act (see **illustration**). To this end, researchers have developed ways to make the normal process of duplicating DNA [called polymerase chain reaction (PCR)] error-prone so that on average each new copy of a gene contains a single random mistake. Another strategy introduces sequence variation along the entire length of each gene by combining large chunks of DNA from multiple parents. Originally this "gene shuffling" technique depended on the parent sequences being similar (homologous), but newer methods allow sequences with much lower homologies to be mixed together, creating even more diversity.

In the next stage, the libraries of DNA sequences are introduced into microbial hosts (typically bacteria or yeast), so that each cell receives a slightly different version of the gene. The microbes contain all the essential supplies needed to produce proteins, and so the introduced genes are turned into enzymes within the cells. In this way, the instructions for mak-

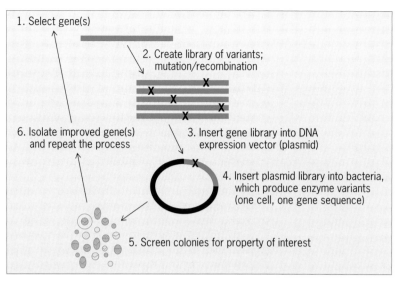

1. Select gene(s)

2. Create library of variants; mutation/recombination

3. Insert gene library into DNA expression vector (plasmid)

4. Insert plasmid library into bacteria, which produce enzyme variants (one cell, one gene sequence)

5. Screen colonies for property of interest

6. Isolate improved gene(s) and repeat the process

Schematic of typical directed evolution experiment, where variant genes have been created by error-prone PCR.

ing the enzyme and the enzyme itself are compartmentalized together.

The next step is to evaluate or screen the mutant enzymes one by one for the greatest improvement in the desired property. The screen is obviously critical because it can reveal only what it has been designed to measure. In the rare cases that the cell's survival can be made to depend on a new enzyme's activity (called in vivo complementation), selection of the best catalysts is straightforward. More typically, the experimenter has to be able to analyze a large number of variants for particular traits without sacrificing the ability to measure even minor improvements. To this end, individual cells are usually grown in multiwell plates to form tiny cultures (as many as 1536/plate), whose properties can then be evaluated using a wide array of high-throughput analytical devices.

To date, most screening for activity has been based on turning substrate into a fluorescent or highly colored product. Cells harboring the most efficient enzymes fluoresce intensely or turn bright yellow, and these signals can be measured by a machine, allowing these cells (and therefore the enzymes inside them) to be selected. As this approach is inherently limited to certain reactions, researchers are creating sophisticated screening methods, which in principle are applicable to any type of chemistry. For example, engineers have devised an assay, called chemical complementation, where the level of catalysis by an enzyme is linked to the activation of a "reporter" gene, which in turn produces an obvious change in the cells' appearance (such as turning the cells from blue to white). The model system is already suitable for a range of enzymes that catalyze both bond forming and breaking reactions, such as glycosyltransferases, aldolases, esterases, amidases, and Diels-Alderases.

The final step in the process is to retrieve the genes encoding the most useful enzymes from the selected cells and use them as parents for the next generation of potential catalysts. The mutation/selection process is then repeated in the hope that beneficial mutations will begin to accumulate. In the best cases, the result is a new enzyme that is highly evolved for a particular function or attribute. It is interesting that sequencing of the resulting mutants often reveals changes that would never have been predicted rationally through using the three-dimensional structure of the protein, validating this random approach to enzyme engineering.

Successes. Over the last decade, directed evolution has demonstrated its ability to both improve enzymes and generate novel biocatalysts (see **table**). A significant focus has been to optimize the properties of enzymes used in industrial processes, particularly their productivity and stability at high temperatures and at a range of pH. Popular targets have included the alkaline proteases, amylases, and phytases. Directed evolution has also been applied to tune the mechanisms and substrate specificities of enzymes, such as the mono- and dioxygenases, for use in fine-chemical synthesis. Catalysts have also been evolved for use as human therapeutics and for detoxification and bioremediation, such as for the degradation of organophosphates and chlorinated aliphatics. Despite these successes, no one has yet managed to elicit a fundamentally new chemistry from an existing enzyme structure, and this continues to be the "Holy Grail" of enzyme redesign.

For background information *see* BIOCHEMICAL ENGINEERING; CATALYSIS; DEOXYRIBONUCLEIC ACID (DNA); ENZYME; GENE; GENE AMPLIFICATION; GENETIC ENGINEERING; MUTATION in the McGraw-Hill Encyclopedia of Science & Technology.

Kira J. Weissman

Bibliography. J. Affholter and F. H. Arnold, Engineering a revolution, *Chem. Brit.*, 35:48–51, 1999; F. H. Arnold, Unnatural selection: Molecular sex for fun and profit, *Eng. Sci.*, 1/2, 1999; E. K. Wilson, Prospecting for proteins—Chemical engineers' meeting highlights the big business of directed evolution, *Chem. Eng. News*, 79:49–52, 2001.

Display manufacturing by inkjet printing

Piezoelectric drop-on-demand (DOD) inkjet technology is most often selected for manufacturing processes such as printing electronic components (resistors and capacitors) and circuit board materials (solder, solder mask, and conductive traces). Piezoelectric DOD printheads offer the critical combination of high productivity, high reliability, and uniform characteristics (such as drop volume consistency, velocity characteristics, and jet straightness) for jetting materials dissolved or dispersed in organic or aqueous media. In recent years, it has proved possible to design and manufacture highly productive and reliable printheads which meet the exacting dispensing requirements in flat-panel display manufacturing.

Piezoelectric DOD printing. In its simplest embodiment, a piezoelectric DOD printhead ejects ink droplets because the piezo element changes its shape or dimensions slightly in response to an applied voltage. A rapid but small displacement of the piezoelectric element generates a pressure wave in the pumping chamber so that a drop is expelled for each voltage pulse (**Fig. 1**). Piezo inkjets can deposit almost any fluid that is both low in viscosity (typically 8–15 mPa · s) and relatively nonvolatile to avoid drying in the nozzle between firings. Printhead materials can be selected so that the printhead is resistant to many organic solvents and even acidic aqueous dispersions. **Figure 2** shows the structure of one such printhead.

The quality and uniformity of drop formation affect wetting around the nozzle opening, and consequently affect both jetting straightness during printing and nozzle maintenance after printing. Jetting reliability is improved for fluids with high surface tension and newtonian behavior. Fluids with high surface tension typically form spherical drops quickly after breakoff, minimizing satellite (small, trailing droplets) formation, as well as the length of the attached ligament (tail). Non-newtonian behavior, which results from the presence of high-molecular-weight materials, influences drop formation and drop breakoff. For these fluids, very long ligaments and delayed drop formation are characteristic.

Methods have been developed to fabricate arrays (containing more than 100 nozzles) which eject uniform drops, allowing complex patterns of drops to be deposited at high rates without sacrificing drop

---

Examples of biocatalyst properties improved by directed evolution, 2001–2003

Altered property	Target enzymes
Increased thermal stability	Lipase B Maltogenic amylase Glucose dehydrogenase Galactose oxidase Subtilisins
Increased activity	α-Arylase Alkaline phosphatase Amidase Subtilisins
Altered substrate specificity	Cytochrome P450s DNA repair enzymes Arsenate reductase Butyrylcholinesterase DNA polymerase Cre recombinase Catalase Lipase
Altered reaction stereospecificity	Lipase Aldolase Esterase
Detoxification/ bioremediation	Cytochrome P450s Toluene *O*-monooxygenase Biphenyl dioxygenase

volume control. Drop volume constancy for all the jets is achieved by using special electronics that permit precise adjustment of the operating voltage for each nozzle. Precise placement of each drop is important in obtaining uniform pixel fill and is necessary to avoid contaminating adjacent pixels.

Almost any operation requiring the precise metering of materials to specified locations on a substrate is a candidate for piezo inkjet printing. In inkjet printing, soluble or dispersible electronic materials (that is, the fluids) are precisely deposited additively only where they are required, reducing the waste of expensive materials and the cost associated with waste disposal. Inkjet printing is a noncontact and noncontaminating method that is compatible with clean room standards. Inkjet printheads are used primarily because digital printing enables the production of variable patterns without retooling and promises significantly lower production costs because of process simplification and reduced capital investment.

Flat-panel display applications. Although inkjet printheads have often been considered as graphic arts printing tools for the home and office, their use in the flat-panel display industry has proved most effective in creating relatively simple images that require precise control of drop placement and drop volume. DOD inkjet printheads operate with high reliability and speed, making piezo technology an ideal match for the precise metering and deposition requirements of flat-panel display manufacturing.

New market and technical opportunities for flat-panel displays have encouraged the development of new manufacturing approaches. As an example, for liquid crystal displays (LCDs) to compete with plasma displays in the large-screen television market, the cost of fabricating LCD panels must be greatly reduced. Most of the cost is in the materials, and major manufacturers are constructing factories to handle substrate of 4 m² (43.1 ft²). Numerous and complex steps are required to construct a LCD panel and some of these steps do not scale well for large substrates. LCD processes may capitalize on inkjet printing as an opportunity to save time, as well as to reduce the associated chemistry waste stream and possibly the thickness of the panels. Another example of display manufacturing that requires new technology is fabricating displays using polymer light-emitting diodes (PLED). The availability of electroluminescent or light-emitting polymers (LEPs) in all colors has moved the key manufacturing step

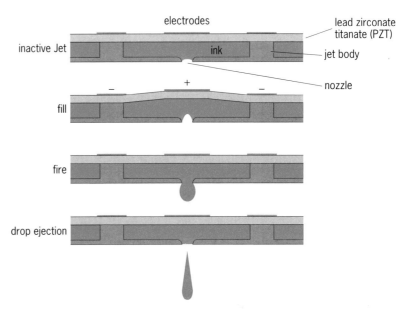

Fig. 1. Schematic for drop ejection from a piezoelectric drop-on-demand inkjet (shear mode).

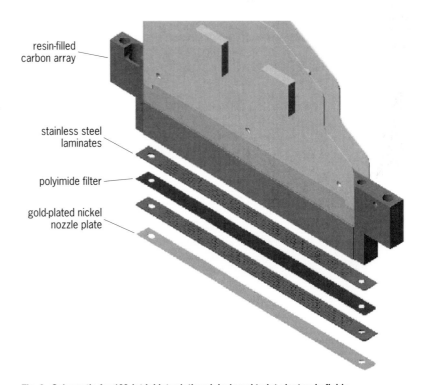

Fig. 2. Schematic for 128-jet inkjet printhead designed to jet electronic fluids. (*Spectra, Inc.*)

from spin coating (for monochrome displays) to jetting patterned films (for full-color displays).

Table 1 lists some of the flat-panel display applications for which inkjets are being evaluated. Serious consideration is being given to inkjet deposition of a uniform coating, for example, the polyimide layer required to align liquid crystals in LCDs. Thus far, fabrication of color filters and inkjet printing of light-emitting polymers have received the most attention. For both of these applications, the ability to print fluid in a well-defined area is required; the dried films

TABLE 1. Examples of flat-panel display manufacturing opportunities for piezo inkjet printheads

Type	Printing application
Full-color display	Light-emitting polymers
Full-color display	Small-molecule light emitters
LCD	Spacers
LCD	Color filters
LCD	Back planes
LCD	Alignment layers

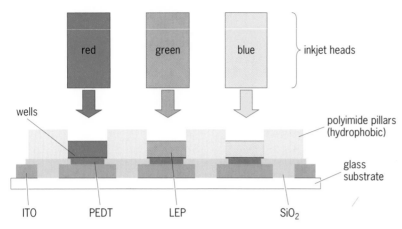

Fig. 3. Concept for inkjet printing full-color displays based on light-emitting polymers, where indium tin oxide (ITO) is the transparent electrode (anode) and PEDT [poly(3,4-ethylenedioxythiophene)] is the base (hole injection) layer. [*N. Beardsley, United States Display Consortium (USDC)*]

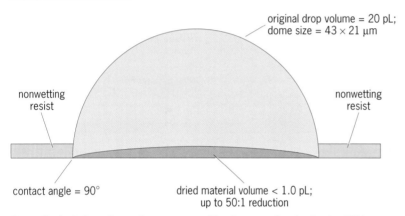

Fig. 4. Pixel printing using surface energy modification to confine the droplet. (*With permission of Litrex Corp., Pleasanton, CA*)

significantly larger than is often desired. Formation of uniform coatings requires a balance of forces. Once the fluid has been optimized for jetting reliability, it may not be sensible to further modify it to improve its coating characteristics. To compensate for fluid flow, surface energy and substrate morphology could be modified. Increasing the surface energy of the substrate would improve wettability. Another approach to improving coating uniformity would be to change the shape of the pixel itself; for example, by rounding the corners to minimize (thin) edge effects. Pixel printing with a dilute solution of light-emitting polymer is illustrated in **Fig. 4**. Quality of the resulting coating depends on control of drop volume and position plus fluids formulated to dry uniformly.

Drop size, uniformity, and placement accuracy are determined by the desired resolution of the target display. A typical pixel size for a high-resolution mobile phone display is 50 μm for each RGB color on a 200-μm superpixel (blocks of the 3 RGB pixels) center. The remaining area is used by a structured polyimide coating that separates the RGB color cells primarily by surface energy forces. Important requirements for material deposition include uniform pixel fill, even pixel wetting, and no cross contamination between color subpixels. These requirements can be translated into printhead and materials specifications as indicated in **Table 2**, providing the target inkjet specification for generation (Gen) 2 [substrate size = 400 × 500 mm (16 × 20 in.)] and Gen 3.5 [substrate size = 600 × 720 mm (24 × 28 in.)] equipment for the mass production of passive and active matrix full-color PLED displays. With a 50-μm pixel size, very few drops are needed to fill the cell with fluid. As the drop size decreases, more drops are used per cell. By adding additional drops, any variation in drop uniformity created by the inkjet head may be decreased. This results in a greater ability to control film thickness.

The specifications for a printhead to jet RGB fluids for a color filter for a high-resolution LCD display are essentially the same as those in **Table 3** except that the printhead materials need to be compatible with the vehicle for color filter inks. However, manufacturers are now beginning to look to inkjets for the manufacture of color filters for LCD TV. The superpixel size grows commensurate to the size of the display. As a result, each pixel increases in size, thus requiring more fluid to fill than does a high-resolution pixel.

must also be thin and uniform. **Figure 3** is a stylized representation of how displays utilizing LEP might be fabricated using an inkjet printhead for each color—red, green, and blue (RGB). Unique fluids have been developed so that RGB color filters can be digitally printed. The size of the color filter or RGB pixel is often defined by photopatterning a thin layer of polymeric material that then is plasma-treated to repel the jetted fluid.

Manufacturing requirements. Direct patterning of electronic materials with inkjets is limited to relatively large features because of drop spread. For example, a 5-picoliter drop may spread to a spot 50 to 100 micrometers, in diameter on a clean substrate,

TABLE 2. Inkjet printhead specification for PLED manufacturing		
Parameter	Gen 2, 130 pixels per inch	Gen 3.5, 200 pixels per inch
Drop placement accuracy at 1 mm stand-off	±10 μm	±5 μm
Drop velocity	4–8 m/s	4–8 m/s
Drop velocity uniformity	±10%	±10%
Drop volume	<15 pL	<5 pL
Drop volume uniformity	±2%	±2%
Maximum jetting frequency	> 5 kHz	>10 kHz
Materials compatibility	LEP solvents, PEDT	LEP solvents, PEDT

TABLE 3. Desired characteristics for inkjet printhead

Parameter	Desired value
Straightness	0.6° (all jets)
Drop volume	5–15 picoliter
Drop velocity	4 to 8 m/s
Volume/velocity uniformity	±2% from all sources; special electronics may be required
Operating temperature	Ambient to 55°C (131°F)
Materials compatibility	1–3 months may be acceptable
Maximum frequency	Up to 10 kHz

Demand for precision placement and drop volume uniformity remains the same, but productivity must increase. One solution is to use inkjet printheads with larger intrinsic drop volume so that only three to six drops will fill the pixel. Many jet nozzles, accurately aligned, are necessary to achieve high throughput for Gen 6/7 substrates. [Gen 6 may have a substrate size of about 1500×1600 mm (59×62 in.), and Gen 7 may be as large as 2000×2100 mm (79×83 in.).]

At present, manufacturing flat-panel displays with inkjet is in its infancy. Some of the possible opportunities are moving from research laboratories to the factory floor, and the success of these early adopters will determine the rate at which new manufacturing approaches become widespread.

For background information *see* ELECTROLUMINESCENCE; ELECTRONIC DISPLAY; FLAT-PANEL DISPLAY DEVICE; INK; LIGHT-EMITTING DIODE; LIQUID CRYSTALS; NEWTONIAN FLUID; PIEZOELECTRICITY; SURFACE TENSION in the McGraw-Hill Encyclopedia of Science & Technology.   Linda T. Creagh

Bibliography. H. Becker et al., Materials and inks for full-color polymer light-emitting diodes, Pap. 21.02, *SID Symposium Digest of Technical Papers*, SID, San Jose, CA, 2002; J. Bharathan and Y. Yang, Local tuning of organic light-emitting diode color by dye droplet application, *Appl. Phys. Lett.*, 72:2660, 1998; P. Dunn and B. Young, The emerging OLED landscape: The OLED emitter, *Display Search* (Austin, TX), vol. 1, no. 1, 2001; M. Fleuster et al., Pap. 44.2, *SID 04 Digest*, SID, San Jose, CA, 2004; P. Hill, *Display Solutions*, December 2003/January, no. 12, 2004; T. Shimoda et al., Inkjet printing of light-emitting polymer displays, *MRS Bull.*, 28(11):821, 2003.

Distributed generation (electric power systems)

Distributed generation (DG) refers to dispersed sources of electric power generation located close to the load they serve. Distributed generation sources can be connected with electric power systems at utility or customer sites at levels ranging from utility subtransmission or distribution voltages to substation or lower voltages (**Fig. 1**). The reliability, power quality, and environmental needs of utilities, businesses, and homeowners, as well as the availability of more efficient, environmentally friendly, and modular electric generation technologies, have stimulated the expanding interest in distributed generation (particularly onsite).

A distributed generation system consists of a number of functional building blocks to meet the requirements of distributed generation subsystems, local loads, and the electric power system (**Fig. 2**). Functional requirements indicate what the system needs to do, or services that the system must offer, but not what equipment should be selected or how it should accomplish the need. Nonfunctional requirements address a specific desired property or constraint on how the system will be implemented.

DG includes devices that convert energy to electricity directly (called prime movers) and energy storage systems. It also encompasses subsystems for electrical integration and interconnection with the electric power system. The distributed generation electrical integration functions are largely intertwined with technology-specific or equipment-specific aspects of subsystems. The interconnection system must be robust enough to accommodate the requirements, constraints, and risks posed by differing electric power system or grid-specific application scenarios.

Distributed generation technology. Numerous distributed generation technologies are available today. Figure 2 shows the major functional components of a generalized distributed generation system, where power flow functions are shown as rectangles and solid lines indicate the power flow path. Similarly, operation control functions are shown as ovals, and dotted lines indicate the control flow paths. The physical components of a distributed generation system may be discrete subsystems or pieces of equipment, but modern components often accomplish more than one function. Not shown in the figure is a functional module for thermal energy use, such as waste heat recovery from diesel engines and very small combustion turbines, called microturbines.

The prime mover or "engine" differentiates distributed generation systems from one another and is typically the most expensive subsystem. Prime movers may be traditional fuel-powered units or renewable-energy-powered units, and can change energy from one form to another. They may use direct physical energy to rotate a shaft (for example, in a hydro turbine or wind energy device), or they may use chemical energy conversion to change thermodynamic energy to physical energy (for example, burning diesel fuel, natural gas, or propane to move a piston or turbine to rotate a shaft).

The electric generator converts prime-mover energy to electrical energy. Historically, most electrical generation has been accomplished via rotating machines. The prime mover drives an electric generator, which is either a synchronous or an induction (asynchronous) machine. The generator has a stator that consists of a set of alternating-current (ac) windings and a rotor with windings, which are either ac

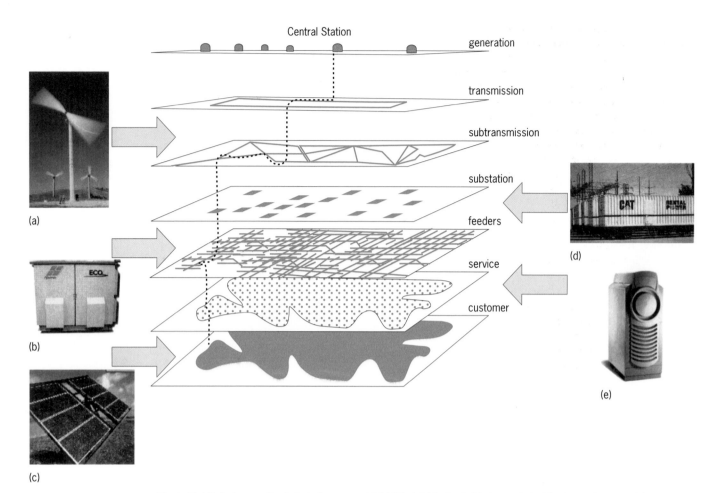

Fig. 1. Distributed generation in today's power grid. (*a*) Wind, (*b*) fuel cell, (*c*) photovoltaic, (*d*) microturbine, and (*e*) diesel power systems.

or direct-current (dc). The stator is stationary, and the rotor is rotated by the prime mover. The means of power conversion is through the interaction of the magnetic fields of the stator and rotor circuits.

Most large utility generators are synchronous generators. Induction generators, on the other hand, are well suited for rotating systems in which prime-mover power is not constant (such as for wind turbines and small hydro turbines), but they can also be used with engines and combustion turbines. The electrical output of the synchronous or induction generator electrical output may then undergo further conversion to achieve standards of power quality that are compatible with utility grid interconnection.

Solid-state converters are static power conversion devices that use hardware and software. They are based on three fundamental technologies: power semiconductor devices, microprocessor or digital signal processor technologies, and control and communications algorithms. Generally, solid-state converters are used with prime movers that provide dc electricity (such as photovoltaic or fuel cell units) or with microturbines, which are small, high-speed, rotating combustion turbines directly coupled to a synchronous-type electric generator. In the case of the microturbine, the generator output is high-

frequency voltage which is well above the utility grid's 50 or 60 cycles per second (Hz). The high-frequency waveform is rectified to dc form, and the dc electricity is then synthesized to a sinusoidal waveform suitable for utilization within the grid.

The remaining power flow functions of distributed generation interconnection are protective relaying and paralleling systems. Protective relaying provides proper response to contingencies (emergency, fault, or abnormal conditions) that arise at the distributed generation unit, on the loads, or on the utility grid. One example response is disconnection from the grid. Contingency conditions include short or open circuits, the utility grid or distributed generation voltage not meeting acceptable standard ranges, and loss of the utility grid.

The operational control function includes monitoring, control, and metering, which include autonomous and semiautonomous functions and operations. Distributed generation and load controls manage the status and operation of the distributed generation and local loads. The status includes on/off and power-level commands. This function can also control hardware to disconnect from the electric power system. Communications allow distributed generation and local loads to interact and operate as part of a larger network of distributed generation

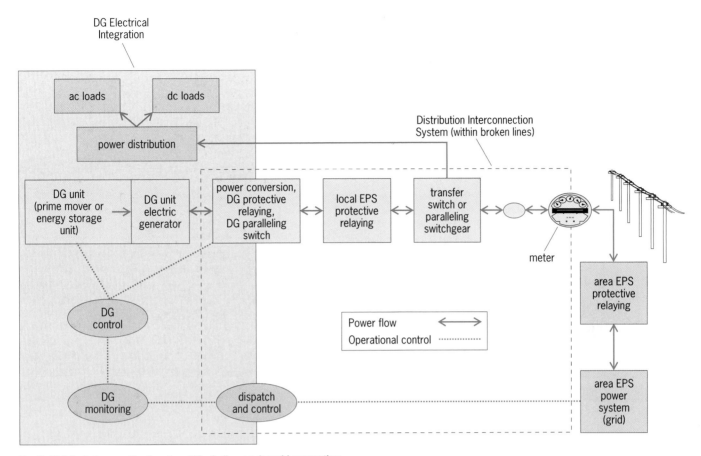

Fig. 2. Distributed generation functional block diagram for grid connection.

systems at end-user sites. The metering function allows billing for distributed generation energy production and local loads.

Distributed generation systems. The attributes and features of some specific distributed generation systems are discussed in the following examples.

Combustion turbines. Combustion turbines are an established technology, ranging from several hundred kilowatts to hundreds of megawatts. They are involved in both distributed generation applications and utility central-station generation. These turbines produce high-quality heat that is used to generate steam for additional power generation or onsite use. They can burn natural gas or petroleum fuels or can have a dual-fuel configuration. Their maintenance costs per unit of power output are among the lowest of combined heat and power technologies. Low maintenance requirements and high-quality waste heat make them an excellent choice for industrial or commercial cogeneration applications larger than 5 MW.

Microturbines. These are small combustion turbines, with typical outputs of 30–200 kW. Microturbines are sized appropriately for commercial buildings or light industrial markets for combined heat and power or power-only applications. Microturbine developers quote an electrical efficiency at the high-frequency generator terminals of 30–33%. The solid-state conversion component introduces about

5% additional losses in the conversion from high-frequency to 60-Hz power. Additional losses from parasitic loads up to 10% of capacity are accrued by the natural-gas fuel compressor required to increase typical gas delivery pressures of 2 psig (14 kPa) or less to 75 psig (500 kPa). These adjustments bring the electrical efficiency to less than 26% for some configurations. Future developments are expected to improve efficiencies and lower costs.

Photovoltaics. A photovoltaic cell is a semiconductor device that converts light (typically sunlight) striking the cell's surface into dc electricity at the cell's output conductors. A photovoltaic array is a collection of photovoltaic cells electrically connected to provide a desired current and voltage. The output of a photovoltaic array depends on the amount of sunlight striking it, its operating temperature, and its photovoltaic-cell semiconductor type. Unlike the output current of rotating machines, batteries, and other generator devices, the output current of a photovoltaic array is limited by photovoltaic-cell characteristics.

An inverter converts the array's dc output into utility-compatible ac. The inverter in a photovoltaic system often provides all utility interconnection functions and features. Unlike other prime movers, photovoltaic modules are not designed around a specific inverter, and inverters are not designed to work with only one photovoltaic type.

Fuel cells. Fuel cells and batteries are sources of electrochemical energy conversion that provide dc electricity. In a simple way, a fuel cell can be thought of as a battery that is continually recharged as it is discharged. A fuel cell has an anode connection that conducts electrons that are freed from pressurized hydrogen molecules. The anode typically has etched channels that uniformly distribute the pressurized hydrogen gas over the surface of a catalyst where hydrogen ions are established. The cathode is the other connection of the fuel cell that carries electrons back from the external circuit to the catalyst, where it combines with hydrogen ions and oxygen to form water, the by-product of the fuel cell. The electrolyte is the proton exchange membrane, which is a specially treated material that allows for the conduction of positively charged ions. It does not allow electrons to pass through. The catalyst is usually made of platinum powder, coated on carbon paper or cloth. The catalyst is porous to maximize the exposure (surface area) to the hydrogen and oxygen. The platinum-coated side of the catalyst faces the electrolyte. The oxygen and hydrogen reaction takes place in one cell, resulting in approximately a 0.7-V cell potential. To bring the operational voltage up to usable levels, many cells are connected in series.

Fuel-cell systems use inverters to condition the dc electric power to grid-compatible 60-Hz power and provide the utility interconnection capabilities and protective functions described previously.

Wind power. In 2003, United States wind power plants operating in 32 states had a total generating capacity of 6374 MW. In addition to large-utility-scale technologies, small wind systems for residential, farm, and business applications are in the range of 5–100 kW.

Wind energy systems are packaged systems that include the turbine blades, rotor, drive or coupling device, and generator. Most systems have a gearbox and a generator in a single unit behind the turbine blades. The generator output typically is fed to an inverter, which typically also performs the protective relaying and paralleling functions to connect with the utility distribution line. Wind-system performance depends on siting and wind conditions. A minimum wind speed is necessary for operation, and coastlines and hills are favorable locations. A major cost is the tower necessary to raise the wind turbine well above the ground or local obstructions.

Internal combustion engines. The reciprocating internal combustion engine is a well-known prime mover. Natural gas is the preferred fuel for power generation; however, these engines can burn propane or liquid fuels. Current internal combustion engines offer low-first-cost, easy startup, proven reliability when properly maintained, and good load-following characteristics. Internal combustion engines are well suited for standby, peaking, and intermediate applications and for packaged combined heat and power in commercial and light industrial applications less than 10 MW.

For background information *see* ELECTRIC DISTRIBUTION SYSTEMS; ELECTRIC POWER GENERATION; ELECTRIC POWER SYSTEMS; FUEL CELL; GAS TURBINE; INTERNAL COMBUSTION ENGINE; PHOTOVOLTAIC CELL; PRIME MOVER; SOLAR CELL; STEAM TURBINE; WIND POWER in the McGraw-Hill Encyclopedia of Science & Technology.   Thomas S. Basso

Bibliography. R. B. Alderfer, M. M. Eldridge, and T. J. Starrs, *Making Connections: Case Studies of Interconnection Barriers and Their Impacts on Distributed Power Projects*, NREL/SR-200-28053, National Renewable Energy Laboratory, Golden, CO, May 2000; Electrical Generating Systems Association (eds.), *On-Site Power Generation: A Reference Book*, 4th ed., 2002; N. R. Friedman, *Distributed Energy Resources Interconnection Systems: Technology Review and Research Needs*, NREL/SR-560-32459, National Renewable Energy Laboratory, Golden, CO, September 2002; L. L. Grigsby (editor-in-chief), *The Electric Power Engineering Handbook*, CRC Press, 2001; J. F. Manwell, J. G. McGowan, and A. L. Rogers, *Wind Energy Explained: Theory, Design and Application*, Wiley, 2002; R. Messenger and G. G. Ventre, *Photovoltaic Systems Engineering*, 2d ed., CRC Press, July 2003.

Document scanning

Paper documents are ubiquitous in our society, but there is still a need to have many of these documents in electronic form. There are large collections of data originating in paper form, such as office data, books, and historical manuscripts, that would benefit from being available in digital form. Once converted to electronic form, the documents can be used in digital libraries or for wider dissemination on the Internet.

To increase efficiency of paperwork processing and to better utilize the content of data originating in paper form, not only must the paper documents be converted to images of the documents in electronic form, but the text also must be converted to a computer-readable format, like ASCII. This will allow editing as well as search and retrieval of information in these documents. The conversion is done by scanning the document and then processing the digital image with an optical character recognition (OCR) algorithm. The operations of scanning, printing, photocopying, and faxing introduce degradations to the images. Humans often notice the degradations present in photocopies of photocopies of text documents. Smaller image degradations, which may not be noticeable to humans, are still large enough to interfere with the ability of a computer to read printed text. OCR algorithms can achieve good recognition rates (near 99%) on images with little degradation. However, recognition rates drop to 95% or lower when image degradations are present. Typical pages of text have more than 5000 characters per page. So an error rate of 5% results in more than 250 mistakes. Before the mistakes can be corrected,

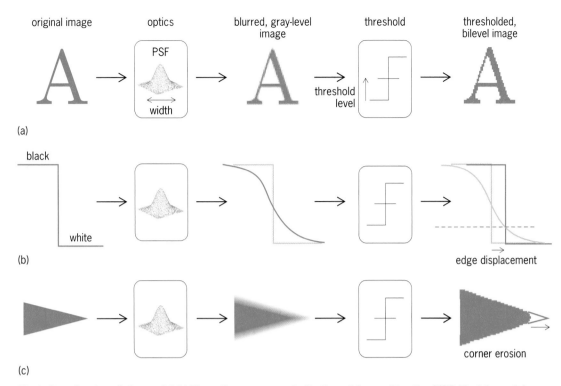

(a)

(b)

edge displacement

(c)

corner erosion

Fig. 1. Scanning degradation model. (*a*) The optics are represented by the point spread function (PSF). The intermediate gray-scale image is then thresholded to form a bilevel image. (*b*) The degradation model has distinct effects on edges and (*c*) on corners.

they must be located, making the correction process even more tedious.

Degradation models. The degradations in the scanned text image occur from many sources, which can be categorized into three areas: (1) defects in the paper (yellowing, wrinkles, coffee stains, speckle); (2) defects introduced during printing (toner dropout, bleeding and scatter, baseline variations); and (3) defects introduced during digitization through scanning (skew, mis-thresholding, resolution reduction, blur, sensor sensitivity noise). Photocopiers and fax machines utilize both a scanning and a printing component.

Multiple models describing various parts of the degradation process have been developed. One major model mathematically describes how images change based on the physics of digitizing images on a desktop scanner (**Fig. 1***a*). The input to the model is the original paper document. It is desired for each pixel value in the digitized image to be based entirely on the amount of light reflected from the paper exactly at the central sensor location. Due to the optics, the scanner forms each digital pixel value by acquiring light from a neighborhood around each sensor. Therefore the pixel value will be determined from image values over a region of space. The light returning to the sensor from the center of the pixel will usually make a larger contribution to the final pixel value, and light from distances farther from the pixel center will make a lower contribution. Each resulting pixel value is determined from a weighted average of the paper image brightness around the sensor through a blurring process called convolution. The

weighting function in this process is called the point spread function (PSF). The PSF is commonly modeled for desktop scanners with the shape of a cube, a cylinder, a cone, or a bivariate gaussian function, although many other shapes are used. The digital image immediate output from the optics will have a range of gray values based on the amounts and locations of black and white varying around each sensor within the support of the PSF. Since OCR is predominantly performed on bilevel images (black and white), the continuous ranged values that the sensor observes are thresholded such that when more light than a threshold value is received, the pixel is assigned a white value, otherwise it is assigned a black value. The width of the PSF and the level of the binarization threshold are parameters in this model.

Degradation categories. Scanning will affect character shapes in text documents by varying amounts dependent on the values of the PSF width and

increasing threshold level

increasing PSF width

Fig. 2. Effects of different PSF widths and binarization thresholds on character images.

binarization threshold (**Fig. 2**). Degradations that predominantly affect character images have been defined and quantified relative to this degradation model for how much edges have been displaced and how corners have been eroded.

Edges. An edge is defined as a sharp transition from black to white. When an edge is input to the scanner, the effects of the convolution with the PSF will blur or smooth the edge (Fig. 1*b*). After thresholding, the image will again have only two values, but the place where the image makes the transition from black to white will change by an amount called the edge displacement. This edge displacement will be constant for all edges in the image that are sufficiently far from all other edges (usually a few pixels). The amount the edge will be displaced is a function of both the threshold level and the width of the PSF. If there is a positive edge displacement the strokes of the characters will become wider, and if there is a negative edge displacement the strokes will become thinner.

Corners. A similar process happens to corners in images. Here the erosion or expansion of the edges happens from all directions, so in addition to the edges being displaced, the sharp tip of the corner will be rounded (Fig. 1*c*). The edge displacement determines how the bulk of the degraded corner relates to the original corner. The corner erosion is defined as the amount of the tip of the resulting corner that is eroded. This erosion will take place on both exterior (black on white) and interior (white on black) corners.

Effects on characters. The amount of edge displacement and corner erosion in an image depends on the width of the PSF and the level of the threshold. Figure 2 shows samples of the letter A under many model parameter possibilities. When the PSF width is larger, the effects of threshold are greater and the character will usually be degraded more. However, it is harder to physically build scanners and camera systems with small PSF widths.

These degradations can lead to broken or touching characters, which are the major causes of OCR errors. For instance, a negative edge displacement will result in thinning of strokes which can cause an "m" to resemble the pair of letters "r n" or for an "e" to resemble a "c." If the edge displacement is positive, the stroke width will increase, which can cause an "r n" to resemble an "m" (**Fig. 3**).

When the degradation process is repeated multiple times (as in multiple-generation photocopies), the effect will be to further break up or fill in characters. Because of the corner erosion and the possibilities of strokes totally vanishing, neither a thinning followed by a thickening nor a thickening followed by a thinning will return the characters to their original states.

Applications. The degradation models allow researchers to improve OCR algorithms and also can be used to develop better scanning technologies. Better experiments on the OCR systems can be performed if a good degradation model is available.

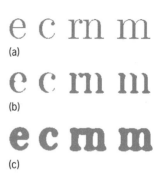

Fig. 3. Examples of characters with different degradations that lead to common OCR errors. (*a*) True characters. (*b*) Characters with negative edge displacement. (*c*) Characters with positive edge displacement.

Degradation models can be used to produce large calibrated data sets where the true label of each character is known. If the OCR algorithm is trained on samples of text with the degradation that will be observed in practice, it will correctly recognize the observed character more easily. If the OCR algorithm knows the degradation model parameters before it sees the degraded image, the recognition rates will be even higher. For instance, if the OCR algorithm knows, based on the degradation model parameters, that the horizontal bar on the "e" will be missing (Fig. 3*b*), it will not have a hard rule saying that the letter "e" must have a closed loop in the upper half of the character. Developers of printers, scanners, and photocopiers can use a degradation model to investigate how design changes will affect a text document so their products would be more likely to produce quality output under a variety of operating conditions.

For background information *see* CHARACTER RECOGNITION; IMAGE PROCESSING in the McGraw-Hill Encyclopedia of Science & Technology.

Elisa H. Barney Smith

Bibliography. H. S. Baird, Document image defect models, in H. S. Baird, H. Bunke, and K. Yamamoto (eds.), *Structured Document Image Analysis*, pp. 546–556, Springer, Berlin, 1992; E. H. Barney Smith and X. Qiu, Statistical image differences, degradation features and character distance metrics, *Int. J. Doc. Analy. Recogn.*, 6(3):146–153, 2004; S. V. Rice, G. Nagy, and T. Nartkar, *Optical Character Recognition: An Illustrated Guide to the Frontier*, Kluwer Academic Publishers, 1999.

Earth Simulator

The Earth system extends from the uppermost ionosphere to the innermost solid core of the planet and exhibits a wide variety of phenomena, such as climate change, ocean circulation, earthquakes, and geomagnetism. A vast number of parameters and precise models are needed to understand these complicated phenomena, which can significantly affect human society. Until recently, even the state-of-the-art supercomputers were not sufficiently powerful to run realistic numerical models of these phenomena. The Earth Simulator has been designed to provide a powerful tool for investigating these phenomena numerically with unprecedented resolution. It was built in 2002 as a collaborative project of the National Space Development Agency of Japan (NASDA), Japan Atomic Energy Research Institute (JAERI), and Japan Agency for Marine-Earth Science and Technology Center (JAMSTEC).

The Earth Simulator is a highly parallel vector supercomputer and consists of 640 processor nodes, which are connected by a high-speed network (**Fig. 1**). Each processor node has 8 vector arithmetic processors and 16 billion bytes of memory. The peak performance of each arithmetic processor is 8 gigaflops. (A gigaflop is a unit of computer speed equal to one billion floating-point arithmetic operations per second.) Because the Earth Simulator has 5120 (640 × 8) arithmetic processors, its entire memory amounts to 10 trillion bytes and its theoretical peak performance is 40 teraflops, enabling 40 trillion arithmetic calculations per second. Immediately after it started operation in March 2002, the Earth Simulator recorded the highest performance on record of 35.86 teraflops with the Linpack benchmark, which solves a dense system of linear equations. In June 2002 it was ranked first among the 500 most powerful supercomputers and presently is the world's fastest and largest supercomputer.

Projects. Thirty-four collaborative research projects have been selected by the Mission Definition Committee of the Earth Simulator Center for large-scale numerical simulation in the fields of (1) ocean and atmospheric science, (2) solid-earth science, (3) computer science, and (4) epoch-making simulation. These projects include research at Japanese institutions and universities, as well as international collaborative research among the Earth Simulator Center and institutions outside Japan.

Ocean and atmospheric science. In 2003 there were 12 research projects categorized as ocean and atmospheric science. These projects covered subjects such as climate change prediction using a high-resolution coupled ocean–atmosphere climate model, development of an integrated Earth system model for predicting global environmental change, and an atmospheric composition change study using regional chemical transport models. The main target of the ocean and atmospheric research is to provide reliable estimates for possible climate change associated with global warming. Typical climate models used on the Earth Simulator divide the atmosphere into 56 layers and the ocean into 46 layers and take into account the coupling of the ocean and the atmosphere. The Earth Simulator enables the use of climate models with unprecedented spatial resolution and provides precise estimates of future climate change.

Computer science and epoch-making simulations. There are 13 research projects categorized as computer science and epoch-making simulations. These projects include geospace (Sun–Earth) environment simula-

Fig. 1. Earth Simulator. It has 320 boxes, each equipped with 16 arithmetic processors, housed in the Earth Simulator Building at the JAMSTEC Yokohama Institute, in Japan. (*Earth Simulation Center*)

tions, biosimulations, nuclear reactor simulations, and large-scale simulation of the properties of carbon nanotubes. For example, it has become possible to perform large-scale simulations of carbon nanotubes to study various physical properties such as their thermal conductivity.

Solid-earth science. There are nine research projects categorized as solid-earth science. These projects include simulations of seismic-wave propagation, the Earth's magnetic field, mantle convection, and the earthquake generation process. The Earth Simulator has shown its powerful potential in simulating seismic waves expected to be generated by powerful earthquakes along the Nankai Trough off the south coast of Japan. This kind of simulation will help in hazard mitigation planning against the strong ground shaking from future earthquakes. Modeling of the Earth's magnetic field using dynamo theory, which is driven by thermal convection within the Earth's fluid core, is also expected to be advanced by large-scale numerical simulation on the Earth Simulator.

Large-scale numerical simulations in seismic studies. The use of seismic waves has been a unique way to probe the Earth, which we cannot directly study inside. Therefore, accurate modeling of seismic-wave propagation in three-dimensional (3D) Earth models is of considerable importance in studies for determining both the 3D seismic-wave velocity structure of the Earth and the rupture process during a large earthquake. Numerical modeling of seismic-wave propagation in 3D structures has made significant progress in the last few years due to the introduction of the spectral element method (SEM), a version of the finite element method. Although SEM can be implemented on a parallel computer, lack of available computer resources has limited its application. For instance, on a PC cluster with 150 processors, it was shown that seismic waves calculated with the SEM were accurate at periods of 18 s and longer. However, these periods are not short enough to capture the important effects on wave propagation due to smaller 3D heterogeneities of the Earth. To examine the shortest period attainable, SEM software was implemented on the Earth Simulator.

In the SEM, the Earth is divided into grid points, where a finer grid interval results in higher resolution in seismic-wave calculation. A typical computation of seismic propagation on the Earth Simulator with 5.467 billion grid points uses 243 nodes (1944 processors). This translates into an approximate grid spacing of 2.9 km (1.8 mi) along the Earth's surface. This number of grid points comprises all the known 3D structure inside the Earth, including the 3D seismic velocity structure inside the mantle, the 3D structure of the crust, and the topography and bathymetry at the Earth's surface. Using 243 nodes, a simulation of 60 min of seismic-wave propagation through the Earth (accurate at periods of 5 s and longer) requires about 15 h of computational time.

The computation of seismic waves on a global scale has been done using a normal mode summation technique (traditional simulation). This technique has computational difficulties in obtaining accurate normal modes at shorter periods and is accurate up to 6 s for seismic waves that propagate in a spherically symmetric Earth model. **Figure 2** compares theoretical seismic waves calculated using the traditional normal mode summation technique for a spherically symmetric Earth model and those calculated using the SEM for a fully 3D Earth model. Because the seismic waves calculated by normal mode summation are accurate up to 8 s and longer, 48 nodes of the Earth Simulator were used to calculate the synthetic seismograms using the SEM. The results clearly demonstrate that the agreement between synthetics and actual seismograms (obtained at seismic observatories) is significantly improved by including the 3D Earth structure in the SEM synthetics. It is remarkable that the Earth Simulator allows us to simulate global seismic-wave propagation in a fully 3D Earth model at shorter periods than a traditional quasi-analytical technique for a spherically symmetric Earth model.

Figure 3 compares the theoretical seismic waves calculated using the SEM on the Earth Simulator and the observed seismic waves for the magnitude 7.9 Denali earthquake that occurred on November 3, 2002, in Alaska. Because hundreds of seismic observatories now record seismic waves generated by earthquakes, it is possible to directly compare theoretical seismic waves with real data from recorded

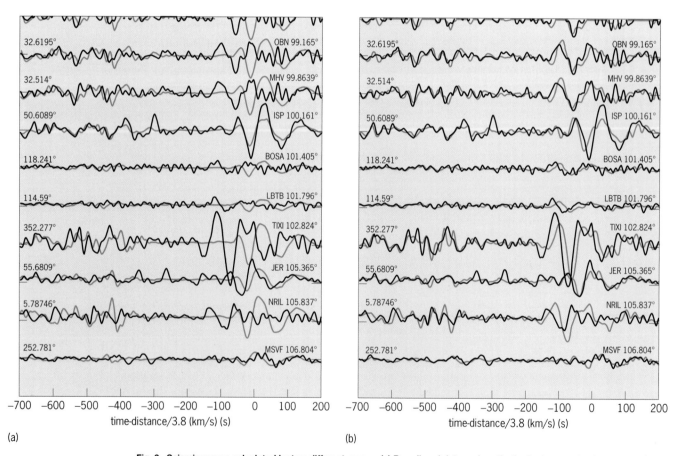

(a)   (b)

Fig. 2. Seismic waves calculated by two different means. (*a*) Broadband data and synthetic displacement seismograms for the September 2, 1997, Colombia earthquake bandpass-filtered between periods of 8 and 150 seconds. Vertical component data (black) and synthetic (color) displacement seismograms aligned on the arrival time of the surface wave (Rayleigh wave) are shown. Synthetics were calculated using a traditional normal mode summation technique. For each set of seismograms the azimuth is printed above the records to the left, and the station name and epicenter distance are printed to the right. (*b*) The same stations are as shown in *a*, but the synthetics were calculated with the SEM using 48 nodes of the Earth Simulator. (*Courtesy of Jeroen Tromp, California Institute of Technology*)

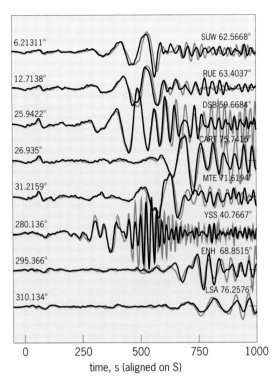

Fig. 3. Theoretical seismic waves and data for the 2002 Denali, Alaska, earthquake, bandpass-filtered between periods of 5 and 150 seconds. Horizontal (transverse) component data (black) and theoretical seismic waves (color) of ground displacement aligned on the arrival time of the secondary wave (S) wave are shown. For each set of traces the azimuth is plotted above the records to the left, and the station name and epicenter distance are plotted to the right. (*From S. Tsuboi et al, 2003*)

earthquakes. The 3D models of the seismic-wave velocity structure of the Earth are traditionally built based on a combination of travel-time anomalies of short-period body waves and long-period surface waves. However, independent validation of such existing 3D Earth models has never been attempted before, owing to the lack of an independent numerical way of computing the seismic response in such models. The agreement between the theoretical seismic waves and observed records is excellent for these stations, which means that the 3D seismic velocity model used in this simulation is accurate. Therefore, this simulation demonstrates that the 3D model represents the general picture of the Earth's interior fairly well. The results demonstrate that the combination of the SEM with the Earth Simulator makes it possible to model seismic waves generated by large earthquakes accurately. This waveform-modeling tool should allow us to further investigate and improve Earth models.

For background information *see* EARTH INTERIOR; EARTHQUAKE; FINITE ELEMENT METHOD; GEODYNAMO; GEOPHYSICS; MICROPROCESSOR; SEISMOLOGY; SIMULATION; SUPERCOMPUTER in the McGraw-Hill Encyclopedia of Science & Technology.

Seiji Tsuboi

Bibliography. F. A. Dahlen and J. Tromp, *Theoretical Global Seismology*, Princeton University Press, 1998; D. Komatitsch, J. Ritsema, and J. Tromp, The spectral-element method, Beowulf computing, and global seismology, *Science*, 298:1737–1742, 2002; D. Komatitsch and J. Tromp, spectral-element simulations of global seismic wave propagation, I. Validation, *Geophys. J. Int.*, 149:390–412, 2002; S. Tsuboi et al., Broadband modeling of the 2003 Denali fault earthquake on the Earth Simulator, *Phys. Earth Planet. Int.*, 139:305–312, 2003.

Electrospinning

Electrospinning is a fiber-forming process capable of producing continuous nanoscale-diameter fibers using electrostatic forces. The combination of high specific surface area, flexibility, and superior directional strength makes fiber a preferred material for many applications, ranging from clothing to reinforcements for aerospace structures. Although the effect of fiber diameter on the performance and processability of fibrous structures has long been recognized, the practical generation of nanoscale fibers was not realized until the rediscovery and popularization of electrospinning technology by D. Reneker in the 1990s.

The ability to create nanoscale fibers from a broad range of polymeric materials and in a relatively simple manner by electrospinning, coupled with the rapid growth of nanotechnology in the recent years, have greatly accelerated the growth of nanofiber technology. Although there are several alternate methods for generating nanoscale fibers, none matches the popularity of electrospinning technology due to the simplicity of the process.

Electrospinning can be done from polymer melt or solution. Most electrospinning has been solution-based due to the greater capital investment required and difficulties in producing submicrometer fibers by melt techniques.

Recognizing the enormous increase in specific fiber surfaces, bioactivity, electroactivity, and the enhancement of mechanical properties, numerous applications have been identified including filtration, biomedical, energy storage, electronics, and multifunctional structural composites.

Process. Electrostatic generation of ultrafine fibers has been known since the 1930s. The rediscovery of this technology has stimulated numerous applications, including high-performance filters and scaffolds in tissue engineering, which utilize the nanofibers' high surface area (as high as 10^3 m²/g). In electrospinning, a high electric field is generated between a droplet of polymer solution at the capillary tip (also known as the nozzle or spinneret) of a glass syringe and a metallic collection screen (**Fig. 1a**). When the voltage reaches a critical value, the electric field strength overcomes the surface tension of the deformed polymer droplet, and a jet is produced. The electrically charged jet undergoes a series of electrically induced deviations from a linear to a curvilinear configuration, known as bending instabilities,

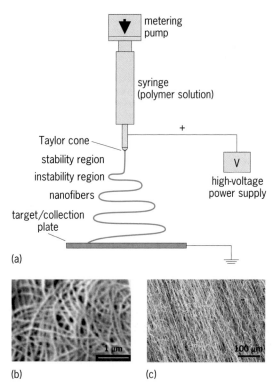

(a)

(b)   (c)

Fig. 1. Electrospinning. (a) Process. (b) Random nano-fibers. (c) Aligned nanofibers.

during its passage to the collection screen, which results in the hyperstretching of the jet. This stretching process is accompanied by the rapid evaporation of the solvent, which reduces the diameter of the jet in a cone-shaped volume called the envelope cone. The "dry" fibers are accumulated on the surface of the collection screen, resulting in a nonwoven random fiber mesh of nanometer- to micrometer-diameter fibers (Fig. 1b). The process can be adjusted to control the fiber diameter by varying the electric field strength and polymer solution concentration. By proper control of the electrodes, aligned fibers also can be produced (Fig. 1c).

Processing parameters. A key objective in electrospinning is to generate nanometer-diameter fibers consistently and reproducibly. Parameters that affect the diameter and influence the spinability and the physical properties of the fibers include the electric field strength, polymer concentration, spinning distance, and polymer viscosity.

The fiber diameter can be controlled by adjusting the flow rate, the conductivity of the spinning line, and the capillary tip diameter. Fiber diameter can be minimized by increasing the current-carrying capability of the fiber by adding a conductive filler, such as carbon black, carbon nanotubes, metallic atoms, or an inherently conductive polymer. Increasing the current-carrying capability of the fiber by 32 times will bring about a 10-fold decrease in the fiber diameter. Alternatively, keeping the current constant and reducing the flow rate by 32 times will bring about a 10-fold decrease in the fiber diameter. Reducing the nozzle diameter also can decrease the

fiber diameter. Increasing the ratio of the initial jet length to the nozzle diameter from 10 to 1000 will decrease the diameter of the fiber by approximately two times.

Experimental evidence has shown that the diameter of the electrospun fibers is influenced by molecular conformation, which is influenced by the molecular weight of the polymer and the concentration of the polymer solution. It was found that the diameter of fibers spun from dilute polymer solutions is closely related to a dimensionless parameter, the Berry number Be, which is a product of intrinsic viscosity $[\eta]$ and polymer concentration C. This relationship has been observed in a large number of polymers. An example of this relationship is illustrated using poly(lactic acid) or PLA. The relationship between the polymer concentration and fiber diameter for solutions of different-molecular-weight PLA in chloroform is shown in **Fig. 2**. The fiber diameter increases as the polymer concentration increases, and the rate of increase in fiber diameter is greater at higher molecular weight. Accordingly, one can tailor the fiber diameter by proper selection of the polymer molecular weight and polymer concentration.

When the fiber diameter is expressed as a function of Be, a pattern emerges, as indicated by the four regions in **Figs. 3** and **4**. In region I, a very dilute polymer solution in which the molecular chains barely touch each other makes it almost impossible to form fibers by electrospinning, since there is not enough chain entanglement to form a

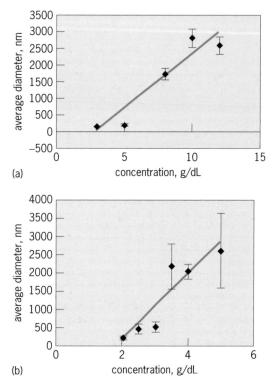

(a)

(b)

Fig. 2. Relationship of fiber diameter to concentrations at different molecular weights. (a) Mol wt = 200,000. (b) Mol wt = 300,000.

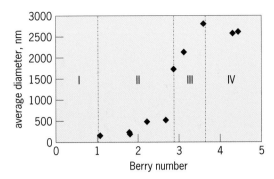

Fig. 3. Relationship between the Berry number and fiber diameter.

continuous fiber and the effect of surface tension will make the extended conformation of a single molecule unstable. As a result, only polymer droplets are formed.

In region II, the fiber diameter increases slowly with Be (from about 100 to 500 nm). In this region, the degree of molecular entanglement is sufficient for fiber formation. The coiled macromolecules of the dissolved polymer are transformed by the elongational flow of the polymer jet into oriented molecular assemblies with some level of inter- and intramolecular entanglement. These entangled networks persist as the fiber solidifies. In region II, some bead formation is observed because of polymer relaxation and surface tension effects.

In region III, the fiber diameter increases rapidly (from about 1700 to 2800 nm) with Be. In this region, the molecular chain entanglement contributes to the increase in polymer viscosity. Because of the increased level of molecular entanglement, a higher electric field strength is required for fiber formation.

In region IV, the fiber diameter is less dependent on Be. With a high degree of inter- and intramolecular chain entanglement, other processing parameters, such as electric field strength and spinning distance, become the dominant factors affecting fiber diameter.

Nanofiber yarn and fabric. The formation of nanofiber assemblies provides a means of connecting the effect of nanofibers to macrostructure performance. Electrospun fiber structures can be assembled by direct, nonwoven fiber-to-fabric formation and by the creation of a linear assembly or a yarn from which a fabric can be woven, knitted, or braided. The linear fiber assemblies can be aligned mechanically or by an electrostatic field. Alternatively, a self-assembled continuous yarn can be formed during electrospinning by proper design of the ground electrode. Self-assembled yarn can be produced in continuous length with appropriate control of the electrospinning parameters and conditions. In this process, the self-assembled yarn is initiated by limiting the footprint of fiber deposition to a small area of the target. The fibers are allowed to build on top of each other until a branched (treelike) structure is formed. Once a sufficient length of yarn is formed, the coming fibers attach themselves to the branches and continue to build up. A rotating drum can be used to spool the self-assembled yarn in continuous length (**Fig. 5**). This method produces a continuous length of partially aligned nanofiber yarn.

Applications. To date, industrial filter media are the only commercial application of electrospun fibers. However, electrospinning has been transformed from an obscure technology to a familiar word in academia and is gradually being recognized by industrial and government organizations. This was evident at the Polymer Nanofiber Symposium held in New York by the American Chemical Society in September 2003, where 75 papers and posters were presented. Based on recent publications, potential applications of electrospun fibers include biomedical products (such as tissue engineering scaffolds,

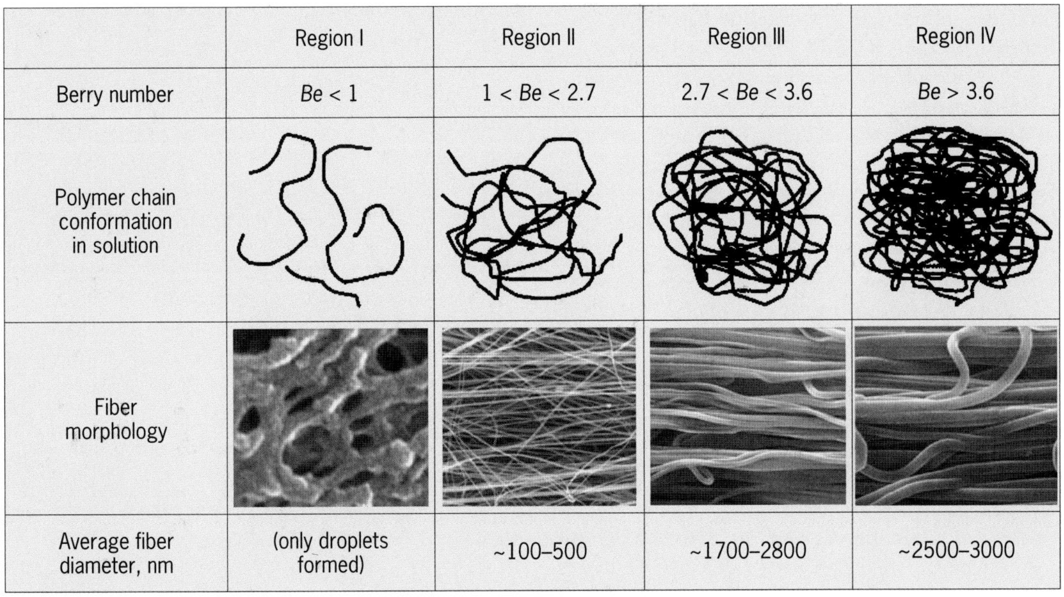

	Region I	Region II	Region III	Region IV
Berry number	$Be < 1$	$1 < Be < 2.7$	$2.7 < Be < 3.6$	$Be > 3.6$
Polymer chain conformation in solution				
Fiber morphology				
Average fiber diameter, nm	(only droplets formed)	~100–500	~1700–2800	~2500–3000

Fig. 4. Polymer chain conformation and fiber morphology corresponding to four regions of Berry number.

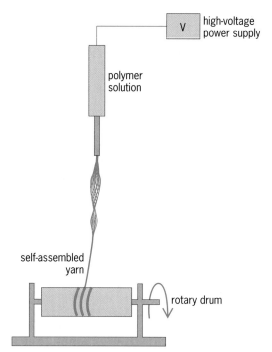

Fig. 5. Continuous electrospinning of self-assembled yarn.

wound care, superabsorbent media, and drug delivery carrier), electronic products (such as electronic packaging, sensors, wearable electronics, actuators, and fuel cells), and industrial products (such as filtration, structural toughening/reinforcement, and chemical/biological protection).

For background information *see* ELECTRIC FIELD; NANOSTRUCTURE; NANOTECHNOLOGY; POLYMER; TEXTILE; TEXTILE CHEMISTRY in the McGraw-Hill Encyclopedia of Science & Technology.   Frank K. Ko

Bibliography. H. Fong and D. H. Reneker, Electrospinning and the formation of nanofibers, Chap. 6 in D. R. Salem and M. V. Sussman (eds.), *Structure Formation in Polymer Fibers*, pp. 225–246, Hanser, 2000; A. Frenot and I. S. Chronakis, Polymer nanofibers assembled by electrospinning, *Curr. Opin. Colloid Interface Sci.*, 8:64–75, 2003; Z. M. Huang et al., A review on polymer nanofibers by electrospinning and their applications in nanocomposites, *Composite Sci. Technol.*, vol. 63, 2003; F. Ko et al., Electrospinning of continuous carbon nanotube-filled nanofiber yarns, *Adv. Mater.*, vol. 15, no. 14, July 17, 2003; F. K. Ko et al., Electrostatically generated nanofibers for wearable electronics, Chap. 2 in X. M. Tao (ed.), *Wearable Electronics*, Woodhead, 2004; A. L. Yarin, *Electrospinning of Nanofibers from Polymer Solutions and Melts*, Lecture Notes 5, Institute of Fundamental Technological Research, Polish Academy Of Sciences, 2003.

Environmental management (mining)

Mineral exploration is the process whereby mineral deposits are found and investigated in sufficient detail to determine whether they can be feasibly developed into a mine. The first stage of exploration commonly involves the review of existing information, such as maps, aerial photographs, and satellite imagery, often followed by airborne surveys. Airborne surveys use various techniques to find changes in the geology or anomalies below the surface. This initial stage of exploration generally results in little physical disturbance and low potential for environmental impacts to flora and fauna, even where ground field work such as tracing back samples to the primary source is involved. This article focuses on subsequent stages of exploration, where activities, such as drilling, trenching, and collection of a bulk sample, potentially result in physical disturbance of the land, animals, and people. Only a very small proportion of the exploration prospects identified ever get to this stage of physical investigation (and an even smaller number eventually become mines).

Exploration activities. Mineral exploration activities, on land or underwater, depend on the characteristics of the mineral deposit (close to the surface versus at a significant depth), as well as the physical nature of the rock under investigation and the environmental conditions at the location.

The three primary means of exploration that are most likely to have an appreciable environmental impact are drilling, trenching, and collection of a bulk sample. These activities may require establishment of supporting infrastructure such as trails or roads, storage areas for mineral samples, a power supply, accommodation for workers, waste management, and an influx of nonlocal workers, all of which may cause environmental impacts.

Drilling. This is the process of sampling representative portions of a mineral deposit underground. It can occur either from the surface (land or ice) or from a location underground once access is made (such as through an exploration shaft or tunnel). A continuous circular core is drilled out of the deposit using a specialized drill often with a diamond-studded head that can grind through bedrock. As the drill moves deeper, the rock core is pushed up and can be extracted and examined.

Trenching. This is a method of exploration on surface, whereby elongated ditches are excavated or blasted through soil to collect samples of a vein or other bedrock feature over a larger area than is exposed naturally (an outcrop).

Bulk sample collection. This is a process whereby a large sample (hundreds of tons) of the deposit is collected either from the surface or from a location underground (once underground access is made). Once the material is released from the ore body by blasting or other means, the bulk sample can be removed using conventional equipment such as loaders and trucks.

Environmental impacts. There is widespread recognition in the minerals industry that exploration may have a significant environmental impact, including a detrimental effect on local communities. Effective environmental management requires an understanding of the potential environmental impacts associated with exploration (see **table**).

Examples of potential physical environmental impacts associated with mineral exploration

Exploration activity	Potential impact
Land and vegetation clearing for access, line cutting, surveys, drilling, and support infrastructure	Erosion and release of sediments into watercourses; loss of habitat; disruption of wildlife; noise and air-quality issues; damage to archeological and heritage features
Vehicle and equipment use	Land disturbance and erosion; impairment of water quality in surrounding watercourses; noise and air-quality issues
Use of water/chemicals required for drilling activities	Impairment of water quality in natural watercourses; soil contamination
Petroleum hydrocarbon fuel and lubricant usage	Soil and water contamination; release of air emissions from power generation and equipment use
Exposure of bedrock to surface	Exposure of certain minerals to air leading to chemical reactions, such as generation of acid which can impair water quality
Management of domestic and other wastes	Impairment of soil and water quality; may attract or injure wildlife; disease
Influx of workers from other locations	Disruption of community life; introduction of disease and incompatible cultural values; potential for conflict; creation of false expectations

These impacts can be controlled or mitigated during exploration or after exploration ceases, with proper planning and implementation of industry-standard environmental management practices, commonly termed best management practices. Best management practices at a minimum comply with the more stringent of local or home-country regulatory requirements, and typically are responsive to site-specific conditions.

Regulatory context. The environmental laws that apply to exploration are commonly of a general nature and would apply to a number of different industries. For example, in a given jurisdiction there may not be exploration-specific laws or regulations governing the storage of petroleum hydrocarbon at exploration sites. However, exploration activities would be required to comply with existing legislation related to hydrocarbon storage and handling. In addition, there may be laws and regulations related to approval of the drilling program, camp construction, access to land, and decommissioning at the end of the exploration program, particularly in developed countries.

General environmental regulations and laws are becoming more consistent worldwide with countries and lending agencies demanding that companies use universal practices particularly at locations where strict regulations are not present. For this reason, many exploration companies have chosen to take an approach of utilizing environmental best management practices at all of their operations worldwide. The management practices are modified in jurisdictions where necessary to meet or exceed regulatory requirements or address particular local concerns or issues.

Environmental best management practices. As companies have come to realize that exploration may have a significant environmental impact, they, their organizations, and government agencies have become proactive in compilation of information to improve environmental performance. The Prospector and Developers Association of Canada (PDAC) recently developed the Environmental Excellence in Exploration (E3) online reference source of best management practices to promote environmental management practices at the earliest stages of exploration. These references provide a source of material based on field-proven information from industry experts, with application to Canadian and international conditions.

In addition, the Australia Environmental Protection Agency in association with a steering committee comprising representatives of the minerals industry, government agencies, and conservation organizations have developed a comprehensive set of booklets describing mining best management practices, including one booklet on onshore minerals and petroleum exploration.

The key aspect of both of these information sources is that they are based on field experience from exploration practitioners. Hence, the reliability of the information and reproducibility of results is high. Both provide an excellent source of information regarding practices to minimize environmental impacts from exploration.

Community engagement. For many communities, the first direct exposure to the minerals industry comes about through an exploration company entering the area to explore for minerals. This first contact can flavor how the people view a future mine and the minerals industry in general. Best management practices require that the exploration company actively engage the community and its membership (such as through employment and provision of support services) to keep them informed and establish trust. The key principles are to make only promises that can be kept, avoid raising false expectations, respond quickly to all concerns, and recognize that there may be cultural differences that need to be considered. Archeological, heritage, and other cultural sites are often protected by local laws and regulations. Even when they are not, the best management practice employed by exploration companies is to determine whether there are potential issues before initiating exploration.

Site access. Exploration sites may require establishment of access routes since they are often located away from existing roads and infrastructure. In some locations, helicopter access may be possible and preferred, rather than development of permanent access roads or trails. Impacts from site access can be minimized by locating new routes in consultation with local communities and government agencies to avoid sensitive areas, and by use of vehicles and equipment less prone to creating erosion, vegetation damage, and development of wheel ruts.

Fuel and oil handling. Petroleum hydrocarbons (fuel and oils) are a necessary supply at remote exploration sites, primarily for power generation and equipment operation. Standard practices include storage of these fluids away from watercourses and areas where inadvertent damage might occur, and provision of secondary containment around the storage area. Spill procedures and spill response equipment should be in place with detailed cleanup procedures to ensure that even when a spill occurs the environmental impacts are minimized.

Drilling procedures. Once access is available and assuming that the site is properly established from an environmental perspective, the key environmental aspect at drilling sites is the management of drilling return fluids. Return fluid is liquid composed of water, mud, and polymers that are used to lubricate the hole and ensure effective coring and removal of drill cuttings (rock from cutting at the bottom of the drill hole). Bentonite mud or polymers may be used to prevent hole collapse. The return fluid is collected on the surface in an often lined, in-ground sump, or portable aboveground tanks that allow the sediments to be removed prior to discharge of the clean water to the environment.

Trenching procedures. Trenching causes primarily surface impact by removal of vegetation and soil cover to expose bedrock. Best practices are to construct trenches across slopes where reasonable to minimize the potential for erosion and ensure that the ends of the trench are sloped so that animals can escape. Trenches must be refilled with previously stockpiled material, once exploration is complete, to ensure the safety of humans and wildlife.

For background information *see* GEOCHEMICAL EXPLORATION; GEOPHYSICAL EXPLORATION; LAND RECLAMATION; MARINE MINING; MINING; PROSPECTING; SURFACE MINING; UNDERGROUND MINING in the McGraw-Hill Encyclopedia of Science & Technology.

Sheila Ellen Daniel

Bibliography. Australia Environment Protection Agency, *Onshore Minerals and Petroleum Exploration: One Module in a Series on Best Practice Environmental Management in Mining*, Commonwealth of Australia, 1996; Environment Australia, *Overview of Best Practice Environmental Management in Mining: One Module in a Series on Best Practice Environmental Management in Mining*, Commonwealth of Australia, 2002; P. A. Lugli, Sound exploration practice in exploration drilling, in *Mining Environmental Management*, pp. 20–22, Mining Journal Limited, July 2001; Prospectors and Developers Association of Canada, E3: *Environmental Excellence in Exploration* e-manual.

Environmental sensors

New sensing technologies detect and analyze, often in real time, pollutants in the water, air, and soil. The information collected by such devices extends environmental scientific and engineering knowledge, protects public health and the environment, and contributes to regulatory monitoring and compliance.

Oil spills, smog, and contaminants in the soil or water require hundreds to thousands of measurements or sensors to determine the extent of the environmental impact. In many cases, the cost of the sensors prohibits taking an adequate number of measurements. In addition to the measurement of the pollutant, its location must be determined, often requiring expensive surveying measurements in hostile environments. To overcome this problem, integrated circuits with Global Positioning System (GPS) technology are attached to the sensor to provide its location. The cost of this circuit is less than the cost of surveying a location. For a complete sensing system, the sensor and GPS location circuit can be connected to a wireless transmitter, similar to a cellular phone. The sensor now can be moved in time and its location still determined.

To efficiently measure environmental pollutants, the next-generation sensing modules will contain a sensing element, a miniature processor for obtaining data from the element, a solar cell or battery for power, a GPS chip for location information, and a wireless transmitter (**Fig. 1**).

Sensing elements. Sensitivity and selectivity define the quality of a sensing element. The sensitivity or detection limit is the lowest level or concentration that can be detected. It is usually specified in parts per billion or parts per million of the media that is being sensed, such as water or air.

Selectivity, also called resolution, is the ability of the sensing element to distinguish between different substances or analytes. The sensor must react specifically to the pollutant it is to sense and not react, or react very little, to other analytes. For example, a sensor that was designed to detect nitrous oxide (N_2O) but also reacts to oxygen (O_2) would not be practical to use in air because it could not "select" between the two gases. For general-purpose sensors, the sensor can react to more than one analyte, but the output would have multiple output indications separated in time or voltage (**Fig. 2**). In most cases, the number of chemicals that the sensing element will see in a field application is extensive, and the difficult task for scientists who are developing sensors is to be able to detect one analyte while ignoring the others.

The classes of sensing elements are: (1) Chemical—senses specific chemicals by a chemical reaction; usually must be combined with an electrical or optical device to produce a practical sensor with a measurable output. (2) Mechanical—identifies analytes by sensing their weight. (3) Electrical—the conduction properties of some exotic materials and structures change when exposed to certain chemicals; this change in electrical properties can be related to the concentration and type of chemical. (4) Ionic—determines an unknown chemical by sensing the transit time of an ion in an electric field. (5) Optical—senses chemicals by measuring their optical properties. Many combinations of these sensing elements also exist.

Chemical sensors. The simplest chemical sensing elements are the metal oxide sensors; including tin oxide (senses NO_2), indium oxide (senses ozone), zinc oxide (senses hydrocarbons), as well as many other exotic materials. Chemicals of environmental interest that can be sensed include nitrous oxide, hydrogen sulfide, propane, carbon monoxide, sulfur dioxide, and ozone. The sensors are thin films of a metal oxide and operate in an oxygen environment. Oxygen draws the electrons out of the thin film and increases its resistivity. When chemicals to be sensed are introduced, they react with the surface oxygen and release electrons (oxidation), which return to the thin film and decrease its resistivity—a change which is easily converted to a measurable electrical signal. The oxidation of the analyte requires a temperature of 250°C (480°F) or more, which is a disadvantage of this type of sensor.

A more recent family of chemical sensors has been constructed from polymers, which are common organic materials. Specifically, polythiophene has been integrated with the most common semiconductor device, the field-effect transistor (FET). The resulting device is called a CHEMFET (**Fig. 3**). As the analyte is captured by the polymer gate material, the work function of the material changes. (The work function controls the conductivity between the source and drain, changing the current through the device, which is measured.) Devices have been constructed in the laboratory to measure carbon tetrachloride, toluene, and methanol.

Mechanical. There are many new sensing elements designed using micro-electro-mechanical system (MEMS) technology. Cantilever MEMS devices are fabricated using silicon semiconductor processing techniques (**Fig. 4**). The beam, which is less than the size of a human hair, has a chemical plated on it that combines specifically with the analyte to be detected. The weight of the captured analyte bends the beam. The resulting strain on the beam can cause a difference in resistance that can be measured. The property of a material, such as quartz or boron, whose resistivity changes under stress is called piezoresistance. In a MEMS device, resistors are formed by implanting boron on the beam and measuring the change in resistance due to the strain. An alternative means of sensing the chemicals is to incorporate the beam as part of a capacitor and to measure the capacitance change. MEMS sensors have been built to detect mercury vapor, lead, and biological agents.

The Environmental Protection Agency (EPA) has defined dangerous levels of small particles in air. These are sensed as a group without regard to the exact chemicals involved. MEMS devices also have been built to sense particles 2.5 micrometers or smaller, and 10 μm or smaller as defined by the EPA standards.

Optical sensors. The advances in laser diodes and semiconductor detectors can be combined in a miniaturized chip, allowing detection of wave-

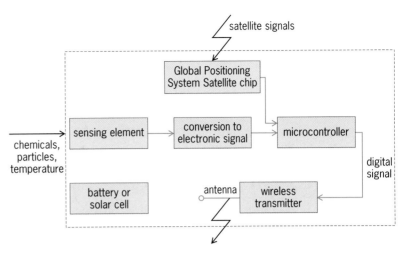

Fig. 1. Block diagram of an environmental sensor module.

lengths from the near-ultraviolet to the near-infrared. Advances from this work should be used to develop sensors for other environmental contaminants in an inexpensive and continuous manner. Prototype devices in this area have been used to detect fluorescence in biological samples of chlorophyll and some aromatic hydrocarbons.

Ion mobility spectrometer sensor. The ion mobility spectrometer (IMS) has been used in laboratories for years to measure a wide variety of chemicals (**Fig. 5**). An unknown sample of gases is injected into a chamber, where it is ionized by a small amount of radioactive material, usually Ni^{63}. (There are a number of alternative methods of ionizing the element.) A voltage-controlled gate is used to allow a small pulse of ions to go down the chamber. The ions are accelerated due to a voltage V being applied along the axis of the chamber. As the ions are propelled

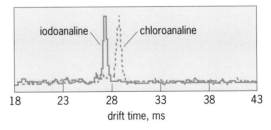

Fig. 2. Multiple output indications separated in time.

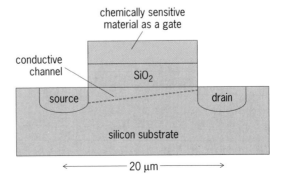

Fig. 3. CHEMFET with a chemical-sensitive material for a gate.

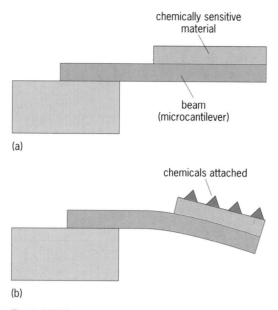

Fig. 4. MEMS structure (*a*) before analytes are introduced and (*b*) after analytes are attached.

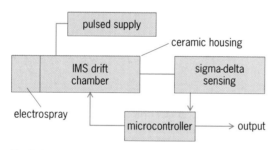

Fig. 5. Ion mobility spectrometer (IMS) sensor.

toward the detector, they continually collide with the "drift" gas in the chamber. The different chemicals travel at different speeds because they have different cross sections. The ions are collected by a metal plate and cause a current, which is then amplified (Fig. 2). The time of the peak indicates the chemical species. Medium-sized IMS sensors are used to detect explosives at airports in carry-on luggage.

Environmental conditions. In field applications, the environmental conditions (pressure, temperature, and humidity) often must be measured, as well as the pollutant. Sensors for these conditions have been miniaturized for years and typically use a change in mechanical or electrical properties to indicate their value.

Significance. Current techniques of taking samples and sending them to a laboratory for analysis are expensive and time-consuming. As a result, the time variance of the environment contaminants is often missed. The significance of these sensors is that they can be deployed in the field and provide data in real time. There will be more data points, resulting in models that are more accurate and provide a better understanding of the environment.

There are applications where the levels of pollutants need to be measured because they may reach a level critical to health. In these cases, an inexpensive sensor can be specifically designed to detect the harmful level at a fraction of the cost of today's techniques. Such sensors can be used as a monitoring or warning system at "smart" waste sites, for gasoline tank leaks, as air quality monitors, or as hazard monitors in factories. They would use wireless remote operation in which no one is endangered by collecting samples. A further advantage of the large number of inexpensive sensors is that they are inherently redundant so that a failure of a sensor or two would not compromise the integrity of the sensing network.

For background information *see* CAPACITANCE; ELECTRICAL RESISTANCE; MICRO-ELECTRO-MECHANICAL SYSTEMS (MEMS); MICROSENSOR; OXIDATION-REDUCTION; SATELLITE NAVIGATION SYSTEMS; TRANSDUCER; TRANSISTOR; WORK FUNCTION (ELECTRONICS) in the McGraw-Hill Encyclopedia of Science & Technology.   Joe Hartman

Bibliography. C. L. Britton, Jr., et al., Battery-powered, wireless MEMS sensors for high-sensitivity chemical and biological sensing, *Symp. Advanced Res. VLSI*, March 1999; P. D. Harris et al., Resistance characteristics of conducting polymer films used in gas sensors, *Sensors Actuators B*, 42:177–184, 1997; J. A. Hartman et al., Control and signal processing for an IMS sensor, *ISIMS (International Conference on Ion Mobility Spectrometry)*, July 27–August 1, 2003; H. Hill et al., Evaluation of gas chromatography···, *J. High Resolution Chem.*, vol. 19, June 1998; J. Hill and D. Culler, MICA: A wireless platform for deeply embedded networks, *IEEE Micro*, pp. 12–24, November 2002; D. C. Starikov et al., Experimental simulation of integrated optoelectronic sensors based on III nitrides, *J. Vac. Sci. Tech. B*, September/October 2001; D. M. Wilson et al., Chemical sensors for portable, hand-held field instruments, *IEEE Sensors J.*, 1(4): 256–274, December, 2001.

Evolutionary developmental biology (invertebrate)

The development of an invertebrate can be described as a pathway of growth and change of form from an egg through embryonic stages and into an adult morphology. As there are over 30 distinctive body plans (phyla) and hundreds of major body types among living invertebrates, all of which have descended from a common invertebrate ancestor, their developmental processes have clearly evolved spectacularly, diverging along a large variety of pathways to lead to their very disparate adult morphologies. The developmental sequence of morphological stages has been described for most invertebrate phyla, although important gaps remain in our knowledge of a few. The invention of powerful tools of molecular biology has enabled great strides in understanding the sorts of genetic mechanisms that underlie the observed morphological sequences in development. Consequently, it has become possible to investigate the evolution of those genetic mechanisms on a comparative basis, and although only a few phyla have been studied in any depth as yet, the findings are providing insights into the evolutionary genetics of the morphological diversity found in both living and extinct animal groups.

Role of developmental regulatory genes. A major finding has been that genetic differences among animal body plans are not based so much on the presence of different genes as they are on different patterns of gene regulation. There is a relatively restricted pool of developmental regulatory genes that mediate the development of major morphological features, and most of those genes are found in all animal phyla; they are involved in production of body plans through regulation of cell multiplication, differentiation, and positioning. Those genes are chiefly transcriptional regulators, together with cofactors and members of their signaling pathways. The regulators bind to target genes at enhancer sites, which are collections of deoxyribonucleic (DNA) sequences adjacent or near to the transcribed portion of the target gene to which numbers of transcription factors attach, providing combinatorial signals to initiate or repress gene transcription. As cell lines proliferate during development, successions of regulatory signals produce differentiation among some daughter cells, founding new cell lines that branch off as distinctive cell types. In many cases, some of the genes expressed in these derivative cells become involved in signaling to other cells—that is, they are regulators themselves—and thus there is a cascade of developmental gene expression that produces a pattern of differentiated cell types within the growing cell population. Differences in body plans among animals are therefore largely owing to differences in regulatory signaling during development that lead to production of distinctive cells, tissues, and the body architectures themselves.

Homeobox and Hox genes. Among the best-known developmental regulatory genes are the homeobox genes. The homeobox is a 180-basepair sequence that encodes a DNA-binding domain found in a large number of transcription factors that are important in animal development. *Hox* genes are an important group of homeobox genes which are largely responsible for mediating anteroposterior morphological differentiation in cnidarians (jellyfish, and so on) and bilaterians (bilateral animals and their descendants); *Hox* genes are not known in sponges. *Hox* genes are usually clustered and are collinear; that is, their order within the cluster is the same as in their domains of expression within the developing organism. Despite this regularity, Hox clusters vary in gene number. Evidence suggests that the most evolutionarily basal bilaterian cluster of Hox-type genes, the ProtoHox cluster, had four *Hox* genes, presumably growing from an ancestral *Hox* gene by duplications. Duplication of this four-gene cluster evidently gave rise to the Hox cluster and to a sister cluster of Hox-type genes, the ParaHox cluster, also widely present in animals. Through gene duplications, Hox clusters have grown in time within many lineages; the largest

Hox cluster yet found is in the invertebrate chordate class Cephalochordata, represented by *Amphioxus*, which has 14 *Hox* genes. However, the history of *Hox* gene duplication and loss has been different in every phylum with a single possible exception, so that generally every phylum has a unique Hox cluster. Many of the duplicated *Hox* genes (paralogs) have acquired unique functions.

Evolution of Hox-type genes. The general body plans of most or perhaps all animal phyla evolved during late Neoproterozoic and Early Cambrian times, from probably just over 600 to about 520 million years ago. By comparing the roles of developmental regulatory genes in their living representatives, it is possible to form a plausible generalized description of the evolution of the developmental genome among invertebrates during that interval, which speaks to some important features of developmental evolution. The Hox cluster serves as the best available example. The earliest animal body fossils appear to be sponges, followed by enigmatic body fossils that may be stem (primitive) cnidarians, and by trace fossils (trails and shallow burrows) indicating the presence of small (chiefly ≥1 mm wide) wormlike bilaterians. Near 543 million years ago the traces begin to be larger and more of them penetrate the sea floor, and near 530 million years ago the body fossils of living bilaterian phyla begin to appear. It is consistent with the fossil record that the body plans of all living phyla had evolved by 520 million years ago.

Hox-type genes must have been present in the common ancestor of the cnidarians and bilaterians. Many of the enigmatic body fossils of the Neoproterozoic are constructed of serial modules, and it is possible that Hox-type genes were responsible for that seriation. Certainly, the minute wormlike bilaterians originally possessed a Hox cluster of at least four genes, which presumably mediated anteroposterior patterning of cell types and tissues, as is true in living worms (such as nematodes) of the same simple structural grade. The fossil evidence suggests an evolutionary scenario involving a radiation of the minute Neoproterozoic worm fauna into a variety of habitats on, within, and above the ancient sea floor, evolving divergent morphological solutions to the distinctive challenges they encountered there. As the Hox gene clusters were directly responsible for morphological changes along anteroposterior axes, it is plausible that Hox clusters underwent divergent evolution during that episode. It was evidently from among these disparate wormlike body plans that the more complex and commonly larger-bodied phyla evolved, producing body plans appropriate to their ancestry and environments, probably accompanied by increases in Hox gene number in many lineages. As complex organs and appendages evolved within Hox expression domains, *Hox* genes came to lie at the head of the increasingly complex cascades of gene expressions required to develop those novel structures. Morphological changes within *Hox* gene domains then began to result from changes within the cascades, while the *Hox* genes themselves were somewhat sequestered from change, and became quite highly conserved within body plans.

Non-Hox genes. Other genes involved in key regulatory functions but not present in clusters, including homeobox genes as well as those with other binding motifs, also occur widely among phyla. Many, perhaps all, of them mediate the development of different features in different groups, not unlike the Hox genes. Commonly they are expressed in nonhomologous cells and tissues, and/or are not expressed in homologous ones, between and within phyla. Such events may sometimes reflect duplication of the regulatory gene and slight divergence of one of the daughter genes (paralogs) to a new function, presumably through mutations that create a binding site for the divergent gene product. In such a case, if the other paralog, which retained its ancestral function, was then lost, the gene would appear to have switched functions. In other cases, functional switching may simply have involved evolution of a novel binding site on a new target gene and loss of the previous one.

An example of the evolution of the expression patterns of non-Hox genes within an invertebrate body plan is provided by the homeobox genes *orthodenticle* (*Otx*) and *distal-less* (*Dlx*), regulatory genes that are expressed multiple times during development in most metazoan phyla. Within the phylum Echinodermata, some of their functions are conserved among classes. However, these genes also function in different tissues and locations in different echinoderm classes: for example, *Otx* is expressed in association with ossicles in the arm tips of brittle stars but not in homologous positions in sea urchins, while *Dlx* is expressed in podia of sea urchins but not in brittle stars. These genes were evidently coopted for different developmental roles as the ancestral echinoderm body plan was modified during evolution of the classes. The seemingly frequent shufflings of regulatory gene activities among targets, changing expression patterns in time and place, are evidently the key processes in the regulatory evolution that underpins the origins of morphological novelty.

Developmental regulation, selection, and biodiversity. Despite differences in invertebrate morphological complexity of orders of magnitude, the sizes of most invertebrate genomes differ only by a factor of 2 or so, and it is likely that the sizes of the fraction of those genomes involved in developmental regulation differ even less. Evidently, when complexity increased, the regulatory genes became involved in increasing numbers of functions, being expressed at more times and places within the developing organism. Such an increase was made possible by evolution of regulatory apparatuses to increase the number of signaling episodes from upstream regulators and to downstream target genes. As new morphologies arose in crown (living) phyla, in response to selection within the mosaic of marine environments, their developmental bases were each underpinned by a complex network of signaling pathways, evolved in large part independently, but with nearly identical toolkits of developmental genes inherited from their small wormlike ancestors. Most of this high-level regulatory genetic activity seems to have occurred during the Neoproterozoic and Early Cambrian. However, on all levels, the marvels of invertebrate biodiversity have evolved, not so much through differences in genes (although these certainly play a role) but through differences in the organization of gene expression during development.

For background information *see* ANIMAL EVOLUTION; ANIMAL MORPHOGENESIS; CEPHALOCHORDATA; CNIDARIA; DEVELOPMENTAL BIOLOGY in the McGraw-Hill Encyclopedia of Science & Technology.

James W. Valentine

Bibliography. S. B. Carroll, J. K. Grenier, and S. D. Weatherbee, *From DNA to Diversity*, Blackwell Science, Malden, MA, 2001; E. H. Davidson, *Genomic Regulatory Systems: Development and Evolution*, San Diego, Academic Press, 2001; B. K. Hall and W. M. Olson (eds.), *Keywords and Concepts in Evolutionary Developmental Biology*, Harvard University Press, Cambridge, MA, 2003; A. H. Knoll, *Life on a Young Planet*, Princeton University Press, 2003; R. A. Raff, *The Shape of Life*, University of Chicago Press, 1996; J. W. Valentine, *On the Origin of Phyla*, University of Chicago Press, 2004.

Evolutionary developmental biology (vertebrate)

The emergence of the new field of evolutionary developmental biology ("evo-devo") has been fueled by the discovery that many very distantly related organisms utilize highly similar genes and molecules to establish important structural patterns during their development (defining, for example, the borders between the head, trunk, abdomen, and tail). The goal of research in this field is to determine how evolutionary mechanisms (such as mutation and natural selection) act on embryonic development, generating diversity in the body plans of various organisms but also constraining the possible directions of evolutionary change. Of particular interest are the so-called macroevolutionary transitions, in which a major change in morphology occurred, such as the emergence of limbed vertebrates (tetrapods) from fish or the evolution of birds from dinosaurian ancestors.

Adult fossils. Although vertebrate embryos occasionally fossilize, the embryos were usually close to birth at the time of fossilization and had already started the process of ossifying their skeleton (replacing the original cartilage with bone). This means that although fossilized embryos can tell much about the reproductive biology of a particular species, they cannot tell us about the early stages of development, when the genes responsible for controlling structure (pattern formation) are expressed. However, advances made by evolutionary biologists in reconstructing phylogeny (evolutionary relationships) have given fossils of adult animals an important role

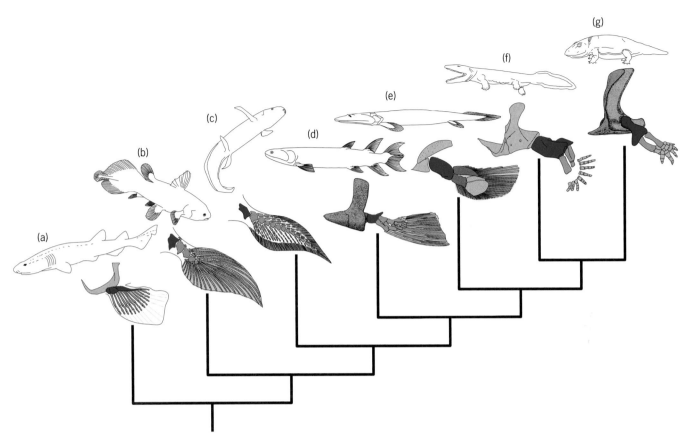

Fig. 1. Phylogenetic tree of fin and limb skeletons, showing the gradual acquisition of limblike characteristics. (*a*) Living sharks; (*b*) living coelacanth; (*c*) living lungfish; (*d*) extinct fish, *Eusthenopteron*; (*e*) extinct fish, *Panderichthys*; (*f*) extinct tetrapod, *Acathostega*; (*g*) extinct tetrapod, *Eryops*.

in evolutionary developmental biology. Without an understanding of their genealogical history, these fossils are simply isolated snapshots of past life. However, once fossils are placed in an evolutionary tree, the morphology can be viewed in a comparative context, and they become route markers of the path of evolutionary history (**Fig. 1**).

The study of adult fossil data can reveal three things: the primitive morphology of structures and organs, the morphological steps in macroevolutionary transitions, and the diversity of ancient vertebrate morphology.

Primitive morphology. Using adult fossil data, developmental biologists can reconstruct the primitive morphology of a particular organ or structure. For example, during the fish-tetrapod transition, what type of fin did the limb evolve from? This approach helps developmental biologists understand which living animals can serve as the best models for elucidating the development of such structures in their long-extinct ancestors (in the case of the fish-tetrapod transition, the living lungfishes are the closest living analogs).

Morphological steps. Developmental biologists can use information from the fossil record to break down major evolutionary transitions into a series of smaller morphological steps. For example, the tetrapod limb did not suddenly appear fully formed; the fossil record reveals many different species bearing fins with increasingly limblike characteristics. Phylogenetic reconstruction helps determine the exact sequence of these changes—for instance, that the wrist evolved after the first appearance of fingers (Fig. 1*f–g*). This process can also reveal unexpected combinations of characteristics, that is, animals very advanced in some respects but primitive in others. For example, the external skull bones of *Panderichthys* (Fig. 1*e*) are very similar to those of early tetrapods (Fig. 1*f*), but the internal braincase is very different, closely resembling that of fishes such as *Eusthenopteron* (Fig. 1*d*).

Phylogenetic reconstruction can also help determine the usefulness of the living species being used as models of development. Because lungfish are difficult to work with in the laboratory, developmental biologists favor fishes that require less care, such as the zebrafish *Danio rerio* (popular in home aquaria). As a ray-finned teleost fish, *Danio* is much farther removed from the origin of tetrapods than lungfish, but its fins still share some important characteristics with limbs—characteristics present in the last common ancestor of the zebrafish and tetrapods. Therefore, the zebrafish can be a useful experimental model for research into the evolution of tetrapod development, as long as attention is paid to which aspects of zebrafish development are likely to be shared primitively with tetrapods, and which evolved after the zebrafish and tetrapod lineages split.

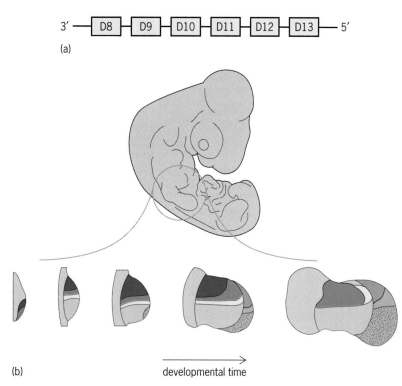

(a)

(b)

developmental time

Fig. 2. *Hoxd* genes in limb development. (*a*) Schematic diagram of part of the *Hoxd* cluster, showing the arrangement of six of the nine *Hoxd* genes along a single chromosome. Although there are many different families of developmental genes, only *Hox* clusters show such close spatial relations. (*b*) Chick embryo, showing the developing limb. As the limb-bud grows, it exhibits a dynamic, overlapping pattern of *Hoxd* expression, which correlates with the spatial arrangement of the genes along the chromosome. This subdivides the developing limb into regions, which then go on to produce distinctive structures. For example, each of the three regions at the apex of the developing limb produce one of the three fingers seen in the adult chicken.

Morphological diversity. The study of adult fossils can also reveal the diversity of ancient vertebrate morphology. Animals alive today show various skeletal specializations, but these are only a small fraction of the range seen in the fossil record. Given that the genetic mechanisms used to pattern various anatomical structures are highly conserved, a survey of the range of morphology occupied over millions of years can give clues about how developmental patterns are modified during evolution. Some features show a great deal of variation—for example, the number of bones making up each finger. Others are conserved—for example, the pattern of bones in the upper arm (humerus, ulna, and radius). This gives an indication of the relative ease with which different parts of the developmental program can be modified (evolutionary constraints). Furthermore, if a similar morphology has evolved several times independently (convergent evolution), it could indicate that a single, specific genetic pathway is being repeatedly selected as a target for evolutionary change (canalized evolution).

Integrating fossil and developmental data. For fossils to be integrated into evolutionary developmental biology scenarios, two practical criteria must be applied. First, the structure or organ under study must be preserved in the fossil record. In vertebrates this generally means that it must be part of the skele-

ton, or in some way leave its mark on the skeleton. This can encompass the bones themselves, as well as softer structures such as muscles, tendons, and parts of the blood supply and nervous system. However, more subtle, and more fundamental, developmental processes can also be observed in the skeleton, such as body segmentation (for example, the way the spine is divided into separate vertebrae) and regionalization (for example, the division of the body into distinct head, neck, trunk, abdomen, and tail areas).

A second criterion is that the developmental genetics of the structure or organ must be sufficiently understood to make meaningful hypotheses about its evolution. There are thousands of different developmental genes, performing many different tasks. In order to understand evolutionary changes, the effects of small networks of interacting genes must be clearly distinguishable.

From fins to limbs. The origin of the tetrapod limb has long been a focus of interest for both evolutionary and developmental biologists. It is not surprising, therefore, that it has become a major topic of evolutionary developmental biology research. Fossils allow us to formulate hypotheses about the timing of the evolution of some of the principal genetic patterning networks (Fig. 1). Fossil data suggest that the genetic mechanisms vital for limb development in living tetrapods emerged gradually, over tens of millions of years. Some, for instance, were present in the last common ancestor of humans and sharks. This undermines traditional definitions of the tetrapod limb and blurs the boundary between fish and tetrapods.

The pentadactyl (five-fingered) limb, considered to be the ancestral condition for all living tetrapods, is in fact the end product of a gradual accumulation of limblike characteristics. Sharks have several cartilages at the base of the fin that articulate with the shoulder girdle; the most posterior of these cartilages (shown to the right in Fig. 1*a*) is elongated and bears several secondary cartilages. The living coelacanth (Fig. 1*b*) and lungfish (Fig. 1*c*) retain only the posterior basal bone (now called the humerus) and have a long axis of secondary bones. More tetrapod-like extinct fish, such as *Eusthenopteron* (Fig. 1*d*), had a shorter axis with more robust secondary bones. The most tetrapod-like extinct fishes, such as *Panderichthys* (Fig. 1*e*), had a flat plate at the apex of the axis. This indicates the evolution of a hand region, although it retained fin-rays. Animals such as *Acanthostega* had no fin-rays (Fig. 1*f*). The hand region consisted of eight digits with a few weak wrist cartilages. This was used as a paddle rather than for walking. It took several millions of years for the wrist to evolve the two rows of bones seen in modern tetrapods, and for the number of digits to reduce to five (Fig. 1*g*).

However, the gradual acquisition of limblike characteristics does not imply that evolution was directed toward the formation of limbs. Each evolutionary change was adaptive for the fin of the species concerned; only in retrospect do these changes seem to be leading to the development of a pentadactyl limb. In fact, some species of ray and teleost fish have independently evolved limblike fins for "walking" on the sea bed, demonstrating that the tetrapod limb was probably only one of many different possible solutions to the problem of locomoting on land.

From limbs to wings. The evolution of limb development is also central in the debate surrounding bird origins. Bird wings are extremely modified forelimbs, bearing just three fingers. However, controversy has surrounded the identification of exactly which three fingers are present. Among evolutionary biologists, there is widespread agreement that birds are the descendants of small, three-fingered dinosaurs. The fossil record of these dinosaurs indicates that digits IV and V (equivalent to the human ring finger and little finger) were gradually reduced and lost, meaning that living birds possess digits I, II, and III (equivalent to the human thumb, index finger, and middle finger). This conflicts with embryological data from birds, which strongly suggest that digits I and V began to form early in development but then degenerated to leave the adult with digits II, III, and IV. Reconciling these apparently incompatible observations has been a challenge for evolutionary developmental biology.

One ingenious hypothesis comes from an understanding of the role of *Hoxd* genes in the patterning of digits. In vertebrates there are nine *Hoxd* genes, clustered closely together on a single chromosome (**Fig. 2a**). During development they are expressed across the limb in a staggered, overlapping pattern that matches their order within the chromosomal cluster. This overlapping pattern divides the developing limb into smaller sections, each with a unique combination of *Hoxd* gene expression and its own developmental "identity" (Fig. 2b). Thus, every digit begins to form in its own *Hoxd* section. However, it has been known for some time that perturbing the combination of *Hoxd* gene expression can alter the identity of a section. Even a small change to a single gene can redirect the development of entire structures; for example, the developing digit III may take on the morphology of the adjacent digit II, and so on. Such homeotic changes can be produced in the laboratory, and are occasionally observed as the result of natural mutations. It has been proposed, therefore, that dinosaurs did indeed lose digits IV and V (as the fossil evidence suggests), but then a homeotic mutation caused the remaining digits to change, leaving them with the identities II, III, and IV (as embryological evidence suggests). While this hypothesis is untested, it remains an intriguing possibility, and currently the only explanation that unites the conflicting data from paleontology and embryology.

Future. Fossil data will continue to make a significant and unique contribution to evolutionary developmental biology research, complementing advances made in our understanding of developmental patterning. A key aim will be to understand why some developmental patterns are modified more frequently than others, and to what extent the requirement for successful embryonic development constrains the possible directions of evolutionary change.

For background information *see* ANIMAL EVOLUTION; AVES; DEVELOPMENTAL BIOLOGY; DEVELOPMENTAL GENETICS; DINOSAUR; FOSSIL; PHYLOGENY in the McGraw-Hill Encyclopedia of Science & Technology.   Jonathan E. Jeffery

Bibliography. S. Chatterjee et al., Counting the fingers of birds and dinosaurs, *Science*, 280:355a (Technical Comments), 1998; J. A. Clack, *Gaining Ground*, Indiana University Press, 2002; M. I. Coates, J. E. Jeffery, and M. Ruta, Fins to limbs: What the fossils say, *Evol. Dev.*, 4:1–12, 2002; G. P. Wagner and J. A. Gauthier, 1,2,3 = 2,3,4: A solution to the problem of the homology of the digits in the avian hand, *Proc. Nat. Acad. Sci. USA*, 96:5111–5116, 1999; C. Zimmer, *At the Water's Edge*, Touchstone, New York, 1999.

Exchange bias

Of considerable scientific and technological importance, exchange bias is the field shift that takes place in the magnetic hysteresis loop of a ferromagnetic material when it is placed in intimate contact with an antiferromagnetic material. Although exchange bias was discovered 50 years ago, it is still a topic of intense interest. This stems partly from the fundamental interest in interface magnetism that the advent of magnetoelectronics has brought about, and partly from its technological application in the magnetic recording industry. Despite extensive research efforts, a full understanding of the origin of exchange bias is still lacking. This article provides a short description of the exchange bias effect, the materials systems in which it is observed, and an intuitive explanation of its origin. This is followed by an explanation of the technological appeal of exchange bias, a description of the theoretical ideas developed to model it, and an outlook on the prospects for future progress.

Basic effect. **Figure 1** shows a schematic of the magnetic hysteresis loop of a ferromagnetic material, with and without an adjacent antiferromagnetic material. Such hysteresis loops, which plot the magnetization (M) against the applied magnetic field (H), exhibit irreversibility due to the energy lost in moving magnetic domain walls. The hysteresis loop of the single ferromagnetic layer is narrow (that is, it has small coercivity), and symmetric with respect to field. The situation is very different for the case of a ferromagnetic system in contact with an antiferromagnetic material. When such a sample is cooled below the ordering temperature of the antiferromagnetic material (Néel temperature) in an applied field (or with the ferromagnet magnetized), the hysteresis loop shifts along the field axis, resulting in an $M(H)$ curve that is asymmetric with respect to $H = 0$. The

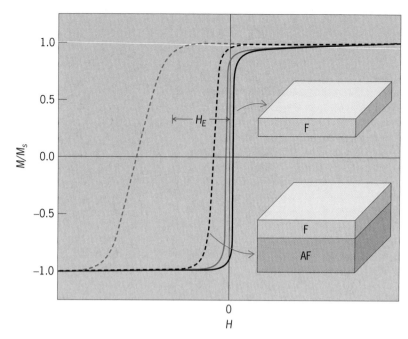

Fig. 1. Schematic hysteresis loop of a ferromagnetic film (F) with and without an adjacent antiferromagnet (AF). The shift of the hysteresis loop along the field axis (that is, the exchange bias, H_E) is shown. M = magnetization. M_S = saturation magnetization (when all the domains are aligned). H = applied magnetic field.

magnitude of this shift is referred to as the exchange bias, H_E. (The ordering temperature of the antiferromagnetic material is the point at which spontaneous alignment of the antiferromagnetic spins occurs, and the antiferromagnetic material begins to influence the properties of the ferromagnetic layer.)

This exchange bias effect is quite general, having been observed at many interfaces between materials with a ferromagnet-like moment and materials that exhibit antiferromagnetic ordering. Antiferromagnets and ferrimagnets as widely varied as transition-metal oxides (for example, CoO, Fe_3O_4) and fluorides (such as FeF_2, $KNiF_3$), perovskite oxides (such as $La_{1-x}Ca_xMnO_3$), and metallic binary alloys (such as $Fe_{50}Mn_{50}$, $Ir_{20}Mn_{80}$) can induce exchange biasing. The effect has been observed in numerous geometries such as bilayer thin films, ferromagnetic layers on bulk antiferromagnetic crystals, ferromagnetic nanoparticles coated with antiferromagnetic layers, ferromagnetic/antiferromagnetic multilayers, and even some naturally occurring biological systems.

Intuitive explanation. Ferromagnetic materials exist in a state where quantum-mechanical exchange interactions between neighboring atoms induce a parallel alignment of electron spins. Conversely, an antiferromagnet is a material where antiparallel orientation of neighboring spins is energetically favorable. To see how a loop shift can occur when these two material types are brought together, consider first the ferromagnetic layer, saturated by a large positive field (**Fig. 2a**). The ferromagnetic spins are aligned, and we assume that the topmost layer of antiferromagnetic spins is aligned with the ferromagnetic layer. Upon reducing the positive magnetic field, the ferromagnetic spins begin to rotate toward

the negative field direction to initiate reversal. The surface antiferromagnetic spins now place a microscopic torque on the ferromagnetic layer, hindering the rotation toward the negative field direction (Fig. 2b). This increases the field required to reverse the magnetization and pushes the left side of the hysteresis loop to larger negative fields (Fig. 2b, c). On the return branch of the loop (Fig. 2d), the antiferromagnetic surface spins now favor the reversal of the ferromagnetic spins back to the positive field direction, meaning that less field is required to induce reversal. The net result is a hysteresis loop shifted along the field axis in the negative field direction. The presence of the antiferromagnetic layer clearly provides unidirectional anisotropy, as the energy minimum is achieved only when the ferromagnetic layer is aligned parallel to the topmost layer of antiferromagnetic spins.

This simple model can be formulated mathematically and is known as the Meiklejohn-Bean model. Although some of its predictions are in qualitative agreement with experimental findings, it fails to account quantitatively for the H_E observed in many systems and also is in direct contradiction with certain experimental facts. More sophisticated models are required to capture the essential physics of the problem. Before discussing these in more detail, along with the rich variety of phenomena that accompany the simple loop shift, a simple explanation for the technological appeal of exchange bias is provided.

Potential applications. Heterostructured devices based on layers of ferromagnetic and nonferromagnetic constituents have revolutionized the magnetic recording industry and led to the birth of magnetoelectronics, or spintronics. The simplest device in spintronics is the spin valve (**Fig. 3a**), which consists of two ferromagnetic layers separated by a thin, nonmagnetic metallic spacer. When a current flows through this device, a fraction of the electrical resistance is determined by the relative magnetization orientations of the two ferromagnetic electrodes. If they are aligned parallel, the resistance is low. If they are antiparallel, spins leaving the bottom electrode experience additional scattering at the interface with the top electrode due to the unfavorable alignment of the magnetization. If the two ferromagnetic layers have different coercivities, this results in the resistance (R) versus H curve shown in Fig. 3a. The resistance is low when the magnetizations are parallel, and high when the magnetizations are antiparallel.

A spin valve is of great importance in magnetic recording, as it is used in read heads to sense the magnetic field emanating from "bits" in the recording medium. As this magnetic field is rather small (about 10 oersteds or 800 A/m), the $R(H)$ curve must be tailored to "switch" at very low fields. Adding an antiferromagnetic layer to cause exchange bias in one of the ferromagnetic layers "pins" the magnetization of that electrode so that its magnetization reversal occurs only in large fields (Fig. 3b). The unbiased or "free" layer can still be reversed in very low fields (its coercivity is low), facilitating antiparallel orientation

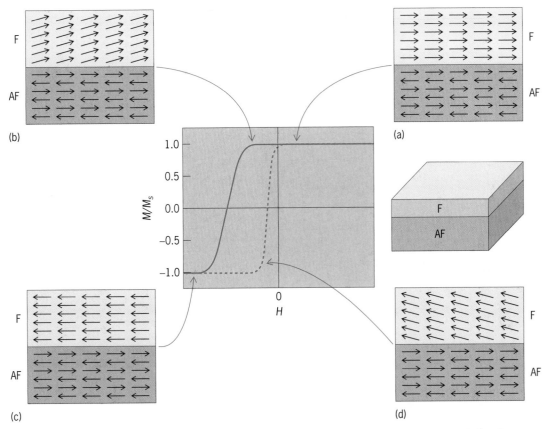

Fig. 2. Schematic of an exchange-biased hysteresis loop along with sketches *a–d* of the antiferromagnetic (AF) and ferromagnetic (F) spin structures at various points around the hysteresis loop. It shows the effect of AF spin or F spin alignment around the loop.

between the two layers and the full magnetoresistance response in small fields. The usefulness of exchange pinning of ferromagnetic layers is not restricted to these hybrid spin valves. In fact, exchange spinning plays a key role in the proposed architectures for future nonvolatile MRAM (magnetic random access memory) applications.

Experimental and theoretical advances. The five decades of experimental research on the exchange bias effect have resulted in the discovery and investigation of numerous supplementary effects in addition to the loop shift, including: (1) a coercivity enhancement accompanying H_E, (2) high field rotational hysteresis, (3) time- and history-dependent effects such as training and "memory" effects, (4) asymmetric magnetization reversals, (5) positive exchange bias, where the $M(H)$ loop shifts in the same direction as the applied field, and (6) even the existence of exchange biasing above the Curie temperature, at which the ferromagnetic layer transitions to the paramagnetic state. These investigations have been accompanied by detailed studies of the influence of structural parameters such as the thickness of the constituent layers, degree of crystalline perfection, interfacial roughness, interdiffusion, crystalline orientation, and so forth. Theoretical developments have taken place at a similar rate. The theoretical models have increased in complexity from the simple Meiklejohn-Bean model. They now embrace the key role of interface roughness, the ten-

dency to form antiferromagnetic domain walls parallel or perpendicular to the interface, the assumption of domain walls in the ferromagnet, orthogonal orientations of the interfacial antiferromagnetic and ferromagnetic spins, as well as the role played by defects in the antiferromagnet.

The vast majority of the models outlined above, and indeed much experimental data, point to the fact that the antiferromagnet and its various imperfections play a key role in exchange bias. Although it is difficult to make "definite" statements, many researchers are converging on the idea that defects in the antiferromagnet layer can generate uncompensated spins (that is, spins in the antiferromagnet that are not compensated by oppositely aligned partners, leading to a net spin inbalance), which couple to the ferromagnet magnetization and give rise to exchange bias. Recently, ingenious (but indirect) experiments, in addition to the application of the first techniques capable of probing both the ferromagnetic and antiferromagnetic interfacial spin structure with high sensitivity, have provided support for this viewpoint. This realization has led to the emergence of several models embracing the idea that realistic models of exchange bias must be developed for specific materials systems and include "real-world" disorder such as interface roughness, defects in the antiferromagnetic layer, and the subsequent uncompensated spin density that it induces. Such models might mark the end of the notion that a universal theory of exchange

(a)

(b)

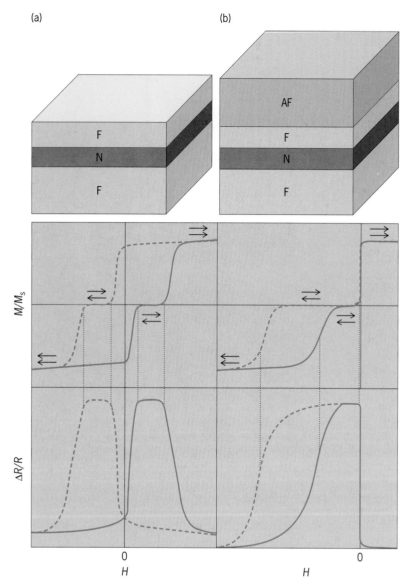

Fig. 3. Schematic explaining the operation of (*a*) a spin valve and (*b*) a hybrid spin valve, which are formed by exchange-biasing one of the ferromagnetic (F) layers with an antiferromagnetic (AF) layer. The graphs show the magnetization hysteresis loops and the magnetoresistance response of the two devices. N = nonmagnetic material. *R* = resistance. Δ*R* = change in resistance.

bias can be developed, but provide hope that in the near future a quantitative understanding of exchange bias can be achieved.

For background information *see* ANTIFERROMAGNETISM; FERRIMAGNETISM; FERROMAGNETISM; MAGNETIC HYSTERESIS; MAGNETIC RECORDING; MAGNETIZATION; MAGNETOELECTRONICS in the McGraw-Hill Encyclopedia of Science & Technology.

Chris Leighton; Michael Lund

Bibliography. A. E. Berkowitz and K. Takano, Exchange anisotropy—a review, *J. Mag. Magnetic Mater.*, 200:552–570, 1999; M. Kiwi, Exchange bias theory, *J. Mag. Magnetic Mater.*, 234:584–595, 2001; W. H. Meiklejohn, Exchange anisotropy—a review, *J. Appl. Phys. Suppl.*, 33:1328, 1962; W. H. Meiklejohn and C. P. Bean, New magnetic anisotropy, *Phys. Rev.*, 102:1413–1414, 1956; J. Nogués and I. K. Schuller, Exchange bias, *J. Mag. Magnetic Mater.*, 192:203–232, 1999; N. Spaldin, *Magnetic Materials: Fundamentals and Device Applications*, Cambridge University Press, 2003; S. A. Wolf et al., Spintronics: A spin-based electronics vision for the future, *Science*, 294:1488–1495, 2001.

Experience and the developing brain

The human brain is an organ that possesses a vast potential for adaptation and change in response to experience. This mutability, which is referred to as brain plasticity, enables humans to solve new problems, learn, and forget. Research reveals two important insights into brain plasticity: (1) Plasticity is influenced by a surprisingly large number of life experiences. (2) The brain retains plasticity into senescence, although the developing brain is likely to have greater capacity for change.

The origins of contemporary research into plasticity derived from two lines of investigation, in rodents and in primates. In the early 1960s, Mark Rosenzweig and colleagues discovered that raising rats in enriched environments produced smarter rats with bigger brains. At about the same time, Harry Harlow was interested in finding out how infant monkeys fared when they were removed from their mothers and raised with surrogate mothers made from wire mesh. He found that monkeys raised with surrogate mothers showed severe emotional and behavioral deficits and had small, underdeveloped brains. The Rosenzweig and Harlow experiments demonstrated that the impact of environment had a profound effect on the brain. The brain was changed not only in size and structure but also in the way that it would subsequently deal with life's challenges.

Historically, the instructions for constructing the brain were attributed to the genetic blueprint found on the chromosomes inherited from parental cells. According to this notion, a brain was predestined to take its adult form solely on the basis of these inherited instructions. It is now known that environment can interact with the genome, a process called epigenetic inheritance, to alter gene expression by either altering patterns of deoxyribonucleic acid (DNA) methylation or by modifying histones (the protein cores around which DNA is wound). These manipulations can silence genes or cause their activation. Altered gene expression changes the expression of proteins. Because proteins are the building blocks of the brain, different proteins build different brains by modifying cell number, cell connectivity, brain size, and, ultimately, behavior. Epigenetic inheritance thus provides an adaptive means for an organism to prepare its brain for the unique environmental challenges that it will face without having to change its genetic blueprint.

In addition to environmental experiences, a surprising number of factors have been identified that influence brain development. The following sections describe four formative causes of brain modification:

environmental influences, nutritional influences, maternal influence, and brain damage.

Environmental influences. Environmental experiences can have a global impact on brain development, and one of the most studied forms of environmental experience is environmental enrichment. Enriched environments are conditions that are stimulating in sensory, motor, and, sometimes, social domains (if animals are housed in groups). They typically consist of larger homes containing motoric challenges and toys that are periodically changed. Laboratory animals reared in enriched environments are faster at completing complex mazes and show increases in brain weight compared with animals raised in standard or impoverished conditions.

The effects of complex housing on brain development and behavior are more pronounced the earlier the animals are exposed to it. Pregnant rats housed in enriched environments give birth to offspring that are able to solve complex maze problems much quicker than offspring born to mothers housed in standard caging. Prenatally enriched animals are also more adept at motor learning and can retrieve food more efficiently in cages designed to assess reaching skills. Simple exposure to classical music during the prenatal and early postnatal period can also improve spatial learning and maze performance of rats.

Nutritional influences. Just as the brain reacts to external sensory stimulation, it responds to the internal stimulation provided by nutrition. Generally, exposure to nutritional elements is accomplished via the maternal blood supply during gestation or ingestion of breast milk during the nursing period. Included among potent nutritional influences are the effects of drugs. Drugs commonly ingested by pregnant women are nicotine, antidepressants such as fluoxetine (Prozac®), and alcohol.

Nicotine. Exposure to low doses of nicotine (equivalent to 1 cigarette smoked daily) throughout the gestation period causes slight cognitive and motor impairments in adult rats. However, this same level of prenatal nicotine exposure is beneficial to perinatally brain-injured rats. When tested as adults, nicotine-treated animals perform better on tests of cognitive and motor behavior than untreated brain-injured animals.

Fluoxetine. Prenatal exposure to fluoxetine causes a reduction in brain and body size in rats and impairments on tests of cognition and motor performance. Brain injury in the perinatal period to animals that were prenatally exposed to fluoxetine is doubly devastating. These animals show severe impairments on behavioral tests and are also smaller in body and brain size than untreated brain-injured animals.

Dietary supplements. Maternal dietary supplements also affect brain development. Choline, an essential nutrient for humans and rats, when supplemented in the maternal diet can afford offspring protection against memory impairments produced by epileptic seizures. Likewise, prenatal choline reduces behavioral impairments associated with early cortical damage in rats.

Vitamin supplements are effective in reducing behavioral impairments in early brain-injured rats. In addition, normal rats show improved cognitive performance when exposed to vitamins present in breast milk of dietary-supplemented mothers and in the feed of the developing pups.

Alcohol. Perhaps the most potent nutritional influence on the brain is alcohol, which if ingested in sufficient quantities at certain times in pregnancy can produce profound alterations not only in physical features but also in brain structure and function. The devastating effects of alcohol in humans is called fetal alcohol syndrome.

Maternal influences. Maternal behavior has also been shown to influence brain development and behavior in offspring. Pregnant female rats are more active than nonpregnant female controls, and their exploratory movements, grooming, and eating and drinking behaviors expose their fetuses to a variety of mechanical pressures and vestibular stimulation. At parturition, mother rats lick and groom their pups and transport them in and out of the nesting area.

Michael Meaney and colleagues have demonstrated that rat dams that show high levels of licking and grooming of their pups have offspring that show reduced stress responses in novel situations and enhanced learning in adulthood. These behavioral effects are accompanied by increased neuronal survival in the hippocampus and increased expression of glucocorticoid receptors (receptors involved in regulating the stress response) in the hippocampus and frontal cortex. Attenuated stress responses are associated with a lower risk of diseases and correlate with successful brain aging (that is, enhanced cognitive skills in senescence).

Females born to dams that display high levels of licking and grooming also spend more time licking and grooming their own offspring. Similarly, female pups born to mothers that engage in low levels of licking and grooming when cross-fostered to high-licking-grooming mothers at birth become high-licking-grooming mothers. Thus maternal behavior is transmitted from generation to generation, not via direct genetic inheritance but via early mother-pup interactions, and so appears to be associated with DNA methylation patterns that alter gene expression to alter adult behavior.

In contrast, prolonged separation from the mother in the early perinatal period causes animals to show exaggerated responses to stress in adulthood. In addition, exposure to stress in pregnant rats can cause greater emotionality and anxiety in their offspring. This heightened reactivity to stress is associated with increases in size of the lateral nucleus of the amygdala (which regulates anxiety) and fewer glucocorticoid receptors in the hippocampus and frontal cortex.

Brain injury. The effects of early brain injury can be devastating when sustained during certain periods of development, while a similar injury can be less severe at other developmental periods. This paradoxical effect is dependent on the type of developmental processes the brain is undergoing at the time of injury. For example, frontal cortex injury in the rat during the first week of life is associated with a poor behavioral outcome in adulthood, whereas the same injury in the second week of life is associated with significant behavioral recovery. Frontal cortex injury at postnatal days 7 to 10 in rat pups causes increases in dendritic branching in the parietal cortex and significant spontaneous regeneration of frontal cortex tissue. During the first week of life, cortical cells are migrating to their final destination in the rat. Astrocytes invade the cortex, dendrites grow, and synapses form during the second postnatal week.

Cortical injury that interrupts neuronal migration seems to have greater negative consequences on the developing brain than does injury that disrupts dendritic outgrowth and synapse formation. The presence of astrocytes in the cortex is also beneficial after injury. Astrocytes are known to produce various proteins that are associated with recovery processes following brain injury, such as basic fibroblast growth factor and glial-derived neurotrophic factor. Thus, the alterations in brain development induced by injury are age-dependent and result in very different behavioral outcomes.

Enriching experience has also been shown to be beneficial after brain injury. Rats that sustain cortical injury in the first week of life perform dismally on tests of both cognitive and motor behavior as adults. If perinatal brain-injured animals are reared in complex environments, the extent of their functional impairments are greatly reduced. Animals that sustain early brain injury also benefit from tactile stimulation. A soft brush is used to stroke infant rats each day following injury until the time of weaning. Tactile stimulation promotes the production of basic fibroblast growth factor in the skin, and this protein is carried via the circulatory system to the damaged brain, where it facilitates reparative processes. Brain-injured animals that receive perinatal tactile stimulation show behavioral improvements as adults on both cognitive and motor tasks.

Environmental enrichment or tactile stimulation of pregnant mothers is also prophylactic for offspring that sustain early cortical injury. Complex housing or gentle stroking of rat dams during their pregnancy produces offspring that are resilient to the damaging effects of early cortical injury and perform at near-normal levels on tests of cognitive or motor behavior as adults.

Conclusion. Although the brain can be modified by these experiental factors, there are times during which particular sensory experiences are important for the induction of appropriate patterns of cell connectivity, particularly in brain sensory areas such as the visual or auditory cortex. These windows of opportunity during brain development are termed critical periods. During critical periods there is a refinement of connectivity, wherein useful connections are induced or maintained and stabilized, and non-functioning connections are pruned away. In other words, one must "use" connections "or lose them." For example, acquisition of language depends on ex-

posure of a child to the sounds of that language. The auditory cortex seems to be maximally tuned to acquiring sound discriminations between the ages of 6 months and 2 years. This does not mean that after age 2 it is impossible to learn to discriminate the phonetics of language; rather, the level of plasticity declines, making it more difficult to learn the discrimination. Thus, the interactions between the many external and internal environmental influences and brain critical periods has a profound effect on the kind of brain that animals, including humans, develop.

For background information *see* BEHAVIOR GENETICS; BRAIN; COGNITION; GENE ACTION; INTELLIGENCE; LEARNING MECHANISMS; NEUROBIOLOGY in the McGraw-Hill Encyclopedia of Science & Technology.

Robbin Gibb; Bryan Kolb; Ian Whishaw

Bibliography.   R. Gibb, Experience, in I. Q. Whishaw, and B. Kolb (eds.), *The Rat*, Oxford University Press, New York, 2004; B. Kolb, R. Gibb, and T. E. Robinson, Brain plasticity and behavior, in J. Lerner and A. E. Alberts (eds.), *Current Directions in Psychological Science*, pp. 11–17, Prentice Hall, Upper Saddle River, NJ, 2004; B. Kolb and I. Q. Whishaw, *Introduction to Brain and Behavior*, Freeman Worth, New York, 2001.

Extensible Markup Language (XML) databases

As the electronic distribution of documents, especially via Web sites, has become more widespread, those responsible for managing document content have looked for new approaches and tools to perform management functions in an efficient and cost-effective manner. Markup languages are useful for organizing and formatting document content for display, exchange, and distribution; however, markup languages are not very helpful with managing and searching large amounts of document content. Consequently, many individuals have realized that the power of database management systems (DBMS) to efficiently store, update, secure, and speedily retrieve data should be applied to the problem of document management as well.

Documents versus data. At first glance, it might seem a contradiction to connect a document markup language, such as can be created with XML, with database technology. Documents and data have historically been processed using quite different software technologies because of their fundamentally dissimilar natures. However, documents and data do share characteristics that allow selected technologies to be used together, so that their best features enhance the way that documents are handled. An explanation of the nature of documents and the added document content of markup languages will show how document markup and database technologies can be combined.

The most common document, known as document-centric type, is principally composed of

```
<SALES>
        <ORDER>
                <OrderNumber>993857322</OrderNumber>
                <CustomerID>3009A</CustomerID>
                <ProductID>QW23UP</ProductID>
                <SpecialInstructions>The order must not be delivered till the end of May.</SpecialInstructions>
        </ORDER>
        <ORDER>
                <OrderNumber>553217858</OrderNumber>
                <CustomerID>2167C</CustomerID>
                <ProductID>AW332X</ProductID>
                <SpecialInstructions>Due to the special nature of the work site this part must be delivered in a
                specific way. These instructions must be followed precisely to ensure safety. First the deliver
                vehicle must be ... </SpecialInstructions>
        </ORDER>
</SALES>
```

Fig. 1. Sample XML markup.

unstructured text. This type of document includes books, advertisements, and magazine articles. Another document, called data-centric type, shares with data the characteristics of a fairly regular structure. Some examples of this type of document include sales orders or flight schedules, where most of the data in the document fall into specific categories such as an order number or a flight number. In practice, many documents are a blend of these two types. These hybrids, called semistructured documents, have structured components and unstructured text. For example, a product catalog would have specific elements containing price, size, and color, but could also include lengthy textual descriptions of the items.

Markup languages. Markup languages are the technology used to identify the various elements of a document and control how the document is displayed. Markup languages add content, called tags or descriptors, to documents to pinpoint the structured and unstructured components of the document. Such added content (data that describe other data) is called metadata.

XML is not a markup language per se; instead, it is a set of flexible markup language rules that can be used to define custom descriptors or tags for specific domains of knowledge such as mathematics, finance, and astronomy. These descriptors are then used to spell out the components, organization, and structure of the document. For example, XML tags developed for sales order documents might identify highly structured components such as order numbers and customer and product identifiers. Each sales order might also contain a quite lengthy unstructured text section with special instructions (**Fig. 1**).

While XML provides a method for identifying and organizing document content, a style markup language, such as the Extensible Stylesheet Language (XSL), must be used to control the manner in which the document will be represented on the Web. XSL tags extract the appropriate data from the document content, and then Hypertext Markup Language (HTML) tags format the document for presentation (**Fig. 2**). This formatting process might include,

among other things, where individual words or images must be placed on a page or how large and in what color the text is to appear.

In contrast, data stored in databases do not have added markup content. The manner in which the data are stored in the rows and columns of a relational database provides an inherent organization and structure. Therefore, all components of the database are well defined. The appearance of the data, as well as what data are retrieved from the database, is determined by instructions written in a database language such as the Structured Query Language (SQL). These controls on data retrieval and formatting are not in the form of tags stored directly with the data in the database, but in completely separate files that the DBMS reads and executes when searching and reporting from the database.

Unstructured textual data can be stored in an ordinary database, but storing unstructured text is not a primary function of a typical database management system that preferably deals with structured data items such as name, address, salary, and so forth. Unstructured text in a database often causes inefficiencies in the physical storage of the data and is

```
<XSL:TEMPLATE MATCH="/">
  <HTML>
  <BODY>
    <H2>Sales Orders</H2>
    <TABLE>
      <TR>
        <TH>Order #</TH>
        <TH>Customer #</TH>
      </TR>
      <TR>
        <TD><XSL:VALUE-OF SELECT="SALES/ORDER/OrderNumber"/></TD>
        <TD><XSL:VALUE-OF SELECT="SALES/ORDER/CustomerID"/></TD>
      </TR>
    </TABLE>
  </BODY>
  </HTML>
</XSL:TEMPLATE>
```

Fig. 2. Sample XSL markup.

certainly much more difficult to search than specific data items such as name and address. Because of these limitations, functionality must be added to a basic DBMS so that it can work properly with document content; otherwise, a completely new type of database system must be developed based on the XML model.

Relational databases and XML documents. Relational database management systems (RDBMS) are by far the most common data management paradigm across all applications in government, industry, and education. Most computer users at all levels of expertise have a reasonable appreciation of the structure and capabilities of the RDBMS. With such a broad base of use, it was only natural that various RDBMS vendors would have "XML-enabled" their database products by adding more universal functionality. This functionality allows document content to be stored in a relational database from which it is extracted and then reformed into a XML document, if necessary.

XML documents can be stored in a RDBMS in several ways. The entire XML document could be stored intact as a single entity in the database. This would be very simple for the RDBMS, since the RDBMS would not need to have any understanding of XML, but would simply serve up the entire document as is. Unfortunately, this approach would not be able to take advantage of primary RDBMS features such as rapid search and retrieval from the unstructured text in the XML documents. On the other hand, this approach would make use of other very useful RDBMS features such as user access permission and backup.

In another approach, the XML document could be broken up into smaller "chunks" such as paragraphs, and these would be stored as specific entities in the relational database. This technique would make search and retrieval faster as entire sections of text, which were not of interest to a specific search, would not need to be read.

In a third approach, all the highly structured portions of the XML document, such as order number, customer ID, and so on, could become specific data items in the database, allowing for extremely fast search and retrieval. Identifying a document's content as structured, unstructured, or semistructured will help in deciding which of these approaches is best.

Given the fundamental nature of an RDBMS, data-centric documents would be able to take best advantage of the capabilities of this type of database management system. The highly structured portions of a data-centric document would fit directly into the basic data model of the RDBMS. Searching through these components, such as name or ID number, would offer a rapid search and retrieval path to the unstructured textual portions of the document. This approach would be preferred, particularly when compared with searching directly through text, which is a very slow process especially when the search criteria are complex.

Native XML databases. Database systems that are specifically designed to store and work with XML documents are called native XML databases (NXD). These database systems are based on the XML document model instead of the relational model used by most other database systems. Consequently, while an XML database system may appear to work just as a RDBMS does, the underlying database storage and retrieval technologies are quite different. An NXD is completely aware of the way XML documents are structured, so queries move through the XML document data model much more efficiently.

While the data storage model is quite different, the native XML database systems provide the same overall feature set, as do most RDBMS. These include, among other capabilities, user access, security, transaction processing, and query languages; however, one feature of an RDBMS for which a native XML database is particularly ill suited is aggregation. It is quite easy for an RDBMS to extract data from a number of places in the database and process them as a group. Examples of such functions include sum, average, minimum, and maximum. However, aggregation is a very difficult operation to perform on the unstructured and semistructured data found in an XML database.

Native XML databases are particularly well suited to working with document-centric content. Since an NXD is completely aware of how XML can describe a document, the XML tags provide the database systems with vital information on how to store, search, and retrieve all structured and unstructured components of any XML document.

For background information *see* DATA STRUCTURE; DATABASE MANAGEMENT SYSTEM; INTERNET; PROGRAMMING LANGUAGES; SOFTWARE; WORLD WIDE WEB in the McGraw-Hill Encyclopedia of Science & Technology.   Robert Tucker

Bibliography. D. K. Appelquist, *XML and SQL: Developing Web Applications*, Pearson Education, Boston, 2002; A. B. Chaudhri, A. Rashid, and R. Zicari, *XML Data Management: Native XML and XML-Enabled Database Systems*, Pearson Education, Boston, 2003.

Extreme programming and agile methods

Extreme programming (XP) is a new software development method proposed as an alternative to traditional approaches. XP is the best known of the half dozen or so agile methods that stand in deliberate opposition to traditional software development methods.

Traditional methods. Digital computers were invented in the 1940s and were programmed by the engineers and mathematicians who created them. These early machines ran relatively short and simple programs, mostly to solve mathematical problems, so programming them was not much of a problem. Computer power and capability grew rapidly, and before long people were using computers to manage large databases and to perform complex computations in business, science, and engineering. These more demanding applications required much larger and more complicated programs, and by the late 1960s a software crisis had emerged. The software crisis was characterized by an inability to deliver correct and useful programs on time and within budget. Many programs were never delivered at all. This crisis precipitated the creation of the discipline of software engineering, whose goal was to figure out how to deliver high-quality software products (programs, data, documentation, and so forth) on time and within budget.

The approach taken by software engineers in the last 30 years is modeled on the way that engineers in other disciplines develop products. These methods emphasize planning both the product to be created and the process used to create it. Process planning includes estimating the time and money needed to create the product, scheduling the work to be done, and allocating human and other resources. Product planning includes carefully specifying product features, capabilities, and properties; the program components and structures to realize them; and the tests for confirming that the product behaves as it should. Traditional methods also include data collection and analysis activities to see if the development process is going according to plan, and to use as a basis for better planning in later projects. Thus, traditional methods devote much effort to formulating, writing, checking, and modifying various plans, as well as to collecting and analyzing data.

The traditional approach has succeeded in relieving (though not entirely eliminating) the software crisis. The success rate for software development projects has improved gradually but steadily for many years. Organizations that make full use of traditional methods usually deliver excellent products on time and within budget. But there are drawbacks too. For instance, organizations using a traditional method usually take a long time to deliver products to customers, and find it hard to change product specifications during development (because all the plans must be changed). As more and more software products need to be delivered on "Internet time," this has increasingly become a problem. The emphasis on planning and data collection in the traditional approach sometimes results in enormous effort wasted on unnecessary or counterproductive work. In addition, some programmers find the culture of caution, order, and control inherent in the traditional approach stifling and dehumanizing.

Agile methods. Around 2000, several prominent software developers began to advocate a different approach in direct opposition to traditional software development methods. These proposals, collectively called "agile methods," emphasize the need to create software products quickly and to respond to product specification changes immediately. Agility is achieved by developing software products in small increments over a short period, typically a month or less. Product increments can be delivered very quickly. Developers do not plan their activities or the product beyond the current increment, so it is fairly easy to react to changing product specifications. Still, the lack of planning (and attendant documentation) in agile methods puts a premium on memory, extensive interpersonal communication, and the ability to quickly adapt software, all of which are likely to break down when large and complicated programs are developed.

Agile methods have been tried successfully on several small projects and advocated in a spate of publications. Many programmers are convinced that agile methods are a revolution in software development, but others regard them as little more than glorified hacking (a pejorative term for poor development practices), and the move to agile methods as a setback to decades of progress in software engineering.

Among the best-known agile methods are Scrum, the Crystal Methods, Lean Development, Adaptive Software Development, the Dynamic System Development Method, and Extreme Programming. The last will be discussed as an example of an agile method.

Extreme Programming. XP, which takes several agile practices to extremes, is the best known and most widely discussed and investigated agile method. It illustrates the contrast between agile and traditional methods.

Although XP includes values, roles, and activities, its central aspect is its 12 core development practices.

Small releases. The product is developed in small increments, which are completed in 2 or 3 weeks. An onsite customer representative (see below) decides if an increment is released to users. Traditional approaches tend to develop product versions over long periods, with major releases to users only every year or two.

Whole team. The development team consists of the programmers and at least one customer representative assigned full time to the development effort. The customer representative writes user stories (see below), elaborates the user stories with the developers, and writes tests for user stories. Traditional approaches develop product specifications with customers early on, but the customers are typically not involved again until development is nearly complete.

Planning game. The customer representative writes user stories on note cards. Each user story describes something about the product that customers want, which can be implemented in a week or two, and can be tested. The customer representative and developers together then decide which user stories to implement in an increment. Product features and the work to be done by programmers are planned only one increment at a time. Traditionally, very detailed and complete product specifications are written before programming begins, and the work to realize the entire program is planned ahead of time.

Test-driven development. Developers are required to write tests first, and then to write code to pass the tests. The traditional practice is to write the code first and then write the tests, though often many tests never are written.

Pair programming. All code is written by two developers sharing a single computer, whereas programmers traditionally work alone.

Continuous integration. Code written by different programming pairs is combined and tested frequently, typically at least once a day. Traditional practice used to be to integrate and test code only after a lot of component development, but nowadays frequent integration has become standard in many traditional methods.

Collective code ownership. Every programmer is responsible for the entire program, and any pair can work on any part of it at any time. Traditionally programmers have been responsible for certain parts of programs.

System metaphor. The team has a shared, metaphorical idea about how the program works. Traditional approaches specify a precise software architecture that guides detailed design and programming.

Simple design. The program is made as simple as possible, while realizing the user stories for the current iteration. Traditional methods usually advocate a design sophisticated enough to support foreseeable program changes and enhancements.

Design improvement. The program is "cleaned up" as it is changed to improve its design. If this were not done, the program would become too convoluted as it evolved. This practice is not needed (as much) in traditional approaches, because the design is determined before the code is written.

Coding standard. The developers must agree on the conventions governing code writing. This practice has been standard in every development method for decades.

Sustainable pace. Developers work at a steady pace with no overtime, so that exhaustion does not degrade productivity. The culture in most traditional development shops is to work very long hours, especially before important deadlines.

Outlook. XP has generated the most controversy between traditional and agile methods supporters. Although the debate goes on, it is likely that agile methods will take their place alongside traditional methods as alternative means of developing software. Agile methods are appropriate when a fairly small team must create a product rapidly, the product specifications are likely to change, and not much will be lost if the product fails. Traditional methods are appropriate when a large team is creating a product with fairly stable specifications whose failure would have serious consequences (such as the loss of life or a lot of money). Emerging methods that combine features of the agile and traditional approaches will be used for products that do not fit neatly into these categories (such as a product with changing specifications developed by a large team).

For background information *see* COMPUTER PROGRAMMING; METHODS ENGINEERING; PROGRAMMING LANGUAGES; SOFTWARE; SOFTWARE ENGINEERING; SOFTWARE TESTING AND INSPECTION in the McGraw-Hill Encyclopedia of Science & Technology.

Christopher Fox

Bibliography. K. Beck, Embracing change with Extreme Programming, *IEEE Comput.*, 32(10):70–77, 1999; B. Boehm and R. Turner, *Balancing Agility and Discipline*, Addison-Wesley, Reading, MA, 2004; A. Cockburn, *Agile Software Development*, Addison-Wesley, Reading, MA, 2002; M. Paulk et al., *The Capability Maturity Model: Guidelines for Improving the Software Process*, Addison-Wesley, Reading, MA, 1995; R. Pressman, *Software Engineering: A Practitioner's Approach*, 6th ed., McGraw-Hill, Boston, 2005.

Firefly communication

The bioluminescent signals produced by the adult beetles in the firefly family Lampyridae are actually species-specific flash patterns used in courtship. Males of many species communicate what species they are as well as how reproductively fit they are, encoded in the timing of the bright flashes. After a species-specific time delay following the flash of a male, a female of the same species might flash back, indicating receptiveness. Then after a short dialog flashed back and forth from one sex to the other, the male is able to locate the position of the female and lands near her to mate. While fireflies, or lightning bugs, are most familiar as the flashing insects chased by children on warm summer nights, not all fireflies are nocturnal nor are they all luminous as adults. Many diurnal species rely solely on pheromonal signals to locate and attract mates, and some species use both pheromonal signals and photic signals. While the adults of some species are not luminous, all of the known firefly larvae are. Past attempts at categorizing the various composite signal systems and recreating the scenarios that explain the sequence in which these systems evolved have been incomplete due to studies focusing only on taxa restricted to specific geographic regions. These studies are too limited in their choice of taxa or too speculative because they did not employ empirical tests of evolutionary scenarios, such as phylogenetic reconstructions. The application of phylogenetic analyses to such questions is the only way to reconstruct historical patterns and test evolutionary scenarios. Such analysis even works for characters that might be evolving at very high rates due to intense evolutionary forces such as sexual selection.

Evolution of bioluminescence. Bioluminescence in beetles is not restricted to the family Lampyridae. When representatives from beetle families closely related to fireflies (such as Phengodidae, Rhagophthalmidae, Omalisidae, Lycidae) were included in a phylogenetic analysis along with firefly taxa, the resulting phylogeny supported the origin of bioluminescence as preceding the evolution of the first firefly. This is significant because the occurrence of bioluminescence in firefly ancestors can provide clues to its use in modern fireflies. The bioluminescent firefly ancestors all have luminescent larvae, while only some of the adults are luminous. This pattern suggests that the first origin of bioluminescence in this lineage was in larvae.

While the function of larval glowing has been the subject of much debate, recent studies have provided excellent support for the idea that the photic emissions of firefly larvae serve as a warning display. Fireflies are protected by defensive chemicals called lucibufagins that cause irritation, gagging, and vomiting in lizards and birds. Since larvae are typically active at night, their sporadic luminous glows are obvious to visual nocturnal predators, which can learn to avoid glowing objects after tasting one or more bioluminescent larvae. The firefly species that are located in the more basal regions of the phylogeny generally possess characteristics that are hypothesized to be ancestral. These characteristics include being diurnally active and using pheromonal sexual signals to locate and attract a mate. Unlike their larvae, the adults of these basal firefly species are nonluminous. However, the ability to be luminous in the adult stage gives some of the more recently derived species use of these photic signals as warning signals, which are then co-opted as sexual signals in courtship. These photic sexual signals have become elaborate, having evolved from a series of sporadic, random glows to a carefully timed species-specific system.

Signal system evolution. Pheromones are the only sexual signals used by the more primitive fireflies found basally in the phylogeny. In this type of signal system, females perch on a stationary object and produce pheromones to which the males are attracted. In other firefly species, pheromones are used in conjunction with photic signals (see **illustration**). In these systems females remain stationary and produce a pheromone while glowing from a paired photic organ on the ventral surface of the eighth abdominal segment. Males at a distance are attracted to the pheromone and, when flying close enough to see her glow, are able to quickly locate a female's position. The phylogeny predicts that pheromones would be lost altogether as photic signaling became the primary mode of communication for attracting and locating mates. In addition, the phylogeny allows the order by which these components arose (or were lost) to be hypothesized. In this analysis, the order of sexual signal evolution in fireflies can be determined as moving from the ancestral condition of sole pheromone use to pheromone use in conjunction with photic signals and then to the sole use of photic signals. It should also be pointed out that the signal modes (pheromonal vs. photic) of these systems change and that the sex of the primary signaler is apparently linked to a given signal mode. A primary signaler is an individual that broadcasts its identity or location to no specific receiver of the opposite sex. In the signal systems where pheromones are used (both with and without photic signals), the female is the primary signaler. This is the ancestral condition in the family. When solely photic signals are used for pair formation, males become the primary signalers, broadcasting their species-specific flash patterns

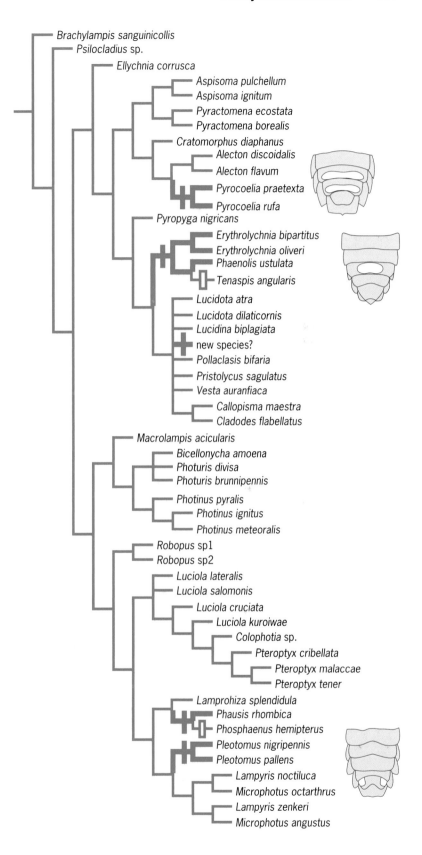

Evolution of signal systems in Lampyridae employing both phermonal and photic components (showing five origins and two losses). Solid boxes indicate origins of combined phermonal and photic signals, open boxes indicate losses. Representative male photic organ morphologies are *Pyrocoelia rufa* (top), *Erythrolychnia oliveri* (middle), and *Pleotomus pallens* (bottom).

while in flight. The male primary signaler is the most derived condition in the family. When females are the primary signalers, they are always sedentary, whereas primary male signalers are almost always active; that is, they fly in search of females while signaling. Therefore, the trend from sedentary females producing pheromonal sexual signals as primary signalers to flying males flashing bioluminescent signals while searching for females may be the effect of intense sexual selection.

Photic organ evolution. Photic organs in adult fireflies are located toward the end of the abdomen. Because male photic organs show more diversity than females, and because males are the primary signalers in most fireflies, the evolution of their photic organs is of primary interest. Adult male photic organs, when present, are located on the ventral surface of abdominal segments six, seven, and eight. The size and shape of these organs vary considerably among the bioluminescent species, from two small photic organs on each side of the eighth abdominal segment (which appears to be homologous with the larval photic organ), to one spot, to a center strip, and then to a photic organ that is expanded to cover the entire ventral surface of the abdominal segment on which it is located. All of these photic organ morphologies appear to have arisen multiple times in the family Lampyridae, with the exception of the one-spot morphology, which arose only once.

In the lineages that use only photic sexual signals, both the photic signals and the photic organs used to produce them have become elaborate. These photic signals have evolved from glows to flashes. In addition to flashed signals hypothetically being able to encode more information than a constant glow, signalers that are flashing luminous signals while in flight may be harder for predators to locate than signalers that are glowing constantly. Flashed signals are produced from photic organs that have increased in size and complexity. These photic organs are enlarged to the point that they completely cover the surface of the abdominal plate they are situated on. Presumably, the larger the photic organ, the brighter the light emission and the easier it is to see. There are also internal changes to these photic organs: they possess reflective layers as well as increased innervation and tracheation, which allow species with such organs to produce rapidly flashed photic signals. These rapidly flashed signals also seem to employ critically timed signal parameters that convey specific information about species identity as well as information about the individual signaler. Current studies suggest that information such as male quality is encoded differently in different firefly species (that is, increased flash rate, increased flash duration, or increased flash intensity). Glowing is generally produced by photic organs that have two larval-like spots or by photic organs that are situated in thin strips covering only a portion of the ventral abdominal plate where they are located. Neither type is as well innervated or tracheated as those of the rapid flashers; therefore, these organs are not as bright

nor do they have the rapid onset and offset of a true "flash." Presumably sexual selection has been the force through which bioluminescent signals and photic organs have become more elaborate, allowing the production of more sophisticated signals and communication systems.

Origins of flashing behavior. The ability to communicate with flashed signals appears to have evolved at least three times in the family Lampyridae. In the New World, flashing arose once in the *Aspisoma-Pyractomena* lineage and once in the *Bicellonycha-Photuris-Photinus* lineage. In the Old World, flashing appears to have evolved only in the *Luciola-Colophotia-Pteroptyx* lineage. "Flashed firefly signals" were previously assumed to have evolved only once. Future studies may focus on differences in the flash communication systems of the above three lineages of fireflies.

It is possible to generate phylogenetic hypotheses and test scenarios about the evolutionary order of specific traits within a lineage. The re-creation in this study of multiple origins for flashed signals and the photic organs that produce them is additional evidence that phylogenetics is a powerful tool for studying the evolution of even rapidly evolving traits.

For background information *see* ANIMAL COMMUNICATION; BIOLUMINESCENCE; CHEMILUMINESCENCE; INSECT PHYSIOLOGY; ORGANIC EVOLUTION; PHEROMONE in the McGraw-Hill Encyclopedia of Science & Technology.   Marc A. Branham

Bibliography. M. A. Branham and J. W. Wenzel, The evolution of bioluminescence in cantharoids (Coleoptera: Elateroidea), *Florida Ent.*, 84:478–499, 2001; R. De Cock and E. Matthysen, Aposematism and bioluminescence: Experimental evidence from glow-worm larvae (Coleoptera: Lampyridae), *Evol. Ecol.*, 13:619–639, 1999; T. Eisner et al., Lucibufagins: Defensive steroids from the fireflies *Photinus ignitus* and *P. marginellus* (Coleoptera: Lampyridae), *Proc. Nat. Acad. Sci., USA*, 75:905–908, 1999; J. E. Lloyd, Sexual selection in luminescent beetles, in M. S. Blum and N. A. Blum (eds.), *Sexual Selection and Reproductive Competition in Insects*, pp.293–342. Academic Press, New York, 1979; J. Sivinski, The nature and possible functions of luminescence in Coleoptera larvae, *Coleopt. Bull.*, 35:167–179, 1981; H. Susuki, Molecular phylogenetic studies of Japanese fireflies and their mating systems (Coleoptera: Cantharoidea), *Tokyo Metro. Bull. Nat. Hist.*, 3:1–53, 1997.

Flapping-wing propulsion

Numerous biological studies confirm that the best hydrodynamic properties in nature belong to dolphins, whereas maximum relative speeds are achieved in air by birds and insects. Actually, flying and swimming creatures employ similar principles for motion. They propel and control their motions using flapping wings or fins.

Wings and fins are flexed by muscles during flight, unlike rigid airfoils. This kind of control flexibility offers some unique benefits. For example, the kinematics of wing motion provides high propulsion and lifting capacity by shedding vortex structures from both the wing and body. Wing motion control is able to utilize energy from these vortex structures to increase propulsion efficiency in a variety of conditions, such as wavy or perturbed flows, or in group motions. Also, winged creatures can fly or glide and turn suddenly because they have muscular control of both wing deformations and amplitudes of oscillation. Swimming creatures have similar capabilities. These abilities offer significant technical potential for human benefit.

The idea of using flapping wings for human flight has been around for a long time. Leonardo da Vinci tried to build flying machines and explain the propulsion principles back in the fifteenth century. Around the beginning of the twentieth century, flapping-wing machines were tried repeatedly (**Fig. 1**) but failed because the available science and engineering was inadequate. Still, the apparent control flexibility, high propulsive force, lifting capacity, and other features of flapping wings warranted scientific study. Moreover, the idea of deriving or borrowing scientific and engineering technology from nature increased. In the 1960s, a separate scientific research field named bionics (the study of systems that function like living systems) was formed, which started a comprehensive study of the principles of flapping-wing propulsion.

Today, intensive theoretical, experimental and engineering studies have demonstrated that this propulsive method possesses high efficiency, low acoustic radiation (for low-frequency oscillations), multifunctionality (ability to operate efficiently at different modes of motion), capacity to combine functions of several devices (propulsor, control device, stabilizer), efficient performance at special modes of motion (for example, hovering), high maneuvering qualities, acceptable cavitation characteristics at the main mode of motion, low mechanical complexity and weight, and simpler control compared to conventional propulsion systems.

This research and development has been underway for some time. Most of the progress has involved merging aerodynamic principles with new scientific fields such as artificial muscle technology (AMT), micro-electro-mechanical systems (MEMS), and microsystems technology. Although some flying laboratory models have demonstrated the feasibility of flapping-wing technology, development of fully functional aircraft has not been achieved. Progress in marine applications, however, is more advanced. This article will review the various types of devices and vehicles under development and the related technologies involved.

Flying machines. In the 1990s, programs supported by defense agencies were directed at research and development of micro air vehicles (MAV) including those with flapping wings. Such systems initially

Fig. 1. Artist's impression of batwings designed by Otto Lilienthal (1848–1896).

were intended for military surveillance, but also are useful in many nonmilitary applications. Some examples are rescue services, traffic control, inspection of vertical building walls, control of state borders, and monitoring of oil and gas piping and electric power lines.

In general, flying vehicles with flapping wings are called ornithopters. They can be divided into two groups differing by size: micro air vehicles and moderate- to full-size ornithopters. The first group includes objects with dimensions less than 6 in. (15 cm); the second group is larger. Vehicles of both groups use flapping wings as a propulsion and lifting device. Micro air vehicles also include devices in the centimeter range (linear dimensions less than 1 in.). Such vehicles are named micro mechanical flying insects (MFI).

The principal problems of an integrated development of micro air vehicles are aerodynamics of the flapping wing, propulsion systems, navigation systems, and energy sources. For micro mechanical flying insects the problems of scaling and miniaturization of the systems and components must be addressed.

Micro air vehicles require integration of five main systems, namely, propulsive-lifting, power supply, flight control, sensors, and communications. The development of such a complex miniature object is a multidisciplinary optimization problem, involving individual systems, integrated systems, and sophisticated aerodynamic designs for a wide range of the input flight parameters.

Modern achievements in new construction materials, micro-electro-mechanical systems (MEMS), microsystems technology (MST), power sources, and energy converters open new horizons for the practical use of flapping wings.

The aerodynamic part of the problem is solved by interconnecting the wing (of complex geometry) and the drive (supporting special kinematics of motion), resulting in a merging of these components into a single module through artificial muscle technology (AMT). Successful work is in progress on the use of different types of integrated-circuit actuator

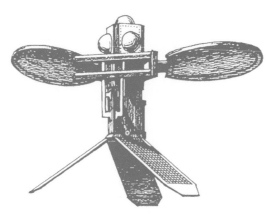

Fig. 2. University of California, Berkeley, micro mechanical insect "Robofly." (*Reproduced with permission from Progress in Aerospace Sciences, vol. 39, pp. 585–633, 2003*)

devices to provide oscillatory motion of the wing. Use of AMT has brought the main aerodynamic module of the micro air vehicle to a new level of development and has stimulated the development of other important systems of the vehicle.

Despite advances, micro air vehicles are still in the development phase. The vehicles designed so far are one-mode devices that can stay in the air a short time only and have very limited application functionality. However, successful steps have been taken toward more advanced technological levels.

Investigations of micro air vehicles with flapping wings are underway at many research centers. Positive results of these projects show that full-fledged micro-electro-mechanical systems–based MAVs with flapping wings can be developed in the near future. This prediction is supported by the existence of integrated projects such as "Robofly" (University of California, Berkeley) [**Fig. 2**], "Microbat" (Aero Vironment), "Entomopter" (GTRI), and others. Some

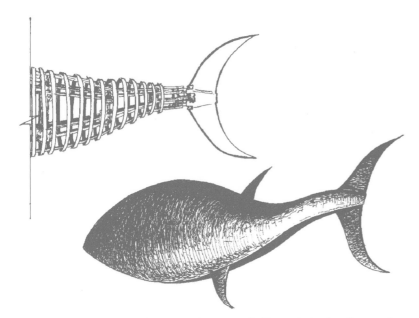

Fig. 3. MIT laboratory robot "Robotuna." (*Reproduced with permission from Progress in Aerospace Sciences, vol. 39, pp. 585–633, 2003*)

of the interest in these technologies is due to their potential for Mars exploration.

Projects also exist for full-size ornithopters. For example, the development of a piloted ornithopter with an engine is underway at the University of Toronto Institute for Aerospace Studies (UTIAS), and a similar project is underway in the RTK Central Research Institute, Russia. *See* ADAPTIVE WINGS; LOW-SPEED AIRCRAFT.

Submersibles and ships. Advances in biorobotics for marine vehicles also started in the 1990s. The objective was to create a variety of crewless underwater vehicles for military purposes, such as reconnaissance, monitoring of water basins, disabling marine mines, and antidiversionary and other missions. Most of the attention was devoted to development of autonomous underwater vehicles (AUVs). Developments of biorobotic AUVs are justified by a reluctance to use marine animals for military goals. In the near future, biorobotic AUVs can become sufficiently inexpensive for practical use. Autonomous underwater vehicles also have commercial applications such as in oceanography, hydrology, and fisheries.

Basically, these AUV designs emulate the swimming of fish and cetaceans (marine mammals such as dolphins) for higher efficiency, lower acoustic radiation, and better maneuverability and steerability as compared to conventional ship propulsion.

Two developmental strategies are underway: fishlike robots having a flexible body–stem–fin structure, and rigid nondeformable hulls with a flapping tail and pectoral fin. Another effort is to combine a nondeformable body with elastic stems and flapping fins. Research centers such as the Massachusetts Institute of Technology, Saint Petersburg State Marine Technical University, Mitsubishi Heavy Industries, Tokyo Institute of Technology, Tokai University, Nekton Research LLC, George Washington University, and the University of Michigan are working on such projects.

Researchers are seeking an integrated solution combining the hydrodynamic design with drive mechanisms to provide the necessary deformations of the body and stem as well as the flapping motion of the fin. However, current mechanical drives for the flapping wings of AUVs are not sufficiently efficient. So, some efforts are focused on artificial muscle technology. Use of artificial muscles as fin drives reduces the number of moving parts, simplifying the fin-drive module and reducing acoustic radiation, thus enhancing stealthiness.

Experimental evidence also proves that flapping-wing propulsion significantly decreases energy consumption of the underwater vehicle. This conclusion has been confirmed by several development projects and mechanical prototypes of fish and cetaceans. One such prototype is MIT's "Robotuna," a laboratory robot comprising eight independently controlled links (**Fig. 3**). Other projects include the "Blackbass Robot" (Tokai University) and "PilotFish" (Nekton Research LLC) projects, aimed at developing fishlike vehicles with nondeforming hulls and equipped with pectoral (rigid and elastic) fins for investigation of precision positioning and maneuvering at low speeds.

In general, up-to-date technology that emulates marine animal movements (called biomimetics) holds particular promise for marine applications. For one thing, developing full-size marine vehicles (ships) is not as difficult as developing aircraft, so progress has been faster. At least two full-size surface ships equipped with flapping propulsors have already been built. One of these ships is from a Russian company, EKIP, another from the Norwegian Fishing Industry Institute of Technology.

Future developments. To reach the goal of broader use of flapping-wing propulsion in micro air vehicles and automatic underwater vehicles, advances in a number of technologies are required. Potential applications in both categories vary widely in size and design to meet different performance requirements. Thus, areohydrodynamic parameters of a variety of wing structures and application conditions need to be better understood. Better microelectric power systems are required to increase operation time because existing power sources for many applications, particularly very small devices, are not sufficient. Alternative power sources need development, perhaps with solar battery technology or laser beam energy transfer. More powerful sensor and control systems are needed to improve control of flight parameters for different motion modes as well as detecting external perturbations.

Thus, the existing flying and swimming technical objects are at a middle stage of active development. However, important steps toward creation of these systems have been achieved. Test prototypes of the micro flying vehicles of the third and fourth generations as well as models of swimming biorobots suggest that full-fledged technical vehicles with flapping wings can appear within 5–7 years.

For background information *see* AERODYNAMIC FORCE; AERODYNAMICS; AIRFOIL; BIOMECHANICS; BIOPHYSICS in the McGraw-Hill Encyclopedia of Science & Technology.

Kirill V. Rozhdestvensky; Vladimir Ryzhov

Bibliography. K. D. Jones and M. Platzer, *Flapping-wing Propulsion for a Micro Air Vehicle*, AIAA Pap. No. 2000-0897, 2000; T. J. Mueller (ed.), Fixed and flapping aerodynamics for micro air vehicle applications, *AIAA Progress in Astronautics and Aeronautics*, vol. 195, 2001; T. J. Mueller and J. D. DeLaurier, Aerodynamics of small vehicles, *Annu. Rev. Fluid Mech.*, 35:89–111, 2003; M. Nagai, *Thinking Fluid Dynamics with Dolphins*, Tokyo Ohmsha-ISO Press, 2002; K. V. Rozhdestvensky and V. A. Ryzhov, Aerodynamics of flapping-wing propulsors, *Prog. Aerosp. Sci.*, 39:585–633, 2003; W. Shyy, M. Berg, and D. Ljungqvist, Flapping and flexible wing for biological and micro air vehicles, *Prog. Aerosp. Sci.*, 359:455–505, 1999; M. S. Triantafyllou, G. S. Triantafyllou, and D. K. P. Yue, Hydrodynamics of fishlike swimming, *Annu. Rev. Fluid Mech.*, 32:33–53, 2000.

Forensic botany

Forensic botany has opened aspects of the field of botany to novel avenues within criminal investigations. Today, several subdisciplines of plant science are being applied successfully in criminal investigations—for example, plant systematics including identification of plant species, plant anatomy, and plant ecology. Often, forensic applications come from combinations of these areas in a given case.

Plant systematics. Plant systematics is the study of evolutionary relationships among plants. Certain knowledge in this field is of considerable use in criminalistics and crime scene work. Being able to attach the proper scientific name to a plant is important in several situations. These items should be gathered carefully and notations made concerning exact location of the plant materials. Collection of plant material evidence is especially important in crime scene investigations, but often is overlooked or collected in an improper manner. The identification of plant materials can involve whole plants, plant fragments, and seeds. There are guides to plant identification called floras. These books are most useful when whole plants or whole flowers are to be identified. In a given setting, the more local the plant identification guide (flora), the better. For example, if dealing with a crime scene at or near Yellowstone National Park, the best source would be the Park's flora.

In some cases, plant fragments may be the only material available, for which there may be published guides. There are several picture guides to plant seeds. These are more difficult to use than the floras because distinctions among seeds can be very subtle. There also are guides for leaf fragments of trees and branch tips from woody shrubs and trees. With seeds and plant fragments, botanical expertise from a local college or university may be needed.

Applications of plant systematics. Sometimes homicide victims may be found in places far from the murder site. In one case, a blanket-wrapped body was found in rural grassland with assorted plants attached to the victim's clothes and the blanket. The plant fragments could not have come from that grassland because they were from ornamental plants associated with formal gardens. The person had been killed in the city and the corpse transported many miles into the country. The ornamental plants were in agreement with those found in a suspect's yard, linking the victim to the suspect, and perhaps to the crime scene.

In another case, a kidnapping victim was grabbed from the street and released somewhere in the mountains above Boulder, Colorado. The investigators suspected useful evidence might be found at the release site. A van used in the crime had been recovered. The victim identified the van, and her DNA was found in it. As investigators inspected the vehicle, they collected plant fragments and seeds from the seats and floor, the floor pads and pedals, the tire treads, the outside window wells, and the undercarriage of the van. The plant samples were collected with forceps and placed in coin envelopes. Once the plants pieces were assembled and identified, the release site was pinpointed, and further evidence was found on site. Investigators are urged to collect plant material with forceps rather than with the special vacuum sweepers found in many departments, because vacuuming shatters dry plant fragments and complicates the identification process.

Plant anatomy. Plant anatomy is the study of plant cells. Applications of plant anatomy in forensic science have been based in large part on a unique characteristic of plants—their cellulose cell walls. Cellulose is a complex carbohydrate that is highly resistant to ordinary decomposition processes. For example, wood is made up of cellulose that has undergone a process known as lignification. Plant cells can pass through the human digestive tract with their walls intact, maintaining the same size and shape they possessed before ingestion. Because the human diet in any culture is made up of a relatively few kinds of plant foods, seldom more than 75 different kinds, and because each food plant has cells of distinctive sizes, shapes, cell arrangements, and cell inclusions such as crystals, it is possible to identify the plants consumed in someone's last meal (see **illustration**). This use of plant anatomy has been employed especially to postmortem the stomach contents.

A further use of stomach contents can be an estimation of time of death because the valve allowing stomach contents to proceed to the small intestine closes at death. If the individual was healthy and had eaten an average-sized meal, the material would remain in the stomach for only 1–3 hours under ordinary circumstances and then pass to the intestines.

Fecal analysis shows identifiable plant cells as well, a technique used with great success by anthropologists in their studies of ancient cultures.

People are sometimes stabbed with wooden instruments. One of the properties of cellulose is that it is birefringent (glows) under light microscopy. This allows determination that the stabbing was carried out with a wooden instrument, and may be used as well to link the instrument with a specific wound.

Plant anatomy applications. Sometimes the contents of the last meal can play a role in criminal investigations. In a "black widow" case, the suspect claimed to have fed the victim his last meal (supper), giving her an alibi for the estimated time of his murder. But analysis of his stomach contents showed his last meal differed greatly from her menu. He was observed earlier in the day eating his true last meal (breakfast) at a restaurant. This occurred at a time for which the "black widow" had no alibi.

In another case, the vegetables that had been selected for a pizza order helped the delivery person recall the individuals in the room where the delivery was made. Several persons in the room were killed, and the investigators thought the perpetrator might also have been present when the pizza arrived, so the deliverer's recall was significant.

Plant ecology applications. Plant ecology is the study of the relationships between plants and their physical and biological environments. The body of a missing person was discovered along a ditch in midsummer in the Great Plains wheat country. An attempt had been made to hide the corpse by covering it with giant sunflowers growing along the roadside. The victim had been reported missing only a few days before the body was discovered, but the crime scene investigators suspected the person had been dead for a longer time due to the state of decay and evidence from forensic entomology. The suspect had left the area about 10 days before the body was discovered and well before the missing person's report was filed. The sunflowers used for covering the body were wilted. In order to determine how many days it took for an uprooted sunflower to reach this particular wilted condition, several living sunflowers were collected from the site where the body was found, exposed to the environmental conditions at the site, and observed daily until the same level of wilting occurred. The appropriate level of wilting took 8 days and was maintained at that state up to 15 days, providing further evidence that the homicide

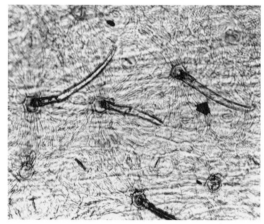

(a)

(b)

Surface of a green bean pod, showing its spines using (*a*) light microscope (*courtesy of David O. Norris*) and (*b*) scanning electron microscope (*courtesy of Meredith Lane*).

had occurred before the person was reported missing and before the suspect had left the area.

A federal agency suspected a kidnapping had occurred based on a single photograph it had received. This photo was circulated widely at forensic science meetings and elsewhere. This photo was all the evidence the agency had that a crime had occurred, and workers were anxious to locate the crime scene. It was concluded that the picture had been taken in the upper montane ecological zone of the southern Rocky Mountains, in either northern Colorado or southern Wyoming, based upon the Engelmann spruce trees and logs, the metamorphic rocks, and certain shrubs shown in the photo. But it was not possible to provide a more specific location.

NecroSearch International Ltd. is a nonprofit organization made up of people from many disciplines that have forensic relevance. Their purpose is to locate clandestine graves. Sequential changes in plant species known as plant succession patterns are distinct over the remains of buried bodies and often remain distinct from surrounding ground cover for many years. Evidence of plant succession patterns has been used in locating such graves.

Summary. Forensic botany is an underutilized tool in criminal investigations. These techniques are simple for the most part and are readily accepted in court testimony because the field is at least as old as the light microscope. For much of the work, qualified consultants can be found at local colleges and universities among the biology faculties trained in plant science.

For background information *see* CRIMINALISTICS; ECOLOGICAL SUCCESSION; ECOLOGY; FORENSIC EVIDENCE; PLANT ANATOMY; PLANT TAXONOMY in the McGraw-Hill Encyclopedia of Science & Technology.   Jane H. Bock; David O. Norris

Bibliography. J. H. Bock and D. O. Norris, Forensic botany: An under-utilized resource, *J. Forensic Sci.*, 42(1):364–367, 1997; J. H. Bock and D. O. Norris (eds.), *Handbook of Forensic Botany*, Humana Press, 2004; M. J. Crawley (ed.), *Plant Ecology*, 3d ed., Blackwell Science, Oxford, England, 1997; W. C. Dickenson, *Integrative Plant Anatomy*, Academic Press, London, 2000; S. Jackson, *No Stone Unturned*, Kensington Publication Corp., Boston, 2002; D. O. Norris and J. H. Bock, Method for examination of fecal material from a crime scene using plant fragments, *J. Forensic Invest.*, 52(4):367–377, 2001; J. S. Sachs, *Corpse: Nature, Forensics, and the Struggle To Pinpoint Time of Death*, Perseus Publishing, Cambridge, MA, 2001.

Fossil microbial reefs

Stromatolites are laminated sedimentary deposits created by benthic (bottom-dwelling) microbial organisms, primarily bacteria and cyanobacteria but also eukaryotic algae, capable of trapping, binding, and cementing sediment grains in thin layers. These kinds of organisms remain an important constituent in many marine communities, including reefs, but are easily overlooked because of their tiny size, unfamiliar morphology, uncertain taxonomic affinities, and obscurity in recrystallized or metazoan-dominated samples. Stromatolites first appeared on Earth 3.5 billion years ago and were the earliest ecosystems to evolve on the planet. Their abundance in the Precambrian geologic record indicates microorganisms were the only common forms of life for more than 3 billion years.

Microbial reef growth. Before the widespread evolution of skeletonized animals in the Early Cambrian period (543 million years ago), stromatolites predominated in most marine settings and grew to enormous sizes as moundlike microbial reefs. Reefs, by definition, are biologically produced, wave-resistant structures that have relief above the seafloor and thereby exert a hydrodynamic influence on adjacent environments. In microbial reefs, sediment trapping, binding, and cementation induced by a variety of microorganisms enhanced mound accretion. Despite stromatolite decline in the Late Precambrian period (Proterozoic Eon), microbes played an important ecologic role in many reefs and associated habitats throughout the Phanerozoic Eon, even facilitating the growth of associated macrobiotas. Microbial communities characterized by distinctive biofabrics (such as laminae, clots, and bushy dendrites) and unique species combinations indicate evolutionary change through time in response to biologic and nonbiologic agents. Environmental conditions that favored microbial reef development and fossilization are uncertain but may have been due to an increase in available nutrients, elevated carbonate saturation levels in seawater, and/or mass extinction of the metazoans (multicellular animals) that competed with them.

Precambrian microbial reefs. Two billion years ago in Canada's Northwest Territories, microbes capable of accreting and cementing sediment produced a stromatolite barrier reef up to 1 km (0.6 mi) thick and 200 km (120 mi) long. These reefs are composed of fine sedimentary laminae (layers) and cavities filled in with microbially produced sediment layers. On the forereef, large stromatolitic debris fell as talus blocks, and in the backreef distinctive microbial communities and sediments were deposited. In those reefs, isolated and semijoined stromatolitic mounds are characterized by distinct, microscopic biofabrics comprising partially linked columns and radiating branches, suggesting the presence of different microbial communities; however, specific microbe types are unknown because they are not preserved.

Laminar, clotted, and sheetlike encrustations characterize kilometer-scale stromatolites in much younger Precambrian reefs (approximately 800–1200 million years old). Three types of calcium carbonate–secreting microbes that formed tubes, clots, and crusts are important evidence that preservable microorganisms had appeared and diversifed by the Late Precambrian. These calci-microbes evolved the capability of fractionating ions in seawater and precipitating calcium carbonate ($CaCO_3$) during cyanobacterial photosynthesis or other processes. The formation of new crystals (nucleation) of calcite on or within organic sheaths that encapsulated the microbes enhanced their preservation. Preservation occurred while the organisms were alive or during postmortem decay and early chemical diagenesis (that is, chemical modifications that take place after initial deposition). In Late Precambrian deposits of Namibia (approximately 550 million years old), microbial reefs comprise lightly mineralized, cnidarian (jellyfish)–like animals fossilized in clotted (thrombolitic) and laminated (stromatolitic) sedimentary structures. Significantly, the presence of calcimicrobes and invertebrate animals in these mounds represents a transitional phase between Precambrian and Phanerozoic reefs. Stromatolite reefs, then, became important environments not only for microbes but for metazoans as well.

Despite the long-term dominance of microbes in Precambrian reefs and other habitats, as the Precambrian came to a close, stromatolites underwent a significant reduction in abundance and diversity. They experienced further decline in the Early Paleozoic, especially after the Early Ordovician. Debates about the reason for their decline continue to focus on whether newly evolved metazoans in the Late Precambrian–Early Cambrian inhibited stromatolite formation by feeding on and burrowing through stromatolites or whether physical and chemical factors in world oceans were less conducive to stromatolite formation and preservation.

Although the underlying causes are controversial, scientists agree that stromatolites and microbial reefs are less prevalent today than they were in the Precambrian. However, the discovery of well-developed stromatolites in many modern and ancient Phanerozoic sites has generated a greater appreciation for—and new debates about—the importance of stromatolites in aquatic ecosystems since the Precambrian. Some scientists have suggested that post-Ordovician stromatolites were restricted to the intertidal zone or were episodic members of normal marine ecosystems, where they functioned primarily as "disaster" species after global ecologic crises, such as mass extinctions. Others have countered that substantial fossil evidence indicates that stromatolites had a persistent presence in many global environments, including reefs, throughout the Phanerozoic. Importantly, debates now focus not only on the microbes within stromatolitic reefs but also on the invertebrate and algal components with which they co-evolved.

Paleozoic microbial reefs. After the Cambrian radiation of life, archaeocyaths (marine calcite sponges) were one of the first invertebrates to "insinuate" themselves into reefs that, similar to their Precambrian precursors, were constructed largely by microbial organisms. Production of excess oxygen by cyanobacteria, use of bacteria as nutrient sources, possible symbioses and biochemical stimuli for larval settlement, as well as sediment stabilization and early lithification induced by microbes were factors that probably favored invertebrate adaptation to microbial reefs. Where macroscopic invertebrates dominate at many, but not all, younger Phanerozoic reef sites, incomplete preservation and other factors—including the uncertain, but possibly microbial, origin of clotted lime mud (micrite)—make it difficult to assess microbial contributions to reef growth. Nevertheless, microbial reefs become especially obvious in the fossil record following archaeocyathan decline in the Early Cambrian, after other mass extinctions, and at geographically disparate sites during many intervals in the Phanerozoic.

Cambrian-Ordovician reefs. In the Cambrian-Ordovician periods, for example, *Renalcis-* and *Epiphyton-*group microbes were the primary builders of microbial reefs comprising dense microbial intergrowths, which formed not only the reef framework but also thick cavity encrustations. Corals, bryozoans, brachiopods, and other reef-dwelling invertebrates had volumetric significance but were subordinate in many reefs to calci-microbes, calcareous algae, and other microbial forms. Sponges show a particularly persistent and ecologically important presence during the Paleozoic Era. Archaeocyaths in the Cambrian, aphrosalpingids (small, chambered sponges) in the Silurian, and other calcified sponges in the Permo-Triassic were densely aggregated in upright positions or entangled together in sheetlike fashion. They acted as rigid traps for loose particles, aided in the binding of internal sediment, and were substrates for microorganismal encrustations and successive layers of marine cement. Multiple generations of bio-encrustations and cements on and within sponge body cavities created structural scaffolding. This elevated reef strength and early lithification, thereby enhancing long-term preservation in the fossil record.

Silurian reefs. Stromatolitic limestones (**illus.** *a–c*) that accumulated along the margins of the Uralian Seaway (proto-Arctic Ocean basin; specifically, present-day Alaska and Russia's Ural Mountains) in the Late Silurian (approximately 420 million years ago) allow comparison with older and younger microbial reefs. Calcified cyanobacteria with associated microorganisms of unknown affinity were the primary frame-building and encrusting constituents in the barrier reefs. Epiphytaceans, such as *Ludlovia* (illus. *d–f*) and *Hecetaphyton* (illus. *g*), had diversified from earlier microbial stocks and, at least regionally, had ecologically replaced the *Renalcis-* and *Epiphyton-*group fossils (illus. *i–j*) as the primary constructors of the Uralian Seaway reefs. Well-developed cavities (illus. *b, h*) occluded by cements, laminae (illus. *c*), and stromatolite talus blocks in deep-water deposits indicate the microbes were capable of constructing seafloor reefs similar to those of the Precambrian. Aphrosalpingids and hydroidlike animals were volumetrically important within these reefs and created an intricate biofabric of intergrown micro- and macroorganisms. These microbial-sponge interactions yield insights into specialized ecologic relationships that evolved in stromatolitic reefs by the middle of the Paleozoic Era. Environmental

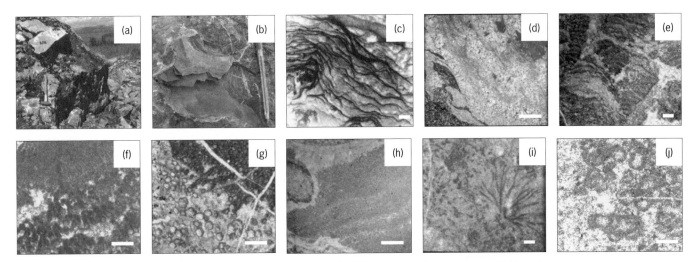

Case example of a microbial reef from the Silurian of southeastern Alaska. (*a–c*) Macroscopic views show typical features in a microbial reef complex, including (*a*) massive limestones, (*b*) well-developed cavities, and (*c*) stromatolitic laminae. (*d–j*) Thin sections reveal microscopic aspects of reefal structures, including laminae comprising bushy growths of probable cyanobacteria (*d–f*) *Ludlovia* and (*g*) *Hecetaphyton*, (*h*) microcavity filled with laminated sediment, and rare evidence of (*i*) *Epiphyton* and (*j*) *Renalcis*. Hammer in *a* is 30 cm long; pen in *b* is 15 cm long; and scale in *c–d* is 4 mm, and in *e–j* is 2 mm. (*Images f and g reprinted with permission, from C. M. Soja, Origin of Silurian reefs in the Alexander terrane of southeastern Alaska, Palaios, 6:111–126, 1991*)

factors, including depth, substrate, water turbulence, and competition, affected the paleoecological distribution of the stromatolitic microbes in ways similar to those of reef invertebrates. These stromatolite reefs became widespread along the Uralian Seaway before the extinction event at the end of the Silurian, possibly in response to ongoing environmental perturbations that inhibited metazoan life and/or promoted stromatolite growth and lithification.

Devonian reefs. Devonian stromatolites (approximately 350 million years old) superbly exposed in the Canning Basin of Western Australia reveal a similar story. At that site, microbes—intergrown with and attached to sponges, brachiopods, and other reef invertebrates—were particularly important participants in extensive reef development. These reef microbes achieved dominance before and after the Late Devonian mass extinction. As in the Silurian example, this suggests that microbes were the ultimate "survivors," capable of weathering even the most severe environmental disturbances, which suppressed the reef-building activities of massive sponges and corals for millions of years.

Comparison with modern reefs. Similarities emerge when comparisons are made between the Paleozoic microbial reefs and modern coral-algal reefs, including their Tertiary predecessors in Europe of Miocene age (approximately 7 million years old). In Spain and Hungary, for example, centimeter-thick microbial encrustations interlayered with calcite crystals occur on platy and upright finger corals. This reflects the age-old processes of microbially induced cementation associated with sediment trapping and binding. In Tahiti, 13,500-year-old stromatolitic crusts, associated with mineralization of biofilms in open cavity systems, are a conspicuous structural component in shelf-margin reefs. Similar laminated microbial crusts are recognized today to have high volumetric significance in high energy coral-algal reefs at modern sites

in the Caribbean Sea, Indian Ocean, and southwestern Pacific Ocean.

Conclusion. Diverse types of microorganisms were especially important agents in reef colonization and stabilization, growth, lithification, and preservation in the past and present. Environmental factors that favored proliferation and preservation of microbial reefs in the Phanerozoic Eon are still debated, including (*a*) increased terrestrial runoff and nutrient concentrations that fueled microbial-algal "blooms"; (*b*) enhanced calcification events during episodes of elevated seawater alkalinity; and (*c*) relaxed ecological landscapes conducive to opportunistic microorganismal growth after the mass extinction of reef-building competitors (that is, metazoans).

Stromatolites represent the world's oldest, continuously forming ecosystems. Important insights into global change on Earth in the geologic past will stem from a greater understanding of why stromatolites and microbial reefs declined in relative abundance over time. Appreciating the circumstances that promoted stromatolite persistence, periodic proliferation, and preservation may also foster future forecasts about environmental change in the marine realm.

For background information *see* ARCHAEOCYATHA; BACTERIA; CYANOBACTERIA; FOSSIL; PALEOECOLOGY; PALEOZOIC; PRECAMBRIAN; REEF; STROMATOLITE in the McGraw-Hill Encyclopedia of Science and Technology.   Constance M. Soja

Bibliography. S. Bengston (ed.), *Early Life on Earth*, Nobel Symp. 84, Columbia University Press, New York, 1994; J. A. Fagerstrom, *The Evolution of Reef Communities*, Wiley, New York, 1987; R. Riding (ed.), *Calcareous Algae and Stromatolites*, Springer, New York, 1991; R. Riding and S. M. Awramik (eds.), *Microbial Sediments*, Springer, New York, 2000; G. D. Stanley, Jr., *The History and Sedimentology of Ancient Reef Systems*, Kluwer Academic/Plenum Publishers, New York, 2001.

Genetically modified crops

Genetic modification is the newest scientific tool for developing improved crop varieties. Such crops can help to enhance agricultural productivity, boost food production, reduce the use of farm chemicals, and make our food healthier. Genetically modified (GM; also called transgenic, genetically engineered, or bioengineered) crops represent the fastest-adopted technology in the history of agriculture, yet they are not universally accepted because of perceived concerns about their safety. Skeptics believe that such crops may pose unrecognized risks to human and animal health and could damage the environment.

Traditional plant breeding. Humans have been modifying crop plants ever since farming began 10,000 years ago, when wild plants were first domesticated to provide food, feed, and fiber. Traditionally, this modification was accomplished primarily through selection of desirable plant types but more recently by plant breeding, which involves the crossing of plants with desirable traits, followed by many generations of selection to eliminate undesirable traits acquired from wild plants. Thus, every crop is a product of repeated genetic adjustment by humans over the past few millennia. While this process has provided the current crop varieties, it is slow and arduous. In addition, the traits for which plants can be bred are limited to those that occur in the wild in closely related crop species; it has not been possible to incorporate characteristics that occur only in nonrelated species.

Genetic modification technology. Recombinant deoxyribonucleic acid (DNA) technology is the most recent tool employed by crop breeders to improve traditional methods of incorporating desirable traits and to eliminate undesirable characters. Genetic modification technology, while providing greater precision in modifying crop plants, enables scientists to use helpful traits from a wider pool of species to develop new crop varieties quickly. Genetic modification involves a clear-cut transfer of one or two known genes into the plant genome—a surgical alteration of a tiny part of the crop's genome compared with the sledgehammer approaches of traditional techniques, such as wide-cross hybridization or mutation breeding, which bring about gross genetic changes, many of which are unknown and unpredictable. Furthermore, unlike traditional varieties, modern GM crops are rigorously tested and subjected to intense regulatory scrutiny for safety prior to commercialization.

Gene transfer. The direct transfer of genes into plants is achieved through a variety of means, but the *Agrobacterium* vector and the "gene gun" methods are the most common.

Argobacterium vector. Agrobacterium tumefaciens is a soil-borne natural pathogen that causes tumors in plants by transferring a piece of its DNA to plant cells. The bacterial cells harbor a large plasmid, called Ti (tumor-inducing) plasmid, which carries the tumor-causing genes in a region of the plasmid called T-DNA. Scientists have modified this bacterium to eliminate disease-causing genes from the T-DNA region, enabling the disarmed bacterium to deliver desirable bacterial genes, such as those for insect resistance, to the plant cells. The resulting plants grown from such cells are healthy but contain the newly introduced gene in every cell expressing the desired trait (**Fig. 1**).

The *Agrobacterium* approach is the most popular method used to deliver genes to plant cells because of the clean insertion and low-copy number of the inserted genes. However, this bacterium does not readily infect monocotyledonous crops, such as wheat, rice, and corn. Scientists have genetically engineered new plasmids to help overcome this problem, and the *Agrobacterium* method is now being employed to transfer genes into cereal crops such as rice.

Gene gun. In the gene gun technique (also called particle bombardment or the Biolistic® approach), microscopic particles coated with the DNA fragment representing the desired gene are shot into the plant cells using a special device. A small proportion of the DNA which enters the cells becomes incorporated into the chromosomes of the plant cell. The gene gun technique helps overcome some of the deficiencies of the *Agrobacterium* method (such as bacterial contamination, low-efficiency transfer to cereal crops, and inconsistency of results).

GM plant development. With any method of gene transfer, the plant cells containing the introduced gene are allowed to develop into full plants under tissue-culture conditions. To ensure that only the modified cells are grown into full plants, scientists include an antibiotic-resistant marker gene along with the desired gene to be introduced into plant cells. When plant tissues are cultured in a medium containing phytotoxic antibiotics, such as kanamycin, only those genetically modified cells containing the marker gene are able to survive and proliferate. This helps scientists to selectively allow a few genetically transformed plant cells (among a mass of millions of untransformed cells) to develop into full plants.

Today's GM crops. GM plants were first developed in the laboratory in 1983. In 1994, the first commercial crop was released for cultivation by farmers and for public consumption. The Flavr Savr® tomato, developed by Calgene in Davis, California, had a delayed ripening trait so that the fruits stayed firm after harvest. This was achieved by the suppression of the enzyme polygalacturonase, which occurs naturally in the cell walls and causes ripe tomatoes to soften. Thus, Flavr Savr® tomatoes, with lower levels of the enzyme than other varieties, can remain on the vine longer before being picked so that full flavor development can occur, because they soften more slowly and remain firm in the supermarket. Fruits and vegetables with improved shelf life are also being developed by silencing other genes related to ripening such as the ethylene receptor gene. In developing countries, nearly half of all the fresh produce perishes because of poor storage and transportation conditions. Thus, biotechnology may help both farmers

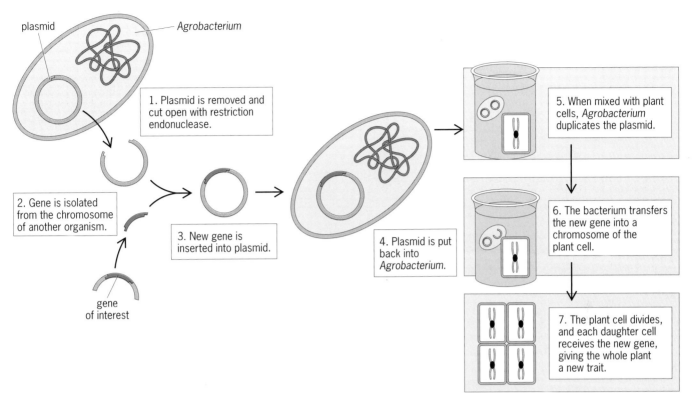

plasmid   *Agrobacterium*

1. Plasmid is removed and cut open with restriction endonuclease.

2. Gene is isolated from the chromosome of another organism.

3. New gene is inserted into plasmid.

gene of interest

4. Plasmid is put back into *Agrobacterium*.

5. When mixed with plant cells, *Agrobacterium* duplicates the plasmid.

6. The bacterium transfers the new gene into a chromosome of the plant cell.

7. The plant cell divides, and each daughter cell receives the new gene, giving the whole plant a new trait.

Fig. 1. *Agrobacterium* vector method. The Ti plasmid of the plant bacterium *Agrobacterium tumefaciens* is used in plant genetic engineering. (*Reprinted with permission from P. H. Raven and G. B. Johnson, Biology, 6th ed., McGraw-Hill, New York, 2002*)

and consumers in these countries through development of fruits and vegetables with longer shelf lives.

Today, GM crops are grown on 70 million hectares across 18 countries. While the United States, Argentina, and Canada dominate the list of countries growing these crops, farmers in less developed countries now plant almost one-third of the world's transgenic crops on more than 20.4 million hectares. Soybeans, corn, cotton, and canola account for most of these crops, and they have been modified for pest resistance, herbicide tolerance, and disease resistance.

Herbicide tolerance. Much of the soybean grown in the United States and Argentina is genetically modified to tolerate the herbicide glyphosate (Round Up®). This herbicide inhibits the production of certain essential amino acids in plants by blocking the action of one enzyme (EPSP synthase) in the biosynthetic pathway of these compounds, and thus is toxic to both weeds and crops. To develop tolerance to this herbicide, scientists introduced into crops a soil bacterium gene encoding an altered EPSP synthase enzyme that eludes binding by the herbicide. Thus, glyphosate will kill the weeds but leave the herbicide-resistant crops unharmed, providing extremely high selectivity between the weed and the crop. Herbicide-resistant crops have proved very popular with farmers, as they simplify the weed management operation, promote no-tillage or low-tillage farming which helps conserve the fertile topsoil, and allow the use of less toxic herbicides.

Pest resistance. GM cotton and corn with enhanced internal pest resistance are also being grown. These were developed through the transfer of genes encoding natural proteins that serve as a natural defense against pests. By transferring the gene for a protein (known as Bt) that blocks digestion in caterpillars from the bacterium *Bacillus thuringiensis* to a plant, scientists have created crops producing novel pesticidal proteins that help control the insect pest when it feeds on the plant. The advent of GM crops with improved pest resistance has led to considerable reduction in pesticide usage on farms and has also increased crop productivity.

Tomorrow's GM crops. Rapid advances in plant genetics, genomics, and bioinformatics are enabling scientists to develop novel GM crops with direct consumer benefits.

Nutritionally enhanced foods. Many nutritionally enhanced foods, such as healthier oils, low-calorie sugars, and vitamin-enriched vegetables and fruits, are under development. A good example of a nutritionally enhanced GM crop is golden rice. This rice is enriched with provitamin A (beta carotene) and, more recently, with elevated levels of digestible iron (**Fig. 2**). The diet of more than 3 billion people worldwide has inadequate levels of essential vitamins and minerals, such as vitamin A and iron. Deficiency in just these two micronutrients can result in severe anemia, impaired intellectual development, blindness, and even death. Many countries, including India, the Philippines, and Vietnam, are

Fig. 2. Golden rice, a GM grain, enriched with beta carotene to address vitamin A deficiency. Light-colored regular rice is interspersed among the grains of golden rice. (*Courtesy of Ingo Potrykus*)

now developing and testing golden rice. Scientists in India have also developed high-protein potatoes by transferring a gene from an amaranth plant.

Other healthful products. Hypoallergenic peanuts, soybean, wheat, and even ryegrass are also on the horizon. Trees requiring fewer chemicals during paper making and even trees that can decontaminate heavy metals from polluted soils are under development. The incorporation of vaccine proteins into commonly eaten foods may provide a way of eliminating diseases in developing countries, where normal inoculation is cost-prohibitive. *See* PHARMACEUTICAL CROPS.

Hardier crops. In many parts of the world, especially in sub-Saharan Africa and South Asia, agricultural productivity continues to be low because of environmental stresses such as drought, heat, soil salinity, and flooding. Hardier crops resilient to these factors are being developed which may make food production possible under marginal conditions and thus may help ensure increased food security.

Benefits. The world population is growing at a rapid pace and will reach 8 billion by 2025. While the available farmland will remain about the same, improving agricultural productivity, especially in the developing world, can ensure food security for all. Improving such farm productivity—while conserving the natural resource base—is a daunting task. The judicious use of genetic modification technology in combination with classical approaches can help improve the food production, reduce the impact of plant diseases or insect pests, tailor crops to harsh environmental conditions, and enhance the nutritive value of foods while reducing the environmental impact of farming.

In 2001, GM soybeans, corn, cotton, papaya, squash, and canola increased U.S. food production by 4 billion pounds, saved $1.2 billion in production costs, and decreased pesticide use by about 46 million pounds. Such crops have also helped lower the pressure on natural resources used in farming: In 2000, U.S. farmers growing transgenic cotton used 2.4 million fewer gallons of fuel and 93 million fewer gallons of water, and were spared some 41,000 ten-hour days needed to apply pesticide sprays. In China, GM cotton varieties have lowered the amount of pesticides used by more than 75% and reduced the number of pesticide poisonings by an equivalent amount.

Safety of GM foods and crops. Despite the success of GM crops, there is still apprehension among some consumers, especially in Europe, about the safety of such foods. However, there has not been any documented case of harm from GM foods since their introduction in 1994. The safety of biotech crops and food has been affirmed by hundreds of studies, including a 15-year study by the European Commission that involved more than 400 research teams on 81 projects. The scientific community has thus repeatedly demonstrated that biotech crops and foods are safe for human and animal consumption.

Antibiotic resistance. Concerns were expressed initially about the safety of antibiotic-resistance marker genes introduced along with desired genes into plants during genetic modification. Drawing from extensive scientific studies, the U.S. Food and Drug Administration and regulators from the European Union have concluded, however, that antibiotic-resistance markers in GM crops do not pose any significant new risks to human health or the environment. Critics have argued that antibiotic-resistance genes might be inadvertently transferred from GM plant cells to bacteria in the guts of animals and humans, making the bacteria resistant to antibiotics and thus rendering some antibiotics less useful for treating bacterial diseases. Research by various world agencies has consistently concluded that these genes have never been shown to be transferred from crops derived through biotechnology to bacteria in nature. Further, there are no antibiotics in these plant foods to cause mutations in the bacteria that would cause them to become resistant to antibiotics, as occurs through excess antibiotic use in much meat.

Environmental toxicity of herbicide-resistant crops. Crop resistance to the herbicide glyphosate enables glyphosate to kill all weeds without harming crops. It makes growing the crop easier for farmers and eliminates the need to use high dosages of far more toxic compounds. Glyphosate has no effect on animals (unlike more toxic compounds often used) and breaks down rapidly (much more than other pesticides) in the environment. One of the main advantages of GM crops is that they enable a decrease in the use of a wide range of pesticides that are damaging to human health and the environment.

Bt pollen toxicity. Researchers at Cornell University found that monarch butterfly larvae that were fed milkweed leaves coated with high levels of pollen from Bt corn (corn genetically modified to produce the Bt protein) ate less, grew slower, and suffered a higher death rate than larvae that consumed milkweed leaves with pollen from non-Bt corn or with no pollen at all. However, field evaluations show that exposure of nontarget organisms, such as monarch larvae, to Bt pollen would in fact be minimal. The Bt pollen toxicity is degraded rapidly by sunlight and is washed off leaves by rain. In addition, due to the enhanced pesticide resistance of the Bt corn, lower amounts of monarch-lethal insecticides can be sprayed.

Other concerns. Other concerns about genetically modified crops include the possibility that pollen from GM "supercrops" may spread to natural environments and become invasive or persistent weeds. However, a team of British scientists has recently found that genetically modified potatoes, beets, corn, and oilseed rape (canola) planted in natural habitats were as feeble at spreading and persisting in the wild as their traditional counterparts. Scientists said research should allay fears that genetic engineering per se would make plants more prone to becoming vigorous, invasive pests.

Another concern is that transferred genes may produce proteins that could cause allergic reactions. However, careful studies are performed to ensure that no such products are put on the market. Recently, a large company stopped developing a genetically engineered soybean that contained a protein from Brazil nuts because research showed that the new soybean triggered the same allergic reactions as does the Brazil nut.

Conclusion. The governments of the United States and other nations have always maintained strict oversight over GM crops. In the United States, they are subject to the extensive, science-based regulations of the Department of Agriculture, the Food and Drug Administration, and the Environmental Protection Agency, which examine the safety of such crops on a case-by-case basis. The American Medical Association believes GM crops have the potential to improve nutrition as well as prevent and even cure disease. Moreover, the World Health Organization believes that GM crops can help developing nations overcome food security problems. Recently the National Academy of Sciences reported that foods from GM crops are as safe as any other foods in the supermarket. While public unease on any new technologies related to food is understandable, continued education on the safety of GM foods and continued oversight of their use may eventually quell such fears.

For background information *see* AGRICULTURAL SCIENCE (PLANT); BIOTECHNOLOGY; BREEDING (PLANT); GENE; GENE ACTION; GENETIC ENGINEERING; HERBICIDE; PESTICIDE in the McGraw-Hill Encyclopedia of Science & Technology.　　C. S. Prakash

Bibliography. M. J. Chrispeels and D. E. Sadava, *Plants, Genes and Crop Biotechnology*, Jones and Bartlett, Boston, 2003; A. Hiatt, *Transgenic Plants: Fundamentals and Applications*, Marcel Dekker, New York, 1993; E. Galun and A. Breiman, *Transgenic Plants*, Imperial College Press, London, 1997; A. McHughen, *Pandora's Picnic Basket: The Potential and Hazards of Genetically Modified Foods*, Oxford University Press, 2000; H. I. Miller and G. Conko, *The Frankenfood Myth; How Protest and Politics Threaten the Biotech Revolution*, Praeger, 2004.

Glaciology

Glaciology emerged from being a descriptive branch of geology to an analytical branch of physics in the years surrounding the International Geophysical Year (1957–1958), as the result of field research in the polar regions. This period, roughly from 1950 to 1970, was dominated by tractor-train (convoy of specially designed vehicles pulling gear-loaded sleds) traverses, mainly over the Antarctic ice sheet, for collecting data on ice elevation and thickness, surface temperature, meteorology, mass balance, and geophysical properties of the bed. The goal was to understand ice sheet flow from the interior ice divides to the sea. This was also the time when the first core holes (ice cores) to bedrock were drilled at Camp Century, Greenland, and Byrd Station, Antarctica, providing the first long-term climate records. Analyzing these climate records became critical for understanding global climate change.

From 1970 to 1980 in Antarctica, the research focus shifted from sheet flow to shelf flow. The Ross Ice Shelf Project (RISP) aimed at drilling through the ice shelf, and the Ross Ice Shelf Geophysical and Glaciological Survey (RIGGS) aimed at collecting tractor-train data sets at a regular grid of stations serviced by airplanes. Shelf flow occurs around Antarctica when the ice sheet becomes afloat in deep water. The Ross Ice Shelf occupies the Ross Sea Embayment between East Antarctica and West Antarctica. It is the size of Texas or France, and includes a rich record of past glaciation preserved in the Transantarctic Mountains to the west and south and in the morphology of floor of the Ross Sea to the north. These glacial and marine geology studies showed that the Ross Ice Shelf formed mostly during the Holocene, when the Antarctic ice sheet underwent gravitational collapse in the Ross Sea sector.

Three foci of glaciological research dominated the period from 1980 to 2000. One focus was to obtain long-term climate records from numerous deep-drilling sites in Greenland, including the two core holes to bedrock at the summit of the Greenland ice sheet, and at Ice Dome C and Vostok Station in interior East Antarctica but 600 km (400 mi) apart, and at small ice domes near the coast. These are Siple Dome and Taylor Dome on the east and west sides of the Ross Ice Shelf, and Law Dome between Totten and Vanderford glaciers on the Budd Coast of East Antarctica. Ice cores recovered from all these sites provided the first evidence that global climate could change radically, even within a few years or less, because the resolution of ice-core chemistry in the Greenland cores revealed seasonal variations in climate. The second focus of glaciological research was the large ice streams that supply the Ross Ice Shelf. In Antarctica, slow sheet flow converges to become fast stream flow, which becomes afloat and merges in embayments to become fast shelf flow. The West Antarctic ice sheet was found to be grounded largely below sea level during the tractor-train traverses and was identified as an unstable "marine" ice sheet by John Mercer, a glacial geologist at the Ohio State University. Since ice streams discharge 90% of Antarctic ice, field research to study the proposed instability was focused on four of the six large West Antarctic ice streams that supply the Ross Ice Shelf.

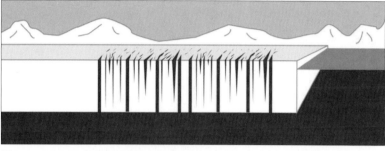

(1)

(2)

(3)

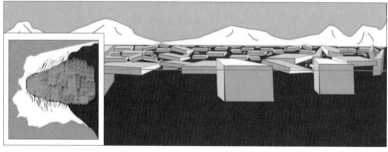

(4)

Fig. 1. Computer model of catastrophic ice-shelf disintegration developed to simulate the sequence of events (1–4) in the Larsen B Ice Shelf, West Antarctica, in 2002 (D. R. MacAyeal et. al., 2003). [*Reproduced with permission from D. R. MacAyeal (University of Chicago) and the International Glaciological Society*]

This began as the Siple Coast Project and matured into the West Antarctic Ice Sheet Initiative (WAIS). The third focus of research was in Greenland, the major area studied under the Program for Arctic Regional Climate Assessment (PARCA). It encompassed a mass-balance assessment of the Greenland ice sheet using satellite remote sensing, airborne remote sensing, and surface glaciology, and included studies of major ice streams that drain the ice sheet. In 2002, results were presented as a special issue of the *Journal of Geophysical Research* (vol. 105, no. D24).

Holistic research. By 2000, glaciologists were aware that a holistic approach was needed in data acquisition and for numerical models of ice dynamics that used these data, as well as in linking glaciological research to parallel research in other dynamic components of the Earth's climate, notably the ocean and the atmosphere. In the 1970s, airborne ice-penetrating radar made it possible to map surface and bed topography, as well as the internal ice stratigraphy. Glaciology made a quantum leap when NASA became involved. The goals of NASA scientists were to develop sensors for satellites orbiting over the ice sheets that could measure temperatures and accumulation and ablation rates at the ice surface, map surface topography with high precision, and chart ice motion. These sensors provided optical, infrared, and microwave imagery, radar sounding, laser profiling, and interferometry. By 2000, the technology had matured to such a degree that remote sensing was used for mapping the distribution of water under the ice sheets, including mapping subglacial lakes, and identifying regions of basal freezing and melting. This included programs to map bed topography from subtle variations in high-resolution surface topography. Parallel developments in drilling technology and downhole instrumentation by glaciologists at the California Institute of Technology made possible the direct sampling of basal till and in-situ studies of till deformation beneath ice streams and the associated subglacial hydrology. Advances in ice-core technology and data analysis led to an understanding of the atmospheric circulation paths over the ice sheet and their changes through time. Remotely operated submersibles were developed to investigate ocean-ice interactions under Antarctic ice shelves and ice streams that become afloat without merging to become ice shelves.

Tractor-train traverses were reactivated for the International Trans-Antarctic Scientific Expeditions (ITASE), with the goal of understanding climate changes from the end of the Little Ice Age through the industrial revolution to the present day. This was accomplished by combining the traditional tractor-train data sets with shallow and intermediate coring along traverse routes linked by radar soundings that mapped internal reflection horizons as timelines in the Antarctic ice sheet. Analyses of the full suite of airborne anions, cations, stable and radioactive isotopes, diatoms, and particulates allowed the origin, direction, and intensity of incoming storm systems to be mapped through time over the ice sheet. Twenty nations are involved in ITASE, with the most ambitious being traverses by the University of Maine in East and West Antarctica inland from Pine Island Bay in the Amundsen Sea and on all three sides of the Ross Sea Embayment. These traverses encompass regions drained by many large ice streams now undergoing rapid lowering, linked to major calving events from the ice shelves they supply. Holistic research investigates these links as responses within a unified dynamic system.

The advent of high-speed linked computers has made possible analyses of these data, as well as incorporation of the data into models for simulating flow in the big ice sheets and the surrounding ocean and atmosphere. This involves nesting of local and regional flow models into models of large-scale flow over a range of resolutions. For holistic ice-sheet modeling, it also involves solving the momentum equation for the full stress tensor, so transitions from sheet flow to stream flow to shelf flow can be simulated realistically, allowing coupling of these models to models of subglacial hydrology and to models of iceberg calving from ice streams and ice shelves. These calving outbursts are capable of suddenly introducing vast storehouses of latent heat into the surrounding oceans, thereby altering patterns of ocean circulation.

Catastrophic events. In 2000, a gigantic iceberg the size of Connecticut was released from the Ross Ice Shelf within hours. On the Antarctic Peninsula, Larsen B Ice Shelf disintegrated catastrophically within days (**Fig. 1**). In Greenland, the floating front of the world's fastest ice stream, Jakobshavn Isbrae, disintegrated within months (**Fig. 2**). Similar calving activity occurred in Pine Island Bay on the Amundsen Sea flank of the West Antarctic Ice Sheet, and large cracks announcing imminent calving appeared on Amery Ice Shelf, which is supplied by Lambert Glacier, the largest East Antarctic ice stream. When combined with earlier major calving events on both sides of the Antarctic Peninsula, and along the calving fronts of Filchner Ice Shelf and Ronne Ice Shelf in the Weddell Sea Embayment (**Fig. 3**), this calving activity may be an early warning of impending gravitational collapse in sectors of the Greenland and Antarctic ice sheets on the scale of Holocene collapse of Antarctic ice to form the Ross Sea Embayment. The primary candidate for a similar large-scale gravitational collapse is the West Antarctic ice sheet inland of Pine Island Bay. With National Science Foundation (NSF) field logistics largely committed to rebuilding South Pole Station, a historic collaboration with the Centro de Estudios Científicos (CECS) in Chile and NASA was inaugurated in 2002, with a program of aerial radar sounding, laser profiling, and photographic mapping over the Pine Island Bay ice drainage system, where rapid inland surface lowering has accompanied high basal melting rates and major calving activity at the floating ends of the two fastest Antarctic ice streams, Pine Island and Thwaites glaciers, both of which enter Pine Island Bay.

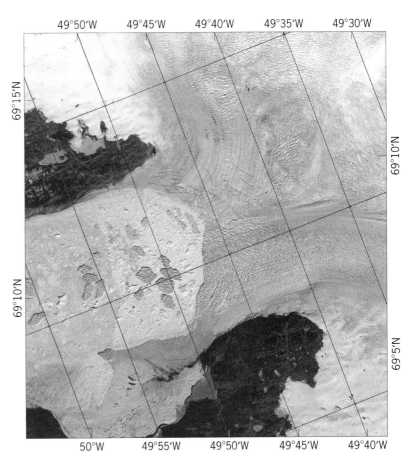

Fig. 2. ASTER image on May 28, 2003, showing ongoing disintegration of the calving front of Jakobshavns Isbrae, West Greenland, that began in 2002. [*Reproduced with permission from Konrad Steffen (CIRES, University of Colorado)*]

Outlook. Antarctic ice streams studied in the last decade as part of WAIS are increasingly seen as representing the last stage of an ice sheet that has been collapsing into the Ross Sea for 10,000 years. Attention is now turning to ice streams entering the Amundsen Sea, where American and British glaciologists see evidence for a new episode of collapse from the rapid downdraw of interior ice as ice streams accelerate in response to rapid disintegration of their buttressing ice shelves. Processes leading to catastrophic disintegration of ice shelves on the Antarctic Peninsula, and downdraw of the supplying mountain glaciers reported by Argentine glaciologists should affect the vast Antarctic ice shelves if climate warming continues. In Greenland, NASA glaciologists have linked downdrawn ice to summer surface meltwater in the ablation zone of the ice sheet that is reaching the bed through crevasses and moulins, thereby increasing basal water pressure that decouples ice from bedrock and converts any intervening till layers into slippery "banana peels." Subglacial hydrology, previously neglected, is now seen as a key to formation and subsequent drainage of large subglacial lakes, producing linked instabilities that propagate up or down ice streams.

Calving of icebergs into the sea has become an exciting part of glaciological research because icebergs

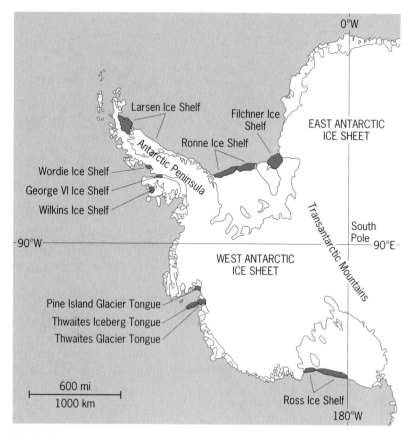

Fig. 3. Recent major iceberg calving events from ice shelves and floating ice streams surrounding the West Antarctic Ice Sheet. Color areas have calved since 1985.

are viewed as entering the Earth's climate system, and massive iceberg outbursts can trigger abrupt climate changes called Heinrich events. All three dynamic components of the Earth's climate—the ice sheets, oceans, and atmosphere—come together most energetically in the calving environment. Calving events illustrated in Figs. 1–3 were accompanied by newly measured rapid surface and especially basal (tens of meters per year) melting rates of floating ice, detected by scientists at Columbia University, Cal Tech's Jet Propulsion Laboratory, the U.S. Geological Survey, and the British Antarctic Survey. This is causing ice-shelf grounding lines to retreat and the ice shelves themselves to disintegrate.

Disintegration of Larsen B Ice Shelf in Antarctica, depicted in Fig. 1 and modeled by Douglas MacAyeal at the University of Chicago, was preceded by ubiquitous crevassing that left ice slabs standing like dominoes and having unstable centers of gravity submerged in water, so the slightest disturbance caused them all to tumble, producing a great outburst of ice. Jesse Johnson, a glaciologist at the University of Montana, introduced at the 2003 WAIS workshop the concept of "jamming" of granular materials flowing through an orifice, commonly observed in grain elevators (or even in a salt shaker). If floating ice is already heavily crevassed, it may flow in part as a granular material. Breaking an ice jam, such that

formed as heavily crevassed ice entering Jakobshavn Isfjord, may have accounted for the rapid disintegration and increased velocity seen in Fig. 2. Calving of gigantic icebergs in Fig. 3 is less understood, but their release is also linked to deep crevasses formed in an ocean environment.

All of these new perspectives represent the vitality and growing importance of glaciology as part of the great scientific enterprise of understanding the history and causes of abrupt climatic change, and the accompanying rapid changes in global sea level. This is perhaps the most urgent scientific concern of our time, with vast economic, social, and political ramifications worldwide if these changes happen.

For background information *see* ANTARCTIC OCEAN; ANTARCTICA; ARCTIC AND SUBARCTIC ISLANDS; GLACIOLOGY; GLOBAL CLIMATE CHANGE; ICE FIELD; ICEBERG; SEA ICE; SOUTH POLE in the McGraw-Hill Encyclopedia of Science & Technology.

Terence J. Hughes

Bibliography. S. A. Arcone et al., Stratigraphic continuity in 400-MHz short-pulse radar profiles of firn in West Antarctica, *Ann. Glaciol.*, 37, in press, 2004; C. D. Clark et al., A groove-ploughing theory for the production of mega-scale glacial lineations, and implications for ice-stream mechanics, *J. Glaciol.*, 49(165):240–256, 2003; H. De Angelis and P. Skvarca, Glacier surge after ice shelf collapse, *Science*, 299:1560–1562, 2003; D. R. MacAyeal et al., Catastrophic ice-shelf breakup by an ice-shelf-fragment-capsize mechanism, *J. Glaciol.*, 49(164): 22–36, 2003; E. Rignot and S. S. Jacobs, Rapid bottom melting widespread near Antarctic Ice Sheet grounding lines, *Science*, 296:2020–2023, 2002; A. Shepherd et al., Inland thinning of Pine Island Glacier, West Antarctica, *Science*, 291:862–864, 2001; R. H. Thomas et al., Investigations of surface melting and dynamic thinning on Jakobshanvs Isbrae, Greenland, *J. Glaciol.*, 49(165):231–239, 2003; H. J. Zwally et al., ICESat's laser measurements of polar ice, atmosphere, ocean, and land, *J. Geodynam.*, 34:405–445, 2002; H. J. Zwally et al., Surface melt-induced acceleration of Greenland ice-sheet flow, *Science*, 297:218–222, 2002.

GPS modernization

The Global Positioning System (GPS) consists of a constellation of 24 or more satellites that transmit navigation signals in two frequency bands called L1 (centered at 1575.42 MHz) and L2 (centered at 1227.6 MHz). The signals are synchronized to within about 10 nanoseconds of a composite time frame known as GPS System Time. The satellites also transmit messages that accurately describe their orbital path in space. GPS users with unobstructed visibility can receive up to 12 of these signals, each one coming from a different direction. The GPS receiver measures the arrival time of each signal. With a minimum of four signals, the receiver can calculate its three physical position coordinates (for example, latitude, longitude, and altitude) plus a precise correction for its internal clock. In general, the more signals used, the better the result.

GPS was declared fully operational on December 8, 1993, after about 20 years of planning, development, testing, and preoperational use. Three years later, the previous navigation satellite system, called Transit, was retired after 32 years of service. A major transition was complete, from Transit, which gave two-dimensional position fixes every 90 min or so, to GPS, which provides continuous time and three-dimensional position and velocity. The advent of GPS has transformed navigation, and GPS has become an integral part of everyday life. Now GPS itself is on the verge of a major transformation. This modernization will introduce more robust signals, improving availability in difficult environments and accuracy for many applications. The coming changes are substantial and will be extremely valuable, but it will take a decade or two before all current satellites reach end of life and are replaced by new versions with the new signals.

Selective availability and antispoofing. The main incentive for the U.S. Department of Defense to develop and operate the dual-use (civil-military) GPS was to enhance military effectiveness. Therefore, it was unacceptable to provide the same advantages to adversaries. To prevent this, the first operational satellites (Block II) were designed with selective availability and with antispoofing features. These were enabled on March 25, 1990. Selective availability reduced the accuracy available to potential adversaries but also to all civilian users. The original GPS signals employ two codes, the C/A (coarse/acquisition) code on L1 and the P (precision) code on L1 and L2. Antispoofing encrypts the P code into a Y code [known as the P(Y) code] to make it unpredictable, so an adversary cannot transmit a false, or spoofed, code to disrupt military navigation. Selective availability limited civilian horizontal accuracy to about 50 m (165 ft) root mean square (RMS) [that is, the horizontal position given by GPS had a 63 to 68% probability of being within 50 m of the actual position]. Antispoofing prevented most civilian access to GPS L2 signals. In one sense, GPS modernization sprang from the civilian reaction to these limitations.

Countermeasures. A consequence of the limitations was that many organizations began developing countermeasures. For example, sending correction signals from one or more well-surveyed reference stations not only defeats the intent of selective availability but also improves the available accuracy. This technique is known as differential GPS (DGPS). A satellite-based augmentation system (SBAS) can provide DGPS corrections over continent-wide areas. The U.S. Federal Aviation Administration began to develop an SBAS called the Wide Area Augmentation System (WAAS). Commissioned in 2003, it supports precision approaches at properly equipped airports

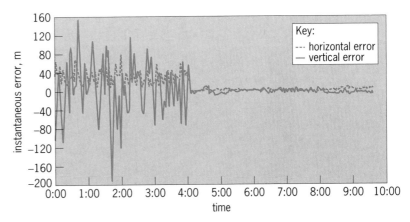

Fig. 1. Change in civilian navigation error when selective availability was discontinued. The horizontal scale is the UTC time on May 2, 2000. (It was May 1 in most of the United States but May 2 in UTC time when the transition occurred.) 1 m = 3.3 ft. (*U.S. Space Command*)

throughout the continental United States. Other nations are developing similar systems. The U.S. Coast Guard transmits DGPS signals from ground-based transmitters throughout the United States, and private organizations also used DGPS corrections to eliminate selective availability errors and achieve superb accuracy. In addition, because dual-frequency measurements had become vital for centimeter-level accuracy in survey applications, companies developed ways to continue making L2 signal measurements in spite of antispoofing.

Deactivation. Two studies published in 1995 pointed out the inadequacy of selective availability to protect national security and recommended that it be deactivated. A presidential directive in March 1996 said that selective availability would be discontinued by 2006, with annual reviews beginning in 2000. Many were surprised when it was turned to zero on May 1, 2000, with "no intent to ever use selective availability again." **Figure 1** shows the dramatic improvement in civilian accuracy when selective availability was discontinued. Instead of continuously degrading civilian accuracy worldwide, the military now depends on denying access by hostile forces to civil GPS signals in a local area of conflict.

New signals. This review of GPS policy triggered plans to enhance civilian GPS performance beyond simply turning selective availability to zero. On March 30, 1998, it was announced that a civilian code would be added to the GPS L2 frequency, giving direct and unhindered access to both the L1 and L2 navigation signals; and on January 25, 1999, the decision to transmit a third civil signal at L5 (centered at 1176.45 MHz), optimized for safety-of-life aviation applications, was announced.

Satellite implementation. Based on the policies noted above, the United States began to develop a fourth-generation Block II satellite called the IIF. (The previous generations were II, IIA, and IIR.) However, to accelerate availability of a new M code military signal and the second civil signal, a limited number of IIR

Current and future GPS signals						
Band: Frequency:	L1 1575.42 MHz		L2 1227.6 MHz		L5 1178.75 MHz	
Signal:	Military	Dual use	Military	Civil	Military	Civil
Legacy	P(Y)	C/A	P(Y)			
IIR-M	P(Y) & M	C/A	P(Y) & M	L2C		
IIF	P(Y) & M	C/A	P(Y) & M	L2C		L5
GPS III	P(Y) & M	C/A & L1C	P(Y) & M	L2C		L5

satellites were retrofitted or "modernized" into IIR-M satellites. Up to eight of these may be launched beginning in 2005, followed by the IIF satellites. A program to design GPS Block III satellites, to be launched no sooner than 2012, also has begun. The **table** shows the legacy C/A and P(Y) navigation signals available today and the signals to be transmitted by subsequent generations of satellites. From only three navigation signals broadcast now, the number eventually will grow to eight.

Signal structure. **Figure 2** shows how four signals will share the L1 band in the GPS III era. It plots the power spectral density of each signal relative to 1 watt. (The vertical scale is dBW/Hz, which means the number of decibels above or below 1 watt of power within a 1-hertz bandwidth.) The legacy C/A code is a binary phase-shift keyed (BPSK) signal with a 1.023-MHz clock rate. (BPSK means the signal carrier switches between only two phase states, one being the inverse of the other, depending on the state of a pseudorandom binary code. Codes change

state, for example, from +1 to −1 or back, at a fixed clock rate. Although a pseudorandom code appears to be random, it is completely predetermined and repeats identically.) The spectrum of such signals has its maximum power spectral density at the center frequency. The power density rolls off to first nulls (zeros) at plus and minus the code clock rate (±1.023 MHz for the C/A code). The legacy P(Y) code is a BPSK signal with a 10.23-MHz clock rate, so its first nulls occur at ±10.23 MHz. Its peak power density is one-tenth (10 dB below) the C/A signal because its energy is spread over a ten times wider bandwidth.

The new L2 Civil (L2C) signal also has a 1.023-MHz clock rate, but with four significant improvements over the C/A code now on L1. First, it consists of two components, one with data and the other with no data. Even with only half the total signal power in the data component, data recovery is more reliable because the data is "convolutionally encoded" with a forward error correction (FEC) algorithm. (The

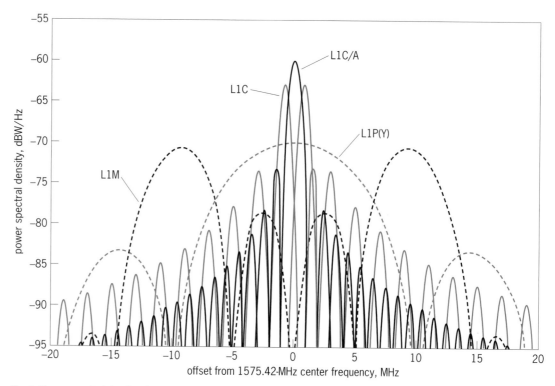

Fig. 2. **Power spectral density of expected GPS III signals in the L1 band (each relative to 1 watt).**

chosen FEC transmits symbols at twice the message data bit rate, where each symbol is defined by encoding a string of data bits. As a result, transmission errors can be corrected when the receiver reassembles the received signals into the actual message.) The signal component with no data is a pilot carrier, unaffected by data modulation. As a result, it can be tracked by a conventional phase-locked loop rather than by a squaring loop, required when data are present. The phase-locked loop has a 6-dB (power factor of 4) threshold tracking advantage over a squaring loop. Thus, even with only half the total signal power in the pilot carrier, the overall signal tracking threshold is improved by at least 3 dB (a power factor of 2). The third improvement is use of longer codes which eliminate cross-satellite interference problems sometimes experienced with C/A. Finally, a new message structure improves resolution and provides more flexibility, which eventually will lead to better navigation accuracy.

The new M codes on L1 and L2 are exclusively for military use. The M code has the same advantages as the new L2C but with much longer codes and with a 5.115-MHz clock rate. Like P(Y), M code is encrypted to prevent spoofing or use by an adversary. The double hump spectrum shown in Fig. 2 results because this BPSK signal also includes a binary offset carrier (BOC). This means that the underlying code is further modulated by a square wave. In this case, the 5.115-MHz code is modulated by a 10.23-MHz square wave. The resulting spectrum has two main peaks at ±10.23 MHz with a null at the peak of the C/A signal. The objective is to spectrally separate the civil and military signals as much as possible so civil signals can be denied to an adversary in a local area by jamming the center of the band with little impact on M code users.

IIF satellites will transmit the new L5 signal at 1176.45 MHz. This signal has the same BPSK spectrum as P(Y), but it includes all the structural improvements of the other modernized signals. Importantly, L5 is in an aeronautical radio-navigation service (ARNS) frequency band, as is L1. Therefore, these two signals will be the backbone of future aviation use of GPS.

Although the first launch of a GPS III satellite will not be before 2012, it is expected that another new civil signal called L1C will be added. It is a BOC signal with the code clock and the square wave both being 1.023 MHz, as reflected in its spectrum (Fig. 2). The design intent is to provide all the benefits of a modernized civil signal on L1 while minimizing interference with legacy C/A signals and maintaining adequate spectral separation from the M code.

Service enhancement. Today civilians have one signal, the C/A code. Over time there will be four civil signals at three different frequencies. The new signals are more robust, so they can be used in more difficult situations, such as in a forest, beside tree-lined roads, or inside some buildings. Access to three fre-

quencies will make scientific, survey, and machine-control applications more efficient and more accurate over longer distances. The accuracy, availability, and continuity provided by SBAS systems like WAAS will dramatically improve when signals at two or more frequencies are available so the receiver can calculate and eliminate ionospheric refraction errors. The ionosphere is the largest source of difficulty for WAAS today, and two-frequency receivers eliminate the problem entirely. These improvements will enhance performance and fuel the growth of many new applications.

New applications. Until recently car navigation was the largest market for GPS receivers, with well over 10 million users. While car navigation continues to grow rapidly, a new and even larger market has emerged for GPS receivers embedded in cell phones. The initial impetus was a U.S. Federal Communications Commission requirement that emergency workers be able to locate a cell phone after it makes an emergency call (E911). Embedding a GPS receiver in cell phones is one way to provide this service, and GPS gives better accuracy than alternative solutions. It also was discovered that a cell phone network can assist GPS receivers to function under extremely difficult signal reception conditions, such as inside a building. These results encouraged other applications, such as providing directions and concierge services to position-enabled cell phone customers. Some expect the total number of cell phone GPS users eventually to approach 1 billion.

Galileo compatibility and interoperability. The European Union has embarked on development of another navigation satellite system called Galileo. The main benefit to civilian users of having two navigation satellite systems is the improved observation geometry that results from doubling the number of visible satellites. Instead of 24 to 28 total GPS satellites, the combination will provide a minimum of 48 satellites in orbit. Many applications suffer from inadequate satellite coverage because terrain or buildings block many of the signals. Doubling the number of satellites will significantly improve the ability to navigate in these difficult environments. For aviation and many other safety-related applications, important integrity evaluation tools are receiver autonomous integrity monitoring (RAIM) and fault detection and excision (FDE). These depend on sufficient redundancy of signals to detect a problem and not use any bad signals. Some experts say a minimum of 36 satellites is required to provide enough signal redundancy to support continuous, worldwide RAIM and FDE. Thus, GPS alone is marginal for this purpose, but GPS combined with Galileo is more than sufficient to assure continuous, worldwide RAIM and FDE.

The digital portion of a satellite receiver is one of the least expensive parts because it is implemented in an application-specific integrated circuit (ASIC). Different signal waveforms, codes, and message structures can be handled easily in such digital electronics. More expensive are the antenna and radio-frequency components, which must be replicated or made more complex for each additional signal frequency. Therefore, commercial receiver manufacturers will gravitate toward using GPS and Galileo signals which have a common center frequency. These provide the largest number of satellite signals at the lowest cost and with the least complexity. Galileo currently plans to overlay GPS signals only on the L1 and L5 frequencies. All other noncommon GPS or Galileo frequencies are likely to be ignored by civil receiver manufactures except for applications mandated by government authorities.

The European Union and the United States have committed to ongoing discussions about compatibility (do no harm) and interoperability (provide the best service). Thus, the opportunity exists to go beyond present plans and further improve the effectiveness and utility of GPS plus Galileo.

For background information *see* MODULATION; PHASE MODULATIONS; SATELLITE NAVIGATION SYSTEMS in the McGraw-Hill Encyclopedia of Science & Technology.   Thomas A. Stansell, Jr.

Bibliography. R. D. Fontana et al., The new L2 Civil signal, *Proceedings of the Institute of Navigation ION GPS-2001*, Salt Lake City, September 2001; R. D. Fontana, W. Cheung, and T. Stansell, The modernized L2 Civil signal: Leaping forward in the 21st century, *GPS World*, September 2001; *The Global Positioning System: A Shared National Asset*, National Academy Press, Washington, DC, 1995; *Global Positioning System: Papers Published in NAVIGATION*, vols. 1–6, Institute of Navigation, 1980–1999; National Academy of Public Administration, *The Global Positioning System: Charting the Future*, 1995; A. J. Van Dierendonck, Understanding GPS receiver terminology: A tutorial, *GPS World*, 6(1):34–44, January 1995.

Harmonics in electric power systems

Alternating current (ac) power generators produce sinusoidal voltage waveforms. Transmission through powerlines and transformers has little effect on the shape of the voltage. Electronic loads, however, produce distortions in the waveform. The amount of voltage distortion is a function of the amount of current and the impedance of the delivery system. Electronic equipment generally controls the current drawn by starting or stopping the current flow within each voltage cycle. The result is a nonlinear current draw and a distorted current flow. Thus, the introduction of modern electronic systems has required changes in the designs of private and utility electric power systems. Commercial buildings and industrial facilities are likely to contain equipment that distorts the voltage of the electricity supply. Modern designers try to restrict that effect to the smallest possible portion of the system.

Origin and effects of harmonics. Harmonics is the name given to waveform distortion in ac electrical systems. Actually an analytical method, harmonics has come to describe any waveform distortion. Jean Baptiste Joseph Fourier (1768–1830) demonstrated that repetitive nonsinusoidal waveforms could be analyzed by breaking down the waveform into an infinite series of sinusoids of different frequencies. Those frequencies are integer multiples of the repetition rate, called the fundamental frequency. The most complicated repetitive waveform can be reduced to harmonic numbers of given magnitudes and relative phase angles by harmonic analysis.

Waveform distortion in the electrical system occurs whenever some electrical threshold event occurs repetitively. The event might be a voltage reaching a particular level or a machine reaching a certain angle. Any change in the use or generation of electricity that occurs on a regular basis within the ac cycle creates harmonics.

What difference does waveform distortion make? Motors, transformers, and other electrical equipment designs are optimized for sinusoidal (50- or 60-Hz) electric input. Any other waveshape makes them less efficient and adds heat. Harmonic analysis helps estimate the losses.

The most common waveform distortion comes from nonlinear electrical loads—loads that draw current at a rate that is not directly proportional to the applied voltage at all times. The current waveform is nonsinusoidal, and the voltage drop in the supply system has exactly the same waveshape.

A simple example of a nonlinear load is a diode and resistor in series. Since the diode blocks the flow of current on half of the cycle, the current is nonsinusoidal. The voltage drop in the supply impedance will occur only on the half-cycle when current flows.

The most common nonlinear load today is the switch-mode power supply, favored in personal computers and other electronic equipment for its efficiency and adaptability to a wide range of input voltages (**Fig. 1**). The incoming ac is immediately rectified and the direct current (dc) is applied to a capacitor. A high-frequency inverter and a small transformer produce the appropriate voltages, and diodes make dc for use in the equipment. This arrangement makes the input a nonlinear load, since the diode-capacitor draws current only from the peaks of the ac sinusoid.

Other types of load, such as lamp dimmers and adjustable speed drives, have similar characteristics. They result in a current that flows only during part of the ac cycle.

In systems experiencing harmonic waveform distortion, the third harmonic is a special case. A three-phase system connected in a wye configuration with single-phase loads will experience third-harmonic distortion, sometimes in great quantities. Balanced single-phase load current on a three-phase system is expected to cause little current in the neutral conductor. Nonlinear single-phase loads draw current at the voltage peaks, which evenly spaces those current peaks through time (**Fig. 2**).

The single-phase load currents come together at the three-phase neutral, but the current peaks do not overlap in time as full sinusoidal currents would. Each current peak travels to the source on the neutral conductor of the three-phase system. The result is that the three-phase neutral has significant current. In many cases, the neutral current is larger than any of the phase currents. Modern office buildings and data centers are constructed with neutral conductors that are larger than the phase conductors, rather than smaller.

Effects of harmonics on wire-wound equipment. In motors, the 5th harmonic (300 Hz on a 60-Hz system) causes losses because the torque produced happens to be opposite to the forward direction of the motor (**Fig. 3**). The 5th harmonic, the 11th, 17th, and others apply a braking action on the rotating machinery, in addition to causing extra heating. If the applied voltage has distortion with these harmonics present, motors operate less efficiently and deliver less power per watt than intended.

In transformers, the extra losses are estimated using the square of the harmonic order. One ampere of 5th harmonic causes losses 25 times the losses of an ampere of fundamental current. As a result, transformers feeding nonlinear loads must be designed to operate efficiently with distorted waveshapes, or their ratings must be changed (derated).

Harmonic frequencies have faster rise times and sharper peaks, and make the magnetic core material swap poles faster. That extra work generates heat, creating additional losses in the transformer.

Third harmonics also create losses in transformer windings connected in delta, because that current circulates in the transformer winding, never reaching into the supply system. A delta-wye transformer can be used to trap 3d harmonic distortion, at the expense of some transformer efficiency.

Equipment derating. Electrical equipment under distorted waveform conditions is sometimes derated. That is, the capacity of equipment using wire windings (motors, transformers) must be reduced to account for the losses caused by harmonics. Transformer heating is approximately proportional to $h^2I_h^2$, where h is the harmonic order and I_h is the current at that harmonic order. The problem increases with higher harmonic orders. Today, transformers designed to supply distorted waveforms to electronic loads are available. The ability to deliver rated capacity at increasingly distorted waveshapes is indicated by the transformer's k factor rating. (The k factor is equal to the sum of the squares of the normalized harmonic currents multiplied by the squares of the harmonic orders. By providing additional capacity—larger-size or multiple winding conductors—k factor–rated transformers are capable of safely withstanding additional winding eddy current losses equal to k times the rated eddy current losses.)

Motor windings are also affected by waveform distortion. Design factors such as the number of poles and the ratio of rotor slots to stator slots determine the efficiency of the motor under distorted voltage

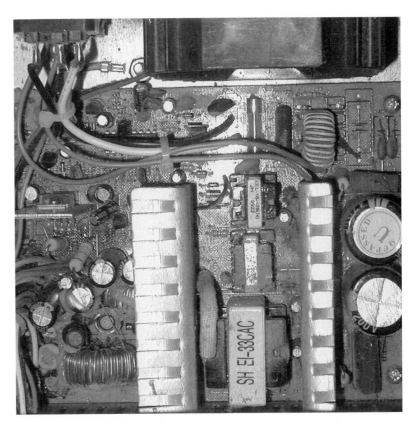

Fig. 1. Switch-mode power supply.

conditions. But in three-phase motors, an applied voltage containing 5th harmonics will cause braking and produce excess heat. The National Electrical Manufacturers Association (NEMA) has supplied derating factors for use with motors in environments with 5th-harmonic voltage present.

Harmonic resonance. Since wires in circuits have inductance, and cables and installed equipment exhibit capacitance to ground, all installations have a natural frequency. Like bridges and turbine blades,

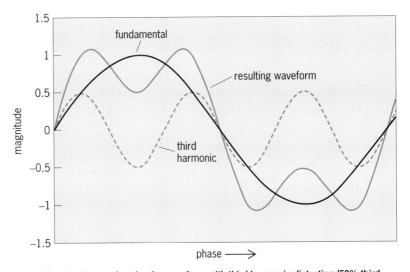

Fig. 2. One fundamental cycle of a waveform with third harmonic distortion (50% third harmonic).

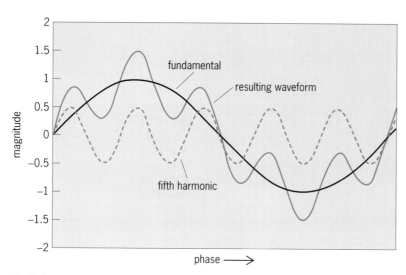

Fig. 3. One fundamental cycle of a waveform with fifth harmonic distortion (50% fifth harmonic).

circuits tend to produce undesirable results when excited at their natural, or resonant frequency. Fortunately, resonance at 50 or 60 Hz is very rare.

When a lineman switches an underground cable system, or when a fuse blows on a utility transformer, a condition called ferroresonance may occur. Identified in the early 1920s yet still little understood today, this condition requires a specific combination of underground cable, a certain transformer connection, and an open phase, but the results can be spectacular. Voltages and harmonic distortion build to very high levels, followed by events such as the triggering of lightning protection systems or equipment failure.

Resonance at higher frequencies is more common, and some powerline harmonics can resonate with line and power factor correction capacitors. The resonant frequency that oscillates in the system may cause telephone interference or other undesirable side effects.

Standards. In the United States, ANSI/IEEE Standard 519-1992 is the harmonic standard for electric systems. Addressing the interface between electric utilities and customers, the standard has been applied in many other ways. As a recommended engineering practice, Standard 519 includes material allowing engineers to make their own guidelines for adding new equipment or solving problems at other locations.

The next version of this standard will include recommended limits for voltages under 1000 V. That will address the present suboptimal situation, in which application engineers apply limits intended for utility connections to panel boards within manufacturing plants.

A new companion document, IEEE Standard 519.1, gives examples and case studies of the application of Standard 519. This book gives deeper insight into the application of limits and explains when the limits might be exceeded. It also gives examples of the three data points needed to apply the standard:

the point of common coupling, the available short-circuit current, and the total demand current.

For background information *see* ALTERNATING CURRENT; ALTERNATING-CURRENT CIRCUIT THEORY; ALTERNATING-CURRENT GENERATOR; DISTORTION (ELECTRONIC CIRCUITS); ELECTRONIC POWER SUPPLY; FOURIER SERIES AND TRANSFORMS; HARMONIC (PERIODIC PHENOMENA) in the McGraw-Hill Encyclopedia of Science & Technology.   Bill Moncrief

Bibliography. R. C. Dugan et al., *Electrical Power Systems Quality*, McGraw-Hill, New York, 2003; IEC, *Testing and Measurement Techniques—General Guide on Harmonics and Interharmonics Measurements and Instrumentation, for Power Supply Systems and Equipment Connected Thereto*, CEI/IEC 61000-4-7, 2d ed., August 2002; *IEEE Recommended Practices and Requirements for Harmonic Control in Electrical Power Systems*, IEEE Stand. 519-1992, 1992; *Planning Levels for Harmonic Voltage Distortion and the Connection of Non-Linear Equipment to Transmission Systems and Distribution Networks in the United Kingdom*, Engineering Recommendation G5/4, February 2001.

Hearing disorders

Approximately 1 in 1000 children is born with a functionally significant, permanent hearing loss, and about 1 in 500 will have a hearing loss by the age of 9. Audiologic testing demonstrates that this hearing loss is attributable to dysfunction of the inner ear or the auditory nerve. Approximately one-half of these cases are caused by environmental factors such as infections, and the other half are genetic in origin. One-third of genetic cases are inherited as part of a syndrome that affects other organ systems. The other two-thirds exhibit hearing loss as the only clinical feature and are termed nonsyndromic. Early observations of relatively high frequencies of normal-hearing offspring born to two parents with nonsyndromic deafness indicated that there were many different genes in which mutations could cause hearing loss. Although hearing loss is increasingly prevalent among older adults, affecting as many as one in two individuals by the age of 80, the genetic basis of this age-related hearing loss, presbycusis, is not well understood. There is a genetic component, but the genetic basis appears to be complex and may involve more than one gene in any individual with presbycusis.

Genetic mapping. The chromosomal locations of over 60 different genes underlying human hereditary hearing loss are now known, and over 30 of these genes have been identified. This progress is attributable to advances in genetic mapping and deoxyribonucleic acid (DNA) sequencing techniques, and the development of sophisticated bioinformatics resources for the entry, organization, extraction, and analysis of vast quantities of mapping and sequence data. These important advances have stemmed, for the most part, from the Human Genome Project.

Mode of inheritance. Hereditary hearing loss genes are mapped and discovered through clinical and molecular genetic studies of large human families with multiple affected members. The mode of inheritance is determined through medical history interviews and audiologic evaluations of all members of the family, followed by construction and inspection of a family tree (pedigree) indicating which members are affected. The majority of nonsyndromic hearing loss is inherited as a dominant or recessive trait. Recessive deafness is typically severe to profound, and its onset is usually congenital, or at least before the development of speech and language at approximately 1-2 years of age. In contrast, dominant deafness usually begins after speech development, and may even start during adulthood. Dominant deafness typically progresses, although the final endpoint of severity varies according to the gene that is mutated.

The majority of recessive deafness genes have been mapped and identified in families from areas where consanguineous matings ("inbreeding") and large family sizes are common; many such families have been studied in southern Asia and the Middle East. In contrast, the majority of dominant deafness genes have been mapped and identified in families from North America and Europe. There are no data to indicate that recessive hearing loss is more common in southern Asia and the Middle East, or that dominant hearing loss is more common in Western societies. This bias probably arises from the decreased severity of dominant hearing loss, which may not come to the attention of physicians and scientists in societies with less accessibility to health care services.

DNA markers and trait linkage. DNA for genetic mapping studies is usually isolated from an approximately 20-mL blood sample. Genetic mapping relies upon a polymerase chain reaction (PCR)–based analysis of anonymous DNA markers that serve as molecular fingerprints to track the inheritance of individual chromosomes among members of a family. Markers that are located near the mutated gene are co-inherited along with hearing loss in the family, and thus indicate the general location of the mutated gene. Analysis of about 300 different markers spanning the human genome is sufficient to provide a rough estimation of the mutated gene's location; this preliminary analysis is supplemented with analyses of additional markers in the region showing linkage to the hearing loss trait. A statistical analysis is performed to determine the level of significance of the linkage, because in small families with only a few affected members the linkage may be coincidental. The linkage analysis defines a genetic map interval that typically contains from a few genes up to several dozen genes that may underlie the hearing loss. Larger families tend to yield smaller genetic map intervals, thus reducing the number of candidate genes and simplifying the search for the causative gene and mutation. In experienced laboratories, mapping a new gene for

hearing loss can currently be accomplished in a few months.

DNA sequencing. After a genetic map interval is defined, its DNA sequence is inspected (using publicly available data derived from anonymous normal human samples) for genes located in the interval. These candidate genes are analyzed, usually by DNA sequencing, for the presence of mutations in samples from affected family members. However, it is not unusual for the predicted effects of DNA sequence changes to be too subtle to permit definitive conclusions about their causality. The strength of the conclusion is increased by the detection of mutations in the same gene in additional families with hereditary deafness linked to the same genetic map interval. The conclusion is also enhanced by a demonstration of specific expression or function of that particular gene in the inner ear, or by one or more mutant mouse strains with hereditary deafness caused by mutations in the corresponding mouse gene. In fact, there are several examples of human deafness genes whose cloning was greatly facilitated by prior identification of the corresponding mouse gene in a deaf mutant strain.

The identification of deafness genes has provided many different entry points for studies of the development, maintenance, structure, and function of the inner ear. Some of the genes have known molecular and cellular functions in other tissues, although it is not usually clear how they integrate into the physiology of the inner ear. The sequence of a deafness gene may offer clues to its probable function, whereas some genes are completely novel with no clues from their sequence.

Animal models. Animal models will be required for studying the pathogenesis of hearing disorders and for developing and testing therapies to prevent, retard, or reverse hereditary hearing loss. Animal models for hearing disorders are especially helpful, since molecular or cellular studies of the human inner ear are usually impossible due to its inaccessibility and susceptibility to permanent hearing loss with any invasive manipulations. The mouse has become the model organism of choice due to the anatomic and physiologic similarity of its inner ear to that of humans. Moreover, the genetics of mice have been thoroughly characterized, and the complete mouse genomic sequence has been determined. Finally, mice are small, are inexpensive, and breed rapidly in comparison with other mammals.

GJB2 mutation testing. An especially important discovery in genetic deafness has been translated into routine clinical practice: 20-80% of nonsyndromic recessive deafness in some populations is caused by mutations of a single gene, *GJB2. GJB2* encodes connexin 26, a protein component of gap junctions in the cochlea and other tissues of the body. Gap junctions are intercellular complexes that allow the passive transfer of small solutes and water between adjacent cells. *GJB2* is a small gene with a simple organization that permits rapid and efficient analyses to detect mutations. In addition, in populations in which *GJB2*-related deafness is common, there is often one highly prevalent mutation that simplifies molecular testing even further. *GJB2* mutation testing has a high yield for a definitive genetic diagnosis and has thus become a routine part of clinical diagnostic evaluations for childhood deafness. The identification of two mutant alleles of *GJB2* enables accurate recurrence risk and genetic counseling for patients and their families, avoidance of additional costly diagnostic evaluations, and more precise prognostic counseling on communication rehabilitation options. For example, a molecular genetic diagnosis of *GJB2* mutations is a positive relative prognostic indicator for rehabilitation of hearing with a cochlear implant.

Treatment. It is possible that many, if not most, forms of genetic deafness will not be amenable to gene replacement therapy approaches. This is especially true for severe congenital forms of deafness since there may be irreversible secondary degeneration and loss of critical inner ear neurosensory tissue at the time of birth. However, other molecular therapies may indirectly evolve from an understanding of the pathogenesis of these disorders as well as normal inner ear biology, including the ability to regenerate neurosensory tissue by other approaches such as stem cell therapy. At a minimum, the ability to pinpoint a specific molecular and cellular lesion in each patient will allow clinicians to stratify the patients for more effective counseling on communication rehabilitation and perhaps will lead to the development of tailor-made auditory rehabilitation strategies according to genetic diagnosis.

For background information *see* GENE; GENETIC MAPPING; HEARING (HUMAN); HEARING IMPAIRMENT; HUMAN GENETICS; MUTATION in the McGraw-Hill Encyclopedia of Science & Technology.

Andrew J. Griffith

Bibliography. T. B. Friedman and A. J. Griffith, Human nonsyndromic sensorineural deafness, *Annu. Rev. Genom. Human Genet.*, 4:341–402, 2003; A. J. Griffith and T. B. Friedman, Autosomal and X-linked auditory disorders, in B. J. B. Keats, A. N. Popper, and R. R. Fay (eds.), *Genetics and Auditory Disorders*, pp. 121–227, Springer, New York, 2002; K. P. Steel and C. J. Kros, A genetic approach to understanding auditory function, *Nature Genet.*, 27:143–149, 2001.

Helper and regulatory T cells

The immune system has evolved complex processes to ensure tolerance to autoantigens (self antigens) while preserving the potential to mount effective humoral (mediated by antibodies secreted by B lymphocytes, or B cells) and cellular (mediated by T lymphocytes, or T cells) immune responses against invading pathogens. The establishment of an immune response to various foreign antigens involves a complex interplay of not only pathogen-related factors that influence infectivity but also host-related factors

that help mount, sustain, and control the necessary adaptive immune responses against the pathogen. Indeed, when a pathogen invades its host, an innate and T helper (Th) cell–dominated adaptive immune response is induced, clearing the infection and, most importantly, establishing life-long immunity to the pathogen. (Note that Th cells are also known as CD4+ cells.)

Th1/Th2 model of T-cell regulation. In order to mount an appropriate T-cell response to a given inflammatory stimulus, the immune system has evolved means of initiating, maintaining, and controlling the timing, magnitude, and quality of these responses. Observations by C. R. Parish and F. Y. Liew in 1972 indicated that cell-mediated and humoral immunity continue to alternate in reciprocal dominance over each other in response to a wide range of antigens. In the mid-1980s, T. Mossman and R. Coffman proposed the Th1/Th2 model of T-cell regulation stemming from experimental observations made in mice, in which two subtypes of Th cells can be distinguished based on their cytokine (intercellular messenger proteins that activate cell division or differentiation) secretion pattern. They suggested that these counterregulatory Th1 and Th2 cells were important regulators of immune responses; each population provides help to different arms of the immune system. The in-vivo relevance of this regulatory Th1/Th2 paradigm was initially demonstrated in murine leishmaniasis and human leprosy, and has now been extended to various types of human immune responses.

A plethora of in-vitro and in-vivo observations have delineated the functional differences between Th1 and Th2 cells. In this model, Th1 cells produce little or none of the cytokine interleukin-4 (IL-4), but they are copious producers of the cytokines interferon-γ (IFN-γ) and lymphotoxin, and are critically dependent on IL-12 for their differentiation. Conversely, Th2 cells produce little or no IFN-γ; are good producers of IL-4, IL-5, IL-10, and IL-13; and generally require IL-4 for their development. Some cytokines such as granulocyte-macrophage colony-stimulating factor, IL-3, and tumor necrosis factor (TNF)-α, are produced by both subsets, while others such as IL-10 are also produced by other cell types including dendritic cells (DC; a type of antigen presenting cell). IL-10 has a variety of immunosuppressive effects, including inhibition of Th1 cytokine production and suppression of antigen-presenting cell (APC) function, thus suppressing local Th1 cell–mediated activity. In addition to T cell–derived cytokines, there is a substantial contribution of cytokines, such as transforming growth factor β (TGF-β), produced by non-T cells to help suppress Th1/Th2 activity. Thus, the two Th-cell subsets produce a unique array of cytokines that seem to cross-regulate each other's development and function. Some studies have also suggested that Th1 and Th2 cells differ in their arrays of cell-surface receptors for cytokines and chemokines (chemoattractant cytokines), endowing them with selective functional characteristics.

The functional differences between Th1 and Th2 cells result in drastically different immunological outcomes. Generally, Th1 cell–dominated responses potentiate cell-mediated immunity and are characteristic of type 1 immunity, which is required for the activation of antigen-presenting cells, mounting effective responses against intracellular pathogens such as viruses and parasites, raising the classic delayed-type hypersensitivity skin response to viral and bacterial antigens, and potentiating antitumor activity. In addition, Th1 cells are implicated in the immunopathology resulting from chronic inflammatory disorders and organ-specific autoimmune diseases, such as rheumatoid arthritis, multiple sclerosis, and type 1 diabetes. In stark contrast, Th2-dominated responses potentiate humoral-mediated immunity and are characteristic of type 2 immunity, which is necessary for B-cell activation and antibody production against allergens and extracellular pathogens such as helminths (parasitic worms). Th2 responses are also implicated in allergy and predisposition to systemic autoimmune diseases such as systemic lupus erythematosus. The ability of cytokines such as IL-4 and IL-10 to inhibit Th1 cells may explain why cell-mediated and humoral immunity are often mutually exclusive, and implicates the Th2 subset as important regulators of cell-mediated immunity.

Differentiation of T helper cells. A number of in-vitro and in-vivo observations have enabled the elucidation of the differentiation process that directs a naive precursor (antigen-inexperienced) Th cell (that is, a precursor Th cell; down either a Th1 or Th2 maturation pathway. The ultimate Th1/Th2 balance is affected by several factors, including the anatomical location of T-cell priming; nature, type, and dose of antigen; the diversity and relative intensity of cell-to-cell interaction with antigen-presenting cells; and the cytokine receptors available on the naive cell. In the face of these varying conditions, the immune system must be constantly adapting, mobilizing, and functionally integrating numerous cellular signals for a rapid response.

In the Th1/Th2 model of T cell regulation, the cytokine environment during T-cell activation by antigen-presenting cells probably plays the most important regulatory role in the differentiation of naive CD4+ T cells. Antigen-presenting cells, particularly DCs, are indispensable for sensing antigen intrusion and initiating a response, including priming and driving the differentiation of noncommitted naive precursor Th cells into Th1 or Th2 cells. Following antigen challenge, DCs are activated and become type 1 (DC1, IL-12-producing) or type 2 (DC2, IL-12-inhibited) cells. The polarized DC1 cells can then promote IL-12-dependent Th1 differentiation, while the DC2 cells promote IL-4-dependent–Th2 cell development. In contrast to DC1, DCs exposed to high IL-10 (a specialized type of dendritic cell called DC3) likely become IL-12-inhibited, are unable to stimulate naive T cells, and permit the emergence of an IL-10-producing regulatory T-cell type (Treg), which can downregulate both the Th1 and Th2 pathways. Dif-

ferentiation is initiated following direct contact of naive Th cells with antigen-presenting cells, during which a major histocompatibility complex bearing an antigenic peptide on the surface binds to the T-cell receptor on the surface of the naive Th cell, forming an immunological synapse. The naive Th cells then pass through a Th0 transient state on their way to becoming Th1 or Th2 cells (see **illustration**).

Thus, the IL-12/IL-4/IL-10 cytokine microenvironment of a Th cell (especially their concentration over time and distance) within the immunological synapse is the determining factor in the Th1/Th2 decision checkpoint. Other soluble factors (including chemokines, prostaglandins such as prostaglandin E2, and various inflammatory mediators) also contribute to the signal and may affect the functional outcome. The commitment to become a Th1 or Th2 cell appears to be irreversible. The local inflammatory environment also plays an important role in the Th-cell differentiation process. For example, in a tumor microenvironment, IL-12 production tends to be decreased, while levels of IL-4 and IL-10 tend to be increased, often resulting in the downregulation of antitumor Th1 activity.

Regulatory T cells. The immune system has evolved numerous mechanisms, including Treg cells, to downregulate immune responses and to assure peripheral immune self-tolerance (which prevents autoreactivity of mature lymphocytes in the peripheral tissues). To this end, a complex network of Treg cells exists to assure timely, efficient, and multilevel control of T-cell responses, depending on the level of inflammation, timing, anatomical location, and nature of the inflammatory insult. Although Treg cells come in multiple types, the general view is that there are two main categories of Treg cells, each differing in their ontogeny and effector mechanism. The first subset includes several types of induced Treg (iTreg) cells, which develop as a consequence of activation of classical naive T-cell populations under particular inflammatory conditions (see illustration). The second Treg subset develops during the normal process of T-cell maturation in the thymus, resulting in the generation of a naturally occurring population of CD4+CD25+ Treg (nTreg) cells that survives in the periphery of normal individuals, poised to control immune responses (see illustration). Thus, peripheral T-cell immunoregulation may, in principle, be assured by a synergistic functional relationship between nTreg and iTreg cells to control immune responses.

iTreg cells. Functional studies have shown that several iTreg populations may help terminate Th1- or Th2-related inflammatory responses. In general, most of these Treg cells arise after deliberate antigen exposure, and their definition has been based on their phenotype and their relative cytokine production capabilities. (Th1 and Th2 cells can be considered regulatory by virtue of their cross-antagonizing actions.) Induced Treg cells also include an antigen-specific, IL-10-producing type 1 regulatory (Tr1) cell, which requires IL-10 during its priming phase (possibly by

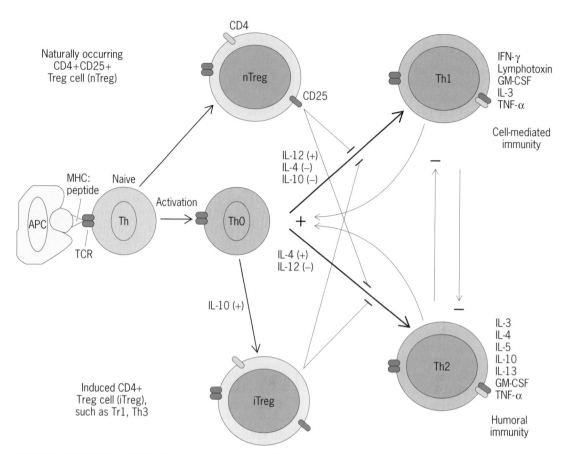

Induction and regulation of CD4+ Th immune responses. Both naturally occurring (nTreg) and induced (iTreg) CD4+ regulatory T-cell populations potentially downregulate the function of activated effector Th1 and Th2 cells in several immunological settings. While CD4+CD25+ Treg cells differentiate in the thymus and are found in the normal naive CD4+ T-cell repertoire, multiple iTreg-cell subsets emerge from conventional CD4+ T cells, which are activated and differentiated in the periphery under unique stimulatory conditions. The relative contribution of each population in the overall regulation of immune responses is unclear, but both conceivably can cooperate to achieve this goal. Each of the Th1 and Th2 cell subsets express a unique profile of cytokines that are counterregulatory in nature and potentiate different arms of the adaptive immune system, cell-mediated and humoral immunity, respectively. APC, antigen-presenting cell; GM-CSF, granulocyte macrophage colony-stimulating factor; IFN-γ, interferon-γ; IL, interleukin; MHC, major histocompatibility complex; TCR, T-cell receptor; TNFα, tumor necrosis factor-α.

a DC3) and mediates its biological effects via IL-10. Oral administration of antigen may also induce in gut mucosal sites antigen-specific, Th3 cells, which produce neither IL-4 nor IFN-γ, and mediate their suppressive effects via TGF-β. Other iTreg cells have also been described. The reason for this diversity is unknown, but there may be developmental or functional overlap between these described Treg subsets. Alternatively, a given cell type can differ according to the stage of activation or the microenvironment in which it develops.

nTreg cells. In recent years, nTreg cells have emerged as the cornerstone component for the generation and maintenance of peripheral self-tolerance. In normal animals, nTreg cells constitute 5–10% of peripheral CD4+ T cells and possess potent immunoregulatory functions. The removal of n-Treg cells from peripheral immune systems, in conjunction with an inflammatory signal, generally increases tumor immunity and pathogen clearance, and reproducibly results in the development of a wide range of organ-specific autoimmune diseases. In normal mice, nTreg cells have a diverse set of T-cell receptors, and could

conceivably recognize a wide array of self antigens; thus nTreg cells may exert their primary function by preventing the activation of autoreactive T cells. Recent studies favor the view that nTreg cells also play critical roles in modulating the immune response to infectious agents and preventing and downregulating excessive inflammatory responses. Currently, a number of fundamental questions regarding their ontogeny, specificity, and mechanism of action are still unanswered. In addition, the signals for the generation, maintenance, and survival of a functional Treg cell pool are incompletely defined.

Future research. The Th1/Th2/Treg paradigm has considerably shaped understanding of the mechanisms that dictate the evolution of normal immune responses and the pathogenic basis for several immunological disorders. This concept illustrates how the overall balance between Th subsets also influences the balance between health (homeostasis) and disease (inflammation). A considerable body of research is focused on the potential application of Th1 or Th2 cytokines as pharmacologic immune response modifiers or the induction/potentiation of

Treg cells in order to influence the clinical course of disease, by tipping the balance away from inflammation and toward protection. Thus, novel immunotherapeutic strategies to modulate the balance between Th1/Th2/Treg cell activity are desirable: potentiating Th1 cells can potentially boost immunity to pathogens and tumors and attenuate allergic responses, while suppression of Th1 cells, via Th2 or Treg induction, may be protective in organ-specific autoimmunity.

For background information *see* ACQUIRED IMMUNOLOGICAL TOLERANCE; CELLULAR IMMUNOLOGY; CYTOKINE; IMMUNE SYSTEM; IMMUNITY; IMMUNOSUPPRESSION in the McGraw-Hill Encyclopedia of Science & Technology.   Ciriaco A. Piccirillo

Bibliography. S. L. Constant and K. Bottomly, Induction of Th1 and Th2 CD4+ T cell responses: The alternative approaches, *Annu. Rev. Immunol.*, 15:297–322, 1997; K. M. Murphy and S. L. Reiner, The lineage decisions of helper T cells, *Nat. Rev. Immunol.*, 2:933–944, 2002; C. A. Piccirillo and A. M. Thornton, Cornerstone of peripheral tolerance: Naturally occurring CD4(+)CD25(+) regulatory T cells, *Trends Immunol.*, 25:374–380, 2004; R. A. Seder and W. E. Paul, Acquisition of lymphokine-producing phenotype by CD4+ T cells, *Annu. Rev. Immunol.*, 12:635–673, 1994; E. M. Shevach, CD4+ CD25+ suppressor T cells: More questions than answers, *Nat. Rev. Immunol.*, 2:389–400, 2002.

High-altitude atmospheric observation

Despite the enormous progress made in atmospheric research during the past 20 years, scientists still lack a basic understanding of a number of issues in the higher regions of the Earth's atmosphere. These issues include the reasons for the dryness of the stratosphere, whether or not there is evidence of the greenhouse effect in the upper atmosphere, and various questions connected with the air motion and transport mechanisms in the transition layer between the troposphere and lower stratosphere. The troposphere extends from the ground to altitudes of 8 km (5 mi) in polar regions and to 19 km (12 mi) at equatorial latitudes. The tropopause separates the troposphere from the stratosphere, which extends above the troposphere to altitudes of roughly 50 km (31 mi).

In addition to the stratospheric dryness problem, the mechanisms for creating and maintaining the stratospheric sulfate aerosol layer, also known as the Junge aerosol layer, are not well understood. This layer spans the globe at altitudes from 15 to 20 km (9 to 12 mi) and consists mainly of sulfuric acid droplets. Other questions remain such as the mechanisms behind the observed but largely unexplained global downward trend in stratospheric ozone, and the extent that the greenhouse effect interferes with the recovery of the lower stratosphere ozone layer. Understanding the various issues will result in more accurate weather, storm, and precipitation forecasts,

as well as improved numerical models of atmospheric dynamics, chemistry, and climate change due to anthropogenic influences.

The atmospheric processes occurring in the tropical regions are of considerable interest. Currently, researchers are studying the anthropogenic emissions (for example, due to biomass burning) in the populated and fast-growing tropical regions to determine what amount is reaching other regions of the globe and the effects of these emissions on the global atmosphere, including cloud formation and regional and global precipitation.

Pollutant gases and aerosol particles emitted from commercial aircraft increase the atmospheric burden in the upper troposphere and lower stratosphere. For example, nitrogen oxides, NO_x [that is, nitrogen monoxide (NO) and nitrogen dioxide (NO_2)], emitted from the jet engines have the potential of destroying ozone in the lower stratosphere or generating ozone in the upper troposphere. Because the aircraft emissions of nitrogen oxides compete with natural processes (for example, NO_x generated by lightning flashes which are most frequent in the tropics), additional studies are required to explain the respective contributions to the atmosphere.

Research platforms. Direct observations in the upper troposphere and lower stratosphere are difficult but essential for understanding atmospheric processes and designing models. In the tropics, the tropopause is between 16 and 19 km (10 and 12 mi), an altitude that regular aircraft cannot reach for direct measurements. As a result, specialized research platforms are needed. Even though recent advances in satellite technology have greatly improved monitoring capabilities, data coverage, and data quality, limitations exist with respect to (mainly vertical) resolution and periods with obstructed views.

Satellites depend on "ground truthing" for direct verification of their readings. This is critical for satellite instrument design and optimizing the mathematical algorithms required to retrieve the atmospheric variables from the remote satellite sensor data. For example, if the satellite measures the extinction of light at wavelengths corresponding to the characteristic spectral lines for a specific atmospheric trace gas, it is necessary to calibrate the detected extinction against in-situ measured abundances of that trace substance. Otherwise, the errors inherent in the satellite measurements remain unacceptably large.

Balloons. While rockets are able to reach the highest atmospheric regions, they record data records for only seconds to minutes. Balloons have proven to be powerful tools for unraveling many questions, such as those concerning the ozone hole. However, balloon launches can be performed only under very specific and limited meteorological conditions. Specific atmospheric events often cannot be actively probed because the flight path of a balloon is largely outside the control of the ground crew.

The first direct balloon-borne mass spectrometric measurements inside polar stratospheric clouds were acquired during flights that were major achievements. During these flights in the arctic stratosphere, the operators succeeded in penetrating these tenuous clouds several times by remotely increasing and decreasing the balloon's buoyancy while the balloon passively drifted in the wind field.

Before these balloon flights, a number of in-situ and remote sensing measurements, as well as laboratory and theoretical studies, had indicated that polar stratospheric clouds consisted of nitric acid hydrate particles or ternary solution droplets of water, sulfuric acid, and nitric acid. However, direct evidence was missing until the balloon-borne mass spectrometer showed molar ratios of water and nitric acid inside these particles to be exactly 3:1, giving proof for the existence of nitric acid trihydrate particles. The detection of such higher molar ratios provides further evidence that ternary solution droplets appear inside polar stratospheric clouds.

These results were key to understanding the heterogeneous chemical processes occurring in the Arctic and Antarctic polar stratosphere, ultimately leading to understanding the phenomenon of the ozone hole. Today, research balloons are capable of carrying instrumental payloads up to 500 kg (1100 lb) to altitudes of 35 km (22 mi).

Research aircraft. High-altitude research aircraft are available for probing the upper troposphere and lower stratosphere. Research aircraft can carry up to 2-ton payloads to altitudes of 20–24 km (12–15 mi), thus reaching the center of the Earth's ozone layer. Although various military aircraft can fly much higher, they usually move at supersonic speeds. For research purposes, air containing trace gases and aerosol or cloud particles has to be representatively sampled, which is possible only at subsonic velocities. Consequently, high-altitude research jet aircraft look much like large sailplanes and sometimes operate under marginal aerodynamic conditions in the sense that they are close to stalling due to slow speeds or their wings are close to mechanically disintegrating due to strong lift forces.

Because of aerodynamic limitations for flight durations of up to 8 hours, the aircraft's maximum altitudes are 16–24 km (10–15 mi), where ambient pressures range from 104 to 30 hectopascals (0.1 to 0.03 atm) and air temperatures can be as low as $-90°C (-130°F)$. Worldwide, there are currently four high-altitude research aircraft equipped as "airborne laboratories," of which three are single-seated (the NASA ER-2, the Australian Egret, and the Russian M-55 Geophysica) and the fourth has two seats (NASA WB57F). Instrument operation is fully automated. The instrument designs reflect some of the world's best engineering capabilities because of stringent safety and operating conditions, high measurement accuracy and precision, and low weight and power consumption requirements.

The instruments onboard the aircraft (such as the ER-2) measure atmospheric gases (such as methane, nitrogen dioxide, carbon monoxide, chlorofluorocarbons, ozone, and various nitrogen-containing gases), highly reactive radicals (such as ClO, OH, HO_2, NO_x), aerosol and cloud particles, water vapor, as well as ambient pressure, temperature, radiative fluxes, and other meteorological variables.

Stratospheric ozone depletion. In 1987, stratospheric in-situ measurements onboard the ER-2 led to an understanding of the mechanisms for the springtime ozone hole over Antarctica. These data, together with numerical models for the ozone-related atmospheric chemistry, established that anthropogenic chlorine-containing gases act together with naturally occurring polar stratospheric clouds to destroy ozone, if certain atmospheric and meteorological conditions prevail.

Additional research flights in the Antarctic and the Arctic stratosphere provided the detailed information needed to improve modeling of these stratospheric processes, such that scientists are able to predict the ozone losses in the Southern Hemisphere and recently Northern Hemisphere to a high degree of confidence and accuracy.

On the surface or inside the volume of the aerosol particles in polar stratospheric clouds, heterogeneous reactions convert passive chlorine-containing species into active forms which are capable of chemically destroying ozone. Such reactions also occur on the microscopic sulfuric acid droplets of the stratospheric sulfate aerosol layer and with enhanced efficiency when sulfate aerosol is injected into the stratosphere by volcanic eruptions.

The existence of similar heterogeneous processes was confirmed directly by in-situ measurements onboard the ER-2 when gases and aerosols of the volcanic plume spreading within the stratosphere from the Mount Pinatubo eruption (June 1991) were sampled. These measurements allowed researchers to understand the role of naturally occurring heterogeneous ozone-depleting reactions involving nitrogen oxides (NO_x), dinitrogen pentoxide (N_2O_5), and nitric acid (HNO_3).

Air traffic. Of increasing environmental concern are the emissions inherent in the growing commercial air traffic. Plans to build and operate a fleet of up to 500 stratospheric supersonic passenger aircraft have raised questions about their impact on air chemistry and aerosol content in the stratosphere. In-situ measurements where made with the ER-2 sampling exhaust from a Concorde cruising in the stratosphere. The data obtained for aerosols, NO_x and nitrogen-containing trace constituents, and other trace gases and meteorological parameters led to improved numerical models for the assessment of such commercial stratospheric air travel, as well as for the impact of growing upper troposphere/lower stratosphere subsonic civil air transport.

Cirrus clouds. Recently, cirrus clouds were discovered that are so thin they cannot be seen from the ground by an unaided observer. They extend over very large areas in the tropical tropopause. Theoretical thermodynamics studies, as well as indirect interpretations of satellite data, indicated that these clouds could consist of nitric acid–containing particles or possibly ice. The chemical composition of the cloud particles was unknown until in-situ measurements with the ER-2 and Geophysica detected particles containing no HNO_3, providing evidence that the cloud particles were frozen water.

This is of major importance for understanding the water budget and partitioning in the tropical upper troposphere and for trying to assess the reasons behind the low water vapor content in the lower stratosphere. In the tropical regions where the Hadley circulation provides a slow upward motion within the upper troposphere and lower stratosphere, these clouds act as "freeze dryers" because moisture contained in the rising air condenses and freezes onto cloud particles, effectively removing water from the gas phase. The dried air then enters the stratosphere.

One important scientific question is to what extent this mechanism is responsible for the globally observed low stratospheric water vapor content. Such mechanisms are directly connected to the effects of global warming, as rising temperatures tend to increase the atmosphere's total water vapor content. In fact, rising water vapor levels have already been observed in the upper troposphere and lower stratosphere above some locations. Increasing water vapor levels interfere with the microphysics of cloud formation and maintenance, air chemistry, and the atmospheric radiative energy budget, among others.

Outlook. The deployment of instrumented high-altitude research aircraft and balloons has been of tremendous importance for unraveling a number of fundamental scientific questions related to the upper atmosphere as well as for understanding of environmental problems vital for life in general. Based on the results of such research, policies could be formulated and implemented, ultimately mediating further intensification of the polar stratospheric ozone destruction. More research is needed, and efforts are underway to equip remotely piloted research aircraft, carrying instrument payloads, to achieve higher altitudes for longer flight durations, while operating under less stringent safety restrictions.

For background information see AEROSOL; ATMOSPHERIC GENERAL CIRCULATION; METEOROLOGICAL ROCKET; STRATOSPHERE; STRATOSPHERIC OZONE; TROPICAL METEOROLOGY; TROPOPAUSE; TROPOSPHERE; UPPER-ATMOSPHERE DYNAMICS in the McGraw-Hill Encyclopedia of Science & Technology.

Stephan Borrmann

Bibliography. B.-J. Finlayson-Pitts and J. N. Pitts, Jr., *Chemistry of the Upper and Lower Atmosphere: Theory, Experiments and Applications*, Academic Press, 2000; D. K. Lynch et al., *Cirrus*, Oxford University Press, 2002; *Scientific Assessment of Ozone Depletion: 2002*, Global Ozone Research and Monitoring Project, Rep. No. 47, World Meteorological Organization, Geneva, March 2003; S. Solomon, Stratospheric ozone depletion: A review of concepts and history, *Rev. Geophys.*, 37:275–316, 1999; R. P. Wayne, *Chemistry of Atmospheres*, 3d ed., Oxford University Press, 2000.

HIV vaccines

Figures from the Joint United Nations Programme on human immunodeficiency virus (HIV)/acquired immune deficiency syndrome (AIDS) suggest that 40 million people are infected with HIV worldwide. Thus the need for an effective HIV vaccine is more pressing than ever. For individuals already infected, potent drugs are available to control the virus. However, these drugs often have significant toxicities and are difficult to afford in many parts of the world, particularly where they are most urgently needed. In addition, noncompliance with drug regimens is believed to contribute to the development of drug-resistant strains of HIV. Hence a vaccine remains the best hope of controlling the HIV pandemic.

Traditional approaches to vaccine development. In determining which kind of vaccine approach might be successful against HIV, it is instructive to consider approaches that have been successful against other pathogens. Many currently available vaccines are designed to induce circulating antibodies that bind to and inactivate bacteria or their released toxins, but antibodies can also be effective against viruses (**Table 1**). Antibodies can neutralize viruses and block or limit their infectivity.

Live attenuated vaccines. For many viruses, the correlate of protection (the presumed mediator of protective immunity) is neutralizing antibodies, which can be induced by inactivated viruses or by live attenuated viruses, which are much weakened versions of the original virus. However, the use of live attenuated HIV vaccines in humans raises formidable safety and liability issues, making it exceptionally unlikely that this approach would ever be used. For viruses which are not directly cytopathic to the cells they infect, including HIV, it is believed that cellular immunity (mediated by T cells) will play an important role in protection. Virus-infected cells can be identified and killed by a subset of T cells called cytotoxic T lymphocytes (CTL). Unfortunately, of the traditional vaccine approaches available, only live attenuated vaccines are able to induce CTL and this approach is inappropriate for HIV. An additional complicating

TABLE 1. Approaches to vaccine development	
Traditional approaches	Successful vaccines
Live attenuated viruses	Mumps, measles, rubella, polio
Inactivated viruses	Influenza, hepatitis A
Bacterial toxoids	Tetanus, diphtheria, pertussis
Killed bacterial cells	Pertussis (not U.S.)
Protein/polysaccharide conjugates	Hib, meningococcus, pneumococcus
Recombinant proteins	Hepatitis B
Novel approaches	Examples
Recombinant attenuated viruses	Adenovirus, vaccinia, canary pox
Nonreplicating viruses	Alphaviruses
DNA vaccines	Bacterial plasmids

factor for HIV is that the virus actually infects and kills T cells, and it is the depletion of these cells that ultimately makes HIV-infected individuals susceptible to many diseases and results in AIDS.

Recombinant protein vaccine. An alternative approach to the development of vaccines, which was successful for hepatitis B virus, has been extensively evaluated for HIV. Recombinant deoxyribonucleic acid (DNA) technology allows the expression of antigenic proteins from a pathogen in flasks of cultured cells. The antigens can then be purified and used as a vaccine to induce antibodies. Unfortunately, the external glycoprotein of HIV (env gp120), when expressed as a monomeric protein, induced antibodies that could neutralize HIV grown under lab conditions but were unable to neutralize primary isolates of HIV that had been recently obtained from patients. HIV expresses a protective "cloud" of sugars on the viral surface, which helps to make it resistant to neutralization. Despite these limitations, the HIV env gp120 recombinant protein approach was evaluated in a phase III efficacy trial, completed in 2003. Unfortunately, the monomer protein vaccine did not offer any evidence of protection against HIV infection.

Overall it has become clear that the traditional approaches to vaccine development are unlikely to be effective for the development of an HIV vaccine. More complex approaches, involving new concepts and technologies, are required to stimulate more potent and diverse immune responses. The protective antigens that need to be included in an HIV vaccine are yet to be determined, as are the components of the immune response which should to be stimulated. Moreover, it is not yet clear if immune protection is actually possible.

Challenges in vaccine development and testing. HIV presents a number of unique difficulties for vaccine development.

Strain variability. HIV shows an extraordinary degree of variability, which is due to the variability inherent in its replication mechanism plus the immune pressure from the infected individual, which promotes viral evolution to escape the immune response. Moreover, while different HIV subtypes have predominated in different geographic regions, there is now increasing genetic mixing of viruses within and between the different subtypes. The degree to which an immune response against one strain of the virus could offer protection against diverse strains is not currently understood. Hence it is not known if it will be possible to develop a single vaccine that would be effective worldwide, or if subtype-specific vaccines will be needed in different geographic areas.

Mucosal transmission and immunity. Most viral infections occur initially at the mucosal sites of the body (in the genital, respiratory, or gastrointestinal tracts). Although some viruses remain confined to these sites, HIV is a virus that spreads throughout the body, usually following infection via the genital tract. Since HIV is mainly a sexually transmitted infection (it can also be transmitted directly into the blood through

contaminated needles), mucosal immunity at the site of infection may be required for protection. Traditional vaccine approaches do not induce mucosal immunity in the genital tract, and few vaccines exist to offer protection against sexually transmitted diseases. After initial infection, HIV briefly circulates at high levels in the blood before it becomes established in tissues. Hence, an effective HIV vaccine may be required to stimulate several arms of the immune system, including cellular and antibody-mediated immunity, in both systemic and mucosal tissues.

Genome integration and latency. While some viruses are cytopathic, causing the death of the cells they infect and resulting in viral clearance, HIV becomes stably integrated into the genome of infected cells. Moreover, HIV establishes an often latent ("silent") infection at immunoprivileged body sites, which may not be susceptible to immunological attack.

Absence of animal model. The absence of correlates of protection in humans is compounded by the absence of a reliable animal model in which to conduct preclinical studies. Although monkeys can be infected with viruses similar to HIV, they often show much more rapid disease progression. In addition, the mechanism of virus transmission in these studies is very different from natural HIV exposure. Unfortunately, it is impossible to mimic the real-life situations which result in transmission of HIV. Moreover, since it is not clear if the immune system can protect humans from HIV infection, a vaccine may be required to induce responses that are not normally induced during natural exposure.

Inadequate testing resources in developing nations. Clinical testing of vaccines is normally undertaken through distinct stages, called phases I, II, and III, involving increasingly large numbers of individuals as safety and efficacy are evaluated. The level of risk associated with phase III efficacy trials may be reduced by expanding phase II trials and by looking closely for correlates of protection. However, since the largest burden of HIV infection is predominantly in underdeveloped nations, future studies will need to focus in these areas, where local expertise and facilities are not always available to allow reliable determinations of immune responses.

Possibilities for HIV vaccine. Despite the many difficulties associated with making an HIV vaccine, a number of observations suggest that a vaccine might be possible. For example, live attenuated vaccines can protect monkeys against a virus closely related to HIV. In addition, passive administration of neutralizing antibodies to chimps protects them against HIV. Moreover, in humans already infected with HIV, cellular immunity correlates inversely with viral load, strongly suggesting that the immune responses can control the virus, often for extended periods. Experimental depletion of cellular immunity in infected monkeys results in reactivation of virus, offering further support for the control of the virus by the immune response. Intriguing observations have also been made in individuals who have been highly ex-

TABLE 2. Examples of novel HIV vaccines currently under clinical evaluation as single approaches or in prime/boost settings

Vaccine	Antigens	Clade*	Phase
DNA and modified vaccinia (strain Ankara)	gag epitopes	A	II
Canary pox and recombinant protein	env/gag/pol and env protein	B	III
Recombinant proteins	env protein	B, E	III (completed)
Adenovirus	gag	B	I
DNA	gag	B	I
Lipopeptides	gag/pol/nef epitopes	B	I
PLG/DNA and recombinant oligomeric protein	gag and env	B	I
Recombinant proteins	env, nef-tat	B	I
Venezuelan equine encephalitis virus (Alphavirus)	gag	C	I

*HIV is genetically diverse and is divided into six major clades that predominate in different geographic areas. The predominant clade in Europe and the United States is clade B, while clade A dominates in central Africa and clade C in southern and western Africa and in Asia.

posed to HIV but have resisted infection. In some individuals, it appears that immune responses in the genital tract induced by repeated virus exposure may offer protection, although this remains controversial.

Overall, although the correlates of protection against HIV have not been established, there is a consensus that an optimal vaccine candidate should induce potent cellular immunity, particularly involving cytotoxic T lymphocytes (which have the ability to kill virally infected cells), neutralizing antibodies (which are effective against wild-type viral isolates from diverse strains), and mucosal immunity in the genital tract. Nevertheless, it is clear that superinfection with HIV can occur, in which an already infected individual can be reinfected with a second HIV strain.

New approaches to vaccine design. A number of novel concepts in vaccinology are in early stages of clinical evaluation for HIV (**Table 2**). In general, these approaches have been shown to induce more potent immune responses than earlier-generation vaccines. These vaccine candidates have often been used in novel prime/boost combinations, which use different vaccine approaches in combination to promote optimal immune responses. The approach first became established using DNA vaccines as a prime, followed by boosting with recombinant live attenuated viruses. These live recombinant viruses are similar to traditional live attenuated vaccines, but employ much safer viruses that have been modified to express HIV antigens.

DNA vaccines. DNA vaccines comprise encoded antigens (DNA), which have been inserted into a bacterial plasmid (a circular stretch of DNA) designed to ensure maximal antigen expression in the body. Following vaccination, the DNA is taken up by cells at the injection site, which then express the encoded antigens. Expression of antigens by cells is thought to closely mimic what happens when an individual is actually infected with a virus. However, since only a few viral antigens are encoded by the plasmid, a whole virus cannot be produced and a true infection cannot result. DNA has significant potential for the development of a successful HIV vaccine because DNA is very potent at inducing cytotoxic T lymphocytes responses. DNA vaccines also have a number

of other potential advantages, as follows:

1. DNA vaccines may allow the induction of CTL activity without the use of a live attenuated virus.

2. They should have an improved safety profile compared with live attenuated vaccines.

3. They should be stable and inexpensive to manufacture.

4. Many antigens can be easily included in the vaccine.

5. Antigens will be produced by cells in the body, so they will be in the correct format to produce an optimal immune response.

6. Using DNA avoids the need to grow dangerous pathogens in culture.

7. DNA vaccines should not be affected by maternal antibodies unlike live attenuated viruses, allowing vaccination early in life.

8. Protective immunity has already been shown for DNA vaccines in many animal models of infectious disease.

9. DNA vaccines appear safe in early human clinical trials.

Nevertheless, in initial clinical trials DNA vaccines have proven only weakly immunogenic (capable of inducing an immune response). Therefore, a number of strategies have been employed to improve their potency, including the addition of adjuvants.

Adjuvants. Adjuvants, which are designed to enhance the immunogenicity of vaccines, can be in the form of delivery systems, which direct antigens into cells of the immune system, or immunopotentiators, which better activate these cells. A novel DNA delivery system that recently entered clinical trials for HIV comprises biodegradable polylactide co-glycolide (PLG) microparticles, which protect the DNA against degradation and promote its uptake into dendritic cells, which control the immune response. The most commonly used immunopotentiators for DNA vaccines are biological molecules called cytokines, including interleukin-2, -12, and -18, which can be administered as recombinant proteins or as plasmids to induce cytokine expression at the injection site.

DNA and modified vaccinia virus. DNA vaccines have commonly been used in conjunction with modified strains of vaccinia virus, which was previously used as a smallpox vaccine. It has been shown in clinical trials that prime/boost with DNA and attenuated viral vaccines was able to exert significant control over virus levels and to protect against disease progression in monkeys infected with viruses closely related to HIV. However, the vaccines did not protect against infection. In addition, follow-up studies in monkeys have shown that viral escape mutants can occur, resulting in a loss of control over the virus and progression to disease. Another drawback with this approach is that many of the viral vaccines can be inactivated by preexisting immunity in individuals who were previously exposed to the viruses through vaccination or natural means.

PLG/DNA and recombinant oligomeric env protein. An alternative prime/boost strategy currently under clinical investigation involves immunization with a DNA vaccine coadministered with an adjuvant (PLG/DNA), followed by boosting with a novel recombinant oligomeric env protein. This vaccine candidate is designed to induce both cellular and antibody-mediated immunity. This vaccine should induce more effective neutralizing antibodies, due to the use of an oligomeric protein with a more native structure, which is similar to how the protein appears on the virus. In addition, the protein has been modified to expose more conserved regions for antibody induction.

Future research. A number of novel vaccine candidates are in late-stage preclinical testing or have recently entered clinical trials. What should emerge from these studies is a clear picture of which technologies should be combined in novel prime/boost vaccine approaches. (Although it would be highly desirable if a single technology were to emerge as a candidate vaccine, this seems unlikely.) Most importantly, the correlates of protection against HIV and related viruses must be sought in preclinical studies. The evaluation of individuals who are repeatedly exposed to HIV but remain uninfected may provide valuable insights.

Potential for HIV immunotherapy. Once it is established that HIV vaccine candidates are able to induce potent immune responses, including cellular immunity, these vaccines may be considered for evaluation as immunotherapy. An immunotherapeutic vaccine would be designed to control disease in already infected individuals. An effective immunotherapeutic vaccine would have a number of advantages over current drug therapies. It would need to be administered only intermittently, unlike the complex daily regimens for current drugs, and it would be significantly less toxic than the current approaches. However, this concept awaits a clear demonstration that HIV can be controlled in humans by an immune response induced by a vaccine.

For background information *see* ACQUIRED IMMUNE DEFICIENCY SYNDROME (AIDS); CELLULAR IMMUNOLOGY; DEOXYRIBONUCLEIC ACID (DNA); PLASMID; RETROVIRUS; VACCINATION; VIRUS in the McGraw-Hill Encyclopedia of Science & Technology.

Derek T. O'Hagan

Bibliography. R.D. Klausner et al., The need for a global HIV vaccine enterprise, *Science*, 300:2036, 2003; N.L. Letvin, Strategies for an HIV vaccine, *J. Clin. Invest.*, 110:15, 2002; A. McMichael et al., Design and tests of an HIV vaccine, *Brit. Med. Bull.*, 62:87, 2002; P. Searman, HIV vaccine development: Lessons from the past and promise for the future, *Curr. HIV Res.*, 1:309, 2003.

Homo erectus

Late in the nineteenth century, the Dutch physician Eugene Dubois traveled as a military doctor to the Dutch East Indies (now Indonesia) in search of the fossil remains of human ancestors. He was following the predictions of Ernst Haeckel, who anticipated a prehuman ancestor that walked upright but did not speak ("Pithecanthropus alalus"), and surmised that the tropics would be the origin of modern humans. His search was rewarded in the 1890s when workers discovered the skull cap of an early human at the site of Trinil along the Solo River of Java. From the same locality, but perhaps younger in age, came upper leg bones as well. Dubois named the fossils *Pithecanthropus erectus*, upright ape-man.

In the 1920s and 1930s, a wealth of fossils similar to the Trinil cranial specimen were recovered in China and Indonesia, but few of these preserved the facial skeleton. The Chinese fossils were called *Sinanthropus pekinensis* (Chinese early man from Peking, now Beijing). Because of their anatomical similarity, *P. erectus*, *S. pekinensis*, and some fragmentary fossils from North and South Africa were merged into the single species *Homo erectus* in the mid-1900s. Beginning in the 1960s, a number of fossils were recovered from East Africa, especially at Olduvai Gorge in Tanzania and Koobi Fora in Kenya, that were likewise added to *H. erectus*. These fossils preserved more of the face and include key specimens, such as crania KNM-ER 3733 and 3883 from Koobi Fora; Olduvai Hominid 9 from Olduvai Gorge; and in the 1980s, the remarkably complete skeleton of a boy from West Turkana (also known as Nariokotome Boy, KNM-WT 15000). From this growing group of fossils and from finds of fossils of the earlier, more primitive genus *Australopithecus*, it became clear that *H. erectus* was much more humanlike than had been appreciated at the turn of the century. Today *H. erectus* fossils from Africa to Asia represent the first fossil humans with modern postcranial proportions. *Homo erectus* is considered to be ancestral to all later species of *Homo*, including *H. sapiens* and *H. neanderthalensis*.

Geological age and distribution. In the past 10 years, radiometric dating techniques, particularly the ^{40}Ar/^{39}Ar technique, have substantially revised the geological age of *H. erectus*. The earliest *H. erectus* fossils appear in Africa about 1.8 million years ago at Koobi Fora, Kenya. They also appear nearly this early in the Republic of Georgia (at Dmanisi about 1.7 million years ago) and in Indonesia (at Sangiran by 1.6 million years ago and at Perning about 1.8 million years ago). Previous claims for early remains from China have been refuted. The species seems to have persisted in Africa until a little less than 1 million years ago, but to have remained longer in Asia, until about 200,000 years ago in China and perhaps as late as 50,000 years ago in Indonesia. Historically important sites in China and Indonesia span large amounts of time; Zhoukoudian probably spans 200,000 to 300,000 years and the Sangiran dome at least 500,000 years. Other sites, such as the early Dmanisi and later Ngandong localities, appear to represent much shorter periods of time.

Anatomy. *Homo erectus* is usually defined on the basis of its skull anatomy. The brain of *H. erectus* is smaller, on average, than that of modern humans, with a mean capacity of about 1000 cm^3 (versus 1300 or 1400 cm^3 in humans), but with a large range from about 700 cm^3 in early Africa and Georgia to over 1200 cm^3 in later Indonesia. However, small-brained individuals exist at all times in all places (see **illus.**). Typically, the *H. erectus* skull has well-developed browridges, a less rounded and lower braincase than modern humans, thickened ridges along the top and back of the skull (that is, sagittal keels and occipital tori, respectively), a thicker skull wall, a receding chin, and relatively large teeth, as well as some peculiarities of the bottom of the skull.

There is both geographical and temporal variation in the cranial anatomy of *H. erectus*, with keels and tori being more strongly developed in Asia than in Africa, resulting in some scholars arguing for the separation of *H. erectus* into two species: *H. erectus* in Asia and a more primitive species, *H. ergaster*, in early Africa and Georgia. This division is complicated, however, by recent finds in Africa and Georgia that show Asian characteristics.

The postcranial skeleton of *H. erectus* is enlarged compared with earlier fossil human species, with longer lower limbs (probably indicating more efficient walking ability) and reduced length of the upper limb. There is also some evidence from the skeleton of *H. erectus* (the juvenile from Nariokotome, Kenya) that early African *H. erectus* may have been the first hominid to exhibit the long, narrow body proportions typical of tropically adapted modern human groups. This may mean that *H. erectus* was maintaining body temperature (thermoregulating) in much the same way that modern humans do, by sweating to lose heat. Although such a mechanism has the disadvantage of making the organism highly dependent on water, it is an effective way of avoiding heat stress during walking or long-distance running and during activity at peak midday temperatures. However, there appears to be a large range of body sizes, with the Georgian fossils being smaller (about 46 kg) than the early African fossils (about 58 kg), and it is currently unclear whether the same body proportions will be seen in all geographical regions.

Life history. There are few remains of *H. erectus* children, so it is difficult to compare their growth patterns with our own. Human development is typified by an extended childhood and an adolescent growth spurt, both of which allow extended time

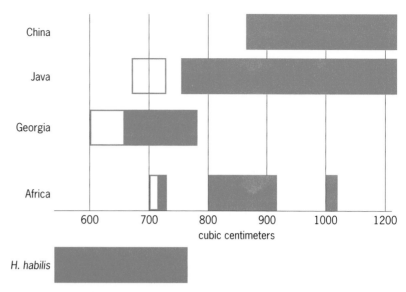

Homo erectus **brain sizes by region compared with** *Homo habilis* **(an early species). Open rectangles represent juvenile fossils, and filled rectangles are adult sizes.**

for learning and training the brain. Recent evidence suggests that the teeth of *H. erectus* developed and erupted faster than those of modern humans. On this basis, the Nariokotome boy, for example, is probably closer to 8 years than 11 years of age. Thus, *H. erectus* seems to have matured a little more quickly than modern humans.

Although average body size appears to have increased for both males and females relative to earlier human species, there is some evidence that the size difference between *H. erectus* males and females decreased. This may have been due to the increased difficulty and energetic costs to females of carrying and birthing large-bodied offspring. It is argued that these births and care of offspring required help from group mates, perhaps grandmothers; however, it is difficult to assess such hypotheses using the fossil record.

Dispersal. Since there are no earlier hominids known from outside Africa, the early appearance of *H. erectus* in Asia suggests a quick dispersal from Africa into western and eastern Asia. Such a quick dispersal is unusual for apes, who are strongly k-selected (have few offspring, in which they invest a lot of resources).

New foraging strategy. Several lines of evidence suggest that the key to the success of *H. erectus* may have been its ability to adopt a new foraging strategy that included larger amounts of protein (probably meat) in its diet. Just before the origin of *H. erectus*, paleoclimatic reconstructions suggest that Africa was becoming less forested and filled with more grasslands. Wooded areas provide most primary plant nutritional resources, such as fruit, leaves, bark, and tubers. In grasslands, primary productivity decreases, but the amount of resources available in the form of animals, so-called secondary productivity, increases. That is, more meat is available in more open areas. Animal products (meat, marrow, fat) are high-quality food items, and any animal that could take advantage of them might live successfully. Alternatively, animals that remained reliant on resources from the wooded areas would need to either reduce their body size to accommodate the decrease in resources or travel farther to obtain the same quantities of resources. *Homo erectus* grew larger, indicating that it had found a way to increase its dietary quality. Both large body size and high diet quality are correlated in primates with large home ranges (the area that an animal ranges over in a given time period, usually a year), and large home ranges are correlated with greater dispersal capabilities.

Tool use. Arguably, *H. erectus* became a habitual carnivore, and this foraging shift facilitated its speedy dispersal from Africa. However, carnivory required that *H. erectus* use tools, since it did not have large cutting and slicing teeth. These tools were the simple stone products of the Oldowan type, that is, cores and flakes. Core and flake tools have been found in the earliest *H. erectus* sites outside Africa, such as Dmanisi in the Republic of Georgia. Although underground tubers have been suggested as an alternative protein source to meat, these require heating to release nutrients, and the use and control of fire appears to have been a later acquisition.

The dispersal of *H. erectus* from Africa was entirely on foot. At the time of its dispersal, sea levels were low due to the presence of large glaciers in northern latitudes. For example, low sea levels exposed the relatively shallow continental shelf that united what are today the islands of southeast Asia with continental Asia. *Homo erectus*, with its long legs and simple tools, could have walked as far as the eastern edge of the island of Bali.

Evolution and fate. *Homo erectus* exists in the fossil record, at least that of southeast Asia, for more than 1.5 million years. During this time, average brain size appears to have increased slowly. In some regional groups, such as *H. erectus* in Indonesia, there is a common anatomical pattern that is persistent throughout the time range of *H. erectus*. There is also regional variation within Asia that differentiates Chinese *H. erectus* from Indonesian *H. erectus*, regardless of age. These differences suggest periodic genetic isolation of the Chinese and Indonesian populations, perhaps attributable to rises in sea level periodically separating Indonesia from the mainland or, possibly, glacial advance separating the Chinese populations from those farther to the south. It is likely that the late survival of *H. erectus* in Indonesia was facilitated by the biogeographic isolation of this group during periods of sea-level rise.

Unlike in Asia, *H. erectus* disappears in Africa by about 1 million years ago. Whether this species is called *H. erectus* or *H. ergaster*, it is commonly thought that this part of the species gives rise to later *H. sapiens*, whereas the Asian portions of the species continue to evolve separately, and are ultimately marginal to later human evolution. Also, unlike later in Asia, after about 1.5 million years ago in Africa, *H. erectus* begins to develop a more complicated stone technology, the Acheulean. This technology is typified by bifacial handaxes and cleavers and is often argued to be associated with big-game hunting behavior.

For background information *see* EARLY MODERN HUMANS; FOSSIL HUMANS; PALEOLITHIC; PHYSICAL ANTHROPOLOGY; PREHISTORIC TECHNOLOGY in the McGraw-Hill Encyclopedia of Science & Technology.
Susan C. Antón

Bibliography. S. C. Antón, Natural history of *Homo erectus*, *Yearbook of Physical Anthropology*, 37: 126–170, 2003; G. P. Rightmire, *The Evolution of Homo erectus: Comparative Anatomical Studies of an Extinct Human Species*, Cambridge University Press, New York, 1993; A. Walker and P. Shipman, *The Wisdom of the Bones: In Search of Human Origins*, Vintage Press, New York, 1997.

Human/machine differentiation

Networked systems are vulnerable to cyber attacks in which computer programs—variously called bots, spiders, scrapers, and so forth—pose as legitimate human users. Abuses have included defrauding financial payment systems, creating vast numbers of accounts for sending unsolicited email ("spamming"), stealing personal information, skewing rankings and recommendations in online forums, and ticket scalping, to name but a few.

Efforts to defend against these "automated impersonation" attacks over the last several years have triggered investigations into a broad family of security protocols designed to distinguish between human and machine users automatically over networks. When such a test authenticates a user as belonging to the group of all humans (and hence excluding all machines), it is called a CAPTCHA: a completely automated public Turing test to tell computers and humans apart.

In 1950 Alan Turing proposed a methodology for testing whether or not a machine effectively exhibits intelligence. He used an "imitation game" conducted over teletype connections in which a human judge asks questions of two interlocutors—one human and the other a machine—and eventually decides which of them is human. Turing proposed that if judges fail sufficiently often to decide correctly, then that would be strong evidence that the machine possessed artificial intelligence. CAPTCHAs are simplified, specialized "Turing tests" in the sense that they are administered by another machine, not by a human as in a real Turing test.

Currently, most commercial uses of CAPTCHAs exploit the gap in reading ability between humans and machines when confronted with degraded images of text. The first reading-based CAPTCHA was developed by a research team at AltaVista. When a potential user—who ought to be human but may be a machine—attempts to access a protected service, the CAPTCHA system generates a challenge which is shown to the user who must read and type the text on the keyboard (**Fig. 1**). It is easy in this case

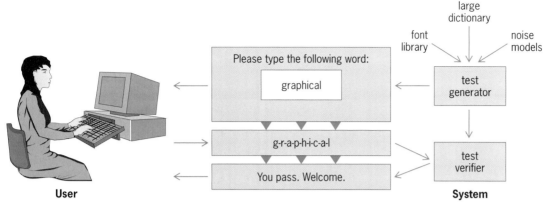

Fig. 1. Overview of a CAPTCHA protocol.

(indeed, trivial) to verify whether the user has passed the test.

With a large dictionary, a library of font styles, and a variety of synthetic noise models, a nearly endless supply of word images can be generated. Despite decades of research and commercial development, optical character recognition (OCR) systems are not robust enough to handle the wide range of degradations that can be created, and it seems unlikely that they will improve sufficiently to do so any time soon.

CAPTCHAs and the problems they are designed to address are not limited to graphical displays. Spoken language interfaces are proliferating rapidly and will soon play important roles in mobile devices and at times when a user wants his or her hands or eyes free for other tasks (for example, when driving a car). Bank-by-phone and telephone reservation systems are good examples in this domain. In such cases, visual tests are not appropriate, but it is still possible to exploit the gap in perception between humans and machines by turning to speech-based CAPTCHAs which combine synthesized speech, audio signal processing, and confusing background noise in much the same way that graphical CAPTCHAs distort word images.

Basic features of CAPTCHAs. The principal desirable properties of any CAPTCHA are:

1. The test's challenges should be automatically generated and graded.

2. The test should be suitable to be taken quickly and easily by human users (that is, the test should not go on too long).

3. The test should accept virtually all human users (even young or naive users) with high reliability while rejecting very few.

4. The test should reject virtually all machine "users" (note that, inevitably, any CAPTCHA can be broken, at some cost, by hiring a human to solve the challenge).

5. The test should resist automatic attack for many years even as technology advances and even if the test's algorithms are known (for example, if the underlying methods are published or released as open source software).

6. The challenges must be substantially different almost every time, otherwise they might be recorded, answered off-line by humans, and then used to answer future challenges (hence, they are typically generated pseudorandomly and not simply stored in a large static database).

Visual CAPTCHAs. The essential principles of visual CAPTCHAs can be illustrated with two recently published examples, PessimalPrint and BaffleText (**Fig. 2**). PessimalPrint is poor-quality but still (just) legible images of machine-printed text synthesized pseudorandomly over certain ranges of words, typefaces, and image degradations. Experiments at the Palo Alto Research Center (PARC) and the University of California—Berkeley showed that judicious choice of these ranges can ensure that the images are legible to human readers but illegible to three of the best present-day OCR software packages. This approach was motivated by a decade of research on the performance evaluation of OCR systems and on quantitative stochastic models of document image quality.

One key insight that made PessimalPrint possible was that within particular parameter settings for image degradation models the accuracy of OCR systems was consistently observed to be a smooth and approximately monotonic function of the parameters. This allowed the mapping (in the parameter space) of the "domain of competency" of machine reading skill, and the location of failure regions where with high reliability all the images that are

EZ-Gimpy

PessimalPrint

BaffleText

Fig. 2. Examples of degraded-text CAPTCHAs.

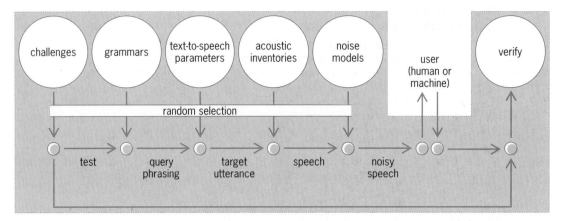

Fig. 3. Pipeline stages for an auditory CAPTCHA.

generated cause OCR machines to fail. Monotonicity also provides a technical basis for roughly grading the difficulty of the images for both human and machine readers.

BaffleText, another reading-based CAPTCHA developed at PARC, uses random masking to degrade images of non-English (but familiar, "pronounceable") character strings. The techniques underlying BaffleText were inspired by:

1. A critical analysis of weaknesses of earlier CAPTCHAs, in particular (*a*) the insecurity of natural language lexica (such as English word lists), which are too small and too well known to defend against a variety of pruning-and-guessing attacks (in which partly correct recognition results are used to generate a short list of possible intrepretations from within the known lexicon, one of which is chosen at random); and (*b*) the ease of restoring, by standard image processing techniques, many varieties of physics-motivated image degradations.

2. Knowledge of the state of the art of machine vision, especially its inability to solve a broad class of segmentation problems where parts of patterns are missing or spurious (requiring Gestalt perception for success).

3. Guidance from the literature on the psychophysics of reading, especially (*a*) the assistance that all forms of familiarity—not simply that an image is of a correctly spelled English word—give to human comprehension, and (*b*) an easy-to-compute image metric, "perimetric image complexity," that tracks human reading difficulty.

Experimental trials confirmed that perimetric image complexity correlates well with actual difficulty (the objective error rate) that people exhibit in reading CAPTCHAs. Also, perceived (subjective) difficulty was observed to correlate well with perimetric image complexity. These facts made possible the automatic selection, from among candidate images generated at random, challenges which are both legible and well tolerated by human users; they also have resisted attack by modern computer vision technology better than competing CAPTCHAs.

Auditory CAPTCHAs. Speech-based services (such as bank-by-phone, airline reservation systems, and customer relationship management) are proliferating because of their ease of use, portability, and potential for hands-free operation. Building a malicious "bot" to navigate a spoken language interface is a tractable problem, especially if there is a fixed sequence of predefined prompts. Hence, it is possible to anticipate attacks on such systems and a similar need to prevent machines from abusing speech-based resources intended for human users. Visually impaired users may also benefit from auditory CAPTCHAs, even when graphical tests are available.

Possible strategies for generating such CAPTCHAs include:

1. Making the dialog difficult for machines to process (exploiting high-level understanding, such as language syntax or semantics).

2. Making the speech signal difficult for machines to process (exploiting low-level understanding, such as recognizing phonemes).

3. Processing the speech to create aural illusions that affect humans but not machines (for a CAPTCHA to be effective, it is important only that humans and machines be distinguishable, not that humans be the ones to get the "right" answer).

As in visual CAPTCHAs such as BaffleText, it is possible to take advantage of the fact that certain pattern recognition tasks are significantly harder for machines than they are for humans. One approach that was investigated by researchers at Bell Labs is to use text-to-speech synthesis (TTS) to generate tests, thereby exposing the limitations of state-of-the art automatic speech recognition (ASR) technology. **Figure 3** presents a general overview and an indication of the wide range of parameters available to the builder of such CAPTCHAs.

An analysis of 18 different sets of distortions demonstrated that there are a variety of ways to make the problem hard for machines. In general, these experiments demonstrated that humans can handle noise levels about 15 dB higher than ASR can. Likewise, humans can understand spoken digit strings when more than half of the signal is missing (for example, replaced with white "noise"), a point at which the machine already has a 100% error rate.

Moreover, when humans make mistakes transcribing such speech, the errors they make are fundamentally different from the kinds of errors a machine would make, a feature that can be employed to improve discrimination.

Ongoing research. To date, CAPTCHAs have been used to protect dozens of commercial Web services, including those offered by AltaVista®, Yahoo!®, PayPal®, Microsoft®, and TicketMaster®. New applications arise on a daily basis. Still, this technology is in its early stages of development. Important questions that have not yet been fully investigated include:

1. How can successful attacks on CAPTCHAs be automatically detected?

2. How can CAPTCHAs be made easy and inoffensive enough to be tolerated by a wider audience?

3. How can CAPTCHAs be parametrized so that their difficulty, for humans as well as for machines, can be controlled?

For background information *see* CHARACTER RECOGNITION; COMPUTER SECURITY; COMPUTER VISION; IMAGE PROCESSING; INTELLIGENT MACHINES in the McGraw-Hill Encyclopedia of Science & Technology.   Daniel Lopresti; Henry Baird

Bibliography. H. S. Baird and M. Chew, BaffleText: A human interactive proof, *Proc., 10th IS&T/SPIE Document Recognition & Retrieval Conf. (DR&R2003)*, Santa Clara, CA, Jan. 23–24, 2003; M. Blum, L. A. von Ahn, and J. Langford, *The CAPTCHA Project: Completely Automatic Public Turing Test To Tell Computers and Humans Apart*, Department of Computer Science, Carnegie-Mellon University, Pittsburgh; A. L. Coates, H. S. Baird, and R. Fateman, PessimalPrint: A reverse Turing test, *Proc., IAPR 6th Int. Conf. Document Analysis and Recognition, Seattle*, pp. 1154–1158, Sept. 10–13, 2001; G. Kochanski, D. Lopresti, and C. Shih, A reverse Turing test using speech, *Proc. 7th Int. Conf. Spoken Language Processing, Denver*, pp. 1357–1360, September 2002; M. D. Lillibridge et al., Method for Selectively Restricting Access to Computer Systems, U.S. Patent No. 6,195,698, issued February 27, 2001; A. Turing, Computing machinery and intelligence, *Mind*, 59(236):433–460, 1950; L. von Ahn, M. Blum, and J. Langford, Telling humans and computers apart automatically—How lazy cryptographers do AI, *Commun. ACM*, 47(2):57–60, February 2004.

Human-powered submarines

A human-powered submarine is typically a wet (flooded) underwater vehicle that is pedal-powered by one or two occupants (**Fig. 1**). The occupants, in most cases, breathe from SCUBA gear mounted inside the submarine. Human-powered submarine competitions are held regularly by two organizations: the International Submarine Races (ISR) held in odd-numbered years in the David Taylor Model Basin at the Carderock Division, Naval Surface Warfare Center in Bethesda Maryland; and the Human-Powered

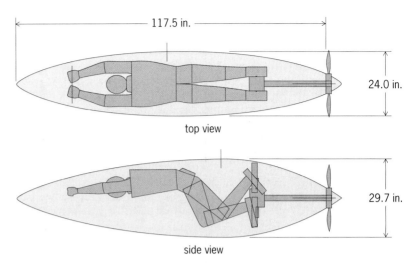

Fig. 1. Typical one-person human-powered submarine, the Virginia Tech *Phantom III*, 1 in. = 2.5 cm.

Submarine Contest (HPS) sponsored by the San Diego section of the American Society of Mechanical Engineers and held at the Offshore Modeling Basin in Escondido, California (**Fig. 2**) during even-numbered years. The first race was held in 1989 by the ISR at Riviera Beach, Florida, with the stated objective of fostering advances in hydrodynamics, propulsion, and life support systems of subsea vehicles. It was held in the open ocean, where submarines raced twice around a closed-circuit course at about 23 ft (7 m) depth. During the ensuing years, battles with the elements, both above and below the surface, and problems with directional control have transformed the race to a straight-line sprint against the clock in an indoor tank at depths of 22 ft or 6.7 m (ISR) or 15 ft or 4.5 m (HPS). The races today are essentially a collegiate competition among engineering students, providing them with an opportunity to turn some of their ideas into reality and learn the practical

Fig. 2. Human-powered submarine at the Off-shore Modeling Basin in Escondido, California. (*Escondido North County Times*)

difficulties involved. There are speed awards for one-person and two-person propeller-driven submarines and one- and two-person alternative-propulsion (nonpropeller) submarines as well as awards for design, innovation, operation, and safety. The fastest submarine to date is *Omer 4* from the École de Technologie Supérieure in Montreal, Canada, which was clocked at 7.192 knots (8.285 mi/h or 13.3 km/h) during the 6th ISR races in 2001.

Overall considerations. Despite the fact that the design point for most such submarines is a steady top speed, today's races place an emphasis on acceleration. The submarines are ballasted to be neutrally buoyant by adding weight in the bottom or foam at the top. This means that, in operation, they are moving a mass equivalent to that of the water displaced by their volume. The smaller the volume of the submarine, the smaller the mass that must be accelerated to speed. These submarines displace a significant amount of water—1400–1500 lb (635–680 kg) for a one-person sub to around 2000 lb (900 kg) or more for a two-person submarine. In the smaller tank of the HPS races, contestants have a distance of only about 165 ft (50 m) to accelerate before entering the 32.8-ft (10-m) long timing zone. Smaller submarines, tightly fitted around the required interior volume, that accelerate more quickly with a given thrust have been the fastest in that tank. A recent trend has been to have both occupants of a two-person submarine provide power. Even though the two-person submarine is larger than the one-person submarine, with two propulsors one can achieve a larger amount of thrust per unit displacement.

A second major consideration is control. Just as the ability to control an aircraft led to the success of the Wright *Flyer*, the ability (or lack of ability) to control direction has a direct bearing on the success of a human-powered submarine design.

Hull construction. Hulls are now usually fabricated from fiberglass using either a foam-core sandwich or single-skin approach. Carbon fiber, Kevlar®, and Nomex® also have been used. Although not nearly as common, wood and Lexan® have been used as hull materials. The construction of a fiberglass or other composite hull begins with the construction of a mold of one-half of the hull. Layers of fiber are laid-up on the mold using a vacuum bagging technique to ensure thorough resin penetration throughout the fiber and close conformity to the mold. Once two halves have been produced, they are joined to form the hull. Female molds produce a much smoother exterior surface that requires much less finishing work. Hatches may be cut from the molded hull or formed in a separate process. The fabrication of a transparent nose is particularly challenging for the student teams. Approaches to this have included a single heat-formed plastic piece, smaller heat-formed pieces joined together, milled and polished domes from larger glued-up pieces, and even flat pieces joined in pyramid type shapes.

Propulsion. The vast majority of submarines use propellers to produce thrust. Both single propellers and counterrotating pairs of propellers have been used. The small efficiency gain of the counterrotating props is of little consequence to the operation of these submarines; however, the elimination of the torque on the submarine (the torque on one propeller is canceled by the opposite torque on the other propeller) can be a great benefit to control. It is difficult to stay on course if the sub rolls over 30 or 40° under the torque applied to the propellers during initial acceleration. Single-propeller submarines need a large amount of ballast weight in the bottom to supply the righting moment necessary to counter this torque. The downside of counterrotating propellers is the added mechanical complexity. Because these submarine propellers are reasonably lightly loaded, the most efficient designs look more like aircraft propellers than marine screws. It is possible to approach 90% design propulsive efficiency at speed.

In an effort to increase efficiency at lower speeds, variable-pitch propellers have been used. *Omer*, mentioned above, used an electronic microprocessor control to vary the pitch during a run in a preprogrammed manner.

Although rotary-pedal motion is usually employed, linear-stroke mechanisms have been used successfully as well. No clear advantage has been shown for one over the other. Early submarines were nearly all chain-driven, but because of the losses inherent in moving a chain through water, most of today's submarines employ a gearbox to turn a long propeller shaft. In 2000, Virginia Tech's *Phantom III* (Fig. 1) introduced a gearbox filled with air and kept dry with the use of a regulator that maintained the pressure slightly above the ambient water pressure. This eliminated even the drag of the gears running through water.

Alternative propulsion submarines have often used articulating fins and waterjets. Most have had very limited success. The entry of the University of California at San Diego at the HPS 2000 races, *SubSonic*, successfully used an oscillating hydrofoil for propulsion. At the 7th ISR races in 2003, Virginia Tech, in an innovative interpretation of the definition of a human-powered sub, introduced *Spectre*. It used a monofin (large SCUBA fin that both feet fit into) on the single occupant whose lower body was enclosed in a fabric skirt that was attached to a rigid cowling covering the pilot's upper body and SCUBA gear. It reached a speed of 3.52 knots (4.05 mi/h or 6.52 km/h).

To make further progress in propulsion system design, more information needs to be known about the effective power output of a human pedaling underwater. Early testing indicated a sustained output of 0.17 hp (127 W) from an extremely fit propulsor. If one optimistically estimates the drag coefficient of a submarine to be 0.1 and the overall propulsive efficiency to be 85%, a simple calculation indicates that the propulsor needs to produce 0.36 hp (269 W) to drive the submarine at 7 knots (8 mi/h or 13 km/h). Informal testing at Virginia Tech has confirmed an output of approximately 0.33 hp (250 W) from a propulsor in a prone position underwater. Instrumentation in future submarines will confirm this power output.

Control. Control is nearly universally accomplished through the use of external fins. Designers size the control surfaces to strike a balance between adequate control forces and the added drag of these appendages. Control difficulties are common at low speeds, where these fins are less effective. It is during the initial acceleration stage, however, when the largest course corrections are often needed. Common fin arrangements have included rear rudder (vertical surface) and elevator (horizontal), fixed midhull stabilizers and, forward canard wings. Actuation of the surfaces is commonly done with mechanical cable systems controlled by the pilot, although pneumatic systems are beginning to appear. In a one-person submarine, this puts significant additional workload on the pilot, who is also trying to concentrate on pedaling as hard as possible. At the HPS 2002 races, the Virginia Tech team introduced an electronic control system that was joystick-activated and included a microprocessor control of pitch and depth employing pressure sensors in the bow and stern. At the HPS 2004 races, the plan is to add directional control, as well the use of an electronic compass. *See* SUBMARINE HYDRODYNAMICS.

Life support and safety. The safety of everyone involved takes the highest priority at every submarine race. SCUBA tanks are mounted in the submarines, and air consumption is closely monitored. Propulsors use high-flow regulators, and all submarine occupants must have spare air canisters that can be carried with them when they exit the submarine. Entry hatches can be operated from either inside or outside the submarine in case rescue is needed. Propulsors need to push against some part of the submarine in order to put force on the pedals. In a one-person submarine, the pilot also needs to have his or her hands free to operate the controls. This necessitates the use of a harness or other pilot restraint with the capability of being quickly and easily released by either the pilot or an outside rescuer. It is a tribute to the many safety measures taken that there has never been a significant accident in a human-powered submarine competition.

For background information *see* PROPELLER (MARINE CRAFT); SHIP, POWERING, MANEUVERING, AND SEAKEEPING; SUBMARINE; UNDERWATER VEHICLE; WATER TUNNEL in the McGraw-Hill Encyclopedia of Science & Technology.   Wayne L. Neu

Bibliography. N. R. Hussey and J. F. Hussey, Fifth International Submarine Races: Encouraging Student Engineers, *Sea Technol.*, 38(9):49–54, September 1997; K. McAlpine, Bobbing along with the sub geeks, *Aqua*, PADI Diving Society Magazine, 2(3): 66–73, July 1999; S. L. Merry, Human-powered submarines: The second generation, *Sea Technol.*, 32(9):35–45, September 1991; *Oceans 2003 Proc.*, San Diego, Sept. 22–26, 2003.

Humans and automation

The term automation originated during the 1950s to refer to automatic control of the manufacture of a product through a number of successive stages. It subsequently was expanded to mean the application of automatic control through the use of electronic or mechanical devices to replace human labor in any branch of industry or science. Replacing physical labor was of principal concern decades ago, but today, at least in the developed nations, the goal is to replace mental labor. Computers and the accompanying sensors that interpret inputs, record data, make decisions, or generate displays to give warnings or advice to humans are now also regarded as automation, though in the strict sense none of these latter functions may be complete automatic control.

Human factors, also known as human engineering, or human factors engineering, is the application of behavioral and biological sciences to the design of machines and human-machine systems. The most relevant behavioral sciences are cognitive psychology (hence the term cognitive engineering) and the broader field of experimental psychology (the study of sensation, perception, memory, and thinking as well as motor skill). To a lesser extent, organizational and social psychology, sociology, and psychometrics are included, but not clinical psychology. The most relevant biological science is human systems physiology (the study of organ functions above the cellular level). Often the term ergonomics (literally "work study") is used as a synonym for human factors, but usually it refers to anthropometry, biomechanics, and body kinematics, as applied to the design of seating and workspaces. In that sense, the term is less appropriate to the interaction of humans with automation.

Supervisory control. It may seem strange to speak of human interaction with automation because of a common belief that systems are either automated or they are not. However, the fact is that people in all walks of life are interacting more and more with automation, often without even knowing it. Typically the human interacts with a machine through some display-control interface, observing a display of computer-generated information of key variables and how they change over time, and choosing actions with control devices such as buttons and knobs. The machine executes the task using its own electromechanical sensors and actuators, displaying its progress and giving the human a chance to interrupt or override its actions. This new role for the human, where he or she is not actually doing the task but is instructing a machine when and how to do it and monitoring the machine's actions, has come to be called supervisory control. The human is a supervisor of a lower-level intelligent machine, in the same sense that a human supervisor instructs and monitors the actions of human subordinates.

Aviation. Aviation, including both aircraft and air-traffic control systems, involves probably the most sophisticated examples of human-machine interactions. First, there were autopilots, enabling the pilot to command the aircraft to maintain a certain heading, speed, altitude, or combination of these. Now, autopilots have evolved to the point where the pilot can instruct the aircraft to take off automatically at a given airport, fly to a given city, and land on a specified runway—all without any further intervention by the human pilot. Moreover, the autopilot has been integrated into a larger computer system, called the flight management system. This system provides navigation advice, weather information and advice on how to fly through it, status information on engines or any other aircraft subsystems, warning and alarm information, and more. The pilot has truly become a manager of a very sophisticated flying robot that can not only obey commands but also give advice when requested (and in some cases unsolicited). Modern military aircraft have become even more robotic. The human pilot of an unoccupied aeronautical vehicle (UAV) stays on the ground and flies one or multiple aircraft by remote (supervisory) control through a radio link.

Air-traffic control uses a variety of automatic systems that collect information about aircraft flight plans, aircraft positions, and weather, then displays this information to human controllers in airport towers and radar rooms. Other automated systems assign "slots" to different flights before they take off and plan spacing of aircraft as they enter the terminal areas around destination airports, but these are only advisory suggestions that the human controller must approve.

In rare instances, automation is the final authority. In some commercial aircraft, if the human pilot commands a combination of speed, pitch angle, and bank at an altitude such that the aircraft would stall, the computer overrides the pilot's command and prevents the stall. Another automated system detects and warns of another aircraft that might be on collision course. If the pilot does not properly respond, the system will automatically force the aircraft to change altitude and direction to avoid a collision.

Space flight. Spacecraft and planetary rovers are telerobots, commanded and monitored from distances so great that even radio communication at the speed of light poses time delays of seconds (to the Moon) to hours (planets in deep space). Because of such delays, continuous control would be unstable. Therefore, the telerobot must function autonomously for the period required for it to execute any subtask, and for data to be fed back to Earth, interpreted, and acted upon.

Other examples. Human-automation interaction is increasingly occurring in many other settings as well. In modern manufacturing plants, for example, it may appear that robots are assembling the automobiles, plugging chips into circuit boards, and performing similar functions, but behind the scenes humans are continuously monitoring the automation and programming new operations. Machines in nuclear and chemical plants and anesthesia and related machines in hospitals are other examples.

Allocation of tasks. Many tasks cannot currently be automated and must be done by humans. Other tasks are done by humans simply because they like to do them. Still others are done by humans because the tasks are trivial, or are embedded in other tasks, and therefore are inconvenient to automate. When a task is boring, fatiguing, or hazardous for a human, it should be automated if possible. Otherwise, when a task can be done by either human or machine, it is worth considering which should do it or whether the two might work together and complement one another.

Currently there are many acceptable techniques for analyzing tasks, but there is no commonly accepted optimal way to allocate tasks between human and machine. There are two main reasons for this: (1) while tasks may indeed be broken into parts, those parts are seldom independent of one another, and task components may interact in different ways depending upon the resources chosen for doing them; and (2) criteria for judging the suitability of various human-machine mixes are usually difficult to quantify and seldom explicit.

Automation need not apply uniformly to entire tasks but can be of benefit to some parts more than other parts. A helpful taxonomy that applies to most complex human-machine systems is the sequence of operations shown in **Fig. 1**: (1) acquire information; (2) analyze and display; (3) decide action; and (4) implement action. Below each box in the figure are some typical variables relevant to that stage. **Figure 2** presents a seven-level scale of degrees of automation. The point of these figures is that

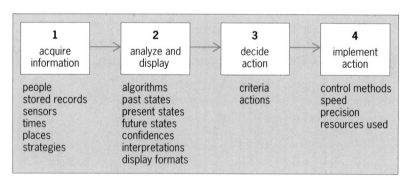

Fig. 1. Four stages of a complex human-machine task.

1. The computer offers no assistance: the human must do it all.
2. The computer suggests alternative ways to do the task.
   The computer selects one way to do the task, and:
3. executes that suggestion if the human approves, or
4. allows the human a restricted time to veto before automatic execution, or
5. executes automatically, then necessarily informs the human, or
6. executes automatically, then informs the human only if asked.
7. The computer selects the way, executes the task, and ignores the human.

Fig. 2. Scale showing degrees of automation.

different kinds of tasks call for different levels of automation at different stages. In air-traffic control, information acquisition (such as radar, weather, and schedule information) is, and should be, highly automated. However, analysis and decision making, except for certain computer-based aids now under development, should be done by human air-traffic controllers. Implementation is in the hands of the pilots, which in turn is largely given over to autopilots and the other parts of the flight management system. In other systems—for example, in market research—the information acquisition is by humans, the analysis and decision making is automated, and the implementation of the marketing itself is a human function. This is almost the exact opposite profile from air-traffic control.

Sometimes different interrelated variables of the same system can be controlled simultaneously by human and machine (called sharing the task). For example, when an automobile is in cruise control the computer controls the car's speed but the human continues steering. Human and computer can also control a single variable alternately (called trading). This is a useful technique if the human needs relief; for example, speed control in an automobile may be handed off to a cruise control system so the driver can rest his or her foot, but the human must regain control when coming too close to the car ahead. Sharing and trading strategies are sometimes called adaptive automation.

Potential problems. Automation may perform a task quicker and more reliably than if a human did it without any help. Yet, it is not true that automation necessarily relieves the human of participation, especially of mental effort. Humans are expected to be safety backups for automated systems when there are significant economic and safety risks to system failure. Even so, humans are not good at monitoring the performance of automatic systems, especially over periods longer than 30 minutes. They get inattentive and bored. Also, it is not easy to understand what complex automation is doing, what it will do next, and whether it is beginning to fail. Machines are not always good at telling human operators what they are "thinking." These problems pose significant challenges as automation is developed to take over more and more important human tasks.

For background information see AIR NAVIGATION; AIR-TRAFFIC CONTROL; AIRCRAFT COLLISION AVOIDANCE SYSTEM; AUTOMATED DECISION MAKING; AUTOMATION; AUTOPILOT; HUMAN-COMPUTER INTERACTION; HUMAN-FACTORS ENGINEERING; HUMAN-MACHINE SYSTEMS; REMOTE MANIPULATORS; ROBOTICS in the McGraw-Hill Encyclopedia of Science & Technology.   Thomas B. Sheridan

Bibliography. A. Degani, *Taming Hal*, Palgrave (Macmillan), 2004; G. Salvendy (ed.), *Handbook of Human Factors and Ergonomics*, Wiley, 1997; N. Sarter and R. Amalberti (eds.), *Cognitive Engineering in the Aviation Domain*, Lawrence Erlbaum, 2000; T. Sheridan, *Humans and Automation*, Wiley Interscience, 2002.

Hurricane-related pollution

Soil on which municipal, industrial, and agricultural waste has been deposited by storms, particularly hurricanes, can adversely affect local and regional water, air, and soil quality. Soil pollution can directly affect public health through human contact with deadly pathogens. In addition, it can lead to air pollution which causes pulmonary congestion, disorientation, altered breathing, and headaches in humans, as well as decreased crop yields. Historically the focus of research and public concern was to understand and control waste-loaded water systems, whereas the focus now is to understand and control waste-loaded soil systems.

There has been motivation to learn as much as possible about waste-loaded soil systems after Hurricane Floyd struck eastern North Carolina in 1999. During this 36-hour storm, massive quantities of untreated municipal and agricultural wastewater were washed into rivers and subsequently deposited onto flood-plain soils in amounts never previously experienced. During the hurricane and its 5-day aftermath, about 2.8 billion gallons (10.6 billion liters) of municipal wastewater and 8 billion gallons (30 billion liters) of swine waste from 500 storage lagoons were released.

Waste-loaded soils. Nitrogen is one of the many pollutants in the hurricane wastewater that is deposited on soil. From within waste-loaded soil, nitric oxide (NO) gas is produced and transported to the lower troposphere where it serves as a precursor to the air pollutant ozone (O_3). The basic chemical, microbiological, and physical factors responsible for the production and transport of the nitrogen oxides (NO_x) in natural and chemically fertilized soils are suggested in the literature. However, nothing is known about the complex nature of the flux of nitric oxide from soil loaded with municipal and agricultural waste as the result of natural storm disasters, and little is known about its contribution to ozone formation.

The composition and characteristics of waste-loaded soils are very complex and different from chemically fertilized soils in terms of (1) soil microbiology including microbial communities and organic content; (2) soil chemistry including form, availability, and transformation of nitrogen; and (3) soil physics including soil porosity, pore size distribution, and NO transport within the soil matrix. Thus nitrogen transformations and transport, particularly NO production and transport through these waste-loaded soils, are highly unpredictable. It is erroneous to assume that this NO source will perturb the biogeochemical cycle in the same manner as NO oxide emissions from ordinary agricultural land.

Case study. A study was undertaken recently to gain a better understanding of the magnitude of NO emissions from waste-loaded soils from hurricanes in eastern North Carolina's Neuse River basin to gain an appreciation of the factors contributing to those emissions, subsequent ozone formation, and potential controls. Prior to Hurricane Floyd, field tests of a variety of agricultural soils in eastern North Carolina indicated that NO moves from the soil to the lower levels of the troposphere and contributes directly to the formation of ozone.

Storm pollution. Hurricane Floyd impacted the Neuse River basin by inducing severe flooding which is estimated to adversely impact water quality parameters. Municipal waste-water treatment facilities serve about 175,000 people in the Neuse River basin, and flooding and power outages are estimated to have added 1.5–9.1 billion gallons (5.7–34.4 billion liters) of untreated waste-water into the river system. Approximately 31,000–180,000 lb of nitrogen (0.4 mg/L), 7000–42,000 lb of phosphorus (0.1 mg/L), and 700,000–4,200,000 lb of total solids (9.2 mg/L) are estimated to have entered the basin from these municipal sources. Hog lagoons for about 2 million swine were inundated with rain and floodwaters allowing an estimated 14.25–85.5 million pounds of untreated hog waste to enter the Neuse River basin. Approximately 720,000–4,300,000 lb of nitrogen (9.4 mg/L), 240,000–1,400,000 lb of phosphorus (3.1 mg/L), and 14,500,000–88,000,000 pounds of total solids (190 mg/L) are estimated to have entered the Neuse River Basin due to hog waste lagoon flooding.

Soil emissions. Waste-flooded soils were found to produce over 30 times greater NO emissions than nonflooded soils, with NO fluxes ranging from 0.1 to 102.5 $ng-N/m^2s$. The NO flux measured from soil samples collected from the downstream regions of the Neuse River basin was observed to be significantly higher than from upstream soils even months after the flooding occurred. The mean (95% confidence) NO flux from the Wake Forest, Kinston, and New Bern soils was 0.5, 17.5, and 77.0 $ng-N/m^2s$, respectively. These findings suggest the increased nitrogen loading from waste-loaded floodwaters enhanced the NO emissions from the downstream areas of the Neuse River Basin. No historical data were available to distinguish the incremental effects associated with flooding due to individual storm events.

Atmospheric chemistry. Whether increased NO emissions lead to increased ozone pollution depends on other important atmospheric variables, notably the NO_x/volatile organic compound (VOC) ratio. Eastern North Carolina and other parts on the United States appear to be NO_x-limited, with VOCs abundant in the lower levels of the troposphere, suggesting that an increase in NO emissions from the soil into the troposphere will increase ozone concentrations. Thus any increase in NO in the lower troposphere that leads to an increase in the concentration of photochemical oxidants, particularly ozone, will adversely affect the health of humans, plants, and animals. In the absence of natural storm disasters, the Office of Technology Assessment estimates that high O_3 concentrations cost the United States $1–5 billion annually in terms of human health care and reduced crop yields. This impact is not limited to the United States, as 10–35% of the world's total grain is produced in regions of the northern midlatitudes, where nitrogen fertilizer is applied to the soil and ozone concentrations are high enough to decrease crop yields. Nitric oxide–induced ozone pollution is global in nature, and local events caused by natural storm disasters are potentially of even greater concern.

Outlook. Continued study is required of the effects of soil characteristics and soil contaminants on nitric oxide flux, the impact of nitric oxide on air quality, and the agricultural value of nitrogen lost to receiving water bodies and to the airshed as nitric oxide. In addition, the influence of varying types of soil on ammonia adsorption and nitric oxide emissions needs further investigation. Improved management of the disposal of animal waste stored in lagoons in hurricane-prone areas is needed.

For background information *see* ATMOSPHERIC CHEMISTRY; BIOGEOCHEMISTRY; ENVIRONMENTAL ENGINEERING; HURRICANE; NITROGEN OXIDES; SOIL CHEMISTRY in the McGraw-Hill Encyclopedia of Science & Technology.   J. Jeffrey Peirce

Bibliography. *Hurricane Floyd Beats Out Fran with Record Flood Levels on N. C. Rivers*, U.S. Geological Survey, Sept. 21, 1999, Raleigh, NC, 1999; J. J. Peirce and V. P. Aneja, Nitric oxide emissions from engineered soil systems, *J. Environ. Eng.*, 127(4): 322–328, 2000; J. J. Peirce et al., *Environmental Pollution and Control*, Butterworth-Heinemann, Boston, 1998; B. E. Rittmann and P. L. McCarty, *Environmental Biotechnology*, McGraw-Hill, New York, 2001; C. L. Sawyer et al., *Chemistry for Engineers*, McGraw-Hill, New York, 1994; R. W. Tabachow et al., Hurricane-loaded soil: Effects on nitric oxide emissions from soil, *J. Environ. Qual.*, 30:1904–1910, 2001; *USGS Scientists Tracking Environmental Damage from Floyd... Heavy Flooding Caused Heavy Pollution*, U.S. Geological Survey, Sept. 23, 1999, Reston, VA, 1999; I. Valiela et al., Ecological effects of major storms on coastal watersheds and coastal waters: Hurricane Bob on Cape Cod, *J. Coastal Res.*, 14(1):218–238, 1998.

Integrated electric ship power systems

Electric propulsion for ships is not new. It has been employed since the early 1900s as a more efficient, lower-maintenance replacement for the traditional mechanical reduction gear used to link the relatively high rotational speed of ships' turbine and diesel engines and the much lower propeller speeds. What is new, however, is the concept of the integrated power system. In this arrangement, all of a ship's power is provided by engines driving generators that feed a common distribution system—as opposed to the more traditional arrangement, in which there are separate engines and transmission systems for propulsion and for auxiliary (lighting, elevators) or hotel (kitchen, housekeeping) loads. This is sometimes referred to as the central power station concept. The integrated power system permits greatly improved efficiency because engines can be optimally

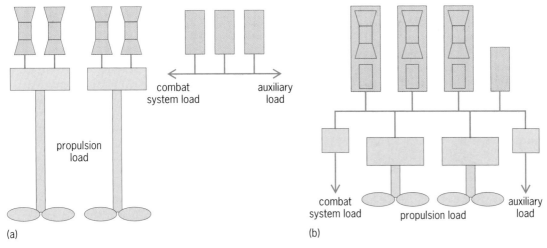

Fig. 1. Comparison of segregated and integrated power systems. (*a*) Segregated power system with seven prime movers. Turbines propel the ship directly, and additional fuel-burning engines generate electricity for combat and auxiliary loads. (*b*) Integrated power system with four prime movers. All engines generate electricity into a common distribution bus.

loaded for any given operating condition, and over the full operational profile (**Fig. 1**). Reliability is also improved because any engine/generator set can be used to power any electrical load on the ship.

All-electric ship. The all-electric ship attempts to optimize the total ship system by eliminating as many steam, pneumatic, and hydraulic systems as possible, replacing them with electrically driven components. In addition to improving operational flexibility, this step promotes full automation, remote manning, and reconfiguration to accommodate fault conditions. Almost all cruise ships built during recent years have employed integrated power systems. These applications exploit the flexibility of the all-electric ship for better space utilization and improved passenger comfort. Naval ships can benefit from these same characteristics both to enhance performance and to add new capabilities that might otherwise be unachievable or too costly.

All-electric naval ships. Electrification of naval ships offers the potential for even greater benefits than those realized by cruise ships, but it presents much greater challenges. Combatant ships require high levels of installed power for propulsion and combat systems, but they are volume- and weight-constrained. This has led to intensive efforts to improve the power density of electrical systems so that the space constraints of the smaller hull forms can be accommodated. Considerable attention has been paid to recent developments in permanent magnet and superconducting materials that offer potential for significant size and weight reductions in rotating machines. Of perhaps greater significance have been the advances in solid-state power semiconductor devices that operate at high voltages with low losses and switch at kilohertz speeds. For example, devices based upon silicon, such as the insulated gate bipolar transistor (IGBT) that can switch several

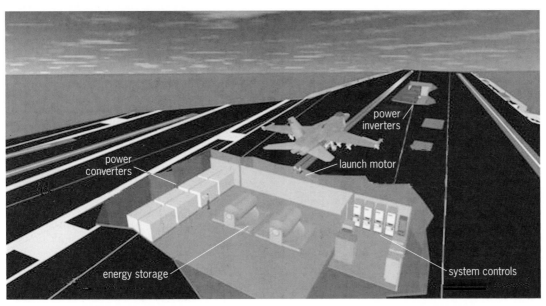

Fig. 2. Electromagnetic Aircraft Launch System (EMALS) as it would appear on an aircraft carrier flight deck.

thousand volts at several kilohertz frequency, are the components employed in most systems today. In the future, devices using wide-band-gap materials such as silicon carbide, and perhaps cryogenically cooled devices, promise to extend these operating ranges by greater than an order of magnitude. These devices, which can be installed in convenient locations, are the keys to creating the high-efficiency, compact power conversion equipment required for future naval applications. In the near term, naval ships require highly survivable and reconfigurable systems that can operate in a variety of mission scenarios. In the long term, directed energy systems for both defensive and offensive application will be employed to meet emerging mission needs.

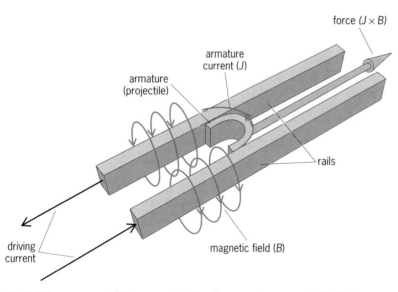

Fig. 3. Rail gun concept. Driving current in the rails generates a magnetic field that provides force to launch the projectile.

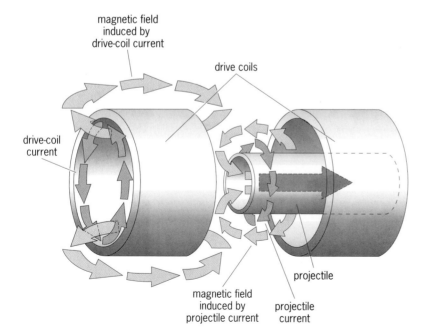

Fig. 4. Coil gun. Currents in the coil induce an opposing current in the projectile. The magnetic fields generated repel each other, which launches the projectile.

Projects in progress. The U.S. Navy is presently designing the first electric ship class for combat duty. These ships will employ an integrated power system with zonal distribution so power can be directed to vital loads from multiple sources. DC distribution is employed when required for these functions, although the main distribution bus is an AC system. In the future, an all-DC system could be employed. This would permit distributed energy storage units to be strategically placed for emergency power when all other sources are lost. These and other power converters would lead eventually to a system, called an integrated fight-through power system, that would automatically reconfigure to permit a ship to continue to defend herself after suffering combat damage, and possibly continue to execute assigned mission duties.

A second major development being pursued by the U.S. Navy is the Electromagnetic Aircraft Launch System (EMALS) that is being built by the General Atomics Company in San Diego (**Fig. 2**). This program is part of an initiative to replace steam and hydraulic launch systems with electrically driven units. EMALS will replace the existing steam catapults with linear motor launchers that will be more efficient and require significantly less maintenance. An added advantage will be the ability to launch larger aircraft than can be accommodated with existing steam-driven units. A companion program, now in the contractor selection stage, is to develop an advanced aircraft arresting gear system to replace the existing hydraulic system used to retrieve aircraft. This system, which may also be electrically driven, will replace the most maintenance-intensive system on today's aircraft carriers.

Directed energy weapons. One of the major reasons that ships are prematurely retired from active service is that they are unable to support combat system modernization because of insufficient electric power generation capacity. It is this limitation that sparked a lot of the Navy interest in the integrated power system. A combatant ship will have on the order of 100 megawatts of installed power. With the segregated systems of mechanical drive ships, less than 10 MW of that power is available to auxiliary and combat system loads. The balance can be used only for propulsion. With the integrated power system, the entire inventory is available to be redirected to any load upon demand. This means that it will be possible to install directed energy weapons as upgrades to operational ships at some time during their life cycles. This is important because the design and construction cycle of a class of ships can span 20 years, and with an expected life of 30 years for each ship, a ship class from cradle to grave can span 50 years. New weapons systems with enhanced capabilities will certainly become available during that long a time span.

Two directed energy weapon concepts involving the ability to launch projectiles to targets hundreds of miles inland, with low latencies (short time durations from launch to target), from safe distances at sea are of interest for land attack missions. These are the rail gun and the coil gun. The rail gun concept is a linear DC motor that uses a Lorentz force to accelerate the projectile (**Fig. 3**). The force is generated by applying a driving current to the rails, creating a magnetic field that interacts with the current in the armature. Relatively high muzzle velocities are possible using this principle. Technical challenges include high current switching and barrel wear resulting from the sliding contact between the armature and the rails. The coil gun functions via the principle that currents circulating in the drive coils induce an opposing current in the projectile, creating magnetic fields that repel each other and accelerate the projectile (**Fig. 4**). Since the projectile is magnetically levitated, barrel wear problems are eliminated. However, this concept has a lower efficiency than the rail gun and may not be capable of achieving comparable muzzle velocities.

Other directed energy concepts of interest include lasers and microwaves. High-energy lasers for offensive purposes operating at sea level present unique challenges and have distinct range limitations. However, lasers are attractive for point defense against threats such as missiles. Microwaves also show promise for close-in self-defense, particularly from small craft. They could conceivably enable a perimeter to be established around a ship that would be difficult to penetrate. A distinct advantage is that this could be a nonlethal form of defense. A second application might be the use of microwaves to beam power to remote surveillance sensors and autonomous vehicles.

Submarine applications. Submarines already employ sophisticated electrical distribution systems. Due to severe volume constraints, power density in submarines is extremely important, and advances that reduce the size of power electronic components will pay huge dividends. These may include further reductions of steam-driven equipment and elimination of the need for hydraulic systems to control actuators. Other major drivers will continue to be safety and acoustic performance (stealth).

Outlook. The advances in materials and power electronics during the last 15 years have caused some to refer to this period as the second electronic revolution, comparable to what the transistor did for communications. The integrated power system is in its first generation, and requires that system architecture and control procedures be established and proven. After this is accomplished, technology insertions that enhance or provide new capabilities will permit full exploitation of the potential of this concept. Some of these even may be applicable to aircraft systems. Electric actuators, if developed for submarines, also may be applicable to the control surfaces of airplanes, thereby contributing to a goal of aircraft with improved safety and reliability.

For background information *see* ELECTRIC POWER GENERATION; ELECTRIC POWER SYSTEMS; LASER; MAGNETISM; MARINE MACHINERY; MICROWAVE; NAVAL SURFACE SHIPS in the McGraw-Hill Encyclopedia of Science & Technology.   Cyril Krolick

Bibliography. N. Doerry et al., Powering the future with the integrated power system, *Nav. Eng. J.*, pp. 278-282, May 1996; *IMarE, SEE AES 2003 Broadening the Horizons: New Ideas, New Applications, New Markets for Marine Electrical Technologies, Proceedings*, Edinburgh, Feb. 13-14, 2003; *IMarE Proceedings of the All-Electric Ship Conference, AES 98*, London, Sept. 29-30, 1998; IPS, The U.S. Navy next-generation power/propulsion system, *Maritime Defence*, pp. 278-279, December 1996; Naval Research Advisory Committee, *Roadmap to an Electric Naval Force*, July 2002; *SEE, IMarE Proceedings of the All-Electric Ship Conference, AES 2000*, Paris, Oct. 27-28, 2000; *SEE Proceedings of the All-Electric Ship Conference, AES 97*, Paris, Mar. 13-14, 1997; *Technology for Future Naval Forces*, Chap. 8: Electric Power and Propulsion, 2001; E. J. Walsh, Surface fleet looks to "all-electric ships," *Sea Power*, pp. 33-34, May 1996.

Intelligent transportation systems

The Intelligent Transportation Systems (ITS) program was established in 1991 by the United States Congress to promote the use of advanced technologies (such as electronics, control systems, telecommunications, and computers) in order to improve the safety, operational, and environmental efficiency of the national transportation system. The creation of ITS has accelerated the pace of innovation and integration of these advanced technologies specifically for transportation applications. The most important functional areas of ITS can be described by the following interrelated and overlapping systems: transportation management, traveler information, emergency management, public transportation, electronic toll, commercial vehicle operations, travel demand management, vehicle technologies, and rural transportation.

Transportation management. The main goal of the Advanced Transportation Management System (ATMS) component of the ITS is to provide safe, efficient, and proactive management of traffic from central control facilities such as Transportation Management Centers (TMCs; **Fig. 1**) within and between large metropolitan areas. Data and information from sensors inside pavements, closed-circuit television (CCTV) cameras, transponder-equipped vehicles, and Global Positioning System (GPS) receivers are collected and analyzed in a timely manner in these facilities. TMC operators then use the information to time ramp metering and intersection signals more efficiently, respond to and manage incidents quickly, and disseminate important information to motorists through media such as dynamic message signs, highway advisory radios, and the Internet.

Traveler information. Advanced Traveler Information Systems (ATIS) use advanced technologies such as the GPS to provide the transportation user the information that is essential in traveling more efficiently from an origin to a destination. Travelers can

Fig. 1. Example of a Traffic Management Center.

access the information before a trip starts and while the trip is in progress. Information may be static (changes infrequently) or real time (changes frequently). Examples of static information include planned construction and maintenance activities, special events such as parades and sporting events, toll costs and payment options, transit fares, schedules and routes, intermodal connections, commercial vehicle regulations (such as hazardous materials, and height and weight restrictions), parking locations and costs, business listings (such as hotels and gas stations), tourist destinations, and navigational instructions. Examples of real-time information include roadway conditions including congestion and incident information, alternate routes,

weather conditions such as snow and fog, transit schedule adherence, parking-lot space availability, travel time, and identification of the next stop on a train or bus. ATIS dissemination platforms can also be divided into pretrip or en-route. Pre-trip ATIS platforms include television and radio, kiosks placed in high foot traffic areas (such as shopping malls and campuses), as well as ATIS software and databases for personal computers that are equipped with communication links for accessing real-time information. The Internet may also be used to access ATIS Web pages in different areas. Audiotext telephone services are also used for traveler information over a wireline telephone connection. En-route ATIS platforms include cell phones, portable computers, and in-vehicle devices such as Route Guidance Systems (RGS; **Fig. 2**).

Emergency management. Incidents contribute to nearly 60% of annual roadway congestion. Different ITS technologies are applied in predicting hazardous conditions and in detecting, verifying, responding to, and clearing incidents as quickly as possible. Predicting hazardous conditions which may lead to incidents are accomplished either through CCTVs or by using sophisticated algorithms that make use of pavement sensors' data on volume, speed, and occupancy. Detecting incidents may be done automatically through the vehicle's automated collision notification system, or by TMC operators through their CCTVs or incident detection algorithms. Verifying the incident and providing detailed information about the nature of the incident are done through CCTVs. Most, if not all, TMCs are members of their regional integrated computer-aided dispatch system.

Fig. 2. Example of a Route Guidance System.

TMC operators provide real-time traffic information to all needed emergency responders. Police, fire, and ambulance vehicles responding to the scene are provided with real-time traffic information to determine the best routes to take to expedite their response. Emergency preemption at signalized intersections may be activated by the TMC operator to expedite the response time of emergency vehicles. TMC operators also coordinate and implement alternative route plans to assist with managing traffic around the accident scene. This information is communicated to the public through dynamic message signs, highway advisory radios, and other mediums discussed previously.

Public transportation. Public transportation agencies use a combination of Automatic Vehicle Location (AVL) devices, Computer-Aided Dispatch (CAD), vehicle performance monitoring, automatic passenger counters, and customer information systems to improve performance and levels of service provided to customers. AVL allows for the determination of the real-time location of the vehicle. CAD centers receive information from AVLs, enabling dispatchers to monitor vehicle performance and transmit instructions to the vehicle operators to improve regular schedule adherence or make unscheduled adjustments in service such as detours. Passenger counters automatically record data on persons entering and leaving the vehicle. Coupled with the route and location data, accurate information is provided on vehicle loading at individual bus stops. Passenger information systems convey information to the customer. Information on system status and vehicle arrival times can be provided to travelers waiting for a vehicle, seeking information on available service, or planning a trip from home or office. Variable message signs at bus shelters, interactive kiosks, dial-in phone services, personal paging systems, and the Internet are some of the platforms used to convey real-time transit information to the customer.

Electronic toll. The two most important technologies used in Electronic Toll and Traffic Management (ETTM) are the Automatic Vehicle Identification (AVI) and the Dedicated Short-Range Communication (DSRC). ETTM offers a range of benefits to the road users, toll operators, and to society in general. These benefits include reduced congestion, less construction, less pollution, improved customer convenience, improved revenue security, better information, and improved safety.

Commercial vehicle operations. The Commercial Vehicle Operations (CVO) portion of ITS consists of technologies that support operations related to both passenger and cargo movement. These include activities related to carrier operations, vehicle operation, safety assurance, credentials administration, and electronic screening. Carrier operations include such functions as customer service, safety management, scheduling, load matching, order processing, dispatching, routing, tracking equipment and shipments, driver management, vehicle maintenance, and communications among commercial fleet managers, commercial drivers, and intermodal operators. Vehicle operation includes functions such as vehicle location and navigation, controlling the vehicle and monitoring vehicle systems, driver monitoring, logging driver activities, and communications with the commercial fleet managers. Safety assurance includes tasks such as roadside safety monitoring, inspection and reporting, providing carrier, vehicle, and driver safety information to roadside enforcement personnel and other authorized users, and compliance reviews. Credentials administration includes applying and paying for credentials, processing credential applications, collecting fees, issuing credentials and maintaining records about credentials, supporting base state agreements, and CVO tax filing and auditing. Electronic screening includes functions such as sorting vehicles that pass a roadside check station, determining whether or not further inspection or verification of credentials is required, and taking appropriate actions.

Travel demand management. Travel Demand Management (TDM) measures include promotion of transit and ride sharing; flexible working arrangements such as staggered work hours, flex time, and telecommuting; traffic calming measures such as bike lanes, speed humps, and police enforcement; and driving restrictions and prohibitions. Specific applications of ITS technologies in TDM include real-time traveler information, interactive ride matching, parking management, dynamic pricing of highway capacity, and telecommuting. Real-time traveler information uses advanced communication technologies to provide commuters and other travelers with accurate information about travel options. Roadside dynamic message signs, highway advisory radios, and the Internet are used extensively for this purpose. Interactive ride matching consists of commuters using the Internet or a specific organization's intranet to enter personal and carpool information into the computer bulletin boards. The information then matches all the travelers who may ride together in car or van pools. This is done either automatically, by software developed specifically for this purpose, or manually. Parking management uses advanced communication technologies to provide the traveler with accurate information about the exact status of parking lots. Dynamic pricing of highway capacity has been used extensively in other countries to manage peak time traffic in major metropolitan areas. Electronic toll collection technology allows variable rate tolls to be deducted from the driver's prepaid stored-value tag while the vehicle is in motion. Telecommuting involves the use of advanced technologies such as video conferencing, voice-data links, wireless transmission, and portable communication to allow individuals to work from home or at strategically located remote telework sites rather than physically travel to congested areas.

Vehicle technologies. The most important vehicle technologies related to ITS include technologies that deal with sensing other vehicles, objects and lane positions, predicting vehicle performance, and sensing absolute position and motion. Sensing absolute position and motion of vehicles is important since it enables TMCs to constantly update their sophisticated route optimization algorithm results and consequently the route guidance information that drivers receive. Sensing other vehicles, objects, and lane positions as well as predicting vehicles' various performance features such as brakes are advanced technologies that enhance the safety of individual vehicles, and automobile manufacturers are increasingly incorporating these features in their products.

Rural transportation. Because congestion is not a major concern in most rural areas, the main purpose of applying ITS technologies there is to aid transportation managers to enhance safety and reliability of the transportation system. In terms of safety, ITS technologies are useful in rural areas where single-vehicle, run-off-the road accidents are overrepresented. Other promising technologies include friction or ice detection and warning systems, intersection crossing detection, animal-vehicle collision avoidance, and horizontal curve speed warning advisory. ITS technologies that have enhanced the reliability of operations of rural transportation systems include pager activation systems, automated highway pavement management systems, permanent or mobile weather sensors, and excessive-speed vehicle-warning systems. These technologies assist rural transportation managers to accurately detect and monitor factors such as dynamic weather conditions, pavement and bridge maintenance conditions, work zone locations and crews, and maintenance fleet activities. Many of the ITS technologies have also been used in rural areas to enhance tourism and trade. Information such as availability of lodging or scheduling of special events can be used in coordination with transportation management information such as status of road construction projects, weather conditions, and traffic congestion to make the traveling public more informed.

Outlook. Because the price of electronics and computer technologies decline over time, many of the ITS technologies described above not only will be cheaper and more readily available but will also be applied in a more integrated fashion to the transportation infrastructure. Many of the overlapping issues related to metropolitan traffic congestion, public transportation systems, commercial traffic operations, and rural transportation will be monitored and controlled from centralized, regional transportation management system units throughout the country. ITS technologies alone, however, will not be the panacea to the worsening of the congestion, safety, and environmental conditions of urban, and to an increasing extent, rural areas of the United States. This is due to an ever-increasing demand for auto travel. A fresh look at the basics of land use in relation to urban design, providing more facilities for walking, biking, and motorcycling, together with an increase in the use of public transportation will be required to potentially curtail the transportation issues of the present and future.

For background information *see* HIGHWAY ENGINEERING; SATELLITE NAVIGATION SYSTEMS; TRAFFIC-CONTROL SYSTEMS; TRANSPORTATION ENGINEERING in the McGraw-Hill Encyclopedia of Science & Technology.   Ardeshir Faghri

Bibliography. American Society of Civil Engineers, *Proceedings of the 5th International Conference on Applications of Advanced Technologies in Transportation Engineering*, Irvine, CA, 1997; *Intelligent Transportation Primer*, Institute of Transportation Engineers (ITS), Washington, DC, 2000; ITS America, *National Program Plan for ITS*, 1995; R. R. Stough and G. Yang, Intelligent transportation systems, in T. J. Kim (ed.), Transportation Engineering and Planning, in *Encyclopedia of Life Support Systems* (EOLSS), EOLSS Publishers/UNESCO, Paris, 2003; U.S. Department of Transportation, *Building the ITI: Putting the National Architecture into Action*, 1996; U.S. Department of Transportation, *Intelligent Transportation Infrastructure Benefits: Expected and Experienced*, 1996; U.S. Department of Transportation, *Intelligent Transportation Systems: Real World Benefits*, Federal Highway Administration, 1998; U.S. Department of Transportation, *National ITS Architecture*, Federal Highway Administration, 1998.

Ionic liquids

In the simplest sense, an ionic liquid is the liquid phase that forms on the melting of an ionic compound (that is, a salt) and consists entirely of discrete ions (**Fig. 1**). More specifically, ionic liquids are the liquid phase of organic salts (organic, complex cations) which melt at, or near, ambient temperature (below 100–150°C; 212–302°F). Ionic liquids chemically resemble both phase-transfer catalysts and surfactants and, in some cases, have been used for both purposes.

The most extensively studied ionic liquids are systems derived from 1,3-dialkylimidazolium (structure **1**), tetraalkylammonium (**2**), tetraalkylphosphonium (**3**), and *N*-alkylpyridinium (**4**) cations.

Derivatives made from these cations can increase or decrease functionality, such as alkyl chains, branching, or chirality. Common anions that allow for the formation of low-melting ionic liquids typically have

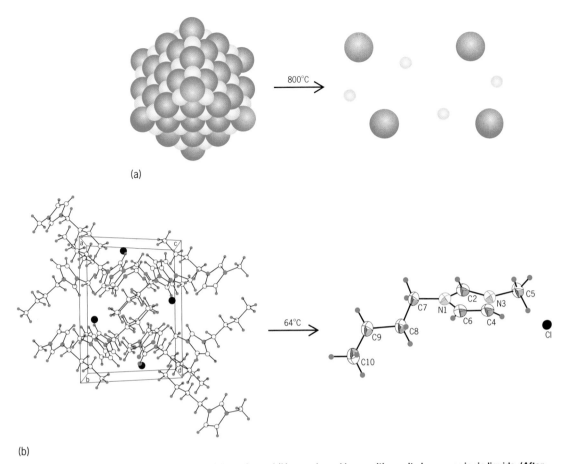

Fig. 1. Examples of (*a*) inorganic and high-melting salts and (*b*) organic and low-melting salts known as ionic liquids. (*After J. D. Holbrey et al., Crystal polymorphism in 1-butyl-3 methylimidazolium halides: Supporting ionic liquid formation by inhibition of crystallization, Chem. Commun., 2003:1636–1637, 2003*)

diffuse (delocalized) charge and can range from simple inorganic anions such as chloride (Cl⁻), bromide (Br⁻), and iodide (I⁻), through larger pseudospherical polyatomic anions, including hexafluorophosphate [PF_6]⁻ and tetrafluoroborate [BF_4]⁻, to larger, flexible fluorinated anions such as bis(trifluoromethylsulfonamide) [(CF_3SO_2)N]⁻ and tri(perfluoroX) trifluorophosphate. A wide range of simple polyhalometallate and halometallate/halide complex anion ionic liquids have also been extensively investigated and have important application in electrochemistry and electrodisposition. The anionic component of the ionic liquid typically controls the solvent's reactivity with water, coordinating ability, and hydrophobicity. Anions can also contain chiral components or can be catalytically active, such as carboranes, polytungstates, and tetrachloroaluminate anions. Manipulation of the rheological properties through mixing of cation-anion pairs to obtain materials that support or enhance reactions makes the use of ionic liquids in organic synthesis an intriguing possibility.

Synthesis. Preparation of the cationic portion of an ionic liquid can be achieved through the quaternization of phosphines or amines with a haloalkane or through protonation with a free acid. Quaternization is generally regarded as the more sound approach because cations prepared through protonation reactions can be degraded easily through deprotonation, leading to the breakdown of the solvent. Reac-

tion time and temperature for typical quaternization reactions depend upon both the haloalkane and the cation "backbone" used. The most widely used ionic liquids in research include 1-butyl-3-methylimidazolium chloride [C_4mim]Cl, [C_4mim][PF_6], and [C_4mim]-[BF_4] (**Fig. 2**). Ionic liquids prepared using this method can be modified through anion exchange reactions, acid treatment, or metathesis reactions to prepare ionic liquids containing a desired anion. (The choice of anion is important, as it is the most influential factor on the physical properties of an ionic liquid.) Through these basic reactions, it is possible to prepare a wide range of ionic liquids possessing varied physical properties that can be used to aid in creating optimal reaction conditions.

Purification of ionic liquids is of utmost importance, specifically those used in organic synthesis, as impurities from preparation could adversely affect the reactions. To synthesize pure ionic liquids, it is necessary to begin with purified starting materials. However, the use of pure starting materials does not ensure pure product. Common impurities in ionic liquids include halide anions from metathesis reactions, the presence of color and water, and acid impurities. Dichloromethane extraction is the most widely accepted method for removing excess halide anions from ionic liquids. Since ionic liquids typically have no measurable vapor pressure, dichloromethane can be removed easily under vacuum. As with

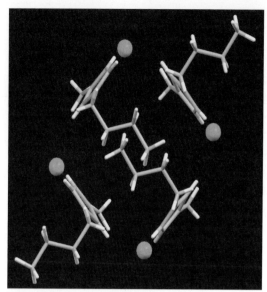

Fig. 2. X-ray crystal structure of 1-butyl-3-methylimidazolium chloride ([C$_{4mim}$]Cl) in its ortho polymorph (its most common conformation). (*After J. D. Holbrey et al., Crystal polymorphism in 1-butyl-3 methylimidazolium halides: Supporting ionic liquid formation by inhibition of crystallization, Chem. Commun., 2003:1636–1637, 2003. Reproduced by permission of The Royal Society of Chemistry.*)

organic solvents, water can also be removed from an ionic liquid under vacuum with heat to drive off unwanted moisture. Ionic liquids normally exist as colorless solvents. Impurities that can cause an ionic liquid to become colored can be avoided by using pure starting materials, avoiding the use of acetone in the cleaning of glassware, and keeping temperatures as low as possible during the quaternization step of its synthesis.

Properties. Interest in ionic liquids can be attributed to a number a unique qualities inherent to these materials. First, ionic liquids are excellent conducting materials that can be used in a number of electrochemical applications, including battery production and metal deposition. In fact, they were first designed to be used in electrochemical applications. Second, ionic liquids typically have no measurable vapor pressure, making them an attractive replacement for volatile organic solvents in synthesis. Third, ionic liquids are composed of ions that can be varied to create materials with vastly different physical and chemical properties, compared with conventional solvents. The ability to "fine-tune" the solvent properties of ionic liquids has been exploited to create solvents ideal for a range of organic syntheses. Lastly, ionic liquids exhibit varying degrees of solvation and solubility in a range of organic solvents, allowing simple separations and extractions.

Use in organic reactions. The replacement of volatile organic solvents with ionic liquids in organic reactions is very exciting. Ionic liquids are polar solvents which can be used as both solvents and reagents in organic synthesis, and the possibility exists that ionic liquids may positively affect the outcome of the reactions.

Certain ionic liquids are suitable for particular classes of organic reactions. For example, neutral ionic liquids are commonly used in Diels-Alder and condensation reactions, as well as nucleophilic displacement. Ionic liquids that possess Lewis acid properties are used in acid-catalyzed reactions, including Friedel-Crafts alkylations and acylations and electrophilic substitutions or additions.

Product isolation from ionic liquids is generally accomplished through extraction with organic solvents. Because most ionic liquids have no measurable vapor pressure, simple vacuum techniques can be used to recover the product in the organic solvent. Biphasic systems composed of ionic liquids and water or organic phases have been used in catalysis reactions exploiting the properties of some ionic liquids to easily recover both product and catalyst. As stated previously, the ionic constituents of an ionic liquid can be "tuned" to exhibit desired physical properties to support such biphasic reactions.

The unique properties of ionic liquids, such as their stability and nonvolatility, make them good candidates for use as solvents in homogeneous catalysis systems. Polymerization reactions in ionic liquids, using transition-metal catalysts and conventional organic initiators, have been studied and have demonstrated increased efficiency compared with traditional polymerization solvents.

Outlook. Sustained interest in ionic liquids can be attributed to their desirable physical properties, such as their electron conductivity, as well as their novelty and the possibility of enhanced reactions. The perceived potential to eliminate volatile organic solvents in synthetic and separation processes has also driven interest and investigation into ionic liquids.

Ionic liquids have many characteristics relevant to a general "green chemistry" approach, including the lack of volatility; however, almost all other properties (such as toxicity, stability, and reactivity) vary with the cation and anion components and cannot readily be generalized. The utility and interest in ionic liquids, as a class of fluids, rests with individual examples displaying new, improved, or different combinations of solvent properties.

For background information *see* ACID AND BASE; DIELS-ALDER REACTION; ELECTROCHEMISTRY; EXTRACTION; FRIEDEL-CRAFTS REACTION; HOMOGENEOUS CATALYSIS; PHASE-TRANSFER CATALYSIS; QUATERNARY AMMONIUM SALTS; SALT (CHEMISTRY); SOLVENT; SUBSTITUTION REACTION; SURFACTANT in the McGraw-Hill Encyclopedia of Science & Technology.

M. B. Turner; J. D. Holbrey; S. K. Spear; R. D. Rogers

Bibliography. R. D. Rogers and K. R. Seddon (eds.), *Ionic Liquids; Industrial Applications to Green Chemistry*, ACS Symp. Ser. 818, American Chemical Society, Washington, DC, 2002; R. D. Rogers and K. R. Seddon (eds.), *Ionic Liquids as Green Solvent, Progress and Prospects*, ACS Symp. Ser. 856, American Chemical Society, Washington, DC, 2003; P. Wasserscheid and T. Welton (eds.), *Ionic Liquids in Synthesis*, Wiley-VCH, Weinheim, 2002; T. Welton, Roomtemperature ionic liquids. Solvents for synthesis and catalysis, *Chem. Rev.*, 99:2071, 1999.

Killer whales

Found in every ocean and sea on Earth, killer whales (*Orcinus orca*) [**Fig. 1**] are the top predator in the marine ecosystem, with no natural enemies other than humans. These carnivorous whales feed primarily upon fish, great whales, sea lions, seals, and sea otters. They are the largest members of the oceanic dolphin family (Delphinidae), suborder Odontoceti (toothed whales).

Within marine ecosystems, the role of killer whales has recently become clear as a result of an inadvertent "experiment" conducted in the North Pacific. From the end of World War II to the mid-1970s, intensive industrial whaling in the North Pacific Ocean resulted in the severe depletion of great whale stocks. Coincident with this decline in great whales, killer whales were noticed preying increasingly upon smaller marine mammals. Current research on this chain of events is now linking increased predation by killer whales to three decades of sequential decline among the northern fur seal (*Callorhinus ursinus*), harbor seal (*Phoca vitulina*), Steller sea lion (*Eumetopias jubatus*) [**Fig. 2**], and sea otter (*Enhydra lutris*) populations in the North Pacific and Bering Sea. Apparently, killer whales have such huge energetic demands and are such effective hunters and killers that they are capable of dramatically depleting their prey populations.

Industrial whaling. In the seventeenth and eighteenth centuries, preindustrial whaling depleted several of the whale species in the Pacific Ocean, including the North Pacific right whales (*Eubalaena japonica*), bowhead whales (*Balaena mysticetus*), humpback whales (*Megaptera novaeangliae*), blue whales (*Balaenoptera musculus*), and gray whales (*Eschrichtius robustus*). Prior to World War II, however, large North Pacific populations still existed of fin whales (*Balaenoptera physalus*), sei whales

Fig. 2. Steller sea lion (*Eumetopias jubatus*). (*Courtesy of Rolf Ream, National Marine Mammal Laboratory*)

(*Balaenoptera borealis*), and sperm whales (*Physeter macrocephalus*). Then, in the late 1940s, whaling vessels equipped with post–World War II maritime technology began to target these last abundant whales. Within 20 years, modern industrial whaling had removed a half million great whales from the North Pacific Ocean and southern Bering Sea. By the mid-1970s, stocks of all great whales of the North Pacific were severely depleted. Although some species have made striking recoveries (gray and humpback whales), great whale populations are estimated to be 14% of prewhaling levels. The speed of the whale decline due to modern whaling was unprecedented and created dramatic ecological changes over a brief time span.

Fig. 1. Killer whales (*Orcinus orca*). (*Courtesy of Brad Hanson, Alaska Fisheries Science Center, Seattle*)

Scientists have only recently begun to appreciate the impacts of industrial whaling on the marine ecosystem. Whaling may have triggered an ecological chain reaction that seriously impacted the feeding habits of killer whales. Although information on the historic feeding behavior of killer whales is limited, it is likely that great whales were an important food source in the North Pacific Ocean.

Natural history. An estimated 30,000 to 80,000 killer whales are found in the oceans. They are easily recognizable, with black bodies and distinct white patches over the eyes and around the belly (Fig. 1). Females and males can grow to approximately 26–28 ft (8-8.5 m), respectively. Maximum estimated ages range 50–60 years for males and 80–90 years for females.

Although adaptable to almost any conditions, killer whales commonly occur within 800 km (500 mi) of major continents, particularly in the colder waters of the Arctic and Antarctic. They can often be found near Iceland, Norway, Japan, and the northeastern Pacific coast, from Washington to the Bering Sea, excluding areas of ice pack because of their dorsal fin.

Highly social, killer whales travel in female-led, socially hierarchical groups known as pods. Killer whales are very vocal, with a repertoire of 7 to 17 unique calls, some of which are used to communicate solely within a pod, others between pods. Two individuals will often echo the same call back and forth many times, precisely repeating each other. This behavior has been observed in other species, although the meaning is not yet understood. Communication among killer whales is considered complex, with different dialects from one pod to another. In addition to slightly different languages, each of the various killer whale populations existing around the world may have evolved distinctive foraging and behavioral traditions, which can include some social insularity, reproductive isolation, and genetic discreteness. Within the jurisdiction of the United States and Pacific Ocean, for example, five distinct killer whale stocks are recognized.

Along the northeastern Pacific coast, killer whales are divided into two types: resident and transient. Although the two types look similar, residents have a small home range within which they travel in pods of 5 to 50 and eat primarily fish and squid. Transients roam over large areas in pods of 1 to 7 and feed on marine mammals (seals, Steller sea lions, and great whales). Thus, it is predation by transients that is implicated in the population collapse of small marine mammals. With the depletion of great whale stocks, and the subsequent creation of a "caloric void," it is now believed that some whale-eating killer whales have compensated by increasing their consumption of small marine mammal species.

Declining small marine mammal populations. Harbor seal, northern fur seal, Steller sea lion, and sea otter populations have fallen drastically over the last three decades. Scientists originally attributed the declines to bottom-up forces, which include nutritional limitation due to climate change and a reduction in food resources due to commercial harvesting of fish. Predation by killer whales, a top-down force, is now thought by many scientists to be the most significant factor in the sea lion and sea otter population crashes. Although predation was formerly considered unlikely to seriously impact prey populations, scientists have determined that if killer whale foraging behavior changed only modestly, it could account for the precipitous fall in abundance of both the sea lion and sea otter.

Steller sea lions. The western stock of Steller sea lions (Fig. 2) has significantly declined throughout the North Pacific Ocean and southern Bering Sea, from hundreds of thousands of individuals in the 1960s to roughly 30,000 individuals in 2001. In Alaskan waters, where approximately 70% of the Steller sea lion population resides, the decline has exceeded 80%. The collapse of these populations has been so dramatic that the National Marine Fisheries Service listed Steller sea lions as "threatened" under the Endangered Species Act in 1990, and due to the continued population plunge west of Cape Suckling, Alaska, the listing was changed to "endangered" in 1997.

Marine mammal population declines have traditionally been attributed to human activities such as commercial harvest and to the mammals' interactions with commercial fisheries. Steller sea lions, however, have not been harvested since 1972, and fatal interactions with commercial fisheries are estimated to be insignificant. Nonetheless, to protect Steller sea lion food sources, commercial fisheries were regulated in Alaska. The possibility that killer whales, not commercial fisheries, may be the primary cause of decline is controversial. If, in fact, Steller sea lions are imperiled because of killer whales and not because of industrial fishing, constraints on commercial fisheries under the auspices of the Endangered Species Act may be misguided. This scientific debate highlights the importance of not just noticing when species are imperiled, but understanding what the root causes of their endangerment might be.

Sea otters. Sea otters are considered a coastal keystone species, on which the balance of the entire coastal ecosystem rests. Keystone species, such as the sea otter, play a disproportionately large role in determining the overall community structure within an ecosystem. The removal, addition, or the fluctuation of populations of sea otters can significantly impact the functioning of ecosystem processes, predatory relationships, and long-term system stability. Sea otters prey heavily on sea urchins, consequently preventing urchins from overgrazing algae and kelp beds. When sea otters were historically removed by trappers and fishermen, sea urchin populations bloomed, decimating kelp beds and consequently resulting in the decline of important commercial fish species that were dependent on these kelp beds.

Predation by killer whales upon sea otters, the least calorically profitable mammal (that is, the sea otter offers the least calories relative to its body mass and the energy output required to hunt it), suggests that they became a substitute food source only as the availability of other small marine mammals decreased. Thus, sea otters have been the most recent small marine mammal population to crash. In 1980, the sea otter population in the Aleutian Sea numbered 55,000 to 100,000 individuals. By 2000 this population had dropped to 6000.

In addition, fewer sea otters resulted in changes to the coastal ecosystem. With fewer predators, sea urchin populations flourished, decimating coastal kelp forests by the late 1990s. There is now concern by some scientists over the effects of an oceanic keystone species (killer whales) feeding upon a coastal keystone species (sea otters) and the consequences of this for the balance of coastal food webs.

Ecological chain reaction. Killer whales are part of a marine ecosystem made up of complex multispecies relationships. If, in fact, a caloric void was left by industrial whaling, those relationships may be significantly changing. The effects upon killer whale predatory habits, small mammal populations, and the balance of marine ecosystems (both oceanic and coastal) are only now being seriously considered. Previously, very little attention was given to the role of predation in structuring marine ecosystems and food webs.

Uncovering the primary cause of these population declines has major implications not only for conservation but for the way we manage our marine ecosystems. If removing one species can trigger a chain reaction of shifts in numerous other species, it is crucial that we focus our efforts on identifying and conserving keystone species.

For background information *see* CETACEA; MARINE ECOLOGY; MARINE FISHERIES; OTTER; PINNIPEDS; PREDATOR-PREY INTERACTIONS; TROPHIC ECOLOGY in the McGraw-Hill Encyclopedia of Science & Technology.   Terra Grandmason

Bibliography. J. A. Estes et al., Killer whale predation on sea otters linking coastal and nearshore ecosystems, *Science*, 282(5388):473-476, 1998; J. K. B. Ford et al., Dietary specialization in two sympatric populations of killer whales (*Orcinus orca*) in coastal British Columbia and adjacent waters, *Can. J. Zool.*, 76(8):1456-1471, 1998; A. M. Hammers, Killer whales—killing other whales, *Nat. Geog. Today*, 2003; K. Heise et al., Examining the evidence for killer whale predation on Steller sea lions in British Columbia and Alaska, *Aquatic Mammals*, 29.3:325-334, 2003; C. O. Matkin, L. B. Lennard, and G. Ellis, Killer whales and predation on steller sea lions, *Steller Sea Lion Decline: Is It Food II*, Alaska Sea Grant College Program, 2002; National Research Council, *The Decline of the Steller Sea Lion in Alaskan Waters: Untangling Food Webs and Fishing Nets*, National Academy Press, Washington, DC, 2003; K. S. Norris, Facts and tales about killer whales, *Pacific Discovery*, 11:24-27, 1958; SeaWeb, Collapse of seals, sea lions and sea otters in North Pacific triggered by overfishing of great whales, *PNAS Release*, 2002; A. M. Springer et al., Sequential megafaunal collapse in the North Pacific Ocean: An ongoing legacy of industrial whaling?, *PNAS*, Vol. 100(21):12223-12228, 2003; USGS, Collapsing populations of marine mammals: The North Pacific's whaling legacy?, *USGS Releases*, 2003.

Lake hydrodynamics

Lakes are valuable natural resources for water supply, food, irrigation, transportation, recreation, and hydropower. They also provide a refuge for an enormous variety of flora and fauna. There are more than 110,000 lakes larger than 1 km^2 (0.39 mi^2) covering a total area of 2.3×10^6 km^2 (0.9×10^6 mi^2, 1.7% of the Earth terrestrial surface). Besides the many millions of smaller lakes (less than 1 km^2), approximately 800,000 artificial lakes and reservoirs have been constructed covering 0.5×10^6 km^2 (0.2×10^6 mi^2). Large cities not located along coasts are typically near freshwater lakes, and rely on these resources. As a result of near-lake development, pollutants and nutrients threaten the ecological integrity of lakes, since the former can poison or kill aquatic organisms and the latter stimulates excessive growth of algae and water plants.

Hydrodynamics is highly variable among lakes because of the many different geometries, surrounding topographies, hydrological and geochemical loadings, and meteorological exposures. Obvious differences are in the surface area, which ranges up to 82,100 km^2 (31,700 mi^2) for Lake Superior, the largest freshwater lake by area, and in depth, which ranges up to 1642 m (5378 ft) for Lake Baikal, whose 23,000 km^3 (5500 mi^3) makes it the largest freshwater lake by volume. Variability in heat fluxes through the lake surface, in addition to chemical and biological properties (such as concentrations of salt, dissolved gases, particles, and algae), also affect the stratification and water movement. The resulting internal hydrodynamics is important in understanding the lake's physical, chemical, and biological structure. The transport and distribution of dissolved and particulate substances have important management consequences for the wise use of lakes.

Density stratification. Crucial for the hydrodynamics as well as the ecosystem functioning is that almost all lakes, at least those deeper than a few meters, experience density stratification and destratification throughout the seasons. Most important for stratification is the temperature dependence of the water density, which reaches a maximum close to 4°C (39°F). During spring and summer, the surface water is heated and usually a quasi-two-layer structure develops, with warmer and lighter water on top of the cooler and heavier water underneath (**Fig. 1**). These two layers are usually separated by a zone of rapid temperature change, the thermocline, which forms anywhere from just below the surface to tens of meters deep. The temperature contour plot in Fig. 1*b* shows such a thermocline at a depth of about 5-10 m (15-30 ft), which is typical for a midlatitude

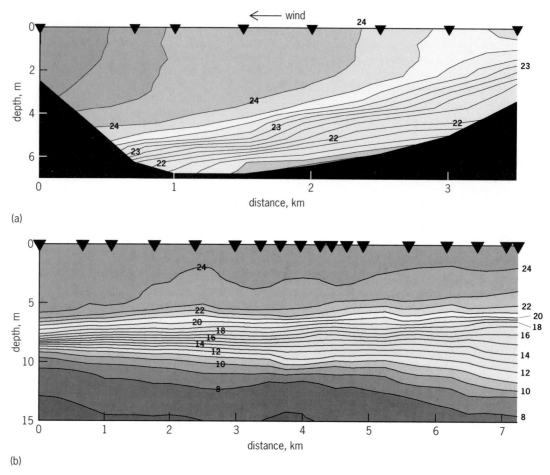

Fig. 1. Temperature contour plots for two lakes. Temperature profiles were obtained along the long axis of the lakes at various locations (inverted triangles on the top axis). Numbers on contour lines are temperatures in °C. °F = (°C × 1.8) − 32. 1 m = 3.3 ft. 1 km = 0.6 mi. (a) Plot for Müggelsee, Germany, on August 27, 1997, demonstrating temperature structure resulting from simple two-layer seiching (*from A. Lorke and A. Wüest, Turbulence and mixing regimes specific to lakes, in H. Baumert, J. Simpson, and J. Sündermann (eds.), Marine Turbulence: Theories, Observations and Models, Cambridge University Press, 2005*). (b) Plot for Lake Hallwil, Switzerland, on August 29, 2001, showing temperature structure resulting from three-layer seiching.

lake during summer. This thermocline implies a large density change [since water at 25°C (77°F) is about 3 kg m^{-3} (0.2 lb ft^{-3}) lighter than water at 4°C (39°F)], which leads to strong stability of the water column. This stability effectively suppresses vertical mixing, and consequently the deep water can remain cold (sometimes near 4°C) for the entire summer.

The water density also depends on pressure (not relevant for shallow lakes and internal motions), salinity (salt content), and particle and gas concentrations in the water. This implies that strong vertical gradients of these properties can also contribute to the stability of the water column stratification. Temperature is usually the most important stratifying agent in roughly the top 50 m (150 ft) in small- and medium-sized lakes, whereas at greater depths it is often salinity. Salinity gradients result from the algal production-decomposition cycle, subaquatic sources, and river inflow. In cold alpine reservoirs, very fine suspended glacial particles can sometimes lead to density stratification, whereas in volcanic crater lakes the vertical gradient of the dissolved gases, particularly carbon dioxide (CO_2), contributes to stability.

In fall and winter—especially at night—the lake surface cools and the uppermost water becomes denser. Subsequently, small parcels of water form plumes sinking at a few millimeters per second. This small-scale convection mixes the entire surface layer, which slowly deepens until the entire lake volume is mixed. In some deep lakes, the convection during the cool season (or dry period in the tropics) may not last long enough for complete mixing and some deep regions stay stratified. If the stratification is due to chemical gradients, the lake may remain permanently stratified (for hundreds or thousands of years). Many small lakes with high algal productivity are permanently stratified, as well as several large freshwater (for example, Tanganyika and Malawi) and saltwater bodies (for example, the Caspian Sea and Lake Van).

Large-scale horizontal motions. During the stratification season, the vertical exchange is greatly reduced and the main motions are almost entirely horizontal, following the contours of equal density (Fig. 1). The two major drivers of horizontal motion are wind and density differences in horizontal directions. As water is 800 times denser than air and as

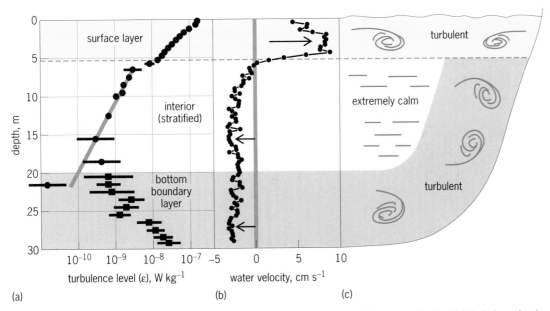

Fig. 2. Vertical profiles in lakes. (*a*) Turbulence profile in Lake Alpnach, Switzerland, demonstrating the high turbulence levels in the surface layer due to wind mixing and in the bottom boundary layer due to bottom (seiche) currents at the sediment-water interface. (*b*) Typical velocity profile for two-layer seiching. Measurements were obtained in Lake Hallwil, Switzerland, which has similar geometry and wind forcing. 1 cm s^{-1} = 0.4 in. s^{-1}. (*c*) Schematic of resulting turbulent zones in a lake enclosing the low-turbulence, quiescent zone in the interior.

momentum is transferred at the surface, the lake receives only about 3.5% of the wind energy from the atmosphere. Surface waves transport and dissipate a portion of this energy, whereas the remaining energy forms large-scale currents, with typical surface water speeds of about 1.5–3% of the wind velocity. For example, in **Fig. 2***b* the water velocity is between 5 and 7.5 cm s^{-1} (2.3 in. s^{-1}) in the surface layer, but it can occasionally reach up to several tens of centimeters per second. In large lakes, surface currents cause a stratified water body to pivot with warm water piling up at the downwind end and deepwater surfacing at the upwind end. After the wind ceases, the water displacement relaxes, and two-layer seiching motions occur; that is, the top and bottom layers oscillate in opposite directions (Fig. 2*b*), resulting in a vertical shifting of the temperature isotherms (Fig. 1*a*). While two-layer seiching is generally observed in most lakes, such as in the example shown in Fig. 2*b*, more complex three-layer seiching patterns can occasionally also be observed; that is, the top and deep layers oscillate in the same direction with the thermocline moving in the opposite direction in between (Fig. 1*b*). Seiching can continue for many days in small-to-medium sized lakes and for several months in very large lakes (for example, Tanganyika) until the energy dies out due to friction. The eigenperiod of the seiche (the time for a complete oscillation) is determined by the depth structure of the basin geometry and the strength of stratification. Seiching is the simplest form of many types of wind-forced waves occurring in stratified lakes. Other types include gravitational surface waves, high-frequency internal waves, as well as inertial, Poincaré, Kelvin, and Rossby waves.

Wind also results in another type of large-scale horizontal current pattern. These currents typically form cyclonic gyres and are often observed in larger lakes (for example, Lake Michigan) ranging in speed from 1 to 10 cm s^{-1} (0.4 to 4 in. s^{-1}; **Fig. 3**). Gyres form mainly as a result of the Coriolis effect and nonuniform wind forcing. The Coriolis effect is the observed deflection of the current direction due to the Earth's rotation, and is to the right (left) in the Northern (Southern) Hemisphere. Factors contributing to the formation of vortices include variable wind stress and stratification. Uniform wind can also result in gyres, but these are typically due to asymmetric lake topography. As a result, the surface water is transported to the shore whereas deeper water surfaces in the center of the gyre. If the gyre persists for long enough, it leads to a curved, convex thermocline, with a thinner top layer in the center and a thicker top layer along the shore, and to a vertical circulation cell, consisting of shoreward flow at the surface and toward the center of the lake in the upper thermocline.

Residual mixing of stratified water. During the warm season (rainy season in the tropics), at least some parts of the lake are usually stratified. However, wind-driven horizontal currents cause vertical shear, resulting in turbulence and vertical exchange within the water column. The direct effect of the wind generates turbulence in the surface layer (Fig. 2*a*). The mixing energy from the wind rapidly dies out with increasing depth, resulting in slow (even approaching molecular diffusion) vertical mixing in the lake interior (Fig. 2*a*). Seiching, other large-scale motions, and the breaking of internal waves also cause shear, particularly in the thermocline and above the sediment (Fig. 2*b*). Shear above the sediment increases

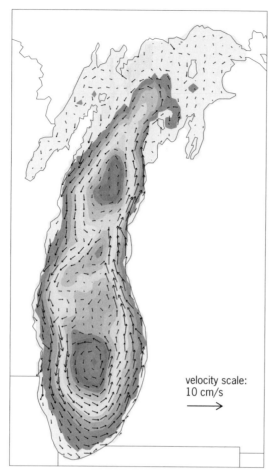

Fig. 3. Two counterclockwise gyres in Lake Michigan. The current speeds were averaged from November 2002 to April 2003. Lightly shaded areas indicate generally clockwise (anticyclonic) rotation, and more heavily shaded areas indicate counterclockwise (cyclonic) rotation. (*From D. J. Schwab and D. Beletsky, Relative effects of wind stress curl, topography, and stratification on large-scale circulation in Lake Michigan, J. Geophys. Res., 108(C2): 26–1 to 26–10, 2003. Copyright 2003 by American Geophysical Union. Reproduced/modified by permission of American Geophysical Union*)

turbulence, creating a well-mixed bottom boundary layer (Fig. 2*a*). The absence of the vertical density gradients in this bottom boundary layer allows much more rapid vertical transport near the sediment compared to the lake interior, where the vertical density gradients are orders of magnitude stronger.

Relevance. Many intriguing hydrodynamic processes govern how substances are transported and distributed in a lake. For example, excessive algal growth due to high nutrient input (agriculture and wastewater) can lead to the depletion of dissolved oxygen in the lake water through the settling of dead algae, which undergo bacterial decomposition, thereby consuming oxygen. As the only significant source of dissolved oxygen is from transfer at the lake surface, oxygen can become depleted in the deep-water, jeopardizing fish habitats. Anaerobic bacteria then take over the decay process, resulting in the production of undesirable substances, for example the greenhouse gas methane. Additionally, drinking

water withdrawn from lakes should be cool and relatively low in organic matter (algae) to avoid disinfection by-products as well as taste and odor problems. Mixing processes in a lake will therefore dictate the location where water is withdrawn and where nutrient-rich effluents should be discharged to avoid additional algal growth that conflicts with the water supply. Understanding naturally occurring mixing processes in lakes also aids in determining the ultimate fate of pollutants, and supports good management strategies and practice.

For background information *see* CORIOLIS ACCELERATION; LAKE; MEROMICTIC LAKE; SEICHE; WAVE MOTION IN LIQUIDS in the McGraw-Hill Encyclopedia of Science & Technology.

Daniel McGinnis; Alfred Wüest

Bibliography. J. Imberger, *Physical Processes in Lakes and Oceans*, Coastal and Estuarine Studies, no. 54, American Geophysical Union, Washington, DC, 1998; A. Lerman, D. M. Imboden, and J. R. Gat (eds.), *Physics and Chemistry of Lakes*, Springer, Berlin, 1995; P. E. O'Sullivan and C. S. Reynolds (eds.), *The Lake Handbook: Limnology and Limnetic Ecology*, Blackwell, 2004; D. J. Schwab and D. Beletsky, Relative effects of wind stress curl, topography, and stratification on large-scale circulation in Lake Michigan, *J. Geophys. Res.*, 108(C2):26-1 to 26-10, 2003; R. G. Wetzel, *Limnology*, 3d ed., Academic Press, 2001; A. Wüest and A. Lorke, Small-scale hydrodynamics in lakes, *Annu. Rev. Fluid Mech.*, 35:373–412, 2003.

Language and the brain

Many species have evolved sophisticated communication systems (for example, birds, primates, and marine mammals), but human language stands out in at least two respects which contribute to the vast expressive power of language. First, humans are able to memorize many thousands of words, each of which encodes a piece of meaning using an arbitrary sound or gesture. By some estimates, during the preschool and primary school years an English-speaking child learns an average of 5–10 new words per day, on the way to attaining a vocabulary of 20,000–50,000 words by adulthood. Second, humans are able to combine words to form sentences and discourses, making it possible to communicate an infinite number of different messages and providing the basis of human linguistic creativity. Furthermore, speakers are able to generate and understand novel messages quickly and effortlessly, on the scale of hundreds of milliseconds.

Linguists and cognitive neuroscientists are interested in understanding what special properties of the human brain make such feats possible. Efforts to answer this question go back at least 150 years. A great deal of attention has been given to the issue of which regions of the human brain are most important for language, first using findings from brain-damaged patients, and in recent years adding a

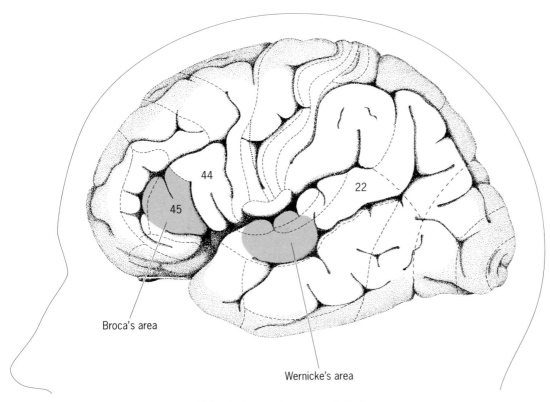

Broca's area

Wernicke's area

View of the left hemisphere of the cortex, with key brain areas for language indicated.

wealth of new information from modern noninvasive brain recording techniques such as positron emission tomography (PET) and functional magnetic resonance imaging (fMRI). However, knowing where language is supported in the human brain is just one step on the path to uncovering the special properties of those brain regions that make language possible.

Classic aphasiology. Nineteenth-century studies of aphasic syndromes, which are selective impairments to language following brain damage, demonstrated the importance for language of a network of left-hemisphere brain areas. Although this work was extremely difficult, requiring neurologists to compare language profiles with autopsy findings sometimes years later, the main findings have been largely confirmed, at least in their broad outlines. It is estimated that around 95% of right-handed people and 70% of left-handed people show left-hemisphere dominance for language.

In terms of more specific brain regions, the model proposed by the 26-year-old German neurologist Carl Wernicke in 1874 has proven to be remarkably accurate for clinical purposes. Wernicke classified language areas of the brain primarily in terms of the tasks that they were responsible for. Damage to the left inferior frontal gyrus is associated with a syndrome in which language comprehension appears to be relatively intact but language production is severely impaired and shows halting speech and difficulty with function words such as determiners (for example, the, a, this) and auxiliary verbs (for example, is, would, can). The brain area is known as Broca's area (also known as Brodmann's areas

44 and 45) and the syndrome as Broca's aphasia, Paul Broca, in 1861, was the first to claim a link between this brain area and language (see **illustration**). Wernicke proposed that Broca's area is specialized for the task of converting mental representations of language (which he assumed to be fundamentally auditory in nature) into speech.

Damage to an area in the superior posterior part of the left temporal lobe, Wernicke's area (also known as Brodmann's area 22, see illustration), is associated with a different syndrome (Wernicke's aphasia), in which language comprehension is seriously compromised and language production is grammatically fluent but often semantically inappropriate or lacking in coherence. Wernicke proposed that this second area was responsible for decoding and storing auditorily presented language. Wernicke's model and an updated account presented by Norman Geschwind in the 1960s made a number of additional predictions about specific kinds of neural damage that should lead to specific language impairments (for example, speech and comprehension problems without impairment to repetition), and stands out as a landmark in efforts to understand mind-brain relations.

Modern aphasiology. Whereas the classic model of aphasia emphasized a division of language areas of the brain based on tasks (for example, speaking and understanding), modern aphasia research suggests that it may be more appropriate to differentiate language areas based on the types of information that they preferentially deal with, such as syntax, phonology, or semantics. The most obvious clinical symptom of Broca's aphasia is labored language production, but careful studies from the 1970s onward have revealed that persons with Broca's aphasia also have comprehension difficulties, particularly in situations in which successful comprehension requires close attention to function words and inflectional morphemes, such as the plural suffix -s and the tense suffix -ed. For example, individuals with Broca's aphasia often misunderstand who did what to whom in a passive sentence such as "The dog was chased by the cat." Such findings have led to the suggestion that Broca's area has a task-independent role in syntactic processing that makes it important for speaking and understanding alike. Similarly, it has been suggested that Wernicke's area is responsible for semantic processes, both in speaking and in understanding.

Functional brain imaging. The advent of modern noninvasive brain imaging technologies has had a major impact on the understanding of brain areas responsible for language. First, it is now straightforward to determine a patient's lesion site shortly after damage occurs, rather than depending on autopsy findings. This has led to a dramatic increase in the database of knowledge available for deficit-lesion correlations. The crucial role of left-inferior frontal regions for language production and syntax has been strongly supported, although important correlational studies by Nina Dronkers suggest that the clinical symptoms of Broca's aphasia may be most strongly associated with a deeper left frontal structure called the insula.

Second, techniques such as fMRI can be used to test for correlations between selective activation patterns in normal adults and selective deficits in patients. In these studies, regular magnetic resonance imaging (or computerized tomography), which produces images of internal body structures, is used to determine the locus of structural damage in the brains of the patients. Using fMRI, which measures and maps neural activity in the brain, images are taken of the brains of normal speakers while they perform the tasks that are impaired in the patients to determine whether normal speakers show activation in the regions that are damaged in the patients. This work has largely confirmed the importance of classic left frontal and temporal language areas but has highlighted a number of additional left-hemisphere language areas, predominantly in the frontal and temporal lobes.

Third, brain stimulation studies using transcranial magnetic stimulation (TMS) have made it possible to noninvasively apply stimulation to create momentary activation or impairment in highly specific cortical regions. Stimulation studies are important, because they can show that a particular area is essential for a specific task rather than merely involved in that task. To date there have been very few TMS studies of language, but their findings largely support the conclusions of deficit-lesion correlation studies in patients.

Finally, it has become apparent that some classic language areas are also implicated in nonlanguage tasks. For example, Broca's area has been implicated in studies of motor planning and short-term memory for verbal items. This raises the possibility that brain areas previously thought to be specialized for specific types of language tasks or linguistic information processing may in fact be specialized for specific types of mental computation, in a modality-independent fashion. For example, both syntactic production and motor planning require the coordination and sequencing of a hierarchically organized plan.

Plasticity and signed languages. Although left-hemisphere dominance is the normal pattern, there is evidence for at least a limited degree of plasticity in the language system. In cases of children whose left hemisphere is removed early in life (for example, before the age of 7–10 years) to control intractable epilepsy, fairly good recovery of language abilities is typically observed. This indicates that the right hemisphere is able to take over many language functions if the left hemisphere is removed. Studies of signed languages indicate that the left hemisphere remains very important for language even when it is conveyed through a different modality. Recent fMRI studies by Helen Neville and colleagues have shown that the processing of American Sign Language recruited cortical areas in both hemispheres of native signers, who had used American Sign Language from early childhood, while the processing of written English was left-lateralized. However, from the neuropsychological point of view, Ursula Bellugi and colleagues have shown that sign language aphasia is due primarily to left-hemisphere lesions.

Temporal dynamics. In contrast to findings about the localization of language in the brain, studies using electroencephalography (EEG) and magnetoencephalography (MEG) measure the scalp voltages or magnetic fields generated by electrical activity in the brain, and provide a detailed record of the temporal dynamics of brain activity related to language. Studies of this kind have provided important clues about the mechanisms that allow language processing to be so fast and efficient. A family of different brain responses that appear within 100–600 ms after the presentation of a linguistic event have been found to be highly sensitive to the predictability of the sound or word, suggesting that prediction of upcoming material plays an important role in rapid language processing.

Outlook. Advances in noninvasive brain recording techniques have led to dramatic improvements in the understanding of the localization and temporal dynamics of human language, but answers remain elusive regarding the underlying question of what special properties of the human brain allow it to support language. Nonhuman primates are able to learn small numbers of arbitrary pairings of symbol and meaning, and researchers seek an explanation for why the human capacity for word learning in particular is quantitatively greater than other primates. In addition, nonhuman primates do not appear able to learn hierarchically organized grammatical systems, and thus researchers are searching for an explanation for how human brains can rapidly encode systematic combinations of words (that is, sentences), organized into recursive hierarchical structures. The current leading ideas on this question focus on the encoding of word combinations using the time structure of neural activity, although it is unclear how this could capture the differences between humans and other primates, given the overall similarities across species in basic neural mechanisms. An important challenge for coming years will be to find whether the brain areas that are implicated in language studies turn out to have distinctive properties at the neuronal level that allow them to explain the special properties of human language.

For background information *see* APHASIA; BRAIN; LINGUISTICS; MEDICAL IMAGING; PSYCHOLINGUISTICS; SPEECH; SPEECH DISORDERS in the McGraw-Hill Encyclopedia of science & Technology.

Colin Phillips; Kuniyoshi L. Sakai

Bibliography. N. Dronkers, B. B. Redfern, and R. T. Knight, The neural architecture of language disorders, in M. S. Gazzaniga (ed.), *The New Cognitive Neurosciences*, pp. 949–958, MIT Press, Cambridge, MA, 2000; R. S. J. Frackowiack (editor in chief), *Human Brain Function*, 2d ed., Academic Press, 2004; N. Geschwind, The organization of language and the brain, *Science*, 170:940–944, 1970; M. D. Hauser, N. Chomsky, and W. T. Fitch, The faculty of language: What is it, who has it, and how did it evolve?, *Science*, 298:1569–1579, 2002; A. C. Papanicolaou, *Fundamentals of Functional Brain Imaging: A Guide to the Methods and Their Applications to Psychology and Behavioral Neuroscience*, Swets & Zeitlinger, 1998; T. Swaab, Language and brain, in M. Gazzaniga, R. Ivry, and G. R. Mangun (eds.), *Cognitive Neuroscience: The Biology of the Mind*, 2d ed., pp. 351–399, Norton, New York, 2002.

Lignin-degrading fungi

The plant cell wall consists of a multilayer structure in which cellulose and hemicellulose are intimately associated with lignin. Lignin provides strength and impermeability, and serves as a barrier against microbial attack across the cell wall. In contrast to other natural polymers composed of regularly interlinked, repetitive monomers, lignin is an amorphous polymer made of randomly distributed phenylpropanoid monomers. Lignin production is estimated at 20×10^6 megatons annually; lignin therefore plays a central role in the global renewable carbon cycle. Its structure imposes unusual restrictions on its biodegradability.

Lignin degradation. Only a few organisms are capable of degrading lignin, the most efficient of which are fungi. The three groups capable of lignin degradation are brown rot, soft rot, and white rot fungi (see **table**).

Brown rot fungi belonging to the basidiomycetes extensively degrade cell-wall carbohydrates and only modify the lignin, causing wood decay. Soft rot fungi taxonomically belong to the ascomycetes and deuteromycetes. They are generally active in wet environments but also decompose plant litter in soils. Both types perforate lignin in order to penetrate the secondary wall of the wood cell, forming cylindrical cavities in which the hyphae propagate. The rot is of limited extent, being closely associated with the fungal hyphae, because the cellulose-degrading enzymes, cellulases, do not diffuse freely through the wood.

White rots, also belonging to the basidiomycete genera, are the most efficient lignocellulose degraders and the only known organisms that can completely break down lignin to carbon dioxide and water. The name white rot derives from the appearance of wood attacked by these fungi, in which lignin removal results in a bleached appearance to the substrate. Lignin cannot be degraded as a sole source of carbon and energy; however, its degradation enables white rot fungi to gain access to cellulose and hemicellulose, which serve as their actual carbon and energy source. The main mechanism of lignin degradation by these fungi involves one-electron oxidation reactions catalyzed by extracellular phenol oxidases (laccases) and peroxidases.

Lignin-degrading enzymes. Lignin-degrading peroxidases are heme-containing enzymes which require hydrogen peroxide (H_2O_2) as an electron acceptor to oxidize lignin and lignin-related compounds. They are monomeric with molecular weights ranging from 35 to 47 kD. Three types of lignin-degrading peroxidases have been reported in white rot fungi: lignin peroxidase (LIP), manganese-dependent peroxidase (MnP), and, more recently, versatile peroxidase (VP). Laccases are blue copper-containing phenol oxidases. Rather than H_2O_2, these enzymes utilize O_2 as the electron acceptor, reducing it by four electrons to H_2O. Laccases have a molecular weight of 50–300 kD. In most white rot fungi species, peroxidases and laccases are expressed as several isozymes. Both types of enzyme are glycosylated (containing sugar units), which may increase their stability.

Lignin-degrading fungi and distribution		
Fungi	Examples	Distribution
White rot	*Phanerochaete* sp. *Pleurotus* sp. *Bjerkandera* sp. *Trametes* sp. *Phlebia* sp.	Predominantly degrade hardwood from deciduous trees. Degrade lignin to completion.
Brown rot	*Serpula lacrymans* *Piptoporus betulinus* *Gloeophyllum trabeum* *Postia placenta*	Preference for coniferous wood which are softwoods. Partially depolymerize lignin.
Soft rot	*Chaetomium* sp. *Ceratocystis* sp. *Phialophora* sp.	Decay both hardwood and softwood. Partially modify lignin.

The ligninolytic system of white rot fungi also includes extracellular H_2O_2-generating enzymes, essential for peroxidase activity. To date, a number of oxidase enzymes have been reported to be involved in the production of H_2O_2. Other extracellular enzymes involved in electron transport, such as cellobiose dehydrogenase (CDH), and some reactive oxygen species also seem to be involved in lignin degradation by white rot fungi.

Applications of white rot fungi. White rot fungi are being used in a number of areas, such as agriculture and environmental protection.

Mushroom production. Several saprophytic fungi, most of them white rots, are cultivated for edible mushroom production on lignocellulosic materials, such as wheat and rice straws and wood residues, or agricultural wastes, such as cotton stalks, sawdust, coffee pulp, flax shive, corn cob, sugarcane bagasse, citronella bagasse, and rice hulls. The most commonly cultivated species are *Agaricus bisporus* (champignon mushroom), *Pleurotus* spp. (oyster mushroom), *Volvariella volvacea* (straw mushroom), and *Lentinula edodes* (shiitake mushroom). Commercial production techniques for these basidiomycetes are well developed (see **illustration**). Since such fungi can decompose lignocellulose efficiently without chemical or biological pretreatment, a large variety of lignocellulosic wastes can be utilized and recycled. The substrates used in each region of the world depend on the locally available agricultural wastes. Bioconversion of lignocellulosic agroresidues through mushroom cultivation offers the potential for converting these residues into protein-rich palatable food, reducing the environmental impact of the wastes. Mushrooms usually have high contents of protein, total carbohydrates, and minerals but low contents of lipids and nucleic acids. The fruiting bodies of edible mushrooms are known for their unique flavor and aroma properties. The volatile fraction and, specifically, a series of eight-carbon aliphatic compounds have been reported to be the major contributors to the characteristic mushroom flavor of each species. Among the volatile compounds that constitute edible mushroom flavor, 1-octen-3-ol is considered to be the major contributor.

Upgrading agricultural wastes for animal feed. The use of lignocellulosic residues as ruminant animal feed represents one of their oldest and most widespread applications, and they play an important role in the ruminant diet. The idea of using white rot fungi to improve the digestibility of lignocellulosic material for ruminants was first developed almost a century ago, suggesting the use of fungi for the improvement of lignocellulosic wastes. Since then, a considerable amount of work has been conducted on the upgrading of lignocellulosics to fodder using white rot fungi.

The lignocellulose complex in straw and other plant residues is degraded very slowly by ruminants because of the physical and chemical barrier imposed by lignin polymers, which prevent free access of hydrolytic enzymes, such as cellulases and hemicellulases, to their substrates. Biological delignification of straw seems to be the most promising way of improving its digestibility. A widely studied fungus for this purpose is *Pleurotus* spp. under different conditions and substrate pretreatments. Wheat straw is the most common substrate studied. Cotton stalk is another. This material poses agrotechnical problems, since the stalks have a fibrous structure similar to that of hardwood.

The main economical barrier to fungal upgrading of straw to fodder is the lack of inexpensive methods for preparation of the fungal growth substrates. A novel process for the preparation of agricultural wastes for fungal fermentation has been developed. This includes pasteurization of the wet straw by solar heat, treatment with detergents, and amendment with wastes from the food industry, such as potato pulp and tomato pomace (residue after pressing). The method has been found to be suitable for application at the farm with low energy consumption.

Biopulping. One of the most obvious applications of white rot fungi and their oxidative enzymes is biobleaching and biopulping in the pulp and paper industry to replace environmentally unfriendly chemicals (such as chlorine), save on mechanical pulping energy costs, and improve the quality of pulp and the properties of paper. The ligninolytic enzymes of white rot fungi selectively remove or alter lignin and allow cellulose fibers to be obtained. Recent data suggest that biopulping has the potential to be an environmentally and economically feasible alternative to current pulping methods.

Soil bioremediation. Bioremediation is the process by which hazardous wastes are biologically converted to harmless compounds or to levels that are below concentration limits. The ligninolytic enzymes of

Closeup view of *Pleurotus* spp. fruiting bodies growing naturally on a dead tree log. The common name oyster mushroom comes from the shell-like appearance of the fruiting body. Note the characteristic eccentric caps.

white rot fungi have broad substrate specificity and have been implicated in the transformation and mineralization (by metabolizing organic compounds) of organopollutants with structural similarities to lignin. White rot fungi have been demonstrated to be capable of transforming or mineralizing a wide range of organopollutants, including munitions wastes, very toxic and recalcitrant pesticides such as organochlorine insecticides and herbicides, polychlorinated biphenyls, polycyclic aromatic hydrocarbons (PAHs), bleach plant effluent, synthetic dyes, synthetic polymers, and wood preservatives. Safe disposal of munitions waste is a constant problem for the military. Not only does the detonation risk of explosives render them hazardous, but the constituent compounds are also toxic and persistent in the environment.

The most frequently studied and applied white rot fungi for bioremediation are *Coriolopsis polyzona*, *Pycnoporus sanguineus*, *Trametes versicolor*, *Bejerkendera adusta*, *Phanerochaete chrysosporium*, *Irpex lacteus*, and *Pleurotus eryngii*. Due to their enhanced proliferation in arid and semiarid environments and their ability to secrete extracellular oxidizing enzymes, white rot fungal bioremediation treatments may be particularly appropriate for in-situ remediation of soils, where recalcitrant compounds (such as the larger PAHs) and bioavailability are problematic. White rot fungi have been applied for large-scale in-situ remediation and biopiling (mixing of contaminated soil with compost in an aerated pile) of heavily contaminated soils. Seeding is performed by introducing fungal preinoculated woodchips, which helps for the fungi's competition with soil microorganisms and also serves as the basis for their long-term proliferation.

Decolorization. One of the most studied applications of ligninolytic fungi is in the treatment of bleach-plant effluents and of synthetic-colorants wastewater. The production of high-quality paper requires a chlorine-mediated bleaching process to remove color associated with the 5–10% residual lignin in pulp. As a result, large aqueous volumes of toxic, low-molecular-mass, halogenated lignin degradation products are released into the environment from bleach plants. These include chlorolignins, chlorophenols, chloroguaiacols, chlorocatechols, and chloroaliphatics.

Synthetic dyes are used extensively in the biomedical, foodstuff, plastic, and textile industries, where it is estimated that 10–14% of dye is lost in effluent during the dyeing process. Over 100,000 commercial dyes are available to textile industries, which consume vast quantities of water and other chemicals during processing, ultimately producing effluents consisting of complex and recalcitrant compounds. Synthetic dyes are designed to be resistant to light, water, and oxidizing agents and are therefore difficult to degrade, and many are also toxic, once released into the environment. Many azo dyes, constituting the largest dye group, are decomposed into potential carcinogenic amines under

anaerobic conditions after discharge into the environment.

The ability of white rot fungi to degrade and decolorize chlorolignins and a wide range of synthetic dyes has been demonstrated and potentially warrants their use in treating pulp, paper, and textile wastewater. The high efficiency of white rot fungi for decolorization of these compounds seems to reside in the involvement of their extracellular ligninolytic enzymes as confirmed by several independent studies using purified cell-free enzymes.

Conclusion. The inherent ability of lignin-degrading fungi to degrade synthetic and natural aromatic macromolecules and the ability to survive in environments generating reduced oxygen species make them a unique living niche. The genome of the most extensively studied white rot basidyomycete, *Phanerochaete chrysosporium*, has been elucidated, which establishes the basis for a deeper understanding of their physiological and metabolic capabilities. In addition to the many known uses of these fungi, many applications of the ligninolytic enzymes are being established in the textile, cosmetic, and laundry industries.

For background information *see* BIODEGRADATION; CELLULOSE; FUNGI; LIGNIN; WOOD ANATOMY; WOOD PROPERTIES in the McGraw-Hill Encyclopedia of Science & Technology.

Carlos G. Dosoretz; Yitzhak Hadar

Bibliography. T. K. Kirk and R. L. Farrell, Enzymatic "combustion": The microbial degradation of lignin, *Annu. Rev. Microbiol.*, 41:465–506, 1987; A. T. Martinez, Molecular biology and structure-function of lignin degrading heme peroxidases, *Enzyme Microb. Technol.*, 30:425–444, 2002; D. Martinez et al., Genome sequence of the lignocellulose degrading fungus *Phanerochaete chrysosporium* strain RP78, *Nat. Biotechnol.*, 22:695–700, 2004; M. L. Rabinovich, A. V. Bolobova, and L. G. Vasilchenko, Fungal decomposition of natural aromatic structures and xenobiotics (a review), *Appl. Biochem. Microbiol.*, 40:5–23, 2004; S. P. Wasser, Medicinal mushrooms as a source of antitumor and immunomodulating polysaccharides., *Appl. Microbiol. Biotechnol.*, 60:258–274, 2002.

Lipidomics

Lipidomics is a new field of research focused on the identification and determination of the structure and function of lipids and lipid-derived mediators in biosystems. As practiced today, lipidomics can be subdivided into architecture/membrane lipidomics and mediator lipidomics.

Cell membranes are composed of a bilayer that contains phospholipids, fatty acids, sphingolipids, integral membrane proteins, and membrane-associated proteins. The membrane composition of many cell types is established. However, membrane organization and how it affects cell function remains an area of interest. Membranes serve as barriers by separating the inside of the cell from the outside environment or from compartments within cells, and by regulating passage of nutrients, gases, and specific

ions. Cell membranes also generate signals to the intracellular environment via their ability to interact with key proteins. Determining the nature of these various interactions and decoding the structure-function information within their organization is one promise of lipidomics. Metabolism of fatty acids is also an important energy source; hence, catabolic breakdown of fatty acids is an important area of "metabolomics" that links to the signaling pathways and roles of lipid mediators. Given the swift advances in genomics and proteomics, it appears that metabolomics, the appreciation of metabolic networks and pathway intermediates, is the next wave in our appreciation of the molecular basis of life.

Advances in computers, software, algorithms, chromatography, and identification of bioactive mediators help to form the basis for lipidomics. Recent advances in the use of liquid chromatography combined with tandem mass spectrometry (LC-MS-MS) permits profiling of closely related compounds without the need for prior derivatization (modification to improve detectability) of samples. The interface of mass spectrometry with liquid chromatography also permits profiling of lipid-derived mediators with reduction of work-up induced artifacts. This overview describes the potential uses of lipidomics in biology.

Membrane architecture. The organization and precise compositions of microdomains surrounding key integral membrane proteins and other lipid-enriched domains within cells remains to be fully elucidated. There is little information on the organization of discrete lipid patches and microdomains in the structure of plasma membranes. Nonetheless, these microdomains and patches (also known as lipid rafts) are of considerable importance in regulating cellular responses. The ability to identify each of the phospholipids (such as phosphatidic acid, phosphatidyl choline, and phosphatidyl ethanolamine), as well as their fatty acids (such as myristic acid, palmitic acid, linoleic acid, linolenic acid, and arachidonic acid) is important because the physical properties of phospholipids with different acyl-chain compositions have dramatically different effects in cell function. Major phospholipid and fatty acid structures are shown in **Fig. 1**. Each phospholipid contains two fatty acids, one each in the 1 and 2 positions, that establish its identity. Saturated fatty acids in the 1 or 2 position—for example, steric acid and oleic acid (with only one unsaturated double bond)—can make membranes more crystalline-like or rigid in regions in which these phospholipids exist. In contrast, fatty acids with an increased number of double bonds or polyunsaturated fatty acids increase membrane fluidity because they have fewer hydrogen-hydrogen interactions and so are less rigid.

Sphingolipids are another form of complex lipid that can also give rise to signaling molecules. The chemical composition and biosynthesis of each of the major classes of lipid structures are known. However, how each class is organized within membrane structures and its dynamics during cell activation and

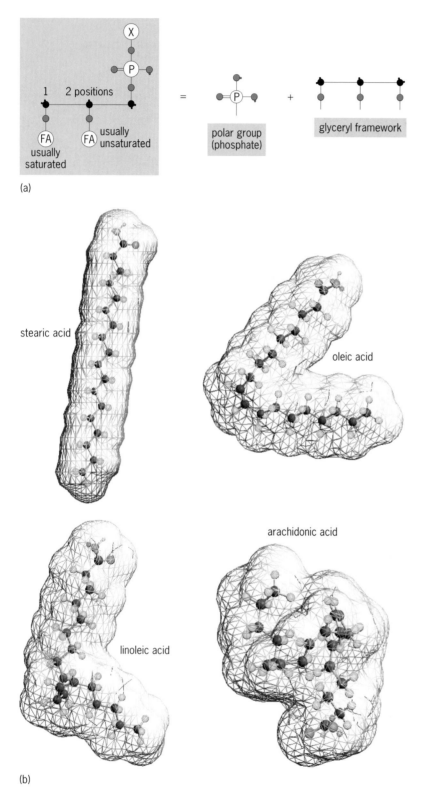

Fig. 1. Components of phosphoinositides. (*a*) General structure of phospholipids, showing positions sn-1 and sn-2, where various fatty acids may attach. X represents a phosphate-linked group; FA represents a fatty acid. (*b*) Space-filling models of four common fatty acids.

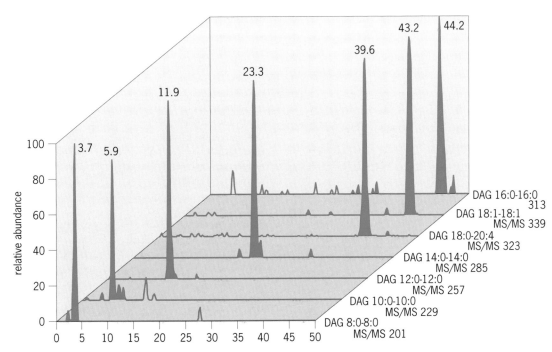

Fig. 2. Selected MS ion chromatograms of synthetic 1,2-diacyl-sn-3-glycerol (DAG) molecular species from PMN of patients with local aggressive periodontal disease. DAG species were resolved and identified by LC-MS-MS using specific retention time, acyl chains, and unique MS-MS signature ion for each molecular species.

generation of intracellular second messengers has yet to be fully appreciated. One major group focusing on these compounds using lipidomics is at the Medical University of South Carolina. Given the considerable diversity in triglyceride structure and the importance of phosphatidylinositol as well as other sugar-linked phospholipids in cell signaling, a systematic analysis is also underway via the National Institute of General Medical Science (NIGMS) Lipid MAPS (metabolites and pathways strategy) consortium. This consortium plans to assemble maps of analytical profiles as cells are activated in experimental settings.

Lipid signals in human disease. Diacylglycerol (DAG) is an intracellular second messenger that helps to illustrate the potential for second-messenger lipidomics. Recently, a genetic abnormality in patients with local aggressive periodontal disease was linked to impaired DAG kinase activity in their peripheral blood neutrophils. The neutrophils in this familial disorder display reduced chemotaxis toward microbes, reduced transmigration, and the ability to generate reactive oxygen species. Using a lipidomics approach to identify molecular species at positions in the 1,2-diacyl-sn-3-glycerol backbone in these patients' neutrophils, alterations in levels and specific molecular species were found (**Fig. 2**). LC-MS-MS-based lipidomics was performed to identify and quantitate individual species of DAGs involved in second-messenger signaling. Profiles of specific DAG species were identified by their physical properties, including molecular ion, specific daughter ions, and co-elution with authentic standards of the major species. Both molecular and temporal differences in DAG signaling species between healthy neutrophils

and those taken from individuals with localized aggressive periodontal disease were demonstrated. Hence, this type of structure-function profiling of intracellular messengers improves our understanding of signaling pathways and their alterations in disease.

Mediator lipidomics. Another powerful use of lipidomics is the characterization of mediators, or mediator lipidomics. For instance, during the release of arachidonic acid and its transformation to bioactive lipids, specific stereoselective hydrogen abstraction leads to formation of conjugated diene-, triene-, or tetraene-containing chromophores (chemical groups that impart color to a molecule), particularly in the eicosanoid group of mediators, such as the leukotrienes and lipoxins (**Fig. 3a**). The presence of both specific ultraviolet (UV) chromophore and characteristic MS-MS spectra, the fragmentation of these compounds, and retention time of each of these related structures permit identification and profiling from cellular environments. Unlike phospholipids or other structural lipids that keep a barrier function, those derived from arachidonic acid, such as prostaglandins, leukotrienes, and lipoxins, have potent stereoselective actions on neighboring cells. As closely related structures may be biologically devoid of actions, accurate structural and functional profiling within a snapshot of a biological process or disease state can give valuable information. Also, when specific drugs are taken, such as aspirin, the relationship between individual pathway products can be changed and altered, and their relationship may be directly linked to the drug's action while in vivo. Hence, mediator lipidomics provides a valuable means toward understanding the phenotype in

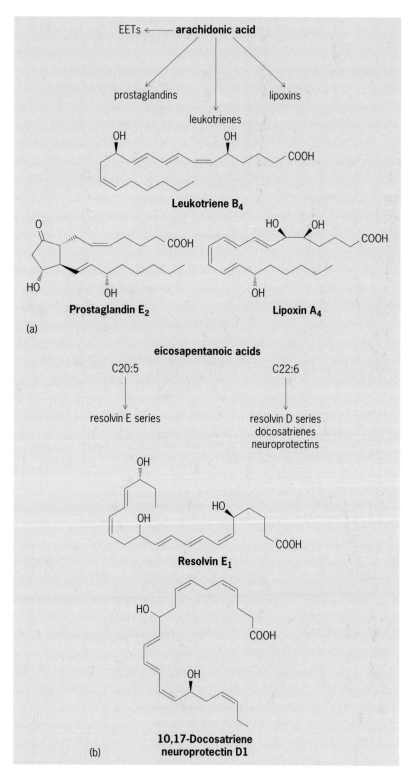

Fig. 3. Families of bioactive lipids. (*a*) Arachidonic acid is the precursor for many of the known bioactive mediators, such as EETs, prostaglandins, leukotrienes, and lipoxins. (*b*) Eicosapentanoic acids (C22:5 and C22:6) are precursors to potent new families of mediators termed resolvins and neuropectins.

many prevalent diseases, particularly ones in which inflammation has an important pathologic basis.

Pathways of inflammation resolution. It is now appreciated that inflammation plays an important role in many prevalent diseases in the Western world. In

addition to the chronic inflammatory diseases, such as arthritis, psoriasis, and periodontitis, it is now increasingly apparent that diseases such as asthma, Alzheimer's disease, and even cancer have an inflammatory component associated with the disease process. Therefore, it is important for us to gain more detailed information on the molecules and mechanisms controlling inflammation and its resolution. Toward this end, new families of lipid mediators generated from fatty acids during resolution of inflammation, termed resolvins and docosatrienes, have been identified (Fig. 3*b*).

Exudates were sampled during inflammation resolution as leukocytic infiltrates were declining to determine whether there were new mediators generated. **Figure 4** schematically represents a functional mediator-lipidomics approach using LC-MS-MS-based analyses to evaluate and profile temporal production of compounds at defined points during experimental inflammation and its resolution. Libraries were constructed of physical properties for known mediators, such as prostaglandins, epoxyeicosatrienic acids (EETs), leukotrienes, and lipoxins, as well as theoretical compounds and potential diagnostic fragments as signatures for specific enzymatic pathways. When novel compounds were pinpointed within chromatographic profiles, complete structural elucidation was performed as well as retrograde chemical analyses that involved both biogenic and total organic synthesis, which permitted scaling up of the compound of interest and its evaluation in vitro and in vivo.

This full cycle of events defines mediator-based lipidomics because it is important to establish both the structure and functional relationships of bioactive molecules in addition to cataloging and mapping their architectural components as in biolipidomics. With this new lipidomics-based approach, a novel array of endogenous lipid mediators were identified during the multicellular events that occur during resolution of inflammation. The novel biosynthetic pathways uncovered use omega-3 fatty acids, eicosapentanoic acid, and docosahexanoic acid as precursors to new families of protective molecules, termed resolvins (Fig. 3*b*). Resolvin E1 is a downregulator of neutrophils and stops their migration into inflammatory loci. Neuroprotectin D1, in addition to stopping leukocyte-mediated tissue damage in stroke, also maintains retinal integrity, and is formed from docosahexaenoic acid by activated neural systems. **Figure 5** gives lipidomic profiles of resolvin E1 and neuroprotectin D1.

Mapping of the local biochemical mediators and impact of drugs, diet, and stress in bionetworks is exciting terrain and will no doubt enable us to appreciate that size of the peak (that is, the relative abundance of a compound) does not always count. Transient, seemingly small, quantitatively fleeting members of lipid mediator pathways and their temporal relationships change extensively during the course of a physiologic or pathophysiologic response. These changes in magnitude and hence

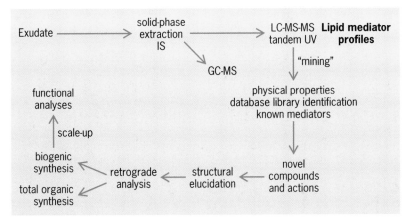

Fig. 4. Functional-mediator lipidomic approach to evaluate and profile compounds during experimental inflammation and its resolution.

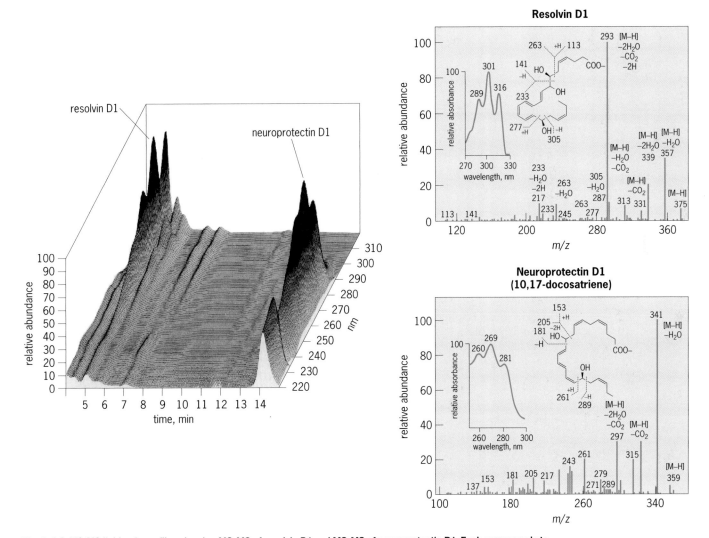

Fig. 5. LC-MS-MS lipidomic profiles showing MS-MS of resolvin D1 and MS-MS of neuroprotectin D1. Each corresponds to the materials beneath the UV chromatophore at the left.

their relationship within a profile of local mediators is a network of events that can be decoded by mediator lipidomics, thus its utility in finding the basis of complex human diseases and developing new therapeutic interventions.

For background information *see* CELL MEMBRANES; EICOSANOIDS; INFLAMMATION; LIPID; LIQUID CHROMATOGRAPHY; MASS SPECTROMETRY; SECOND MESSENGERS in the McGraw-Hill Encyclopedia of Science & Technology.   Charles N. Serhan

Bibliography. K. Gronert et al., A molecular defect in intracellular lipid signaling in human neutrophils in localized aggressive periodontal tissue damage, *J. Immunol.*, 172:1856–1861, 2004; S. Hong et al., Novel docosatrienes and 17S-resolvins generated from docosahexaenoic acid in murine brain, human blood and glial cells: Autacoids in anti-inflammation, *J. Biol. Chem.*, 278:14677–14687, 2003; Y. A. Hannun and L. M. Obeid, The ceramide-centric universe of lipid-mediated cell regulation: Stress encounters of the lipid kind, *J. Biol. Chem.*, 277:25847–25850, 2002; B. D. Levy et al., Lipid mediator class switching during acute inflammation: Signals in resolution, *Nat. Immunol.*, 2:612–619, 2001; Y. Lu, S. Hong, and C. N. Serhan, Mediator-lipidomics databases, algorithms, and software for identification of novel lipid mediators via LC-UV-MS3, in *52d American Society for Mass Spectrometry Conference*, Nashville, 2004; P. K. Mukherjee et al., Neuroprotectin D1: A docosahexaenoic acid-derived docosatriene protects human retinal pigment epithelial cells from oxidative stress, *Proc. Nat. Acad. Sci. USA*, 101:8491–8496, 2004; C. N. Serhan et al., Novel functional sets of lipid-derived mediators with antiinflammatory actions generated from omega-3 fatty acids via cyclooxygenase 2-nonsteroidal antiinflammatory drugs and transcellular processing, *J. Exp. Med.*, 192:1197–1204, 2000; C. N. Serhan et al., Resolvins: A family of bioactive products of omega-3 fatty acid transformation circuits initiated by aspirin treatment that counter proinflammation signals, *J. Exp. Med.*, 196:1025–1037, 2002.

Location-based security

In numerous applications it is desirable to restrict access to plaintext data based on the recipient's location or the time. For instance, in digital cinema distribution, the distributor would like to restrict access to the location of the authorized theater and at the authorized times. Hospitals might want to restrict access to patient records to the locality of the hospital in order to better comply with regulations of the Health Insurance Portability and Accountability Act (HIPAA). Banks and financial institutions could use location-based security techniques to authenticate senders and recipients based on their locations.

In the military arena, navigational waypoints obtained from a captured Global Positioning System (GPS) receiver could reveal the mission's objective. By encrypting them with location and time constraints, a captured navigation set would not easily reveal mission parameters. Orders could be encrypted so as to be undecipherable until in the neighborhood of the target. Weapons systems could be made active, inactive, or booby-trapped based on their location and time.

A prototype location-sensitive encryption-decryption system has been implemented using GPS as the location sensor. Plans call for development of a location-based security module for digital cinema applications where the decoder is actually made part of the projector.

With conventional cryptography, once the decryption keys are known, there is no restriction on access to the data. Geo-encryption builds on established cryptographic algorithms and protocols in a way that allows data to be encrypted for a specific place or a broad geographic area, and supports constraints in time as well as velocity.

The term "location-based encryption" is used here to refer to any method of encryption wherein the cipher text can be decrypted only at a specified location. If an attempt is made to decrypt the data at another location, the decryption process fails and reveals no information about the plaintext. The device performing the decryption determines its location using some sort of location sensor, for example, a GPS receiver or some other satellite or radio-frequency positioning system.

Encryption algorithms. Encryption algorithms can be divided into symmetric algorithms and asymmetric algorithms. Symmetric algorithms use the same key for encrypting and decrypting plaintext (**Fig. 1a**).

Numerous, very fast symmetric algorithms are in widespread use including the Data Encryption Standard (DES), Triple-DES, and the Advanced Encryption Standard (AES), released in 2001. Keeping the key private is essential to maintaining security and

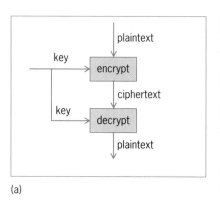

(a)

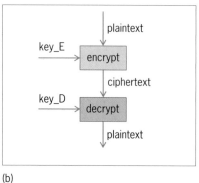

(b)

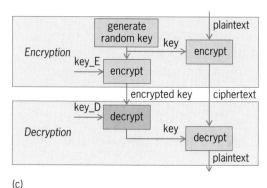

(c)

Fig. 1. Encryption algorithms. (a) Symmetric algorithm. (b) Asymmetric algorithm. (c) Hybrid algorithm.

therein lies the crucial question of how to share keys securely. Numerous techniques have been developed.

Asymmetric algorithms are comparatively new, with the first published description in 1976. Also known as public-key algorithms, they have distinct keys for encryption and decryption (Fig. 1b). Here, key_E can be used to encipher the plaintext but not to decipher it. A separate key, key_D, is needed to perform this function.

In principle, to securely convey the plaintext, the intended recipient could generate a key pair (key_E, key_D) and send key_E, also known as the public key, to the originator via unsecured channels. This would allow the originator (or anyone else) to encrypt plaintext for transmittal to the recipient who uses key_D, the private key, to decrypt the plaintext. Clearly, it is important that the recipient never reveal the private key for this system to be secure.

One major drawback with asymmetric algorithms is that their computational speed is typically orders of magnitude (about 1000 times) slower than comparable symmetric algorithms. This has led to the notion of hybrid algorithms (Fig. 1c). Here, a random key, sometimes called the session key, is generated by the originator and sent to the recipient using an asymmetric algorithm. This session key is then used by both parties to communicate securely using a much faster symmetric algorithm. The hybrid approach has found wide application, notably on the Internet where it forms the basis for secure browsers and secure e-mail.

Location and time constraints. In principle, one could attach location and time specifications to the ciphertext file and build devices that would decrypt the file only when within the specified location and time constraints. There are several potential problems with such an approach:

1. The resultant file reveals the physical location of the intended recipient. The military discourages this type of practice, at least for their own forces. Furthermore, it provides vital information to someone who wants to spoof the device (that is, make it appear that his or her signals originate from the device when they do not actually do so).

2. If the device is vulnerable to tampering, it may be possible to modify it so as to completely bypass the location check. The modified device would decrypt all received data without acquiring its location and verifying that it is correct. Alternatively, an adversary might compromise the keys and build a modified decryption device without the location check. Either way, the modified device could be used anywhere and location would be irrelevant.

Another possibility is to use location itself as the cryptographic key to an otherwise strong encryption algorithm such as AES. This is ill-advised in that location is unlikely to have sufficient entropy (uncertainty) to provide strong protection. Even if an adversary does not know the precise location, there may be enough information to enable a rapid brute force attack analogous to a dictionary attack. For example, suppose that location is coded as a latitude-longitude pair at the precision of 1 cm, and that an adversary is able to narrow down the latitude and longitude to within a kilometer. Then there are only 10^5 possible values each for latitude and longitude, or 10^{10} possible pairs (keys). Testing each of these would be easy.

A guiding principle behind the development of cryptographic systems has been that security should not depend on keeping the algorithms secret, only the keys. This does not mean that the algorithms must be made public, only that they be designed to withstand attack under the assumption that the adversary knows them. Security is then achieved by encoding the secrets in the keys, designing the algorithm so that the best attack requires an exhaustive search of the key space, and using sufficiently long keys that exhaustive search is infeasible.

The GeoEncryption algorithm addresses these issues by building on established security algorithms and protocols by modifying the previously discussed hybrid algorithm to include a GeoLock (**Fig. 2**). On the originating (encrypting) side, a GeoLock is computed based on the intended recipient's PVT block. The PVT block defines where the recipient needs to be in terms of position, velocity, and time for decryption to be successful. The GeoLock is then eXclusive ORed (XORed) with the session key (key_S) to form a GeoLocked session key. (XORing is a bit-by-bit logic operation where, if the two input bits are the same, the resultant is a 0, else it is a 1.) The result is then encrypted using an asymmetric algorithm and conveyed to the recipient, much as in the case of the hybrid algorithm (Fig. 1c). On the recipient (decryption) side, GeoLocks are computed using an antispoof GPS receiver for PVT input into the PVT→ GeoLock mapping function. If the PVT values are correct, the resultant GeoLock will XOR with the GeoLocked key to provide the correct session key, key_S. This is because XORing with the same

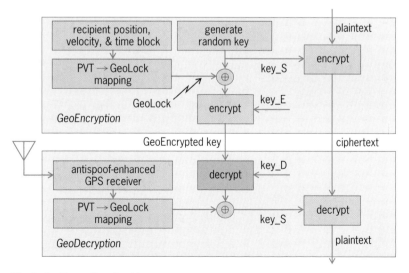

Fig. 2. GeoEncryption algorithm.

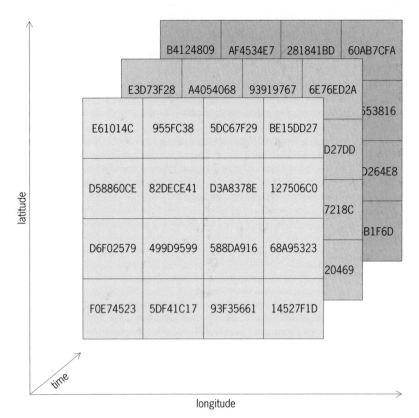

Fig. 3. PVT → GeoLock mapping function.

value, twice in succession, yields the original (A⊕B⊕B=A).

PVT→GeoLock mapping function. The way in which GeoLocks are formed is indicated in **Fig. 3**, which shows a notional diagram of a PVT→GeoLock mapping function where latitude, longitude, and time constitute the inputs. Here, a regular grid of quantized latitude, longitude, and time values has been created, each with an associated GeoLock value. The PVT→GeoLock mapping function itself may incorporate a hash function or one-way function with cryptographic aspects in order to hinder using the GeoLock to obtain PVT block values. Additionally, the algorithm may be deliberately slow and difficult, perhaps based on solving a difficult problem.

Grid spacing must take into account the accuracy of the GPS receiver at the decrypting site; otherwise, erroneous GeoLock values may result. It makes no sense to have 1-cm grid spacing if a standalone GPS receiver with 5-m accuracies is being used. Conversely, if an interferometric GPS receiver capable of 2-cm accuracy is used, 10-m grid spacing is overly conservative. Grid spacing may also be wider in the vertical direction to account for the poorer vertical positioning accuracy typical of most GPS receivers because of satellite geometries.

A more complete PVT→GeoLock mapping function could actually have eight inputs: East, North, and Up coordinates of position; East, North, and Up coordinates of velocity; time; and coordinate system parameters. The velocity inputs might actually map

into a minimum speed requirement so as to ensure that the recipient is actually underway.

For background information *see* COMPUTER SECURITY; CRYPTOGRAPHY; SATELLITE NAVIGATION SYSTEMS in the McGraw-Hill Encyclopedia of Science & Technology.   Logan Scott

Bibliography. R. Anderson, *Security Engineering: A Guide to Building Dependable Distributed Systems*, Wiley, 2001; B. Schneier, *Applied Cryptography*, 2d ed., Wiley, 1996; L. Scott and D. Denning, A location based encryption technique and some of its applications, in Institute of Navigation, *2003 National Technical Meeting Proceedings*, ION-NTM-2003, pp. 734–740, 2003.

Low-speed aircraft

It might be thought that low-speed small-scale aircraft are simply smaller versions of full-scale designs and that textbook aerodynamics apply to them. However, low-speed aircraft have very special lift and drag characteristics that distinguish them from their higher-speed counterparts. To a large measure, this is determined by a key nondimensional parameter called the Reynolds number (Re). It was derived from water-pipe experiments by Professor Osborne Reynolds at the University of Manchester over 100 years ago, but the concept has found crucial application in aerodynamics. For such a powerful parameter, the Reynolds number has a very simple form, as seen below, where ρ is the density of the

$$\mathrm{Re} = \frac{\rho V L}{\mu}$$

fluid, V is the flow velocity, L is the length of the object in the flow, and μ is the fluid's viscosity.

An insightful physical interpretation is to look at the Reynolds number as the ratio of inertial force to viscous force. For example, if we consider two fluids of similar density, such as water and syrup, the Reynolds number for the syrup will be much lower for given values of L and V. In other words, viscous forces will dominate in the flow. Likewise, if we consider an object traveling through air, the Reynolds number equation states that if V and L are small enough, viscous forces will dominate. As a result, we may conclude that, for a small insect, air would seem like a very viscous medium.

A tiny insect is an extreme example of a low-Reynolds-number flying object. This is not a regime approached by even the smallest aircraft [although it is the goal for micro air vehicles, which are typically defined as less than 15 cm (6 in.) in length, width, or height]. Nonetheless, Reynolds number effects are still important. For example, since early wind tunnels were fairly small and low-speed, when wing models were tested the Reynolds numbers were much lower than those for their full-scale counterparts. As a consequence, the best-performing airfoils from these tests were thin and highly curved (cambered), like that on the Wright Flyer, and aircraft

(a)

(b)

Fig. 1. Airfoils from the early twentieth century. (a) Thin airfoil (Göttingen 342). (b) Thick airfoil (Göttingen 387).

designers incorporated such airfoils well into World War I. What was not realized was that at the full-scale Reynolds number achieved by these aircraft a very different type of airfoil would be optimal, one with greater thickness (**Fig. 1**). This was discovered at Göttingen University, where a larger, higher-speed wind tunnel allowed the development of a new family of airfoil sections. The Fokker Company used such airfoils in their famous designs, including the D-7 and D-8 fighters (**Fig. 2**).

The use of thin airfoils by other designers had important nonaerodynamic consequences. Such sections have a fairly small internal volume for structure and thus suffer low torsional rigidity. Consequently, a biplane configuration requires that the wings form the top and bottom of a truss structure, resulting in drag from the bracing wires and struts. By comparison, thick airfoils give significant interior volume for the incorporation of structure. Thus the D-7 and D-8 were very "clean" designs with no external bracing wires.

From that time on, aerodynamicists developed a keen appreciation of Reynolds number effects, and a major activity during the interwar years was the construction of larger, higher-speed, and even variable-density wind tunnels. Research focused on high Reynolds number flow, and the realm of low-speed aircraft was left to model airplane designers and builders whose knowledge of Reynolds number effects was virtually nonexistent. By cut-and-try methods, as well as the motivation of competition, fairly efficient designs were developed. There was also a certain amount of serendipity, in that the multispar construction used by some builders provided a flow turbulation that enhanced performance.

Boundary-layer flow. The earliest low-speed research was performed by F.W. Schmitz in Germany during the early 1940s. His systematic series of wind-tunnel tests confirmed the low-Reynolds-number superiority of thin, cambered sections, as compared to thick airfoils. He also showed how purposeful turbulation could increase the performance of thick sections. All flow has a thin layer that moves slowly relative to the surface of the object over which it is flowing. This is called the boundary layer. Its speed is zero at the surface and increases to full exterior-flow speed a short distance away from the surface. For example, a dusty car cannot be cleaned by driving fast, because the dust is within the boundary layer. Even though the boundary layer is usually very thin compared with the size of an airfoil, it determines

how the exterior flow will behave. That is, controlling the boundary layer controls how the airfoil performs. This is what turbulation does.

The boundary layer may be categorized into two types: laminar and turbulent. A laminar boundary layer consists of very smooth flow. It also produces the lowest skin friction, and has been sought after for decades by full-scale airplane designers. The turbulent boundary layer incorporates small-scale turbulence in its flow and produces more skin friction. However, it is the more stable of the two at higher Re values, and the laminar layer readily transitions to that state. From a drag consideration, this is a nuisance for full-scale aircraft, and special measures have to be taken to achieve any degree of laminar flow (such as polished surfaces, special airfoils, boundary-layer suction, and so on). By comparison, laminar boundary layers are fairly stable at low Re values, where ironically they are not desired. This is because flow separation is a major consideration at low Reynolds numbers, and turbulent layers are better able to resist this. That is, the small-scale turbulence energizes the layer and allows it to move farther against an adverse pressure gradient (such as on the leeward side of bodies and airfoils). An example is golf balls, where the flight of early smooth balls was found to improve with dings and scratches. Hence, dimples were incorporated to "trip" the boundary layer from laminar to turbulent.

Such tripping was also accomplished by the multi-spar construction mentioned earlier, where the wing's covering was stretched over a series of surface spars, giving ridges perpendicular to the flow (**Fig. 3a**). Schmitz identified this phenomenon and expanded on the idea by purposely attaching fine wires to the wing's upper surface or mounting a wire ahead of the leading edge (Fig. 3b). His discoveries were carefully noted by European model-airplane builders, and their designs dominated international competition throughout the 1950s. Since that time, awareness of fundamental aerodynamic effects has been a part of model-airplane design, and this has motivated research and produced results that now have application for the wings and propellers of small

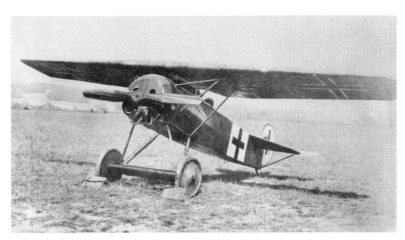

Fig. 2. 1918 Fokker D-8 fighter plane. (*National Air and Space Museum*)

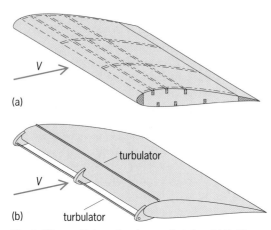

Fig. 3. Wings with boundary-layer turbulation. (a) Multispar construction. V = air velocity relative to wing. (b) Wing with turbulators mounted on upper surface and ahead of leading edge.

unmanned air vehicles (UAVs). Computer codes now exist for which optimized low-Re airfoils may be designed with extraordinary accuracy.

Laminar separation bubble. A crucial low-Re aerodynamic phenomenon that has been identified and modeled is the laminar separation bubble. This effect, when properly used, can provide a natural transition from a laminar to a turbulent boundary layer, without the need for a tripping device. For example, a cigarette smoldering in an ashtray in an absolutely calm room produces a thin, steady laminar stream of smoke that rises straight up. However, no matter how calm the atmosphere, there is a height at which the stream will transition to a turbulent plume, broadening out in a chaotic fashion. This is similar to Reynolds' experiments, in which a thin stream of dye was injected into a smooth flow of water in a glass pipe. At some combination of length and speed, the dye stream would transition like the smoke stream. These values of length and speed, combined with the water's density and viscosity, gave a single nondimensional value which is now known as the transition Reynolds number.

In a simplistic way, **Fig. 4** illustrates the laminar separation bubble effect. When the flow first en-

counters the airfoil, it establishes a laminar boundary layer. However, as it rounds the leading edge and begins to encounter the adverse pressure gradient on the airfoil's upper side, the flow separates from the surface. This action causes the stream to transition and plume, spreading and reattaching to the surface as a turbulent boundary layer. The actual behavior of a laminar separation bubble is more complicated than this, and is the subject of considerable research. However, the key point is that if the bubble is small and tight, it can be beneficial to the airfoil's performance. On the other hand, a large bubble causes early stall and high drag. It is a scientific art form among the designers of low-Re airfoils to tailor the airfoil's pressure distribution to produce the desired bubble behavior.

Low-speed flight. Much of this research is finding application in modern small-scale UAV designs. That is, even though a low-speed UAV may have a configurational similarity to a full-scale counterpart, the differences are subtle and crucial in that its airfoils and propellers must be specialized to this flow regime. This importance grows exponentially as the size decreases toward micro air vehicles.

Nature. In nature, solutions to low-speed flight almost universally use flapping wings. It had been imagined that this was a case of making a virtue out of the necessity of cyclic muscle action. However, recent research on bird, bat, and insect flight has shown that such unsteady motion may offer some unique aerodynamic advantages. For example, at the larger scale of flapping flight (which would include geese and fruit bats), there is strong evidence of propulsive efficiencies matching the best that propellers can produce at that scale. Moreover, the unsteady motion may augment the aerodynamic characteristics through an effect known as dynamic-stall delay. The way this works is that a steady wing will experience flow separation (stall) when its angle to the flight direction (angle of attack) is slowly increased beyond a certain value, whereas if that angle is rapidly increased, the stall will be delayed to a higher angle. Thus, if the angle is quickly reversed before the stall occurs, one may oscillate with attached flow at angles beyond those for steady-state stall. This effect has been experimentally observed for a scaled ornithopter wing.

At the other end of the size spectrum (insects and small hummingbirds), the situation digresses even further from established aerodynamics. Here, purposeful flow separation and vortex generation have important roles. It has always been a goal of aircraft designers to achieve nonseparated flow. This is the condition for minimum drag and thus minimum power for flight. Therefore, the situation becomes very strange for a classically trained aerodynamicist when small flying creatures are studied. For example, Tokel Weiss-Fogh at Cambridge University observed extraordinary lift behavior from hovering tiny wasps and hypothesized that the pair of wings were flapping in a unique fashion. That is, the wings would clap together at the end of each flapping cycle. It

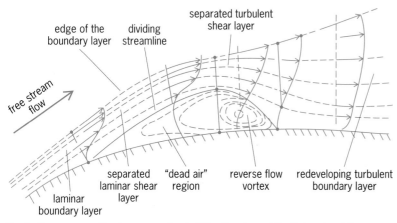

Fig. 4. Time-averaged laminar separation bubble. (*Courtesy of the University of Notre Dame*)

Fig. 5. Micro air vehicle, *Mentor*. The total span (tip-to-tip distance on the wings) is 14 in. (36 cm).

was found that in the course of pulling apart, strong bound vortices were formed on the leading edge of each wing. This gave a boost in lift beyond that explained by any steady-state model. Later, it was found that this "clap-fling" behavior is used by other creatures, such as pigeons making a panicked takeoff. In this instance, it is working as "nature's afterburner," producing emergency thrust at the cost of rapid energy expenditure.

Micro air vehicle. The clap-fling effect has also been mechanically implemented in a micro air vehicle, called the *Mentor*, where two pairs of wings, moving opposite to one another, provide dynamic balance (**Fig. 5**). As with pigeons, this is not the most energy-efficient way to sustain flight, but it is a way to provide considerable thrust within a size-limited wingspan. The bench-top experiments and flight of *Mentor* have clearly confirmed Weiss-Fogh's hypothesis. This also shows how nature provided a useful solution for a very small-scale aircraft. It is anticipated that other such insights will result from ongoing animal flight research. Motivated by micro air vehicle funding, zoologists and aerodynamicists have been collaborating. This is an extraordinarily rich topic because clap-fling is only one of numerous ways in which small creatures fly. For example, the dragonfly is capable of hovering as well as quick translational flight. Dragonflies have been mounted in a wind tunnel to measure forces and observe streamline patterns. This has provided a database for analytical modeling, which promises to be extraordinarily challenging because not only is the flow separated throughout most of the flapping cycle, but also the shed vortices from the front wing evidently interact with the rear wing in an aerodynamically

beneficial way. *See* ADAPTIVE WINGS; FLAPPING-WING PROPULSION.

For background information *see* AERODYNAMIC FORCE; AERODYNAMICS; AFTER BURNER; AIRFOIL; BOUNDARY-LAYER FLOW; FLIGHT; FLUID FLOW; LAMINAR FLOW; REYNOLDS NUMBER; TURBULENT FLOW; WIND TUNNEL; WING in the McGraw-Hill Encyclopedia of Science & Technology.   James D. DeLaurier

Bibliography. J. A. D. Ackroyd, The United Kingdom's contribution to the development of aeronautics, Part 1: From antiquity to the era of the Wrights, *Aeronaut. J.*, Royal Aeronautical Society, January 2000; M. Dreala, *XFOIL: An Analysis and Design System for Low Reynolds Number Airfoils*, Conference on Low Reynolds Number Airfoil Aerodynamics, University of Notre Dame, June 1989; R. Eppler, *Airfoil Design and Data*, Springer, 1990; M. M. O'Meara and T. J. Mueller, Laminar separation bubble characteristics on an airfoil at low Reynolds numbers, *AIAA J. Aircraft*, 25(8):1033–1041, August 1987; M. S. Selig and M. D. Maughmer, Generalized multipoint inverse airfoil design, *AIAA J.*, 30(11): 2618–2625, November 1992; F. W. Schmitz, *Aerodynamics of the Model Airplane*, N70-39001, National Technical Information Service, Springfield, VA, November 1967.

Macromolecular engineering

Macromolecular engineering refers to the process of designing and synthesizing well-defined complex macromolecular architectures. This process allows

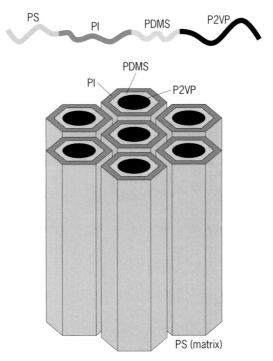

Fig. 1. Schematic illustrating the hexagonal triple coaxial cylindrical structure observed in PS-*b*-PI-*b*-PDMS-*b*-P2VP found by transmission electron microscopy. (*Reprinted with permission from Macromolecules, 35:4859–4861, 2002. Copyright 2002 American Chemical Society.*)

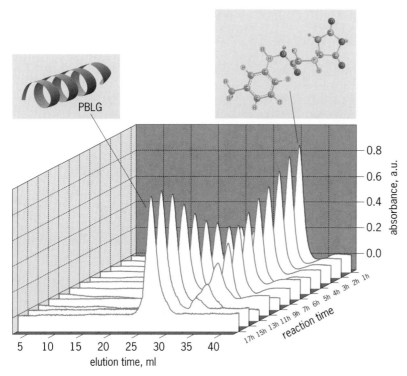

Fig. 2. Monitoring the living polymerization of γ-benzyl-L-glutamate NCA with *n*-hexylamine by size exclusion chromatography with a UV detector. The monomer is transformed 100% to polypeptide. (*Reprinted with permission from Biomacromolecules, 5(5):1653–1656, 2004. Copyright 2004 American Chemical Society.*)

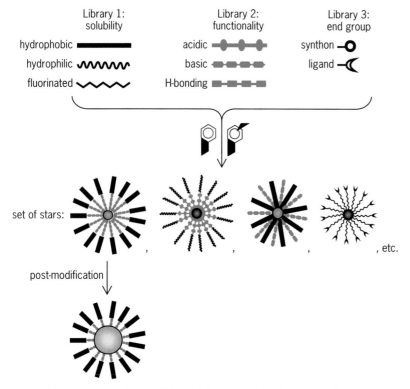

Fig. 3. Schematic representation of the modular approach to star polymers by nitroxide-mediated living radical polymerization. (*Reprinted with permission from J. Amer. Chem. Soc., 125:715–728, 2003. Copyright 2003 American Chemical Society.*)

for the control of molecular parameters such as molecular weight/molecular weight distribution, microstructure/structure, topology, and nature and number of functional groups. In addition, macromolecular engineering is the key to establishing the relationships between the precise molecular architectures and properties of polymeric materials. The understanding of the structure-property interplay is critical for the successful use of these elegantly tailored structures in the design of novel polymeric materials for hi-tech applications, such as tissue engineering, drug delivery, molecular filtration, micro- and optoelectronics, and polymer conductivity.

The year 2003 was characterized by a continuous stream of developments in macromolecular engineering. Although new polymerization techniques were not announced, novel synthetic routes for preparing complex architectures were reported, utilizing living polymerization methods (for which there is no termination step to stop chain growth) and suitable postpolymerization reactions. Novel block copolymers, including star-shaped, branched, grafted, and dendritic-like polymers, were prepared by a variety of living polymerization methods, such as anionic, cationic, living radical, metal-catalyzed polymerization, or combinations of these methods. Only a few examples will be given.

Anionic polymerization. The first organometallic miktoarm (mikto is from a Greek word meaning mixed) star copolymer of the AB$_3$ type, where A is polyferrocenyldimethylsilane (PFS) and B is polyisoprene-3,4 (PI), was synthesized by I. Manners and coworkers by anionic polymerization. The synthetic approach involved the anionic ring-opening polymerization of ferrocenophane, followed by reaction with excess silicon tetrachloride. After the removal of the excess of the volatile linking agent, the resulting macromolecular linking agent PFS-SiCl$_3$ was reacted with excess of living polyisoprenyllithium. The desired product (PFS)Si(PI)$_3$ was separated from the excess PI arm by preparative size exclusion chromatography to yield a near-monodisperse miktoarm star copolymer. The presence of a metallopolymeric block can give redox activity, as well as semiconducting and preceramic properties, to these materials.

The first tetrablock quaterpolymers of polystyrene (PS), polyisoprene (PI), polydimethylsiloxane (PDMS), and poly(2-vinylpyridine) [P2VP] having different total molecular weights and compositions were synthesized by N. Hadjichristidis and coworkers. The synthetic strategy relied on recent advances in the living anionic polymerization of hexamethylcyclotrisiloxane and the use of a heterofunctional linking agent. This heterofunctional linking agent contains two functional groups, one of which reacts selectively with the polydimethylsiloxanyl lithium. The synthetic approach involved the sequential polymerization of styrene, isoprene, and hexamethylcyclotrisiloxane, followed by reaction with stoichiometric amount of the heterofunctional linking agent 2-(chloromethylphenyl)ethyldimethyl chlorosilane. A slight excess of living poly(2-vinylpyridine) lithium was then added to react with the functional group of the resulting macromolecular linking agent PS-PI-PDMS-Ph-CH₂Cl. The excess P2VP was removed by fractional precipitation to give the pure tetrablock quaterpolymers. Due to the incompatibility of the four different blocks, these new quaterpolymers gave a unique four-phase triple coaxial cylindrical microdomain morphology never obtained before (**Fig. 1**). These structures could be used, for example, as multifunctional sensors or multiselective catalysts for sequential or simultaneous chemical reactions.

New ways of living ring-opening polymerization of α-amino acid-*N*-carboxy anhydrides (NCA) to yield polypeptides have been presented, thereby resolving a problem which had existed for more than 50 years. H. Schlaad and coworkers synthesized polystyrene-*b*-poly(γ-benzy-l-glutamate) block copolymers by using the hydrochloric salt of an ω-functionalized —NH₂ polystyrene as the macroinitiator of the ring opening polymerization of γ-benzy-l-glutamate NCA. The use of the hydrochloric acid, instead of the free amine, suppressed the termination reactions. Hadjichristidis and coworkers synthesized well-defined di- and triblock copolypeptides, as well as star-shaped homo- and copolypeptides, by extensively purifying the solvent (dimethylformamide), the monomers (NCAs), and initiator (primary amines) in order to maintain the conditions necessary for the living polymerization. A variety of NCAs, such as γ-benzy-l-glutamate NCA, ε-carbobenzoxy-l-lysine NCA, glycine NCA, and O-benzyl-l-tyrosine NCA, were used to synthesize several combinations of block copolypeptides, showing that this is a general methodology (**Fig. 2**). The combination of the self-assembly of block copolymers and the highly ordered three-dimensional structures of proteins is expected to give novel supramolecular structures.

Living radical polymerization. A modular strategy for the preparation of functional multiarm star polymers by using nitroxide-mediated, living radical polymerization was proposed by J. M. Fréchet, C. Hawker, and coworkers. The approach involved the use of a variety of alkoxyamine-functional initiators for the polymerization of several vinyl monomers. These linear chains, containing a dormant chain end, were coupled with a crosslinkable monomer, such as divinylbenzene, resulting in a star polymer. The ability of this initiator to polymerize a great variety of vinyl monomers, along with the great diversity of the block sequence, led to the synthesis of a myriad of functionalized three-dimensional star polymers. A few examples are given in **Fig. 3**. The diversity of the method becomes even higher when performing postpolymerization reactions on specific blocks of the stars. These unique structures are useful in a range of applications such as supramolecular hosts, catalytic scaffolds, and substrates for nanoparticle formation.

K. Matyjaszewski and coworkers synthesized a series of three- and four-arm stars of poly(bromopropionylethyl-methacrylate-*g-n*-butyl acrylate) by using

atom transfer radical polymerization (ATRP). The synthetic approach involved the use of tri- and tetrafunctional initiators for the ATRP of a 2-(trimethylsiloxyethyl)ethyl methacrylate (HEMA-TMS), which is a monomer containing protected hydroxyl groups. After the polymerization, the hydroxyl groups were deprotected and transformed to new bromine-containing groups, which are initiating sites of ATRP. The three- and four-arm star multifunctional macroinitiators were subsequently transformed to the densely grafted molecular brushes of poly(bromopropionylethyl-methacrylate-*g-n*-butyl acrylate), after the ATRP of poly(*n*-butyl acrylate) [**Fig. 4**]. The researchers were able to see the actual stars by atomic force microscopy (AFM). The synthesis of 4-arm star brushes is given in Fig. 4.

A universal iterative strategy for the divergent synthesis of dendritic macromolecules was presented by V. Percec and coworkers. They used living radical polymerization of conventional monomers, such as methyl methacrylate (MMA), to synthesize the polymeric blocks of the dendrimers, followed by reaction with a specially designed compound, called TERMINI (from TERminator Multifunctional INItiator). TERMINI acts both as a terminator of the living

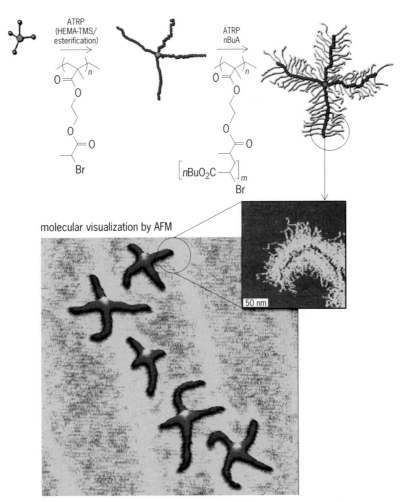

Fig. 4. Reaction sequence for the synthesis of four-arm star brushes, and AFM images. (*Reprinted with permission from Macromolecules, 36:1843–1849, 2003. Copyright 2003 American Chemical Society.*)

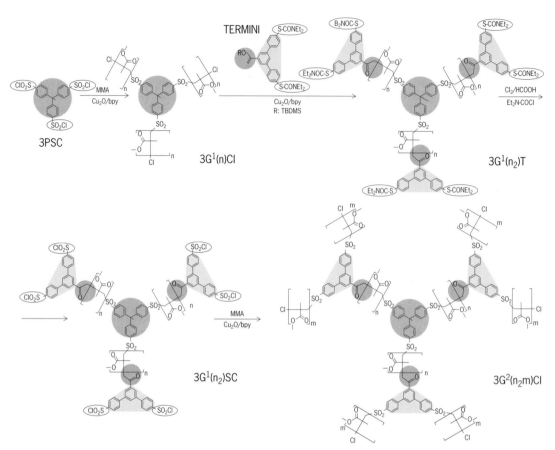

Fig. 5. Divergent iterative synthetic strategy for synthesis of dendritic PMMA. (*Reprinted with permission from J. Amer. Chem. Soc., 125:6503–6516, 2003. Copyright 2003 American Chemical Society.*)

polymer chains and as a masked difunctional initiator. Consequently, after terminating the living polymer and demasking, TERMINI is able to reinitiate the polymerization of MMA by ATRP, generating two new PMMA branches per TERMINI unit with nearly 100% efficiency. Using this methodology, dendrimers with up to four generations were synthesized (**Fig. 5**). The dendrimers were extensively characterized, showing that they exhibited high molecular weight homogeneity.

Cationic polymerization. Benzoic acid 2-methylsulfanyl-4,5-dihydro-oxazolinium-4-ylmethyl ester triflu-

oromethanesulfonate was employed as the initiator by T. Endo and collaborators for the cationic ring-opening polymerization of 1,3-oxazolidine-2-thione. The unique feature of this initiator is that it is stable under air and water, thus promoting the living cationic polymerization and leading to monodisperse polymers.

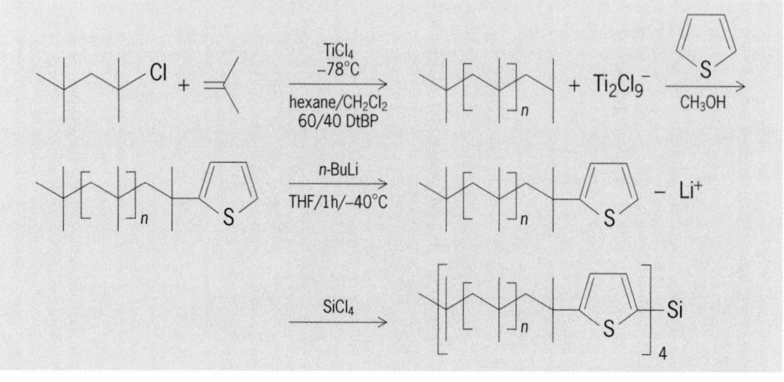

Catalytic polymerizations. Living polymerization of propylene was achieved by T. Fujita and coworkers using a series of titanium (Ti) complexes featuring fluorine-containing phenoxy-imine chelate ligands in the presence of methylaluminoxane (MAO). Despite the fact that the Ti catalyst possesses C_2 symmetry, the polypropylene produced is syndiospecific, through a chain-end control mechanism. Highly syndiotactic and monodisperse polypropylenes were synthesized at the temperature range between 0 and 50°C (32 and 122°F).

Well-defined poly(*n*-butyl methacrylate-*b*-methyl methacrylate) block copolymers were prepared by Hadjichristidis and coworkers, using zirconocene catalysts, fluorophenyl borate cocatalysts, and sequential addition of monomers, starting from the polymerization of *n*-butyl methacrylate. The success of the procedure was based on previous kinetic investigations that have revealed the exact experimental conditions (such as the temperature, nature of catalytic system, catalyst and cocatalyst concentration, and time for the completion of polymerization)

Fig. 6. Synthesis of 4-arm PIB stars by site transformation of thiophene end-capped PIB chain ends.

under which the polymerization is controlled and the termination reactions are minimized.

Postpolymerization reactions. Miktoarm star copolymers PS$_3$PtBA$_3$, where PS is polystyrene and PtBA is poly(t-butyl acrylate), were prepared by J. Teng and E. Zubarev using a rigid hexafunctional core and postpolymerization reactions between functional groups. Carboxy-terminated PS and PtBA chains were linked to a silyl-protected 3,5-dihydroxybenzoic acid in a two-step procedure. Demasking of the carboxyl group from the silyl functions was followed by the reaction with the hydroxyl groups of the hexafunctional core leading to the formation of the miktoarm star. Subsequent hydrolysis under mild acidic conditions produced the corresponding star with poly(acrylic acid) arms.

Combinations of techniques. Specific polymerization techniques are applicable to a limited number of monomers. Therefore, the combination of different polymerization methods is expected to lead to new and unique macromolecular architectures. A new and efficient procedure for the synthesis of polyisobutylene (PIB) stars and PIB-poly(t-butyl methacrylate) block copolymers was developed by A. Müller, R. Faust, and collaborators by combining living carbocationic and anionic polymerizations. Monoaddition of thiophene to living PIB leads to the formation of 2-polyisobutylene-thiophene macroinitiator, which was subsequently metalated with n-butyllithium to transform the living cations to the respective anions. The stable carbanionic species were then used to initiate the polymerization of t-butyl methacrylate. PIB four-arm stars were also prepared via coupling of the living macrocarbanions with SiCl$_4$ as a coupling agent (**Fig. 6**).

Y. Gnanou and coworkers prepared double hydrophilic star-block (PEO$_3$-b-PAA$_3$) and dendrimer-like copolymers (PEO$_3$-b-PAA$_6$) consisting of three inner poly(ethylene oxide) [PEO] arms and either three or six outer poly(acrylic acid) [PAA] arms. Three arm PEO stars, having terminal —OH groups, were prepared by anionic polymerization, using a suitable trifunctional initiator. The hydroxyl functions were subsequently transformed to either three or six bromo-ester groups, which were used to initiate the polymerization of t-butyl acrylate by ATRP. Subsequent hydrolysis of the t-butyl groups yielded the desired products (**Fig. 7**). These double hydrophilic block copolymers exhibit stimuli-responsive properties with potential biotech applications.

Polystyrene/poly(ethylene oxide), star-block, (PS-b-PEO)$_n$ [n = 3, 4] and dendrimer-like, PS$_3$-b-PEO$_6$ copolymers were synthesized by ATRP and anionic polymerization techniques. Three- or four-arm PS stars were prepared using tri- or tetrafunctional benzylbromide initiators. The end bromine groups were reacted with the suitable hydroxylamine in order to generate the same or twice the number of hydroxyl end groups. These functions were then used for the polymerization of ethylene oxide to afford the desired products (**Fig. 8**).

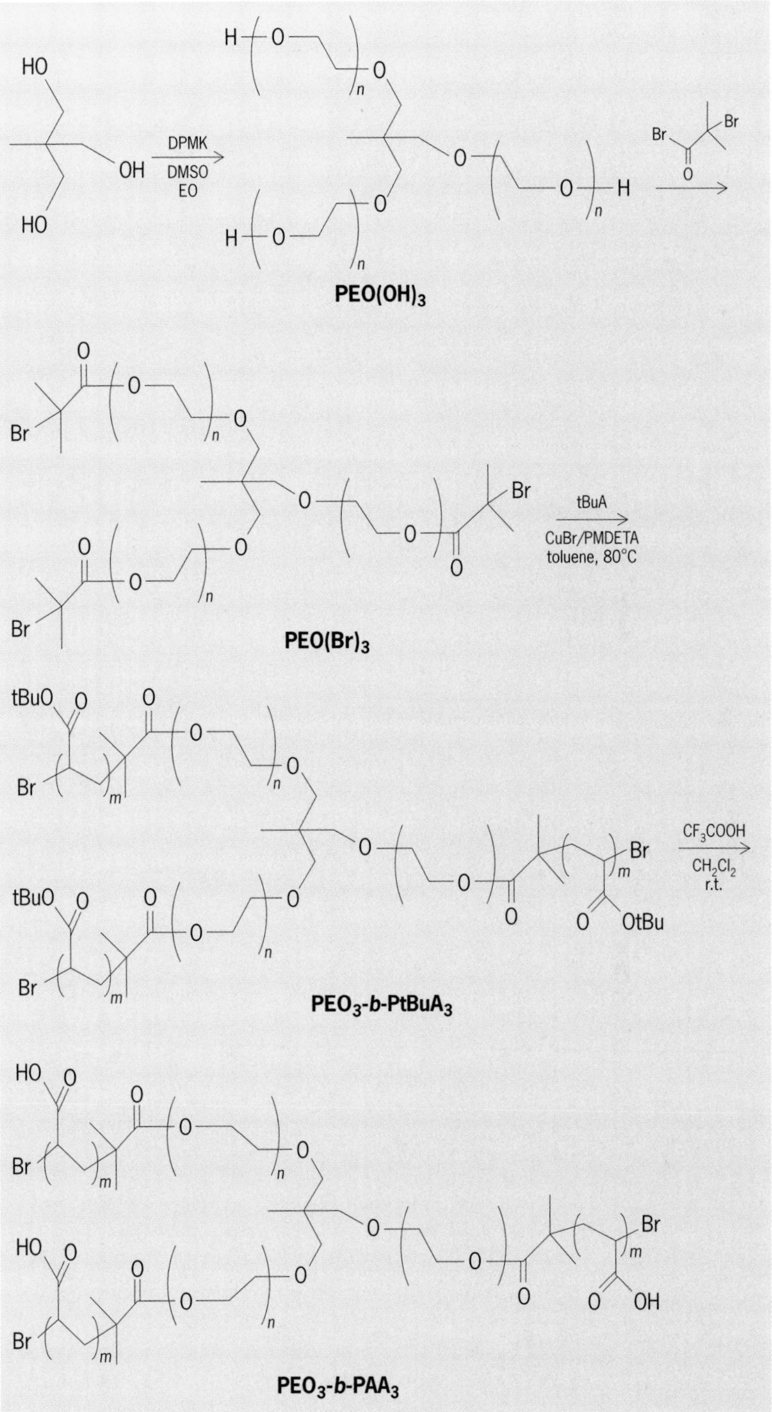

Fig. 7. Synthesis of PEO$_3$-b-PAA$_3$ star block copolymers.

Graft copolymers having poly(methyl methacrylate) [PMMA], backbone and polystyrene [PS], polyisoprene [PI], poly(ethylene oxide) [PEO], poly(2-methyl-1,3-pentadiene) [P2MP] and PS-b-PI branches were prepared by Hadjichristidis and coworkers using the macromonomer methodology. The methacrylic macromonomers were synthesized by anionic polymerization, whereas their homopolymerization and copolymerization with MMA were performed by metallocene catalysts. Relatively high

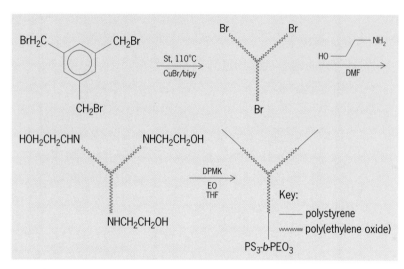

Fig. 8. Synthesis of PS₃-*b*-PEO₃ star block copolymers.

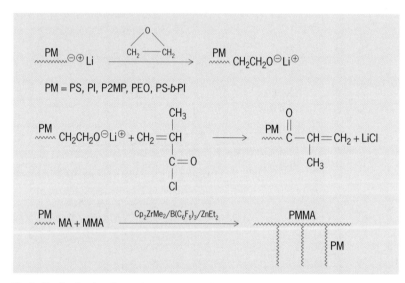

Fig. 9. Synthesis of graft copolymers by combination of anionic and metallocene-catalyzed polymerizations.

macromonomer conversions were obtained in all cases. Parameters such as the flexibility, structure, and molecular weight of the macromonomer as well as the nature of the catalytic system were found to affect the copolymerization behavior (**Fig. 9**).

For background information *see* BRANCHED POLY-MER; COPOLYMER; DENDRITIC MACROMOLECULE; ORGANIC SYNTHESIS; POLYMER; POLYMERIZATION in the McGraw-Hill Encyclopedia of Science & Technology.

Nikos Hadjichristidis; Hermis Iatrou; Marinos Pitsikalis

Bibliography. T. Endo et al., Living cationic ring-opening polymerization by water-stable initiator: Synthesis of a well-defined optically active polythiourethane, *Chem. Commun.*, 24:3018–3019, 2003; J. M. Fréchet, C. Hawker, et al., A modular approach toward functionalized three-dimensional macromolecules: From synthetic concepts to practical applications, *J. Amer. Chem. Soc.*, 125:715–728, 2003; T. Fujita et al., Syndiospecific living propylene polymerization catalyzed by titanium complexes having fluorine-containing phenoxy-imine chelate ligands, *J. Amer. Chem. Soc.*, 125:4293–4305, 2003; Y. Gnanou et al., Synthesis and surface properties of amphiphilic star-shaped and dendrimer-like copolymers based on polystyrene core and poly(ethylene oxide) corona, *Macromolecules*, 36:8253–8259, 2003; Y. Gnanou et al., Synthesis of water-soluble star-block and dendrimer-like copolymers based on poly(ethylene oxide) and poly(acrylic acid), *Macromolecules*, 36:3874–3881, 2003; N. Hadjichristidis et al., *Block Copolymers: Synthetic Strategies, Physical Properties and Applications*, Wiley-Interscience, 2003; N. Hadjichristidis et al., Complex macromolecular architectures utilizing metallocene catalysts, *Macromolecules*, 36:9763–9774, 2003; N. Hadjichristidis et al., Living polypeptides, *Biomacromolecules*, 5(5):1653–1656, 2004; N. Hadjichristidis et al., Synthesis and characterization of linear tetrablock quaterpolymers of styrene, isoprene, dimethylsiloxane and 2-vinylpyridine, *J. Polym. Sci., Polym. Chem.*, 42:514–519, 2004; N. Hadjichristidis, H. Hasegawa, et al., Four-phase triple coaxial cylindrical microdomain morphology in a linear tetrablock quarterpolymer of styrene, isoprene, dimethylsiloxane and 2-vinylpyridine, *Macromolecules*, 35:4859–4861, 2002; I. Manners et al., Synthesis of the first organometallic miktoarm star polymer, *Macromol. Rapid Commun.*, 24:403–407, 2003; K. Matyjaszewski et al., Effect of initiation conditions on the uniformity of three-arm star molecular brushes, *Macromolecules*, 36:1843–1849, 2003; A. Müller, R. Faust, et al., Polyisobutylene stars and polyisobutylene-block-poly(*tert*-butyl methacrylate) block copolymers by site transformation of thiophene end-capped polyisobutylene chain ends, *Macromolecules*, 36:6985–6994, 2003; V. Percec et al., Universal iterative strategy for the divergent synthesis of dendritic macromolecules from conventional monomers by a combination of living radical polymerization and irreversible terminator multifunctional initiator (TERMINI), *J. Amer. Chem. Soc.*, 125:6503–6516, 2003; H. Schlaad et al., Synthesis of nearly monodisperse polystyrene–polypeptide block copolymers *via* polymerisation of *N*-carboxyanhydrides, *Chem. Commun.*, 23:2944–2945, 2003; E. Zubarev et al., Synthesis and self-assembly of a heteroarm star amphiphile with 12 alternating arms and a well-defined core, *J. Amer. Chem. Soc.*, 125:11840–11841, 2003.

Madagascan primates

The fauna and flora of Madagascar, Africa's Great Red Island, can be described as fitting what botanist Quentin C. B. Cronk recently called the "diversity and stability paradox" of island biogeography. The diversity paradox is that, despite the existence of numerous unique island taxa (whose loss would drastically impact the taxonomic diversity of represented groups), the actual species diversity (or richness) of island biota is typically low in comparison with that of like-sized continental landmasses. The stability paradox is that, although island flora and fauna may appear to be stable (as many of the taxa are ancient in origin), they are highly susceptible to rapid change and decimation, particularly when subjected to human colonization.

Extant lemurs. Lemurs are Madagascar's outstanding species. They are native only to Madagascar and the Comores (small islands off the coast of Madagascar); they constitute about one-third of the Earth's families of living primates.

Taxonomy. Living lemurs belong to five families: Daubentoniidae, Cheirogaleidae, Megaladapidae, Lemuridae, and Indriidae (**Fig. 1**) [see **table**]. Additional families (Palaeopropithecidae and Archaeolemuridae) have been erected to accommodate some of the recently extinct, "subfossil" lemurs.

Diversity. The diversity of lemurs has diminished dramatically since the advent of humans on the island (as evidenced by the dating of *Palaeopropithecus* bones with human-induced cutmarks and the pollen of introduced plants, such as marijuana) a little over 2000 years ago. About one-third of lemur species that occupied the island during the Pleistocene and early Holocene have disappeared completely. At least half of the remaining lemur species are threatened with extinction, about one-third critically so. It is likely that fewer than 5000 individuals remain for some lemur species (for example, the greater bamboo lemur, *Hapalemur simus*), and sharp decreases in the geographic ranges of still-extant species can be traced over the past several thousand years, and even the past 100 years. There are about 500 threatened animal species living on Madagascar today (according to the World Conservation Union), and many species of plants are also threatened.

Behavior and ecology. The living lemurs have been the subjects of intensive research over the past 25 years, and much is now known about their social behavior, dietary adaptations, reproduction, demography, and activity patterns (see table). Lemurs are highly arboreal, although some (notably, ringtailed lemurs) spend considerable time on the ground. Social organization is quite variable, ranging from troops of several dozen individuals, to small groups, to solitary foragers (though these individuals may form fairly large sleeping clusters). Activity patterns also vary, from strictly nocturnal to cathemeral (active day and night) to strictly diurnal. Some consume gums, insects, and fruit, whereas others are highly folivorous (specialized leaf eating). Some are skilled vertical clinger-and-leapers, whereas others are arboreal quadrupeds.

Physiological adaptations. Primatologist Patricia C. Wright argued that living lemurs have adaptations to conserve energy or to maximize exploitation of scarce or difficult-to-process resources. Such adaptations may help lemurs to cope with periodic resource shortages in their island environment. Adaptations for energy conservation include physiological tolerance of low water consumption (in some species); behavioral adaptations, such as huddling or

Fig. 1. Coquerel's sifaka (genus *Propithecus*) belongs to the family Indriidae, the sister taxon to the sloth lemurs. Coquerel's sifaka lives in the dry forests of northwest Madagascar. (*Photograph courtesy of Laurie Godfrey, 2004*)

sunning to conserve body heat (in some species); and a low basal metabolic rate (in many species). Lemurs are energy minimizers, lowering activity levels to decrease water and caloric requirements. The smallest species (mouse lemurs) experience daily torpor; that is, they lower their internal body temperature during the coldest time of day. Some cheirogaleids enter seasonal hiberation, apparently to conserve energy or water during the season of scarce resources.

Adaptations to maximize the exploitation of scarce resources include cathemerality (the ability to shift periods of activity from day to night or both). This behavioral pattern, which occurs in some species, may facilitate access to a wide variety of resources. Female dominance over males (which is exhibited by many lemur species) allows females to gain the body mass they require to support gestation and lactation. Special feeding adaptations are exhibited by some lemur species: aye-ayes use their incisors to bore holes and their specialized fingers to extract beetle larvae from dead wood. Many species have the digestive adaptations to process tough, fibrous foods, including mature leaves. *Hapalemur aureus*, for example, thrives on a poisonous species of bamboo.

New species. The number of lemur species remaining on Madagascar today is unknown. Deforestation continues at an alarming rate; meanwhile, new species of lemurs (particularly small-bodied, nocturnal ones) are being discovered in small pockets of remaining forest. In some cases, new species have been proposed on the basis of museum specimens (usually, individuals captured at some time during the past 200 years); the status of some of these has not been confirmed in the wild. In other cases, new species have been recognized on the basis of

Lemur families and genera		
Family and included genera	Common names	Characteristics
Palaeopropithecidae *Palaeopropithecus* *Archaeoindris* *Mesopropithecus* *Babakotia*	Sloth lemurs	This diverse family is entirely extinct. Called sloth lemurs because of similarities to sloths, these lemurs are related to living indriids. The largest of all lemurs, *Archaeoindris*, belongs to this family. Sloth lemurs had long, curved digits, and most were specialized hangers. They fed on a combination of leaves, fruit, and seeds. Sloth lemurs survived the advent of humans to Madagascar by at least 1500 years. There is evidence of human butchery of sloth lemurs in southwest Madagascar more than 2000 years ago, shortly after humans first colonized the island.
Indriidae *Indri* *Propithecus* *Avahi*	Indris, sifakas, and woolly lemurs	This family includes the largest of living lemurs (*Indri indri* and *Propithecus diadema*) as well as a smaller-bodied form (*Avahi*). The larger-bodied taxa are diurnal, but the smaller-bodied taxon is nocturnal. Indriids move by leaps in the trees and hops on the ground. They eat leaves, fruit, and seeds.
Megaladapidae *Megaladapis* *Lepilemur*	Koala and sportive lemurs	The extinct koala lemurs (*Megaladapis*) were much larger than their extant relatives, the nocturnal sportive lemurs (*Lepilemur*), but they resembled the latter in diet (leaves) and commitment to life in the trees. The largest *Megaladapis* species was the size of a male orangutan. *Megaladapis* was still extant when humans arrived on Madagascar, and well into the last millennium.
Archaeolemuridae *Archaeolemur* *Hadropithecus*	Monkey lemurs	These robust, baboon-sized lemurs may have been among the last of the giant lemurs to become extinct. Called monkey lemurs because of convergences to baboons and macaques, these were likely the most terrestrial of the giant lemurs. Archaeolemurids were able to break open hard objects (such as nuts) with their teeth; there is also some evidence for omnivory in *Archaeolemur*.
Daubentoniidae *Daubentonia*	Aye-ayes	This family includes the living aye-aye and its giant extinct relative, *Daubentonia robusta*. The latter was still extant when humans arrived on Madagascar; its incisors were collected, drilled, and probably strung on necklaces. Aye-ayes are the largest nocturnal lemurs of Madagascar.
Lemuridae *Lemur* *Hapalemur* *Eulemur* *Varecia* *Pachylemur*	Bamboo, brown, mongoose, red-bellied, ruffed, and ringtailed lemurs	This family of quadrupedal lemurs has one extinct member, *Pachylemur* (three times the mass of the largest living members of this group). Bamboo lemurs prefer bamboo; all other lemurids prefer fruit. Extant lemurids are either cathemeral or diurnal.
Cheirogaleidae *Cheirogaleus* *Microcebus* *Mirza* *Phaner* *Allocebus*	Dwarf, mouse, Coquerel's, fork-marked, and hairy-eared lemurs	These small-bodied lemurs are entirely nocturnal. Some weigh only 30 grams as adults. Mouse and dwarf lemurs experience daily or seasonal torpor. There are no known extinct cheirogaleids on Madagascar.

behavioral, ecological, and genetic studies of wild animals. During the past two decades, new species of the genera *Microcebus, Cheirogaleus, Avahi, Hapalemur,* and *Propithecus* have been named, and still others have been tentatively proposed. Although the number of recognized lemur species (particularly nocturnal ones) is growing, the future looks grim for many of these.

Extinct lemurs. The extinct lemurs comprise eight genera (17 species, including one in the process of being described). Their skeletal remains have been found alongside those of still-extant species at marsh, stream, and cave sites throughout Madagascar. They belong to the island's modern primate fauna; some were alive as recently as approximately 500 years ago. Paleoecologist David Burney has developed a chronology of events associated with the disappearance of the giant lemurs. In geologic terms the extinctions were catastrophic and coincident (occurred at the same time), following the arrival of humans on Madagascar. Within a modern human time frame, however, that extinction window was prolonged, spanning a period of at least 1500 years (from approximately 2000 to 500 years ago). Numerous latest-known occurrences of now-extinct lemurs as well as nonprimate megafauna (gigantic ratites called elephant birds, couas, giant tortoises, an aardvark-like termite feeder called *Plesiorycteropus,* several species of hippopotami, and giant rodents) fall in this period.

Natural history. Modern tools of analysis have enabled researchers to reconstruct the lifeways of extinct species more thoroughly than ever before. The diets of extinct lemurs have been reconstructed based not only on dental morphology and craniofacial architecture, but also on dental enamel prism structure and direct evidence, such as scratches on the teeth (or microwear), the chemical composition of the bone, and the colon contents (fecal pellets) found in association with an *Archaeolemur* skeleton in a cave. Paleontologists have been able to reconstruct virtually complete skeletons of extinct lemurs. Molecular biologists have been able to extract and amplify deoxyribonucleic acid (DNA) from the bones. It is now possible to explore the molecular and anatomical evidence for extinct lemur phylogeny.

All of the extinct lemur species were large in body size. They ranged in size from under 10 kg (22 lb) to over 200 kg (440 lb)—that is, from little over the size of an indri to about the size of a large adult male gorilla (**Fig. 2**). Most exhibit skeletal adaptations for arboreality (tree climbing, though in this case not leaping). This suggests a commitment to wooded, though not necessarily closed-forest, habitats. Among the extinct lemurs, the most celebrated candidates for terrestriality (and, therefore, open-habitat exploitation) are *Archaeolemur* and *Hadropithecus*. Tooth use-wear analysis suggests that most of Madagascar's extinct lemurs were tree-foliage or mixed fruit-and-tree-foliage browsers. There is some evidence for omnivory, as well as hard-object exploitation (possibly fruit with hard pericarps, hard seeds, or gastropods), in the Archaeolemuridae.

Megafaunal extinction hypotheses. A variety of explanations have been proposed for the megafaunal extinctions, including:

1. Paul Martin's blitzkrieg or overkill hypothesis—Overhunting by humans killed the megafauna.

2. Joel Mahé's climate change hypothesis—Increasing aridification resulted in the drying of water sources essential to the survival of megafauna.

3. Henri Humbert's great fire hypothesis—Fires set by humans transformed the landscape.

4. Ross MacPhee's hypervirulent disease hypothesis—Pathogens introduced by humans killed the megafauna.

5. Robert Dewar's biological invasion hypothesis—Endemic species were replaced through resource competition with livestock, particularly cattle, introduced by humans.

6. Daniel Gade's deforestation hypothesis—Human-induced forest destruction was responsible for the disappearance of the megafauna.

7. David Burney's synergy hypothesis—A variety of natural and human impacts interacted to endanger many species and finally to exterminate some.

Each of these hypotheses attempts to explain the particular susceptibility of large-bodied species to endangerment and extinction. The blitzkrieg model argues that large-bodied species make easy hunting targets for humans and provide the best rewards per unit hunting effort. The hypervirulent disease hypothesis argues that large-bodied species are least

Fig. 2. Gorilla-sized *Archaeoindris fontoynontii* (bottom) and its relative, chimpanzee-sized *Palaeopropithecus ingens* (top) belong to the extinct family, the Palaeopropithecidae, known as the sloth lemurs due to their remarkable convergences to sloths. (*Photograph courtesy of Ian Tattersall, American Museum of Natural History.*)

able to rebound from the effects of an epidemic, given their slow reproductive rates and relatively large number of particularly vulnerable (that is, immature or elderly) individuals. The biological invasion hypothesis sees introduced livestock as ecological vicars (and thus competitors) of large-bodied, terrestrial, endemic megafauna. Deforestation (including forest fragmentation) is known to threaten large-bodied before smaller-bodied species. In general, species with slow reproductive rates and larger habitat-area requirements are more vulnerable to extinction than species with fast reproductive rates and smaller habitat-area requirements, and large-bodied species are likely to have slow reproductive rates and large habitat-area requirements.

Recent evidence on cause of extinctions. Recently, an elegant study has shed considerable light on megafaunal extinctions in Madagascar. Looking at datable sediment cores, Burney and his colleagues examined the concentration of spores of a dung fungus called *Sporormiella*. Fossil spores of this fungus provide an excellent indirect measure of megafaunal biomass, because *Sporormiella* can complete its life cycle only on large dung. Burney found a sharp reduction in the *Sporormiella* spore concentration immediately following human arrival. This decline is followed centuries later by a sharp increase in microscopic charcoal (indicative of an increased frequency or severity of fire). *Sporormiella* spore concentrations do not rise again until 1000 years later, in all likelihood with the proliferation of introduced livestock.

Burney hypothesizes that human overhunting of open-habitat, grazing species (giant tortoises, hippos, and elephant birds) would have initially disrupted the producer/consumer equilibrium on savannas and bushlands, effecting a decrease in the biomass of the megafaunal consumers and an increase in the terrestrial herb biomass, which in turn would have impacted the fire ecology of Madagascar. The combination of increased fire and continued hunting pressure would have then threatened the forests and the species (including lemurs) that depended on them. The late apparent proliferation of livestock speaks against the competitive exclusion hypothesis, and the drop in megafaunal biomass prior to the microscopic charcoal spike speaks in favor of hunting as providing the initial (but by no means the only) human insult to the system—triggering a domino process that continues today.

Conservation. Madagascar was recently designated a biodiversity hotspot to underscore its importance to biodiversity conservation as well as the ongoing loss of critical habitat. Despite conservation efforts, the lemurs of Madagascar continue to decline under the combined impacts of deforestation, forest fragmentation, and hunting. Few lemurs survive anywhere in the central portion of Madagascar, where they once thrived. There are today approximately 50 legally protected areas on Madagascar comprising over 1,500,000 hectares. Reserve management depends on designated conservation priority levels, which in turn depends on the biodiversity represented at each site. Ecological monitoring is essential, and an effort is being made to expand both the monitoring and the physical network of protected areas.

For background information *see* BIODIVERSITY; ENDANGERED SPECIES; FOSSIL PRIMATES; POPULATION ECOLOGY; PRIMATES in the McGraw-Hill Encyclopedia of Science & Technology.   Laurie R. Godfrey

Bibliography. S. M. Goodman and J. P. Benstead (eds.), *The Natural History of Madagascar*, University of Chicago Press, 2003; R. A. Mittermeier et al., *Lemurs of Madagascar*, Conservation International, Washington, DC, 1994; J.-J. Petter and F. McIntyre, *Lemurs of Madagascar and the Comores* (CD-ROM), ETI (Expert Center for Taxonomic Identification), Amsterdam, 2000; K. Preston-Mafham, *Madagascar, a Natural History*, Facts on File, Inc., New York, 1997; I. Tattersall, *The Primates of Madagascar*, Columbia University Press, New York, 1982.

Magnetorheological finishing (optics)

Magnetorheological finishing (MRF) is revolutionizing the way that optics are manufactured. The process utilizes polishing slurries based on magnetorheological fluids, which exhibit an increase in viscosity in the presence of a magnetic field. These fluids are mixed, pumped, and conditioned in their liquid states, then converted to a semisolid state to create a stable and conformable polishing pad when applied to the workpiece surface. Material removal rates can be adjusted over a wide range by changing various MRF process parameters, enabling the removal, with each iteration, of more than 10 micrometers of material, or as little as 10 nanometers. MRF offers a deterministic and flexible polishing and final figuring process that can simultaneously improve surface form, roughness, and integrity (by minimizing subsurface damage and residual stress in the crystal lattice of the lens from grinding), and it can be used on a wide variety of materials and part geometries.

Advantages. For centuries, optics have been polished using abrasive slurries and dedicated polishing laps: A desired spherical surface (denoted by its radius of curvature, R) is polished by creating pressure and relative velocity between the optical surface and a full-aperture spherical lap with precisely the opposite radius, $-R$. The traditional technique has a number of significant drawbacks. Convergence of the surface to the desired shape can be slow and may be unpredictable for a number of reasons, such as lap wear and changes in the slurry over time. Part geometry can also strongly influence final quality. For example, high-aspect-ratio optics (diameter greater than ten times the thickness) are challenging because it is difficult to get uniform pressure over the entire surface. Large tooling inventories may be required because a dedicated tool is needed for every desired radius. Finally, it is not possible to conventionally polish nonspherical optics—aspheres—which are an increasingly popular class of optics that improve imaging quality and reduce overall system size and weight. The constantly changing curvature of aspheric optics prevents a full-aperture lap from continuously contacting the surface of the optic during polishing.

The MRF process overcomes many of these limitations. The polishing tool is fluid-based and therefore conforms to the surface being polished. This allows the same machine to polish convex, concave, and plano (flat) surfaces, both spherical and aspheric, without the need for dedicated tooling. Short setup

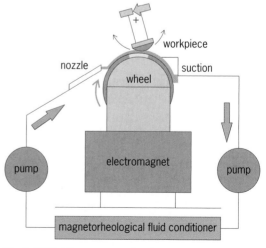

Fig. 1. MRF delivery system showing the main components and basic motion of the workpiece.

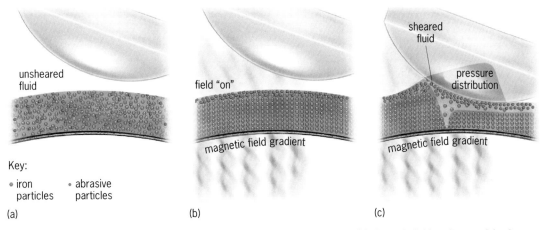

Key:

• iron
  particles
• abrasive
  particles

(a)   (b)   (c)

Fig. 2. Magnetorheological fluid flow in the polishing zone. (*a*) No magnetic field. (*b*) Magnetic field on; iron particles form chains. (*c*) Part immersed in stiffened fluid creates subaperture polishing zone.

times also facilitate cost-effective processing of small batches of high-precision optics.

MRF process. An MRF polishing machine consists of a fluid delivery system (creating the MRF tool) and a computer-numerical-controlled (CNC) machine for controlling the position and velocity of the part relative to the tool. The MRF fluid delivery system circulates about 1 liter (0.26 gallon) of magnetorheological fluid in a closed loop, actively controlling the properties of the fluid that affect the removal rate. A delivery pump forces the magnetorheological fluid through a nozzle onto a rotating wheel, forming a thin ribbon of fluid approximately 2 mm (0.08 in.) thick and 5 mm (0.2 in.) wide (**Fig. 1**). An electromagnet creates a strong magnetic field gradient at the top of the wheel, stiffening the fluid where the optical surface will be immersed. The wheel carries the fluid out of the magnetic field to a suction/scraper unit that removes the now-liquid fluid from the wheel surface. A pump returns the fluid to a conditioning vessel that maintains the fluid's temperature and prevents sedimentation. The result is a dynamic polishing tool that is extremely stable—the removal rate is maintained to within a few percent over an entire day.

Magnetorheological fluid is essentially a mixture of water, iron, and polishing abrasive. When this fluid passes through a magnetic field, the iron particles form chains along the magnetic field lines, giving the fluid a stiff "structure" and increasing its viscosity by more than 4 orders of magnitude (**Fig. 2**). The stiffened fluid is compressed in the converging gap between the rotating wheel and the surface of the optic. The iron particles are attracted downward toward the wheel, and the liquid and polishing abrasive are forced up into a thin (100–150 μm) sheared fluid layer, creating a high tangential shear force on the optical surface. This shear-dominant removal mechanism is the key to MRF's ability to produce extremely low surface roughness (less than 0.5 nm root-mean-square), remove subsurface damage, and relieve surface stresses induced in prior grinding or polishing steps. It also leads to peak removal rates that are very

high relative to other polishing processes (as high as 5–10 μm/min on most optical materials).

The MRF process creates a polishing tool (called a spot) with a unique shape (**Fig. 3**). There is a "soft" leading edge as the fluid first comes into contact with the part. As the converging gap is reduced, the pressure and shear forces increase, as does the removal rate. As the fluid passes by the apex of the wheel (the minimum gap), the removal rate reaches its maximum, and then rapidly falls off because the fluid remains stiff and compressed against the wheel. This shape offers many advantages, as long as the shape—especially the sharp edges—is well characterized. The sharp trailing edge of the spot allows features even smaller than the spot itself to be corrected or induced. MRF can simultaneously combine high removal rates with high-resolution correction capability to enable short cycle times while performing high-precision figure correction (convergence on the optimal shape of the optic).

To figure-correct an optical surface, the dwell time of the MRF spot is varied as it moves across the

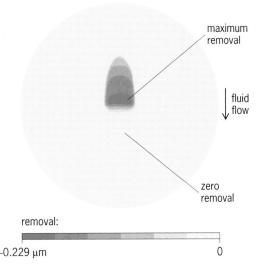

maximum
removal

fluid
flow

zero
removal

removal:

−0.229 μm   0

Fig. 3. Measurement of an MRF polishing spot, illustrating the unique three-dimensional shape.

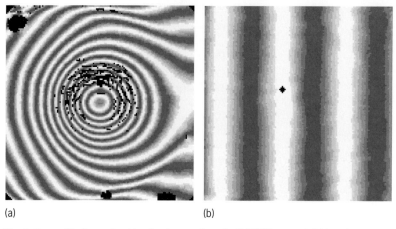

(a)   (b)

Fig. 4. Transmitted wavefront (system) correction of a Nd:YLF laser rod. (*a*) Interference fringes showing pre-MRF condition (∼8 λ). (*b*) Interference fringes showing post-MRF condition (∼0.15 λ).

surface (the longer the spot spends over an area, the more material is removed). Dwell time is manipulated by varying the velocity of the part along a spiral or raster-based toolpath produced by the CNC ma-

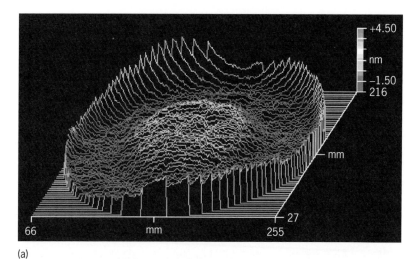

(a)

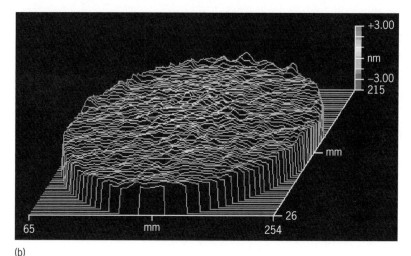

(b)

Fig. 5. Thickness uniformity correction of a thin silicon film on a silicon-on-insulator (SOI) wafer. (*a*) Pre-MRF condition (PV ∼ 6 nm). (*b*) Post-MRF condition (PV ∼ 1.7 nm).

chine. The precise dwell time required for each point on the optical surface is determined using advanced deconvolution algorithms that take into account an empirical characterization of the MRF spot and measurements of the optical surface to be corrected. For the dwell time to be accurate, the spot shape must be well characterized. Accurate characterization is achieved by using a high-resolution interferometer to measure the actual removal rate on a sample part (same material, same curvature). This process generates an accurate three-dimensional representation of the actual removal rate and alleviates the need for a complicated model that would depend on fluid formulation, process parameters, and the material and shape of the part to be polished.

Applications. MRF has many applications in the manufacture of high-precision optics, four of which are given here.

High-precision figure correction. Figure correction begins with a measurement of the surface to be corrected, typically obtained through interferometry, but profilometry or other surface metrology techniques can be readily applied. Surface measurements determine the desired "hitmap" for the dwell/velocity calculation. The velocity schedule is calculated and converted into CNC instructions. Improvements in root-mean-square (RMS) error on the order of a factor of 5–10 are typical in one iteration. This high rate of convergence typically leads to finished optics in one or two iterations. Metrology is often the limiting factor—the reproducibility of most interferometers is sensitive to environmental conditions and mounting, and bottoms out at around λ/20 peak-to-valley (PV), where λ is the wavelength. MRF can continue to correct beyond that limit if accurate metrology can be obtained (as the final example below will illustrate).

Projection optics in microlithography tools demand the most precise optical elements that can be manufactured today: ever-shrinking linewidths are needed to pack more circuitry into tighter spaces. MRF is applied after conventional (pitch) or CNC polishing, to bring surfaces from ∼λ/4 PV to better than λ/20 PV.

Aspherizing. A second common application of MRF is to induce a prescribed aspheric departure into a conventionally polished best-fit sphere (a preliminary sphere from which the least amount of material has to be removed to obtain the desired aspheric surface). This is optimal when the departure is relatively small (less than 10 μm) but has been applied to aspheres in excess of 40 μm of aspheric departure. The desired aspheric prescription is entered into the control software as well as the starting spherical radius, and the MRF system calculates and removes the difference. The figure correction process can then continue as in the first example as long as there is a way to measure the error from the desired asphere (such as using a profilometer or an interferometer plus a null lens).

System correction. A less obvious application of MRF is in system correction: correcting the transmitted wavefront quality of an optical system by inducing a prescribed error into only one surface. MRF's deterministic, subaperture approach makes it ideal for this type of correction. For example, **Fig. 4** shows a Nd:YLF laser rod measured in transmission before and after system correction. The total error is a combination of the front and back surface flatness as well as the index of refraction variation (inhomogeneity) of the rod itself. MRF correction is applied to one surface only, creating the "perfectly bad" figure error that will compensate for the total transmitted wavefront errors. This correction led to a focal spot near the diffraction limit, which allowed a larger area of the rod to be utilized as a gain medium, thus obtaining higher energy output.

Thin-film uniformity correction. A final application of MRF illustrates the relationship between metrology capability and figuring capability. Silicon-on-insulator (SOI) wafers have a thin (~150 nm) insulating layer of silicon dioxide (SiO_2) sandwiched between a silicon substrate and a thin (50–100 nm) silicon layer. The thickness uniformity of the top silicon layer is crucial to the performance of the circuitry that will be created on it. Ellipsometry and spectral reflectometry are two optical metrology techniques that can measure the absolute thickness of a thin film across its surface with subnanometer precision. MRF can use one of these thickness maps—instead of the typical interferometric surface map—to define the desired hitmap for an MRF correction. The machine then removes the prescribed amount of material on the wafer surface to improve the thickness uniformity, not the flatness. For example, the PV of a silicon layer 200 nm in diameter and 70 nm thick can be reduced from ~6 nm to 1.7 nm (**Fig. 5**). The corrected PV is 10 times smaller than the PV (~20 nm) of the high-precision figure correction example above, showing that MRF can typically correct up to the accuracy of the available metrology.

For background information *see* COMPUTER NUMERICAL CONTROL; CONFORMAL OPTICS; GEOMETRICAL OPTICS; INTERFEROMETRY; OPTICAL SURFACES; RHEOLOGY in the McGraw-Hill Encyclopedia of Science & Technology.   Paul Dumas

Bibliography. V. Bagnoud et al., *High-Energy, 5-Hz Repetition Rate Laser Amplifier Using Wavefront Corrected Nd:YLF Laser Rods*, presented at CLEO, Optical Society of America, 2003; D. Golini et al., MRF polishes calcium to high quality, *Optoelectr. World*, 37(7):85–89, July 2001; D. Golini et al., Precision optics fabrication using magnetorheological finishing, *Proc. SPIE*, vol. CR67, Optical Manufacturing and Testing Conference II, SPIE Annual Meeting, San Diego, July 1997; W. I. Kordonski and S. D. Jacobs, Magnetorheological finishing, *Int. J. Mod. Phys. B*, 10(23–24):2837–2848, 1996; J. T. Mooney et al., Silicon-on-insulator (SOI) wafer polishing using magnetorheological finishing (MRF) technology, *Proc. 2003 ASPE 18th Annual Meeting*, 2003; A. B. Shorey et al., Experiments and observations regarding the mechanisms of glass removal in magnetorheological finishing, *Appl. Opt.*, 40:20–33, 2001.

Mantle transition-zone water filter

The Earth's mantle is the 2900-km-thick (1800-mi) layer of rock between the crust and the iron core. Although solid, the mantle flows very slowly through solid-state creep (similarly to how glaciers flow), moving about as fast as human fingernails grow. What makes the mantle move is convection, the process of hot buoyant material rising and cold heavy material falling. Convection in the mantle is powered by both heat production from decay of radioactive isotopes (such as uranium and thorium) and the loss of primordial heat left over from the Earth's accretion (that is, impact-generated heat). Mantle convection drives plate tectonics (continental drift) and all its attendant phenomena such as earthquakes and volcanoes. Over geologic time scales, mantle convection is vigorous and should thoroughly stir the mantle. However, considerable evidence exists to suggest that the mantle is poorly mixed, and even layered. This paradox of the well-stirred but unmixed mantle remains one of the unsolved mysteries of the Earth's interior and has been an area of active debate for decades. Its solution is at the heart of

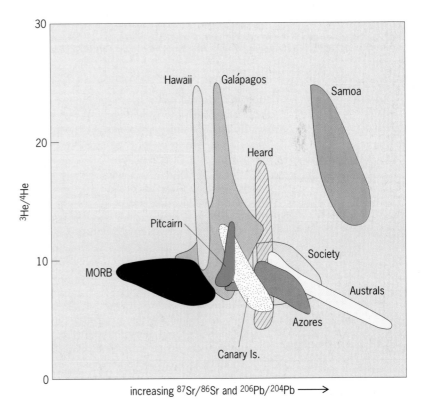

Fig. 1. Isotopic ratios show the distinction between mid-ocean-ridge basalts (MORB) and ocean-island basalts (OIB; regions are labeled by specific island provinces). The isotopic ratios are radiogenically produced (^{4}He is produced in alpha-particle decay of heavy elements such as uranium and thorium, while ^{3}He is nonradiogenic; ^{206}Pb is produced after many steps by ^{238}U decay, while ^{204}Pb is nonradiogenic; ^{87}Sr is produced by decay of ^{87}Rb, while ^{86}Sr is nonradiogenic). MORB is relatively depleted in radioactive elements, which is why the daughter products (the radiogenic isotopes displayed) are in relatively low abundance. This suggests that the MORB source has been cleaned of these elements, while the OIB source has not. Helium is opposite to this trend, usually interpreted to mean that primordial ^{3}He was degassed from the MORB source region but not the OIB source region. (*Adapted from P. E. van Keken, E. H. Hauri and C. J. Ballentine, Mantle mixing: The generation, preservation, and destruction of chemical heterogeneity, Annu. Rev. Earth Planet. Sci., 30:493–525, 2002*)

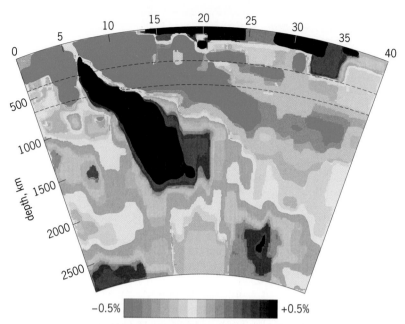

Fig. 2. Seismic tomographic cross section through a subduction zone beneath Mexico shows where compressional seismic waves (acoustic waves) are relatively fast (black) or slow (green). Fast seismic velocities are interpreted as cold, while slow is hot. The figure shows a "cold" current or slab of material extending from the surface subduction zone itself well into the lower mantle and toward the core-mantle boundary. (*After H. Bijwaard, W. Spakman, and E.R. Engdahl, Closing the gap between regional and global travel time tomography, J. Geophys. Res., 103:30055–30078, 1998; via an adaptation by P. E. van Keken, E. H. Hauri, and C. J. Ballentine, Mantle mixing: The generation, preservation, and destruction of chemical heterogeneity, Annu. Rev. Earth Planet. Sci., 30:493–525, 2002*)

understanding the nature and structure of mantle convection, the history of cooling of the entire globe, and the chemical differentiation of the Earth (for example, formation of continental crust and oceans from the mantle).

Upwellings and lavas. The mantle rises to the surface at two main sites: mid-ocean ridges (for example, the East Pacific Rise) where the tectonic plates spread apart and draw mantle up from shallow depths, and ocean islands or hotspots (for example, Hawaii) which are thought to arise from hot narrow plumes of fluid upwelling from deep within the mantle. As these upwellings approach the surface, and they melt, producing lavas and volcanic edifices. Although the lavas produced at these two sites ostensibly come from the same mantle, their chemical signatures are signficantly different (**Fig. 1**). For example, the oceanisland lavas seem to be enriched in heavy trace elements such as uranium and thorium, while the mid-ocean ridge lavas seem to be depleted or cleaned of these elements. This and a host of other largely geochemical observations suggest that these upwellings come from different reservoirs in the mantle that have somehow remained isolated from one another; otherwise, if the mantle were well mixed, the upwellings coming to the surface would all look the same. One simple possibility is that the mantle is divided into two layers, with the deeper layer being the source for the enriched mantle plumes that cause ocean islands, while the upper layer provides depleted mid-ocean ridge lavas.

Sinking slabs and mantle stirring. Although a layered mantle could explain most geochemical evidence, seismological observations suggest that the entire mantle is well stirred by subducting slabs. Subducting slabs are the result of tectonic plates growing relatively cold and heavy, eventually sinking into the mantle through subduction zones, which are usually manifested as oceanic trenches and regions of deep-earthquake seismicity (called Wadati-Benioff zones). Such subducting slabs are essentially the cold downwellings of mantle convection, and are thought to be the major convective currents in the mantle. Seismological images of subducting slabs indicate that while they might linger at certain boundaries in the mantle [for example, at 660 km (410 mi) depth], they usually descend into the lower mantle, even all the way to the core-mantle boundary at 2900 km (1800 mi) depth (**Fig. 2**). Such stirring by slabs sinking across the entire mantle would likely destroy any layering.

Paradox. These two major pieces of contradictory evidence—the geochemical one for a poorly mixed or layered mantle, and the seismological one for a vigorously convecting well-mixed mantle—have provided the Earth science community with one of its greatest paradoxes: How is a well-stirred but unmixed mantle maintained?

Many models and theories have been put forward to explain this mystery. Most concern how to keep the mantle layered or poorly mixed, even in the face of stirring by subducting slabs. However, these models are themselves often fraught with problems and tend to fail various observational tests. For example, in order to isolate the mantle layers, these models prescribe that the layers are separated by a boundary that is just permeable enough to allow relatively weak (both mechanically and buoyantly) hot plumes to leak out of, or entrain pieces of, the enriched lower layer, but is also impermeable enough to resist disruption by strong slabs. There is no clear observational evidence for such a boundary; the best candidates for observed mantle boundaries, such as the 660-km-deep seismic discontinuity, have been shown to be quite permeable to slabs. Layering, of course, is not the only choice for isolated reservoirs; it is possible and even highly likely that the whole-mantle stirring would not entirely homogenize the mantle but would give it a marble-cake appearance. It is then possible that mid-ocean ridges and mantle plumes that lead to ocean islands have access to the same overall mantle, but preferentially extract from differently "flavored" parts of the marble cake. Why plumes or mid-ocean ridges would do such selective sampling of marble-cake components is not understood, although progress within this approach is ongoing.

Filter hypothesis. The recently proposed mantle transition-zone water-filter hypothesis offers an alternative model to explain this mantle paradox (**Fig. 3**). The theory proposes that rather than being divided into isolated unmixed reservoirs, the mantle is merely filtered as it passes through a boundary at a depth of 410 km (255 mi). The filtering occurs primarily to the "background" or ambient mantle which is forced up by the downward flux of subducting slabs. As this ambient mantle rises up through a region called the transition zone [involving most of the mantle's major mineralogical phase changes between 660 km and 410 km depths], it gets somewhat hydrated; this occurs because transition-zone minerals are able to absorb relatively large quantities of water. As this slowly upwelling mantle passes out of the transition zone at the 410-km boundary, it turns from the transition-zone mineral wadsleyite into the upper-mantle mineral olivine, which has a much smaller capacity for absorbing water. Thus, what was only a modest or small amount of water for the wadsleyite below 410 km causes the olivine above 410 km to be saturated with water. Since water-saturated minerals have a relatively low melting point (much lower than dry minerals), and since these upwelling materials are already at hot ambient mantle temperatures, they partially melt. Melt is very good at dissolving water and only a little bit of it will pull out most of the water and dry the remaining solid minerals. Melt will also pull out the incompatible trace elements, which by definition do not fit well in crystal lattice structures and are thus readily extracted to melts (which have more flexible and disordered structure). The small amount of melting thus cleans or filters the solid minerals and leaves them depleted of trace elements as well as dry. These solid minerals continue to rise slowly as part of the ambient upwelling and eventually provide the source for depleted lavas at mid-ocean ridges. Meanwhile, the water-rich melt, which cleaned the mantle and is analogous to a dirty filter, is presumed heavy because the high mantle pressures compress the melt so easily; it therefore pools at the 410-km boundary until it is refrozen and dragged back down by subducting slabs into the deeper mantle, where the enriching incompatible elements (the "dirt") can be restirred into the enriched lower mantle.

Unfiltered mantle plumes. Although most of the mantle is filtered through the above process, mantle plume material would not undergo cleaning or filtering since it is too hot [being as much as 300°C (572°F) hotter than the rest of the mantle] and moves too fast to have absorbed much water while in the transition zone. Moreover, laboratory experiments suggest that transition-zone minerals absorb water more poorly when hotter. Therefore, when plume material exits the transition zone and transforms into olivine-dominated upper-mantle material, it would not be water-saturated and would thus undergo neither melting nor filtering. Plumes rising out of the enriched lower mantle would thus remain unfiltered and eventually deliver enriched material to the surface at ocean islands.

Conclusion. The transition-zone water-filter theory explains the well-stirred but unmixed mantle paradox by means quite different from the classical layering or poor mixing approach. In particular, the model allows for the bulk of the mantle to undergo vigorous, slab-driven whole-mantle convection, while the

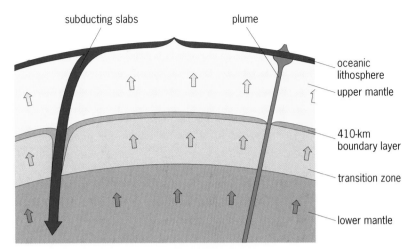

Fig. 3. Schematic of the transition-zone water-filter model. Subducting slabs descend from the cold oceanic lithosphere into the lower mantle and force up the ambient background mantle (arrows). The upwelling passes from the lower mantle through the transition zone where it is hydrated because of the high water-solubility of transition-zone minerals. Upon rising out of the transition zone, the hydrated material transforms to upper-mantle material that cannot dissolve as much water as transition-zone material; hence the water carried out of the transition zone causes it to be saturated. Since saturated minerals have lower melting temperatures (and the ambient upwelling is already at hot ambient mantle temperatures) the material partially melts. The melt extracts both water and incompatible elements leaving the solid phase cleaned and depleted. The rising solid phase (upper-mantle material) provides the depleted mid-ocean ridge basalt source that is tapped by mid-ocean ridges (the ridge shape in the oceanic lithosphere). The melt is assumed heavy (due both to the high pressures at that depth and the high compressibility of melt) and thus it does not rise but instead pools above the 410-km boundary until it is entrained by the slabs back into the deeper mantle. Plumes are too hot and move too fast to get hydrated in the transition zone, hence are not saturated when crossing into the upper mantle and thus do not melt, do not get filtered, and therefore deliver enriched material to the surface to create ocean-island basalts. (After D. Bercovici and S.-i. Karato, Whole mantle convection and the transition-zone water filter, Nature, vol. 425, pp. 39–44, 2003)

trace incompatible elements are decoupled from this whole-scale flow and undergo a separate circulation that is held by the water filter below 410 km. The hypothesis, drawing on well-known observations of the mineralogy and chemical behavior of the transition zone, provides a framework for explaining a wide range of geochemical observations of mantle reservoirs as well as geophysical observations of whole-mantle convective stirring. It also predicts the presence of melt at the 410-km boundary; indeed, several different seismological studies have detected zones of very low seismic velocity sitting atop the 410-km boundary in different areas of the world, and these are usually interpreted as due to partial melting. Encouraging evidence aside, the water-filter hypothesis is still a new idea; and inasmuch as it offers several testable predictions, there is also a considerable amount of work remaining to test it.

For background information *see* EARTH, CONVECTION IN; EARTH, HEAT FLOW IN; EARTH INTERIOR; ELEMENTS, GEOCHEMICAL DISTRIBUTION OF; GEODYNAMICS; HOTSPOTS (GEOLOGY); MID-OCEANIC RIDGE; PLATE TECTONICS; SUBDUCTION ZONES in the McGraw-Hill Encyclopedia of Science & Technology.

David Bercovici; Shun-ichiro Karato

Bibliography. D. Bercovici and S.-i. Karato, Whole mantle convection and the transition-zone water filter, *Nature*, 425:39–44, 2003; G. F. Davies, *Dynamic Earth*, Cambridge University Press, 1999; A. Hofmann, Just add water, *Nature*, 425:24–25, 2003;

L. H. Kellogg, B. H. Hager, and R. D. van der Hilst, Compositional stratification in the deep mantle, *Science*, 283:1881–1884, 1999; G. Schubert, D. L. Turcotte, and P. Olson, *Mantle Convection in the Earth and Planets*, Cambridge University Press, 2001; P. E. van Keken, E. H. Hauri, and C. J. Ballentine, Mantle mixing: The generation, preservation, and destruction of chemical heterogeneity, *Annu. Rev. Earth Planet. Sci.*, 30:493–525, 2002; Q. Williams and R. J. Hemley, Hydrogen in the deep earth, *Annu. Rev. Earth Planet. Sci.*, 29:365–418, 2001.

Mass extinctions

In the 1830s, geologists realized that there were major differences between the trilobite-coral-crinoid-brachiopod-dominated fossil assemblages early in geologic history, and the mollusk-echinoid-dominated faunas that populate the sea floor today. They named the Paleozoic, Mesozoic, and Cenozoic eras for these major changes in marine fossils. In recent years, paleontologists have come to realize that the boundaries between the Paleozoic and Mesozoic eras (or the Permian and Triassic periods) and the Mesozoic and Cenozoic eras (or the Cretaceous and Tertiary periods) are marked by mass extinctions, when more than 30% of the genera of animals become extinct in less than a million years. These mass extinctions are more than just the accumulation of extinction at the normal "background" rate, because they often wipe out long-lived groups of animals that were successful during normal times and resisted extinction over long time scales but did not survive the catastrophic events of a mass extinction.

There are five recognized intervals (the "Big Five") where the extinction rate was well above the background level. These intervals have become accepted as the major mass extinction events in life's history. In recent years, much attention and research has focused on these events, and many claims have been made for their causes. However, as these ideas have undergone scientific testing, some of the more outrageous claims have been discredited.

Permian-Triassic extinction. The biggest mass extinction of all occurred at the end of the Paleozoic Era, 251 million years ago (Ma). This was "the mother of all mass extinctions" when possibly 95% of all marine species on the sea floor vanished and many land animals died out as well. The victims included most of the groups that dominated the Paleozoic Era, from the last surviving trilobites, graptolites, tabulate and rugose corals, and blastoid echinoderms, to the majority of the groups that were thriving in the Permian, including the fusulinid foraminiferans, the common groups of brachiopods, crinoids, and bryozoans, and all but two species of ammonoids. The clams and snails, on the other hand, suffered much less extinction, and during the ensuing Triassic Period they came to dominate the sea floor (as they

still do today). On land, many groups of "protomammals," archaic amphibians, and land plants vanished. Recent detailed studies showed that the Permian-Triassic extinction was fairly abrupt (less than 10,000 years in duration).

In past years, scientists blamed the Permian-Triassic extinction on the loss of shallow-marine habitat on the sea floor when the supercontinent of Pangaea formed (but this occurred gradually, starting at the beginning of the Permian), or on the global cooling caused by south polar glaciation (but this also occurred early in the Permian). Recently, several important historical phenomena have emerged. The oceanographic evidence from carbon isotopes suggest that oxygen levels were severely depleted, and that huge amounts of carbon were released into the world's oceans; this would have led to carbon dioxide–oversaturated waters (hypercapnia) that may have poisoned the sea floor, and low oxygen levels on land that made it hard for most land animals to breathe. In addition, other evidence points to a runaway greenhouse warming and rapidly fluctuating climates at the end of the Permian, which may have raised global temperatures beyond the tolerance of many organisms. The end of the Permian was also marked by one of the biggest volcanic eruptions in Earth history, the Siberian basalts, which released over 1.5×10^6 km^3 (3.6×10^5 mi^3) of lava, and may have also released huge quantities of carbon dioxide and sulfur into the atmosphere that could have triggered the hypercapnia and global warming. Some scientists have tried to blame the Permian extinctions on some kind of extraterrestrial impact, but the evidence for impact has always been questionable, and does not explain the geochemical signals of carbon, oxygen, and sulfur found in the oceans and on land.

Cretaceous-Tertiary (KT) event. The second biggest extinction event, when perhaps 70% of species on Earth vanished, occurred 65 Ma. This is the most well studied of all mass extinctions, because it marked the end of the dinosaurs (except for their bird descendants), as well as the coiled ammonites (like the chambered nautilus) that ruled the seas for over 300 million years and survived many previous mass extinctions. This period also marked major extinctions of the plankton, the clams and snails, and all of the marine reptiles that had once ruled the seas. On land, the nonavian dinosaurs and some land plants died out, but most other land animals and plants were unaffected.

For decades, scientists speculated about the extinction of the dinosaurs and suggested various ideas about their demise. They thought that climates might have become too hot or too cold, or that mammals ate their eggs (but mammals and dinosaurs had coexisted for over 150 million years), or that dinosaurs could not eat the new flowering plants (but these plants appeared at the beginning of the Cretaceous), or disease. But all of these explanations fail because they focus exclusively on the dinosaurs, when the KT event was a mass extinction that affected everything from the bottom of the food chain (plankton, land plants) to organisms such as clams and ammonites, to the top land animals like the dinosaurs and the top predators of the sea, the marine reptiles.

In 1980, the KT extinction debate was galvanized when evidence of a huge impact of a 10-km-diameter (6-mi) asteroid was discovered. This led to more than 20 years of fierce debate as the evidence was evaluated and much of it rejected. Almost 25 years later, it is now clear that there was a major impact that hit the northern Yucatan Peninsula at a site known as Chicxulub. The end of the Cretaceous was also marked by huge volcanic eruptions, the Deccan lavas, which covered over 10,000 km^2 (3860 mi^2) of western Indian and Pakistan with as much as 2400 m (7875 ft) of lava flows, and released huge quantities of mantle-derived gases into the atmosphere. The Cretaceous ended with a global drop in sea level, or regression, that affected both land and marine communities.

Although the impact advocates have dominated the discussion and the popular views, the data are much more complex than has been portrayed. Many of the organisms (especially the rudistid and inoceramid clams, most of the plankton, the marine reptiles, and possibly the dinosaurs and ammonites) were dying out gradually in the latest Cretaceous, and may not have even been alive during the Chicxulub impact. This evidence cannot be explained by a sudden impact but only by a gradual cause, such as a regression or global climatic change triggered by volcanic gases. More importantly, many marine organisms, such as the majority of clams, snails, fish, and echinoids, some plankton, and many land organisms (such as the crocodilians and turtles, the amphibians, and most of the mammals) lived right through the KT extinctions with no apparent ill effects. If the KT impact were so severe that it caused a global "nuclear winter" of cold and darkness, and huge clouds of acid rain (as some have suggested), none of these organisms would still be alive on Earth. Amphibians, in particular, are sensitive to small changes in the acidity of their ponds and streams, so the "acid rain" hypothesis is discredited. Currently, most scientists agree that some sort of gradual cause such as regression and/or the Deccan eruptions might explain the demise of most Cretaceous organisms, and the Chicxulub impact was only a minor coup de grâce that may have finished off the last of the nonavian dinosaurs or ammonites.

Other extinctions. The remaining "Big Five" extinctions were much less impressive than the two that began and ended the Mesozoic Era. The third biggest was the Late Devonian extinction (about 365 Ma), when much of the tropical coral and sponge reef community was decimated along with many other marine invertebrates and archaic fish groups that had flourished in the earlier Devonian. Although impacts have been blamed for the event as well, current evidence from marine isotopes favors a severe global cooling. The fourth biggest is the Late Ordovician extinction (about 445 Ma), when about 60% of the marine genera died out. This, too, has been blamed on a short-term global cooling event, with no evidence for impact. The fifth event, the Triassic-Jurassic extinction (about 208 Ma), affected some marine organisms (like the ammonoids, conodonts, snails, brachiopods and crinoids), and many archaic land reptiles, amphibians, and protomammals. Although this event had also been blamed on impact the evidence has been discredited, and currently the huge eruptions caused by the opening of the North Atlantic are considered a more likely cause.

Mass extinctions are major events that completely rearrange the ecology of life on Earth. Over the past 20 years they have been intensively studied, and some scientists have gone so far as to blame them all on extraterrestrial impacts. However, more recent analyses have shown that only one event (the KT extinction) is clearly associated with impact, and even in this case it is not the major cause of the mass extinction. Instead, massive volcanic eruptions and/or major climatic changes played a more important role.

For background information *see* CENOZOIC; CRETACEOUS; DATING METHODS; EXTINCTION (BIOLOGY); FOSSIL; GEOLOGIC TIME SCALE; MESOZOIC; PALEONTOLOGY; PALEOZOIC; PERMIAN; STRATIGRAPHY; TERTIARY; TRIASSIC in the McGraw-Hill Encyclopedia of Science & Technology.   Donald R. Prothero

Bibliography. J. D. Archibald, *Dinosaur Extinction and the End of an Era: What the Fossils Say*, Columbia University Press, New York, 1996; M. J. Benton, *When Life Nearly Died: The Greatest Mass Extinction of All Time*, Thames & Hudson, London, 2003; D. Erwin, *The Great Paleozoic Crisis: Life and Death in the Permian*, Columbia University Press, New York, 1993; A. Hallam and P. B. Wignall, *Mass Extinctions and Their Aftermath*, Oxford University Press, 1997; N. MacLeod and 21 others, The Cretaceous-Tertiary biotic transition. *J. Geol. Soc. London*, 154:265–292, 1997; D. M. Raup, *Extinction: Bad Genes or Bad Luck?*, Norton, New York, 1991; D. M. Raup and J. J. Sepkoski, Jr., Mass extinctions in the marine fossil record, *Science*, 215:1501–1503, 1982; P. D. Ward, *Gorgon: Paleontology, Obsession, and the Greatest Catastrophe in Earth's History*, Viking, New York, 2004.

Mass spectrometry (carbohydrate analysis)

The functional significance of carbohydrates in biological systems is quite diverse. For example, carbohydrates cover the surface of some bacteria, such as *Haemophilus influenzae*. Carbohydrates in the outer lipid coat may aid in disguising these bacteria by mimicking substances naturally present in humans. They are also key components of cartilage, where they form polymeric chains of negatively charged monosaccharides. The negative charges provide electrostatic repulsion when the cartilage undergoes compression. Carbohydrates also are present on the surface of many proteins. Their presence is sometimes necessary to assist in protein folding or receptor binding. The protein that covers the surface

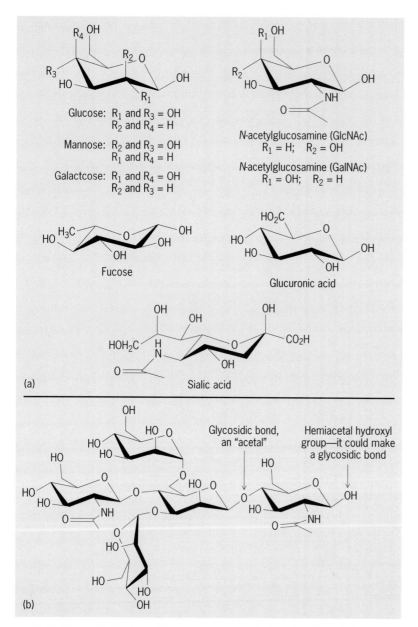

Fig. 1. Carbohydrate structures. (a) Common monosaccharides found in carbohydrates (oligosaccharides). (b) Example of a carbohydrate with considerable branching. The central monosaccharide is connected to four other monosaccharides.

The carbohydrates are linked through glycosidic bonds; the hemacetal hydroxyl group on one monosaccharide is replaced with a hydroxyl group from another monosaccharide to form an acetal, a glycosidic bond. Because the only requirement for glycosidic linkage is that an acetal bond be formed, it is possible to attach one to four monosaccharides to any given monosaccharide unit (Fig. 1b). This branching is unique to carbohydrates, as neither deoxyribonucleic acid (DNA) nor proteins contain this element of complexity. Since there is considerable variation in the type of building blocks and in how they can be linked, carbohydrates have been identified as the most information-dense biological polymer (or oligomer) in existence.

In addition to the variability in the building blocks and the way they are arranged, carbohydrates have one more level of structural complexity: their microheterogeneity. Microheterogeneity is a phenomenon in which the carbohydrate structure at any given site of carbohydrate attachment (to a protein or lipid) may be quite varied, even when the protein sequence (or lipid) is not varied. For example, the carbohydrate attached to one site of the glycoprotein ovine luteinizing hormone may have more than 20 different structural forms (**Fig. 2**). This structural variation is present because, unlike proteins or DNA, carbohydrates are assembled by enzymes in a process that does not follow a rigid building plan.

Analytical approach. Carbohydrate samples are available only in limited quantities because they must be isolated from biological sources and cannot be amplified by the polymerase chain reaction (PCR) used to amplify DNA or by recombinant techniques used to amplify proteins. In addition, the structural characterization of carbohydrates is challenging because of the variation in types and arrangement of the groups present and the microheterogeneity of carbohydrate samples. To characterize carbohydrates, the appropriate analytical technique must be sensitive enough to analyze small samples, and it should provide as much information as possible about the structure of the molecules.

Mass spectrometry is an analytical technique that is naturally suited to carbohydrate analysis because of its sensitivity and ability to discriminate among all compounds of different masses. A mass spectrometer is an instrument that identifies molecules, often in very small quantity, according to their masses. It has an ion source, a mass analyzer that separates ions according to their mass/charge (m/z) ratios, and a detector.

Carbohydrate analysis by mass spectrometry can be performed on less than 1×10^{-12} mol of sample, and all nonisomeric carbohydrates are distinguishable using high-resolution mass spectrometers. In addition, current mass-spectrometric studies of carbohydrates have demonstrated that this technique can be used to characterize mixtures of carbohydrate samples, rich in microheterogeneity, and it can be used to obtain some structural information about the carbohydrates as well. These achievements

of the human immunodeficiency virus (HIV), gp120, is heavily glycosylated; these carbohydrates are believed to shield the protein from immune attack.

Structural complexity. Carbohydrates have a considerable amount of structural variation to match their diverse functional roles. All carbohydrates contain monosaccharide building blocks, which may be linked in a variety of ways. **Figure 1a** shows some of the common building blocks present in carbohydrates. The monosaccharides, such as glucose, galactose, and mannose, can be stereoisomeric, in which case only the arrangement of the hydroxyl groups differs. Nonisomeric monosaccharides are also possible; fucose, sialic acid, N-acetylhexosamines, and glucoronic acids are examples. These monosaccharides may also contain sulfate or phosphate groups.

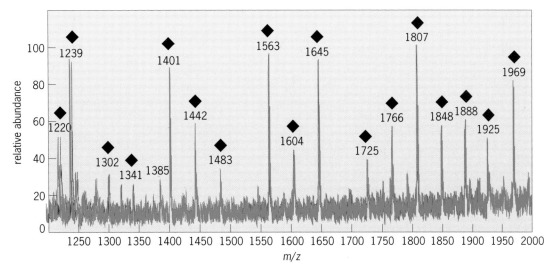

Fig. 2. Mass spectrum originating from ovine luteinizing hormone, a glycoprotein that has a high degree of microheterogeneity. Each peak with a diamond corresponds to a glycopeptide with a unique carbohydrate composition. The composition can be determined from the *m/z* value obtained.

would not have been possible without the recent developments of soft ionization sources and tandem mass-spectrometric experiments.

Soft ionization sources. Soft ionization sources, especially electrospray ionization (ESI) and matrix-assisted laser desorption ionization (MALDI), are heavily relied upon for mass spectral analysis of carbohydrates. Developed within the last 20 years, they have enabled the analysis of many different biological materials. Both techniques make ions in a "soft" process that does not force the compound to fragment. Instead, molecules in solution can be directly ionized. In addition, these ionization techniques can be used for nonvolatile compounds.

Soft ionization is important for suppressing fragmentation. If the carbohydrates fragmented extensively during mass spectral analysis, fragment ions could be confused for additional components. It is very difficult to purify a single carbohydrate sample due to the microheterogeneity of most samples, so often mass spectral analysis is used to identify many components present in a mixture simultaneously. Extensive fragmentation would prohibit this.

In addition to being a soft ionization process, ESI and MALDI share the characteristic that they can be used for nonvolatile compounds. Since carbohydrates are not volatile, mass-spectrometric analysis for these compounds was very difficult using older ionization techniques, such as chemical ionization (CI-MS) or electron impact (EI-MS.)

The mass spectra of carbohydrates can be used to determine compositional information about the sample.

Tandem mass spectrometry. Structural information may also be obtained using tandem mass-spectrometric experiments (**Fig. 3**). In these experiments, a precursor ion is selected from the mass spectrum, isolated inside the mass spectrometer, and then fragmented via low-energy collisions with a neutral gas (usually helium or argon). The fragmen-

tation products are displayed in another mass spectrum (Fig. 3). The resulting spectrum is known as a collision-induced dissociation (CID) spectrum. It may also be called a collisionally activated decomposition (CAD) spectrum or an MS/MS spectrum. The CID spectrum contains the masses of smaller segments of the carbohydrate, and this information can be used to obtain structural information about the original compound.

Screening of lipooligosaccharides and glycoproteins. Mass spectrometry is very useful for obtaining compositional information about oligosaccharides (carbohydrates) attached to lipids or proteins. Typically, the carbohydrates are cleaved chemically or enzymatically from the protein or lipid, and the carbohydrate fraction is injected directly into the mass spectrometer, without derivatization. By acquiring a mass spectrum of the oligosaccharide fraction, all the nonisomeric carbohydrates are identified by their *m/z* values. Alternatively, glycoproteins can be digested enzymatically to produce glycopeptides, which can be analyzed in the same fashion (Fig. 2).

Using mass spectrometry to "screen" the composition of the sample is advantageous because of the

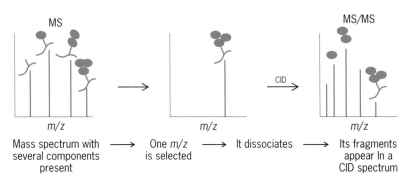

Mass spectrum with → One *m/z* → It dissociates → Its fragments several components   is selected   appear In a present   CID spectrum

Fig. 3. Collision-induced dissociation (CID) spectra are acquired by isolating one *m/z* from a mass spectrum, fragmenting it, and then displaying the fragmentation products.

method's speed. Any other technique would require first separating all the different carbohydrate forms prior to analysis, which can be difficult or impossible depending on the complexity of the sample. While mass-spectrometric screening of carbohydrate mixtures is very useful, its key limitation is that only an m/z ratio is determined for each carbohydrate. As a result, the types of monosaccharide building blocks are known, but no information is (necessarily) available about their arrangement. This limitation can be mitigated to some extent using CID experiments.

Structural analysis. Several examples exist where CID spectra have been used to identify structural features of carbohydrates. One example involves determining the stereochemistry of unknown *N*-acetylhexosamine monosaccharides using CID data. In this experiment, CID data from derivatized monosaccharide standards are compared to CID data from a derivatized, unknown monosaccharide. The identity of the unknown is determined by "matching" its CID spectrum to that of a standard. (The derivatization is necessary because all the underivatized monosaccharide isomers give identical CID spectra.) This approach is useful for monosaccharide analysis. However, the method cannot be applied to stereoisomeric monosaccharides within an oligosaccharide without first hydrolyzing the oligosaccharide to release the monosaccharides.

Another example of CID experiments for carbohydrate analysis is their use in discriminating among isomeric carbohydrates that contain the same monosaccharide units but different branching patterns. In these isomers, the arrangement of the monosaccharide building blocks differs. For these compounds, CID spectra of carbohydrate standards are not used to "match" the CID spectra of unknown compounds, because relatively few carbohydrate standards are available. Instead, the CID spectrum of the unknown carbohydrate may be solved in a process similar to putting together a puzzle. Like puzzle pieces, the ions in the CID spectra can sometimes be used to construct a unique structure. While CID experiments cannot always be used to determine the branching pattern of carbohydrates, this approach has been used successfully several times by different research groups in the analysis of many types of carbohydrates.

Alternative approaches. Mass spectrometry is very useful for compositional analysis of complex mixtures of oligosaccharides, but currently it cannot be used to provide complete structural information about carbohydrates. Enzymatic methods and nuclear magnetic resonance (NMR) analysis provide more structural information. These techniques can be used to identify how the monosaccharides are arranged (branching), how each monosaccharide is connected to the next (linkage), and which stereoisomeric monosaccharides are present at what position in the oligosaccharide. The chief limitation of these techniques is that they require purification of a single carbohydrate prior to analysis. In addition, NMR may require considerable sample quantities.

For background information *see* ANALYTICAL CHEMISTRY; CARBOHYDRATE; GLYCOPROTEIN; MASS SPECTROMETRY; MASS SPECTROSCOPE; MONOSACCHARIDE; OLIGOSACCHARIDE in the McGraw-Hill Encyclopedia of Science & Technology.

Heather Desaire

Bibliography. R. A. Dwek, Glycobiology: Toward understanding the function of sugars, *Chem. Rev.*, 96:683–720, 1996; H. Geyer and G. Geyer, Strategies for glycoconjugate analysis, *Acta Anat.*, 161:18–35, 1998; Y. Merchref and M. V. Novotny, Structural investigations of glycoconjugates at high sensitivity, *Chem. Rev.*, 102:321–369, 2002.

Medicinal mushrooms

Certain mushrooms have long been valued as flavorful and nutritious foods by many societies worldwide. Furthermore, some of these societies, especially in the Far East, have long recognized that extracts from certain mushroom species possess health-promoting benefits and, consequently, use them as ingredients in many herbal medicines, such as traditional Chinese medicine (TCM). While approximately 14,000 species of mushrooms are believed to exist worldwide, at least 600 have been shown to possess various therapeutic properties, and the term "medicinal mushroom" is gaining worldwide recognition for such species. (The term medicinal mushroom was derived in the Far East but is now accepted, in broad terms, by Western medicine standards.) The most commercialized edible mushrooms that demonstrate principal medicinal or functional properties include species of *Lentinus*, *Hericium*, *Grifola*, *Flammulina*, *Pleurotus*, and *Tremella*, while others known only for their medicinal properties, such as *Ganoderma* spp., *Schizophyllum commune*, and *Trametes* (*Coriolus*) *versicolor*, are nonedible because of their coarse texture and bitter taste.

For traditional therapeutic purposes, medicinal mushrooms were used not necessarily as fresh whole mushrooms but as hot-water extracts, concentrates, or powdered forms which could be incorporated into health tonics, tinctures, teas, soups, and herbal formulas. Over the last few decades new analytical biochemical techniques have enabled the separation of many of the principal active compounds, especially those displaying anticancer activities. The availability of these partially purified extracts has allowed extensive screening in animal models of cancer and human cancer tissue cell lines, and an increasing number of successful clinical trials. Several of these compounds are now registered as anticancer drugs in Japan, China, and Korea.

Principal medicinal compounds. Of the many different compounds isolated from medicinal mushrooms, the most significant are various polysaccharides and

TABLE 1. Polysaccharide antitumor agents developed in Japan (immunotherapeutical drugs as biological response modifiers, BRM)

	Krestin	Lentinan	Sonifilan
Abbreviation	PSK	—	SPG*
Common name	Krestin	Lentinan	Schizophyllan
Fungus (origin)	*Trametes versicolor* (mycelium)	*Lentinus edodes* (fruit body)	*Schizophyllum commune* (medium product)
Polysaccharide	β-glucan-protein	β-glucan	β-glucan
Structure	-1,6-branching -1,3: 1,4-main chain	-1,6-branching -1,3-main chain	β-1,6-branching β-1,3-main chain
Molecular weight	100,000	500,000	450,000
Route of administration	Oral	Intraperitoneal; intravenous	Intraperitoneal; intravenous
Cancer treated	Cancer of digestive tract, lung, and breast	Cancer of stomach	Cervical cancer

*SPG, *Schizophyllum* polysaccharide.

polysaccharide-protein complexes which have shown important immunomodulating and anti-cancer effects in animals and humans (**Table 1**). Such bioactive polysaccharides have been isolated from mushroom fruit bodies, submerged culture mycelial sources, and liquid culture broths, and are water-soluble β-D-glucans, β-D-glucans with heterosaccharide side chains, or β-D-glucans in protein complexes, that is, proteoglycans. The main medically important polysaccharides that have undergone extensive anticancer clinical trials in Asia in the last two decades include lentinan (β-D-glucan) [*Lentinus edodes*], schizophyllan (β-D-glucan) [*Schizophyllum commune*], grifron-D/maitake-D (β-D-glucan) [*Grifola frondosa*], PSK, and PSP (polysaccharide peptides) [*Trametes versicolor*]. PSK differs from PSP only in its saccharide makeup, lacking fucose and containing arabinose and rhamnose. The polysaccharide chains are true β-D-glucans. All of these biopolymers are currently produced by Asian pharmaceutical companies.

More recently, purified polysaccharides derived from *Agaricus braziliensis*, *Phellinus linteus*, *Ganoderma lucidum*, and active hexose correlate compound (AHCC) [derived from liquid fermentation of several medicinal mushrooms] are being proposed in the Far East as producers of a new generation of anticancer compounds.

Possible modes of action. Various polysaccharides and proteoglycans have been shown to enhance cell-mediated immune response in vivo and in vitro in animals and humans. Their ability to function as immunomodulators relies on several concurrent factors, including dosage, route and frequency of administration, timing, specific mechanisms of action, and the site of activity. Several of these anticancer polysaccharides have been able to potentiate the host's innate (nonspecific) and acquired (specific) immune responses, and to activate many kinds of immune cells and substances that are essential for the maintenance of homeostasis, such as host cells (cytotoxic macrophages, neutrophils, monocytes, and natural killer cells) and chemical messengers (cy-

tokines such as interleukins, colony stimulating factors, and interferons) that trigger complement and acute-phase responses. They have also been considered as multicytokine inducers capable of modulating gene expression of various immunomodulatory cytokines via specific cell-membrane receptors.

Several of the proprietary mushroom polysaccharides (for example, lentinan, schizophyllan, PSK, and PSP) have proceeded through phase I, II, and III clinical trials mainly in China and Japan and to a limited extent in the United States. In these clinical trials the polysaccharides have mainly been used as adjuvant treatments with conventional chemotherapy or radiotherapy for many types of cancer. In many cases there have been highly encouraging improvements in quality of life together with the observations that their incorporation into treatment regimes significantly reduced the side effects often encountered by patients. For example, PSK has been in clinical use in Japan for many years with no reports of any significant short- or long-term adverse effects. Such compounds are not miracle drugs but have been shown to increase the quality of life of cancer patients and may offer increased survival rates for some types of cancer when used in an integrative manner with conventional therapies.

Chemoprevention. There is increasing evidence with experimental animals that regular feeding with powdered medical mushrooms can have a cancer-preventing effect, demonstrating both high antitumor activity and restriction of tumor metastasis. A recent epidemiological study conducted among Japanese mushroom workers in the Nagano Prefecture implied that the regular consumption of the edible medicinal mushroom *Flammulina velutipes* was associated with a lower death rate from cancer compared with other people in the Prefecture (**Table 2**). This has prompted a national survey by the Japanese Ministry of Health and Welfare.

Medicinal mushrooms with clinical trial support. *Lentinus edodes* is a fungus indigenous to Japan, China, and other Asian countries with temperate climates. In Japan it is called shiitake and has been renowned

TABLE 2. Comparison of cancer death rates in Nagano Prefecture, Japan*

Average cancer death rate in the Prefecture	Total	160.1
	Men	90.8
	Women	69.3
Average cancer death rate of farmers producing and consuming a medicinal mushroom	Total	97.1
	Men	57.5
	Women	39.7

*Cancer death rate: rate per 100,000 age-adjusted rates. Total population is 174,505. Years investigated are 1972–1986.

for thousands of years both as a food and as a medicine. There is a substantial corpus of literature on preclinical data for lentinan (the medically active extract), including antitumor activity, prevention of metastasis (cytostatic effect), and prevention of chemical and viral-induced oncogenesis by lentinan in animal models. There have been many clinical trials in Japan, although many of these carry possible methodological flaws by Western standards. Despite this, lentinan has been found to alleviate tumors and conventional treatment-related symptoms, and to demonstrate efficacy. Lentinan has been approved for clinical use in Japan since the early 1980s and is widely used in hospitals as an adjuvant treatment for certain cancers in Japan and China, especially gastric cancer.

Few adverse reactions to lentinan have been noted, and it is well tolerated. Perhaps the most intriguing aspect of lentinan use in conjunction with chemotherapy or radiotherapy is its apparent ability to greatly reduce the debilitating effects of chemotherapy, such as nausea, pain, hair loss, and depressed immune function. Although there have been few formal quality-of-life studies, this observation has been noted as a feature of most of the mushroom polysaccharides when used in cancer therapy.

Schizophyllum commune is a very common fungus that has worldwide distribution and is found growing on dead logs. Schizophyllan is an exopolysaccharide excreted into the liquid fermentation medium when the fungus is grown in the filamentous mycelial form. It has been approved for clinical use in Japan. Schizophyllan increased the overall survival of patients with head and neck cancers and significantly increased the overall survival of stage II cervical cancer patients but not stage III.

The fungus *Trametes versicolor* has worldwide distribution and grows as multicolored, overlapping flaps on dead logs. PSK and PSP are relatively similar polysaccharopeptides produced by deep-tank liquid fermentation cultivations of the fungal mycelium. PSK is a Japanese product, and PSP is a more recent Chinese product. Both products have been registered as pharmaceutical drugs in their respective countries. In 1987 PSK accounted for more than 25% of total national expenditure for anticancer agents in Japan. There have been extensive successful clinical trials over the past two decades against a variety of solid tumors including head and neck, upper gastrointestinal tract, colorectal, and lung cancers, al-

ways as adjuvants to chemo- or radiotherapy. Patients also showed improvement in appetite and fatigue, together with a stabilization of hematopoietic parameters such as changes in white blood count. Recent studies with PSP in combination with chemotherapy have shown improved disease-free survival in gastric, esophageal, and non-small-cell lung cancers with a concomitant reduction in treatment-related side effects.

While the role of medicinal mushrooms in immunomodulations and anticancer activities have been highlighted in this article, it is pertinent to note that many of the medical mushroom extracts have been highly valued for other medicinal properties, including treatment of hypercholesterolemia, high blood pressure, and diabetes as well as for their antimicrobial, antioxidant, and free-radical-scavenging activities. Indeed, the latter two aspects are now considered to be an integral part of the anticancer effects of specific medicinal mushrooms.

Mainstream medicine is beginning to recognize health care values in both herbal and traditional Chinese medicine. Complementary or integrative medicine is now being actively pioneered in the United States and Europe in parallel with intensive reappraisal of standards of herbal medicines in the Far East. Western pharmaceutical companies are slowly becoming aware of the considerable health benefits that may accrue from traditional Chinese medicine concepts but are concerned over patentability problems and possible negative interaction with prescribed drugs. Undoubtedly, some complex human diseases, such as cancer and cardiovascular disturbances, may well benefit from a closer interaction of conventional medicine and quality-controlled and authenticated traditional Chinese medicine.

For background information *see* CHEMOTHERAPY; FUNGAL BIOTECHNOLOGY; IMMUNOTHERAPY; MUSHROOM; MYCOLOGY in the McGraw-Hill Encyclopedia of Science & Technology.   John E. Smith

Bibliography. M. Fischer and L. X. Yang, Anticancer effects and mechanism of polysaccharide-K (PSK): Implications of cancer immunotherapy, *Anticancer Res.*, 2:1737–1754, 2002; P. M. Kidd, The case of mushroom glucans and proteglycans in cancer treatment, *Altern. Med. Rev.*, 5:4–27, 2000; I. Ikekawa, Beneficial effects of edible and medicinal mushrooms on health care, *Int. J. Med. Mushrooms*, 3:291–298, 2001; J. E. Smith, N. J. Rowan, and R. Sullivan, *Medicinal Mushrooms: Their Therapeutic Properties and Current Medicinal Usage with Special Emphasis on Cancer Treatments*, Special Report Commissioned by Cancer Research UK, 2002; J. E. Smith, R. Sullivan, and N. Rowan, The role of polysaccharides derived from medicinal mushrooms in cancer treatment programs: Current perspectives (review), *Int. J. Med. Mushrooms*, 5:217–234, 2003; A.-T. Yap and M.-L. Ng, Immunopotentiating properties of Lentinan (1,3)-β-D-glucan extracted from culinary medicinal Shiitake mushrooms *Lentinus edodes*, *Int. J. Med. Mushrooms*, 5:339–358, 2003.

Menopause

Human females are unusual among mammals in experiencing menopause, a nonvoluntary and irreversible cessation of fertility long before the senescence of other bodily systems and the end of average lifespan. Two main hypotheses have been proposed to explain the origin of menopause. Some evolutionary biologists and anthropologists suggest that menopause is an adaptation—the result of natural selection for a postreproductive lifespan to permit increased maternal investment in rearing offspring. At some point in human evolution, females who ceased reproducing before the end of their lives gained a genetic fitness advantage from dedicating their remaining reproductive effort to enhancing the reproductive success of their existing progeny. Others hypothesize that menopause is an epiphenomenon (by-product)—either the result of a physiological trade-off favoring efficient early reproduction or merely an artifact of increased lifespan and life expectancy.

Adaptation hypotheses. Menopause may have arisen 1.9–1.7 million years ago during a time in hominid evolution of increasing bipedality and rapid encephalization (increase in brain size). Changes in pelvic anatomy to allow for more efficient bipedal locomotion and increases in fetal head size eventually made the birth process so difficult that further rapid brain growth and organizational development had to occur after birth. Hominid infants were born in an altricial (less developed) state, requiring intensified and prolonged maternal investment. Due to the increasing risk of maternal death in childbirth, the production of babies later in life jeopardized the survival of existing offspring.

There are two adaptation explanations for menopause: the grandmother and mother hypotheses. In the grandmother hypothesis, the benefit of menopause derives from increasing the fertility of adult daughters and survival of grandchildren; whereas in the mother hypothesis, the benefit derives from increasing the survival and potential fertility of one's own offspring. The grandmother hypothesis ignores the fact that increased infant helplessness and prolonged juvenile dependence primarily required a change in young females' reproductive strategies. The fitness-enhancing strategy was to switch from serial to overlapping childcare, leaving some postreproductive time to finish raising the lastborn.

Epiphenomenon hypotheses. The assumption underlying the epiphenomenon explanations is that evolution is constrained by phylogenetic history, developmental limitations, and physiological trade-offs resulting from genes that have a positive effect on one component of fitness and a negative effect on some other component of fitness. Given semelgametogenesis (the trait of producing all one's gametes at one time), premature reproductive senescence in human females is the result of a physiological trade-off favoring efficient and intensive early reproduction to maximize reproductive output before the dwindling supply of eggs jeopardizes hormonal sup-

port for ovulation and the negative consequences of old eggs predominate. In the longer lifespan and life expectancy hypothesis, menopause is simply the result of women outliving their supply of eggs or ability to sustain ovulatory cycles. However, the notion that postreproductive life is the result of recent increases in life expectancies due to improved sanitation and medical care is belied by biblical references to menopause. Nor is the postreproductive lifespan the result of increases in maximum lifespan, since there is fossil evidence indicating that hominid maximum lifespan exceeded the current age of menopause well before the appearance of *Homo sapiens*.

Support for adaptation and physiological trade-off hypotheses. The adaptation and physiological trade-off hypotheses are supported by evidence showing that both menopause and high early fertility are universal among and unique to human females, by the dynamics of follicular depletion in human females, and by estimates of the heritability of age of menopause in women.

Universality and uniqueness of menopause. The lack of true menopause in other mammals supports the notion that a period of postreproductive maternal investment is the response to a unique set of socioecological, anatomical, and physiological pressures of critical importance to how humans evolved.

The front-loaded human fertility pattern (high fertility early in life) is universal among human females and is different from fertility patterns observed in other mammals, which supports the physiological trade-off hypothesis. Instead of fertility starting to decrease when women are in their midtwenties, age-specific fertility functions for macaques, olive baboons, lions, and East African elephants, for example, remain relatively constant over a period of time, then terminate quite abruptly only a few years before maximum age at death. However, the universal and unique fertility pattern in human females—with its early peak, subsequent decline, and relatively long posreproductive lifespan—also supports the adaptation hypothesis, in that selection for efficient early reproduction facilitates prolonged offspring dependence, overlapping childcare, and postreproductive maternal investment.

Reports of menopauselike physiological phenomena in nonhuman primates reveal that the reproductive changes observed represent something different from human menopause. These changes are idiosyncratic and far from species-wide, while age at reproductive cessation is variable and postreproductive lifespans are relatively short. Comparison of the life histories of Ache women, former foragers from eastern Paraguay, and common chimpanzee females (*Pan troglodytes*), illustrate this well. Approximately half of all chimpanzee mothers never outlive their reproductive capacity, whereas approximately half of all reproductive-age Ache women live at least 18 years after reproductive cessation. Even longer-lived species, such as elephants and most whales, retain reproductive capacity until very old age relative to their maximum lifespans. The only real exception to the general mammalian fertility pattern are female short-finned pilot whales (*Globicephala macrorynchus*), which have a mean postreproductive lifespan of 14 years.

Follicular depletion. Additional support for the adaptation and physiological trade-off explanations comes from the dynamics of follicular depletion in human females. Histological investigation of human ovaries indicates an acceleration in the rate of atresia (the process of follicular depletion) around age 40. Without this acceleration, women would have enough eggs to last about 70 years. The acceleration provides the proximate physiological cause of premature reproductive cessation and is consistent with selection for efficient early reproduction and a postreproductive lifespan.

Heritability in the age of menopause. Though not conclusive evidence, Greek, Roman, and medieval texts suggest that age of menopause has remained roughly constant for a few thousand years, despite tremendous socioeconomic and demographic changes. With estimated heritabilities as large as 40–60% and no appreciable upward movement in the age of menopause, it is reasonable to hypothesize that age of menopause is under some degree of stabilizing selection; this in turn suggests that there are costs to prolonging fertility and benefits to reproductive cessation in modern industrialized and agricultural societies. The fact that there are presently costs and benefits means that there is nothing exclusive to the foraging way of life of our ancestors that makes premature reproductive senescence adaptive. This should not be surprising, given that reproductive patterns in human females are remarkably consistent across populations and time regardless of mode of subsistence. However, the relatively large amount of variation in age of menopause within and across populations suggests the stabilizing selection on age of menopause is weak. This could be the consequence of stronger selection for some other trait that results in menopause, such as efficient early reproduction or a postreproductive lifespan. Front-loaded fertility and the universal mean age at last birth of approximately 40 years are consistent with both.

Fitness costs and benefits. Evidence of fitness costs of prolonged fertility and benefits of reproductive cessation is not overwhelming. In historical data, there is evidence of fitness costs of prolonged fertility, stemming from increased maternal mortality and elevated fetal and infant mortality associated with pregnancies late in life. In modern industrialized societies, mortality rates for older mothers and their infants are very low, but late pregnancies still carry high risk of fetal loss, stillbirth, and birth defects.

To discover the fitness benefits of reproductive cessation, researchers have analyzed the work performed by postreproductive women in extant traditional populations. There is evidence from the Hadza of Tanzania that grandmothers' foraging time positively correlates with grandchildren's weight. A recent study in rural Gambia demonstrated that postreproductive maternal grandmothers significantly increase survival of grandchildren by improving their nutritional status. However, models designed to test fitness benefits associated with reproductive cessation do not show significantly positive effects among the Ache of Paraguay.

Conclusion. The universality of menopause among human females implies that front-loaded fertility and the postreproductive lifespan are both the response to some unvarying constellation of selective pressures. Infant altrativity, prolonged offspring dependence, and the expedience of overlapping childcare are major components of that constellation. If menopause is indeed the result of selection for efficient early fertility to accommodate overlapping childcare, it is also the result of changes in the reproductive strategies of young females, which provides compelling support for the mother hypothesis.

Thus, menopause is both the result of a physiological trade-off favoring intensive early fertility and an adaptation for more effective maternal investment during and after reproduction. Subsequent increases in lifespan or life expectancies would have resulted in longer postreproductive lifespans for human females and eventually opportunities for grandmothering. The current very long postreproductive lifespan characterizing human females is undoubtedly the result of large recent increases in life expectancies.

For background information *see* ANIMAL EVOLUTION; ANTHROPOLOGY; MENOPAUSE; ORGANIC EVOLUTION; PHYSICAL ANTHROPOLOGY in the McGraw-Hill Encyclopedia of Science & Technology.

Jocelyn Scott Peccei

Bibliography. K. Hawkes et al., Hardworking Hadza grandmothers, pp. 341–366 in V. Standen and R. A. Foley (eds.), *Comparative Socioecology*, Blackwell Scientific, Oxford, 1989; K. Hill and A. M. Hurtado, The evolution of premature reproductive senescence and menopause in human females: An evaluation of the grandmother hypothesis, *Human Nat.*, 2:313–350, 1991; C. Packer, M. Tatar, and A. Collins, Reproductive cessation in female mammals, *Nature*, 392:807–811, 1998; S. J. Richardson, V. Senikas, and J. F. Nelson, Follicular depletion during the menopausal transition: Evidence for accelerated loss and ultimate exhaustion, *J. Clin. Endocrinol. Metab.*, 65(6):1231–1237, 1987; G. C. Williams, Pleiotropy, natural selection, and the evolution of senescence, *Evolution*, 11:398–411, 1957; J. W. Wood, *Dynamics of Human Reproduction: Biology, Biometry, Demography*, Aldine De Gruyter, New York, 1994.

Micro hydropower

Traditional water-power technology has been used extensively throughout the world for small-scale processing of agricultural produce for many centuries. Micro hydropower has developed from this technology and has been adapted to meet modern requirements for electricity and mechanical power. It has become a low-cost option for rural electrification, particularly for remote communities in developing countries. There is already widespread use of this technology in Nepal and significant potential in many other countries.

Micro hydro usually refers to schemes of up to 100 kW output. Installations below 5 kW are usually referred to as pico hydro. The available power (in watts) from a hydro plant may be calculated from the equation below, where ρ is the density of water

$$P = \rho\, gQH$$

(1000 kg/m³), g is the gravitational acceleration (9.81 m/s²), Q is the flow rate or discharge in m³/s, and H is the head in meters. (Head is the vertical height between the water level at the intake and the level at the turbine limit.) **Figure 1** shows the useful ranges of head and flow for various turbine types. The output power is reduced by energy losses in the pipes and channels bringing water to the power house, in the turbine, and in the generator and distribution system. For a micro hydro scheme the total losses are typically 30–60% of the available power.

Site surveys. Designing a micro hydro scheme is time-consuming because each site has different characteristics in terms of head and flow available and the relative position of intake, power house, and consumers. The layout of a hypothetical micro hydropower site is shown in **Fig. 2.** Carrying out thorough site surveys and designing equipment for each site can increase the engineering costs out of proportion to the size of the scheme. One focus of recent research has been the reduction of engineering time through standardization and the use of new technology to reduce costs while maintaining performance and reliability. These techniques are especially valuable for pico hydro installations.

Hydrological databases are being developed that engineers can access to estimate flow rates in rivers. Such databases are gradually becoming available for even more remote areas in the world. However, for pico hydro schemes it is difficult to obtain accurate data from maps or databases as the catchment areas are too small. In this case, flow measurements must be made at the driest time of year in order to design the scheme so that the power can be continuously available.

For high-head schemes (H greater than 30 m or 100 ft), height measurements can be made using a hand-held digital altimeter with sufficient accuracy (±1 m) to carry out the scheme design. All the site survey data can be collected during one site visit using a standard Global Positioning System (GPS) unit. This data can later be downloaded to a computer and used to calculate lengths of pipes and cables. Software is becoming available that can optimize the pipe sizes and cable layouts, leading to economic selection of materials while saving many hours of skilled engineering time.

Selection of equipment. Larger hydro schemes have all of the equipment custom-designed. For micro

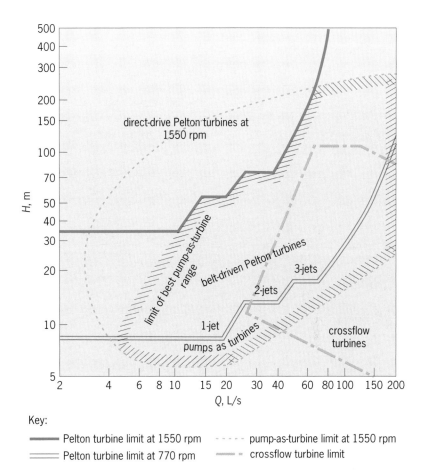

Fig. 1. Head-versus-flow diagram, showing useful ranges for various turbine types. Head (*H*) in meters is plotted against flow (*Q*) in liters per second (*After A. Williams, Pumps as Turbines: A Users Guide, 2d ed., 2003*)

hydro, this approach is not cost-effective. Instead, turbines are often made in a range of standard sizes and adapted to site conditions by changing the operating speed. Local manufacture of turbines can keep costs down, but the designs have to be appropriate for available materials and manufacturing equipment. Pico turbines are often produced in small workshops, so the designs are simplified still further by eliminating components such as those that control flow rates. Often a direct drive to a fixed-speed generator is used, in which case the site layout may be designed to fit the closest available turbine option, rather than the other way around. Turbine costs can be further reduced if batch production methods are introduced.

For high-head sites, Pelton turbines are used, having a nozzle that directs the water onto specially shaped split buckets. These buckets have been scaled down for use in the smallest turbines, with some adaptations so that they can be locally cast from bronze or aluminium, which is sometimes melted down from scrap. Small Pelton turbines are often connected directly to the generator (as in **Fig. 3**) so they run at 1000–3800 revolutions per minute (rpm). Further development of small Pelton turbine designs has resulted in a unit that has an efficiency of 70% for only 2 kW output. This is a significant improvement

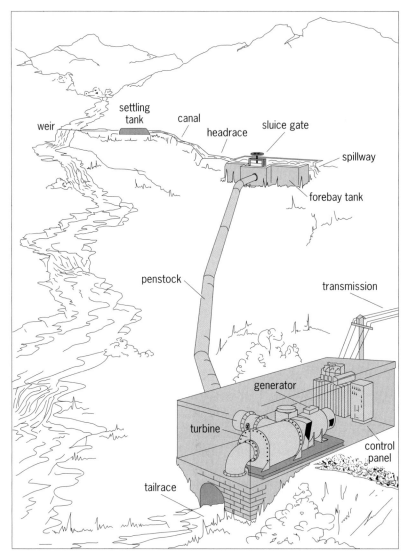

Fig. 2. Layout of a hypothetical micro hydropower site, showing placement of major components. (*After R. Holland, Micro Hydro Electric Power, ITDG Publishing, 1986*)

on most previous designs that could be made in developing countries.

For medium-head schemes (H between 10 and 30 m, or 30 and 100 ft), Mitchell-Banki or "crossflow" turbines are more commonly used. Water is directed through a cylindrical rotor fitted with a set of curved blades. These turbines have simpler geometry and can be manufactured easily from welded steel in developing countries. Standard designs have been developed by the Swiss development agency SKAT. Operating speeds are usually between 250 and 800 rpm. Another option for medium-head schemes is to use a standard centrifugal pump as a turbine. Although the standard pump geometry is not optimized for turbine operation, in practice the efficiency as a turbine is usually within 1% of the pump efficiency, and is occasionally better. Because pumps are mass-produced, they tend to be cheaper than equivalent turbines and easier to install and maintain. However, pump manufacturers do not normally have any information on the head and flow of their pumps when operated as

turbines. Methods have therefore been developed for predicting turbine performance from pump data, but these are not always accurate. Further investigation is being carried out, using laboratory tests and computational fluid dynamics (CFD) modeling to assist in the selection of appropriate pumps and to identify any modifications that can be made to optimize efficiency.

For low-head sites (H less than 10 m or 30 ft), axial-flow propeller turbines are commonly used. Again, these have been simplified for the smallest schemes. To lower costs, the designs eliminate variable guide vanes and variable-pitch runner blades. Current research on small propeller turbines includes development of design parameters that will allow fish to pass through the turbines with much lower chance of injury. This follows from designs of large "fish-friendly" turbines that have mainly been developed in the United States. Computational fluid dynamics is being used as a research tool to investigate turbine designs with regard to energy losses and scaling effects, so that in the future even pico turbines may be easily designed with predictable performance.

Water wheels are coming back into use for low-head sites, particularly where the main use of power is for mechanical equipment rather than generating electricity. In Nepal and in the northern parts of India, the traditional wooden waterwheel (*pani ghatta*) has a vertical axis. They run typically at speeds between 100 and 200 rpm. Improved designs using steel have been successfully implemented. They are cost-effective as they require less maintenance and produce more power from the same head and flow so that modern processing machinery, such as rice hullers, can be driven. In Germany a number of small manufacturers have been installing wheels of the traditional European design, with a horizontal axis. The new wheels use modern components and materials, which result in a better-engineered product that requires less maintenance.

For micro hydro sites that generate electricity, there has been a trend away from mechanical control of water turbines to electronic load controllers. These sense the generator output voltage and use ballast or dump loads to maintain the generator speed as required. These are usually air or water heaters that absorb electric power when other appliances are switched off. Electronic control is also used in cases where micro hydro schemes are connected to a grid network. Electronic circuits which generate electricity of fixed frequency (50 Hz or 60 Hz) for a range of generator speeds are also being developed.

For pico hydro projects, standard industrial three-phase motors have been adapted for use as induction generators to supply single-phase loads. They have no slip rings or brushes and are more reliable as well as cheaper than small alternators. An electronic induction generator controller (IGC) has been developed to maintain motor stability and is now being manufactured in several countries in Asia, Africa, and Latin America.

Compact turbine units, such as the Pico Power Pack (Fig. 3), have been developed with the turbine runner attached to a shaft extension from the motor running as generator. An additional shaft extension has been fitted at the other end of the generator to drive mechanical equipment. Where pumps are used as turbines, close-coupled units (**Fig. 4**) are available in which the pump impeller is fitted to the motor shaft, and they run in reverse, with the pump running as a turbine and the motor running as a generator.

Technology dissemination. For small-scale rural electrification projects, there has gradually been a move away from projects funded purely by outside agencies such as regional governments or development charities. Many successful projects are now being implemented through local entrepreneurs, and a market is being developed for micro hydro equipment. For successful dissemination of the technology, manufacturers need to be capable of producing a reliable product, and consumers need to have access to small-scale finance. Development

Fig. 3. Installing a pico power pack a Kathamba in Kenya. Pelton-style buckets are used on the shaft. (*Phillip Maher*)

Fig. 4. Installing a centrifugal pump as a turbine at Thima in Kenya.

organizations are taking on the role of enablers within this process. In Peru, for example, a successful "revolving fund" has been set up by Intermediate Technology Development Group to assist entrepreneurs or community organizations to set up micro hydro projects. The loans have been paid back because most schemes are designed to include productive end-uses as well as supplying electricity for lighting.

For pico hydro schemes, the cost per household has been reduced by the use of compact fluorescent lamps, which are now widely available. Only 20 W is enough power to light a typical rural house, so a 2-kW generator is enough to supply up to 100 households with electricity, with power available during the daytime for charging batteries or driving agro-processing equipment. The cost per household can be significantly less than a grid connection or the cost of a solar home system. Even light-emitting-diode (LED)-based lamps may soon become a cost-effective option for rural lighting. The future for micro hydro looks brighter than ever.

For further background information *see* COMPUTATIONAL FLUID DYNAMICS; GENERATOR; HYDRAULIC TURBINE; IMPULSE TURBINE; MOTOR; PUMP; REACTION TURBINE; TURBINE; WATERPOWER in the McGraw-Hill Encyclopedia of Science & Technology.

Arthur A. Williams

Bibliography. T. Anderson et al., *Rural Energy Services: A Handbook for Sustainable Energy Development*, ITDG Publishing, 1993; A. Harvey, with A. Brown, P. Hettiarachi and A. Inversin, *Micro-Hydro Design Manual: A Guide to Small-Scale Water Power Schemes*, ITDG Publishing, 2002.

Microcombustion

The physicist Richard Feynman gave a prophetic lecture more than 40 years ago in which he predicted much of the miniaturization that would occur in the ensuing years. To date, miniaturization has been demonstrated in many fields, such as electronics, information storage, and micro-electro-mechanical systems (MEMS) in the range of scales that Feynman paradoxically described in his lecture "Plenty of Room at the Bottom." Perhaps deliberately, he did not mention power generation. The power generating component is often the factor determining the size and weight of a number of portable consumer electronics (for example, mobile phones) and hampering further miniaturization.

Power generation at small scale has been funded in recent years mostly by the Defense Advanced Research Projects Agency (DARPA), the research arm of the U.S. Department of Defense, which embarks in high-risk/big payoff initiatives. A critical parameter to compact power generation, regardless of the strategy that will ultimately deliver such a power, is gravimetric energy density, in terms of which fuels fare much better than conventional batteries. The **illustration** shows in a logarithmic scale the energy density of a typical hydrocarbon, an alcohol, and

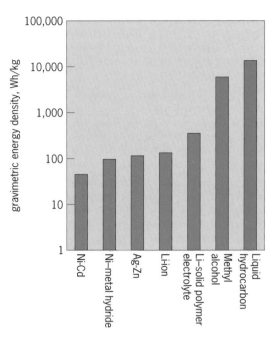

Comparison of the gravimetric energy density of some common batteries and liquid fuels.

some batteries in the common and somewhat awkward units typically used for batteries (1 Wh/kg = 3.6 kJ/kg). Even operating with a chemical-to-electric conversion efficiency as low as 5%, fuel-based devices can in principle be several times "denser" than conventional batteries, provided that their structure is sufficiently light. Moreover, they can be recharged easily and quickly by refilling the tank.

To harness the chemical power of fuels, small-scale combustion for thermal power generation enters the picture as an intermediate step for subsequent conversion to electricity. In addition, microcombustion is relevant to applications in which thrust or mechanical power is required—probably a less challenging task since, once thermal power is generated, a simple expansion through a nozzle in a suitable microthruster may achieve the goal.

Systems and scales. Technological goals that have spurred researchers' interest in microcombustion include micro heat engines as readily rechargeable batteries for electric power generation; micro combustors to power actuators in robotic applications; miniaturized internal combustion engines for the propulsion of micro air vehicles; and micro thrusters to power small satellites impulsively or to launch and power "smart dust" acting in a wireless network as microsensors for monitoring weather and air quality. The power requirements of all these applications are on the order of tens or hundreds of watts, and the longest dimension of the device perhaps a few centimeters. Combustion at this scale, although small by conventional combustion standards, is more appropriately called mesoscale combustion. At an even smaller overall scale is the concept of localized power generation for MEMS, to avoid complications associated with "bussing around" electric power. This goal would require making thermochemical power sources sized at the microscale (submillimeter), so that every micro device would carry its own power source to satisfy requirements on the order of perhaps only tens of milliwatts. The microscale has interesting manufacturing prospects through the possible use of the same microfabrication techniques that have been developed for the electronics industry. These techniques easily lend themselves to mass production, which has been a key factor in the extraordinary growth of the electronics industry in the last few decades. For our purposes, both mesoscale and microscale combustion will be considered as microcombustion, since both are much smaller than conventional combustion systems.

Limitations. Microcombustion has some fundamental limitations. As size decreases, the combustor surface-to-volume ratio increases. This in turn is a rough indicator of the ratio between heat loss and heat release rate. At a sufficiently small scale, losses will be too high for combustion to be sustained, and flame extinction or quenching will occur. Interesting fundamental work at the University of Illinois has shown recently that flames can propagate in micro passages, with characteristic dimensions on the order of 100 micrometers, provided that the surface is hot. The nature of the surface under these high-temperature conditions plays a role, which suggests that, in addition to thermal quenching, chemical quenching by surface scavenging of highly active chemical species (such as radicals) becomes significant.

Energy management. Control of flow/surface interaction requires a careful design and a judicious selection of operating conditions and wall materials, and ultimately appropriate thermal management of the entire device. Moreover, even if steady combustion is sustained, the anticipated efficiency of energy conversion modules will be at best about 20%, and more likely in the single-digit percentages. Since the majority of the fuel's chemical energy is transformed to heat, proper energy management is crucial. One reason is that the dissipation of thermal energy is a major practical impediment in some applications, especially for certain consumer electronics and in military applications in which thermal signature could give away the presence of someone wearing such a battery. At a minimum, thermal management must recover the bulk of the energy contained in the exhaust to preheat the incoming reactant stream under conditions of modest temperature differences across the combustor, as compared with conventional systems. In this way, the total reactant energy, including chemical and thermal components, can be higher than it would without recovery, which has prompted the labeling of these systems as "excess enthalpy" burners. The Swiss roll combustor is a compact combustor design in which the exhaust heat energy (enthalpy) is in part recovered and used to preheat the reactant streams of fuel or oxidizer. As a result, the overall burner efficiency can be dramatically increased. Proper energy management is also crucial because the efficient operation of an energy conversion system depends on maintaining a large temperature difference between the high-temperature and the low-temperature sides of some kind of thermodynamic cycle. Achieving this objective in a small volume is a major challenge. For that reason, micro heat engines should be more difficult to realize than microthrusters should.

Fuel. Complete conversion of the fuel to carbon dioxide (CO_2) may be difficult due to the likely short residence time in the combustor, and may create significant emission of carbon monoxide (CO) and unburned hydrocarbons, both of which are health hazards. Although pollutant emission at first may not appear to be significant for devices which would handle modest flow rates, their use indoors should still dictate maximum fuel conversion, which suggests the use of catalytic techniques. The increase in the surface-to-volume ratio with miniaturization favors the catalytic approach, especially at the microscale where surface effects are very significant since, if the surface is chemically active, quenching losses are minimized.

Additional complications may result from the need to store the fuel in the liquid phase, an inevitable choice if the energy density of the fuel is to be exploited. This restricts the fuel choice to highly volatile hydrocarbons, such as propane or butane, which can be stored in the liquid phase and vaporized readily at room temperature. Hydrogen, which has been a convenient choice in the early steps in microcombustion, is totally impractical since the high-energy-density objective would be defeated by the taxing requirements of cryogenic fuel storage. Alternatively, one could reform conventional hydrocarbons for in-situ generation of hydrogen. But this is a challenging task at the microscale. If successful, however, direct energy conversion via fuel cells may prove more attractive than the combustion route. Heavier fuels, such as logistic fuels used in the military, need to be dispersed by a suitable atomization technique to ensure full evaporation and mixing in small volumes—a nontrivial task when small size is a premium. A promising approach in this context relies on the use of an electrostatic spray to disperse the fuel into small, rapidly vaporizing droplets of uniform size, with negligible pressure drop.

Even if microcombustion is feasible, the combustor is but one element of the energy converter in which it is employed. For any system to find practical application, ancillary equipment (such as pumps, blowers, compressors, valves, and recuperators) of commensurately small size and weight would need to be developed, and in some cases with good tolerance for potentially harsh environments. To date, hardly anything is available off-the-shelf in this category.

Electric power generation. There are several strategies for producing electric power.

Miniaturization of conventional engines. This approach involves moving parts, that is, the direct miniaturization of conventional engines, using classical thermodynamic cycles such as those adopted in internal combustion engines and gas turbines. Examples include the gas turbine Brayton cycle (MIT), rotary-piston Wankel cycle (U. C. Berkeley), swing engine (U. Mich.), free piston engines (Georgia Tech. and Honeywell). Problems, especially at the microscale, include sealing, balancing rotating components, separating the "hot" from the "cold" side of the thermodynamic cycle, and difficulties in microfabrication of the complex geometries of the necessary parts. The MIT micro gas turbine attracted a lot of attention as the pioneering effort in the field but has not yet delivered positive power. The swing engine at the mesoscale appears to have the best prospects. It is based on a rotationally oscillating free-piston engine in which combustion occurs in four chambers separated by a single rotating swing arm. It has an energy density on the order of 1000 Wh/kg, and an (fuel-to-electric) efficiency currently at 8% and projected to improve to 15%.

Thermoelectric generator. This approach entails no moving parts. A typical example adapts the combustor to a thermoelectric generator, for example, as in ongoing projects at the University of Southern California, Penn State, and Princeton. Thermoelectrics are notoriously inefficient. Parasitic losses of peripherals and some inevitable combustor losses would lower further the overall fuel-to-electric conversion efficiency to single-digit percents. Short of breakthroughs in the materials properties (for example, figure of merit) of high-temperature thermoelectrics, the low conversion efficiency of this route may prove to be a major drawback.

Free-piston Stirling engine. This is an interesting prospect at the mesoscale. The free-piston Stirling engine has been the darling of thermodynamicists because of its high thermodynamic efficiency. Its practical appeal is that it is essentially an external combustion engine in which the cycle working gas is hermetically sealed. Its linear free-piston motion is simple and virtually immune from wear and tear. The engine is maintenance-free, with a lifetime of several years. For that reason, the free-piston Stirling engine can be considered intermediate in complexity between the two categories discussed above. Its mechanical energy can be easily converted to electrical energy via a linear alternator. Current projections at the mesoscale include an energy density on the order of 1000 Wh/kg and an (fuel-to-electric) efficiency greater than 20%.

Mechanical power generation. For mechanical power generation, microrocket projects have been pursued at TRW, U. C. Berkeley, Penn State, and MIT, among others. In this context, solid-fueled systems, such as the 100-W unit developed at TRW, may be simpler to implement since their application would be less dependent on the concomitant realization of pumps, valves, and miscellaneous mechanical systems that liquid-fueled counterparts would necessitate. The drawback is that solid-fueled rockets can be fired only once, which may limit the type of missions for which they are suited.

Outlook. Microcombustion is an intriguing field, the practical realization of which in standalone energy conversion systems presents many challenges, despite considerable progress in its relatively short life. Currently, the best feasibility prospects remain at the mesoscale.

For background information *see* COMBUSTION; MESOSCOPIC PHYSICS; MICRO-ELECTRO-MECHANICAL SYSYTEMS (MEMS); STIRLING ENGINE; THERMODYNAMIC CYCLE; THERMOELECTRICITY in the McGraw-Hill Encyclopedia of Science & Technology.

Alessandro Gomez

Bibliography.   W. J. A. Dahm et al., Micro internal combustion swing engine (MICSE) for portable power generation systems, *AIAA Aerospace Sciences Meeting*, Jan. 14–17, 2002; A. C. Fernandez-Pello, Micro-scale power generation using combustion: Issues and approaches, *Proc. Combust. Inst.*, 29:883–889, 2002; R. P. Feynman, in H. D. Gilbert (ed.), *Miniaturization*, pp. 282–296, Reinhold, New York, 1961; C. M. Miesse et al., Sub-millimeter scale combustion, *AIChE J.*, 50:3206–3214, 2004; F. J. Weinberg et al., On thermoelectric power conversion from heat re-circulating combustion systems, *Proc. Combust. Inst.*, 29:941–947, 2002.

Mississippi River degradation

The Mississippi River ranks as the third longest river in the world, and it is eighth in terms of average water volume discharged. The river and its tributaries collect runoff and sediment from 31 states, making it the second largest watershed in the world. The sediment-laden Mississippi has, over thousands of years, built a vast wetland system along the lowland coast of Louisiana. Due to the delta-forming processes of the river, the nutrient-rich salt-water and fresh-water marshes and estuaries are nursery grounds for one of the most productive fisheries in the world. Louisiana's wetlands also buffer New Orleans and other coastal towns from erosion and high-velocity hurricane winds, and serve as important habitat for 15 million migratory birds and many other wildlife species.

The Mississippi is also classified as one of the world's most endangered rivers. Over the last century, particularly the last 50 years, humans have dramatically altered the river and surrounding areas. The consequences are severe. For example, largely due to hydrological engineering projects (such as extensive canals), the Louisiana coastal area has lost 1900 mi^2 of wetlands since 1930, and is predicted to lose another 500–700 mi^2 (1295 to 1813 km^2) by 2050. In addition, in the Gulf of Mexico, contamination from farming practices upstream is creating a massive "dead zone," in which dissolved oxygen concentrations are too low to support fish and invertebrate life.

Channelization. The Mississippi River has been dammed and leveed extensively to keep it from flooding its borders and from changing course. Over the past 6000 years, the river has changed course six times; without water control projects, scientists project that the Mississippi River would have abandoned its path long ago in favor of the Atchafalaya

River basin. The Atchafalaya branches off to the west 170 mi (274 km) upstream from New Orleans and currently captures 30% of the Mississippi's flow. Its route to the ocean is only 142 mi (229 km), about half as long as the Mississippi's. Many predict that some-day soon runoff from a large storm will cause the Mississippi to overpower its human-engineered diversions and make the switch, leaving New Orleans and Baton Rouge behind. If this happens, ocean salt water will rush in to fill the empty riverbed and wreak havoc with local water supplies.

Canalization. Coastal wetlands are constantly un-dergoing the processes of accretion (soil buildup) and subsidence (loss of soil volume). The Mississippi River slows as it enters the Gulf of Mexico, and the sediments it has picked up along the way begin to drop out. When the sediments pile up above the tide level, plants take hold, creating diverse habitat for coastal marine life. Older wetlands usually show more signs of subsidence, while newly formed wet-lands show more accretion.

The rate of wetland subsidence in new and older wetlands increased dramatically in recent decades, particularly during the 1960s and 1970s. This cor-responds to a period when extensive networks of canals were built to aid in navigation for oil and gas extraction (see **illustration**). The subsidence rate has slowed in recent years but continues to be faster than historical levels. Current rates threaten the homes of about 2 million people, as well as the wildlife and fishing industries, which supply about 40% of the seafood in the United States. Canals ac-count directly for about 12% of land loss, but canal edges, or spoil banks, severely alter the flow of water

through the ecosystem even decades after they have been abandoned. Natural wetland soils are mostly water (92%), sediment (4%), and peat or root matter (4%), whereas spoil bank soils are heavily compacted and support trees and shrubs instead of native marsh plants. Complicated networks of spoil banks impede the seasonal migrations of juvenile fish, as well as water and nutrient flow through the system. Canals also allow for longer-than-natural dry periods in wet-lands, which contributes to marsh grass mortality. As wetland plants die, organic material in the soil is lost, and subsidence increases. Canals can also in-crease salt-water intrusion, impacting sensitive wet-land species.

Estimates for restoring the coast using a variety of methods, including intensive engineering projects, exceed $14 billion over the next 30 years. How-ever, simple restoration projects involving in-filling of old canals are among the cheapest methods for rebuilding wetlands. Preventing wetland loss in the first place may be the most effective way to protect the coast, since restoration is still such an experimen-tal science, and in many cases restoration will not be possible.

Dead zone. Levels of nitrogen in the Mississippi have doubled in recent years, primarily due to fertil-izer runoff from midwestern farms, as well as cattle manure and auto emissions. Every spring and on through the summer, algae in the Gulf of Mexico flourish from the extra dose of nitrogen nutrients from upstream. The algae die after a few days and sink to the bottom of the continental shelf. As the algae decompose, they use up all available oxygen, causing fish and invertebrates to rush to get out to

Example of extensive oilfield navigation canal networks in Terrebonne Bay, which are contributing to rapid coastal land loss along the Louisiana coast.

avoid suffocation. As the surface water cools in the fall, the stratified layers mix and oxygen levels are restored until the following spring. This recurring hypoxic (low-oxygen) area has grown in recent years to 7000–9000 mi^2 (18,130–23,310 km^2).

The degradation of the Gulf of Mexico due to nitrogen pollution from the Mississippi River is a harbinger of environmental damage expected to spread worldwide. Globally, humans have doubled the total amount of nitrogen added to terrestrial systems, and the inputs are still increasing due to exponentially growing fertilizer use. Since nitrogen and fertilizer are required for the high agricultural productivity of the United States Great Plains, reducing nitrogen input into the Mississippi will require innovations in how fertilizer is applied, and perhaps the breeding of plants that use fertilizer more efficiently. Unlike many forms of pollution, nitrogen performs an invaluable service in terms of promoting crop growth, and it cannot easily be replaced. Nonetheless, scientists and government agencies have plans to cut the size of the dead zone in half over the next 15 years by replanting trees and shrubs along tributary streams and rivers to absorb and transform the nitrogen runoff from farms. Some researchers are also looking to reduce fertilizer application while maintaining crop production levels, and others are discussing the merits of reducing emissions.

Industrial pollution reserves. Southern states receive and store more toxic waste than other areas of the country. For example, Robert Bullard reported in the book *Dumping in Dixie* that in 1987 Louisiana was one of five southern states that harbored 60% of the nation's hazardous wastes by volume.

The Department of Environmental Quality (DEQ) in Louisiana has issued health warnings, citing high levels of hexachlorobenzene, hexachloro-1,3-butadiene, polychlorinated biphenyls, dichlorodiphenyltrichloroethane, dioxin, and mercury in major waterways. The DEQ warns against contact with sediment or water in some rivers and lakes, and it cautions that the seafood is too toxic to eat more than once or twice a month. The DEQ also warns of phenols, metals, oil and grease, and pathogens in the Mississippi and of several "unknown toxicities" in many coastal streams. Local health studies have found high levels of these types of chemicals in blood, hair, and breast milk samples. Coal-burning power plants, refineries, and chemical manufacturing plants are major sources of this pollution. Higher levels of pollution correlate with higher levels of illness, asthma, and overall mortality. In addition, studies indicate that some chemicals, even at very low levels, can mimic the body's natural hormones, and could be contributing to rising cancer rates and to reproductive behavioral changes. Since pollution can cross continents on wind and ocean currents, its negative impacts are not limited to local human and wildlife communities.

Conclusion. The Mississippi River and Louisiana coastal wetlands have been severely altered over the last several decades. Many of the functional aspects of the land (including support for diverse species, protection of inland areas from storms, and provision of clean water for drinking, recreation, and fishing) have been severely diminished. The most widespread symptoms of eutrophication (the deterioration of the life-supporting and esthetic qualities of a water body caused by excessive fertilization) and environmental degradation in the Mississippi and Louisiana estuaries are low concentrations of dissolved oxygen due to nitrogen pollution and loss of submerged aquatic vegetation due to canalization. These, in turn, result in impaired recreational and commercial fishing and shellfishing. Recent analyses by the National Oceanic and Atmospheric Association predict that conditions are expected to noticeably worsen by 2020.

Even if nitrogen inputs are reduced, the wetlands will need considerable restoration. It is much cheaper and simpler to keep the land and water quality high than to go back and try to restore it. Restoration efforts are still very experimental, and it will take many decades to prove if they have worked. Estuaries and salt-water wetlands, such as those supported by the Mississippi River, are estimated to produce services to humans (including food production, flood control, and storm protection) that have an annual value ranging between $5000 and $10,000 per hectare per year. Given that value, the cost of pollution abatement and habitat restoration may well be a bargain.

For background information *see* EUTROPHICATION; ENVIRONMENTAL TOXICOLOGY; RIVER; RIVER ENGINEERING; WATER CONSERVATION; WATER POLLUTION; WETLANDS in the McGraw-Hill Encyclopedia of Science & Technology.   Stacey Solie

Bibliography. S. Bricker et al., *National Estuarine Eutrophication Assessment*, NOAA, Special Projects Office, Silver Springs, MD, 1999; R. D. Bullard, *Dumping in Dixie: Race, Class, and Environmental Quality*, Westview Press, Boulder, CO, and United Kingdom; Committee on Hormonally Active Agents in the Environment, National Research Council, *Hormonally Active Agents in the Environment*, 1999; Environmental Protection Agency, *Action Plan to Reduce the Size of the "Dead Zone" in the Gulf of Mexico*, EPA 841-F-01-001, January 2001; J. Kaiser, Panel cautiously confirms low-dose effects, *Science*, 290:695–697, 2000; A. Mosier et al., Policy implications of human-accelerated nitrogen cycling, *Biogeochemistry*, 57/58:477–516, 2002; N. Nosengo, Fertilized to death, *Nature*, 425:894–895, 2003; B. Streever, *Saving Louisiana? The Battle for Coastal Wetlands*, University Press of Mississippi, Jackson, 2001; U.S. Geological Service. *100+ Years of Land Change for Coastal Louisiana*, Jimmy Johnston, Lafayette, LA, 2003.

Mobile satellite services with an ancillary terrestrial component

Mobile satellite service (MSS) systems provide voice and data communications service to users over a wide area. MSS systems can serve a whole continent

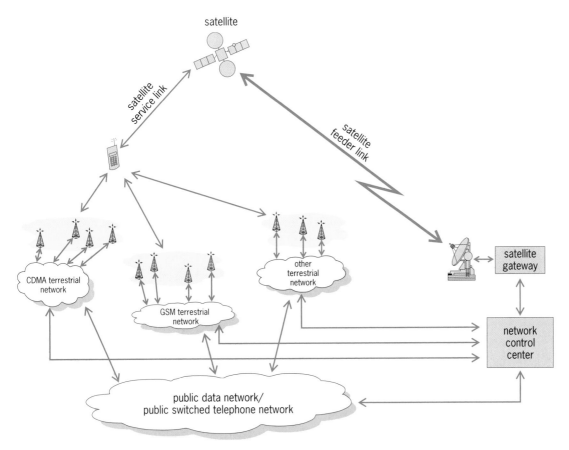

Fig. 1. Satellite-based/ancillary terrestrial component architecture.

and its coastal waters via one or several Earth-orbiting satellites. In contrast, cellular communications systems use thousands of base transceiver stations to provide communications in desired areas.

Two major factors have restricted the success of MSS systems. (1) Users have been required to use expensive and physically large terminals because of satellite technical limitations and the distance between the satellite and the service area. Even the smallest terminals have been unsuitable for carrying in a shirt pocket, for example. (2) Despite their size and expense, these terminals do not work in buildings or even on the street in dense urban areas. The struggling MSS industry recently received a much-needed stimulus, as these factors have been overcome by technical advances and a decision by the Federal Communications Commission (FCC).

The availability of much larger satellite antennas, as well as other innovations, will provide the satellite power and sensitivity to enable the use of satellite terminals similar in size and cost to existing terrestrial cellular terminals. Following a lengthy technical debate, the FCC authorized MSS providers to use ancillary terrestrial components (ATC), a network of terrestrial base transceiver stations similar to those used by the cellular industry but integrated with the satellite network in such a way that the terrestrial network reuses the satellite spectrum and the user terminal operates on either network. The MSS operator would provide service on its ATC where large build-

ings or other obstructions block the line of sight to the satellite. Such shared use of spectrum has never been approved before. This authority will allow the MSS provider to use its spectrum to provide additional services more efficiently and to offer better coverage in crowded urban areas, as well as along highways and waterways, and in the rural or relatively remote areas that have long been the hallmark of MSS.

System architecture. The system architecture will consist of two integrated subnetworks: a space-based network (SBN) and an ATC. **Figure 1** shows the interoperation of the SBN and the ATC system architecture. In the SBN, a satellite communicates with its gateway via feeder links and provides service to its mobile terminals via service links. In the ATC, the terrestrial network serves the same mobile terminals utilizing any of the standard cellular system protocols, such as the Global System for Mobile (GSM) Communications and Code Division Multiple Access (CDMA). Since the service footprint of the SBN is large, covering all of Central and North America, ATC would be deployed within the SBN footprint in urban areas, where buildings block satellite signals from the streets and indoor coverage is required.

The satellite will have a large reflector, greater than 20 m (65 ft) in diameter, to achieve high downlink power density, as well as high sensitivity to signals transmitted by user terminals. The satellite will provide approximately 200 small-diameter beams over

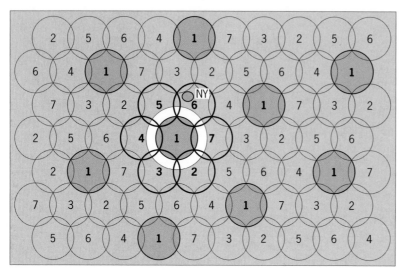

Fig. 2. Beam frequency reuse pattern.

Central and North America, instead of one large beam. **Figure 2** shows the beam pattern conceptually, with each small circle representing one beam from the satellite. To reduce interference, different portions of the available radio frequency spectrum will be used in adjacent beams. Figure 2 shows a seven-beam spectrum reuse pattern in which all beams numbered 1, for example, will reuse the same portion of the available spectrum, but the same portion of the spectrum is never used in adjacent beams.

The SBN and the ATC both share and reuse the same radio-frequency spectrum, and therefore intrasystem interference between them must be considered in the overall design. The same issue is faced by existing terrestrial cellular networks since they use and reuse the same spectrum many times within a given service area. Unique to the SBN/ATC architecture is that the SBN is, in effect, adjacent to every ATC cell in the network. In addition, an SBN beam is much larger than an ATC cell and will cover the same geographical region that is covered by a large number of ATC cells. The critical balance of spectrum reuse and the control of intrasystem interference will take place in the network control center shown in Fig. 1.

If a satellite beam numbered 6 in Fig. 2 is using a given frequency range, that same frequency range may not be reused by any ATCs within that beam numbered 6. Actually, the exclusion region for frequencies used in any given beam is somewhat larger than the nominal coverage area of the beam, and is illustrated in Fig. 2 by the white ring around one of the beams numbered 1. Each such exclusion region represents a spatial guard band that allows the encircled satellite beam pattern to develop an average discrimination of 10 dB relative to an ATC area that may reuse the same frequency set.

In Fig. 2, New York is being served by a satellite beam that uses frequency range 6 and by the ATC within that beam. Spectrum planning and control by the network control center would allow the New York ATC to reuse all satellite band frequencies that

are being used by satellite cells 1, 2, 4, and 7, but not 6. (It also perhaps may use to a lesser extent frequencies of cells 3 and 5, since in this example New York may be within the exclusion zone for beams 3 and 5).

ATC-induced interference potential. There is always a certain amount of interference or unwanted energy (noise) that will affect the reception of a signal. As an ATC initiates terrestrial reuse of the frequencies of a neighboring satellite beam, the neighboring satellite beam will inadvertently receive additional ATC generated cochannel interference. The key question is how much and whether the affected satellite return link is robust enough to accommodate the increased interference level. All radio communications systems are limited by many factors, including how much signal power is transmitted, how much signal power is lost on the path to the receiver, and how much interference (referred to as noise) is generated by random processes within the receiving system. The effect of the cochannel interference induced by the ATC into a satellite receiver may be characterized by an equivalent percentage increase in noise, and can be expressed in terms of an equivalent receiver effective temperature differential increase.

Consider the intrasystem interference potential from a user terminal operating with the ATC system. The spatially averaged radiated power of an ATC terminal in the direction of the satellite is −4 dBW (based on real antenna measurements of cellular terminals). That quantity is reduced before reaching the satellite receiver by several factors. These factors include the path spreading loss over the path between the terminal and the satellite; satellite antenna discrimination in the direction of an ATC terminal that, to be cofrequency, must be no closer than an adjacent satellite beam; and reduction in ATC terminal radiated power due to the use of closed-loop power control.

Despite the factors that reduce the interference at the satellite, with a 400-fold cochannel reuse by the ATC in the vicinity of a satellite cell, a more than 100% rise in the noise power spectral density at a satellite receiver can occur. This performance degradation may be unacceptable since the satellite link margin (power at the receiver in excess of that required to establish the link) that would have to be expended to accommodate the effect (3 dB) may be unavailable. However, ground-based signal processing does relieve the situation.

Ground-based interference cancellation. Figure 3 shows the key elements of the receiver architecture of an interference canceller. The receiver is configured for canceling the ATC-induced interference on a cochannel receiver of satellite beam 1 (Fig. 2). The signals that are intercepted by the "desired" satellite beam (beam 1) and by the neighboring satellite beams over the frequency span of the desired signal are transported to the satellite gateway, where they are linearly combined (via fractionally spaced transversal filters) to form an optimum decision variable in accordance with a least-squares-error

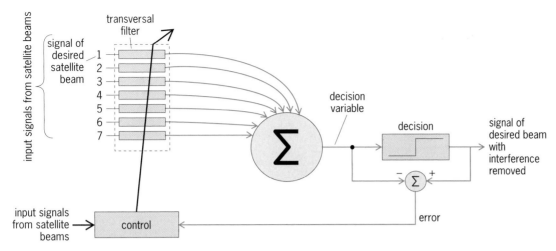

Fig. 3. Adaptive receiver processing at the network control center.

criterion. The performance of a satellite receiver equipped with the interference canceller of Fig. 3 has been evaluated by Monte Carlo computer simulation. The results show that very large ATC-induced increases in noise power values (of the order of 600%) can be reduced by 2 orders of magnitude.

Conclusion. Large efficiencies in spectral reuse can be attained by configuring a geostationary satellite system with ancillary terrestrial components to reuse the available satellite band frequencies over both satellite and terrestrial links. Besides providing significant additional frequency reuse and therefore additional system capacity, this architecture greatly improves the system's service reliability since the ATC overcomes service interruptions that would otherwise occur to the mobile satellite terminal in populous, urban environments. Negative effects to the system's satellite links stemming from terrestrially reusing the satellite-band frequencies can be controlled by appropriate signal processing. The use of these techniques will facilitate the construction of an integrated SBN-ATC system that will allow users with small cellular-type terminals to obtain the benefits of both cellular and mobile-satellite communications.

For background information *see* COMMUNICATIONS SATELLITE; DATA COMMUNICATIONS; ELECTRICAL COMMUNICATIONS; MOBILE RADIO in the McGraw-Hill Encyclopedia of Science & Technology.
Peter D. Karabinis

Bibliography. Comtek Associates Inc., *Use of Mobile Satellite Spectrum To Provide Complementary Terrestrial Mobile Service To Improve Satellite Coverage*, Final Report Prepared for Industry Canada, November 5, 2002; G. Maral and M. Bousquet, *Satellite Communications Systems*, 4th ed., Wiley, West Sussex, 2002; Report and Order and Notice of Proposed Rulemaking, FCC 03-15, *Flexibility for Delivery of Communications by Mobile Satellite Service Providers in the 2 GHz Band, the L-Band, and the 1.6/2.4 Bands*, IB Docket No. 01-185, adopted Jan. 29, 2003, released Feb. 10, 2003; M. Richaria, *Satellite Communications Systems*, 2d ed., MacMillan, Hampshire, Ltd., 1999.

Modern traditional sailing ships

In the exploratory and commercial heyday of sailing ships, there were many tragic accidents resulting in the loss of ships and, in many cases, great loss of life. The responses of governments to these tragedies generally took the form of increasingly tighter regulations of commercial vessels (cargo and passenger alike) to try to prevent recurrences of such accidents. Therefore, rehabilitated original ships and accurate replica sailing ships now are required to reconcile historical accuracy with modern practices and government regulations. If one were to apply today's regulations, especially those relating to stability, to ships of a hundred years ago, the great majority of those ships would not meet requirements. Designers of replica ships, if these ships are to be true to the appearance, performance, and experience of their forebears, are therefore faced with a very challenging set of parameters for a given project, with fire-retardancy and stability issues being two of the most important and often the most difficult. Fortunately, modern designers of traditional ships now have new technologies available which not only can facilitate the solutions to these design problems but in many cases can be hidden within the structure so as not to adversely effect the overall goals of the client and the ship (**Fig. 1**).

Modern requirements, modern methods. We now see many departures from, and improvements to, the design and equipment of ships of old; most are required for safety, but a few are expected for convenience and comfort. For example, most sailing ships now have engines which, in addition to adding a safety factor, improve maneuverability and schedule keeping. New requirements have resulted in new safety equipment being carried aboard or installed, such as lifeboats or lifefloats, modern fire-fighting equipment, safety harnesses, emergency position indicating radio beacons (EPIRBs), as well as communications radios, radar, and electronic navigation equipment. Other systems not seen in years past are now standard, such as electrical systems, modern galley appliances, and modern sanitation. Lifeline

Fig. 1. *Simon Bolivar.* Launched in 1979, it has become a familiar participant in Tall Ship events, and is an excellent example of a modern traditional sailing ship. It is exclusively used as a training vessel for the Venezuelan Navy.

tion appeared in the nineteenth century, first in Europe and later in America (Europe suffered wood shortages much sooner than did America), and allowed sailing ships to be built bigger, stronger, and faster. A number of new materials are available to build with, some as replacements to wood, such as aluminum, steel, and fiber-reinforced plastics (FRP; fiberglass), and some which are used in conjunction with wood, such as fiberglass and its associated resins (polyesters, vinylesters, and epoxies). These latter forms of construction, which over the years have combined a variety of materials, are referred to as composite construction. Indeed, many good vessels have been built of wood planking over steel frames, an early form of composite construction. More recently, especially with replica ships, we are seeing traditional wood construction with the various wood components set in resins (usually epoxies) to improve strength, watertightness, and resistance to fire. Further departures from traditional construction include modified wood construction utilizing FRP skins to form sandwich hull panels and other panels, often obviating the need for frames (**Fig. 2***a*, *b*). A method called cold molding uses very thin wood sheets or veneers laid over temporary frames and set in epoxy (the wood fibers essentially replacing the fiberglass in FRP).

Even traditional wood construction has benefitted from new methods and materials, especially with new types of fasteners and preservatives, and with much more efficient tools than were available years ago. We also have at our disposal the means to procure a greater variety of tropical hardwoods, something not normally possible to the old builders, who had to rely on local woods. These special woods provide a much broader array of choices to the builder, who can now use, for example, a very heavy, dense,

requirements have altered vessel profiles by requiring higher minimum rail heights and prescribing their methods of construction. Belowdecks accommodations, especially for passenger-carrying vessels, require minimum measurements and construction for passenger comfort, safety, and ease of emergency egress, resulting in spaces bearing no resemblance to those of a century ago (although ships involved strictly in sail training are allowed somewhat rougher accommodations).

We now have new methods of construction not available in past centuries. Iron and steel construc-

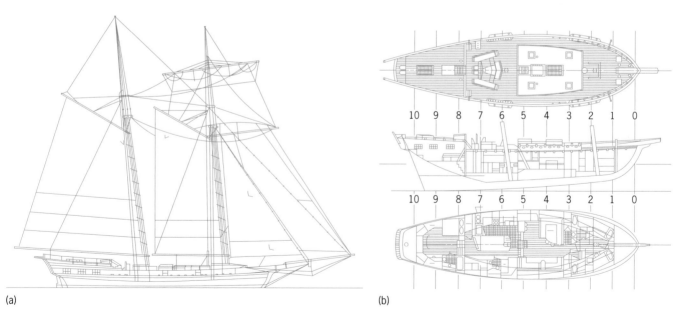

(a)   (b)

Fig. 2. *Berbice.* The author's design of a replica of the original ship, a late-eighteenth-century American topsail schooner captured by the British. A private vessel, it is of fully composite construction, specifically "sheathed strip plank," laying wood in epoxy and sheathing both inside and out in FRP (fiberglass). It also incorporates integral cast-lead ballast. (*a*) Sail plan profile. (*b*) Top: deck plan; middle: inboard profile/longitudinal section; bottom: accommodations.

and strong wood (for example, greenheart, cortez, purpleheart) in the lower structures, the "backbone," of the ship. Lighter, more bendable woods such as silverballi and courbaril might be used in planking. Even the temperate zones provide an excellent assortment of woods. White oak and Douglas-fir are both considered staples of good shipbuilding: They are durable, straight-grained, and rot-resistant, and take well either to single-piece timber construction or to the process of creating preshaped timbers by laminating numerous thinner layers of these woods. This lamination practice not only creates very strong and rot-resistant members but also allows the use of smaller trees and timbers with less waste, an important consideration with dwindling wood supplies and increased reliance on farm-raised woods. Frames and deck beams are usually the prime candidates for lamination, although more and more ships are being built in which almost all of the structural members are laminated. All these methods allow present-day builders to economically produce very strong, lighter-weight vessels.

Significant advances have also been achieved in the art and science of keeping ships' rigs intact and vertical. In the old days, most standing rigging (that rigging which is fixed in place and serves to support the masts) was of natural fibers, mostly hemp, and had to be tended to and adjusted daily, especially with large changes in temperature or humidity. Sails were also made of natural fiber and required constant attention. All of these materials were constantly susceptible to rot. With many ships of old having mediocre stability at best, it was often considered advantageous to have the sails blow out, and even have the rig come down, in extreme circumstances, before the ship itself capsized or foundered.

Today we have metal wires and cables in a variety of alloys and configurations to keep the rigging in place. Steel and galvanized steel wire rigging became prevalent about the same time that iron and steel hulls were developing in the mid and late nineteenth century. More recently, various stainless steels have been developed to further the reliability of this rigging. In an effort to keep the proper appearance and feel of a traditional sailing ship, a number of tricks have been used to essentially hide the wire, mostly by wrapping fiber rope around it. Fiber ropes have been replaced with synthetic ropes in a variety of compositions, such as polypropylene, dacron, and nylon. Some of these synthetic ropes, such as Spunflex™, a polypropylene derivative, have a very similar look and feel to traditional natural fiber ropes. (Rope can be called rope as long as it is still on the spool; after that it is called line.) Different synthetics have different degrees of stretch, so care must be taken to use the right line for a particular function. Indeed, some are very low stretch and have been used in place of stainless wire for standing rigging on some ships.

Sails also are now made almost exclusively of synthetic materials. The most prevalent material today for recreational sailboats is dacron. However, sails on a sailing ship have to be "handed"; that is, crew go aloft, climb out the yards, and manhandle the sails into submission to furl them. Most shiny, stiff synthetic sails do not lend themselves well to that, so a number of synthetics, mostly polyester-based materials and weaves, have been developed such that they handle and behave much like traditional, natural fiber, soft sails. Duradon™, from Europe, is probably the best example of this type of sailcloth, although there is a United States company making a very successful cloth called Oceanus™, using a polyester-based fiber made from recycled soda bottles.

A concern about these new, much stronger synthetic sails and rigging is that the "safety valve" that olden ships had when their sails blew out might be lost.

Stability issue. A ship is considered initially stable if it remains upright without external help. It must also be able to return to this upright position when those forces trying to push it over (heeling forces, wind and waves) are removed. Furthermore, (and this is one of the tests that the original ships would have a very hard time with), a federal regulation pertaining to vessels carrying passengers known as 46 CFR 171.055 (c) (2) states that a sailing vessel must have a positive righting arm up to $90°$ of heel. In other words, it must be able to be heeled over to $90°$, essentially knocked over fully on its side ("on her beam ends"), and still be able to return to its upright position when the heeling forces are removed.

There are also a number of other calculations associated with the above mentioned that include other variables such as height of center of gravity, height of metacenter (the point about which a vessel rotates in the early stages of heeling), vessel displacement, area of lateral waterplane (the vessel's underwater profile), area of windage (the profile area of a vessel subjected to the wind, including sails, spars, rigging, and the like), and so on. All of these result in different quantifications of the vessel's behavior in different situations, so as to give a complete picture of the vessel's overall intact stability, and its survivability in a damaged condition. A vessel must satisfy all these calculations before government approval to carry passengers or cargo for hire. Many of these regulations have resulted in the requirement for additional internal structures never built into the original vessels, for example, watertight bulkheads. The correct number and positioning of these allow any one compartment (or two for larger vessels) to completely flood, without sinking the vessel beyond a line 3 in. (7.6 cm) below the main deckedge. This feature alone would have saved many ships in the past.

Since a primary component of good stability is low center of gravity, we look first at the vessel's hull and interior structures. Today we can utilize the various composite and alternative construction methods discussed above to lighten the upper part of the ship's structure and rig while increasing strength. In the meantime, while the original ships carried their ballast in the bilges, often in the form of bricks and

Fig. 3. Launch of *Pride of Baltimore II*. Designed by Thomas Gillmer, with assistance from the author, it is a replica of early-nineteenth-century Baltimore clippers, which normally had much slacker (more steeply sloped) bilges. Note the firmer turn of the bilges. A significant portion of the visible keel is actually cast-lead ballast.

rigs falling down. But this means higher loads will be carried from the rig into the ship's hull (the sails do not blow out at an early point, and less heeling potential means greater forces on the rig). So the hull has to be stronger as a unit (known as the "hull girder"), while being lighter, especially in the upper portions of the structure, but with a lower and often heavier ballast, to achieve the stability requirements. It becomes clear, then, that without new methods and materials, sturdy, stable, safe, regulation-compliant, and well-performing traditional modern sailing ships would be nearly impossible to achieve.

For background information *see* SHIP DESIGN; SHIP POWERING, MANEUVERING, AND SEAKEEPING in the McGraw-Hill Encyclopedia of Science & Technology.

Capt. Iver C. Franzen

Bibliography. L. A. Aguiar (comp. and ed.), *Sail Tall Ships!: A Directory of Sail Training and Adventure at Sea*, American Sail Training Association, Newport, RI, 2003; N. J. Brouwer, *International Register of Historic Ships*, 2d ed., Sea History Press, Peekskill, NY, 1993; *Code of Federal Regulations, Title 46, Shipping*, GPO, Washington, DC, 2004; Gougeon Brothers, Inc., *The Gougeon Brothers on Boat Construction*, McKay Press, Bay City, MI, 1985; T. Trykare, *The Lore of Ships*, Holt, Rinehart and Winston, New York, 1963; J. A. Wills, *Marine Reinforced Plastics Construction*, Tiller Publishing, St. Michaels, MD, 1998.

stones (the source of "cobblestones" for some seaport streets), we now have the capability of casting specific shapes of lead which can be incorporated directly into the keel structure (**Fig. 3**). By getting a much denser material that much lower in the structure, we achieve a much lower center of gravity of the ballast which, when coupled with the lighter structure above it, results in a significantly lower center of gravity of the entire vessel. The righting moment is therefore increased and, when opposed to heeling moments unchanged from the original vessel, results in a greater range of positive stability to comply with the new regulations.

Another tool available to designers is to increase what is known as form stability. This involves a change in the hull shape in section by subtly flattening the bottom somewhat and creating a "firmer turn of the bilge." This essentially gives the vessel more to "lean on" when heeled. This has been applied to a number of notable vessels, with little, if any, sacrifice of their performance. Indeed, a number of recent replicas have used this approach to a dramatic degree, which, when coupled with very lightweight construction, has resulted in ships which are remarkably stable and fast. These are often referred to as "waterline up" replicas, where the underbodies bear little if any resemblance to the parent vessels.

Conclusion. There is a circular relationship within the design process of a modern traditional sailing ship. Modern safety regulations require a more stable ("stiffer") ship. Those same regulations also frown on

Moisture-resistant housing

For increased energy efficiency, modern homes tend to be "tighter," with less loss of heating and cooling. As a result, moisture generated inside a home may not find a way out and moisture from outside, coming in through a leak, may end up trapped in the home. Moisture is involved in many housing durability, health, and performance problems. Excessive moisture in a house can lead to rot, mold, swelling, staining, dissolution, freeze-thaw damage, and termite attacks that damage building materials and impede building performance. An understanding of moisture content and movement in a building is critical if one is to provide good housing performance, and is particularly important for walls, roofs, windows, and basements. Owners, builders, subcontractors, inspectors, designers, regulatory agencies, and product manufacturers require such knowledge.

Moisture problems. Moisture in houses causes a wide range of problems. The most serious tend to be structural damage due to wood decay, steel corrosion or freeze-thaw, unhealthy fungal growth on interior surfaces, and damage to moisture-sensitive interior finishes.

For a moisture-related problem to occur, at least four conditions must be satisfied: a moisture source must be available; there must be a route or means for this moisture to travel; there must be some driving force to cause moisture movement; and the materials and/or assembly must be susceptible to moisture damage. Elimination of even one condition can prevent a moisture-related problem. This is, however, often practically or economically difficult. In practice it is impossible to remove all moisture sources, to build walls without imperfection, to remove all forces causing moisture movement, or to use only materials that are not susceptible to moisture damage. Hence, designers take the approach of reducing the probability of having a problem by addressing two or more of these prerequisites. Controlling or managing moisture and reducing the risk of failure by judicious design, assembly, and material choices is the approach taken in the design of moisture-tolerant housing.

Moisture balance. If a balance between the amount of wetting and drying is maintained, moisture will not accumulate over time and moisture-related problems are unlikely. The rate and duration of wetting and drying must, therefore, always be considered together with the ability of a material or assembly to safely store moisture (that is, without damaging the material).

Most moisture control strategies of the recent past attempted to reduce the amount of wetting, for example, by increasing air tightness and vapor resistance of the enclosure, or by reducing the volume of rainwater penetration. However, it has become generally accepted that most building construction will not be perfect, and thus wetting will occur—for example, water tends to leak into walls around windows, cladding stops only some of the rain, and some amount of air leakage is inevitable. Old homes often allowed a significant amount of wetting but compensated for this with a large amount of drying and safe storage. Drying was accomplished by allowing heat energy to flow freely through assemblies and by using large volumes of durable materials such as solid timber and masonry. The drawback to this approach is excessive energy consumption and material use. Therefore, modern approaches emphasize balancing a reduced wetting potential with the provision of greater drying potential and safe storage.

The moisture tolerance of houses can be dramatically improved through intelligent design of building location, orientation, geometry, and HVAC (heating, ventilation and air conditioning) design, for example. Sloping the site away from the building encourages surface drainage away; the addition of overhangs, drip edges, and other surface features reduces the amount of rain penetration; and air conditioners that are not oversized ensure better control of interior humidity. This is often the least expensive and most effective approach, although it must be considered early in the design stage.

Moisture sources and wetting mechanisms, moisture sinks and drying mechanisms, and moisture storage and material response will be considered in turn below.

Moisture sources and wetting. The four major sources of moisture, and the wetting mechanisms involved, for a home's enclosure are (1) precipitation, especially driving rain; (2) water vapor in the air transported by diffusion and/or air movement through the wall (from either the interior or exte-

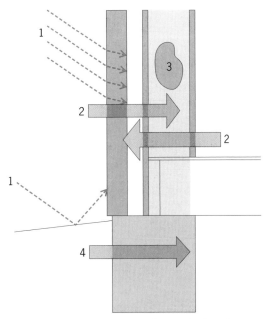

Fig. 1. Moisture sources and wetting mechanisms for enclosures. See text for 1–4.

rior); (3) built-in and stored moisture; and (4) ground water, in liquid and vapor form. These sources are shown in **Fig. 1**.

Driving rain is generally the largest source of moisture for a building, and must be controlled by the enclosure. Various strategies exist for resisting water penetration, but accepting and managing (by storage and drying) some leakage is often the most successful and practical.

Water vapor, generated by occupant activities (such as bathing, cooking, or laundering) by drying out of materials, and from the outdoor air in warm humid conditions, often causes problems in conditioned homes. Vapor transport by air movement is by far the most powerful mechanism, and all homes with conditioned space require an air barrier system. Vapor transport by diffusion is usually much less important, and low-permeance vapor diffusion barriers are often not necessary (but are often confused with air barriers, which are).

Moisture is also often built into new buildings. For example, construction lumber may contain well over 25% moisture by weight, and concrete contains large quantities of water when poured. This source of moisture may be controlled by limiting the use of wet materials or by allowing drying before closing the building in.

Ground water is present in vapor form almost everywhere, even in deserts, whereas liquid is present in areas with high water tables and during rainfalls and snow melts. Low-vapor-permeance ground covers are required between the soil and the house enclosure in almost all climates to prevent diffusion into the home from the soil. Similarly, a capillary break (hydrophobic or water-resistant material) prevents liquid water from wicking into the enclosure from the soil. Another important source of moisture can be plumbing leaks and surface water floods. These can be managed by the use of disaster pans below

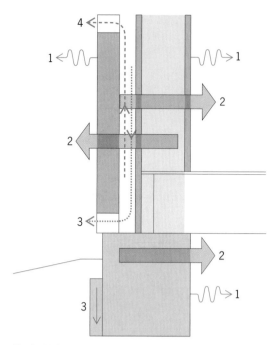

Fig. 2. Moisture sinks and drying mechanisms for enclosures. See text for 1–4.

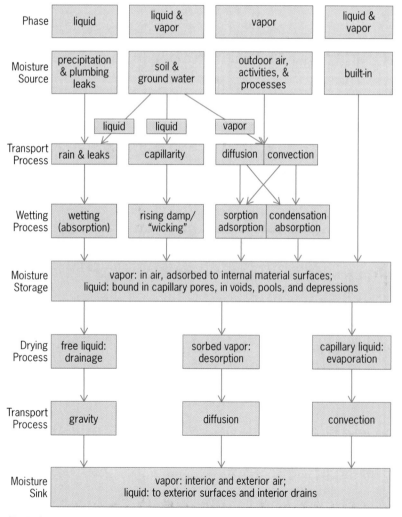

Fig. 3. Summary of moisture sources, sinks, and transport.

appliances, the location of drains to remove plumbing leaks, and the siting of the home above flood levels. Wetting events from these sources usually require immediate attention and special drying techniques.

Moisture sinks and drying. Moisture is usually removed from an enclosure by (1) evaporation of water transported by capillarity to the inside or outside surfaces; (2) vapor transport by diffusion, air leakage, or both, either outward or inward; (3) drainage, driven by gravity; and (4) ventilation (ventilation drying). These methods are illustrated in **Fig. 2**.

Wet materials such as historic solid masonry or thatch will dry directly to their environment, but this mechanism is less important for modern multilayer enclosure assemblies. Water vapor, whether generated internally or the result of evaporation from wet materials, can be removed by diffusion and air movement. Some materials (low-vapor-permeance materials like steel and plastic) resist diffusion drying. In hot weather drying can occur to the interior and the exterior, and in cold weather it tends to occur toward the outside. (Moisture moves from higher concentration to less, and in most cases this means it moves from warm to cold.) In most climates, houses should be designed to allow drying in both directions. The actual direction will vary with weather conditions.

Drainage is the fastest and most powerful means of removing water that may penetrate cladding or leak into a basement. Drainage layers outside basements and drainage gaps behind cladding and below windows have been shown to be very effective. When drainage stops, materials are still saturated and surfaces wet, so other drying mechanisms must be employed.

Ventilation behind cladding and roofing can be an effective mechanism of drying in many climates and situations. If dry cool spaces (such as crawl spaces) are ventilated during warm humid weather, wetting will occur, and so ventilation is not recommended for crawl spaces.

Moisture storage and material response. A building material's or assembly's ability to store moisture is important as it represents the amount of time that can separate wetting and drying events. Moisture can be stored in a variety of ways in enclosure assemblies—as vapor, water, or solid. The volume of water that is stored in an enclosure can be large, in the order of a few to tens of kilograms per square meter. This moisture can be: (1) trapped in small depressions or in poorly drained portions of assemblies; (2) adhered by surface tension as droplets, frost, or even ice, to materials and surfaces; (3) adsorbed within hygroscopic building materials (especially brick, wood, fibrous insulation, and paper); (4) retained by capillarity (absorbed) in porous materials; and (5) stored in the air as vapor.

Time, temperature, and relative humidity are the most important environmental variables affecting durability. Fungal mold and mildew growth can begin on most surfaces when the stored moisture results in a local relative humidity of over about 80% for some time. Corrosion and decay require higher levels of humidity (well over 90%) to proceed at dangerous rates. Both damage mechanisms require temperatures above 5°C (40°F) and increase dramatically with increasing temperature. Freeze-thaw damage and dissolution require the material to be at or near capillary saturation.

Summary. Controlling moisture in housing is fundamental to providing durable, healthy, high-performance housing. A summary of the major wetting and drying processes and the moisture transport mechanisms involved in the movement of moisture into and out of the enclosure is shown in **Fig. 3**. Moisture can be successfully managed in practice by balancing the duration and rates of wetting and drying mechanisms with the available safe storage capacity.

For background information *see* ARCHITECTURAL ENGINEERING; BUILDINGS; FUNGI in the McGraw-Hill Encyclopedia of Science & Technology.

John F. Straube

Bibliography. *ASHRAE Handbook of Fundamentals*, American Society of Heating, Refrigeration, and Air Conditioning Engineers, Atlanta, 2001; J. Lstiburek and J. Carmody, *Moisture Control Handbook*, Van Nostrand Reinhold, New York, 1993; *Moisture Analysis and Condensation Control in Building Envelopes*, ASTM Man. 40, ed. by H. Trechsel, American Society of Testing and Materials, Philadelphia, 2001; J. F. Straube and E. F. P. Burnett, *Building Science for Building Enclosure Design*, Building Science Press, Westford, MA, 2004.

Molecular motors

Movement is a fundamental property of life. Most forms of movement that we encounter in the living world—be it the transport of a tiny vesicle or the swimming of a whale—have a common molecular basis. They are generated by motor proteins that use the energy derived from the hydrolysis of adenosine triphosphate (ATP) to take nanometer-scale steps along a cellular track. Their actions power the movements we see, involving possibly just a single motor molecule in the case of the vesicle and phenomenally large arrays of billions and billions of motors in the case of the whale. Understanding the molecular basis of the behavior of these molecules is a prerequisite to understanding cellular and organismal motion.

Classes of motors. Three classes of motor proteins, each comprising several families of motors of different makeup and function, are known to generate linear movement along cellular tracks composed of polymeric molecules: myosins, which move on actin filaments; and kinesins and dyneins, which use microtubules. The intrinsic molecular polarity of the respective tracks specifies the direction of movement of a given motor. Most myosins move toward the plus end of actin filaments (the end where subunits are added to the polymer track at a high rate). Likewise, most kinesins are plus-end-directed microtubule motors, whereas all dyneins known so far move toward the microtubule minus end. One class of myosin mo-

tors and one large family of kinesins also move toward the minus end. Intermediate filaments, a third form of cytoplasmic filamentous polymers, are not known to support the movement of a motor; these fibers do not possess an intrinsic molecular polarity.

Each motor molecule possesses a roughly globular motor domain, referred to as the head, that harbors the binding site (catalytic site) for ATP and the binding site for the track (**Fig. 1**). The nonmotor domains vary widely in structure and function and include components involved in mechanical amplification, motor self-association, regulation, binding of associated proteins, and attachment to cargo.

The atomic structures of several kinesin and myosin motor domains have revealed an unexpected structural relationship between these two classes of motors. In essence, the central portion of the myosin head has the same structural fold as the core of the kinesin motor domain, suggesting that the two motors share a common evolutionary origin. Atomic structures of dyneins are not yet available, but sequence analyses show dynein to be a member of the AAA (ATPases associated with various cellular activities) protein family, which include proteins with widely divergent cellular functions.

Mechanochemistry. Initial events in force generation involve the transmission of small structural changes triggered by ATP hydrolysis to mechanical elements that translate these initial events into a large conformational change. In myosin, the mechanical amplifier is an α-helix stabilized by two associated polypeptides termed light chains. This "lever arm" domain swings through an angle of up to 70°, which translates into a "power stroke distance" whose magnitude depends on the length of the lever (usually several nanometers). In kinesin, a short, flexible element at the C-terminus of the motor domain, termed neck-linker, undergoes a shift in position from "docked" along the motor domain in the ATP state to "mobile" in the ADP state, again allowing for a power stroke of several nanometers. In dynein, six AAA domains form the ringlike structure of the head, with an N-terminal stem near AAA unit 1 and a stalk carrying the microtubule binding site protruding from the ring between AAA units 4 and 5 at the opposite side. ATP hydrolysis, which occurs only in the first AAA module, somehow triggers a rotation of the ring of AAA units that leads to a shift in the position of the microtubule binding stalk of several nanometers. In all three classes of motors, these structural changes are coordinated with a cycle of ATP hydrolysis and are coupled to a change between tight and weak binding to the track (**Fig. 2**).

Processivity and step size. A single motor domain undergoing a mechanochemical cycle is likely to detach from the track during a phase following the force-producing conformational change. Productive movement would require an ensemble of several motor domains so that at any given time at least one of them would be in a strong-binding state. The smallest unit to achieve that is a dimer of two motors. In these dimeric motors the two heads cooperate in such a way that their ATPase cycles are kept out of

Myosin

Kinesin

Dynein

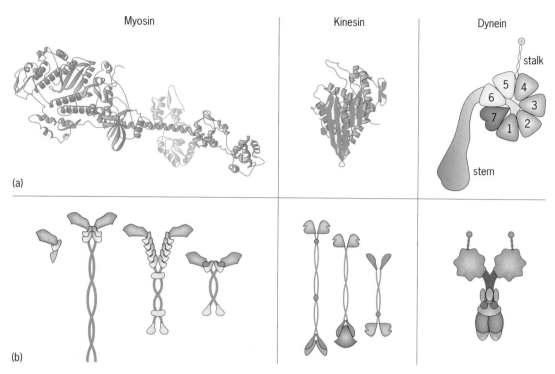

(a)

(b)

Fig. 1. Classes of molecular motors. (a) Structure of the motor domain. The motor domain of the myosin II heavy chain is shown in green, and light chains in light green and gray. The structure of the dynein motor domain is based on high-resolution electron microscopy since a crystal structure of dynein is not yet available. AAA domain 1 hydrolyzes ATP; AAA domains 2–4 bind but do not hydrolyze ATP; AAA domains 5 and 6 do not bind ATP. The stalk with its microtubule-binding site at the tip extends between AAA units 4 and 5. (b) Examples of motor architectures. Heavy chains are shown in gray, light chains in dark/light green (myosins) and dark green (myosins and dynein). Myosins from left to right: monomeric myosin I, muscle myosin II, myosin V, and minus-end-directed myosin VI. Kinesins: conventional, heterotrimeric, and minus-end-directed kinesin. Dynein: cytoplasmic dynein with two heavy chains.

phase. This assures that at any given time at least one motor domain is bound to the track while the other moves toward the next binding site. In this way a single dimer can take several hundred steps without dissociating, a property referred to as processivity. However, not all dimeric motors move in such a hand-over-hand fashion. If the two heads are not co-ordinated, the motor is not processive. Examples are found among both the myosins and the kinesins.

The step size, which is the distance from one point of attachment to the next along the track, may not be identical to the power stroke distance since additional factors may contribute to complete a step, such as diffusion of the unattached head to find the next binding site on the track. The step size is uniformly 8 nm (the distance between adjacent tubulin dimers in a microtubule) for kinesins and presumably dyneins as well. In myosins, it depends in the length of the lever arm and other factors, and may vary from ∼4 nm (the distance between subunits in an actin filament) up to 36 nm (the pitch of the actin helix).

Directionality and force. One motor family in both the myosin and the kinesin class of motors moves in the opposite direction, toward the minus end. Sequence analyses and structural studies do not reveal any major changes in the motor core. Instead, the linkage to the mechanical amplifier is changed. Dyneins, so far, all are minus-end-directed motors. Moreover, no natural motor is known that is able to change the direction of movement along its track.

Remarkably, the forces that the three classes of motors can develop all are in the same range, around 5 piconewtons. On the other hand, the velocity of movement varies widely: 0.1–3 μm/s for different kinesins, 0.1–60 μm/s for myosins, and 1–10 μm/s for dyneins. Some motors do not produce any movement and have adopted other functions, for example, in microtubule depolymerization (some kinesins) or as tension clamps (some myosins).

Cellular functions. As mentioned above, the three classes of cytoskeletal motors comprise superfamilies of a rich collection of motors differing in structural and functional properties. Thus the genome of the plant *Arabidopsis* harbors 61 kinesin genes, while humans possess 40 myosins, 45 kinesins, and 16 dyneins. A classic example of motor function is during muscle contraction, where myosin II molecules (Fig. 1b) bundled in a myosin filament interdigitate with arrays of actin filaments to bring about shortening of a muscle fiber. But the function of myosin is not restricted to contraction. Along with its cousins kinesin and dynein, myosin motors are involved in a wide variety of motile (and some non-motile) processes in cells (briefly summarized in the **table**). It should be kept in mind, though, that many motors have not yet been characterized functionally, so additional tasks may be uncovered.

All functions of motors, at least as studied so far, require anchorage to a substrate or cargo. Interaction is mediated by the nonmotor domain(s) and/or

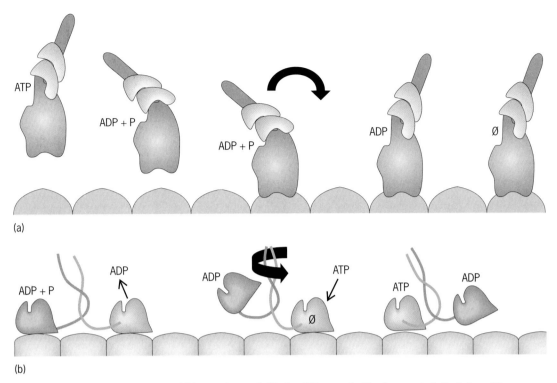

Fig. 2. Postulated structural changes. (*a*) In myosin coupled to the ATPase cycle. The "power stroke" rotation of the light-chain-binding domain is indicated by a curved arrow. Light chains are shown in green and the actin filament surface in light gray. (*b*) In dimeric, processive kinesin coupled to the ATPase cycles in the two heads. One-half of the full cycle is shown, demonstrating how the two heads exchange places.

accessory proteins. Thus, conventional kinesin or dynein can bind directly to certain integral membrane proteins via one of their light chains. In most cases, however, attachment to cargo requires a more complex machinery composed of several polypetides. A prime example is dynein, which interacts, via one of its associated light chains, with a large activator complex called dynactin, itself composed of several different proteins. Often the attachment complex contains regulatory molecules such as small G

Cellular activities of motors	
Activities	Motors*
Organelle and vesicle transport	M K D
Endocytosis	M
Muscle contraction	M
Ciliary and flagellar movement	D
Ciliary and flagellar biogenesis	K D
Transport of RNA, macromolecular complexes	M K D
Intermediate filament interactions	K D
Actin interactions	K
Microtubule polymerization, bundling	K D
Cell movement, cell shape	M (K, D indirect)
Left/right asymmetry	K D
Binding to other motors	M K
Sensory functions	M K D
Signaling	M K
Virus transport	M D
Mitosis/meiosis	M K D
Cytokinesis	M K

*M, myosins; K, kinesins; D, dyneins.

proteins, thus offering a means for the direct regulation of motor-cargo association.

Certain forms of organelle transport, such as pigment granule movements in melanophores or vesicle transport in neurons, may require the action of several different motors. Organelles can move on both microtubules and actin filaments, with kinesin and myosin motors acting sequentially and the organelle switching tracks in the process. In other instances an organelle may change direction of movement along the same track, implying a switch between plus-end- and minus-end-directed motors. Myosin V and conventional kinesin have been shown to interact directly, but it is unclear whether physical interaction is a prerequisite for coordination.

Their role in a wide variety of transport processes also implicates motors in a number of diseases, including defects in muscle systems, respiratory failure, hearing loss, and possibly neurodegenerative conditions. In addition, motors are crucial for important steps in development. Thus, the deployment of certain messenger ribonucleic acid (mRNA) species in the *Drosophila* oocyte establishes the anterior-posterior axis of the embryo. In vertebrates, defects in ciliary biogenesis impair the function of cilia in the node, an organizing structure in the developing embryo. The action of these cilia is important for the determination of the left-right axis and the positioning of internal organs. Thus motors are the driving force behind not only essentially all forms of cellular movement but also many higher-order processes of multicellular organisms.

For background information *see* CELL MOTILITY; MUSCLE; MUSCLE PROTEINS in the McGraw-Hill Encyclopedia of Science & Technology.   Manfred Schliwa

Bibliography.   A. Kamal and L. S. Goldstein, Principles of cargo attachment to cytoplasmic motor proteins, *Curr. Opin. Cell Biol.*, 14:63–68, 2002; M. Schliwa (ed.), *Molecular Motors*, Wiley-VCH, Weinheim, 2003; M. Schliwa and G. Woehlke, Molecular motors, *Nature*, 422:759–764, 2003; J. A. Spudich, The myosin swinging crossbridge model, *Nat. Rev. Mol. Cell Biol.*, 2:387–392, 2001; R. D. Vale, The molecular motor toolbox for intracellular transport, *cell*, 112:467–480, 2003.

Multiple ionization (strong fields)

One of the most important processes that occur when light interacts with matter is the photoelectric effect, in which the light is absorbed and electrons are set free. This process led Albert Einstein to the conclusion that electromagnetic radiation consists of energy packages known as photons. When a photon from the light field is absorbed, an electron is emitted with a kinetic energy given by the photon energy minus the binding energy of the electron in the material. As early as 1931, Maria Göppert Mayer argued that this is not the full story. She predicted that photons are also able to combine their energy to facilitate electron ejection. Such multiphoton processes can be studied with modern short-pulse laser systems. In the focus of such lasers, energy densities beyond 10^{15} W/cm^2 can be routinely reached. This is achieved by combining very short pulses (typically shorter than 10^{-13} s) and tight focusing. A laser beam intensity of 10^{15} W/cm^2 at a wavelength of 800 nanometers corresponds to a density of almost 10^{11} photons in a box the size of the cube of the wavelength. The interaction of matter with such superintense radiation can be best studied by putting a single atom in the focus of the laser beam and examining how it responds. It was soon found that frequently many more photons were absorbed than were necessary to overcome the binding of the electron. This multiphoton process leads to ejection of fast electrons. In addition, an unexpectedly high probability for multiple ionization was observed, that is, ejection of more than one electron from the atom. The key question here is what are the microscopic mechanisms by which many photons are coupled to many electrons in a single atom or molecule. *See* TWO-PHOTON EMISSION.

COLTRIMS. These problems can be addressed with unprecedented detail and accuracy by a powerful multiparticle imaging technique called cold target recoil ion momentum spectroscopy (COLTRIMS). For very low energy charged particles (microelectronvolts to several hundred electronvolts) that are typically found in atomic and molecular processes, this technique is comparable to the cloud chamber and its modern successors in nuclear and high-energy physics. The technique makes it possible to trace the momentum vectors (that is, the energy and angles) of all the charged particles arising from a fragmenting atom or molecule (**Fig. 1**). In the experiment, the laser is focused into a dilute gas jet (which is vertical in Fig. 1). A homogeneous electric field sweeps the ionic core of the fragmented atom to the left and the one or more emitted electrons to the right. The position of impact and the time of flight of the particles to the position-sensitive detectors are measured. From these quantities the initial momenta can be reconstructed.

Further discussion of the ionization processes will be restricted to the adiabatic regime. Here the change of electric field is slow compared to the time scale of the electronic transition. The adiabaticity is quantified by the use of the Keldysh parameter, given by Eq. (1). Here, I_P represents the ionization poten-

$$\gamma = (I_p/2U_p)^{1/2} \qquad (1)$$

tial and U_p is the ponderomotive potential, given in atomic units by Eq. (2), where the intensity and fre-

$$U_p = I/4\omega^2 \qquad (2)$$

quency of the laser light are represented by I and ω, respectively. For $\gamma < 1$, ionization can be quantitatively described by tunneling. The situation is depicted in **Fig. 2a**. The laser wavelength is much longer than the size of the atom. The sum potential of the laser field and the attractive atomic potential creates a barrier. The electron is set free by tunneling through this barrier. The resulting electron wave packet has a very small momentum. Once free, the electron wave packet is driven by the oscillating electric field of the laser beam. The mean energy in this

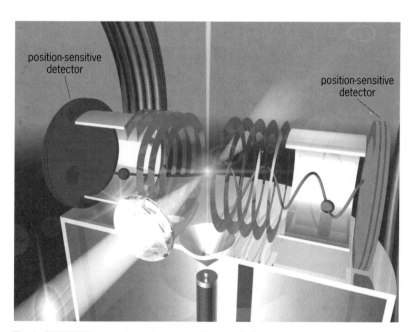

Fig. 1. COLTRIMS momentum microscope. The atom is fragmented in the laser focus. Electrons and ions are detected by position-sensitive detectors. (*From Th. Weber et al., Atomic dynamics in single and multiphoton double ionization: An experimental comparison, Opt. Express, 8:368–376, 2001*)

quiver motion is given by the ponderomotive potential U_p. The momentum of the ion and electron, observed long after the laser pulse has been attenuated, is the residual of this quiver motion. The final drift momentum P_{drift} is a function of the time t_0 representing the birth of the free electron. In the case of an electric field of the form $E(t) = E_0 \sin(\omega t)$, P_{drift} is given by Eq. (3), where q is the charge on the

$$P_{\text{drift}}(t_0) = (q/\omega)E_0 \cos(\omega t_0) \qquad (3)$$

particle and ωt_0 is the tunneling phase.

The key point here is that in COLTRIMS a momentum measurement is effectively a time measurement. This relationship has been used to disentangle the mechanisms of double ionization, which is discussed below. For single ionization, the momentum of the ion and electron peak close to zero, meaning that $\cos(\omega t_0) = 0$, which occurs only if $\omega t_0 = 90°$ or $\omega t_0 = 270°$. This shows that tunnel ionization is most probable at the maximum amplitude of the electric field, $E(t_0) = E_0 \sin(\omega t_0)$, since $\sin(\omega t_0)$ also reaches its maximum amplitude of 1 when $\omega t_0 = 90°$ [$\sin(90°) = 1$] or $\omega t_0 = 270°$ [$\sin(270°) = -1$].

Ionization mechanisms. In its simplest form, multiple ionization in an intense laser field can occur sequentially by successive independent single-electron processes. The first electron can tunnel at one maximum of the electric field and the second electron at some later oscillation of the field. This sequential ionization does not require any interaction between the electrons.

Alternatively, double ionization might be mediated by the electron-electron interaction. One process of this kind is shake-off (Fig. 2b). It is known to be responsible for double ionization initiated by the absorption of a single very high energy photon. In this case, one electron is removed rapidly, and the second electron no longer finds itself in an eigenstate of the altered potential resulting from the vacancy. One or more electrons can be shaken off to the continuum. A further process is termed rescattering (Fig. 2c). Here the first electron is ejected and accelerated by the laser field. Upon reversal of the field direction, it is driven back to its parent ion. When it recollides with the ion, it can knock out a second electron, leading to double ionization.

Ionization rates. The rate of production of singly ionized atoms as a function of the peak laser intensity shows a steep and smooth rise and then saturates (**Fig. 3**). This saturation is reached when the ionization probability in the focus of the laser approaches unity. In contrast, the double ionization rate shows a prominent shoulder. The rates expected for single ionization via tunneling (given by the solid line labeled Ne^{1+} in Fig. 3) agree well with the observations (given by the open circles). For double ionization, however, the rates expected for sequential tunneling of independent electrons (given by the solid line labeled Ne^{2+} in Fig. 3) describes the data (given by the open squares) only at high laser intensities. At lower

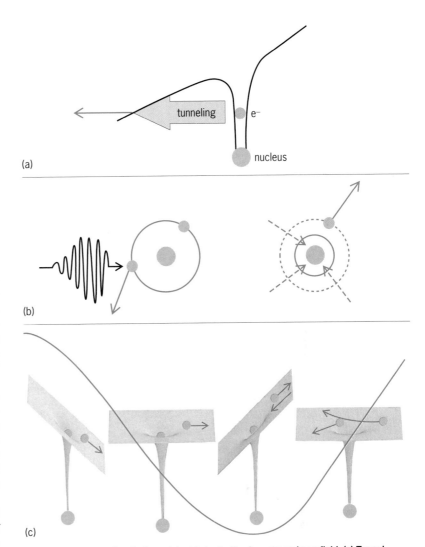

(a)

(b)

(c)

Fig. 2. Mechanisms for single and double ionization in a strong laser field. (a) Tunnel ionization. (b) Shake-off. (c) Rescattering. (*From R. Dörner et al., Wenn Licht Atome in Stücke reisst: Elektrononkorrelationen in starken Feldern, Physik. Blätter, 57(4):49–52, Wiley-VCH, 2001*)

intensities, this process underestimates the observed production rate by more than an order of magnitude, indicating the importance of a nonsequential mechanism that incorporates electron correlation.

Ion momenta. These processes have been elucidated in great detail by investigating the ion momentum and the correlation among the momenta of the two electrons. For both single ionization and double ionization at very high laser intensities, the ion momentum peaks at zero and most of the ions move very slowly (**Fig. 4a, b**). As seen in Eq. (3), this observation directly proves that the ions are created at the peak of the laser field. At lower laser intensities, however, the doubly charged ions show a double-peak structure with a pronounced minimum at the origin (Fig. 4c). This observation rules out the shake-off process as the double ionization mechanism. Since shake-off is an instantaneous process, the second electron would be emitted immediately after the tunneling of the first one at the field maximum. This would lead to a peak at zero ion momentum. In

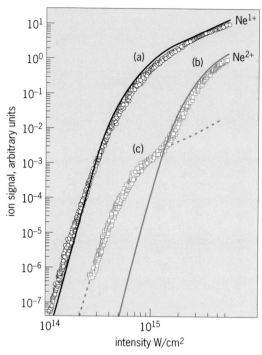

Fig. 3. Ionization of neon in a strong laser field, showing rates of production of singly and doubly charged ions as a function of the laser intensity. The parenthetical letters refer to the parts of Fig. 4. Solid curves labeled Ne^{1+} and Ne^{2+} give rates expected for, respectively, single ionization via tunneling and double ionization via sequential tunneling of independent electrons. The broken curve gives the rate expected for double ionization via mechanisms that involve electron correlation. Open circles and open squares give observed rates of single and double ionization, respectively. (*From R. Dörner et al., Wenn Licht Atome in Stücke reisst: Elektronenkorrelationen in starken Feldern, Physik. Blätter, 57(4):49–52, Wiley-VCH, 2001*)

turn, the double peak at higher momenta is a direct proof of the time delay introduced by the rescattering of the primary electron. (In Figs. 4 and 5, the momentum is expressed in atomic units. The atomic unit of momentum is $\hbar/a_0$, where $\hbar$ is Planck's constant divided by 2π and a_0 is the Bohr radius.)

Electron momentum correlation. The time evolution can be seen in more detail in the momentum correlation between the two electrons, shown in **Fig. 5** for the double ionization of argon. The horizontal axis shows the momentum component of one electron in the direction of polarization of the laser (p_{ez1}), and the vertical axis shows the same momentum component of the second electron (p_{ez2}). The data are integrated over the transverse momenta of the first and second electron. The momentum components of the electrons parallel to the laser field are related to the time when the electrons are set free, according to Eq. (3). At a laser intensity of 3.8 $\times$ 10^{14} W/cm^2, a strong correlation between the momenta of the two electrons is observed (Fig. 5a). There is a clear maximum for both electrons that are emitted with the same momentum component of about 1 atomic unit. Emission to the opposite half planes, however, is strongly suppressed. The laser intensity of 3.8 $\times$ 10^{14} W/cm^2 is right on the bend in the curve of the double ionization rate for argon.

At an intensity of 15 $\times$ 10^{14} W/cm^2 (Fig. 5b), in the regime where sequential double ionization dominates, the correlation between the electrons is completely lost.

The peak in the first quadrant of Fig. 5a corresponds to both electrons being emitted on the same side of the atom with about 1 atomic unit of momentum. The peak can be understood in terms of the following rescattering scenario. The first electron is set free by tunnel ionization, accelerated, and subsequently driven back to the Ar$^+$ ion. In order to eject the second electron, the returning electron must have sufficient energy to at least excite the ion. The phase of tunneling is the corresponding return phase determined by this process. In the recollision, the electron loses most of its energy, and both electrons are set free with little energy at the time of return. For the particular case shown in Fig. 5a, this return phase is about 35° off the maximum field and indeed leads to a momentum of about 1 atomic unit. At the higher laser intensity of 15 $\times$ 10^{14} W/cm^2 (Fig. 5b), multiple ionization is much more likely to come about as a result of the sequential process than the rescattering mechanism. The random time variations between the two independent tunneling events leads to the loss of momentum correlation.

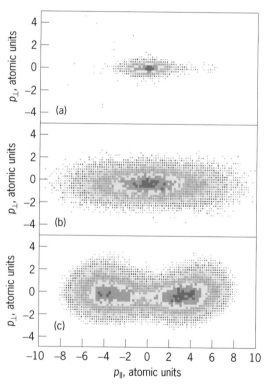

Fig. 4. Ion momenta for singly and doubly charged ions emitted in ionization of neon in a strong laser field at the laser intensities indicated in Fig. 3. Shadings show relative frequency of ion momenta as a function of the momentum components parallel ($p_\parallel$) and perpendicular ($p_\perp$) to the laser polarization. (*a*) Single ionization (Ne^{1+}). (*b*) Double ionization (Ne^{2+}) at very high laser intensity. (*c*) Double ionization (Ne^{2+}) at lower laser intensity. (*From R. Dörner et al., Wenn Licht Atome in Stücke reisst: Elektronenkorrelationen in starken Feldern, Physik. Blätter, 57(4):49–52, Wiley-VCH, 2001*)

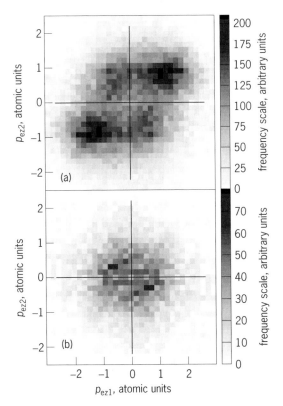

Fig. 5. Correlated electron momenta for double ionization of argon (*a*) at a laser intensity for which rescattering dominates (3.8×10^{14} W/cm²) and (*b*) at a laser intensity for which sequential ionization dominates (15×10^{14} W/cm²). Shadings show relative frequency of double ionizations as a function of the momentum components of electron 1 (p_{ez1}) and electron 2 (p_{ez2}) parallel to the polarization of the laser. (*From Th. Weber et al., Correlated electron emission in multiphoton double ionization, Nature (London), 405:658–661, 2000*)

Open questions. While some of the observed features are qualitatively understood, still many open questions remain. A satisfactory quantitative agreement between experiment and theory has not been reached, and the role of excited states and electron repulsion in the presence of strong fields is unclear. An exciting prospect is the possibility of using the correlated momentum measurement of several particles in a laser field to achieve time resolution in the attosecond domain without the need for attosecond pump-probe pulses.

For background information *see* ATOMIC STRUCTURE AND SPECTRA; LASER; OPTICAL PULSES; PHOTOIONIZATION in the McGraw-Hill Encyclopedia of Science & Technology.   Reinhard Dörner

Bibliography. P. B. Corkum, Plasma perspective on strong field multiphoton ionization, *Phys. Rev. Lett.*, 71:1994–1997, 1993; R. Moshammer et al., Momentum distributions of Ne^{n+} ions created by an intense ultrashort laser pulse, *Phys. Rev. Lett.*, 84:447–450, 2000; Th. Weber et al., Correlated electron emission in multiphoton double ionization, *Nature (London)*, 405:658–661, 2000; Th. Weber et al., Recoil-ion momentum distributions for single and double ionization of helium in strong laser fields, *Phys. Rev. Lett.*, 84:443–447, 2000.

Natural optical fibers

Although optical fiber technology is usually considered to be an excellent example of human innovation and engineering, recent investigations have revealed that silica fibers capable of guiding light are manufactured as well by a simple deep-sea sponge. Indeed, several other "primitive" organisms have already been shown to exhibit sophisticated solutions to complex optical problems. The skeleton of the "Venus's flower basket" (*Euplectella aspergillum*), a

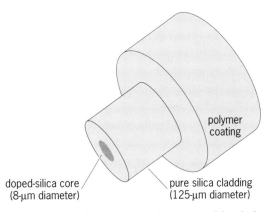

Fig. 1. Schematic illustration of typical commercial optical fiber. Not to scale.

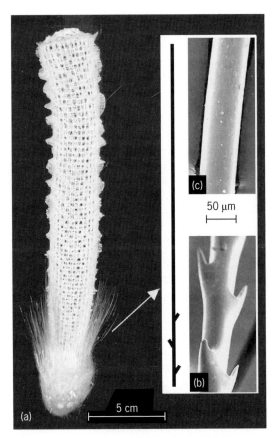

Fig. 2. Typical specimen of *Euplectella aspergillum.* (*a*) Photograph of entire skeleton showing many basalia spicules emanating from the base. (*b*) Barbed middle region. (*c*) Smooth distal region.

hexactenellid sponge that lives at ocean depths ranging from 35 to 5000 m (115 to 16,400 ft), contains silicious spicules (spiny fibers) exhibiting optical and mechanical properties that are comparable, and in some cases superior, to manmade optical fibers. As such, these spicules may serve the sponge as a fiber optic network for the collection or the distribution of light. A better understanding of these natural silica fibers may lead to improvements in the design, fabrication, and performance of commercial optical fibers.

Commercial optical fibers. Commercial optical fibers are fabricated via high-temperature chemical deposition of silica soot to create a preform that is on the order of 1 m (3 ft) long and 10 cm (4 in.) in diameter. The resulting glass preform is drawn (stretched) by a factor of about 10^5 into a 125-micrometer-diameter fiber inside a furnace at approximately 2000°C (3630°F). After cooling, the fiber is coated with a polymer jacket for chemical and mechanical protection. Chemical dopant species are added during silica soot deposition to create variations in the local refractive index of the resulting fiber. Typically, the optical signal is confined by total internal reflection to an 8-μm-diameter germania-doped core region, which exhibits a slightly higher refractive index than the surrounding pure silica cladding (**Fig. 1**).

Modern optical fibers are a marvel of engineering: more than 95% of the original signal power launched into a typical optical fiber remains even after traveling through 1 km (0.6 mi) of glass! Recent technological breakthroughs, including all-optical signal amplification and simultaneous transmission of multiple channels using different wavelengths, permit simple terabits (10^{12} bits) of information per second to be transmitted thousands of kilometers over a single strand of commercial optical fiber. Optical fiber cables are the main links of modern telecommunications networks, spanning the globe in cables laid on ocean bottoms and across continents.

The optical properties of an optical fiber are determined by its refractive index profile, which is the variation of the refractive index across the fiber's diameter. Various schemes for measuring the refractive index profile of commercial optical fibers have been developed. The transverse interferometric method (TIM) is particularly useful for measuring the refractive index profile of commercial fibers and deep-sea sponge spicules. In this approach, the emission of a light source is divided into two beams. One beam is passed transversely across a fiber sample immersed in a bath of refractive index oil, while the other beam traverses an identical bath of refractive index oil without a fiber sample. Due to the wavelike nature of light, interference fringes appear when the two beams are recombined. The appearance of these interference fringes is mathematically related to the refractive index profile of the fiber sample. In this way the refractive index profile of an optical fiber, including spicules of *Euplectella*, can be accurately measured in a nondestructive manner.

Spicule characteristics. The skeleton of the Venus's flower basket consists of a cylindrical cagelike arrangement of spicules (**Fig. 2**) that is typically inhabited by a pair of symbiotic shrimp. The optical properties of the basalia, which are spicules between 5 and 15 cm (2 and 6 in.) in length and between 40 and 70 μm in diameter that secure the sponge to soft sediments in its natural environment, are remarkably similar to commercial optical fibers.

TIM measurements of basalia spicules reveal that their refractive index is very close to that of silica, which is not surprising since chemical analyses show that the spicules are primarily composed of silica. However, the central region of the spicules contains a small-diameter (1–2-μm) core exhibiting a higher refractive index compared with the surrounding portion of the spicule, perhaps because the core contains more sodium. This core/cladding refractive index profile is strikingly similar to commercial optical fibers, and recent experiments have demonstrated that this central core region can guide light much like the core region of a conventional optical fiber (**Fig. 3**).

Since the refractive index of silica (about 1.46) is significantly higher than that of seawater (about 1.33), light can also be guided within the entire spicule by total internal reflection at the spicule's surface. This relatively large refractive index difference and the large diameter of the spicule (relative to its

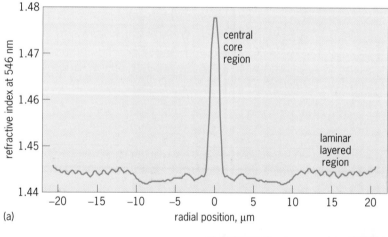

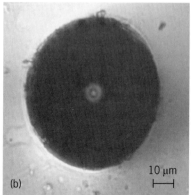

Fig. 3. Optical characteristics of basalia spicules from *Euplectella aspergillum*. (*a*) TIM measurement (at 546 nm) of a typical 45-μm-diameter spicule. (*b*, *c*) Microscopic views of light emerging from the fiber. Depending on the launch conditions and the medium surrounding the fiber, light can be guided (*b*) only in the central core region of the fiber or (*c*) in the entire fiber cross section.

core) make light guidance within the entire spicule more efficient than light guidance within the core alone. Consequently, if the spicule actually guides light in its natural environment, the entire diameter of the spicule is probably used as a light guide.

The middle section of the basalia spicules exhibits many barblike structures (Fig. 2b), which have been shown to act as points of illumination when light is guided in the fibers. Light is able to leak out at the barbs because the geometry of the barb causes the angle between a light ray and the spicule surface in the vicinity of the barb to fall below the critical angle required for total internal reflection. Indeed, the creation of a large number of localized points of illumination may yet be shown to be an important biological function of the basalia spicules of E. aspergillum.

High-purity silica glass is a remarkably strong material, and the failure strength of commercial optical fibers can exceed that of high-strength steel. However, silica glass is a brittle material whose mechanical strength is greatly reduced in the presence of cracks. Commercial optical fibers are coated with a polymer jacket to protect the fiber's surface from scratches that could develop into cracks leading to fracture. Nature has apparently found a different solution to this problem: the basalia spicules are not composed of a continuous piece of glass, like commercial optical fibers, but an assembly of 50–200-nanometer-diameter silica spheres formed into laminar layers held together by organic material. A crack cannot readily propagate from one region to another because it is arrested at the boundaries between the layers. Thus, the spicules are considerably more robust than uncoated commercial optical fibers. Because the fine structural features are much smaller than a wavelength of light (about 500 nm), the spicules are optically transparent.

Implications. Although the basalia spicules have not yet been shown to guide light in their natural environment, this possibility is a tantalizing one. Almost no sunlight penetrates below the top 200 m (656 ft) of the ocean; consequently bioluminescence is the main source of light in the ocean depths. Many deep-ocean organisms are known to be bioluminescent, including the shrimp inhabiting the Venus's flower basket. The basalia spicules may serve as a fiber-optic network for the distribution or collection of bioluminescent light, and may play a role in attracting the shrimp's prey or repelling its predators, or attracting the shrimps themselves.

Nature's solutions to engineering challenges can provide important insights for human technology. For example, the composite structure of E. aspergillum's basalia spicules makes them more mechanically robust than uncoated commercial optical fibers. In addition, certain chemical dopants, such as sodium, that are found in the central core of the basalia spicule may reduce optical attenuation in fibers; however, sodium cannot be readily incorporated into commercial optical fibers because they are fabricated at high temperatures which prevent uniform distribution of the sodium. Diffusion through

high-temperature silica during processing can also preclude the use of other dopants in commercial optical fibers. Furthermore, when a commercial fiber cools at the conclusion of the draw process, undesirable residual mechanical stresses can arise within. Nature has avoided all of these difficulties by evolving an ambient temperature solution for silica fiber fabrication. What is learned from research on such natural mechanisms may someday improve technological routes and lead to commercial fibers of better quality.   Andrew D. Yablon; Joanna Aizenberg

For background information see HEXACTINELLIDA; OPTICAL FIBERS in the McGraw-Hill Encyclopedia of Science & Technology.

Bibliography.   J. Aizenberg et al., Biological glass fibers: Correlation between optical and structural properties, Proc. Nat. Acad. Sci., 101:3358-3363, 2004; A. Ghatak and K. Thyagarajan, Introduction to Fiber Optics, Cambridge University Press, 1998; J. Hecht, Understanding Fibre Optics, 4th ed., Prentice Hall, Upper Saddle River, NJ, 2002; D. Marcuse, Principles   of   Optical   Fiber   Measurements, Academic Press, New York, 1981; V. C. Sundar et al., Fibre-optical features of a glass sponge, Nature, 424: 899-900, 2003.

Neoproterozoic predator-prey dynamics

A century and a half ago, in On the Origin of Species Charles Darwin presented the notion that animal evolution was governed by natural selection, by which fitness in the face of competition, predation, and environmental vicissitudes determined the continued survival of a species or its extinction. New species evolved when variations in morphology, reproductive success, or behavior led to improved adaptation in the struggle for existence and these new traits were passed on to offspring.

Modern marine communities are a densely tangled web of predator-prey interactions that govern the energy flow in ecosystems and regulate their composition. Predation is, therefore, considered to be a prime driver of evolutionary change. However, did it manifest itself from the very beginnings of animal evolution? Did predation provide the motivation for the appearance of shells and exoskeletons in the latest Neoproterozoic and earliest Cambrian (**Fig. 1**)? Or was it not really important until much later in the Paleozoic?

Fossil record of predation. Certainly the fossil record of the last 250 million years suggests an escalating "arms race," beginning with the so-called Mesozoic marine revolution when levels of predation seem to have increased dramatically and, consequently, attack and defense innovations arose in tandem. Did predator and prey species coevolve by reciprocal adaptations? Unfortunately, the fossil record is inadequate for demonstrating most types of biotic interactions which are so readily observable in modern communities. The majority of these interactions involved soft-bodied animals and so go

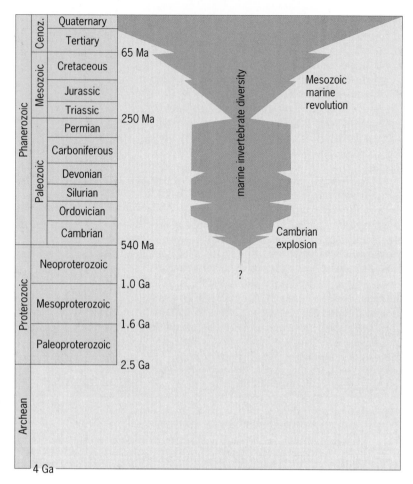

Fig. 1. Chart of geologic time with approximate ages of key boundaries in millions of years (Ma) and billions of years (Ga). The spindle shows diagrammatically an approximation of the relative diversity of marine invertebrate animals. Outward inflections represent diversification, whereas inward inflections represent mass extinction events.

Neoproterozoic animals, large and small, although many of them are peculiar and seem unrelated to younger taxa. At this time there appeared the oldest biomineralized animals, that is, those in which calcium carbonate was precipitated deliberately by enzymatically mediated cellular reactions. These took the form of tubes which may have housed suspension-feeding worms or cnidarians, stalked goblets which may have supported polyps, and primitive skeletons of colonial corals or sponges.

By contrast, Lower Cambrian marine strata yield a bonanza of shells, skeletons, spicules (spiky, knobby, and latticelike supporting structure), and sclerites (hardened plates), mostly calcareous but also siliceous and phosphatic, representing a wide variety of animals belonging to groups familiar to paleontologists and biologists, along with markings and burrows made by soft-bodied animals that lived in the sediment. The Early Cambrian fossil record shows increasing invertebrate diversification over a time span of several tens of millions of years. The complexity of these early ecosystems is indicated by sponges, corals, mollusks, trilobite arthropods, ostracode crustaceans, brachiopods, echinoderms, various wormlike animals, as well as other animals that occupied suspension- and deposit-feeding, grazing, browsing, scavenging, and predatory niches. Newly evolved animals are found in strata representing various environmental settings where they crawled, grubbed, nestled, stuck, swam, and dug, taking advantage of evolving food sources and many different inorganic and organic substrates. In turn, these cascading ecological developments coincided with, and may have been triggered by, momentous environmental shifts involving the breakup of a supercontinent, sea-level rise, and possibly changes in atmospheric and seawater composition. The biosphere then, as now, was a complex interrelated system with feedback from its biological and geological components at many scales.

Cloudina and Sinotubulites. The oldest calcium carbonate shells in the fossil record are millimeter-to-centimeter-sized tubes belonging to *Cloudina*, which have been reported from localities geographically disparate in terminal Neoproterozoic time: Namibia, Oman, China, and western North America. These strata are around 550 million years old. In the Dengying Formation of the Yangtze gorges region of central China, *C. hartmannae* is found with another type of tubular shell, *Sinotubulites cienegensis*. It is not known what the animals themselves looked like. However, the original calcium carbonate (calcite or aragonite) shells of many specimens in these strata were replaced early, in the soft sediment, by micrometer-sized apatite (calcium phosphate) which led to preservation of extraordinary fidelity. Indeed, this kind of phosphatization, so very rare in the rock record, is the most important taphonomic window available onto soft tissues and delicate shell microstructure. It reveals that *Sinotubulites* and *Cloudina*, while producing superficially similar tubes variably adorned with concentric

largely undetected because of taphonomic reasons, that is, the biological, physical, and chemical barriers to their entry into the rock record. The kinds of predation that are potentially preservable in strata are those that led to damage to hard parts—chiefly shells and exoskeletons—in the form of broken specimens, healed injuries, and boreholes. However, in the case of shells that were originally composed of aragonite rather than calcite (both forms of calcium carbonate), the chance of preservation is reduced further because of the metastable nature of this mineral, which is prone to dissolution on the sea floor and subsequent loss during the lithification (rock-forming) process.

Cambrian explosion. The Cambrian explosion refers to the geologically rapid diversification of animal groups between 545 and 500 million years ago (Fig. 1). Animals arose from multicellular ancestors perhaps half a billion years before that, in the latest Mesoproterozoic or earliest Neoproterozoic. Here, the fossil record is unclear, and rates of genetic change extrapolated from living animals and their close ancestors may not be applicable. Certainly by around 565 to 545 million years ago, there is evidence for a fairly substantial assortment of late

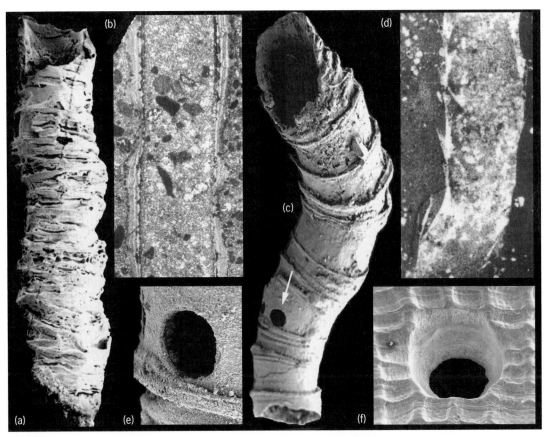

Fig. 2. *Cloudina* and *Sinotubulites* species. (*a*) Scanning electron photomicrograph of phosphatized tubular shell of *S. cienegensis*. Tube width is 1500 μm. (*b*) Optical photomicrograph of thin section of shell of *S. cienegensis* showing laminated walls due to successive accretion. Tube width is 1600 μm. (*c*) Scanning electron photomicrograph of phosphatized tubular shell of *C. hartmannae* showing borehole (arrow). Tube width at aperture is 530 μm. (*d*) Optical photomicrograph of thin section of shell of *C. hartmannae* showing nested thin-walled collars. Tube width is 500 μm. (*e*) Closeup scanning electron photomicrograph of borehole in *C. hartmannae* visible in lower part of specimen shown in c. Borehole diameter is 85 μm. (*f*) Scanning electron photomicrograph of borehole in a Recent pelecypod mollusk (*Codakia* sp., Quintana Roo, Mexico) drilled by the radula (rasplike feeding structure) of a predatory gastropod. Borehole diameter at inner surface of shell is 700 μm.

ridges, wrinkles, and flanges (**Fig. 2**), contrast in how they grew and biomineralized: the walls of *Sinotubulites* are thick and laminated from successive outward accretion along with lengthening of the tubes, while *Cloudina* consists of nested, thin-walled conical collars from successive growth at the aperture. The two taxa, therefore, biomineralized by fundamentally different mechanisms. *Sinotubulites* must have had an epithelial or mantlelike covering that secreted the shell, whereas secretion in *Cloudina* probably took place in an organic rim that occupied just the apertural region. However, it is impossible to know how much of the older tubes was occupied by the living organisms, and exactly which kind of organisms they were. It is presumed that both taxa poked upright in the sediment and extended a feeding apparatus into the water column similar to modern tubiculous taxa.

Evidence of predation. In a collection of nearly 100 specimens of *Cloudina* and over 60 specimens of *Sinotubulites* liberated from their dolomite (calcium magnesium carbonate) host rock by gentle etching in weak acetic acid, one-fifth of *Cloudina* tubes exhibit perforations 15–85 μm in diameter, clustering in the 30–50-μm range (Fig. 2), whereas none of the *Sinotubulites* is so affected. The holes in *Cloudina* are mostly single and are consistently located only on smooth areas of the shell wall, three to four times the tube diameter away from the aperture. One specimen exhibits three incomplete holes. The holes are interpreted to be boreholes made from the shell exterior by a predatory animal, although probably soft-bodied and thus likely never to be preserved in the fossil record.

The fact that the predator selectively targeted *Cloudina* over *Sinotubulites* points to the ability to sense which is which, probably in part because of the presence or absence of an organic covering. The predator chose smooth areas of the shell and seemingly used chemical means, rather than a rasping mechanism, to penetrate the thin shell. Attacks were not always successful initially, but the incomplete holes suggest that the predator was capable of resuming after interruptions.

The location of the boreholes away from the aperture of *Cloudina* argues that the tube occupant was able to keep the predator at bay for some distance by means of an organic covering, chemical defense, or physical sweeping from a flexible apparatus protruding from the aperture. In turn, the predator responded to the defense mechanism.

Complex predator-prey relationship. It can be concluded from the *Cloudina* and *Sinotubulites* fossil evidence that these were two different but coexisting types of small animals with tube-secreting capabilities but contrasting calcium carbonate biomineralization mechanisms. Theoretically, then, these calcareous shells served to thwart any existing predators, but a new animal evolved around the same time with the ability to bore through them. However, both tubular taxa possessed adaptations that gave some degree of protection against this novel style of predation: *Sinotubulites* had an organic covering over its tubular shell, whereas *Cloudina* was able to protect only the growing end. The predator, on the other hand, had an advanced degree of neural complexity that caused it to exhibit stereotypic behavior in that it selected not only its prey but also the specific attack site. These characteristics are surprisingly modern, fundamentally no different than those of extant invertebrate predator-prey interactions.

Evolutionary implications. The fossil specimens of *Cloudina* and *Sinotubulites* demonstrate that the marine food web already encompassed durophagous (shell-destroying) predation from the very appearance of shell secretion. Perhaps their soft parts were also vulnerable, but to a different kind of predator, and perhaps predator-prey interactions existed among soft-bodied animals prior to this biomineralization event. However, evidence for either of these possibilities is elusive. Did predation actively provoke evolutionary novelty in the Cambrian explosion? This is hard to say, but it does stand to reason. Yet, a predation intensity of some 20% in this sample is remarkably high and may even be anomalous, given that individual shell collections for most of the subsequent Paleozoic show typically much less, if any, evidence of predation. Be that as it may, new research and new discoveries continue to bring life to paleoecology, and permit the timing and impact of evolutionary developments in relation to biotic diversity to be reconstructed ever more rigorously.

For background information *see* ANIMAL EVOLUTION; CAMBRIAN; EDIACARAN BIOTA; FOSSIL; ORGANIC EVOLUTION; PALEOECOLOGY; PALEONTOLOGY; PHYLOGENY; PRECAMBRIAN; PROTEROZOIC; SPECIATION; TAPHONOMY in the McGraw-Hill Encyclopedia of Science & Technology.   Brian R. Pratt

Bibliography. S. Bengtson (ed.), *Early Life on Earth: Nobel Symposium No. 84*, Columbia University Press, 1994; H. Hua, B. R. Pratt, and L.-Y. Zhang, Borings in *Cloudina* shells: Complex predator-prey dynamics in the terminal Neoproterozoic, *Palaios*, 18:454–459, 2003; P. H. Kelley et al. (eds.), *Predator-Prey Interactions in the Fossil Record*, Kluwer Academic/Plenum Publishers, 2003; A. H. Knoll, *Life on a Young Planet: The First Three Billion Years of Evolution on Earth*, Princeton University Press, 2003; G. Vermeij, *Evolution and Escalation: An Ecological History of Life*, Princeton University Press, 1987.

Network security and quality of service

Our dependence on computers, networks, and the Internet has grown at a fast pace, as evidenced by the increasing number of applications, as well as computer supervisory control of power, telecommunication, and transportation systems. In their current state, computers and networks are neither secure nor dependable for protecting stored or transmitted information or delivering a desired level of quality.

Computers and networks were not originally designed with security in mind. Many security loopholes or vulnerabilities exist, especially in software systems that manage computer and network resources, including operating systems to control computer resources such as the central processing unit (CPU) and memory, network protocols to coordinate information communication and transmission, and various applications running on operating systems and using network protocols to provide services directly to users.

Security problems caused by cyber attacks exploit these vulnerabilities to gain unauthorized access to confidential information, vandalize Web sites, and corrupt databases. Cyber attacks have taken many forms, including viruses, worms, password cracking, and denial of services.

A cyber attack compromises the availability, confidentiality, or integrity state of one or more computer/network resources, the three major aspects of security. A compromised security state may in turn lead to quality of service (QoS) problems. For example, the QoS of a database can be measured as the timeliness, precision, and accuracy in responding to a query. The compromised availability state of a database may degrade QoS in terms of slow or no response to a query. The compromised confidentiality state of a database may result in leaking unauthorized information to a database query. The compromised integrity state of a database may lead to an inaccurate output for a query due to corrupted, changed, or even lost information.

Cyber attacks are not the only cause of QoS problems. The timeliness of receiving information from a Web site often is not guaranteed. Computer and network resources have limited capacity in terms of CPU time for processing service requests and bandwidth for transmitting data. When multiple service requests compete for service, the resource schedules the service in a certain order. Most service requests are scheduled in the order of their arrival times, using the first-come-first-service (FCFS) rule. So when a resource is serving one request, other requests are placed in a queue or buffer. Since there is no admission control, if the queue is full, a request for service is rejected.

Many denial-of-service attacks have taken advantage of this lack of admission control by sending a large number of service requests. All these service requests are admitted into the resource, flooding the queue and service capacity and leaving no capacity for legitimate requests.

The Internet has evolved from a traditional data transmission network to a shared, integrated platform to carry a broad range of applications with varying traffic characteristics and different QoS requirements. Some applications, such as World Wide Web and e-mail, come with no hard time constraints and QoS requirements. Others, such as teleconferencing and telephony, are time-dependent, placing strict QoS requirements on the carrier network. Data from different applications compete for common computer and network resources such as routers. Using the FCFS scheduling rule, a router treats all application data equally by their arrival times rather than their priorities for QoS.

Technologies and quantitative methods for security. Protecting against cyber attacks requires layers of prevention, detection, and reaction technologies. Common prevention technologies include firewalls, cryptography, authentication, authorization, and access control. These cannot block all attacks, so technologies are required for monitoring, detecting, and stopping attacks as early as possible.

Signature recognition. These techniques identify and store signature patterns of known attacks, match observed activities with signature patterns, and detect an attack when a match is found. Most antivirus software and commercial attack detection tools use this approach. The main drawback of signature recognition is that novel attacks cannot be detected since their signature patterns are not known.

Anomaly detection. These techniques establish a profile for the normal behavior of a user, process, host machine, or network. An attack is detected when the observed behavior deviates largely from the norm profile. An advantage of anomaly detection techniques is that they can detect both known and novel attacks if those attacks demonstrate large deviations from the norm profile.

Statistical methods. In addition to the computer science methods, many quantitative methods from mathematics/statistics and engineering have been employed in signature recognition and anomaly detection techniques to deal with the data complexity in cyber attack detection. For example, the Markov chain model from mathematics/statistics and operations research in industrial engineering has been applied in anomaly detection to capture the stochastic properties of event transitions in the norm profiles. Statistical process control (SPC) techniques from industrial engineering also have been applied in anomaly detection to build the statistical norm profiles on such features as the intensity and frequency distribution of computer/network events. Nong Ye and colleagues have developed a multivariate SPC technique, called the chi-square distance monitoring method (CSDM), to overcome the difficulties of conventional SPC techniques in handling large amounts of messy and complex computer/network activity data. In CSDM, first a data vector representing the exponentially weighted frequency counts for different-type events occurring in a sequence is obtained, including the current event and the recent events. Next, the chi-square distance of this data vector is calculated from the norm profile of the event frequency distribution, which is statistically characterized by the mean vector of the event frequency distributions observed in the normal activities. If the distance is greater than a threshold value (indicating a significantly large deviation of the data vector from the norm profile), the current event is detected as an attack.

Except for antivirus software, no other attack detection tools have gained wide acceptance. Existing attack detection tools mostly monitor activity data, including network traffic data (raw data packets) and data of audited or logged events on computers. Large amounts of activity data can be accumulated quickly, including irrelevant information, which presents a considerable challenge for an attack detection tool to process efficiently. Both signature recognition and anomaly detection techniques incorporate characteristics of either attack data or normal data (but not both), resulting in a high rate of false alarms, and missed attacks. Models of attack signatures do not require the comparison of attack signatures with normal data. As a result, attack signatures, especially those identified manually by examining only attacks or attack data, may also appear in normal data, leading to false alarms as shown in **Fig. 1**. In addition, the anomaly approach may detect as an attack a normal activity (that is, irregular behavior with a large deviation from a norm profile but still different from attack activities), resulting in a false alarm (Fig. 1). A single norm profile can hardly cover all kinds of normal behavior. Most existing signature recognition and anomaly detection techniques are built without a solid theoretical foundation to validate their detection coverage and accuracy.

Signal-noise separation. This method recently has been proposed by Nong Ye to overcome the above problems. It consists of four elements: data, features, characteristics, and signal detection models (Fig. 1). Attack data are considered as cyber signals, and normal data are considered as background or cyber noise. The data variables are first collected at observable attack paths to capture not only the activities (such as computer and network events) but also the availability, confidentiality, and integrity states of computer/network resources (such as the network bandwidth used at an interface and data bytes transmitted through that interface) and the timeliness, precision, accuracy of performance of QoS. The features that enable the distinction of cyber signals from cyber noises (such as the ratio of access frequency to the regularly configured network ports, to the rarely used network ports) are then extracted from the data variables. The characteristics of cyber signals and cyber noises are built into models to separate signals and noises and detect the cyber signals. The scientific understanding of the data, features, and signal/

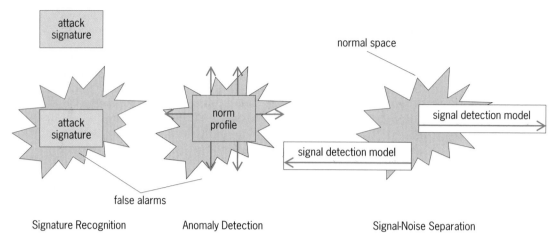

Signature Recognition   Anomaly Detection   Signal-Noise Separation

Fig. 1. Cyber attack detection approaches.

noise characteristics allows the development of specialized cyber sensors to monitor a small amount of highly relevant activity/state/performance data for the efficiency of the detection performance. The goal is to reach the accuracy as seen, for example, in radar detection of hostile air objects.

The reaction to a detected attack usually includes a series of steps to stop it; to diagnose the origin, path, and cause; to assess the impact; to recover computers and networks; and to correct the vulnerabilities exploited to prevent similar attacks. In practice, system administrators often carry out these steps manually. Few technologies exist for the automated attack reaction.

Technologies and quantitative methods for QoS assurance. Integrated service (IntServ) and differentiated services (DiffServ) are the two major architectures proposed for QoS assurance on the Internet.

IntServ. In this architecture, an end-to-end (source to destination computer) bandwidth reservation is required to guarantee the QoS of an individual data flow. The implementation of IntServ requires a range of complex mechanisms for the path reservation, admission control, and QoS monitoring through traffic policing and shaping. IntServ is not scalable or practical for a large-scale network, such as the Internet, due to the management requirements to maintain the state of each data flow.

DiffServ. This architecture divides the Internet into domains with edge and core routers. A domain's edge routers classify and mark data packets based on certain administrative policies. Inside the domain, core routers provide QoS based on the type of data traffic. Multiple queues are maintained to differentiate the services for data packets with different classes of QoS requirements or service priorities. Packets are placed into the queues corresponding to their classes. Each queue services the data packets using the FCFS rule. In DiffServ, there is no need to reserve the bandwidth on an end-to-end connection path and no absolute guarantee for the end-to-end QoS of each individual data flow.

Industrial engineering approach. Nong Ye and colleagues develop QoS models at the local and regional levels of the Internet to minimize the waiting time variance of jobs going through individual or multiple computer/network resources. Mathematical theories from operations research have been applied to develop the batch scheduled admission control (BSAC) method to admit batch service requests and then schedule the service requests (jobs) in a given batch through the Yelf (Ye, Li, and Farley) spiral scheduling method by the following steps: (1) Place the largest job (the job with the longest processing time) last, the second largest job last-but-one, the third largest job first, and the smallest job before the second largest job. (2) Find the next largest job and place it either before or after the smallest job, depending on which placement produces a smaller variance of waiting times for the scheduled jobs up to this point. (3) Repeat the above steps for the remaining jobs.

The arrows in **Fig. 2** indicate the consecutive steps and the resulting schedule of placing the 10 jobs with the processing times of 2, 3, 4, 5, 6, 7, 8, 9, and 10 time units using the Yelf spiral schedule method.

The QoS model using the BSAC method and the Yelf spiral scheduling method stabilizes the service of jobs on a local-level computer or network resource and allows each job to have approximately the same waiting time regardless of when the job comes to the resource for the service. This service stability implies the predictability of the QoS performance in timeliness. Thus, each local-level computer or network resource becomes a "standard part" with stable and predictable QoS performance, ultimately enabling and simplifying the proactive selection and assembly of the "standard parts" on the Internet by an end user to achieve end-to-end QoS. The researchers

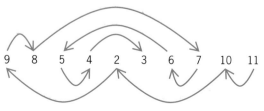

Fig. 2. Yelf spiral scheduling method.

are also developing the QoS models at the global level of the Internet to assure the end-to-end QoS of each individual data flow while maintaining the overall health of traffic flows on the Internet by reducing traffic congestion.

For background information *see* COMMUNICATIONS; COMPUTER SECURITY; COMPUTER-SYSTEM EVALUATION; DATABASE MANAGEMENT SYSTEM; INDUSTRIAL ENGINEERING; INTERNET; OPERATIONS RESEARCH; PRODUCT QUALITY; STOCHASTIC PROCESS; WORLD WIDE WEB in the McGraw-Hill Encyclopedia of Science & Technology.   Nong Ye

Bibliography. E. A. Fisch and E. B. White, *Secure Computers and Networks: Analysis, Design and Implementation*, CRC Press, Boca Raton, FL, 2000; D. C. Montgomery, *Introduction to Statistical Quality Control*, Wiley, New York, 2000; W. L. Winston, *Operations Research: Applications and Algorithms*, Duxbury Press, Belmont, CA, 1994; N. Ye (ed.), *The Handbook of Data Mining*, Lawrence Erlbaum Associates, Mahwah, NJ, 2003.

Neurotrophic factors

Trophic factors are proteins that stimulate growth, activity, and survival. Neurotrophic factors, such as brain-derived neurotrophic factor (BDNF), were first noted for their ability to promote growth and survival in neurons during early development. However, neurotrophic factors are now known to play an active role in the adult brain, and to have an effect on a variety of more subtle cellular functions aside from cell survival. BDNF itself is now assumed to be involved in a variety of brain functions, including memory formation and affective state.

Since its discovery in 1982, appreciation for BDNF's role in the brain has evolved considerably. Once thought to be active only in the developing peripheral nervous system (PNS), BDNF is now known to be a trophic support molecule in the mature central nervous system (CNS) as well. Recent data suggest that BDNF is involved in changing the shape, or morphology, of neurons, and that it can function as a rapidly acting molecule that provides a local "history" of neuronal activity. Evidence also links abnormal BDNF processing and function in the adult brain to a variety of psychiatric disorders, such as depression. This range of roles underscores that BDNF is a critical participant in normal brain functioning, such as in the formation of memories. Instead of merely mediating cell survival, BDNF is now thought to participate in the constant adaptation that brain cells undergo in response to their environment—a neuroremodeling that scientists refer to as neuroplasticity.

Role in cell survival, cell death, and neuroplasticity. BDNF is a small molecule (about 12 kDa), and the mature BDNF protein appears to be identical in all mammals; such evolutionary conservation is typical of proteins that are important to cell survival. Unlike the PNS, where inactivation of individual neurotrophic factors may cause cell death, the CNS appears to have an overlap of neurotrophic factor actions in development such that omission of one does not directly correspond to death of a certain cell type. The messenger ribonucleic acid (mRNA) and protein for both BDNF and its major receptor, TrkB, are made by discrete populations of neurons in the adult brain. Like other neurotrophic factors, BDNF can be retrogradely or anterogradely transported. BDNF and its receptor have a wide distribution in the mammalian brain, but notably high levels of both are seen in regions, such as the hippocampus, which exhibit a high degree of neuroplasticity.

When released from cells, BDNF primarily binds to the TrkB receptor and activates intracellular signaling pathways that promote cell survival and inhibit cell death. Interestingly, two isoforms (molecular variants) of the TrkB receptor are evident in the normal brain: full-length TrkB, which can couple to intracellular signaling pathways linked to cell survival, and truncated TrkB, which lacks the intracellular aspect (that is, an amino acid residue) of the receptor that drives the signaling pathways. Therefore, truncated TrkB appears to be a dominant negative inhibitor of the BDNF signaling cascade (that is, it competes with full-length TrkB for binding of BDNF), providing another way to regulate BDNF signaling aside from changes in the amount of ligand or number of available receptors. In some instances, BDNF will also bind to the low-affinity pan-neurotrophin receptor (p75). Recent studies suggest that while BDNF binding to full-length TrkB may promote cell survival and neuroplasticity and inhibit cell death, BDNF binding to p75 may actually promote cell death. Understanding the fine balance between BDNF's influence on cell survival and cell death is an area of extremely active research.

Role in memory. Neuroscientists are only beginning to understand the molecular and cellular underpinnings of memory. It is clear that a memory is not made or stored in a given cell or synapse. However, researchers know that repeated activation of discrete synapses plays a role in memory formation and consolidation, the "cementing" of a recently acquired memory for longer-term storage. BDNF appears to be well positioned to influence the repeated activation of synapses necessary for processing memories. For example, although BDNF was initially identified as a cell survival factor, it has acute effects on the release of neurotransmitters, including glutamate, gamma-amino butyric acid (GABA), dopamine, and serotonin. Therefore, local release of BDNF can feed back to influence local release of neurotransmitters, which can act to stimulate or inhibit recently active synapses. Conversely, neurotransmitter release can influence BDNF release. For example, glutamate receptor activation increases the expression of BDNF. Such findings emphasize that BDNF can be released from "active" synapses. The activity-induced regulation of BDNF, and its subsequent regulation of synapses, has led researchers to propose a role for BDNF in the "synapse stabilization" thought to be important for memory processes.

Several lines of evidence support the theory that BDNF is important in learning and memory. If an animal learns a hippocampal-dependent task that involves spatial learning, hippocampal BDNF expression will increase. Conversely, animals with reduced BDNF expression show deficits in learning and memory, specifically in the cosolidation of memory. As a clinical corollary to this basic research, scientists have identified a variation in the human *bdnf* gene that results in an alteration in an amino acid sequence (a valine to methionine) that is linked to memory impairment and vulnerability to mental disorders.

Given the role of BDNF in memory formation, it is interesting that manipulations that increase BDNF also improve performance on spatial learning tasks. For example, voluntary exercise is known to both increase BNDF levels in the hippocampus as well as improve memory of spatial learning tasks. In contrast, a diet high in fat and refined sugar decreases BDNF levels in the hippocampus and diminishes performance on spatial learning tasks. While correlative and therefore not conclusive, these studies emphasize an intriguing link between BDNF levels and learning and memory that warrants further exploration.

Role in depression. The etiology of mood disorders, such as depression, is complex. For example, depression is likely due to an interaction of genetic vulnerability and environmental perturbations. However, several pieces of evidence from both clinical and basic research into depression suggest that BDNF may play a role in the etiology and/or the treatment of depression.

The potential involvement of BDNF in depression first came from a series of studies using laboratory animals. Researchers found that animals exposed to stress, a predisposing factor in clinical depression, showed decreased levels of hippocampal BDNF. (Interestingly, voluntary exercise prevents the stress-induced decrease in BDNF.) Stress also increased levels of the BDNF receptor TrkB in the hippocampus, suggesting a compensatory increase in the receptor in response to the decreased amount of ligand available. In contrast, administration of antidepressant medications to laboratory animals increased the levels of BDNF in certain brain regions. Chronic but not acute administration of agents such as fluoxetine (Prozac®) to laboratory animals increased levels of BDNF mRNA as well as levels of BDNF itself in the hippocampus. Another effective clinical intervention for depression, repeated electroconvulsive treatment, was also effective at increasing levels of BDNF in the hippocampus.

These findings were interesting for several reasons. First, the hippocampus has long been a central region in depression research. The volume of the hippocampus is decreased in patients with depression, and researchers hypothesized that reversal of this volume decrease was integral to treatment of depression. Perhaps an increase in BDNF, as mediated by chronic antidepressant treatment or repeated electroconvulsive treatment, could enhance cell survival and ameliorate the volume decrease seen in the hippocampus in depression. Second, while antidepressant medications and electroconvulsive treatment are clearly effective clinically, researchers were still unsure of the mechanism of action. For example, fluoxetine is a blocker of the serotonin reuptake site and thus acts acutely to increase serotonin in the synapse. However, chronic but not acute fluoxetine administration is clinically effective. Since chronic but not acute fluoxetine administration was effective in increasing hippocampal BDNF levels, understanding the neural mechanisms underlying the BDNF increase had some promise in revealing information about how chronic antidepressants were clinically effective.

These early studies with laboratory animals suggested that depression may be marked by dysregulation of BDNF and its signaling cascades, and that effective treatments of depression may normalize levels of BDNF in the brain. Recent clinical evidence supports this theory. Humans with major depression appear to have low serum BDNF levels, and antidepressant treatment appears to reverse that decrease. These findings support the hypothesis of neurotrophic factor involvement in mood disorders, such as depression.

Taken together, the basic and clinical research on BDNF encouraged researchers to pursue the theory that depression is an impairment of neuroplasticity, and that BDNF dysregulation may underlie certain aspects of depression. Much progress has been made in understanding both the "upstream" and "downstream" factors related to BDNF and depression. For example, upstream of BDNF, chronic administration of antidepressants leads to an increase in intracellular signaling molecules, such as $3',5'$-cyclic adenosine monophosphate (cAMP) and cyclic AMP-responsive element binding protein (CREB), which can result in an increase in BDNF. Downstream, increased BDNF itself can lead to a variety of intracellular responses, such as increased birth of new neurons (adult neurogenesis), which may ameliorate the neuropathology thought to occur in clinical depression. The research on BDNF and depression and these upstream and downstream aspects that may contribute to BDNF signaling has resulted in the identification of multiple new targets for antidepressant treatment. For example, enhancers of the signaling upstream from BDNF, positive mediators of adult neurogenesis, and activators of select aspects of glutamatergic signaling are just a few of the targets being actively examined for their potential efficacy in treating depression.

Future research. Given that BDNF's actions at the full-length TrkB receptor and at the p75 receptor may have opposite actions, it is not surprising that BDNF's actions do not result in positive neuroplasticity everywhere in the brain. For example, recent work with BDNF suggests that its antidepressant and proneuroplasticity actions may be brain region–specific. Since many researchers and pharmaceutical agencies are very interested in utilizing BDNF and its related signaling cascades to ameliorate depression, such brain region–specific changes will be very important to examine prior to global manipulation of BDNF levels in the brain.

For background information *see* AFFECTIVE DISORDERS; BRAIN; MEMORY; NEUROBIOLOGY; SIGNAL TRANSDUCTION; SYNAPTIC TRANSMISSION in the McGraw-Hill Encyclopedia of Science & Technology.

Amelia J. Eisch

Bibliography. R. S. Duman, Role of neurotrophic factors in the etiology and treatment of mood disorders, *Neuromol. Med.*, 5(1):11–25, 2004; V. Lessmann, K. Gottmann, and M. Malcangio, Neurotrophin secretion: Current facts and future prospects, *Prog. Neurobiol.*, 69(5):341–374, 2003; H. K. Manji et al., Enhancing neuronal plasticity and cellular resilience to develop novel, improved therapeutics for difficult-to-treat depression, *Biol. Psychiat.*, 53(8):707–742, 2003; E. J. Nestler et al., Neurobiology of depression, *Neuron*, 34(1):13–25, 2002; W. J. Tyler et al., From acquisition to consolidation: On the role of brain-derived neurotrophic factor signaling in hippocampal-dependent learning, *Learn Mem.*, 9(5):224–237, 2002.

Neutral-atom storage ring

Storage rings are usually associated with kilometer-long accelerators and particle colliders, but there is one miniature storage ring. The Nevatron, the first neutral-atom storage ring, is slightly larger than a quarter (25 mm or 1 in. in diameter) and weighs only 200 g (7 oz). While it cannot create antiprotons with teraelectronvolts (TeV) of energy, the Nevatron does bring together laser cooling and magnetic trapping to produce a storage ring for neutral rubidium-87 (^{87}Rb) atoms. This ring has allowed observation of multiple atomic orbits, dual cloud loading, and atomic cloud shaping. It also provides the basis for an atomic interferometer, a device that may one day provide the world's most sensitive gyroscope.

Control of neutral atoms. Creating a storage ring for neutral atoms requires the ability to manipulate their positions. Unlike charged particles, neutral atoms are unaffected by homogeneous electric and magnetic fields. Instead, other weaker interactions must be employed to control atomic motion. While neutral atoms do not have a net charge, they do possess a magnetic dipole moment, similar to a tiny magnet, and this magnet can be used to move the atoms around. If this dipole moment is oriented in the direction opposite to the magnetic field, there will be a force on the atom pushing it toward areas of low field strength. There have been a number of experiments using this effect to guide atoms along straight lines for relatively short distances.

Laser cooling. Although effective, the magnetic dipole force is relatively weak. Neutral atoms at room temperature simply have too much kinetic energy to be controlled in this fashion. The secret to the success of the Nevatron is the tremendous advances made in laser cooling during the end of the twentieth century. This technology makes possible the preparation of a cloud of rubidium-87 atoms that has been laser-cooled to 3 millionths of a degree above absolute zero. The cooling takes place at ultrahigh vacuum in a magneto-optical trap (MOT) where near-resonant laser light is shone upon atoms from all directions. The light is detuned below the resonant atomic transition such that stationary atoms will not interact with the light on their own. However, a moving atom will see Doppler-shifted light and consequently will scatter photons from the direction in which it is moving. These photons impart momentum to the atom and force it to slow down. If this effect, known as optical molasses, is combined with a magnetic field gradient, a cloud of millions of neutral atoms is created with a temperature of about $100 \, \mu$K. Further techniques using the internal structure of the atom reduce the temperature of the cloud to about $3 \, \mu$K. Laser cooling has been performed on most alkali-metal atoms and some metastable noble gases as well. Rubidium was chosen for this experiment because its strongest optical transition is conveniently accessible by the same inexpensive diode lasers used to burn compact disks (CDs).

Ring loading. Once cooled, the atoms must be loaded into the ring. Two parallel current-carrying wires produce the magnetic guiding field used to load the Nevatron. Atoms are attracted to the central area between the wires where the field strength is smallest. **Figure 1** shows the cloud of atoms in the MOT trap being transferred to the magnetic guide. Here each wire has a diameter of 280 μm and carries a current of 8 A. This configuration allows the ring to hold atoms with temperatures of 1.5 mK, or about 100 neV (hence the name Nevatron). After being loaded into the guide, the atomic cloud falls 4 cm (1.6 in.) under gravity down to the overlapping area with the ring (**Fig. 2**). The ring consists of two more wires mounted on a thin aluminum substrate.

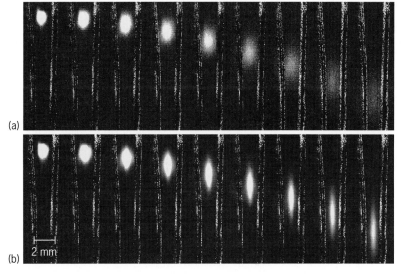

Fig. 1. Time sequences showing evolution of a cloud of rubidium-87 atoms that has been cooled to 3 μK in a magnetooptical trap located between two parallel wires. (*a*) With no current in the wires, the atoms fall away under gravity. (*b*) Current flowing through the wires produces a guiding field that traps the atoms in the region between them. These atoms are then loaded into the storage ring.

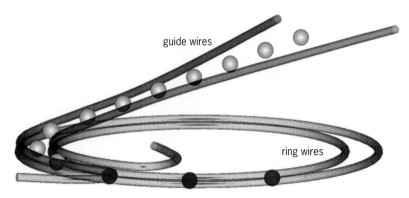

Fig. 2. Structure of the Nevatron. The overlap region, where the atoms are transferred into the storage ring, is shown.

The substrate provides tracking for the wires and also acts as a heat sink. Once the atoms are in the overlap area, the current is transferred from the guide wires to the ring, and the cloud enters the storage ring at about 1 m/s (3 ft/s or 2 mi/h).

Loading into the ring originally proved to be a difficult technical challenge. In an earlier experiment, the ring was not separate from the loading guide. It was found that a time-varying potential (that is, the changing currents described above) must be used to effectively load the ring. Otherwise, atoms that enter the ring will be scattered out again at the entrance point, causing extreme losses. With the new scheme, the atoms make about seven revolutions before being ejected by background gas collisions or imperfections in the ring geometry. The ring can be loaded more than once by collecting more atoms in the magnetooptical trap while the other atoms are orbiting. The second cloud can then be introduced 180° out of phase with the first. Both clouds maintain a constant phase to one another while orbiting. However, this technique cannot be used to arbitrarily increase the number of atoms in the ring, because they are lost too quickly.

Data collection. All data collected from the storage ring have been taken using a low-noise charge-coupled-device (CCD) camera and selective illumination of the atomic cloud. Resonant laser light is focused between the two wires at one of several observation holes in the aluminum substrate located around the ring. The rubidium atoms scatter the photons from the focused laser beam at a known rate. Using this rate, the number of atoms and their location in the ring can be determined from images of the atomic cloud (**Fig. 3**). The imaging process is destructive, so the experiment must be repeated many times using a time advance at each step to view the trajectory of the stored atoms. Each step in the experiment takes about 6 s, dominated by the loading time into the magnetooptical trap. About 200 steps are needed to form a complete trajectory, which takes about 20 min to complete.

Atomic motion in the ring. The cloud expands along the azimuthal direction of the ring as it propagates. From this expansion, the relative velocity and thus the temperature of the atoms in the ring can be cal-

culated. Notably, the temperature does not change after loading into the ring. At 3 μK, the atoms' relative velocity inside the cloud is about 2 cm/s (0.8 in./s), while the cloud itself moves at 1 m/s (3 ft/s). A single orbit in the Nevatron takes about 80 ms to complete. The atoms could be forced to orbit slower or faster by changing the distance from the MOT to the ring, which would impart either more or less kinetic energy to the atoms during their fall.

The atomic trajectory in the ring is not perfect. In particular, there is a small bump in the field where the current is fed in and out. This transfers rotational motion to radial motion and causes atoms to flee the ring. A smoother ring could be constructed from multiple wires, which would suppress the feed-in effect on the field. However, size is very important to these magnetic guides. The amount of force supplied by the ring is inversely proportional to the spacing between the wires. If this spacing were made larger to facilitate more wires, the effectiveness of the guide would decrease. One possible future construction technique would involve etching very closely spaced wires on the surface of a chip. This has since been demonstrated by creating a magnetic conveyor belt for atoms.

Applications. The most attractive application of the Nevatron is in atom interferometry. Light interferometers operate by observing the interference of electromagnetic waves. Due to the small wavelength of light (about 500 nm), these devices can measure very small displacements. If the light is propagated along a ring in both directions, a gyroscope can be created. A rotating ring creates a longer path in one direction and a shorter path in the other. This varying path length induces an interference fringe through the Sagnac effect which can be observed on a photodetector. Here, the amount of shift and hence the sensitivity of the interferometer is proportional to the enclosed area of the loop. According to quantum mechanics, matter too is made up of waves, and these waves can also interfere with one another.

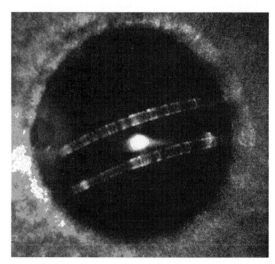

Fig. 3. Image of a cloud of neutral atoms guided in the Nevatron.

Atom interferometers rely on the interference of matter waves, which are much smaller than light waves (about 1 nm) at relevant temperatures. Similarly, atomic gyroscopes have been created with great success and sensitivity. The enclosed area of these gyroscopes is currently only a few square millimeters. The Nevatron offers the possibility of a large enclosed area and multiple orbits that could easily produce an enclosed area of 4400 mm² (6.8 in.²), 200 times larger than the most sensitive gyroscope demonstrated to date.

For background information *see* ATOM OPTICS; BOSE-EINSTEIN CONDENSATION; CHARGE-COUPLED DEVICES; GYROSCOPE; INTEFERENCE OF WAVES; INTERFEROMETRY; LASER COOLING; PARTICLE ACCERATOR; PARTICLE TRAP in the McGraw-Hill Encyclopedia of Science & Technology.   Jacob A. Sauer

Bibliography. P. R. Berman (ed.), *Atom Interferometry*, Academic Press, New York, 1997; H. Haken and H. C. Wolf, *The Physics of Atoms and Quanta*, 6th ed., Springer, 2000; H. J. Metcalf, P. Van Der Straten, and H. E. Stanley, *Laser Cooling and Trapping*, Springer, New York, 1999.

Nobel prizes

The Nobel prizes for 2003 included the following awards for scientific disciplines.

Chemistry. The chemistry prize was awarded for the discovery of the structure and function of proteins, known as channels, which span cell membranes and selectively regulate the movement of water or ions. Peter Agre of the Johns Hopkins University shared the prize with Roderick MacKinnon of the Howard Hughes Medical Institute at Rockefeller University.

The cell membrane is composed largely of lipids and proteins arranged such that internally the membrane is hydrophobic, or impermeable to water and water-soluble ions and molecules. The discoveries made for water and ion channels are important in biology and medicine for understanding physiological process such as nerve impulses and diseases such as kidney disorders.

In 1988, Agre isolated from red blood cells and kidney cells the first water channel, a membrane protein now called aquaporin 1, or AQP1. He confirmed that only cells containing aquaporin absorb water and that chemically blocking aquaporin prevented cells from absorbing water. Since his discovery many other aquaporins have been found in plant and animal cells for specific functions, including root water uptake, fluid balance within the eye and brain, and the production of tears and saliva, to name a few. By 2000, the first three-dimensional structure of AQP1 was determined and used to show how the channels allow only water molecules to pass through while blocking everything else.

In 1998, MacKinnon determined by x-ray crystallography the structure of the potassium ion channel, called KcsA, from the bacterium *Streptomyces livi-*

dans. The structure showed that the channel contained an ion filter at one end which selectively allows potassium ions to cross the cell membrane but not other ions or water, and a gate at the other end which opens and closes in response to a nerve or muscle impulse, or a signaling molecule.

Maintaining the correct potassium concentration inside and outside cells is important for generating nerve signals, regulating the heartbeat, and controlling the release of insulin in response to blood sugar levels. MacKinnon's contribution to the understanding of ion channels should lead to new drugs that function by not just blocking ion channels but by controlling ion channels.

For background information *see* BIOPOTENTIALS AND IONIC CURRENTS; CELL (BIOLOGY); CELL MEMBRANES; CELL PERMEABILITY; PROTEIN in the McGraw-Hill Encyclopedia of Science & Technology.

Physics. Alexei A. Abrikosov (Argonne National Laboratory), Vitaly L. Ginzburg (P. N. Lebedev Physical Insitute in Moscow), and Anthony J. Leggett (University of Illinois at Urbana-Champaign) were awarded the physics prize for pioneering contributions to the theory of superconductors and superfluids.

At very low temperatures, close to absolute zero, quantum physics gives rise to phenomena that are not ordinarily observed in the macroscopic world. In superconductivity, the electrical resistance of certain metals vanishes; in superfluidity, certain liquids flow without viscosity and display other remarkable properties such as flowing over the walls of beakers in which they are placed. These phenomena are related, since the electrons in a metal may be considered to form a fluid, with the electrons that contribute to superconductivity forming a superfluid component.

In the early 1950s, Ginzburg and Lev Landau proposed a theory that describes superconductivity with a complex-valued function called the order parameter, which may be spatially varying, and whose squared value indicates the fraction of electrons in a metal that have condensed into a superfluid. The order parameter obeys an equation similar to the equation for the wave function in quantum mechanics. Although this equation was basically an inspired guess, it was highly successful in describing the behavior of superconductors and was later shown to follow from the Bardeen-Cooper-Schrieffer (BCS) theory (1957), which is based directly on quantum mechanics.

In addition to their absence of electrical resistance, superconductors exclude an external magnetic field (the Meissner effect). However, they can display different behaviors as the field strength is increased. In type I superconductors, superconductivity disappears when the field strength exceeds a certain limit. However, while the magnetic field, above a certain strength, begins to penetrate a type II superconductor in filaments called vortices, the rest of the material remains superconducting, allowing superconductivity to persist at much larger fields. Type II superconductors have important applications such as magnetic resonance imaging and superconducting particle accelerators.

In the 1950s, Abrikosov (then at the Kapitza Institute for Physical Problems in Moscow) extended the Ginzburg-Landau equations, which had been developed to describe type I superconductors, to explain the behavior of type II superconductors as well. He showed how the order parameter can describe the vortices, how an external magnetic field can penetrate the material through the vortices, and how the vortices form a lattice. He also predicted in detail how the vortices grow in number as the field strength increases until superconductivity is lost. This work has continued to be used, with greater frequency in recent years, in developing new superconductors and magnets. *See* COLLECTIVE FLUX PINNING.

The two isotopes of helium, helium-4 (with two protons and two neutrons in its nucleus) and the rare isotope helium-3 (with two protons and one neutron), display very different behaviors at low temperatures, reflecting their different quantum properties. Helium-4 atoms are bosons, and there is no restriction on how many such atoms may occupy a single quantum state; helium-3 atoms are fermions, which obey the Pauli exclusion principle. Superfluidity occurs in liquid helium-4 at about 2 K. Discovered in the late 1930s, it was explained by Landau as an example of Bose-Einstein condensation. Superconductivity in helium-3 can occur only if its atoms form Cooper pairs, similar to the Cooper pairs of electrons which are the basis of superconductivity in the BCS theory. It was discovered in 1972 at a temperature about 1000 times lower than in helium-4.

Helium-3 differs from a superconductor in that the two helium-3 atoms in a Cooper pair have parallel spins and relative orbital momentum. This gives superfluid helium-3 very complex properties, including anisotropy and magnetism, and it displays three different phases, depending on the temperature, pressure, and magnetic field strength. Within a few months of its discovery, Leggett (then at the University of Sussex, England) provided a systematic explanation of these properties. This work remains the basis for research in the field and has also been used in other fields as diverse as liquid-crystal physics and cosmology.

For background information *see* LIQUID HELIUM; SUPERCONDUCTIVITY; SUPERFLUIDITY in the McGraw-Hill Encyclopedia of Science & Technology.

Physiology or medicine. Paul C. Lauterbur of the University of Illinois, Urbana-Champaign, and Peter Mansfield of the University of Nottingham, U.K., were awarded jointly the Nobel Prize for Physiology or Medicine for their discoveries concerning magnetic resonance imaging (MRI).

The development and application of MRI represented a major advance in the ability to noninvasively visualize internal body organs. It has become an important diagnostic tool in the evaluation of diseases of the brain and spine as well as joint, bone, and soft tissue abnormalities. MRI is based on the phenomenon of nuclear magnetic resonance, which was first demonstrated in 1946 by Felix Bloch and Edward Mills Purcell, and for which they shared the Nobel Prize for Physics in 1952.

The nuclei of most atoms possess an intrinsic angular momentum, which can be envisioned as resulting from a nucleus spinning about an axis, analogous to the way the Earth rotates about its axis. Nuclear angular momentum is expressed as a series of quantized levels. Nuclei with an odd atomic mass number are said to possess half-integer spin; nuclei with even number atomic mass possess integer spin. Some nuclei may have a spin value of 0. The nuclei with nonzero spin possess a magnetic moment and can therefore interact with external magnetic fields. If such nuclei are subjected to a strong static magnetic field, an applied radio-frequency signal of proper wavelength will cause a transition in spin state; upon release from the signal, the nuclei will return to their earlier state and emit energy in the form of a radio-frequency signal that can be detected. This is referred to as nuclear magnetic resonance (NMR). Highly sophisticated techniques for applying signals at different wavelengths and measuring resulting signals (NMR spectroscopy) have been developed, enabling the study of molecular structure and dynamics and making NMR spectroscopy an invaluable tool in chemistry and physics.

Magnetic resonance imaging as used in medical diagnosis takes advantage of the amenability of hydrogen nuclei (protons) to NMR detection, their widespread occurrence in the molecules of bodily tissues, particularly in water, and changes in water content in pathological tissues that make such pathology visible through MRI. In essence, the patient is placed within a strong static magnetic field, the body is scanned, and the resulting radio signals are digitally processed to produce a two-dimensional anatomic image of a particular organ.

The Nobel Committee recognized Paul Lauterbur for his work in creating two-dimensional NMR images by introducing gradients in the magnetic field. By analyzing the radio waves emitted by the material under examination, he was able to determine their origin and thus build up a two-dimensional image of the structure. Peter Mansfield was recognized for developing methods of mathematical analysis of the NMR signals produced within magnetic field gradients, thus enabling useful imaging techniques. He also showed how fast imaging could be accomplished.

For background information *see* MEDICAL IMAGING; NUCLEAR MAGNETIC RESONANCE (NMR) in the McGraw-Hill Encyclopedia of Science & Technology.

Olfactory system coding

Detecting and discriminating various chemicals in the environment is critical for the survival of most animals on this planet. They depend on their sense of smell (olfaction) to locate food, mates, and predators. For a compound to be smelled by air-breathing animals, it needs to be volatile or semi-volatile. In general, odors are nonionic, hydrophobic, organic molecules with molecular weights of no more than 350 daltons. The odor quality of a molecule is largely determined by its size, shape, and functional groups. Mammals are able to distinguish among thousands of odors, and this amazing task is performed by the olfactory system.

Organization of the mammalian olfactory system. **Figure 1** shows the anatomy of the mammalian olfactory system. The primary olfactory sensory neurons, also called olfactory receptor neurons, are located in the main olfactory epithelium in the nose. Their axons form the olfactory nerve and project to the olfactory bulb of the brain. The axon terminals of the sensory neurons synapse with the dendritic arbors of the mitral/tufted cells, the projection neurons (that is, neurons with long axons that extend to distant parts of the brain) in the bulb. These neuropils (dense networks of neuronal processes) and glia form special structures called glomeruli, which are distributed in a layer under the surface of the bulb. The mitral/tufted cell axons form the lateral olfactory tract and project to five olfactory cortical regions: anterior olfactory nucleus, olfactory tubercle, pyriform cortex, amygdala, and entorhinal area. Olfactory information is further sent to the frontal cortex from some of these areas (regions 2–5 in Fig. 1)

either directly or through the relay of the thalamus, a key structure for transmitting sensory information to the cerebral hemispheres. The pyriform cortex pathway plays a critical role in olfactory perception, and the amygdala pathway is presumably involved in the emotional effects caused by odors. From the entorhinal area, the olfactory information is sent to the hippocampus, a structure critical for memory consolidation.

In addition to the main olfactory system, most mammals have an accessory olfactory system, which starts from the sensory neurons in the vomeronasal organ located on top of the vomer bone, between the nose and mouth. These neurons mainly sense the chemical signals emitted by other individuals within the same species. Through the relay of the accessory olfactory bulb and subsequently the amygdala, such information is sent to the hypothalamus, which integrates the functions of the autonomic nervous system and hormone release. In general, the accessory olfactory system plays a critical role in breeding and aggressive behaviors. Humans have a structure that looks similar to a vomeronasal organ, but it is generally believed to be a nonfunctioning vestigial organ.

Olfactory coding in the epithelium. Within the olfactory neuroepithelium, there are three cell types: olfactory sensory neurons, supporting cells, and

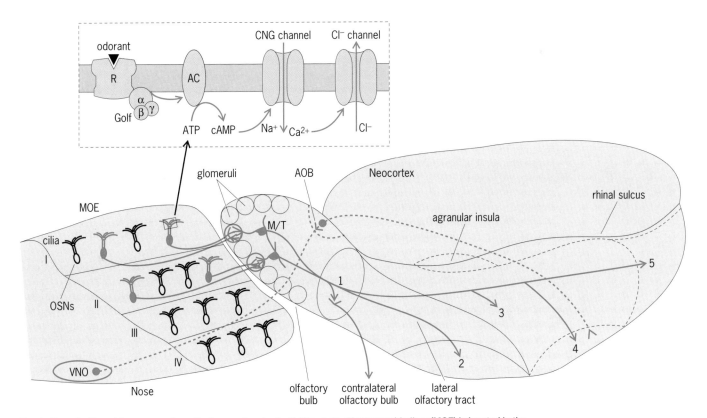

Fig. 1. Organization of the mammalian olfactory system (rodent). The main olfactory epithelium (MOE) is located in the dorsal, posterior part of the nose. I to IV indicate the different receptor gene expression zones. The inset shows the major olfactory signal transduction pathway in the olfactory sensory neurons (OSNs). The axons of the mitral/tufted (M/T) cells in the olfactory bulb form the lateral olfactory tract and project to five regions of the brain (shown here in side view): (1) anterior olfactory nucleus, (2) olfactory tubercle, (3) pyriform cortex (dotted area), (4) amygdala, and (5) entorhinal area. AC, adenyl cyclase; AOB, accessory olfactory bulb; ATP, adenosine triphosphate; cAMP, 3',5'-cyclic adenosine monophosphate; CNG, cyclic nucleotide gated; R, receptor; VNO, vomeronasal organ.

basal cells (which give rise to the other two cell types throughout an animal's life). An olfactory sensory neuron is a bipolar neuron, with a thin axon and a thick dendrite with a swelling at the end (the dendritic knob) bearing 10–15 cilia. The olfactory signal transduction machineries are located in the cilia and the knob of a sensory neuron.

Olfactory signal transduction. A series of events, triggered by binding of an odor molecule to a specific receptor on a sensory neuron, transform chemical energies into electrical signals (inset in Fig. 1). Such binding activates an olfactory-specific G protein (a guanine nucleotide-binding protein) known as Golf. The α subunit of Golf activates the enzyme adenylyl cyclase (type III), generating the major second messenger, 3′,5′-cyclic adenosine monophosphate (cAMP), which directly opens the cyclic nucleotide gated channel, allowing Na^+ and Ca^{2+} to flow in and depolarize the cell. An additional outward Cl^- current (olfactory sensory neurons maintain relatively high intracellular Cl^- concentrations), activated by Ca^{2+} ions, also contributes to depolarization. When the cell reaches the threshold level of depolarization, it fires action potentials (nerve impulses) and sends the information into the olfactory bulb. The nose expresses approximately 1000 different types of receptors, enabling the main olfactory system to use this common pathway to encode thousands of odorants.

Olfactory receptors. The molecular basis for peripheral coding has emerged recently from a series of studies, starting with the cloning of the mammalian olfactory receptor genes, which form the largest gene superfamily in the genome. Olfactory receptors are G-protein-coupled, seven-transmembrane-domain proteins located on the surface of the dendritic cilia of olfactory neurons. A single sensory neuron manages to express only one receptor from a repertoire of approximately 1000 similar ones. The sensory neurons expressing a particular receptor are selectively distributed in one of the few stripes or zones (these different receptor gene expression zones are roughly arranged from the dorsal to ventral epithelium with each zone extending in the anterior-posterior axis; Fig. 1) where they are widely dispersed and intermingled with other sensory neurons. The axons of the neurons expressing the same receptor converge onto a pair of glomeruli in the olfactory bulb (Fig. 1).

Combinatorial coding at the epithelial level. Even though a single sensory neuron expresses only one olfactory receptor, it can respond to multiple odorants with different thresholds and dose-response relations. This is, presumably, because multiple odorants (having a common motif) can bind to a single receptor with different affinities. Conversely, a single odorant can be recognized by multiple receptors, which detect different molecular features (**Fig. 2**). As a consequence, both odor quality and intensity are encoded by combinations of multiple receptors (neurons). Increased odorant concentration will induce stronger activity in individual cells and recruit more cells (receptors with lower affinities for that

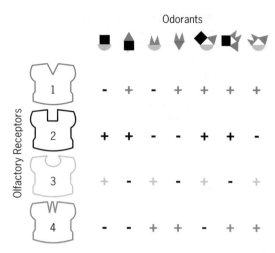

Fig. 2. Olfactory information is encoded by combinations of various receptors at the epithelial level. Representative receptors (1–4) are listed in the left column and odorants in the upper row. A "+" means activation of a receptor by an odorant, and "−" means no response.

odorant). A combinatorial scheme based on approximately 1000 receptors permits the olfactory system an almost unlimited power of detection and discrimination. In other words, any odorant at a given concentration will generate a unique combination of activated olfactory sensory neurons, whose axons carry the information into the olfactory bulb.

Olfactory coding in the bulb. The olfactory bulb has three cell types: (1) mitral/tufted cells—projection neurons; (2) periglomerular cells, which apparently contain multiple subtypes; and (3) granule cells. The periglomerular and granule cells are inhibitory local neurons, whereas a mitral/tufted cell has a single primary dendrite receiving direct excitatory inputs from the olfactory sensory neurons. This signal is modified and processed by local circuits, mainly through dendrodendritic (dendrite to dendrite) reciprocal synapses. For instance, firing of a mitral cell will cause release of the neurotransmitter glutamate from its secondary dendrites, which excites the granule cells that in turn release the neurotransmitter γ-aminobutyrate (GABA) to inhibit the same mitral cell (self-inhibition) or nearby mitral cells (lateral inhibition) [**Fig. 3***a, b*].

Spatial vs. temporal coding in the bulb. The glomeruli serve as the anatomical and functional units in the olfactory bulb. At the input level, a single glomerulus receives convergent inputs from 2000–5000 sensory neurons expressing the same receptor. At the output level, a mitral/tufted cell sends its primary dendrite into a single glomerulus (Fig. 3*a*), which is innervated by approximately 50 mitral/tufted cells. As a result, the olfactory bulb inherits the combinatorial coding strategy used in the nose, that is, a single odorant activates multiple glomeruli and a single glomerulus can be activated by multiple odorants. This lays a molecular and functional foundation for the spatial coding hypothesis, which proposes that an odor stimulus is represented as a unique spatial activity pattern in the olfactory bulb.

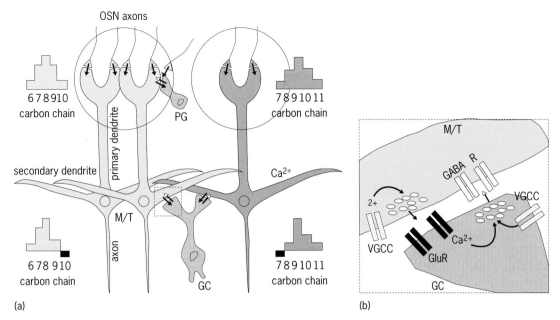

Fig. 3. Olfactory information is processed in the olfactory bulb. (*a*) Synaptic connections within the olfactory bulb. Filled arrows stand for excitatory connections mediated by glutamate, and hollow arrows for GABA-mediated inhibitory connections. The histograms represent the response spectra at the input and output levels for the mitral/tufted cells (M/T) innervating the two glomeruli. OSN, olfactory sensory neuron; PG, periglomerular cells; GC, granule cells. (*b*) Reciprocal synapses between the secondary dendrites of mitral/tufted cells and dendritic spines of the granule cells. GluR, glutamate receptors; GABA R, GABA receptors; VGCC, voltage-gated calcium channels.

An alternative/complementary hypothesis, the temporal coding hypothesis, has been proposed, in which odor information is encoded by the timing of mitral cell firing. The olfactory systems (from insects to mammals) usually show odor-evoked rhythmic activities, caused by oscillatory synchronization of cell populations. Disruption of such synchronization leads to impaired odor discrimination in insects. Even though comparable evidence is still lacking for mammals, it is plausible that both spatial and temporal components contribute to olfactory coding in the olfactory bulb.

Sharpening of the receptive field by lateral inhibition. The mitral cells do not simply relay the sensory input they receive to the next level; instead, their molecular receptive fields (the range of odor molecules that a neuron responds to) can be substantially different from their inputs, mainly due to lateral inhibition by interneurons. In the example shown in Fig. 3*a*, the sensory neurons projecting to the first glomerulus respond to aliphatic aldehydes with a carbon chain length from C6 to C10 (peak at C8). However, the mitral cells innervating the same glomerulus show a narrower spectrum, from C6 to C9 with inhibitory responses to C10. This is due to lateral inhibition from the interneurons excited by the mitral cells innervating a neighboring glomerulus which respond strongly to C9 and 10.

In summary, the projection neurons in the olfactory bulb carry the odor information into the next stage with distinct spatial and temporal activity patterns and modified response spectra compared with the primary sensory neurons.

Olfactory coding in the cortex. The mitral cells receiving inputs from the same receptor type project diffusely to the olfactory cortex, which enables individual cortical neurons to assemble divergent and convergent inputs from distinct mitral cell populations. Consequently, new response features will emerge. Unfortunately, we know very little about how neurons in the olfactory cortex and in higher brain centers encode and decode odor information.

The olfactory sensory neurons and the mitral cells serve as feature detection devices, which suggests that odor perception is the result of analytical processing by the olfactory system. However, psychophysical experiments strongly suggest that olfaction is synthetic; that is, odors are perceived as unitary, irreducible entities, since mammals have only limited ability to identify elements in a simple mixture. In addition, learning and memory plays a critical role in odor perception. One possibility is that odor-specific spatial-temporal patterns generated in the olfactory bulb are synthesized and stored in the cortex. When an odor is encountered, its neural representation is matched to all previously stored encodings, leading to odor perception.

For background information *see* CHEMORECEPTION; NEUROBIOLOGY; OLFACTION; SENSATION; SIGNAL TRANSDUCTION; SYNAPTIC TRANSMISSION in the McGraw-Hill Encyclopedia of Science & Technology.

Minghong Ma

Bibliography. S. Firestein, How the olfactory system makes sense of scents, *Nature*, 413:211–2118, 2001; G. Laurent, Olfactory network dynamics and the coding of multidimensional signals, *Nat. Rev.*

Neurosci., 3(11):884–895, 2002; N. E. Schoppa and N. N. Urban, Dendritic processing within olfactory bulb circuits, *Trends Neurosci.*, 26:501–506, 2003; D. A. Wilson and R. J. Stevenson, The fundamental role of memory in olfactory perception, *Trends Neurosci.*, 26:243–247, 2003.

Optical coherence tomography

Red and near-infrared light with wavelengths between 700 and 1550 nanometers can penetrate a few millimeters into the human body before it is either absorbed by the tissue or scattered (that is, reflected in a different direction). Scattering occurs at locations in the tissue where boundaries between materials, such as vessel walls, give rise to a change in the index of refraction. As a result, some of this light is backscattered (reflected 180°) and exits the tissue in the same place it entered. Optical coherence tomography (OCT) is a system for delivering light and collecting backscattered light to produce images with micrometer resolution.

Presently the most widespread commercial and clinical application of OCT is ophthalmic imaging. Because the eye is relatively transparent, it is possible to observe features from the front of the eye (such as the iris and lens) to the back (such as the retina and optic nerve). In particular, OCT can identify most of the known layered structures within the cornea and retina (**Fig. 1**) and identify many ophthalmic diseases. The capability for gauging macular and nerve fiber thickness with OCT is highly useful for diagnosing macular edema and glaucoma, respectively.

Imaging technique. OCT uses broadband light (a large range of wavelengths $\Delta\lambda$) so that its coherence length l_c is small (l_c is proportional to $1/\Delta\lambda$). Sources of broadband light include broadband lasers, superluminescent diodes, or tungsten lamps. Low-coherence light only interferes with itself if the light paths traveled within two arms of a Michelson interferometer are identical to within l_c. OCT uses this effect to optically range (as in radar) the locations of backscattering within tissue. The OCT interferogram is obtained by changing the optical path length in one arm of the interferometer (the reference arm), collecting backscattered light from a biological specimen in the other arm (the sample arm), and measuring the intensity of the recombined beam at the output with a photodetector. Within this interferogram, the amplitude of the interference fringes (periods of light and dark arising from constructive and destructive interference, respectively) is therefore a measure of the backscattering intensity as a function of depth within the specimen. The frequency of these interference fringes is $2v/\lambda_c$, where v is the speed of the changing optical delay and λ_c is the center wavelength of the broadband light.

The resolution of an OCT image is given in the axial (depth) direction by l_c and in the transverse direction by the spot diameter of the imaging beam (Fig. 1). Two-dimensional OCT image sections are typically rendered by acquiring individual depth scans (through scanning the optical delay) and subsequently translating the imaging beam laterally across the tissue. One can also build up three-dimensional images by acquiring adjacent two-dimensional sections.

Recently an alternate scanning technique called spectral domain OCT (SD-OCT) has been developed in which a grating and line scan camera are used at the output of the interferometer. The interference resulting from an entire depth scan is collected simultaneously on the line camera without the need to scan the optical delay in the reference arm. Because the grating acts to separate wavelength components of light, the inverse Fourier transform of the line camera data results in the time- (and correspondingly depth-) dependent interference amplitude. Instantaneous acquisition has a signal-to-noise advantage because the entire line is insensitive to time-dependent fluctuations (for example, arising from the light source). SD-OCT therefore requires less averaging to achieve a comparable signal, so that images are acquired more rapidly, typically around 30,000 lines per second. This is particularly useful in clinical applications, where fast imaging is required to reduce motion artifacts or to scan large areas of tissue.

Clinical applications. Fiber-based catheters for OCT beam delivery have been designed for endoscopic imaging of the esophagus, other mucous membranes, and blood vessel walls. In recent years there has been much interest in cardiovascular imaging with OCT because it provides superior resolution (better than 10 micrometers) and penetration of the vessel wall. Acute myocardial infarctions (heart attacks) are linked to the disruption of an atherosclerotic plaque, which consists of a fibrous cap overlying a lipid pool on the vessel wall. It has recently been demonstrated that vulnerable plaques can be distinguished in patients using OCT, which provides

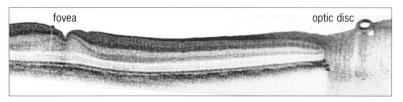
(a)

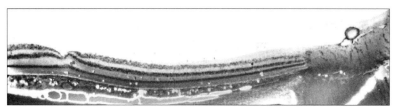
(b)

Fig. 1. Images of monkey (*Macaca fuscicularis*) retina. (*a*) OCT images at 1 × 3 μm (axial × transverse) resolution and (*b*) corresponding histology. (*Reproduced with permission from W. Drexler, Ultrahigh-resolution optical coherence tomography, J. Biomed. Opt., 9:47–74, 2004*)

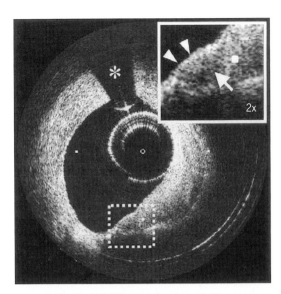

Fig. 2. Endoscopic OCT image of a plaque in the coronary artery of a living patient. (*Reproduced with permission from I.-K. Jang et al., Visualization of coronary atherosclerotic plaques in patients using optical coherence tomography: Comparison with intravascular ultrasound, J. Amer. Coll. Cardiol., 39:604–609, 2002*)

more details of the plaque microstructure than alternative techniques such as intravascular ultrasound. The presence of macrophage cells in these plaques, which degrade the fibrous cap, are correlated with increased plaque instability. Because macrophage cells are large cells characterized by heterogeneities in the refractive index, the macrophage content in plaques has been correlated with the standard deviation of the OCT backscattering signal (**Fig. 2**). OCT therefore shows much promise as a tool for the screening of life-threatening plaques.

Endoscopic OCT has also been used for imaging the gastrointestinal (GI) tract, including the esophagus, stomach, small intestine, colon, and rectum, where it is capable of delineating the mucosa and submucosa and identifying various structures such as glands, vessels, and villi. The emphasis in clinical GI OCT imaging has been on the early-stage diagnosis of cancer. In the esophagus, a disorder known as Barrett's esophagus is a precancerous condition distinguishable in OCT by the presence of gland- and cryptlike morphology and the lack of layered structures. This condition, which is typically treated by frequent screening and random biopsy, may be treated more effectively if OCT imaging algorithms are further developed for identifying dysplastic (advanced precancerous) tissue. In the colon, an algorithmic approach to OCT imaging has demonstrated the accurate identification of cancerous tissue. Colon polyps that are adenomatous (cancerous) exhibit significantly less scattering than hyperplastic (benign) polyps, and also lack morphological structure or organization in clinical OCT images. These properties may aid in diagnosing polyps that are too small to be identified by conventional endoscopy.

Molecular contrast. For over a century, histologists have been using microscopy to visualize the chemical properties of preserved, sectioned tissue by staining with tissue-specific dyes. Recently, molecular imaging—the ability to image the chemistry within living tissue—has become a very powerful tool for clinical diagnosis of disease and for research into disease progression and the benefits of pharmaceuticals. Molecular contrast (the ability to distinguish the presence and location of molecules) is obtained with OCT, either by directly imaging endogenous tissue or by adding exogenous contrast agents which have a high affinity for a specific chemical and exhibit a distinguishable signature in the OCT image.

One of the earliest demonstrations of endogenous molecular contrast with OCT used spectroscopic OCT (SOCT) to associate the spectrum of the backscattered light with the presence of melanocytes within an African frog tadpole. Melanocytes contain melanin, which absorbs more strongly at shorter wavelengths. This measurement is easy with OCT because the interference fringe frequency f is proportional to the wavelength of backscattered light, so that the backscattered light spectrum is proportional to the frequency spectrum of the fringes, which is computed by taking the Fourier transform of the signal. In this way, wavelength-dependent absorption within the bandwidth of the OCT light source is observed as a shift in the spectrum of backscattered light from objects below the absorber. More recently, SOCT has been used to detect exogenous dyes having a wavelength-dependent absorption profile.

Several classes of multiphoton effects have been investigated using OCT-type systems. Multiple-wavelength pump-probe illumination schemes have been developed where the absorption of a probe beam within a specific molecule can be affected by the presence or absence of a pump beam at a different wavelength. Therefore, the difference between the absorption measured with the pump beam on and off is related to the concentration of the target molecule. Nonlinear wave-mixing techniques such as second-harmonic generation and coherent anti-Stokes Raman scattering can also be used to generate molecular contrast in OCT-type interferometers, since the efficiency and wavelength dependence of nonlinear wave-mixing processes are highly dependent on the chemistry within the specimen.

Exogenous contrast agents can be applied topically or injected interstitially or intravenously. Possibly the simplest contrast mechanism is one which enhances the backscattering by using agents which have a large scattering efficiency. Plasmon-resonant gold nanoparticles, protein microspheres incorporating metal nanoparticles into their shell (**Fig. 3**), and liposomes (lipid bilayer micro- and nanospheres) have been used for this purpose. One advantage to protein microspheres and liposomes, which are commonly used as ultrasound contrast agents, is that they can be loaded with therapeutic agents to treat the disease to which they have been targeted, minimizing

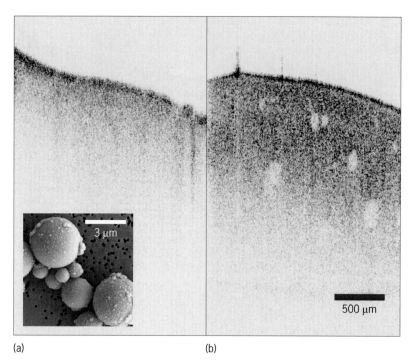

(a)   (b)

Fig. 3. OCT images of a mouse liver (a) before and (b) after intravenous injection of protein microspheres with a surface monolayer of gold nanoparticles. Inset shows scanning electron micrograph of protein microsphere with surface monolayer of silica nanoparticles. (*Reproduced with permission from T.-M. Lee et al., Engineered microsphere contrast agents for optical coherence tomography, Opt. Lett., 28:1546–1548, 2003*)

the harm to healthy tissue. A difficulty with these backscattering agents arises in distinguishing them from the normal tissue backscattering. One way to achieve higher specificity is to use magnetic agents, which are mechanically toggled by an external electromagnet, so that the magnetic-specific signal can be measured by difference imaging (much like the pump-probe imaging described above).

Future challenges lie in surface modification of OCT contrast agents with monoclonal antibodies or receptor ligands that have a high affinity for overexpressed factors present on the surface of the targeted cell types, such as growth factor receptors commonly found on early-stage cancer cells.

For background information *see* COHERENCE; COMPUTERIZED TOMOGRAPHY; INTERFEROMETRY; MEDICAL IMAGING; NONLINEAR OPTICS; OPTICAL COHERENCE TOMOGRAPHY in the McGraw-Hill Encyclopedia of Science & Technology.

Amy L. Oldenburg; Stephen A. Boppart

Bibliography. S. A. Boppart et al., In vivo cellular optical coherence tomography imaging, *Nat. Med.*, 4:861–865, 1998; B. Cense et al., Ultrahigh-resolution high-speed retinal imaging using spectral-domain optical coherence tomography, *Opt. Express*, 12:2435–2447, 2004; W. Drexler, Ultrahigh-resolution optical coherence tomography, *J. Biomed. Opt.*, 9:47–74, 2004; D. Huang et al., Optical coherence tomography, *Science*, 254:1178–1181, 1991; U. Morgner et al., Spectroscopic optical coherence tomography, *Opt. Lett.*, 25:111–113, 2000.

Origin of vertebrates

The fossil record for vertebrates is one of the best of any group of major organisms, enabling the delineation of their evolutionary development from primitive jawless fish through more advanced fish to tetrapods and ultimately humans. Despite this, it is often not clear when major groups diverged from each other, particularly when the first vertebrates separated from other chordates. Since the earliest vertebrates do not appear to have had mineralized skeletons, their preservation potential is low and their fossil record is very sparse. However, in recent years material from Cambrian-age Konservat-Lagerstätten (fossil deposits of exceptional soft-tissue preservation), particularly that of Chengjiang, South China, has pushed the fossil record of vertebrates down into the Early Cambrian [525 million years ago (mya)], indicating that vertebrate origins must lie even earlier. In this context, it is interesting that a possible chordate has recently been reported from the Ediacaran fauna (600–545 mya) of the Flinders Ranges of South Australia. Although as yet undescribed, this material would fit with divergence times based on molecular data suggesting that chordates had evolved in the upper Precambrian and that vertebrates may have diverged not long afterward.

Chordates and early vertebrates. Vertebrates constitute the subphylum Vertebrata within the phylum Chordata, which also includes two other subphyla, the marine cephalochordates and tunicates (also known as urochordata).

Evolutionary development. Vertebrates have the basic chordate characters (a notochord, a tough flexible rod that runs down the back; a dorsal longitudinal nerve cord; gills; and V-shaped muscle blocks, or myotomes, along the length of their bodies). They also have elements of the vertebral column and the ability to secrete phosphatic hard tissue (bone)—characters not shared by the other chordates. The tunicates are small attached filter feeders that have a mobile larval stage, during which they closely resemble a primitive fish. In this stage, the animal has a notochord and a longitudinal nerve cord, but these are lost once it anchors itself and metamorphoses into the adult form. The cephalochordates (or lancelets), small eellike filter feeders exemplified by amphioxus, retain the chordate features as adults and live in sediments as mobile filter feeders. It is generally believed that the cephalochordates are the closest relatives of the vertebrates while the tunicates are less closely related.

Within the vertebrates, the significance of the development of the jaw is recognized by their classification into gnathostomes ("jaw mouths") and agnathans ("without jaws"). The jawless vertebrates are considered to be the most primitive and were separated as the Agnatha, which includes the modern lampreys and hagfishes as well as a number of fossil groups often termed ostracoderms ("shell skin"). The modern forms share a primitive eellike appearance and similar feeding habits.

Fossil record. The earliest fossil chordates are known from the Chengjiang and Burgess Shale Konservat-Lagerstätten, which are Early Cambrian (525 mya) and Middle Cambrian (505 mya) in age, respectively. The best known is *Pikaia gracilens*, which is a leafshaped organism with clearly defined V-shaped muscle blocks and a notochord. There is some doubt about its cephalochordate affinities, however, as there is no evidence for any gill-like structures. A similar animal from the Chengjiang fauna, *Cathaymyrus*, is also thought to be a cephalochordate, and both have been compared with the modern cephalochordate amphioxus. Thus, there may be evidence of the closest relatives of vertebrates as far back as the Early Cambrian. Very recently, however, two additional species have been reported from Chengjiang, *Myllokunmingia* and *Haikouichthys*, both of which are described as vertebrates. Both show sigmoidal muscle blocks, a relatively complex and presumably cartilaginous skull, gill arches, a heart, and fin supports, and have been compared to lampreys. The apparent presence of serially arranged gonads and a dorsal fin with forwardly tilted radials has been cited to cast doubt on their inclusion within vertebrates, however.

Prior to the discovery of the Chengjiang vertebrates, the evidence for their presence in the Cambrian rested on the preservation of fragments of phosphatic hard tissue. Of these, *Anatolepis* has been reported from the Late Cambrian, although it was initially described from the Early Ordovician of Spitsbergen, Norway. It consists of microscopic plates and spines with scalelike ornamentation and a layered internal structure. Although there has been some controversy over the vertebrate attribution of this material, recent histological studies show the presence of dentine, the characteristic vertebrate hard tissue. Although fragments of purported vertebrate hard tissue have also been recently reported from the Late Cambrian of Australia, there is some doubt about their affinity as thin sections show a resemblance to some arthropod cuticles.

Thus, both soft and hard tissue fossil evidence points to the presence of vertebrates within the Cambrian and possibly as far back as the Early Cambrian.

Ediacaran fauna. The occurrence of possible fossil vertebrates as far down as the Early Cambrian has resulted in speculation about their presence in even older rocks and has focused attention on the upper Precambrian Ediacaran fauna, an assemblage of organisms that existed from about 600 to 545 mya (the base of the Cambrian). Originally recognized in the Flinders Ranges of South Australia, the Ediacaran fauna are now known from localities worldwide, where the fossils represent both shallow- and deep-water ecosystems and evoke a marine life very different from that of today's oceans. These representatives of the first metazoans, or complex life, had no hard parts and are preserved mostly as disk- and frond-shaped impressions that may be up to several feet in length. They were originally thought to represent the precursors to animal families present today, such that disk shapes represented ancestral jellyfish and frond shapes represented sea pens (relatives of sea anemones). More recently, it has been proposed that they were not relatives of modern organisms but an unsuccessful natural experiment to test a metazoan body plan utilizing a hydraulic architecture that could be swollen with fluid. This fauna was thought to have become extinct at the end of the Precambrian and to have been replaced by the proliferation of organisms in the basal Cambrian, an important evolutionary event referred to as the Cambrian Explosion.

However, recent work has shown that some Ediacaran organisms did survive into the Cambrian and may indeed have been ancestral to some modern forms. Within the last year, a possible chordate has also been reported from the Precambrian of the Flinders Ranges. Although a scientific description has yet to be published, the photographs show an animal that is certainly fish-shaped but preserved as an impression that does not allow the chordate characters to be easily identified. However, regardless of whether this is a vertebrate or even a chordate, it is possible to explore the timing of the origin of chordates and the divergence of vertebrates from them by utilizing molecular data.

Molecular data. Although fossil evidence continues to refine the picture for the divergence of fossil groups, the lack of fossil data for the earliest history of chordates in general, and vertebrates in particular, requires that other techniques be tried. Studies on organic molecules have shown that in deoxyribonucleic acid (DNA) and proteins, the amino acid sequences are accurately reproduced through the generations with only minor differences caused by mutations followed by natural selection or genetic drift, and that comparison of these sequences among a range of living taxa can be used to provide evidence of interrelationships. Additionally, molecular distances (a measurement of cumulative mutations) can be used to date divergence events between taxa if the molecular clock is first calibrated using organisms which have well-dated divergence events (that is, which have a good fossil record). Recent studies on the divergence time for cephalochordates and vertebrates using molecular data estimate that this event may have happened as far back as 750 mya (see **illustration**), while the divergence between agnathans and gnathostomes may have occurred about 500 mya. These divergence dates await corroboration from the fossil record, but it is interesting that the date for the origin of chordates is about the same as that for the first major Neoproterozoic glaciation event, which is suggested to have led to extensive evolutionary changes due to a reduction of habitat and genetic isolation.

Given the pace at which new information about the earliest vertebrates has appeared recently, it is expected that new material will come to light in the next few years that will substantially clarify our understanding of both the timing and causes of vertebrate origins.

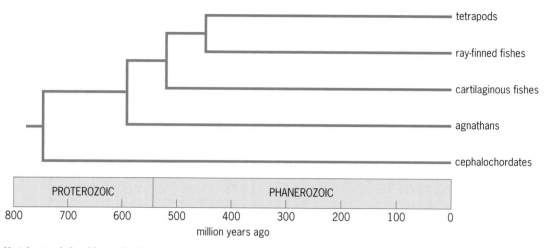

Vertebrate relationships and estimated divergence times using molecular data.

For background information *see* ANIMAL EVOLU-
TION; CHORDATA; EDIACARAN BIOTA; FOSSIL; PRE-
CAMBRIAN; VERTEBRATA in the McGraw-Hill Encyclo-
pedia of Science & Technology.   David K. Elliott

Bibliography. M. J. Benton, *Vertebrate Paleontology*,
Chapman & Hall, 1997; P. Janvier, *Early Vertebrates*,
Oxford Science Publications, 1998; J. A. Long, *The
Rise of Fishes*, Johns Hopkins University Press, 1995;
J. G. Maisey, *Discovering Fossil Fishes*, Henry Holt,
1996.

Orographic precipitation (meteorology)

Mountains may induce precipitation through orographic lifting, sensible heating, instability release,
or impinging weather system modification. An orographic precipitating system may form in various
ways, depending upon the ambient flow speed, vertical stability and structure of the wind, mountain
height, horizontal scale and geometry, and mesoscale
and larger (synoptic) environments. The mechanisms
of orographic precipitation for small hills and for
large mountains are quite different. Over large mountains, the highest rainfall is located on the windward
slope of the prevailing wind, while for small hills the
highest rainfall is on the hilltops. One revealing example of the rainfall contrast between the upslope
and lee slope is the annual precipitation over the Andes in South America, in which the rainfall maximum
is located on the eastern slope in northern Andes (to
the north of 30°S) and on the western slope in southern Andes.

The formation mechanisms of orographic rain may
be classified as the following types: (1) stable (not rising) air ascending (mechanically) over a large mountain and cooling, condensing, and precipitating;
(2) precipitating system triggered by convective instability (warmer air rising into a cooler air mass);
and (3) orographic rain over small hills by a seeder-
feeder mechanism. In a seeder-feeder mechanism,
precipitation from an upper-level (seeder) cloud falls
through a lower-level (feeder) cloud covering a hill,

where it picks up additional moisture, resulting in
more precipitation on the hill than would occur from
the seeder cloud alone (**Fig. 1**). The orographic convective precipitating system may be formed by
three different mechanisms: thermally initiated convection, orographically lifted convection, and leeside enhancement of convection.

Observational, theoretical, and numerical studies.
The following factors have been proposed to enhance orographic precipitation: (1) relative humidity

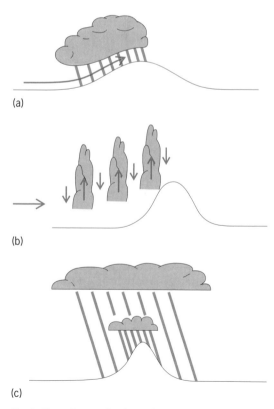

Fig. 1. Formation mechanisms of orographic precipitation.
(*a*) Stable ascent over large mountains. (*b*) Convective
precipitating system. (*c*) Orographic rain over small hills by
seeder-feeder mechanism. (*After C.-M. Chu and Y.-L. Lin,
J. Atmos. Sci., 57:3817–3837, 2000*)

and temperature of low-level air, (2) slope of the hill perpendicular to the wind direction, (3) strength of the wind component normal to the mountain, (4) depth of the feeder cloud, (5) precipitation rate from the feeder cloud, (6) rate of production of condensate in the feeder cloud, (7) cloud water content, and (8) rate of accretion or washout by the precipitation emanating from the seeder cloud.

Strong, low-level wind has a significant impact on the location of maximum rainfall associated with the seeder-feeder mechanism over small hills. Two factors important to wind drift of precipitation are (1) the horizontal drift of a raindrop L_{adv} [Eq. (1)] of

$$L_{adv} \approx Ub_c/V_r \qquad (1)$$

radius r falling at a speed V_r through a cloud of depth b_c by a wind speed U and (2) the scale of horizontal variation of liquid-water content q_c [Eq. (2)], where

$$L_c \approx q_c/(dq_c/dx) \approx L \qquad (2)$$

L is the half-width of the hill. If $L_c \leq L_{adv}$, such as over a narrow hill, wind drift will decrease the magnitude of maximum orographic enhancement because some of the seeder drops entering the feeder cloud summit would fall downwind in the region where both the feeder cloud droplets and seeder drops will evaporate.

With radar observation, it was found that equally vigorous upward and downward motion associated with convection are embedded in the precipitating orographic clouds.

The release of conditional instability (saturated air) and convective instability has been proposed to explain the formation of convective orographic precipitating systems. Occasionally the convection is associated with symmetric instability, that is, slantwise convection in an area of strong vertical shear. With more advanced and accurate observational instruments, such as Doppler radar, polarimetric radar, lidar, wind profiler, dropsonde, mesonet, research aircraft, and satellite, more field experiments have been proposed to investigate orographic precipitation.

A number of researchers have built up theoretical models to study the problem of moist airflow over mountains and apply their results to explain the formation and propagation of the orographic precipitating systems. The most straightforward approach for moisture effects is to assume that the latent heating is everywhere proportional to the vertical velocity [Eq. (3)], where w is the vertical velocity, q is the

$$q = \alpha w \qquad (3)$$

latent heating rate, and α is a constant. This particular parametrization of q in terms of w requires that the heat added to an air parcel lag the vertical displacement by a quarter cycle. It follows that the buoyancy forces can do no work and the heat can only modify, but not generate, internal gravity waves. Based on this theoretical approach, it has been found that (1) mountain waves are strengthened (weakened)

by low-level sensible cooling (heating) in a study of airflow over a two-dimensional mountain ridge when the basic wind is strong; (2) a convergence of the vertical momentum flux at the heating level tends to accelerate the flow there; (3) prescribed heating (cooling) tends to produce downward (upward) displacement, thus reducing (enhancing) the drag on the mountain; and (4) special treatment is required to avoid a net heating problem. Some important dynamical processes have been obtained by the above theoretical studies, leading to improved parametrization or representation of physical processes in numerical models. However, almost all of the theories have some constraints. In order to have a more realistic representation of the complicated orographic rain processes for improving the quantitative precipitation forecasting, numerical modeling appears to be the natural choice.

Compared to some approximations made in theoretical studies—such as small-amplitude (linear), two-dimensionality, idealized basic states, and simple representation of microphysical processes—and the coarse data used for observational analysis, numerical modeling studies provide an alternative, powerful tool for studying the dynamics of orographic precipitating processes. Improvements in computers and numerical techniques make possible detailed simulations of orographic precipitation. However, although some significant advancement in numerical modeling has been made in the last several decades, quantitative precipitation forecasting of orographic rain remains challenging.

Effects of synoptic and mesoscale environments. Severe orographic precipitation events are often associated with weather systems, such as troughs, fronts, storms, and tropical cyclones. Some common synoptic and mesoscale environments shared by heavy orographic precipitation were recognized as early as the 1940s. Based on the analysis of several severe orographic precipitation events in the United States, Europe (Alps), Taiwan, and Japan, the following common features are observed: (1) a conditionally or potentially unstable airstream impinging on the mountains, (2) the presence of a very moist and moderate to intense low-level jet [stream], (3) the presence of steep orography to help release the conditional or convective instability, and (4) the presence of a quasistationary synoptic scale system to impede or slow the progress of the orographically forced convective system over the threatened area. **Figure 2** shows the synoptic and mesoscale environments, conducive to heavy orographic precipitation, which could lead to flooding on the southern side of the Alps.

When a midlatitude cyclone or a frontal system approaches a mesoscale mountain, conditional or convective instability may be triggered due to orographic lifting of the lower layer of the incoming air. The warm, forward-sloping air mass ascending ahead of a front is often considered to be responsible for providing the low-level jet and moisture needed to generate heavy orographic rainfall. A differential advection mechanism has also been proposed to

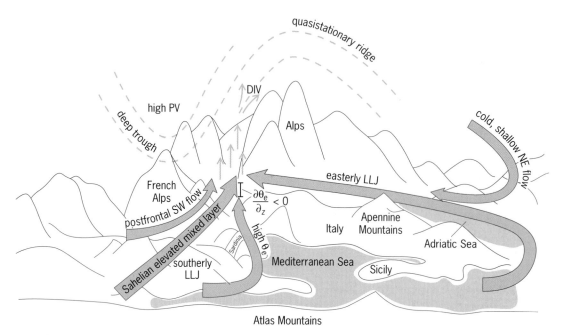

Fig. 2. Schematic for the synoptic and mesoscale environments conducive to Alpine orographic rain.

explain the orographic precipitation associated with deep convection during the passage of fronts.

Moist flow regimes and common ingredients. For a conditionally unstable airflow over a mesoscale mountain, three moist flow regimes may be identified: (1) upstream propagating convective system, (2) stationary convective system, and (3) both stationary and downstream propagating systems. For a three-dimensional flow, it is found that heavy upslope rainfall might be produced in the presence of a low-level jet under regimes 2 and 3, more in line with observations. In addition to the presence of a low-level jet, it appears that some other synoptic and mesoscale features may also play important roles in controlling the generation of heavy orographic rain. Flow regimes are also controlled by other parameters, such as the convective available potential energy (CAPE), steepness of the orography, and vertical wind shear.

Heavy orographic rainfall requires significant contributions from any combination of the following common ingredients: (1) high precipitation efficiency of the incoming airstream, (2) an intense low-level jet, (3) steep orography, (4) favorable (that is, concave) mountain geometry and a confluent flow field, (5) strong synoptically forced upward vertical motion, (6) high moisture flow upstream, (7) presence of a large, preexisting convective system, (8) slow (impeded) movement of the convective system, and (9) conditionally or convectively (potentially) unstable low-level flow.

For background information *see* AIR MASS; CLOUD PHYSICS; CONVECTIVE INSTABILITY; GEOPHYSICAL FLUID DYNAMICS; MESOMETEOROLOGY; PRECIPITATION (METEOROLOGY); RADAR METEOROLOGY; SATELLITE METEOROLOGY in the McGraw-Hill Encyclopedia of Science & Technology.   Yuh-Lang Lin

Bibliography. R. M. Banta, The role of mountain flows in making clouds, *Atmospheric Processes over Complex Terrain*, Meteorol. Monogr. 45, American Meteorological Society, pp. 229–283, 1990; T. Bergeron, Studies of the orogenic effect on the areal fine structure of rainfall distribution, *Meteorol. Inst. Uppsala U.* Rep. no. 6, 1968; P. Binder and C. Schär, MAP design proposal, *MeteoSwiss*, 1996; P. Bougeault et al., The MAP special observing period, *Bull. Amer. Meteorol. Soc.*, 82:433–462, 2001; R. R. Braham and M. Dragnis, Roots of orographic cumuli, *J. Meteorol.*, 16:214–226, 1960; K. A. Browning, Structure and mechanism of precipitation and the effect of orography in a wintertime warm sector, *Quart. J. Roy. Meteorol. Soc.*, 100:309–330, 1974; C.-M. Chu and Y.-L. Lin, Effects of orography on the generation and propagation of mesoscale convective systems in a two-dimensional conditionally unstable flow, *J. Atmos. Sci.*, 57:3817–3837, 2000; Y.-L. Lin et al., Some common ingredients for heavy orographic rainfall, Weath. Forecasting, 16:633–660, 2001; R. B. Smith, The influence of mountains on the atmosphere, *Adv. Geophys.*, 21:87–230, 1979; R. B. Smith et al., Local and remote effects of mountains on weather: Research needs and opportunities, *Bull. Amer. Meteorol. Soc.*, 78:877–892, 1997; R. B. Smith and Y.-L. Lin, The addition of heat to a stratified airstream with application to the dynamics of orographic rain, *Quart. J. Roy. Meteorol. Soc.*, 108:353–378, 1982.

Pentaquarks

Of the hundreds of known subatomic particles, many interact via the strong force. The theory of the strong force, known as quantum chromodynamics (QCD), predicts attraction between quarks. Also, quantum chromodynamics is responsible for the attraction between protons and the neutrons in the nucleus.

Until recently, all strongly interacting particles could be categorized into two groups, baryons and mesons: the baryons are made from three quarks, and mesons are made from a quark-antiquark pair. The theory of quantum chromodynamics forbids some combinations of quarks, such as a single free quark. Other combinations, such as the pentaquark, made from four quarks and one antiquark, are allowed by the rules of quantum chromodynamics. However, experiments in past decades failed to find pentaquarks, although the reason for this failure was a mystery.

Credible evidence for a pentaquark was published by several independent experimental groups in 2003. However, further evidence is necessary to establish pentaquarks as the first new type of "quark-matter" particle to be discovered in decades. If the properties of the pentaquark can be established, it will open a new door to theoretical models of how quarks can be combined into matter. See TETRAQUARKS.

Baryon structure. Before further discussion of pentaquarks, the three-quark baryons will be briefly reviewed. In the simplest description, the proton is a baryon made from two "up" quarks and one "down" quark. The terms up and down are often abbreviated by u and d. (The terms have no relationship with any direction in space—they are just traditional.) Baryons are distinguished by their quantum numbers. For example, the proton is made up of uud quarks, and the neutron has a udd structure. Each quark carries a baryon number of $+1/3$, so the proton and the neutron have a baryon number $B = 1$. Similarly, each antiquark carries a baryon number of $-1/3$.

Exotic and nonexotic pentaquarks. Pentaquarks theoretically come in two types, both of which have baryon number $B = 1$. First is the nonexotic variety, which has the same quantum numbers as a 3-quark baryon. The second, known as an exotic baryon, has a different type of antiquark than any of the quarks, and hence can be distinguished from 3-quark baryons. In order to form an exotic pentaquark, it is necessary to go beyond the ordinary world of u and d quarks and introduce the "strange" or s quark. Some known baryons contain s quarks, but a baryon with an s antiquark would be truly exotic. Experimentally, it is easy to count the number of s and anti-s quarks produced when particles collide by identifying the outgoing particles. So a pentaquark with an s antiquark, as was reported recently, can be distinguished based on its strangeness quantum number.

Analysis of scattering experiments. The measurement of new particles is accomplished by so-called scattering experiments. Subatomic particles are accelerated into beams and directed onto simple nuclei, such as protons or deuterons, at rest in the laboratory. The collisions produce new particles which travel outward, much as billiard balls do in a game of pool, carrying momentum and energy. The total energy is expressed in terms of the momentum p, the rest mass m_0, and the speed of light c by Einstein's equation, $E^2 = (pc)^2 + (m_0c^2)^2$, which makes it possible to calculate the mass of the particles.

Decades ago, experiments recorded the paths of particles by photographs of tracks left by ionization bubbles in liquid hydrogen. The analysis of these pictures was labor-intensive, but the results produced exciting discoveries of new particles. Nowadays, the tracks of scattered particles are recorded electronically and processed by computers. This has enabled experimental physicists to investigate huge volumes of data for rare scattering processes that could not be seen before. This is one of several reasons why the pentaquark could have eluded earlier experiments.

Theoretical prediction. Part of the art of making new discoveries comes from knowing where to look. At the heart of the scientific method is the interplay between experiment and theory, where hypotheses are tested and new measurements lead to advances in theoretical models. In 1997, a theoretical prediction by Dmitri Diakonov, Victor Petrov, and Maxim Polyakov suggested that a pentaquark might exist with a mass of about 1.65 times the proton mass. Furthermore, they predicted that it would have a long lifetime (as compared to a typical baryon of similar mass) before it decayed, making it easier to measure. These predictions arose from a particular theory of the structure of baryons known as the chiral soliton model. This mathematical model had been ignored by many theoretical physicists because of the significant approximations involved. Still, it gave a definite prediction whereby experimenters could localize their pentaquark search and test the theory.

Use of archival data. By 1990, most experimental physicists had given up the search for pentaquarks because of the lack of evidence in earlier searches. Other experiments that were considered more likely to succeed than pentaquark searches were being done. Fortunately, however, some of the experiments that had been done for other reasons could also be used for the pentaquark search. The data were available and just needed to be processed (or analyzed) in a slightly different way in order to look for the predicted pentaquark particles. One such experiment was at the SPring-8 facility in Japan.

SPring-8 experiment. The SPring-8 facility is an electron storage ring where the electrons are accelerated to a high energy of 8 gigaelectronvolts (GeV) and then circulate continuously in a ring of magnets. This facility is used primarily to produce strong x-ray beams for experiments in biology, chemistry, and condensed-matter physics. One beamline is dedicated to nuclear physics, where a laser scatters from the electron beam producing a powerful beam of high-energy photons up to 2.4 GeV in energy. These photons are directed at nuclear targets such as liquid hydrogen or carbon. After the photons strike the nucleus, new particles are produced such as K mesons, or kaons.

The negatively charged kaons contain an s quark and the positively charged kaons contain an s

antiquark, which is ideal for a pentaquark search because the pentaquark was predicted to decay into, say, a neutron and a positive kaon. These particles were in the data from the SPring-8 experiment. Diakonov convinced Takashi Nakano, the leader of the nuclear physics experiment at SPring-8, to look in the Japanese data for evidence of the pentaquark. Although the prospects for a successful outcome appeared doubtful at the outset, Nakano's group found evidence in favor of the pentaquark with a mass near the value predicted by theory.

The evidence from the SPring-8 experiment is in the form of an excess of scattering events producing kaons with momentum and energy consistent with that expected for a new particle with the strangeness and baryon quantum numbers of the pentaquark. When each scattering event was plotted as one count in a mass spectrum, the SPring-8 experiment found 36 counts in the predicted mass region, where only 17 were expected from background sources. The statistical accuracy of this measurement is not overwhelming, but estimates (from gaussian statistics) of the likelihood that it is a statistical fluke give less than 0.01%.

Other experiments. Supporting evidence quickly followed from other experiments once the results of the Japanese experiment were announced. Old data were reanalyzed by physicists from Russia, where a kaon beam was directed at a liquid xenon target, and their results are consistent with the SPring-8 pentaquark. At the Thomas Jefferson National Accelerator Facility (Jefferson Lab) in the United States, data for a photon beam on a deuterium target gave results with better statistical significance, this time using a more convincing data analysis procedure. A similar experiment by the same group, with a photon beam incident on protons, gave evidence with the best statistical significance of any measurement yet.

By the spring of 2004, eight independent experiments from laboratories in Japan, Russia, the United States, and Germany had published statistically significant evidence for the pentaquark. It has since been given the name Θ^+ (theta). [Capital Greek symbols are typically used to name baryons, and the plus sign indicates a positive charge.] However, confirmation of discovery requires stronger evidence, and experiments to gather better statistics are needed.

Jefferson Lab has been actively pursuing a high-statistics measurement of the pentaquark, with photon beams on both deuteron and proton targets. The analysis of these high-statistics data will provide much more information about the detailed properties of the pentaquark, if it truly exists, such as its spin quantum number from the angular distribution of its decay. In turn, theorists will use this information to develop models of how quarks interact to form new particles.

The quest to understand the nature of matter depends, in part, on the understanding of interactions between quarks. The pentaquark provides a new testing ground for theoretical models of quark matter.

For background information *see* BARYON; ELEMENTARY PARTICLE; MESON; PARTICLE ACCELERATOR; PARTICLE DETECTOR; QUANTUM CHROMODYNAMICS; QUARKS; STATISTICS; STRANGE PARTICLES in the McGraw-Hill Encyclopedia of Science & Technology.
                                            Kenneth H. Hicks

Bibliography. F. E. Close, *The Cosmic Onion: Quarks and the Nature of the Universe*, Heinemann Educational Books, 1983; G. D. Coughlan and J. E. Dodd, *The Ideas of Particle Physics: An Introduction for Scientists*, 2d ed., Cambridge University Press, 1991; G. Fraser (ed.), *The Particle Century*, Institute of Physics Publishing, 1998; M. Riordan, *The Hunting of the Quark: A True Story of Modern Physics*, Touchstone Books, 1987.

Persistent, bioaccumulative, and toxic pollutants

Persistent, bioaccumulative, and toxic pollutants (PBTs) are long-lasting substances that are able to build up in the food chain; they are known to cause, or strongly suspected of causing, adverse health effects and ecosystem damage. Numerous PBTs have been associated with effects on the nervous system, endocrine dysfunction, reproductive and developmental problems, cancer, and genetic abnormalities. PBTs can travel long distances, moving among environmental compartments (that is, biota, air, water, and land) and can persist for decades. For example, polychlorinated biphenyls (PCBs) are still found in sediment, soil, water, and organic tissues several decades after they were banned.

PBTs may be organic, inorganic, or organometallic and other metallic complexes (**Fig. 1**). Lead (Pb), mercury (Hg), and their compounds comprise the greatest mass of PBTs released into the United States environment (**Fig. 2**). This is due to the large volume and surface areas involved in metal extraction and refining operations. However, the organic PBTs are important because they are highly toxic, with congeners that are carcinogenic, endocrine-disruptive, and neurotoxic. For example, polycyclic aromatic hydrocarbons (PAHs) are a family of compounds containing two or more fused benzene rings. Their structure renders them highly hydrophobic and difficult for an organism to eliminate. In addition, such structures are able to insert into deoxyribonucleic acid (DNA) and interfere with transcription and replication. Some of the physical and chemical characteristics of PBTs are shown in the **table**.

Persistence. The most widely used metric for chemical persistence is the half-life, the time it takes to degrade one-half the mass of a given chemical. The U.S. Environmental Protection Agency considers a compound persistent if its half-life in water, soil, or sediment is greater than 60 days, and very persistent if its half-life is greater than 180 days. In air, a compound is considered persistent if its half-life is greater than 2 days.

Partitioning. Another factor in determining a substance's persistence is its ability to move in the environment. If a substance is likely to leave the water, it is not persistent in water. However, if the compound moves from the water to the sediment, where it persists for long periods, it is considered environmentally persistent. Such movement among phases and environmental compartments is known as partitioning. PBTs are usually semivolatile, with vapor pressures of 10^{-5} to 10^{-2} kPa (10^{-8} to 10^{-5} atm) at 20°C (68°F) and atmospheric pressure (101 kPa). The low vapor pressure and limited water solubility of most PBTs means they will have low fugacity (fugacity refers to the tendency of a chemical to move from a compartment or phase).

Henry's law states that the concentration of a dissolved gas is directly proportional to the partial pressure of that gas above the solution [Eq. (1)], where p_a is the partial pressure of the gas, K_H is Henry's law constant, and $[c]$ is the molar concentration of the gas; or where C_W is the concentration of the gas in water [Eq. (2)].

$$p_a = K_H[c] \qquad (1)$$

$$p_a = K_H C_W \qquad (2)$$

Henry's law is an expression of the proportionality between the concentration of a dissolved compound and its partial pressure in the atmosphere at equilibrium. In air and water per Eq. (3), K_{AW} is the air/water partitioning coefficient, where C_A is the concentration of a gas in the air.

$$K_{AW} = C_A/C_W \qquad (3)$$

The relationship between the air/water partition coefficient and Henry's law constant for a substance is Eq. (4), where R is the gas constant (8.21×10^{-2} L atm mol^{-1} K^{-1}) and T is the temperature on the

$$K_{AW} = K_H/RT \qquad (4)$$

Kelvin scale. Under environmental conditions, most PBTs have very low K_H values.

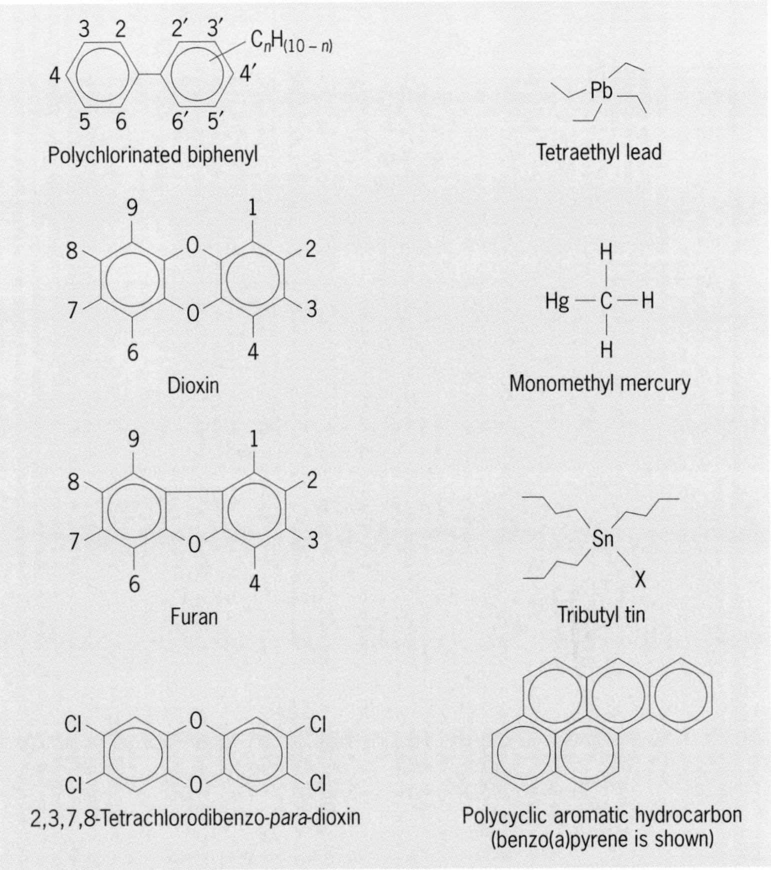

Fig. 1. Some PBT structures.

Another expression of partitioning, the octanol-water coefficient K_{ow}, indicates a compound's likelihood to exist in the organic versus aqueous phase. If a PBT is dissolved in water and the water encounters an organic solvent (such as octanol), the PBT will have a tendency to move from water to the organic solvent. Its K_{ow} reflects how much of the substance will move until the aqueous and organic

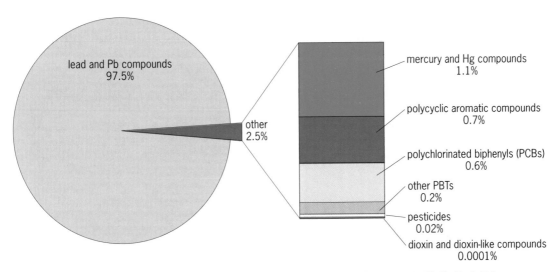

Fig. 2. Total releases of persistent bioaccumulating toxic substances (PBTs) in 2001, as reported in the Toxic Release Inventory (TRI). Total releases = 206 million kilograms (227,000 tons). (*U.S. Environmental Protection Agency*)

Physicochemical properties affecting the persistence, bioaccumulation, and toxicity of environmental contaminants*

Property of substance or environment	Chemical importance	Physical importance
Molecular weight	Contaminants with MW > 600 may not be bioavailable because they are too large to pass through membranes.	Heavier molecules have lower vapor pressures, so are less likely to exist in the gas phase and more likely to remain sorbed to soil and sediment particles.
Chemical bonding	Ring structures are generally more stable than chains. Double and triple bonds add persistence compared to single-bonded molecules.	Large, aromatic compounds have affinity for lipids in soil and sediment. Solubility in water is enhanced by the presence of polar groups in the structure. Sorption is affected by presence of functional groups and ionization potential.
Stereochemistry	Neutral molecules with cross-sectional dimensions greater than 9.5 angstroms have been considered to be "sterically hindered" in their ability to penetrate the polar surfaces of the cell membranes.	Lipophilicity (solubility in fats) of neutral molecules generally increases with molecular mass, volume, or surface area. Solubility and transport across biological membranes are affected by a molecule's size and shape. Molecules that are planar, such as polycyclic aromatic hydrocarbons, dioxins, or certain forms of polychlorinated biphenyls, are generally more lipophilic than are globular molecules of similar molecular weight.
Solubility	Lipophilic compounds may be very difficult to remove from particles and may require highly destructive remediation techniques (such as combustion). Insoluble forms may precipitate out of the water column or be sorbed to particles.	Hydrophilic compounds are more likely to exist in surface water and in solution in interstices of pore water of soil, vadose zone, and aquifers underground. Lipophilic compounds are more likely to exist in organic matter of soil and sediment.
Co-solvation	If a compound is hydrophobic and nonpolar but is easily dissolved in acetone or methanol, it can still be found in water because these organic solvents are highly miscible in water.	This is an important mechanism for getting a highly lipophilic and hydrophobic compound into water, where the compound can then move by advection, dispersion, and diffusion. PBTs like PCBs and dioxins may be transported as co-solutes in water by this means.
Vapor pressure or volatility	Volatile organic compounds (VOCs) exist almost entirely in the gas phase since their vapor pressures in the environment are usually greater than 10^{-2} kPa, while semivolatile organic compounds (SVOCs) have vapor pressures between 10^{-2} and 10^{-5} kPa, and nonvolatile organic compounds (NVOCs) have vapor pressures less than 10^{-5} kPa.	Higher vapor pressures mean larger fluxes from the soil and water to the atmosphere. Lower vapor pressures cause chemicals to have a greater affinity for the aerosol phase.
Fugacity	This is often expressed as Henry's law constant (K_H), the vapor pressure of the chemical divided by its solubility of water. High-fugacity compounds are likely candidates for remediation using the air (for example, pump-and-treat and air-stripping).	Compounds with high fugacity have a greater affinity for the gas phase and are more likely to be transported in the atmosphere than those with low fugacity.
Octanol-water coefficient (K_{ow})	Substances with high K_{ow} values are more likely to be found in the organic phase of soil and sediment complexes than in the aqueous phase. They may also be more likely to accumulate in organic tissue.	Transport of substances with higher K_{ow} values is more likely to be on particles (aerosols in the atmosphere are sorbed to fugitive soil and sediment particles in water), rather than in water solutions.
Sorption	Adsorption (onto surfaces) dominates in soils and sediments low in organic carbon (solutes precipitate onto soil surface). Absorption (three-dimensional sorption) is important in soils and sediments high in organic carbon.	Strong sorption constants indicate that soil and sediment may need to be treated in place. Phase distributions favoring the gas phase indicate that contaminants may be off-gassed and treated in their vapor phase.
Substitution, addition, and elimination	Dehalogenation of organic compounds by anaerobic treatment processes often renders them much less toxic. Adding or substituting a functional group can make the compound more or less toxic. Phase 1 metabolism by organisms uses hydrolysis and redox reactions to break down complex molecules at the cellular level.	Dehalogenation can change an organic compound from a liquid to a gas or change its affinity from one medium (such as air, soil, and water) to another. New species produced by hydrolysis are more polar and, thus, more hydrophilic than their parent compounds, so they are more likely to be found in the water column.
Dissociation	Molecules may break down by hydrolysis, acid-base reactions, phytolysis, dissociation of complexes, and nucleophilic substitution.	Dissociation of compounds may occur via acid-base equilibria among hydroxyl ions and protons and weak and strong acids and bases. Dissociation may also occur by photolysis "directly" by the molecules absorbing light energy, and "indirectly" by energy or electrons transferred from another molecule that has been broken down photolytically.
Reduction-oxidation	Toxic organic compounds can be broken down to CO_2 and H_2O by oxidation processes, including the reagents ozone, hydrogen peroxide, and molecular oxygen (that is, aeration). Reduction is also used in treatment processes. For example, hexavalent chromium is reduced to the less toxic trivalent form in the presence of ferrous sulfate: $2CrO_3 + 6FeSO_4 \rightarrow 3Fe_2(SO_4)_3 + Cr_2(SO_4)_3 + 6H_2O$.	Reductions and oxidations (redox reactions) occur in the environment, leading to chemical speciation of parent compounds into more or less mobile species. For example elemental or divalent mercury is reduced to the toxic species mono- and dimethylmercury in sediment and soil low in free oxygen. The methylated metal species have greater affinity than the inorganic species for animal tissue.
Diffusion	A compound may move by diffusion from one compartment to another (for example, from the water to the soil particle) depending on its concentration gradient.	Diffusion is a very slow process in most environmental systems. However, in rather quiescent systems ($< 2.5 \times 10^{-4}$ cm s^{-1}), such as aquifers and deep sediments, the process can be very important.
Isomerization	Isomers of toxins have identical molecular formulas but differ in atomic connectivity (including bond multiplicity) or spatial arrangement.	The fate and transport of chemicals can vary significantly, depending upon the isomeric form. For example, the rates of degradation of left-handed chiral compounds (mirror images) are often more rapid than for right-handed compounds.
Biotransformation	Many of the processes discussed in this table can occur in or be catalyzed by microbes. For example, many fungi and bacteria reduce compounds to simpler species to obtain energy. Biodegradation is possible for almost any organic compound, although it is more difficult in very large molecules, insoluble species, and completely halogenated compounds.	Microbial processes will transform parent compounds into species that have their own transport properties. Under aerobic conditions, the compounds can become more water-soluble and are transported more readily than their parent compounds in surface and ground water.

*These properties exist at environmental conditions in ambient temperatures near 20°C (68°F) and atmospheric pressures around 101 kPa (1 atm).
SOURCE: Adapted from D.A. Vallero, *Engineering the Risks of Hazardous Wastes*, Butterworth-Heinemann, Boston, 2003.

solvents (phases) reach equilibrium. For example, if at a given temperature and pressure a chemical's concentration in octanol is 100 mg/L and its concentration in water is 1000 mg/L, its K_{ow} is 100/1000 or 0.1.

Since the range is so large among various environmental contaminants, it is common practice to express log K_{ow} values. For example, in a spill of equal amounts of the polychlorinated biphenyl decachlorobiphenyl (log K_{ow} of 8.23) and the pesticide chlordane (log K_{ow} of 2.78), the decachlorobiphenyl has much greater affinity (more than five orders of magnitude) for the organic phases than does the chlordane (**Fig. 3**). This does not mean that a greater amount of either compound is likely to stay in the water column, since both compounds are hydrophobic, but it does mean that the time and mass of each contaminant moving between phases will vary. The kinetics is different, so the time it takes for two compounds with differing K_{ow} values to reach equilibrium will differ.

Compounds with low K_H and K_{AW} values can be transported long distances in the atmosphere when sorbed to particles. Fine particles can behave as colloids and stay suspended for extended periods, explaining in part why low-K_H PBTs are found in remote locations (such as in the Arctic regions) relative to their sources.

Sorption and K_{oc}. Sorption is an important predictor of a chemical's persistence. If the substrate has sufficient sorption sites, as do many clays and organic matter, the PBT may become tightly bound and persistent. Another frequently reported liquid-to-solid-phase partitioning coefficient is the organic-carbon-partitioning coefficient K_{oc}, which is the ratio of the contaminant concentration sorbed to organic matter in the matrix (soil or sediment) to the contaminant concentration in the aqueous phase. Thus, the K_{oc} is derived from the quotient of a contaminant's soil partition coefficient K_d and the fraction of organic matter (OM) in the matrix [Eq. (5)]. The soil

$$K_{oc} = K_d/\text{OM} \qquad (5)$$

partition coefficient K_d is the experimentally derived ratio of a contaminant's concentration in a solid matrix to the contaminant concentration in the liquid phase at chemical equilibriium.

PBTs are expected to be strongly sorbed, but K_{oc} varies from substrate to substrate. **Figure 4** shows that the log K_{oc} values for the PAH (phenanthrene) can range nearly three orders of magnitude, depending on the substrate. The figure shows some promising materials for collecting and treating PBTs. For example, granulated activated carbon has a strong affinity to PAHs.

Bioaccumulation. Physicochemical characteristics of the contaminant affect its bioaccumulative nature. If a substance is likely to be sorbed to organic matter (that is, high K_{oc} value), it will a have affinity for tissues. If a substance partitions from the aqueous phase to the organic phase (that is, high K_{ow}

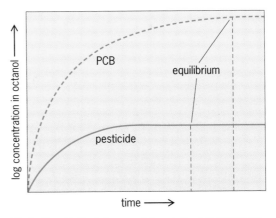

Fig. 3. Hypothetical diagram of the relative concentrations of a polychlorinated biphenyl (log $K_{ow} = 8$) and an orgaochlorine pesticide (log $K_{ow} = 3$) in octanol with time. In this instance, the PCB is more soluble than the organochlorine pesticide in octanol. This is an indication that the PCB's affinity for the organic phase is greater than that of organochlorine pesticide.

value), it is likely to be stored in fats of organisms at higher trophic levels (such as carnivores and omnivores). The bioconcentration factor (BCF) is the ratio of the concentration of the substance in a specific genus to the exposure concentration. The exposure concentration is the concentration in the environmental compartment (usually surface water). The bioaccumulation factor (BAF) is based on the uptake of the substance by an organism from water and food. The BCF is based on the direct uptake from the water only. A BCF of 500 means that an organism takes up and sequesters a contaminant to concentrations greater than 500 times the exposure concentration. Generally, any substance that has a BAF or BCF greater than 5000 is considered highly bioaccumulative, although the cutoff point can vary.

Genera will vary considerably in reported BCF values, and the same species will bioaccumulate

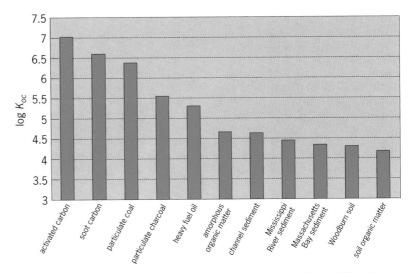

Fig. 4. Comparison of reported values of K_{oc} for phenanthrene sorption on different types of organic matter found in soils and sediments. (*From S. W. McNamara, R. G. Luthy, and D. A. Dzombak, Bioavailability and Biostabilization of PCBs in Soils: NCER Assistance Agreement Final Report, U.S. Environmental Protection Agency, Grant No. R825365-01-0, 2002*)

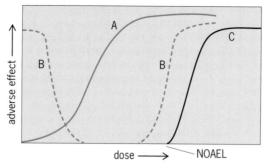

Fig. 5. Three prototypical dose-response curves. Curve A represents the no-threshold curve, which expects a response (for example, cancer) even if exposed to a single molecule (this is the most conservative curve). Curve B represents the essential-nutrient dose-response relationship, and includes essential metals, such as trivalent chromium or selenium, where an organism is harmed at the low dose due to a "deficiency" (left side) and at the high dose due to "toxicity" (right side). Curve C represents toxicity above a certain threshold (noncancer). This threshold curve expects a dose at the low end where no disease is present. Just below this threshold is the "no observed adverse effect level," or NOAEL. (*After D. Vallero, Environmental Contaminants: Assessment and Control, Academic Press, Burlington, MA, 2004*)

different compounds at various rates. In an organism, the amount of contaminant bioaccumulated generally increases with size, age, and fat content, and decreases with increasing growth rate and efficiency. Bioaccumulation also is often higher in males than females and in organisms that are proficient in storing water. Top predators often have elevated concentrations of PBTs.

The propensity of a substance to bioaccumulate is usually inversely proportional to its aqueous solubility, since hydrophilic compounds are usually more easily eliminated by metabolic processes. In most

cases, the larger the molecule, the more lipophilic it is. Generally, compounds with log $K_{ow} > 4$ can be expected to bioaccumulate. However, this is not always the case. Some very large and lipophilic compounds (log $K_{ow} > 7$) have surprisingly low rates of bioaccumulation. For example, very large molecules (with cross-sectional dimensions greater than 9.5 angstroms and molecular weights greater than 600) are often too big to pass through organic membranes.

Toxicity. The toxicity criteria for PBTs include acute and chronic effects. These criteria can be quantitative. For example, a manufacturer of a new chemical may have to show that there are no toxic effects in fish exposed to concentrations below 10 mg/L. If fish show effects at 9 mg/L or lower, the new chemical would be considered toxic.

Dose-response curve. A contaminant is acutely toxic if it can cause damage with only a few doses. Chronic toxicity occurs when a person or ecosystem is exposed to a contaminant over a protracted period, with repeated exposures. The essential indication of toxicity is the dose-response curve. The three curves in **Fig. 5** represent those generally found for toxic chemicals. The curves are sigmoidal because toxicity is often concentration-dependent. As the doses increase, the response cannot mathematically stay linear (that is, the toxic effect cannot double with each doubling of the dose). Instead, the toxic effect will continue to increase, but at a decreasing rate (slope).

Curve A is the classic cancer dose-response curve, in which any amount of exposure to a cancer-causing agent may result in cancer at the cellular level. There is no safe level of exposure. Thus, the curve intercepts the x axis at 0. Metals and nutrients can be toxic at high levels, but are essential to the development and metabolism of organisms. Curve B represents a compound that will cause dysfunction at low levels (below the minimum intake needed for growth and metabolism) and toxicity at high levels. Curve B is a composite of at least two curves where the segment that runs along the x axis is the "optimal range" between deficiency and toxicity. Curve C is the classic noncancer dose-response curve. The steepness of the three curves represents the potency or severity of the toxicity. For example, curve C is steeper than curve A, so the adverse outcome (disease) caused by the chemical in curve C is more potent than that of the chemical in curve A.

In animal studies, uncertainties in dose response may be due to high doses and short durations (at least compared to a human lifetime). When environmental exposures do not fall within the observed range, extrapolations must be made to establish a dose relationship.

Bioaccumulation dynamics. Following uptake and entry into an organism, a contaminant will move and change. Toxico-kinetic models predict the dynamics of uptake, distribution, and elimination of contaminants in organisms. In the kinetic phase, a contaminant may be absorbed, metabolized, stored

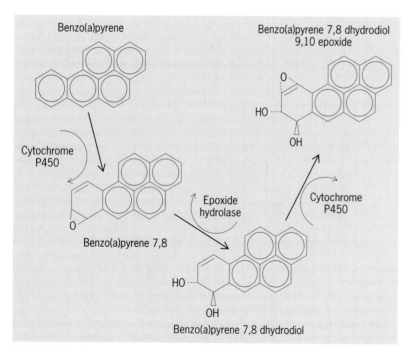

Fig. 6. Biological activation of benzo(a)pyrene to form the carcinogenic active metabolite benzo(a)pyrene 7,8 dihydrodiol 9,10 epoxide.

temporarily, distributed, and excreted. The amount of contaminant that has been absorbed in the form of the original contaminant is known as the active parent compound. If the parent contaminant is metabolized, either it will become less toxic and will be excreted as a detoxified metabolite, or it will become a more toxic active metabolite. In the dynamic phase, the contaminant or its metabolite undergoes interactions with body cells, causing a toxic response. The PBT body burden can be a reliable indicator of time-integrated exposure and absorbed dose, because a PBT is usually easily absorbed, making its kinetic phase straightforward.

Metabolites. Since PBTs are generally lipophilic, their toxic effects increase as they build up in the organism, so the organism needs mechanisms to eliminate them. To transfer a contaminant to bodily fluids, which are predominantly water, the molecule must be transformed into a more hydrophilic metabolite by the attachment of polar groups. Most of these reactions are catalyzed in the endoplasmic reticulum of the cells. Reactions where new polar groups are added include oxidation, reduction, hydrolysis, and hydration. Removing halogens, such as in the metabolism of chlorobenzene, can be quite difficult since they lend so much stability to the compound.

A PBT may become less toxic in the body, in which case the new compound is known as the detoxified metabolite, which is easier to excrete than the parent contaminant. The second possibility is that the PBT becomes biologically activated to a more toxic form. A dramatic example is the epoxidation of benzo(a)pyrene. Scientists hypothesize that the carcinogenic active metabolite is formed by three enzyme-catalyzed reactions (**Fig. 6**). First, benzo(a)pyrene is epoxidized to benzo(a)pyrene 7,8 epoxide via cytochrome P450 catalysis. Later, another epoxide is formed, this time at the 9,10 position, producing the actively carcinogenic metabolite, (+(anti))benzo(a)pyrene 7,8 dihydrodiol 9,10 epoxide.

Many countries are addressing these and several other PBTs, including reducing the exposures and lessening risks, reducing sources, screening new chemicals, and developing alternative chemicals and applying green chemistry to develop alternative chemicals that are less persistent, less bioaccumulative, and less toxic.

[Disclaimer: The U.S. Environmental Protection Agency through its Office of Research and Development funded and managed some of the research described here. The present article has been subjected to the Agency's administrative review and approved for publication.]

For background information *see* ENVIRONMENTAL ENGINEERING; ENVIRONMENTAL TOXICOLOGY; HAZARDOUS WASTE; MUTAGENS AND CARCINOGENS; ORGANIC CHEMISTRY; POLYCHLORINATED BIPHENYLS; TOXICOLOGY; TROPHIC ECOLOGY in the McGraw-Hill Encyclopedia of Science & Technology.   Daniel Vallero

Bibliography. Agency for Toxic Substances and Disease Registry, *Toxicological Profile for Lead: Final Report*, 1993; A. Baccarelli et al., Immunologic effects of dioxin: New results from Seveso and comparison with other studies, *Environ. Health Perspect.*, 110:1169–73, 2002; J. Duffus and H. Worth, The science of chemical safety: Essential toxicology–4: Hazard and risk, *IUPAC Educators' Resource Mater.*, IUPAC, 2001; Environment Canada, *Envirofacts: Toxic Chemicals in Atlantic Canada—Lead*, EN 40-226/1-1995, 1995; B. Hellman, *Basis for an Occupational Health Standard: Chlorobenzene*, National Institute for Occupational Safety and Health, DHHS (NIOSH) Pub. No. 93-102, 1993; H. F. Hemond and E. J. Fechner-Levy, *Chemical Fate and Transport in the Environment*, Academic Press, San Diego, 2000; S. S. Suthersan, *Remediation Engineering: Design Concepts*, pp. 143-144, CRC Press, Boca Raton, 1997; U.S. Environmental Protection Agency, Office of Pesticide Programs, *List of Chemicals Evaluated for Carcinogenic Potential*, Washington, DC, 1999; U.S. Environmental Protection Agency, *Workshop Report on Developmental Neurotoxic Effects Associated with Exposure to PCBs*, EPA/630/R-02/004, Research Triangle Park, NC, 1993.

Pharmaceutical crops

Due to advances in molecular biology, it is now commonplace for scientists to transfer genes from one species to another, with the intention of designing novel organisms with attributes that can benefit humans. This new technology has been most successfully applied to plants. In fact, over the last two decades, scores of different crop species have been genetically engineered, meaning that the crops have been altered by the introduction of foreign genes. To date, the genetically engineered crops that have been approved for commercial production in the United States have been designed primarily to make farming easier and more cost-effective. For example, bacterial genes have been transferred into corn and cotton to produce insect-resistant varieties.

In the last few years, crop genetic engineering has entered a new realm. Biologists from both industry and academia have begun looking at genetic engineering of crops as a possible means of quickly and cheaply producing large quantities of drugs and vaccines. The production of pharmaceuticals in genetically engineered crops is often referred to as pharming.

Pharming overview. The process of pharming typically involves three steps: first, scientists identify particular genes that code for pharmaceutically active proteins, isolate those genes, and incorporate them into the genetic material of the crop. These transferred genes (or transgenes) can potentially come from a different plant species, an animal (even from a human), a bacterium, or any other organism. Second, farmers grow these genetically engineered crops, and the plants themselves act as minifactories, synthesizing the pharmaceutical proteins. Finally, the crop is harvested, and in most cases the drug must be extracted and purified before it can be distributed commercially. In a few instances, crops are actually being engineered so that the vaccine can be delivered simply by having someone eat the food itself, without the cost and inconvenience of extracting the drugs and delivering them via pills or injections.

Some of the pharmaceuticals targeted for pharming have previously been produced by cultures of transgenic animal, bacterial, or yeast cells in large vats, but using plants to produce the drugs could mean higher drug yields. Higher yields are needed because, for some drugs, current production methods cannot keep up with growing demand. Moreover, faster and less expensive production could lead to reduced prices for consumers of these drugs. However, it is unclear whether drug prices would actually decline, because the research and development of pharmaceutical crops will remain very expensive.

Research and development. Approximately 400 different types of pharmaceutical crops are currently in the research and development stage. These crops have been genetically engineered to produce a remarkable variety of enzymes, hormones, vaccines, and antibodies. Potential products of pharming include blood thinners, coagulants, insulin, growth hormones, cancer treatments, and contraceptives. One company is developing genetically modified corn to produce lipase, a digestive enzyme used to treat patients with cystic fibrosis. Another company has developed genetically engineered rice that produces lactoferrin and lysozyme, two compounds used to treat severe diarrhea in infants. Research has been done toward developing plant-derived vaccines for hepatitis B, cholera, rabies, human immunodeficiency virus (HIV), malaria, and influenza, among others. In this case, the transgenes cause the plant to produce an antigen that will induce production of antibodies in people who consume or are injected with the plant-derived antigens. Researchers have focused on the banana as a possible mechanism for oral delivery of vaccines because bananas are widely available in developing nations and can be eaten raw, which avoids the denaturing of proteins caused by cooking. If drugs are to be delivered directly by food products, it is crucial that the food remain uncooked.

Field and clinical trials. Drugs produced by pharmaceutical crops have not yet appeared on the U.S. market, but several are currently making their way through field and clinical trials. In 2002, 130 acres of pharmaceutical corn were cultivated in the United States in field trials. Two-thirds of all pharmaceutical plantings in the United States are corn, but soy, rice, potatoes, alfalfa, wheat, tobacco, and other crops are also being used for transgenic drug production. The first drugs derived from pharmaceutical crops could be on the market in the next 2–3 years.

Risks. Intense public anxiety exists regarding the development and use of pharmaceutical crops. Although earlier methods of pharmaceutical production often involved cultures of genetically engineered cells, these cells were kept under strict confinement in laboratories. Pharming, on the other hand, is done outdoors, where it will be impossible to ensure 100% containment of the transgenes and the crops.

Contamination of food supply. One of the biggest concerns is that the public might someday find unwanted medicines showing up in the general food supply. Contamination of food supplies can occur due to a breakdown in containment of the transgenes or due to the accidental commingling of pharmaceutical foods, intended only for medicinal uses, with nonpharmaceutical varieties headed for dinner tables. Genes coding for pharmaceutical products can escape when pollen from pharmaceutical crops drifts and fertilizes fields of nonpharmaceutical crops. Additionally, seeds from pharmaceutical crops could be dispersed outside their intended locations. Finally, if seeds are left behind in the soil, "volunteer" pharmaceutical plants can establish themselves in the years following the initial plantings, possibly in mixture with nonpharmaceutical crops.

Despite limited experience with pharmaceutical crops, there has already been a dramatic example of transgene "escape" involving field trials of pharmaceutical corn. In November 2002, the U.S. Department of Agriculture discovered that a corporation had failed to comply with federal regulations in two field trials, which were conducted in Nebraska and Iowa, to test corn genetically modified to produce a pharmaceutical protein. In both locations, the company failed to destroy volunteer corn plants in the subsequent growing season. In Nebraska, the volunteer corn had been shredded and mixed among soybeans at a grain elevator, necessitating the destruction of 500,000 bushels of soybeans. In Iowa, 155 acres of corn surrounding a test site were destroyed because of possible contamination via pollen from volunteer plants. The corporation was fined $250,000 for these violations, in addition to the costs of the cleanup.

The possible escape of pharmaceutical products from these engineered crops is of great concern, and an editorial in a 2004 issue of the journal *Nature Biotechnology* offered two suggestions that could help industry to avoid the foreseeable problems. First, geographic isolation can be used to reduce the chances of contamination of the general food supply. For example, pharmaceutical crops might be cultivated on islands from which the crop is otherwise absent. Second the editors recommend that food crops should not be used for the production of pharmaceutical proteins. For example, tobacco might be a wiser choice of species to target for these activities. Industry has not yet embraced this second risk-management recommendation because protein yields from seeds are much higher than from vegetative parts, and tobacco produces very little seed. Industry has instead favored the production of "pharm-grain" and "pharm-oil seed" crops from which large quantities of protein can be more easily recovered.

Detrimental effects on wildlife and soils. Even if all crops engineered to produce drugs were nonfood crops, there could still be risks. For example, no matter where pharmaceutical crops are grown or in what plant species they are produced, there will still exist the possibility that birds, insects, and other wildlife may consume these crops. Another concern is how the cultivation of pharmaceutical crops may affect soils and the diverse species that live in and improve soils. The possible impacts of pharmaceutical crops on wildlife and soils have received scarce attention, but surely will vary greatly from one variety to the next, depending on what drug is being produced by the plant. These are very serious concerns because many of the drugs being worked on are effective at very low doses and can be toxic at higher concentrations. In addition, some of the compounds are highly persistent in the environment or can bioaccumulate, meaning that they could become more concentrated in organisms higher up on the food chain. One possible strategy to avoid these problems would be to engineer drugs in such a way that they do not become biologically active until after they are harvested and purified.

Outlook. Like many new technologies, the genetic engineering of crops to produce pharmaceutical products has great promise. Bananas that could cheaply produce vaccines and be easily delivered to children throughout the tropics seem to be a wonderful invention. But there are downsides to the technology as well—avoiding contamination of food and poisoning of wildlife will be difficult if drugs are widely produced in food crops grown outdoors. The future course of this technology will require thoughtful inputs from ecologists, public health experts, and medical researchers, as well as those who genetically engineer the crops in the first place.

For background information *see* AGRICULTURAL SCIENCE (PLANT); BIOTECHNOLOGY; BREEDING (PLANT); GENETIC ENGINEERING; PHARMACOLOGY in the McGraw-Hill Encyclopedia of Science & Technology.   Michelle Marvier

Bibliography. Editors of Nature Biotechnology, Drugs in crops—the unpalatable truth, *Nat. Biotechnol.* 22:133, 2004; G. Giddings et al., Transgenic plants as factories for biopharmaceuticals, *Nat. Biotechnol.*, 18:1151–1155, 2000; J. K.-C. Ma, Genes, greens and vaccines, *Nat. Biotechnol.*, 18:1141–1142, 2000; Pew Initiative on Food and Biotechnology, On the pharm, *AgBiotech Buzz*, vol. 2, issue 7, July 29, 2002; Pew Initiative on Food and Biotechnology, Minding the pharm, *AgBiotech Buzz*, vol. 3, issue 3, May 14, 2003; Union of Concerned Scientists, *Food and Environment: Pharm and Industrial Crops—The Next Wave of Agricultural Biotechnology*.

Phosphoinositides

The phosphoinositides (PIs) are important building blocks of the lipid bilayer of biological membranes, representing about 10% of the total phospholipids in the membrane. They also have essential roles in cell signaling, membrane trafficking, and cytoskele-

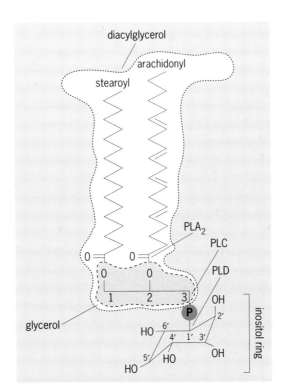

Fig. 1. Structure of the phosphatidylinositol (PtdIns) molecule, showing the sites of hydrolysis by various phospholipases.

ton dynamics. Serious diseases are associated with disorders in PI metabolism.

PI metabolism and regulation. The parent molecule of phosphoinositides is phosphatidylinositol (PtdIns), composed of an inositol 1-phosphate molecule (the "inositol headgroup") connected via a phosphodiester linkage to the sn-3 position of a glycerol backbone (**Fig. 1**). Phosphorylation at several sites of the inositol moiety determines the identity and properties of the various PIs. Fatty acids are linked to the sn-1 and sn-2 positions of the glycerol moiety (in the PIs of mammalian cells, these are most frequently stearic and arachidonic acid, respectively). The presence of nonpolar fatty acid components confines PIs to the nonpolar environment of the lipid bilayer. Under basal conditions, PIs are localized mainly in the inner leaflet of the cell membrane, where steady-state levels are maintained by continuous cycles of phosphorylation by PI kinases, producing PI phosphates (PIPs), and dephosphorylation by PIP phosphatases.

Synthesis of PtdIns takes place in the endoplasmic reticulum (ER); subsequent phosphorylations can occur in several cell compartments, including the ER, nucleus, Golgi complex, endosomes, and at the plasma membrane, depending on the localization of the relevant PI kinases. Translocation of PtdIns between cellular compartments is mediated by PI transfer proteins (PITPs), a class of proteins that binds and transports phospholipids.

Phosphorylation. All polyphosphorylated PIs originate from PtdIns, which is the most abundant PI species, representing ~90% of the total PIs. Phosphorylation is carried out by specific PI kinases that act on the 3, 4, and/or 5 positions of the inositol ring. The resulting polyphosphorylated PIs are named according to the site(s) of phosphorylation: for example, phosphatidylinositol 4,5-bisphosphate (PtdIns4, 5P$_2$), with phosphoryl groups attached at the 4 and 5 positions.

The PI kinases include three classes of PI 3-kinases (distinguishable by their primary structures, regulation, and substrate specificities); two classes of PI 4-kinases that phosphorylate only PtdIns (distinguished by their different sensitivities to specific inhibitors); and two classes of PI phosphate kinases that are able to phosphorylate PtdIns4P to produce PtdIns4,5P$_2$ (distinguished by their different sensitivities to phosphatidic acid).

Dephosphorylation. Dephosphorylation of the phosphorylated PIs is carried out by PIP phosphatases, which also have discrete cellular localizations (see **table**), and are divided into three major categories based on their ability to hydrolyze the 3-, 4- or 5-phosphates of the different PIs. PI 3-phosphatases include the tumor suppressor PTEN (phosphatase and tensin homolog), its related protein TPIP (transmembrane phosphatase with tensin homology),

Intracellular distributions and classifications of the known PI kinases and PIP phosphatases		
Type	Subtype	Intracellular distribution
PI3K	*Class IA*	Plasma membranes, nucleus
	p110α	
	p110β	
	p110δ	
	Class IB	Plasma membranes, nucleus
	p110γ	
	Class II	
	PI3K-C2α	Golgi complex, endosomes
	PI3K-C2β	Plasma membranes
	PI3K-C2γ	Golgi complex
	Class III	
	Vps34	Golgi complex, endosomes
PI4K	*Type II*	
	PI4Kα	Golgi complex, endosomes
	PI4Kβ	Golgi complex, plasma membranes
	Type III	
	PI4Kα	Golgi complex, endoplasmic reticulum
	PI4Kβ	Golgi complex, nucleus
PI5K	PIKfyve	Late endosomes
PIPK	*Type I*	
	PIP5Kα	Golgi complex, endosomes
	PIP5Kβ	Golgi complex, plasma membrane
	PIP5Kγ	Plasma membranes
	Type II	
	PIP4Kα	Plasma membranes
	PIP4Kβ	Plasma membranes
	PIP4Kγ	Endoplasmic reticulum
3-phosphatases	PTEN1,2	Plasma membranes, Golgi complex, nucleus
	TPIPα,β	Endoplasmic reticulum, Golgi complex
	MTM1	Plasma membranes
	MTMR1-8	Plasma membranes, not determined
4-phosphatases	*Type I, II*	Not determined
	Sac1	Endoplasmic reticulum, Golgi complex
	Synaptojanin 1, 2	Clathrin-coated vesicles, synaptic vesicles, mitochondria
5-phosphatases	*Type I*	
	5-phosphatase	
	Type II	
	5-phosphatase2	Not determined
	Ocrl	Golgi complex, endosome lysosomes
	Synaptojanin 1, 2	Clathrin-coated vesicles, synaptic vesicles, mitochondria
	PIPP	Plasma membranes
	Sac2	
	72-kDa 5-phosphatase/pharbin	Golgi complex
	SKIP	Endoplasmic reticulum, plasma membranes
	Type III	
	SHIP 1,2	Plasma membranes, nucleus
	Type IV	
	5-phosphatase	Not determined

myotubularin (MTM), and myotubularin-related proteins (MTMRs). The type I and type II PI 4-phosphatases use PtdIns3,4P_2 as their major substrate, while the PI 5-phosphatases are divided into four types according to their substrate specificities.

Localization in organelles. The localizations of PI-specific kinases and phosphatases produce characteristic PI pools in the different organelles of the cell. The PI-metabolizing enzymes, with rare exceptions, are cytosolic proteins targeted to specific organelles by mechanisms that are not entirely clear but often involve small guanosine triphosphatases (GTPases). The GTPases recruit the necessary enzymes and, in some cases, stimulate the enzymes' activities.

PI-binding proteins. The key roles of the PIs in signal transduction, membrane trafficking, and cytoskeleton dynamics are mediated through proteins that possess domains that bind preferentially to specific PI species. By virtue of their recognition of the different PIs, these PI-binding modules mediate the recruitment of their host proteins to specific cell membranes, within the framework of combinatory membrane-protein interactions that can involve additional, PI-independent binding sites. In this manner, the PIs not only participate in the generation of specific membrane domains but also contribute to the determination of the identities of the different cell organelles that result from these unique combinations of lipids and proteins. The interactions between the PIs and the proteins that surround them can have different effects, such as the recruitment and assembly of protein complexes and the modulation of protein functions.

PI-dependent signaling. Phosphorylated PIs are involved in the regulation of virtually all cellular functions, including cell growth and differentiation, and specialized cell functions such as regulated secretion, cell polarity, and cell movement. PtdIns4,5P_2 is the inositol lipid that has to date received the most attention, due to its dual activity as a precursor for second-messenger molecules, and as a crucial messenger itself for the localization and assembly of protein machineries involved in membrane trafficking and actin polymerization. In the former role as a messenger molecule precursor, PtdIns4,5P_2 is the major substrate of the phospholipase C family of signaling enzymes, which are activated by hormonal stimulation and which form two independent second messengers: inositol 1,4,5-trisphosphate, which causes the release of calcium from intracellular stores, and diacylglycerol, which activates enzymes in the protein kinase C family. These pathways can be activated by several classes of receptors (receptors coupled to GTP-binding proteins; and tyrosine kinase/growth factor receptors) that are usually coupled to specific phospholipase C subtypes. Other phospholipases, which act on all PIs, are phospholipase A_2, which specifically hydrolyzes the sn-2 fatty acid of the glycerol backbone, thus forming the lysoPIs; phospholipase A/lysolipase, which release both of the fatty acids linked to the glycerol backbone, forming glycerophosphoinositols; and phospholipase D, which leads to the formation of phosphatidic acid (**Fig. 2**).

Receptor activation can lead not only to the hydrolysis of PtdIns4,5P_2 but also to its phosphorylation by type I PI 3-kinase, to form PtdIns3,4,5P_3. This latter lipid is instrumental in the recruitment and activation of signaling enzymes, such as Akt/protein kinase B and the PI-dependent kinases, which can then phosphorylate substrates that are essential in the control of cell proliferation and survival.

PIs and membrane traffic. The PIs have roles in membrane trafficking along both the endocytic and exocytic pathways (Fig. 2). At the plasma membrane, PtdIns4,5P_2 is required for the invagination of coated pits, while PtdIns3P and PtdIns3,5P_2 regulate subsequent transport steps at early and late endosomes. PtdIns4,5P_2 is also generated during the early steps of the phagocytic process (the engulfing of large extracellular particles, including microorganisms), while PtdIns3P is required for phagosomal maturation and delivery of phagocytosed materials to lysosomes. At the synapses of neurons, the PIs have a fundamental role in the recycling of synaptic vesicles, since PtdIns4,5P_2 is required for the fusion of secretory granules with the plasma membrane. Finally, PtdIns4P, previously considered merely an intermediate in the synthesis of PtdIns4,5P_2, has recently been shown to have the major role among PI species in the Golgi complex (the main sorting station of the secretory pathway), where it is generated by two different PI 4-kinase isoforms.

PIs and cytoskeleton remodeling. A key event in cell polarization and movement is the reorganization of the cortical actin cytoskeleton (the cytoskeletal network just beneath, and supporting, the cell membrane). Cytoskeleton remodeling is regulated by a complex signaling cascade, which is itself governed by the small GTPases of the Rho family (including RhoA, Rac, and Cdc42). The PIs, and especially PtdIns4,5P_2, directly bind a variety of regulatory proteins involved in the maintenance of the cortical actin cytoskeleton and modulate their functions.

In general, PtdIns4,5P_2 exerts multiple actions, all of which favor the growth and stabilization of actin filaments (Fig. 2). Usually, the barbed ends of actin filaments are capped by large proteins that prevent elongation of the filaments in resting cells; the caps have to be released to allow rapid actin polymerization. By interacting with these capping proteins (such as CapZ and gelsolin), PtdIns4,5P_2 promotes their dissociation from the barbed ends of actin filaments. PtdIns4,5P_2 also inhibits actin-severing proteins and interacts with actin-bundling proteins; these actions stabilize actin filaments and promote cross-linking to form stable actin fibers. In addition, PtdIns4,5P_2 promotes de novo actin polymerization by activating proteins of the WASP (Wiskott-Aldrich syndrome protein) family. When PtdIns4,5P_2 binds WASP, regions of WASP that bind to the actin-nucleator complex Arp2/3 are exposed. The association of WASP with the Arp2/3 complex allows the subsequent nucleation and polymerization of actin.

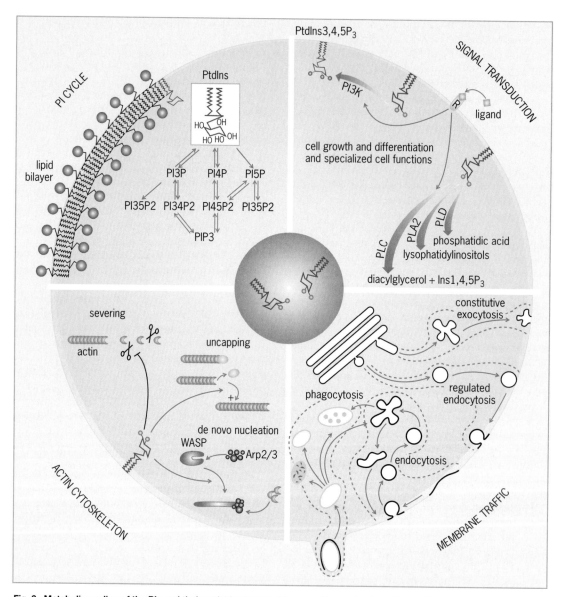

Fig. 2. Metabolic cycling of the PIs and their activities in signal transduction, actin dynamics, and membrane traffic.

Finally, the PIs, PtdIns4,5P$_2$ in particular, regulate the strength of the adhesion between the actin cytoskeleton and the plasma membrane, and they contribute to actin remodeling in the formation of filopodia and lamellipodia (extensions of the cell body that arise during cell movement).

PIs and disease. The importance of the PIs in cell physiology is highlighted when PI metabolism is hijacked by pathogens to favor their entry into and survival within host cells, and by the severe consequences of deficiencies in the PI kinases and PIP phosphatases. In particular, in many human cancers the PIP 3-phosphatase PTEN is mutated; mutations in MTM1 (another PIP 3-phosphatase) cause myotubular myopathy (a severe X-linked congenital disorder characterized by hypotonia and respiratory insufficiency); mutations in MTM-related proteins (MTMR2 and MTMR13/SBF2) cause severe demyelinating neuropathies, such as Charcot-Marie-Tooth disease; and mutations in the PIP 5-phosphatase OCRL-1 are responsible for the severe and rare X-linked oculocerebrorenal (Lowe) syndrome, which is characterized by congenital cataracts, Fanconi syndrome of the proximal renal tubules, and mental retardation.

For background information *see* CELL (BIOLOGY); CELL MEMBRANES; ENZYME; PHOSPHOLIPID; SECOND MESSENGER; SECRETION in the McGraw-Hill Encyclopedia of Science & Technology.

Daniela Corda; Maria Antonietta De Matteis

Bibliography. M. A. Lemmon, Phosphoinositide recognition domains, *Traffic*, 4:201–213, 2003; C. Pendaries et al., Phosphoinositide signaling disorders in human diseases, *FEBS Lett.*, 546:25–31, 2003; A. Toker, Phosphoinositides and signal transduction, *Cell. Mol. Life Sci.*, 59:761–779, 2002; M. R. Wenk and P. De Camilli, Protein-lipid interactions and phosphoinositide metabolism in membrane traffic: insights from vesicle recycling in nerve terminals, *Proc. Nat. Acad. Sci. USA*, 101:8262–8269, 2004; H. L. Yin and P. A. Janmey, Phosphoinositide regulation of the actin cytoskeleton, *Annu. Rev. Physiol.*, 65:761–789, 2003.

Photonic crystal devices

Electronic devices based on the crystalline properties of semiconductors—transistors and integrated circuits—have unleashed immense computing power and transformed disciplines across science and industry. This success has spurred development of optical analogs. The fundamental principles of transistors can be traced to the interaction of the wavelike properties of electrons with the repetitive spacing of atoms found in semiconductor crystals. An optical device would rely instead on the interaction of the wavelike properties of photons with a periodic spacing of dielectric constants. Devices based on this principle are known as photonic crystal (PhC) devices. They offer for optics the same potential for miniaturization that has occurred for electronic integrated circuits.

This article offers a brief overview of the physics and applications of photonic crystal devices. In addition to planar devices to be used in photonic integrated circuits, it looks at the use of photonic crystal fibers for sensing, sources, and high-power transport.

Theory. In the common PhC device shown in **Fig. 1**, light is guided through a channel and around a sharp bend by the periodically spaced holes surrounding it. To understand how this occurs, consider a situation where the electromagnetic wavelength is much larger than the hole spacing. In this case, an incident wave coherently scatters off the holes in the forward direction, and "sees" only the average absorption and dielectric constant (or refractive index) of the PhC. As a result, the wave is not guided by the bend. Shortening the incident wavelength by a small amount increases the number of waves that fit in the crystal, but not the interaction with the bend.

As the incident wavelength is shortened even further, however, a point is reached for a triangular hole pattern where the hole spacing is on the order of one-

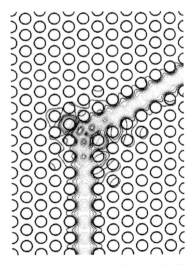

Fig. 1. Photonic crystal waveguide fabricated using round holes in a triangular pattern. Light is injected into the waveguide from the bottom and turns through a sharp bend, exiting to the right with low bend loss. (*Courtesy of Lumerical Solutions Inc., www.lumerical.com*)

half of a wavelength, and the holes and the spacers between them each span approximately one-quarter of a wavelength. Individual Fresnel reflections due to the difference in refractive index between the holes and spacers then interfere constructively in the direction opposite that of the incident wave, reflecting most of the incident wave backward. This process is identical to that found in multilayer thin-film reflectors; an important advantage of the PhC device is that its two-dimensional hole/spacer geometry deters the peak reflection wavelength from shifting with angle of incidence, a shift which is a severe problem for many applications of the thin-film reflector.

The range (or band) of wavelengths that is reflected by the PhC depends on the difference in refractive index between holes and spacers. To see why, consider a one-dimensional thin-film analogy, where the half-wavelength peaks created by the interference of the incident and reflected waves could just as easily center on the holes as on the spacers. With the wavelength in the PhC device fixed by the quarter-wave layers needed for constructive interference to occur, the wave that peaks in the higher-index spacers must have a lower frequency (or photon energy), and that which peaks in the lower-index holes must have a higher energy. These two energies then define the edges of the photonic band gap (PBG). No propagation is possible in the PhC when the incident photon energy is between these extremes (or "in the band gap"). Any such photons will be strongly reflected over a range of incident wavelengths depending on the differences in refractive index and the size and spacing of the holes, expanding our intuitive picture of constructive interference in the backward direction to include a range of wavelengths.

Returning to Fig. 1, we see that light of the correct wavelength would be strongly reflected from a PhC structure that did not have a channel. To deflect an electromagnetic wave around a sharp bend, we must first create the channel by filling in one set of air holes. Light can then be coupled into the channel. As it propagates, it reflects strongly off the channel walls, which consist of the correct hole/spacer pattern for any wavelength in the band gap of the PhC structure. In this way, the PhC device "guides" light around a sharp bend with low loss, by a mechanism quite different from conventional waveguiding based on total internal reflection. Although the details and calculations are considerably more complex than this description, it is this basic principle that governs most of the PhC devices described in the following sections.

Applications. It is one thing to successfully demonstrate a basic principle, quite another to design a commercial device that meets essential criteria, such as low loss, low power consumption, high speed, long-term reliability over a wide temperature range, and low cost. This section reviews the status and limitations of emerging PhC devices.

Photonic crystal fiber. The most mature of the PhC devices are photonic crystal fibers (PCFs), now commercially available. There are two types: solid-core

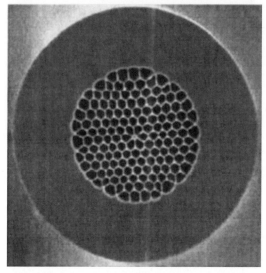

(a)

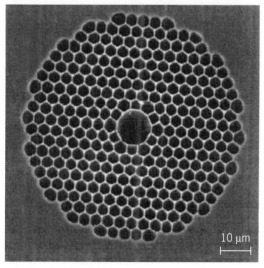

10 μm

(b)

Fig. 2. Scanning electron microscope images of photonic crystal fiber waveguides using (*a*) a small glass core approximately 2 μm in diameter, optimized for large nonlinear index changes, and (*b*) a hollow air-core approximately 12 μm in diameter, optimized for very low nonlinearities. (*Courtesy of Blaze Photonics Ltd., www.blazephotonics.com*)

and hollow-core, each with its own unique properties (**Fig. 2**). Because manufacturing imperfections such as scratches, contamination, surface roughness, and dimensional variations lead to relatively high losses for both types, the primary applications at this point are in sensing, sources, and high-power transport, rather than the application first imagined, long-distance fiber-optic communications.

Solid-core PCF (Fig. 2*a*) consists of a small glass core (such as silica, SF_6) approximately 2 micrometers in diameter, suspended in air by a glass spiderweb structure running down the length of the fiber—cladding that is more hole than glass. As with conventional fiber, waveguiding occurs along

the length of the PCF fiber based on total internal reflection (TIR), with the average index of the air holes and their surrounding silica establishing the lower cladding index required for TIR. The large index difference between the solid core and webbed cladding tightly constrains the intensity profile to the very small core area, increasing the core's nonlinear index change—that is, the change in refractive index with intensity—by a factor of 50 or more over conventional telecommunications fiber.

The large nonlinear index change in the core results in the self-modulation of a pulse's phase by its own intensity profile which, in combination with Raman scattering and four-wave mixing, results in a spectral broadening of the pulse on the order of 1000 nanometers. This has allowed development of broadband ("supercontinuum") PCF sources 10,000 times brighter than sunlight, with applications in spectroscopy and optical coherence tomography. The long interaction length and high intensity at the surface of the core also makes these fibers useful for sensing chemical changes at the air/glass interface. Erbium-doped PCFs with larger cores and much lower nonlinearities are proving useful as high-power fiber lasers and amplifiers. *See* OPTICAL CO-HERENCE TOMOGRAPHY.

Hollow-core PCF (Fig. 2*b*) has an air core approximately 12 μm in diameter. Since the refractive index of air is lower than the surrounding material, waveguiding by TIR is ineffective, so the PBG effect is required. The advantage over conventional fiber is that most of the overlap of the intensity profile is with air, so nonlinearities can be reduced to very low levels, making these fibers useful for high-power transport. In principle, attenuation should also be small, and it is hoped that manufacturing improvements can reduce hollow-core PCF loss to less than that of conventional telecommunications fibers.

Planar devices. Planar PhC devices are typically envisioned as advanced components for fiber-optic networks, primarily due to the possibility of miniaturization and planar (two-dimensional) integration of multiple photonic devices. Using building blocks such as very short cavity lasers and waveguides that can turn sharp bends without significant loss (Fig. 1), PhC concepts may allow the highest component density for photonic and opto-electronic integrated circuits. Some of the devices under development are described below.

Nanocavity semiconductor lasers, where the size of the gain region is only a wavelength or so on each side, surrounded by very high reflectivity PhC mirrors to balance the cavity gain and loss. A number of telecommunications devices have been proposed using these lasers, but the most promising application at this point is chemical sensing. Specifically, by introducing a "defect" (or different-sized air hole) into the laser cavity, which can in turn be filled with a higher-index reagent, the change in index also shifts the lasing wavelength, potentially allowing identification of analytes; sensitivities as small as femtoliters of reagent volume have been reported. In contrast with PCF chemical sensors, it is hoped that these lasers will allow miniaturization and photonic integration of multiple devices, leading to an extremely compact "laboratory on a chip."

Miniaturized, low-power telecommunications components, culminating in various types of photonic integrated circuits. An example considered for future development is the integration of low-threshold nanocavity lasers with PhC wavelength-add filters for a highly miniaturized transmitter/multiplexer that can add individual wavelengths to a multiple-wavelength data stream. This process can be reversed on the receiver side of a fiber-optic network using PhC wavelength-drop filters in combination with cavity-enhanced photodetectors.

Outlook. Given that the price of commercial products generally depends more on packaging efforts than on individual chip cost, the miniaturization enabled by PhC devices will find useful applications only if benefits unique to photonic crystal devices are discovered. In some cases, the high integration density (number of devices that fit on a given wafer) of PhC devices can even be a detriment, for example if temperature or current change in one component affects the performance of a neighboring device. Furthermore, components demonstrated to date have not met all criteria for commercial success described earlier: low loss, low power, low cost, high speed, and long-term reliability. In PhC manufacturing, high yield and low cost remain difficult challenges, and even the seemingly simple task of low-loss coupling of planar PhC devices to fiber on a commercial scale is still problematic.

Conventional devices that compete with PhC devices (microring resonators, for example) are also being improved, so the eventual acceptance of PhC devices will ultimately depend on a comparison of performance and cost. In only a few cases have the unique properties of PhC devices demonstrated commercial importance. Longer term, unique photonic crystal phenomena will open up a number of exciting applications in lighting, spectrometry, and imaging. Examples are metallic PhCs that use a complex dielectric constant to obtain highly efficient incandescence, superdispersion (an extremely large change in refractive index with wavelength) at the band edge, and superresolution (subwavelength) lensing. Perhaps the most progress will occur when PhC devices are no longer thought of as electronic analogs, but as photonic devices in their own right, with a long list of potential applications to be developed over the coming years.

For background information *see* DIFFRACTION; INTEGRATED CIRCUITS; INTEGRATED OPTICS; INTERFERENCE OF WAVES; LASER; NONLINEAR OPTICS; OPTICAL FIBERS; REFLECTION OF ELECTROMAGNETIC RADIATION; REFRACTION OF WAVES in the McGraw-Hill Encyclopedia of Science & Technology.   K. J. Kasunic

Bibliography. F. S. Crawford, Jr., *Waves*, McGraw-Hill, New York, 1968; J. D. Joannopoulos, P. R. Villeneuve, and S. Fan, Photonic crystals: Putting a new twist on light, *Science*, 386:143–149, Mar. 13, 1997; J. C. Knight, Photonic crystal fibres, *Nature*, 424:847–851, Aug. 14, 2003; E. Yablonovitch, Photonic crystals: Semiconductors of light, *Sci. Amer.*, 285:47–55, December 2001.

Photoprotection in plants

The light-harvesting pigment chlorophyll is found in great abundance wherever conditions are conducive for photosynthetic organisms to live. Although chlorophyll facilitates very efficient absorption of visible light (wavelengths between 400 and 700 nanometers) by these organisms, no organism can utilize all of the absorbed energy for photosynthesis under full sunlight. In addition plants are often exposed to environmental stress (for example, temperature extremes in winter and summer, limiting water or nutrients, and high salinity) that reduces photosynthesis further.

Excess absorption of light by a pigment such as chlorophyll, especially when the excitation energy cannot be used for photosynthesis, is potentially very hazardous for two reasons. First, chlorophyll molecules that have become excited through the absorption of a photon of light could pass that energy on to oxygen, resulting in the formation of singlet excited oxygen (in which one electron is promoted to a higher energy level). Second, excitation energy that would normally be used for reducing carbon in photosynthesis may instead be passed on from chlorophyll to oxygen in the form of an electron, leading to the formation of superoxide (O_2^-). If not dealt with appropriately, both of these modified forms of oxygen (known as reactive oxygen species) have the potential to damage various macromolecules [for example, lipids in membranes, deoxyribonucleic acid (DNA), and proteins] and can even result in cellular death. Since photosynthetic organisms, and plants in particular, absorb excess light regularly on a daily basis and continuously during times of stress, it is not surprising to find that they also possess an efficient mechanism for protecting themselves against the threat posed by this situation.

Types of radiation. Solar radiation that reaches the Earth's surface includes wavelengths from the ultraviolet (UV) to the infrared (heat) portion of the spectrum. Ultraviolet radiation is of special concern because of its high energy content and its potential for direct absorption by certain macromolecules (for example, DNA and proteins), resulting in immediate damage. Since ultraviolet radiation is not used for photosynthesis, plants typically accumulate large amounts of UV-absorbing compounds in the epidermis, the cuticle, and the hairs on their leaves to intercept the UV radiation before it can cause damage. Infrared radiation provides only heat, through the excitation of molecules such as water in the plant cells, and poses no danger to the plant per se. Thus visible light, which is actively absorbed by the photosynthetic pigments, represents the greatest threat to plants.

Three levels of intervention. To prevent damage to macromolecules resulting from the absorption of visible light by chlorophyll and the formation of reactive forms of oxygen, there are three broad approaches that plants utilize. The first is to prevent or reduce the absorption of light; the second is to safely channel the excess excitation energy into a dissipative process and release it harmlessly as heat; and the third is to employ antioxidant enzymes and metabolites to safely detoxify singlet excited oxygen and superoxide before they can interact with, and pass on their energy or electrons to, other molecules. All plants possess the third mechanism, but it is far better to prevent the formation of reactive oxygen species than to try and detoxify them once they have been formed. In addition, there are some situations (exposure to low temperatures during winter stress) in which detoxification by antioxidant enzymes is severely hampered or even suppressed entirely.

The mechanisms to reduce the absorption of light include elimination (senescence) of leaves during periods of extreme stress, growth or movement of leaves to minimize absorption during maximal exposure, utilization of leaf hairs and/or substances (for example, waxes and pigments) on or in the epidermal layer of leaves to intercept or reflect light, and changes at the chloroplast level (reorientation and positioning to minimize light absorption or reductions in chlorophyll content). Most of the mechanisms to reduce light absorption are employed by only some species, and none of the mechanisms prevent the absorption of excess light by full-sunlight-exposed leaves.

Fortunately, all plants possess a mechanism to divert excess excitation energy into heat before it can be passed on to oxygen. This photoprotective process, known as the xanthophyll cycle, is highly regulated so that it is engaged when required and is disengaged, so as not to compete with photosynthesis, when light is no longer excessive.

Xanthophyll cycle. As photons of light are absorbed by chlorophyll molecules, the excitation energy migrates to reaction centers, where the energy is used to boost an electron that is passed from carrier molecule to carrier molecule until it can be used to fix CO_2 to an acceptor molecule, collectively leading to the net production of sugars. During this process, protons accumulate on the inside of the internal membranes (the lumen) of the chloroplast. These protons move out of the lumen through a protein complex, and the energy of this movement is captured in the formation of adenosine triphosphate (ATP). At low light levels, protons move out of the lumen just as rapidly as they accumulate inside. For every increase in light intensity, there is a proportional increase in photosynthesis. At higher light levels, however, the utilization of the electrons and ATP generated from light absorption reaches a point of saturation, and further increases in light do not result in further increases in photosynthesis (**Fig. 1**). As the protons accumulate in the lumen, they activate an enzyme that converts a carotenoid pigment called violaxanthin into another called zeaxanthin (through

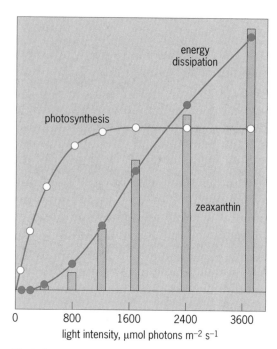

Fig. 1. Increases in photosynthesis (open circles), zeaxanthin content (tinted bars), and energy dissipation activity (solid circles) with increases in incident light in a sunflower *(Helianthus annuus)* leaf. Among plants, crops (such as sunflower) typically have the highest rates of photosynthesis. As shown in the figure, for sunflower, almost all of the light above 1000 μmol photons m^{-2} s^{-1} (about half of full sunlight), is in excess. (*Adapted from B. Demmig-Adams et al., Light response of CO_2 assimilation, dissipation of excess excitation energy, and zeaxanthin content of sun and shade leaves, Plant Physiol., 90:881–886, 1990*)

an intermediate, antheraxanthin). These carotenoids are xanthophylls (oxygen-containing carotenoids), and the interconversion of the three is known as the xanthophyll cycle.

At higher light levels, increasing amounts of zeaxanthin form in proportion to the level of excess light that is absorbed (Fig. 1). Accumulated protons also cause a protein in the chloroplast to change position, bringing the zeaxanthin molecules in proximity to chlorophyll, allowing the excess excitation energy to be transferred from chlorophyll to zeaxanthin, which can easily shed the excess energy as heat. This energy dissipation process increases proportionally with the level of excess light (Fig. 1), just as zeaxanthin levels do, thereby preventing the formation of singlet excited oxygen through the transfer of that excess excitation energy to oxygen.

Energy dissipation rates. In fully exposed leaves of plants growing under favorable conditions, zeaxanthin levels (and low levels of antheraxanthin) increase through midday and then decrease through the end of the day (**Fig. 2***h, l*). Compared with crop species such as sunflower, plants with lower rates of photosynthesis but equivalent exposure to radiation, such as the evergreen ornamental *Euonymus kiautschovicus*, absorb more excess excitation energy, typically form more zeaxanthin (Fig. 2*b, l*), and exhibit higher levels of energy dissipation (Fig. 2*f, j*). As an alternative to photosynthetic utilization, this energy dissipation process lowers the efficiency of

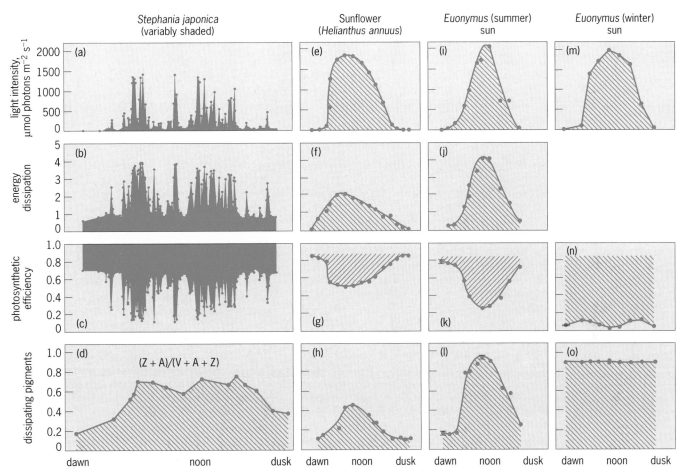

Fig. 2. Factors in photoprotection in there plant species over a day's course. (*a, e, i, m*) Diurnal changes in light intensity. (*b, f, j*) Photoprotective energy dissipation activity. (*c, g, k, n*) Photosynthetic efficiency. (*d, h, l, o*) Level of zeaxanthin and antheraxanthin (Z + A) as a fraction of the total xanthophyll cycle pool [violaxanthin (V) + A + Z]. Note that due to the sustained engagement of zeaxanthin in energy dissipation, it is not possible to accurately assess the level of energy dissipation activity during the winter. (*Adapted from W. W. Adams et al., Rapid changes in xanthophyll cycle-dependent energy dissipation and photosystem II efficiency in two vines, Stephania japonica and Smilax australis, growing in the understory of an open Eucalyptus forest, Plant Cell Environ., 22:125–136, 1999; B. Demmig-Adams and W. W. Adams, The role of xanthophyll cycle carotenoids in the protection of photosynthesis, Trends Plant Sci., 1:21–26, 1996; B. Demmig-Adams et al., Physiology of light tolerance in plants, Hort. Rev., 18:215–246, 1997; and A. S. Verhoeven et al., Two forms of sustained xanthophyll cycle-dependent energy dissipation in overwintering Euonymus kiautschovicus, Plant Cell Environ., 21:893–903, 1998*)

the photosynthetic process to a greater extent in the ornamental species (Fig. 2*k*) compared with sunflower (Fig. 2*g*).

In highly variable light conditions (for example, a partly cloudy day or understory of a forest), in which light may shift from nonexcessive to full sunlight and back again within a matter of seconds, photoprotective dissipation increases rapidly during exposure to high light and then decreases rapidly to permit efficient photosynthetic utilization of low light (Fig. 2*a, b, c*). Although the dissipating pigments are formed and maintained at elevated levels throughout the day (Fig. 2*d*), they can be rapidly engaged and disengaged (Fig. 2*b, c*) through changes in the accumulation of protons in the lumen as light levels shift from excessive to nonexcessive.

In addition to a responsiveness on the order of seconds, photoprotection through the xanthophyll cycle exhibits acclimation over days and even seasons. Compared with shade leaves (that is, leaves that are acclimated to shaded conditions), leaves that develop in full sunlight possess a greater quantity of the

xanthophyll cycle carotenoids and higher capacities for photoprotective energy dissipation. When faced with even more excess light, such as during the winter when the utilization of the absorbed light through photosynthesis becomes inhibited by low temperatures and in some species (such as conifers) is even downregulated by the plants themselves, zeaxanthin can enter into a locked configuration that provides maximal photoprotection throughout the day (Fig. 2*n, o*). Although it is not possible to accurately calculate the levels of energy dissipation in such leaves, high levels of zeaxanthin (and antheraxanthin) are retained throughout the day and night (Fig. 2*o*), and the efficiency of the photosynthetic process is maintained in a minimal state (Fig. 2*n*).

Some annual and biennial species, such as winter cereals and the weed *Malva neglecta*, have high capacities for photosynthesis during the winter and continue to grow during warmer periods. Many evergreen species, however, cease growth and downregulate the capacity for photosynthesis during the winter. This involves the disassembly of a key protein in the first reaction center of photosynthesis, thereby removing the source of electrons for not only photosynthesis but also superoxide formation. The combination of removing the source of electrons that could lead to superoxide formation and the sustained engagement of zeaxanthin in energy dissipation to prevent the formation of singlet excited oxygen thus provides protection against the danger posed by light absorption under the extreme conditions of winter stress. This permits the maintenance of the photosynthetic apparatus in a state that can be rapidly reactivated in the spring through a few minor changes.

For background information *see* ABSORPTION OF ELECTROMAGNETIC RADIATION; CAROTENOID; LIGHT; PHOTOCHEMISTRY; PHOTOSYNTHESIS in the McGraw-Hill Encyclopedia of Science & Technology.

William W. Adams III

Bibliography. W. W. Adams et al., Photoprotective strategies of overwintering evergreens, *BioScience*, 54:41–49, 2004; B. Demmig-Adams and W. W. Adams, The role of xanthophyll cycle carotenoids in the protection of photosynthesis, *Trends Plant Sci.*, 1:21–26, 1996; C. Külheim et al., Rapid regulation of light harvesting and plant fitness in the field, *Science*, 297:91–93, 2002; X.-P. Li et al., A pigment-binding protein essential for regulation of photosynthetic light harvesting, *Nature*, 403:391–395, 2000; Y. Z. Ma et al., Evidence for direct carotenoid involvement in the regulation of photosynthetic light harvesting, *Proc. Nat. Acad. Sci. USA*, 100:4377–4382, 2003; G. Öquist and N. P. A. Huner, Photosynthesis of overwintering evergreen plants, *Annu. Rev. Plant Biol.*, 54:329–355, 2003.

Polymer recycling and degradation

Polymers are used in numerous applications because of their low weight, high adaptability, and ease of processing. The use of low-weight plastics as packaging material and as a replacement for metal parts in cars reduces transportation costs and environmental impact by lowering fossil fuel consumption and greenhouse-gas emissions.

The most widely used plastics are synthetic polymers made from oil. Although only 4% of the world's oil is used for the production of plastics, sustainable production and use of oil-based polymers is needed because oil resources are slowly running out. In addition, plastic waste accounts for approximately 20% by volume of municipal solid waste, much of which is disposed of in landfills. In recent years, environmental awareness and a reduction in landfill capacity have put much focus on techniques for sustainable use of plastics, reuse of plastic waste, and development of plastics from biological sources. Production and use of biodegradable polymers from renewable resources may significantly contribute to the global goals of sustainable development. **Figure 1** shows some common polymers which will be discussed.

Resource and waste management. Different strategies are available for resource and waste management, including reduction of resource use, reuse (also known as primary recycling), mechanical and chemical recycling (secondary recycling), incineration (tertiary recycling), and landfill disposal. Reduction of resources means designing products that use as little material as possible and generate a minimal amount of waste. Primary and secondary recycling turns waste into resources, reducing environmental damage and conserving natural resources. Incineration is considered recycling only if the calorific value of the plastic waste is recovered and used to generate heat or electricity. In the waste management hierarchy, reducing resource use is the most desirable option and disposing of waste in landfills is the least desirable option (**Fig. 2**).

Fig. 2. Waste management hierarchy with the most desirable option at the top.

According to the Association of Plastics Manufacturers in Europe (APME), from 2001 to 2002 the total consumption of polymers in Western Europe increased 4.1% to 47.7 million metric tons (52.6 million tons). In 2002, the total plastic waste in Western Europe was 20.4 million metric tons (22.5 million tons), of which 2.7 million metric tons (3 million tons) were mechanically recycled, 298,000 metric tons (319,000 tons) chemically recycled, and 4.7 million metric tons (5.2 million tons) incinerated

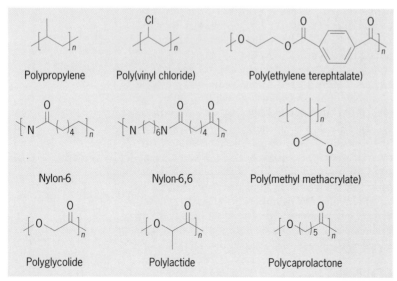

Fig. 1. Polymer structures.

with energy recovery. In total, 38% of the plastic waste was recovered.

Mechanical recycling. Mechanical recycling involves grinding waste and extruding it into new products or pellets for further processing. Waste from plastic processing equipment is routinely ground up and put back into the raw-material feed. Many studies have shown that plastics can be mechanically recycled repeatedly without losing their mechanical properties. Mechanical recycling is best suited to clean, homogenous waste streams. For the mechanical recycling of mixed plastic waste, sorting is highly important as small amounts of contaminating polymer may result in low recyclate quality and limit its use.

Chemical recycling. Chemical recycling, also known as feedstock recycling, transforms waste plastics into original monomers or other basic chemicals. Several processes with widely differing characteristics are classified as chemical recycling. Poly(ethylene terephtalate) is depolymerized by reaction with water, methanol, or diethylene glycol, while nylon-6 treated with steam depolymerizes to its monomer, ε-caprolactam.

Waste plastics can be chemically recycled to compounds other than monomers. Mixed or soiled waste can be treated at high temperatures in the absence of oxygen (pyrolysis), with low amounts of oxygen (gasification), or in an atmosphere of hydrogen (hydrogenation). During pyrolysis, polymers decompose to monomers and other organic substances that can be used as feedstock or for energy generation. Gasification mainly produces a mixture of carbon monoxide (CO) and hydrogen (H_2), known as synthesis gas, which can be used to produce methanol or as fuel. Hydrogenation produces a high-value synthetic crude oil that can be used in a variety of applications.

Energy recovery. Because plastics are derived from oil, they have a calorific value approximately equal to coal. Energy can be recovered from plastic waste through combustion, through co-combustion with traditional fossil fuels, or as a substitute for coal in cement kilns. Energy recovery may provide the most suitable recycling route for plastic waste that cannot be mechanically or chemically recycled due to excessive contamination, separation difficulties, or poor polymer properties. Mechanical-recycling small, lightweight plastic packaging items may be economically and environmentally costly. These items are preferably recycled by incineration as mixed municipal solid waste with energy recovery. In addition, energy recovery can be considered controlled detoxification, as harmful substances present in medical plastics waste are destroyed by incineration.

Bioplastics. Bioplastics are a promising approach to the sustainable use of plastics. The concept of bioplastics is wide and not strictly defined. It can include polymers synthesized by natural sources (biopolymers), plastics made from renewable resources, and petroleum-based polymers blended with polymers made from renewable resources (**Fig. 3**). Bioplastics that readily degrade to environmentally acceptable compounds provide a solution to the increasing amount of plastic waste. Bioplastics made from renewable resources address the problem of oil supply depletion. There are also bioplastics that are both biodegradable and made from renewable resources.

Synthetic. One well-known bioplastic is poly(lactic acid), or polylactide, which is polymerized from the monomer lactic acid. Lactic acid is produced by fermentation of corn and other starch-containing plants by bacteria, yeast, or fungi. Although it can be directly polymerized to poly(lactic acid), in most cases the cyclic dimer of lactic acid, lactide, is produced first and then polymerized to polylactide via ring-opening polymerization. Lactide is readily copolymerized with other monomers to yield polymers with mechanical properties ranging from soft and elastic to stiff and high-strength. Recently, large investments have been made for large-scale production of lactic acid and lactic acid-based polymers in the United States, Europe, and Japan. Traditionally, lactic acid-based polymers have been used in medical applications since the material shows good biocompatibility. With the increased supply of lactic acid,

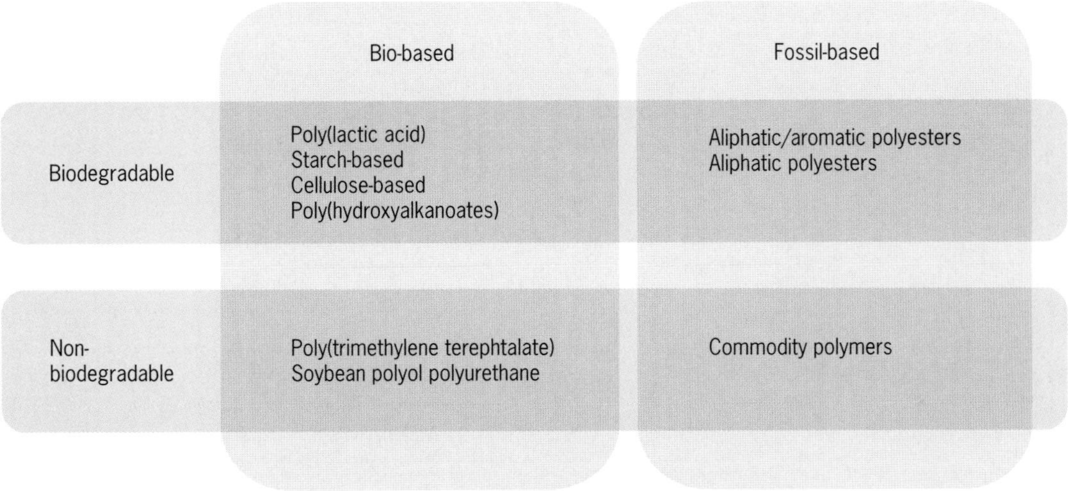

	Bio-based	Fossil-based
Biodegradable	Poly(lactic acid) Starch-based Cellulose-based Poly(hydroxyalkanoates)	Aliphatic/aromatic polyesters Aliphatic polyesters
Non-biodegradable	Poly(trimethylene terephtalate) Soybean polyol polyurethane	Commodity polymers

Fig. 3. Polymers classified according to their source and biodegradability.

these polymers have found applications as packaging materials and fibers.

Filled. In some cases, recycled petroleum-based polymers are considered as bioplastics. Many recent recycling studies have evaluated the use of waste material from different sources as filler material in polymeric composites. The successful use of glass-fiber-reinforced phenolic prepreg (resin-preimpregnated and partially cured) waste as reinforcing filler in polypropylene and polyamide has been reported. The prepreg is usually partially cured immediately after impregnation to facilitate handling. In the United States, thermoplastics filled with wood fibers, called wood plastics composites, have been available for some years. Good mechanical properties can be achieved by controlling the interfacial properties between the filler materials and the polymer matrix using silane coupling agents.

Long-term properties. The increasing recovery and reuse of plastic waste requires the qualified characterization of recycled plastics' long-term properties and environmental impact. In recent years, the development of techniques, such as solid-phase microextraction (SPME), for extracting chemical compounds from a variety of matrices has greatly facilitated studying the formation and release of low-molecular-mass compounds during service life. Emission studies are important environmentally because recycled plastics are used as packaging materials, as furnishings, and in car interiors.

Although many different materials can be used as filler in plastics, the fillers may be very easily degraded, making the composite a high-emitting material and unacceptable for use. For composites using various new and, in some cases, thermally sensitive fillers, it is important to study the formation and release of low-molecular-mass compounds during service life, as the fillers may form additional degradation products that significantly affect the composite's emission profile.

It has been shown that nylon-6,6 can be mechanically recycled up to six times without significant deterioration of its mechanical properties. However, repeated processing makes the material more susceptible to oxidation. After short exposures to relatively moderate temperatures, more low-molecular-mass compounds are released from recycled nylon-6,6 than virgin nylon-6,6. At very low levels of oxidation, SPME revealed more volatile degradation products in recycled material, compared to virgin material, while traditional tensile testing failed to detect any differences.

Degradation. Polymers can degrade by several different mechanisms, such as thermal and mechanical degradation during processing, oxidative degradation during processing and service life, and hydrolytic degradation during service life (**Fig. 4**). Degradation negatively affects material properties and ultimately leads to material failure.

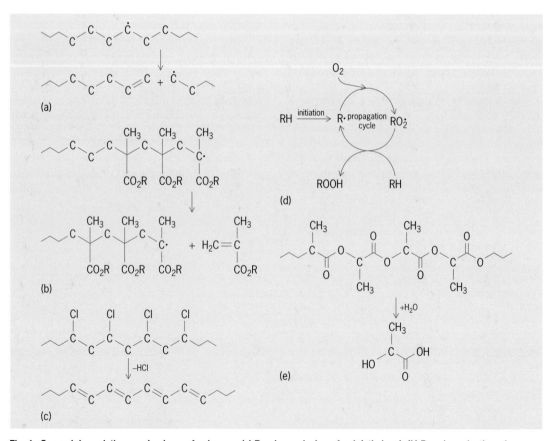

Fig. 4. General degradation mechanisms of polymers. (*a*) Random scission of poly(ethylene). (*b*) Depolymerization of poly(methyl methacrylate), R = CH. (*c*) Side-group elimination of poly(vinyl chloride) yielding hydrochloric acid. (*d*) Auto-oxidation cycle of polymers. (*e*) Hydrolysis of poly(lactid acid) yielding the monomer lactid acid.

For some applications, rapid material degradation is desirable. After completing their function, mulch films, carrier bags, and drug delivery matrices should degrade rapidly to environmentally acceptable low-molecular-mass compounds. For other applications, such as construction materials or pipes, the materials are required to remain unaffected for many years. The lifetime of plastics can be controlled by using additives that promote or prevent degradation.

Thermal degradation. All polymers can be degraded by heat. When heated to the extent of bond rupture, polymers degrade by random scission [for example, polyolefins], depolymerization [for example, poly(methyl methacrylate)], or side group elimination [for example, poly(vinyl chloride].

Oxidative degradation. This reaction is initiated by energy, either heat or sunlight, which breaks bonds in the polymers, yielding polymer radicals. The radicals react with atmospheric oxygen to form peroxides, which decompose and start an auto-oxidative degradation. All plastics react with oxygen during service life, but the reaction is slow at ambient temperature. At elevated temperature, oxidation proceeds quite rapidly. Antioxidants are used to achieve reasonable lifetimes for plastic products.

Hydrolytic degradation. Polymers such as polyesters, polyamides, polycarbonates, and polyacetals can be degraded through hydrolysis by moisture in the environment. Hydrolytic cleavage ultimately degrades the polymers into simple compounds resembling the original monomers.

Outlook. The degradation routes of common polymeric materials have been studied thoroughly and are well known. At present, most research on degradation focuses on tailoring the rate of degradation of polymeric materials. This can be achieved by compounding the polymer with pro- and antioxidants.

Another approach is macromolecular design (advanced polymer synthesis) to control durability and long-term performance. For example, polymer degradation may be tailored by incorporating selected functional groups in the polymer backbone, and by controlling cross-linking density and crystallinity. An interesting application of macromolecular design is the production of polymers for medical applications such as tissue engineering, artificial tendons, nerve guides, or controlled drug delivery. A highly controlled polymer synthesis produces a material that, after completing its performance, dissolves into nontoxic, low-molecular-mass compounds which can be assimilated by the human metabolism. Among the biodegradable polymers, aliphatic polyesters have a leading position. Typical examples are polylactide, polyglycolide, and poly(ε-caprolactone), which also can be produced from renewable resources. These aliphatic polyesters are degraded by hydrolysis to hydroxy carboxylic acids, which in most cases are metabolized.

For background information *see* ANTIOXIDANT; BIOPOLYMER; PHOTODEGRADATION; POLYMER; RECYCLING TECHNOLOGY; STABILIZER (CHEMISTRY) in the McGraw-Hill Encyclopedia of Science & Technology.

Mikael Gröning; Ann-Christine Albertsson

Bibliography. A.-C. Albertsson (ed.), *Degradable Aliphatic Polyesters in Advances in Polymer Science*, vol. 157, Springer, Berlin, 2002; A. Azapagic, A. Emsley, and I. Hamerton (eds.), *Polymers, the Environment and Sustainable Development*, Wiley, Chichester, U.K., November 2003; J. Brandrup (ed.), *Recycling and Recovery of Plastics*, Carl Hanser, Munich, Germany, 1996; J. Scheirs (ed.), *Polymer Recycling, Science, Technology and Applications*, Wiley, Chichester, U.K., 1998.

Polymer solar cells

Conjugated polymers and molecules with semiconducting properties exhibit a high potential for use in fabricating efficient and low-cost flexible photovoltaic devices or solar cells, possibly for large-area applications. One important advantage of polymer semiconductors over their inorganic counterparts is that they can be dissolved and applied as a solution by techniques such as spin coating, doctor blading, and screen printing.

Charge transfer process. Photovoltaic applications require a high yield of photogenerated charge carriers (that is, free carriers generated on light absorption). This is not the case in pure conjugated polymers, which primarily generate electron-hole pairs with low charge mobility and undergo radiative decay resulting in photoluminescence. By blending the polymer with a strong electron acceptor, the light-induced generation of free charge carriers can be stimulated.

Figure 1*a* shows the photoluminescence spectrum of the pristine polymer poly(3-hexylthiophene) [P3HT; see structure] excited at a wavelength

P3HT

of 532 nanometers. The photoluminescence originates from the radiative decay of neutral photoexcited states. The photoluminescence signal is efficiently quenched upon blending P3HT with the strong electron acceptor [6,6]-phenyl-C_{61} butyric acid methyl ester (PCBM; see structure), a methano-

PCBM

fullerene. The quenched photoluminescence indicates that either photogenerated electron-hole pairs have been dissociated via charge transfer from the polymer to the fullerene or they have lost their energy by nonradiative energy transfer. The proof of the charge transfer can be provided by light-induced electron spin resonance (ESR), which detects spin-carrying charge carriers generated by light. No ESR signals were detected from the polymer-fullerene blend in the dark, reflecting that only relatively few paramagnetic species are present in the ground state. Instead, two strong light-induced ESR line groups have been observed, as in Fig. 1*b*, which clearly shows the formation of two independent paramagnetic species (both spin $^1/_2$). Thus, the light-induced ESR studies unambiguously indicate the occurrence of photoinduced electron transfer in polymer-fullerene composites. The charge transfer process is found to be ultrafast; that is, it takes less than 40 femtoseconds (40×10^{-15} s). Moreover, the charge-separated state turns out to be metastable; that is, the hole remains on the polymer backbone, and the electron is localized on the acceptor for some microseconds at room temperature. An efficient generation of free charge carriers, approaching 100%, can be expected.

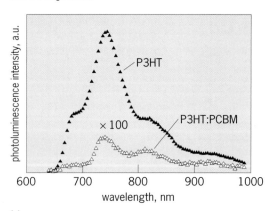

(a)

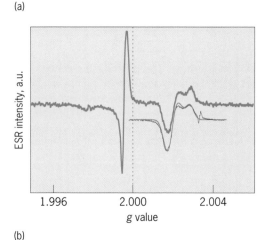

(b)

Fig. 1. Charge transfer process. (*a*) Photoluminescence spectra of pristine P3HT and a P3HT:PCBM blend. (*b*) Light-induced ESR spectrum of a PPV:PCBM blend [PPV = poly(*p*-phenylene vinylene)]. The line group with *g* < 2 is assigned to the fullerene anion, and the signal with *g* > 2 is attributed to the polymer cation. The latter assignment has been confirmed by ESR measurements (dark spectrum) of the iodine-doped PPV, which displays the same *g* values and line shape (see inset curve).

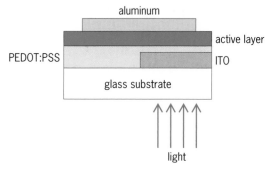

Fig. 2. Device configuration of a polymer-fullerene bulk heterojunction solar cell. The active layer (thickness 100–200 nm) results from a blend of donor (for example, conjugated polymer) and acceptor (for example, fullerene) type materials. The window electrode consists of indium tin oxide (ITO) covered with a thin (50-nm) film of hole-conducting PEDOT : PSS [poly(ethylenedioxythiophene) : poly(styrene sulfonate). For the cathode (100 nm), a low-work-function metal, such as calcium or aluminum, is used.

Converting solar radiation to electricity. One of the most effective and extensively studied photovoltaic device concepts relies on the donor-acceptor bulk heterojunction, where acceptor molecules are dispersed in a donor material (**Fig. 2**). A thin composite film of the two semiconductors is sandwiched between two electrodes with asymmetric work functions (for example, low-work-function metal on one side and higher-work-function metal on the other side). In the above configuration, each electrode can be expected to form an ohmic contact with the *p*- or *n*-type semiconductors (polymer and fullerene, respectively), while simultaneously blocking minority charge carriers. The separated transport paths for electrons and holes are an essential feature of such type of devices. For both the polymer and the fullerene, the electronic transport properties, as well as the optical absorption with respect to the solar spectrum must be optimized (Fig. 1). The latter can be realized by using organic semiconductors with relatively low-band-gap energy, in order to better exploit the red/infrared part of the solar spectrum. (The lower the band-gap energy, the more photons with low energy—that is, high wavelength—can be absorbed by the semiconductor.)

Figure 3*a* shows the external quantum yield of charge carrier generation in a P3HT:PCBM solar cell. The external quantum yield, representing the number of charges extracted by the electrodes to the outer circuit per incident photon, reaches a peak value of 65% ($\lambda = 520$ nm). Within the wavelength range 300–670 nm, the external quantum yield follows the P3HT:PCBM absorption spectrum indicated by the solid line. In contrast, a strong mismatch of the active layer absorption to the solar spectrum, especially in the red (above 625 nm) part of the spectrum, is obvious.

The current-voltage (*J-V*) characteristics of a P3HT:PCBM solar cell with 100 nm absorber thickness are shown in Fig. 3*b*. A photocurrent density J_{SC} of 8.2 mA/cm², an open-circuit voltage V_{OC} of

550 mV, and a fill factor of 50% are typical values for P3HT:PCBM–based devices. Those result in an overall power conversion efficiency η of 2.3%. An important peculiarity of devices based on the above absorber material combination is illustrated in the inset to Fig. 3*b*, where J_{SC} is plotted as a function of temperature at different illumination intensities. At low temperatures (below 270 K), J_{SC} turns out to be strongly temperature-dependent and reflects poor charge transport properties of the composite due to carrier recombination on traps (impurities or defects which can capture electrons or holes). At elevated temperatures, J_{SC} saturates and becomes nearly temperature-independent, clearly indicating that the charge carriers traverse the active layer without significant losses. If so, the active layer thickness can be increased in order to improve the overall absorption and, accordingly, to generate more charge carriers. **Figure 4** shows the current-voltage characteristics of two solar cells based on the P3HT:PCBM absorber with different thicknesses, namely, 100 and 350 nm. As the absorber thickness increases up to 350 nm, a strong increase of the photocurrent density from 8 to 15 mA/cm² can be observed. At the same time, the fill factor decreases from 50 to 37%, at identical open-circuit voltage. The overall power conversion efficiency of the 350-nm-thick device was calculated to $\eta = 3.1\%$.

The low fill factor definitely represents the main drawback of the device configuration. Analysis shows that such deficiency is neither due to a series resistance, which increased from 1 to 3.1 Ω-cm², nor due to the influence of the parallel resistance on the *J-V* characteristics. In the latter case, the total current in the reverse direction would decrease continuously with voltage. Instead, the photocurrent density saturates at about 18 mA/cm², as shown in Fig. 4. The electrical-field-dependent carrier recombination seems to provide a limiting mechanism for the short-circuit current and, thus, the fill factor. In relatively thick devices, the separation of photogenerated charges due to the built-in electrical field is no more efficient, and the charge carrier drift length is smaller than the active layer thickness.

Outlook. To improve the efficiency of polymer solar cells, currently ranging between 2.5 and 3.5%, it is vital to understand which physical mechanisms control the current-voltage characteristics of a given device. New material combinations with a larger mobility lifetime become necessary. Donor and acceptor polymers with higher charge carrier mobility are an important prerequisite. At the same time, in order to suppress recombination or to improve the lifetime of the carriers, the density of defects and impurities in the absorber layer must be reduced. Closely related to this issue is morphology optimization. So far, the solar cell absorbers investigated are three-component systems (that is, donor, acceptor, and solvent); hence, the phase separation of even purified components becomes a nontrivial challenge. The choice of the solvent and the donor-acceptor ratio are only two parameters out of many that need

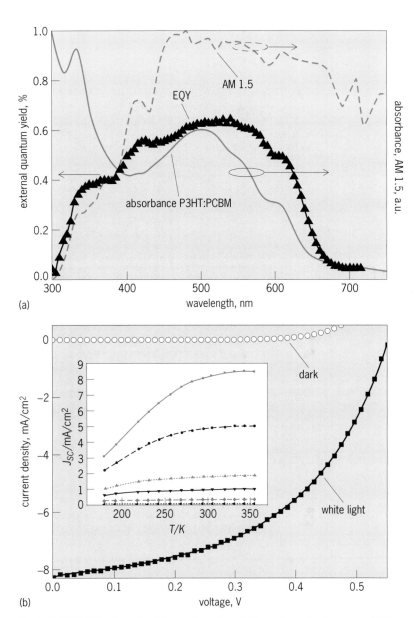

Fig. 3. **P3HT:PCBM solar cell. (*a*) External quantum yield spectrum of a the solar cell. The solid line shows the optical absorption spectrum of the composite, compared with the terrestrial solar spectrum, AM 1.5. (*b*) Current-voltage characteristics of the solar cell (100 nm absorber thickness) under 100 mW/cm² white light illumination and in the dark. Inset shows temperature dependence of the photocurrent density at different illumination intensities (0.07–100 mW/cm²).**

to be optimized. The origin of the open-circuit voltage is a matter of debate in the literature. The data available point at the importance of the absorber-electrode interface as being extremely sensitive to the choice of electrode materials. The analysis of the donor-acceptor energetics, setting the thermodynamic limit for V_{OC}, is not unambiguous, as it follows from electrochemistry data.

Agreement in the scientific community has been achieved on the issue of importance of better exploitation of the solar spectrum, which implies use of donor or acceptor systems with a lower band gap. The combined efforts of organic chemists, device engineers, and spectroscopists concerning band-gap optimization are in progress.

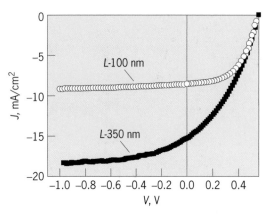

Fig. 4. Current-voltage characteristics of ITO/PEDOT:PSS/ P3HT:PCBM/Al solar cells with 100 and 350 nm absorber thickness (illumination intensity is 100 mW/cm²). [ITO = indium tin oxide; PEDOT/PSS = poly(ethylenedioxy-thiophene)/poly(styrene sulfonate); Al = aluminum.]

For background information *see* CONJUGATION AND HYPERCONJUGATION; CURRENT DENSITY; ELECTRON-HOLE RECOMBINATION; FULLERENE; HOLE STATES IN SOLIDS; ORGANIC CONDUCTOR; PHOTO-CONDUCTIVITY; PHOTOVOLTAIC EFFECT; SEMICON-DUCTOR; SOLAR CELL; SPIN LABEL; TRAPS IN SOLIDS; WORK FUNCTION (ELECTRONICS) in the McGraw-Hill Encyclopedia of Science & Technology.

Ingo Riedel; Jürgen Parisi; Vladimir Dyakonov

Bibliography. C. J. Brabec et al. (eds.), *Organic Photovoltaics: Concepts and Realization*, Springer, Berlin, 2003; C. J. Brabec et al., Tracing photoinduced electron transfer process from a polymer chain to a fullerene moiety in real time, *Chem. Phys. Lett.*, 340:232–236, 2001; C. J. Brabec, J. C. Hummelen, and N. S. Sariciftci, Plastic solar cells, *Adv. Mater.*, 11:15–26, 2001; J. De Ceuster et al., High-frequency (95 GHz) electron paramagnetic resonance study of the photoinduced charge transfer in conjugated polymer-fullerene composites, *Phys. Rev. B*, 64:1952061–1952066, 2001; I. Riedel et al., Effect of temperature and illumination on the current-voltage characteristics of polymer-fullerene bulk heterojunction solar cells, *Adv. Func. Mater.*, 14:38–44, 2004; N. S. Sariciftci et al., Photoinduced electron transfer from a conducting polymer to buckminster-fullerene, *Science*, 258:1474–1476, 1992.

Probiotics

Probiotics are defined by the Food and Agriculture Organization (FAO) of the United Nations as "live microorganisms which when administered in adequate amounts confer a health benefit on the host." The premise of probiotics is that healthy humans (and animals) harbor a large and diverse array of microorganisms, and at times the commensal microbial population needs replenishment, perhaps due to food sterilization or the use of antimicrobials in medications or as preservatives and livestock feed. Such propagation of selected species of the body's microbial content (typically, bacteria of the genera *Lacto-*

bacillus and *Bifidiobacterium*) is often directed at preventing or even treating disease or illness. The belief is that by administering bacteria that are naturally found in a niche (such as the mouth, throat, gastrointestinal tract, vagina, urethra, and skin), the host will be better able to restore and maintain good health.

Mechanisms of probiotic action. How probiotic strains are able to confer a health benefit on the host is still not fully understood; however, various possible mechanisms of action have been proposed.

Acute diarrhea. A number of studies have shown that the duration of acute diarrhea, caused by microorganisms such as rotaviruses in children and enteropathogenic bacteria in adults, can be significantly reduced with rehydration and the administration of probiotics. It is not clear how this works. Although receptor site blockage or competitive exclusion has been shown to take place in vitro (that is, probiotics have exhibited these mechanisms in the laboratory when exposed to pathogens that cause diarrhea), it may not be the main mechanism of action in the resolution of diarrhea in vivo, because pathogens often invade the epithelium or release a toxin before the probiotics are administered. In vitro activity is often cited as proof of probiotic activity; however, there are few cases in which such activity has been proven to be responsible for, or implicated in, resolution of disease.

Another theory is that the immune response of the host is triggered by the probiotics, which helps to eradicate the pathogens and stop the diarrhea. There is evidence that exposure of T cells to lactobacilli can rapidly induce T-cell activity and that ingestion of lactobacilli probiotics can quickly stimulate expression of secretory IgA, an antibody that may prevent antigens from adhering to gastrointestinal mucosal membranes. In addition, one study suggests that deoxyribonucleic acid (DNA) from probiotic strains can downregulate (reduce) inflammation. Why the indigenous lactobacilli of the host are not able to function in this way remains unknown but could be due to immune tolerance, their reduced numbers in the host, or their presence within dense biofilms, limiting their exposure to the immune modifiers. An untested concept is that probiotic bacteria signal the host to downregulate secretory and motility functions of the gut as well as inflammatory cytokine production which have collectively caused the diarrhea.

It has been suggested that probiotic organisms could kill the offending pathogens. However, this seems an unlikely primary mechanism of action in diarrheal intervention, since probiotic strains require 24–48 h to grow and produce sufficient metabolic products to kill other microorganisms, while the symptoms of disease are ameliorated within one day of ingestion of the probiotics.

Inflammatory bowel disease. The ability of probiotics to improve the well-being of patients with inflammatory or irritable bowel problems is being tested. The most encouraging documentation has come from the use of a high-dose, multispecies product, VSL#3, for patients with inflammatory bowel disease. No data exist as to presumptive mechanisms of action. Although modulation of immunity, including downregulation of inflammation, appear to be an end result, the biochemical or immunological pathway is far from clear, as are the roles of the eight constituent strains in the product. Studies are needed on interactions between the ingested organisms and the host and involve receptors (a family of cell-surface proteins that detect and signal an immune response against microbes) and neurotransmitters.

Vaginal and urinary tract infections. Lactobacilli are believed to play an important role in maintaining vaginal health, because they predominate in the normal microbial flora of the vagina and a decrease in the number of lactobacilli is associated with bacterial vaginosis and urinary tract infection. Lactobacilli are believed to maintain vaginal health through the production of bacteriocins (proteins that inhibit or kill competing bacterial species), hydrogen peroxide, and lactic acid, which appear to be important in killing or inhibiting growth of pathogenic organisms and maintaining a healthy microbial balance (approximately 10^7–10^9 bacteria per milliliter of vaginal fluid). However, the presence of these factors in healthy women, as well as their reduced concentrations in patients with bacterial vaginosis, yeast vaginitis, and urinary tract infection, has not been documented. Other possible mechanisms of probiotic activity include production of biosurfactants, or proteins and peptides contained within the biosurfactant mixture, which reduce pathogen colonization (presumably by interference with pathogen adhesion) and signal the downregulation of virulence factors produced by viruses and pathogens, such as *Escherichia coli* and streptococci.

Evaluating probiotics for human use. The simple existence of bacteria in a product, such as yogurt, so-called acidophilus powders, and fermented meat, does not make it a probiotic. Guidelines for the evaluation of probiotics have been proposed by the United Nations and the World Health Organization: In order for a product to be designated a probiotic, it should meet certain criteria. For example, the product should have a designated species and strain, be prepared in a way that ensures viability, and have demonstrated clinical efficacy. At present, however, there are no regulatory agencies to enforce these guidelines; therefore, it is the responsibility of the manufacturer, consumer, referring physician, or scientist to evaluate the health and nutritional properties of probiotics.

Species and strain designation. The bacterial speciation and strain numbers (for example, *Lactobacillus rhamnosus* GG) should be accurate and allow tracking via a science citation search. References should be available that document clearly what the organisms can do and the dose and form proven to confer health benefits.

Preparation to ensure viability. The term "live" in the FAO definition of probiotics is critical, and it means live at the time of use, not at the time of manufacture. Unless products are prepared with specific excipients, coatings, or tableting or encapsulation technologies that preserve viability (usually by blocking access of air and moisture), the contents are unlikely to survive longer than a few weeks or months. In some cases, refrigeration acts to preserve the viable count, such that there are still sufficient living organisms in the product at expiration.

Clinical efficacy. In terms of conferring a health benefit, this requires proving that the probiotics have a physiological effect on the host beyond simple provision of nutrition or placebo activity. Examples could be reducing the effects or risk of diseases or ailments, such as infection, lactose intolerance, inflammation, allergic reactions, and pathogen colonization in food such as poultry and beef. Two independent studies should be performed in which the probiotics are tested against a placebo in a sample size study that reaches statistical significance. If further clinical efficacy is desired, the probiotics can be compared with standard therapy, such as drug treatment for a disease.

In recent years, substantial clinical evidence has been accumulated from randomized, double-blind, placebo-controlled studies that show that certain probiotic strains (particularly *L. rhamnosus* GG, *L. acidophilus* La5, *L. rhamnosus* GR-1, *L. fermentum* RC-14, *L. reuteri* SD2112, *L. plantarum* 299, *L. casei* Shirota, *L. casei* DNN 114001, and *Bifidobacterium lactis* BB12) do indeed fulfill these guidelines. In livestock, there is mounting evidence that probiotic strains can lead to beneficial weight gain and improve the health of the animal as well as potentially reduce colonization of pathogens, such as salmonella in chickens. However, translation of research findings into proven product formulations has, for the most part, not yet been done. Notably, some animal probiotics contain multiple strains and species, such as *Bifidobacterium pseudolongum* or enterococci, that are not normally used in human products.

Summary. The past 5 years has seen an unprecedented interest in probiotics. Sophisticated molecular biology, immunology, microbial genetics, and biochemistry studies are elucidating the physiological roles of indigenous microbiota of the host, candidate probiotic strains, and existing probiotic organisms. The application of the United Nations/World Health Organization Guidelines to the production and marketing of strains, if achieved, will create a new level of reliability that will take probiotics into mainstream medical care.

[The Canadian Research and Development Centre for Probiotics is supported by a grant from the Ontario Research and Development Challenge Fund. Studies are funded by NSERC.]

For background information *see* BACTERIA; CLINICAL MICROBIOLOGY; DIARRHEA; INFLAMMATORY BOWEL DISEASE; MEDICAL BACTERIOLOGY; MICROBIOTA (HUMAN); VAGINAL DISORDERS in the McGraw-Hill Encyclopedia of Science & Technology.

Gregor Reid

Bibliography. D. Goossens et al., Probiotics in gastroenterology: Indications and future perspectives, *Scand. J. Gastroent. Suppl.*, 239:15–23, 2003; *Guidelines for the Evaluation of Probiotics in Food*, Food and Agriculture Organization of the United Nations and World Health Organization Working Group Report, 2002; J. A. Patterson and K. M. Burkholder, Application of prebiotics and probiotics in poultry production, *Poultry Sci.*, 82(4):627–631, 2003; G. Reid et al., New scientific paradigms for probiotics and prebiotics, *J. Clin. Gastroent.*, 37(2):105–118, 2003; G. Reid et al., Potential uses of probiotics in clinical practice, *Clin. Microbiol. Rev.*, 16(4):658–672, 2003.

Protein networks

The primary information carried from generation to generation by an organism's genome resides in the deoxyribonucleic acid (DNA) sequences of genes that code for proteins. The recent determination of genome sequences from a number of organisms including humans has revealed thousands of genes encoding proteins that have yet to be functionally characterized. Proteins play diverse roles in a wide variety of cellular activities, and these roles usually involve specific interactions with other proteins. A view that has become useful in trying to understand the role of proteins in biological systems is to consider each protein as a member of a functional network of interacting proteins. Individual proteins make specific contacts with one or more other proteins, and together form a network that mediates a particular biological process. The protein network perspective, inspired in part by the recent acquisition and analysis of large datasets of protein interactions, is proving invaluable in assigning functions to uncharacterized proteins and in modeling pathways. The recognition of protein networks as important modules of biological function also has contributed to a movement toward systems-level studies of biological processes.

Biologists generally categorize cellular protein networks into two types, those that involve transient interactions among their protein components and those that involve stable multiprotein complexes. The distinction is instructive even if real protein networks actually populate the entire spectrum from highly transient to highly stable. Networks of transiently interacting proteins are typified by the signal transduction pathways involved in sensing and responding to environmental signals. In the canonical signal transduction pathway, a ligand binds to a protein receptor embedded in the cell membrane, leading to a change in the receptor's conformation or enzymatic activity, which in turn promotes its interaction with a second protein in the cytoplasm. This results in modification or activation of the cytoplasmic protein, leading to its interaction with yet another protein, and so on. Ultimately, the original signal (extracellular ligand) is transduced to some readout, such as activation of a transcription factor

and expression of a set of genes that modify the cell's behavior. Stable protein networks, on the other hand, are typified by the protein complexes that form molecular machines to carry out processes such as DNA replication, ribonucleic acid (RNA) transcription, or protein degradation. Other proteins stably assemble into cytoskeletal structures responsible for cell morphology and mobility. In each of these and other systems, information or energy moves through a network of protein-protein interactions, whether the interactions are stable or transient.

A first and essential step toward gaining a full understanding of a functional network is to catalog all of its constituent proteins and all of the binary interactions among them. Such catalogs have begun to emerge, largely through efforts to understand the functions of the huge number of uncharacterized proteins identified in genome-sequencing projects. Clues about the function of an uncharacterized protein can be obtained by identifying other proteins with which it interacts. If a new protein under study is found to interact with an already characterized protein, for example, the new protein is likely to be involved in the same biological process as the characterized protein. The power of this approach, sometimes referred to as "guilt by association," has inspired the development of high-throughput technologies for detecting protein-protein interactions.

Building proteome-scale protein networks. To date, the most widely used approach for large-scale protein interaction studies is the yeast two-hybrid system. In a yeast two-hybrid assay, two proteins are tested for binary interactions within the nucleus of yeast cells. The experimenter first constructs hybrid genes that encode each protein linked or "fused" to a domain from a transcription factor. The hybrid genes are then introduced into yeast where they express the two hybrid proteins. One protein is fused to a DNA-binding domain (DBD) that binds to operator sequences upstream of reporter genes engineered into the yeast strain. The second protein is fused to a transcription activation domain (AD). If the two test proteins interact when coexpressed, transcription of the reporter genes will occur. The two-hybrid assay is simple and requires no protein purification or special instrumentation. Moreover, the assay is independent of native protein expression levels and does not require a system for expression of affinity-tagged proteins in the native organism. With the availability of entire genome sequences and high-throughput cloning techniques it has become possible to generate AD and DBD clone sets representing the majority of the predicted open reading frames (ORFs, the DNA sequences that encode proteins) for an organism. Systematic high-throughput interaction screening is facilitated by introducing the AD and DBD clones into haploid yeast strains of opposite mating types. The resulting strains are combined under conditions promoting mating, resulting in the formation of diploid strains that express both the AD and DBD proteins, which can then be assessed for reporter activity.

Large-scale yeast two-hybrid analyses have been performed to characterize proteins from yeast, *Drosophila*, the roundworm *Caenorhabditis elegans*, and the bacterium *Helicobacter pylori*. These studies have mapped out thousands of protein interactions. In the *Drosophila* study, for example, more than 20,000 interactions were detected among over 7000 proteins, representing a third of the proteome of this insect. These maps are rich with new biologically relevant protein interactions and have helped predict the functions of many uncharacterized proteins by placing them into characterized networks and pathways.

Protein interaction maps and functional networks. Experimentally derived protein interaction maps inevitably include false negatives and false positives, and the maps generated by high-throughput yeast two-hybrid are no exception. Biologically relevant interactions are missed (false negatives) for a variety of reasons, including potential problems with protein folding, the absence of certain posttranslational modifications in yeast, or interference from the fused transcription factor domains, which could obscure the interaction surfaces of one or both proteins. Trafficking of the proteins from where they are synthesized on the ribosomes to the nucleus must also occur for the assay to work, precluding detection of interactions that require some other local environment such as the cell membrane. It is also well known that the yeast two-hybrid system results in the detection of some false positives, protein interactions that do not occur naturally. These are of concern as they cannot be readily distinguished from real interactions simply by inspection of the interaction map. The presence of both false positives and false negatives, therefore, means that the interaction maps are not precise reflections of the functional protein networks within a cell, but only provide useful guides to finding functional networks. To derive functional protein networks, additional data and information from other sources are required.

Data generated from different technologies can help deduce functional protein networks in two ways. First, since false negatives are often specific to certain techniques, the combination of data from multiple technologies can create more comprehensive interaction maps. Second, it has been shown that interactions detected by two or more different high-throughput techniques are more reliable than those detected by just one technique. Thus, the overlapping data from different technologies has a lower false-positive rate. Besides yeast two-hybrid, other high-throughput technologies currently being used to generate information about protein interactions include protein microarrays and the systematic identification of protein complexes via in-vivo pull-downs (precipitation with antibodies of one of the proteins in the complex) and mass spectrometry. A number of in-silico approaches (that is, using computer modeling) that predict protein interactions have also been developed. Additionally, information about genetic interactions and mRNA expression provide potentially useful correlative information for analysis of protein networks.

Network biology. The flood of protein interaction data coming from yeast two-hybrid and other technologies applied at the proteome level has begun to provide insights that go beyond the functions of individual proteins and protein complexes. One surprising finding is that most of the proteins in the interaction map for a particular organism are linked together into one large network component. Of the proteins involved in the ~20,000 *Drosophila* protein interactions identified by yeast two-hybrid, for example, 98% are connected together into a single network (see **illustration**). In this network, some of the proteins are highly interconnected and others are only sparsely interconnected. The finding of similar results in different organisms and with data from different technologies suggests that this phenomenon is not an experimental artifact, but has profound biological significance. Apparently, most of the functional networks and pathways encoded by a genome are linked together into a higher-order structure, which may help the cell coordinate and integrate different processes. Remarkably, further study of protein networks is beginning to benefit from research on other, nonbiological networks.

The study of networks has revealed common properties between such diverse systems as protein networks, the Internet, and social networks. These properties appear to be independent of the nature of the nodes (such as proteins, computers, or people) and the interactions between the nodes, suggesting the possibility of being able to develop some

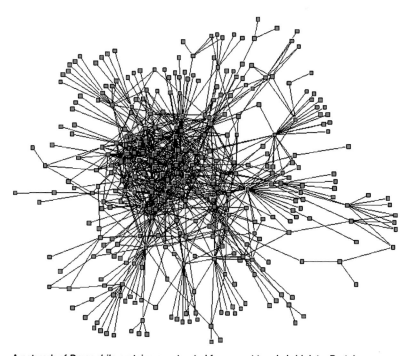

A network of *Drosophila* proteins constructed from yeast two-hybrid data. Proteins are represented as squares and interactions as lines. All proteins shown are connected into one large network. Some proteins have many interactions while others have few. Biologists can use such an interaction map to discover and study biological pathways that operate within cells.

understanding of a system without detailed knowledge of its individual parts. Protein networks and many nonbiological networks, for example, have "scale-free" topologies. One consequence of this topology is that a small number of highly connected nodes contribute significantly to holding the network together. Studies of protein networks in yeast have shown that the more highly connected proteins are the most evolutionarily conserved and that they serve more essential functions. Examination of protein network topology has also revealed that proteins within highly interconnected clusters are often functionally related, providing another useful example of correlation between network topological properties and biological function. Protein networks also have the "small-world" characteristic of interconnectivity that results in relatively short average path lengths between any two proteins. In social networks, this property explains the surprisingly small number of "degrees of separation" between any two individuals, but its importance to protein networks has yet to be fully understood. Additional studies may reveal how topological properties influence the mechanisms by which energy and information flow through functional protein networks. These and other dynamic properties of protein networks must be studied to develop a full understanding of how proteins work together to mediate biological processes.

For background information *see* FUNGAL GENETICS; GENE; PROTEIN in the McGraw-Hill Encyclopedia of Science & Technology.

Jodi R. Parrish; Russell L. Finley, Jr.

Bibliography. A. L. Barabasi and Z. N. Oltvai, Network biology: Understanding the cell's functional organization, *Nat. Rev. Genet.*, 5:101–113, 2004; T. R. Hazbun and S. Fields, Networking proteins in yeast, *Proc. Nat. Acad. Sci. USA*, 98:4277–4278, 2001; E. Phizicky et al., Protein analysis on a proteomic scale, *Nature*, 422:208–215, 2003; C. von Mering et al., Comparative assessment of large-scale data sets of protein-protein interactions, *Nature*, 417:399–403, 2002.

Quorum sensing in bacteria

In its simplest form, quorum sensing refers to the ability of bacteria to communicate. It describes the capacity of certain bacteria to exhibit coordinated behavior in response to a particular population density. These bacteria usually rely on the production, accumulation, and subsequent response to diffusible signal molecules in order to sense population. These signal molecules accumulate in environments that can sustain a sufficiently dense population, or quorum, of the signal-producing bacteria. When the concentration of the signal molecule reaches a critical level, the quorum-sensing bacterial population responds through the concerted expression of specific target genes.

Historically, bacteria were thought of as solitary and self-contained individuals, each growing independently of the population. Implied in this model was the notion that microorganisms lack the ability to organize and communicate. This view has changed dramatically in the last two decades with the discovery that bacteria can use chemical signals to communicate and to coordinate group activities. The pioneering work, from the laboratories of J. W. Hastings and Alex Tomasz, showed that when the population of a particular bacterium reached a critical density, the bacteria triggered the expression of specific genes in a concerted fashion. Hastings' work revealed that the production of luminescence by the gram-negative bacterium *Vibrio fischeri* was controlled through the production and accumulation of a diffusible signal molecule termed an autoinducer. Tomasz demonstrated that the gram-positive bacterium *Streptococcus pneumoniae* controlled its ability to take up deoxyribonucleic acid (DNA) from the environment by responding to another diffusible chemical, later referred to as the competence factor. The work of these two pioneers was initially met with skepticism, but quorum sensing has since been demonstrated in multiple bacterial systems, many of which establish symbiotic or pathogenic relationships with animal or plant hosts.

Gram-negative bacteria. One of the best-studied examples of quorum sensing can be found in *V. fischeri*, a symbiotic marine bacterium that produces light when it is associated with its animal host (**Fig. 1a**). In their seawater environment, free-living *V. fischeri* are found in low population densities and do not emit light. During their symbiotic association, the bacteria inhabit the light organs of several marine fishes and squids at very high densities—where they emit light. The onset of luminescence is regulated by the environmental accumulation of a specific chemical signal. This signal molecule was shown to be a specific acylated derivative of homoserine lactone or AHL (**Fig. 2a**). As the cell density increases during the symbiotic association with the animal host, the diffusible AHL quorum-sensing signal molecule accumulates in and around the cells. Upon reaching a threshold level (10 nanomoles), the AHL interacts with and subsequently activates a regulator protein known as LuxR. The LuxR/AHL complex then modulates the expression of the genes necessary for the synthesis of the light-producing proteins. In addition, the LuxR/AHL complex activates the expression of a gene that encodes the LuxI protein, which is responsible for the synthesis of the AHLs. This leads to a rapid rise in the levels of autoinducer and creates a positive feedback circuit that sustains the production of light. Bioluminescence provides the animal host with a defense mechanism. The host in turn rewards its symbiotic partner with nutrients and a sheltered environment.

In many gram-negative bacteria, the basic AHL structure consists of a homoserine lactone head group attached to an acyl side chain varying in length from 4 to 18 carbon atoms. This variability provides the specificity of quorum-sensing signals. To add complexity, most organisms produce more than one

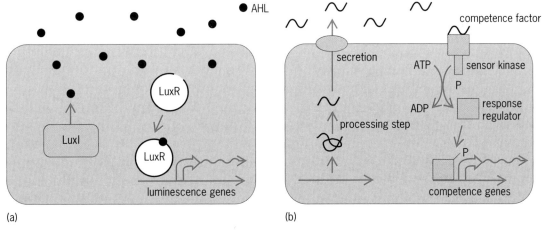

Fig. 1. Quorum-sensing circuits in bacteria. (*a*) In gram-negative bacteria (such as *Vibrio fischeri*), acyl-homoserine lactones (AHLs; filled circles) are produced by the LuxI proteins and detected by LuxR. AHLs, which can freely diffuse into and out of the cell, increase in concentration as the bacterial population rises. The interaction between the LuxR protein and the AHL molecule results in the specific binding of the complex to promoter DNA elements and the transcriptional activation of the luminescence genes. (*b*) Gram-positive bacteria (such as *Streptococcus*) synthesize peptides (wavy lines) that are usually modified and then actively secreted. Detection occurs via a two-component signal transduction circuit, leading to the ATP-driven phosphorylation of a response regulator protein, which then binds to promoter DNA and regulates transcription of the competence genes.

type of AHL, and different organisms can produce the same molecules. There is therefore some overlap in the production and recognition of AHLs by different organisms.

In the gram-negative organisms that have been characterized so far, the quorum-sensing mechanisms are similar to the *V. fischeri* paradigm, but the activated target genes are diverse. For example, in the plant pathogen *Agrobacterium tumefaciens*, quorum sensing regulates the transfer of a plasmid necessary for virulence. This ensures that all of the agrobacteria are properly equipped for the pathogenic plant invasion process. Quorum sensing also regulates the production of virulence factors in the human opportunistic pathogen *Pseudomonas aeruginosa* and the plant pathogen *Erwinia carotovora*, making certain that virulent processes only occur once the bacterium has achieved a particular population density inside the host. Many other important bacterial events are also regulated through quorum sensing such as the synthesis of exopolysaccharide, antibiotic production, and the control of motility.

Gram-positive bacteria. To date, AHL production has not been demonstrated in any gram-positive bacterium. Quorum sensing in gram-positive bacteria is instead commonly associated with the production of

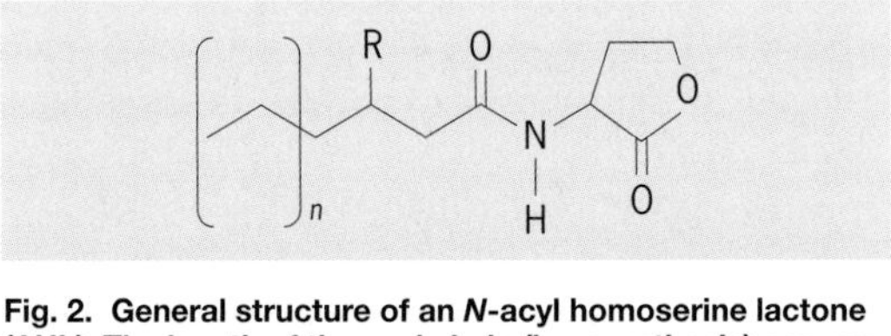

short amino acid chains or peptides as signal molecules. Certain species of *Streptococcus*, *Enterococcus*, and *Bacillus* produce linear peptides, which can be modified after their synthesis. The production and release of cyclic peptides has also been reported in species of *Staphylococcus* and *Enterococcus*.

Most peptide signals in gram-positive quorum-sensing systems are synthesized as an unprocessed translation product and are later modified and secreted into the external milieu. One of the best-described examples of gram-positive quorum sensing can be found in the streptococci, which are able to absorb DNA from their environment and integrate it into their own genetic material in a process known a competence (Fig. 1*b*). High population densities stimulate this process. The peptide signal for *Streptococcus* competence is generated as a 40-amino-acid precursor that is then processed and exported by various bacterial proteins. The final peptide signal is only 17 amino acids in length and is designated the competence-stimulating peptide, or CSP (Fig. 2*b*). CSP acts through a two-protein signaling relay system. In this system, a sensor protein transduces the environmental signals into metabolic changes by transfering a phosphate group to a regulator protein which, once activated, induces the expression of genes. The CSP-regulated genes endow the cells with the ability to bind environmental DNA, to import this genetic material into the cell, and to mediate its integration into the bacterial genome.

Other gram-positive bacteria employ similar quorum-sensing systems that regulate important events such as sporulation and competence in *B. subtilis*, control of toxins and virulence factor production in *S. aureus*, and plasmid conjugation in *E. faecalis*.

Intra- and interspecies cell-cell communication. As described, quorum sensing allows bacteria to

Fig. 2. General structure of an *N*-acyl homoserine lactone (AHL). The length of the acyl chain (in parenthesis) can vary from 4 to 18 carbons ($n = 0$ to 14), and the AHL chain can be modified also at the third carbon position, where R can be —H (fully reduced), —OH (hydroxyl), or ═O (carbonyl).

engage in species-specific recognition of self in a mixed population. It would seem that another important component of bacterial communication should be the ability to detect the presence of other species in a particular environment. This could allow the bacteria to sense the ratio of self to others and to specifically adapt their behavior to optimize survival. Recent work suggests that indeed bacteria use a separate quorum-sensing system for interspecies communication. In addition to the classical AHL-producing quorum-sensing system, *V. harveyi* possesses a second circuit based on a signal originally described as autoinducer-2 (AI-2) and later identified as a furanosyl borate diester. Synthesis of AI-2 requires a protein designated LuxS, homologs of which have been identified in a wide variety of gram-negative and gram-positive bacteria. Analogous to the system already discussed in gram-positive bacteria, two-component signal transduction proteins carry out autoinducer-mediated signal relay in *V. harveyi*. The widespread presence of this AI-2-mediated signaling system suggests that it is a more universal signal that could promote interspecies bacterial communication.

Irrespective of the type of quorum-sensing signal used, intra- and interspecies quorum sensing allows a population of bacteria to coordinate its behavior—thereby allowing a population of individual bacteria to acquire some of the characteristics of a multicellular organism. This type of behavior occurs in the formation of biofilms. Biofilms are highly ordered bacterial communities that allow bacteria to live adhered to surfaces. They are often responsible for the ability of bacteria to cause persistent infections, including some of those seen in the urinary tract, middle ear, heart valves, and implanted medical devices. The mechanism of biofilm formation, and the bacterial communication involved in this type of organized behavior, promises to be a major focus of future quorum-sensing research.

Conclusions. The last two decades have witnessed the development of the exciting new field of bacterial communication. It is now clear that most, if not all, bacterial species have sophisticated communication systems that allow them to coordinate their activities. The identification of species-specific as well as universal intercellular signaling molecules suggests a mechanism by which bacteria can determine the proportions of self and other in mixed-species environments, and respond to this information by appropriately modulating gene expression. Quorum sensing plays a major role in the regulation of many of the genes necessary for successful establishment of pathogenic and symbiotic interactions. Future development of strategies that interfere with intra- and interspecies cell-cell communication could therefore provide a new paradigm of treatment for certain bacterial diseases. The next few years will hopefully provide a wealth of information on how these quorum-sensing systems participate in the insufficiently understood prokaryotic-eukaryotic cell-signaling systems.

For background information *see* BACTERIA; BACTERIAL PHYSIOLOGY AND METABOLISM; BIOFILM; BIOLUMINESCENCE; CHEMORECEPTION; VIRULENCE in the McGraw-Hill Encyclopedia of Science & Technology.

Juan E. González

Bibliography.   C. Fuqua and P. Greenberg, Listening in on bacteria: Acyl-homoserine lactone signaling, *Nature Rev.*, 3:685–695, 2002; C. Fuqua, M. R. Parsek, and P. Greenberg, Regulation of gene expression by cell-to-cell communication: Acyl-homoserine lactone quorum sensing, *Annu. Rev. Genet.*, 35:439–468, 2001; J. González and M. Marketon, Quorum sensing in the nitrogen fixing rhizobia, *Microbiol. Mol. Biol. Rev.*, 67:574–592, 2003; P. Greenberg, Tiny teamwork, *Nature*, 424:134, 2003; N. A. Whitehead et al., Quorum-sensing in Gram-negative bacteria, *FEMS Microbiol. Rev.*, 25:365–404, 2001; K. B. Xavier and B. L. Bassler, LuxS quorum sensing: More than just a numbers game, *Curr. Opin. Microbiol.*, 6:191–197, 2003.

Radiation-curable inks and coatings

Radiation-curable inks and coatings, also known as energy-curable inks and coatings, are based on reactive acrylate chemistry. They can be applied by any standard printing process and then cured (polymerized) in-line to a tough, dry film with a brief exposure to either ultraviolet (UV) or electron beam (EB) energy. The food packaging market is a good example of where radiation-curable inks and coatings have become the dominant printing technology. The resulting ready-to-ship printed packages are a revolutionary change in an industry in which warehouses full of pallets of cartons printed with slow-drying, oil-based inks were the norm.

Composition. Energy-curable inks and coatings are made up of many standard ingredients, including pigments, waxes, and defoamers. However, the predominant part of the formula consists of a mixture of materials, all having pendant, reactive acrylate ($H_2C{=}CHCOOR$) functionality. When this mixture of materials is exposed to a suitable energy source, the acrylate double bonds polymerize and cross-link adjacent polymer chains to form a tough, three-dimensional polymer network on the printed substrate.

Energy-curable materials fall into two classes, loosely based on their molecular masses. The low-molecular-mass materials, usually less than 1000 daltons, are known as monomers. These tend to be fluids and serve as colorant carriers, as does oil in a conventional lithographic ink or water or solvent in a fluid flexographic or gravure ink. In conventional ink drying, the fluids must evaporate or be absorbed by a porous substrate, whereas the fluids (reactive monomers) in UV/EB inks are incorporated into the cured (dried) film. Monomers often have one to four reactive acrylic groups, although some have higher functionality (**Fig. 1**). A functionality of greater than two is required for cross-linking.

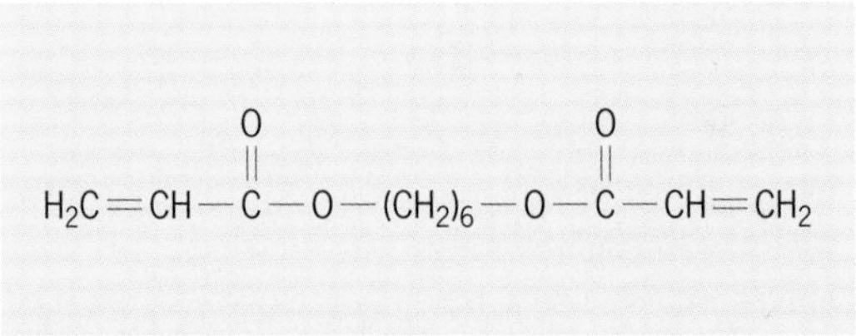

Fig. 1. Hexanediol diacrylate, a typical monomer.

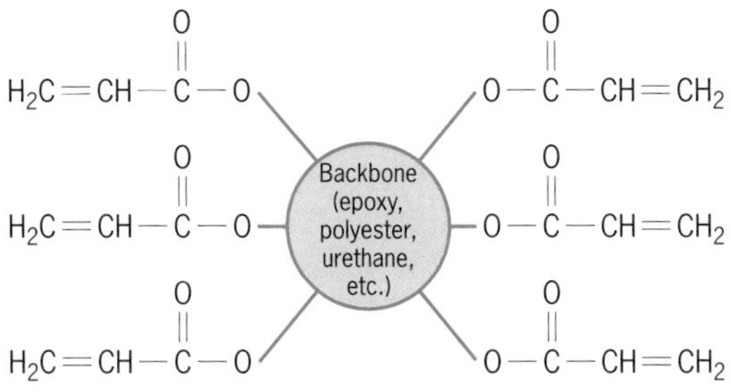

Fig. 2. General structure of an oligomer.

The other class of curable materials has higher molecular masses, often from 1000 to several thousand daltons. The materials are not large enough to meet most definitions of a polymer and are known as oligomers or prepolymers. Oligomers are epoxy, polyester, or urethane resins that have been esterified with acrylic acid (**Fig. 2**). Most of the performance properties of the ink, such as chemical resistance, rub resistance, gloss, and coefficient of friction, come from the oligomers.

UV versus EB. A critical concept is the difference between UV- and EB-curable products. A curtain of high-energy electrons falling on a thin film of energy-curable ink or coating has enough energy to initiate and complete the curing process. The energy output from a UV lamp, however, is much less. Thus, UV-curable inks require a photoinitiator, a chemical that can absorb UV energy, decompose, and have its decomposition fragments pass the energy to the curable components of the ink or coating. Photoinitiators are designed to have a chemical structure that absorbs energy at the peak emission wavelength of the light source, and to have a labile bond that will undergo homolytic cleavage to create relatively stable free radicals. These free radicals initiate the curing process.

UV inks can be used on presses that print individual sheets (sheet-fed presses) or on presses that print on a continuous roll of paper (web presses). EB inks can be printed only on web presses because the curing must be done in an nitrogen-inerted (that is, oxygen-free) zone. This requires that the EB curing device have narrow slits for the web to enter and exit, in order to minimize nitrogen loss. This limits EB to web processing since slits big enough to admit a sheet-gripping device would allow an unacceptable amount of nitrogen leakage, inhibiting cure or requiring a costly amount of nitrogen to be used.

Curing mechanism. In both processes, the curing mechanism is normally free-radical polymerization, with the activated double bonds of an acrylate ester undergoing a typical head-to-tail polymerization. Since the individual monomers and oligomers usually contain multiple acrylate groups, significant cross-linking occurs.

There are a smaller number of products known as cationic UV inks and coatings. These also are formulated with monomers and oligomers containing reactive double bonds, but the photoinitiator decomposes upon exposure to UV energy to give a strong protic acid that initiates the polymerization. The acidic nature of the process limits somewhat the pigments and other ingredients that can be used. Cationic inks are for UV only, as the EB process does not require any photoinitiator.

Curing equipment. In EB curing, accelerated electrons are used to cure inks and coatings. The basic design of an electron-curing unit consists of a tungsten filament, chamber, vacuum pump, high voltage, thin titanium foil, and nitrogen atmosphere. The curing of inks and coatings takes place by forcing the electrons to leave the filament or wire. The electrons are drawn to areas of lower voltage, accelerating to speeds of more than 100,000 mi/s. This acceleration is made possible by using a vacuum, which results in virtually no resistance to movement. Once the electrons have been accelerated to high speeds, they are guided toward a thin titanium-foil window. The electrons pass through the thin foil window and penetrate the ink or coating, initiating polymerization.

UV light is electromagnetic radiation, ranging 180–400 nanometers. Sources of UV light include sunlight, fluorescent lamps, and mercury vapor lamps. UV light is energetic enough to break the chemical bonds of the photoinitiator. This energy starts the photochemical reaction necessary to polymerize UV-curable inks and coatings. The primary wavelengths needed for curing inks and coatings are around 250 nm and 365 nm.

Finished print characteristics. A significant advantage of UV/EB inks is that they are fully dry as they come off the printing press. This means that cartons are ready to die-cut, fill, and ship immediately after printing. Other printing processes can require drying for hours or days before they can be processed. UV/EB inks have good adhesion to a wide range of plastic substrates, as well as high print quality. In addition, cured UV/EB inks and coatings have a high degree of chemical and moisture resistance.

EB offers some advantages over UV in that the degree of cure achieved is higher. Although there are no good quantitative measures for degree of cure, it is known that the energy input of the EB curing unit is higher than a UV curing unit. An EB print has no residual photoinitiator fragments in the ink film and a lower degree of residual extractables and volatiles in the cured ink film, compared to a UV print. These factors encourage the use of EB over UV for food packaging.

UV ink is fully satisfactory for food packaging, and it is widely used in Europe and well known in the United States. However, since test results show that EB residuals in the ink film are even lower than the very low levels from UV printing, EB is more common for longer-run food-packaging work. Given optimal curing conditions for both UV and EB processes, the total residuals in a cured-ink film as measured by a purge-and-trap, gas chromatography/mass spectrometer method are usually less than 100 ppm. The low residuals in the cured films, in addition to the inks low volatile organic compound (VOC) content, make UV/EB inks environmentally friendly (green).

One concern may be the odor of UV/EB inks before printing. The odor is not significantly stronger but it is different from that of all other inks, and pressroom workers may react. After curing, the odor is extremely low. This is important since the major use of UV/EB inks is for printing food packaging.

A disadvantage of UV/EB inks is that they are more expensive than conventional inks. This is because the chemical raw materials are difficult to make and more expensive than conventional raw materials. As a result, UV/EB inks are used only where performance advantages outweigh their added cost.

Markets for energy-curable inks. The lithographic market is about 80% UV and 20% EB. All EB lithographic printing is done on web presses. Around 90% of the EB lithographic market is for food packaging, and most of the rest is for labels that need high chemical resistance.

In UV lithographic printing, the sheetfed-to-web press ratio is about 1:1. Sheetfed UV lithography can be for folding carton, commercial printing, forms, advertising, metal cans, and many submarkets. The substrates used vary from uncoated forms of paper, to board, to coated commercial stock, to many plastics, vinyls, films, and foils. Web UV lithographic printing is mostly for food packaging on board or polyboard. UV lithographic printing for metal beverage cans is a large market but requires specialized presses.

The three largest markets for UV flexographic printing are tags and labels, folding cartons, and flexible packaging. UV flexographic folding carton printing is a rapidly growing area. Most of the business is in either personal care products or other consumer goods.

For background information *see* CHARGED PARTICLE BEAMS; FREE RADICAL; INK; PHOTOCHEMISTRY; POLYMERIZATION; PRINTING; ULTRAVIOLET RADIATION in the McGraw-Hill Encyclopedia of Science & Technology.   Don P. Duncan

Bibliography. C. J. Rechel (ed.), *UV/EB Curing Primer 1*, RadTech International North America, Northbrook, IL, 1997; J. Hess, Signs of a bright future, *Ink World*, p. 26, October 2000; P. K. T. Oldring (ed.), *Chemistry & Technology of UV & EB Formulation for Coatings, Inks & Paints*, vols. 1–5, SITA Technology, London, 1991; S. Oller, UV cures all, *Amer. Printer*, p. 40, June, 2003; S. Oller, UV web presses, *Amer. Printer*, p. 38, April, 2002; K. Tomlinson, EB's bright future, *packagePRINTING Mag.*, p. 61, October, 2002.

Radio-echo sounding

Radio-echo sounding (RES) is a technique used by glaciologists to measure the internal structure, ice thickness, and subice morphology of ice masses. It has been used extensively in Greenland and Antarctica to determine the volume of ice stored within large ice sheets, and in numerous locations to build an understanding of the sizes of smaller ice caps and glaciers. Such information has been central to quantifying the rise in global sea level that might occur through future climate warming and glacier melting.

Data acquisition is highly time efficient, and only a moderate degree of postprocessing is required to make the data usable. Analysis of these data allows the bed configuration to be established, the conditions at the ice-bed interface to be evaluated, the flow history of the ice sheet to be identified, and former rates of ice accumulation to be reconstructed. In addition, RES has been used to identify and measure lakes located underneath ice sheets, the largest and best-known Antarctica subglacial lake being Lake Vostok.

RES was developed by British and American scientists in the early 1960s. Previously, ice thickness was measured using seismic sounding, which is a very time consuming exercise. Consequently, relatively little was known about the configuration of large ice masses. RES, especially if mounted on an aircraft (airborne RES), is a far more efficient means of obtaining information on the depths of ice sheets (see **illustration**). It is now established as the most important technique used to measure the subsurface of large ice masses.

Technique. The RES technique works by firing very high frequency (VHF) radio waves downward into a glacier, where they are reflected off boundaries separating material of contrasting electrical properties (see illustration). Such contrasts occur at the ice-air and ice-bed interfaces of the ice sheet surface and base, respectively. In addition, radar reflections occur off internal layering within the ice sheet. RES works well in ice sheets because ice at low temperatures ($<-10°C$ or $14°F$) is virtually transparent to VHF radio waves. The two-way travel time of the reflected waves, received close to their source, allows the thickness of ice (in which the waves have traveled) to be calculated, as the speed of radio waves in ice is known reasonably well (168.5 m/μs or 552.8 ft/μs).

The radar power P_R received at the surface of an ice mass is related the power reflected at a boundary of dielectric contrast by the equation below, where

$$P_R \propto \frac{P_T q R}{64\pi^2 z^2 L}$$

P_T is the transmitted power, q is the signal gain due to refraction at the air-ice interface, R is the power

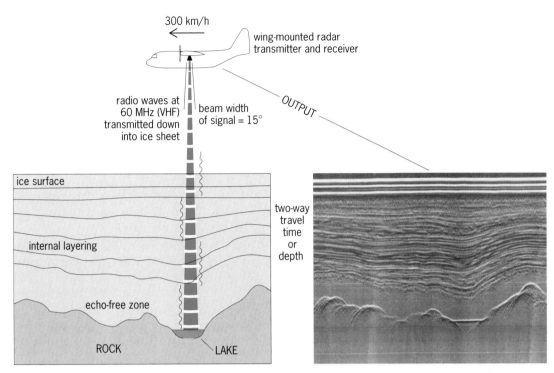

The technique of RES, with a U.S. Navy C130 Hercules aircraft as used in the 1970s to obtain approximately 400,000 km (250,000 mi) worth of flight track in both east and west Antarctica. The example shows how subglacial lakes may be distinguished from other ice-bed reflections. The data also illustrate the nature of internal ice sheet layering measured by RES.

reflection coefficient (related to the electrical properties of the reflector), L is the loss factor due to energy absorption, and z is the distance between the reflector and the transmitter. Factors L and z mean that less energy will be returned to the ice surface the deeper the reflections occur. Although these loss terms can be offset by increasing the power of the transmitted radio wave, they often conspire to limit the ice depth that can be sounded to around 4 km (2.5 mi). Even so, the thickest ice ever measured using RES was recorded at 4776 m (3 mi) in the Astrolabe Subglacial Basin in East Antarctica.

Large ice masses are measured by RES through a series of linear transects, often arranged in a grid. Both of the world's great ice sheets, Greenland and Antarctica, have been measured in this way. In Greenland, RES transects cover the entire region, allowing a full appreciation of the form of this ice sheet. In Antarctica, however, the coverage is less complete. The single most comprehensive RES survey in Antarctica was undertaken in the 1970s by a consortium of the Scott Polar Research Institute (University of Cambridge), the U.S. National Science Foundation, and the Technical University of Denmark. The data collected in this survey were used to construct the Antarctic Geophysical Folio, and remain the basis of subglacial maps of the continent. Subsequently, M. B. Lythe and coworkers compiled all the available ice thickness data in Antarctica to produce the BEDMAP (Bedrock Mapping) database, depicting the continent's subglacial topography.

The accuracy of ice thickness measurements made by RES is of the order of 1.5% of the total ice thickness, and is dependent on the pulse length of the transmitted radio wave. Far greater errors can occur in the "gridded" datasets generated from RES surveys in data-free regions and in between the data transects. Such errors can be significant in Antarctica, where substantial data-free regions exist today. As the bed topography is a vital input to numerical ice sheet models, predictions of the ice sheet response to climate change may be restricted by these errors.

Ice sheet morphology. The power of the subglacial reflection, in conjunction with the general subice morphology, has been used to reveal important information relating to landscape evolution, ice sheet dynamics, and the presence of lakes beneath the ice sheet. The roughness of the reflector will influence heavily the power reflected from it. If the surface is rough (at the wavelength of the radio wave), scattering will occur. Analysis of the returned power has been used to identify an association between enhanced ice flow and smooth subglacial conditions in ice streams. Conversely, rough terrain is often found at the center of ice sheets, where ice flow speeds are low and subglacial erosion is limited. The dielectric constant ε of the media on either side of the reflecting layer will also influence the power reflected. The dielectric contrast (which is related to the reflection coefficient R) may vary considerably beneath an ice sheet. For rock, ε is between 4 and 9, whereas for water it is 81. Reflections that are stronger than that generated by an ice-rock interface may indicate the presence of subglacial water. It is essential to glaciologists to identify if the ice base is "wet," as this is an important control on the flow of ice. Indeed, RES has revealed a connection between ice streams and wet-based subglacial conditions.

In some circumstances, extremely strong subglacial reflections have been obtained in association with a planar morphology. In order to account for the strength of the reflection, a dielectric constant representative of water is required. Such reflections are thought to be from the surface of a subglacial lake (see illustration). So far, over 100 subglacial lakes have been discovered in Antarctica. In six, reflections from just beneath the lake surface have been identified. Radio waves do not penetrate water very well (hardly at all if the water is somewhat saline). Consequently, to receive radio-wave reflections from 10–15 m (33–49 ft) beneath the lake surface means that the lake water must be pure.

Ice sheet history. As radio waves travel through the ice column, they can be reflected off boundaries of contrasting ice conductivity and permittivity. The latter can be caused by ice density changes (that occur in the upper few hundred meters of the ice sheet) and by alterations in the ice crystal fabric. Ice conductivity reflections are thought to be caused by an aerosol product from past volcanic events, which formed an acidic layer on the ice sheet. These internal layer reflections can provide information in three ways about the flow and history of the ice sheet.

First, internal layers reflect the flow of the ice sheet. In slow-flowing regions near the ice sheet center, internal layers are often clearly defined, draped over subglacial topography, and can be traced for several hundred kilometers. Conversely in ice streams, layers are distorted so much that they are often unidentifiable. Between the fast-flowing ice streams and the slow-flowing ice sheet interior, internal layers may buckle slightly, making there appearance rougher than the bed beneath. Such layers occur in the tributaries feeding ice streams. Thus by examining internal layer types, one can ascertain the flow configuration of the ice sheet.

Second, internal layers portray the flow history of the ice sheet. In some situations, buckled internal layers have been identified in slow-flowing regions of the ice sheet. Such layering is evidence of a former ice flow configuration, and demonstrates that ice sheets are susceptible to changes over periods of hundreds to thousands of years in response to environmental change.

Third, internal layers can be used to predict the former rate of ice accumulation. If layers across the slow-flowing interior of an ice sheet can be dated (through connection to an ice core), they can be used as input to simple models to predict the spatial and temporal variation in the rates of ice accumulation. Such knowledge is essential to understanding ice sheet history and the feedbacks between climate and ice sheet growth and decay.

For background information see ANTARCTICA; ARCTIC AND SUBARCTIC ISLANDS; CLIMATOLOGY; GLACIOLOGY; RADAR; RADIO-WAVE PROPAGATION in the McGraw-Hill Encyclopedia of Science & Technology.

Martin J. Siegert

Bibliography. V. V. Bogorodsky, C. R. Bentley, and P. E. Gudmandsen, *Radioglaciology*, Reidel, Dordrecht, 1985; D. J. Drewry, *Antarctica: Glaciological and Geophysical Folio*, Scott Polar Research Institute, University of Cambridge, 1983; M. Fahnestock, et al., Internal layer tracing and age-depth-accumulation relationships for the northern Greenland ice sheet, *J. Geophys. Res.*, 106(D24): 33789–33798, 2001; M. B. Lythe and D. G. Vaughan, BEDMAP: A new ice thickness and subglacial topographic model of Antarctica, *J. Geophys. Res.*, 106: 11335–11351, 2001; G. de Q. Robin, S. Evans, and J. T. Bailey, Interpretation of radio echo sounding in polar ice sheets, *Phil. Trans. Roy. Soc. London A*, 265:437–505, 1969; M. J. Siegert et al., An inventory of Antarctic subglacial lakes, *Antarc. Sci.*, 8:281–286, 1996.

Radio-frequency spectrum management

To accommodate new wireless technologies and promote competition, a new approach to spectrum management has been adopted by the U.S. Federal Communications Commission (FCC). In its early years, the FCC would decide who would use the available radio-frequency spectrum to provide service in a location, and even what technology would be used. In recent years, the FCC has changed the way it operates by delegating more of the decision making to market forces.

Origins of regulation. The first radio communication devices were spark gap transmitters which emitted signals composed of a series of damped oscillation waves that could be turned on and off (pulses) at the command of a telegraph key operator. Spark gap transmitters were good only for transmitting Morse code. Given their limitations, spark gap transmitters were never widely deployed and operated only in controlled situations. This was fortunate because the few spark gap transmitters that were deployed blasted their signals in an almost uncontrolled fashion over a broad range of frequencies. If they had been widely deployed, the interference would soon have been intolerable and regulation would quickly have been required. Shortly after G. Marconi and others accomplished the feat of transatlantic wireless communication, the vacuum tube found its way into radio technology. The use of the vacuum tube allowed transmitters to emit continuous-wave signals which led to the ability to transmit speech.

The ability to transmit speech wirelessly led to radio broadcasting. The first stations often operated at very high power levels, and there were no effective regulations for using radio frequencies. The federal government stepped in as more broadcasters came on the air and the inevitable radio-frequency interference occurred. Congress enacted the Radio Act of 1927, thereby creating the Federal Radio Commission whose purpose was to regulate broadcasting. Not long after, a need was seen for federal regulation of interstate and foreign communication by both wireless and wired communications; thus the

Communications Act of 1934 was passed which created the FCC. Among its responsibilities, the FCC was charged to "Assign bands of frequencies to the various classes of stations, and assign frequencies for each individual station . . . ." To manage the spectrum, the FCC adopted a mechanism where frequency bands are identified and reserved for certain uses (such as television broadcasting or land mobile radio services). Once the bands are reserved for a certain use, other uses are not permitted.

Spectrum management model. For most of its existence, the FCC allocated a defined spectrum band for a defined service. It then established technical rules and often specified the technology that could be used to provide the service. Finally, it selected the licensee to provide that service (usually in a defined geographical area).

The FCC usually embarked upon this process only after a formal request to do so was received from a technology or service provider. On occasion, the FCC began the process on its own (though usually after receiving less formal suggestions from potential spectrum users) or perceiving a public policy need to do so (such as encouraging wireless service in rural areas). In either case, the process of allocating spectrum, crafting spectrum usage rules, and issuing service licenses was an extraordinarily time-consuming process—sometimes taking a decade or more. This is because each step of the process involved issuing notices to the public of the FCC's intentions, receiving written and oral comments, replying to comments making a decision and, often, defending that decision in court. Even the very last stage of the process, issuing the licenses, could take years.

This approach to spectrum management is basically the same approach used by regulators around the world. Traditionally, foreign spectrum regulators allowed less public input, while identifying the spectrum, the service, the technology, and the licensee. Critics began to question this process as changes in technology and service, or even service provider, could not be quickly accommodated. As a result, wireless services and technology always lagged the market—often by a decade or more.

This approach also relied on decisions by government officials about what technology and services should be placed in what spectrum. In effect, the FCC was trying to determine the technology of the future. If the FCC guessed incorrectly, the end result would stifle technological innovation with too little spectrum available for high-demand services. In addition, unused spectrum would be "warehoused" by operators of low-demand services. Moreover, a lack of spectrum access due to incumbents holding prime spectrum "real estate" could stifle innovation by locking out new entrants. The FCC realized just how difficult it was to forecast rapidly evolving technology and markets.

Recent changes. For a number of reasons, the FCC in recent years has worked steadily to change its approach to spectrum management. First, the FCC saw that it could not act quickly enough in making decisions about how the spectrum should be used and by whom. It came to the understanding that only the market could act quickly and efficiently enough to make that determination. Second, the FCC increasingly believed that there are technological ways of solving many of the problems that previously were thought to be exclusively solvable by rules. (For example, "cognitive radios" that can sense the surrounding radio environment and adjust their power levels and operating frequencies to avoid interference with other users.) Third, the FCC became ideologically opposed to making substantive decisions about how the spectrum should best be used or by whom, in the belief that markets would embrace worthwhile new technologies that could better define the "public interest" than could a regulator. The goal of bringing market forces into the process, then, began to drive changes in the FCC's approach to spectrum management.

The FCC was granted auction authority by Congress in the mid-1990s. This permitted the FCC to allow market forces to choose the service providers: the service provider willing to pay the most was granted the license. This produced a quicker and more market-oriented approach to the last steps of the spectrum management process. All the other steps (the allocation of frequencies to one of the broad service categories, and the establishment of technical and operational rules) remained basically the same, and once the auction winner was selected, that provider remained bound by the same restrictions that limited other licensees.

The FCC also broadened its experiment with an "unlicensed" spectrum—such that unlicensed use can occur in virtually any of the radio-frequency spectrum, except in what are deemed restricted bands hosting sensitive services or critical government radio operations—giving market forces almost wholly free rein. Since no licenses were required, anyone can use the spectrum with almost any technology to provide any service. In terms of technology and service, the FCC watched the astonishing success of the unlicensed spectrum, confirming its belief in market forces as a spectrum management tool.

The FCC forged ahead, bringing additional market forces in other parts of the spectrum management process. In the late 1990s it issued wireless communications services (WCS) licenses. More recently, the concept of secondary markets has been introduced—so that licensees can lease their spectrum to others, with minimal FCC involvement. So too was the ability to aggregate and disaggregate spectrum. This means that licensees can add to their licensed spectrum by acquiring additional spectrum or subtract from the spectrum by selling it off. They can also partition their spectrum, that is, sell off part of their licensed service area. Thus, the spectrum used to provide service will be determined in large measure by the market.

Outlook. Currently, the FCC is considering a general practice of issuing non-service-specific licenses that allow license holders the freedom to offer any services they want, and allow those who lease secondary spectrum to offer any service they want that the licensee will agree to. For example, what most people think of as spectrum for third-generation mobile phone service is actually designated for "advanced wireless services"—meaning almost anything the licensee wants to do. Most recently, the FCC has proposed the concept of underlay spectrum, in which the rights of license holders would be specifically defined in technical terms—above the interference threshold (the maximum level of extraneous radio signals a receiver can tolerate and still function properly), the spectrum would in effect belong to the licensee. The public would then be granted an easement to use that same spectrum below the interference threshold for unlicensed services.

Clearly the FCC is in the process of a "paradigm shift" in the way it manages the airwaves. Looking ahead, there will be further attempts to use market forces to manage spectrum and fewer decisions made by regulators.

For background information *see* AMPLITUDE MODULATION; BANDWIDTH REQUIREMENTS (COMMUNICATIONS); ELECTRICAL COMMUNICATIONS; ELECTRICAL NOISE; ELECTROMAGNETIC RADIATION; ELECTROMAGNETIC WAVE TRANSMISSION; FREQUENCY MODULATION; MODULATION; RADIO; RADIO SPECTRUM ALLOCATIONS; RADIO-WAVE PROPAGATION; WAVELENGTH in the McGraw-Hill Encyclopedia of Science & Technology.   S. B. Harris; D. Ladson

Bibliography. R. H. Coase, *The Federal Communications Commission*, Journal of Law & Economics, University of Chicago, 1959; S. B. Harris, *Who's in Charge Here?*, *Space News*, December 2002; W. Lehr, *Should Unlicensed Broadband be Restricted to "Spectrum Siberia"?: The Economic Case for Dedicated Unlicensed Spectrum Below 3 GHz*, New American Foundation, July 1, 2004; On the same wavelength, *The Economist*, Aug. 12, 2004; K. Werbach, *The End of Spectrum*.

Red yeast rice

Red yeast rice, produced by fermentation of red yeast (the filamentous fungus *Monascus*) on rice grains, has traditionally been used in Asian countries both as a food colorant and as a medicinal substance. It is also known as angkak, beni-koji, hong qu, hungchu, monascus, red koji, red leaven, red rice, zhitai, and zue zhi kang. The first mention of red yeast rice occurred in Li Shih-chun's *Pen Chaw Kang Mu*, a monograph of Chinese medicine, in 1590. *Monascus* red rice has been used in Chinese folk medicine for indigestion, dysentery, and anthrax, as well as for relieving bruised muscles, promoting blood circulation, and invigorating spleen function. Recently, red yeast rice was found to contain cholesterol-lowering compounds. In addition, dried red yeast rice is used to color foods such as fishmeal, cheeses, soybean products, alcoholic beverages, and sausages.

Although there are six species of *Monascus*, only two—*M. purpureus* (also known as *M. anka* or *M. pilosus*) and *M. ruber*—are important for the preparation of red yeast rice. *Monascus* species pro-

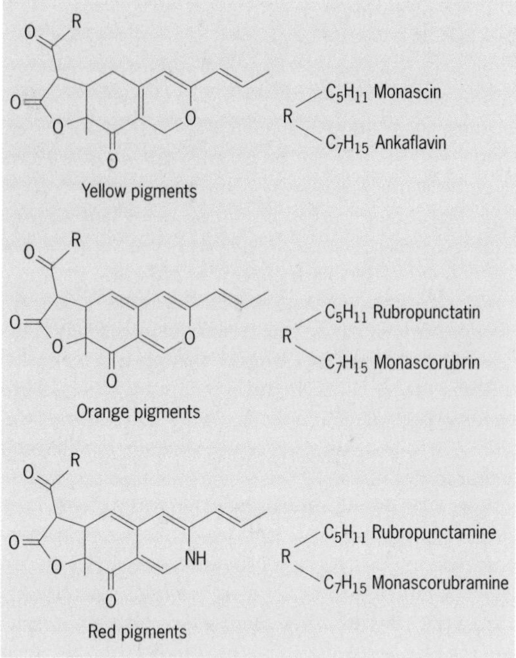

Fig. 1. Major *Monascus* pigments.

duce six major oligoketide pigments (yellow, orange, and red) as secondary metabolites (**Fig. 1**). The yellow and orange pigments are synthesized mostly from acetyl coenzyme A subunits in a manner similar to fatty acid synthesis. The red pigments result from the reaction of the orange pigments with ammonia or compounds containing a primary amino group, such as amino acids, peptides, proteins, amino sugars, and nucleotides. The ability of orange pigments to react readily with such compounds is the reason why the red yeast rice contains mostly yellow and red pigments. In addition, some *Monascus* strains can produce other yellow (such as yellow II, xanthomonasin A, and xanthomonasin B) and red (such as monascopyridine A and monascopyridine B) pigments. Some pigments, especially the orange ones, also have biological effects.

Other well-known secondary metabolites of the fungus *Monascus* found in red yeast rice are the monacolins, the active compounds in a class of blood lipid–lowering drugs known as statins. Among other monacolins, *M. purpureus* and *M. ruber* produce monacolin K (see structure), also known as lova-

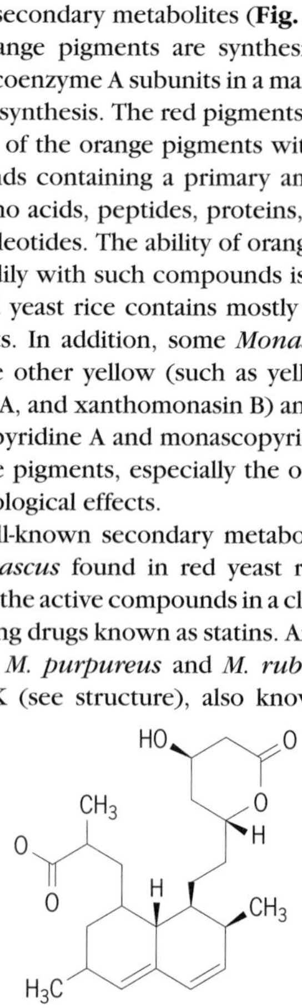

statin or mevinolin, which is converted in the human body to the active form, that is, the β-hydroxy acid of monacolin K. Of all the statins found in red yeast rice, mevinolin, together with related compounds that are either precursors or by-products of monacolin synthesis, are produced in the highest amount. Statins inhibit 3-hydroxy-3-methylglutaryl coenzyme A (HMG-CoA) reductase, a key enzyme in cholesterol synthesis, and therefore can serve as anticholesterol (hyperlipidemic) drugs. In addition, statins can potentially suppress tumor growth due to their ability to inhibit the synthesis of nonsteroidal isoprenoid compounds.

Recently, it was found that some strains of *Monascus* can produce the mycotoxin citrinin, previously reported as monascidine A. Citrinin, which is known to be nephrotoxic and mutagenic, was detected in amounts varying from 0.2 to 17 ppm in samples of commercially available red yeast rice. Although no mutagenic effects were observed at testing of red rice samples, it is recommended to use non-citrinin-producing *Monascus* strains for red yeast rice production.

Production. Traditional manufacturing of red yeast rice usually includes washing, soaking, draining, steaming, sterilizing, and inoculating the rice with *Monascus* spores or mycelia, followed by cultivation and drying. Production methods differ in various regions, and the exact procedure is usually a secret of the producer. In the Philippines the traditional cultivation takes place in bamboo trays covered with banana leaves, while in Taiwan inoculated rice is heaped in a bamboo chamber until its temperature rises to 42°C (108°F) and then the rice is spread on plates and cultivated.

The secrets of success in red yeast rice cultivation are a relatively low substrate humidity (ranging 25–50%, depending on the strain) maintained at a constant level, low substrate pH, favorable rice variety, and substrate pretreatment. Both low humidity and low pH limit glucoamylase activity in favor of pigment production. Glucoamylase activity increases with increasing moisture, resulting in ethanol production from glucose released from the rice starch and less pigment formation. The colonization of the substrate by the fungus is influenced by the rice variety and its pretreatment. The better the colonization, the better the flavor and color of the product. The choice of a poor rice variety, together with high humidity or a bad pretreatment, can cause agglomeration of the rice grains and reduced oxygen supply to the fungi. At the end of the cultivation, the rice grains should have a deep purple-red color and the pigments should be distributed evenly across a broken grain.

Although a low a_w (water activity) value in the rice helps to reduce the risk of product contamination, it is even better to work in aseptic conditions. Because of the long duration of fermentation (typically 7–15 days), in semiaseptic conditions, there is an especially high risk of contamination by *Bacillus* or *Penicillium* species. Industrial cultivation of red yeast rice usually takes place in a fermentor suitable for solid substrate cultivation—for example, a moving bed fermentor. In laboratory cultivation, trays used in the traditional manufacture can be replaced with autoclavable plastic bags that are plugged with cotton stoppers (**Fig. 2**).

Applications. The manufacture of red yeast rice varies according to product use, and the product itself differs in the amount of pigments, enzymes, and monacolins produced. There are three possible uses of the product: food coloring, red koji (used for fermentation), and monacolins production.

Food colorant. When red yeast rice is to be used for coloring foods, it is dried typically at 40, 45, or 60°C (104, 113, or 140°F) for 24 h after cultivation and then ground into powder. The coloring of foods by *Monascus* red yeast rice is not limited to traditional foods. Western-like meat products, such as "Chinese hot dogs," colored with red rice have become very popular in China. Many patents from Asia, Europe, or the United States describe the use of red rice as a food dye. There are also studies demonstrating that the use of red rice can replace in meat products, at least to some extent, nitrate/nitrite salts, which can form carcinogenic nitrosamines. Nitrate/nitrite salts are added to meat products not only for color but also for prevention of bacterial growth. Red yeast rice also exhibits an antibacterial effect (especially against *Bacillus, Listeria monocytogenes*, and *Staphylococcus aureus*). The use of nitrate/nitrite in meat, when supplemented with red yeast rice, can be reduced by at least 60%. However, red yeast rice is generally not allowed as a food additive in the West.

Fermentation. When red yeast rice is used as red koji, the cultivation is shorter and the product contains less pigment in comparison with red yeast rice used for food coloring. Koji represents both an inoculum and a source of hydrolytic enzymes. A common type of koji (white koji) is prepared by solid substrate cultivation of rice with the fungus *Aspergillus oryzae* for 40–48 h at 35–38°C (95–100°F). Red koji is

Fig. 2. Rice grains after 7 days' cultivation of *M. purpureus* at 30°C (86°F).

prepared similarly to the usual koji but with the fungus *Monascus*, and is used for various food fermentations, with the best-known examples being the production of red rice wine (kaoliang) and fermentation of tofu (soybean cheese) for the production of sufu (fermented red tofu).

Monacolins. Mevinolin formation by *M. ruber* in submerged liquid cultivation, discovered by A. Endo in 1979, was initially believed to be the best method of mevinolin production by *Monascus*. However, it was later found that rice could be a suitable substrate for mevinolin production by *Monascus* as well as by *M. ruber* or *M. purpureus* under special conditions. In many studies, it was proved that such a red yeast rice decreases elevated serum triglycerides—that is, fats and low-density lipopolysaccharide (LDL) cholesterol—leaving levels of high-density lipopolysaccharide (HDL) cholesterol unchanged in the human body. It seems that the effect of red yeast rice on hyperlipidemia and high blood pressure treatment are caused not only by mevinolin and related compounds but also by other compounds contained in the rice, such as β-sitosterol and campesterol (which interfere with cholesterol absorption in the intestine), unsaturated fatty acids, B-complex vitamins, fiber, and trace elements. It is probable that all these compounds act in concert at lowering serum triglycerides. Currently, several companies in Asian and European countries offer dried and powdered red rice in capsules as a dietary supplement. In 2001, an American court stated that red yeast rice could not be a dietary supplement because it contained monacolin K, a natural compound closely related to the prescription drug lovastatin.

For background information *see* CHOLESTEROL; FOOD FERMENTATION; FUNGAL BIOTECHNOLOGY; FUNGI; LIPOPROTEIN in the McGraw-Hill Encyclopedia of Science & Technology.

Petra Patáková

Bibliography. A. Endo and K. Hasumi, Mevinic acids, in *Fungal Biotechnology*, pp. 162–172, Chapman & Hall, Weinheim, 1997; P. Jůzlová, L. Martínková, and V. Křen, Secondary metabolites of the fungus *Monascus*: A review, *J. Ind. Microbiol.*, 16:163–170, 1996; L. Martínková and P. Patáková, *Monascus*, in *Encyclopedia of Food Microbiology*, pp. 1481–1487, Academic Press, London, 2000; K. H. Steinkraus, Chinese red rice: Anka (ang-kak), in *Handbook of Indigenous Fermented Food*, p. 547, Marcel Dekker, New York, 1983.

Resveratrol

The last two decades have seen an unprecedented interest in understanding the medicinal properties of naturally occurring compounds. A wide variety of these agents are under scrutiny for their clinical potential in disease prevention and treatment. Among these compounds is a family of plant antibiotics (phytoalexins) named viniferin whose members exhibit strong antifungal properties. One remarkable compound in this family is resveratrol (RSV).

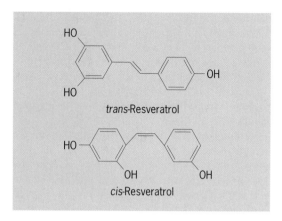

Structure of trans and cis isomers of resveratrol.

Occurrence and Chemistry

RSV (see **illustration**) was first isolated from the roots of the oriental medicinal plant *Polygonum cuspidatum* (Ko-jo-kon in Japanese), and is known for its beneficial effects against a host of human afflictions. Its occurrence has also been documented in trees, including eucalyptus and spruce; in a few flowering plants, such as species of lily; in peanuts and groundnuts; and in grapevines. Since the first reported detection of *trans*-RSV in grapevines in 1976 and in wine in 1992, most research has focused on RSV in grapevines. This is mainly because compounds found in grapevines were implicated in epidemiological data demonstrating an inverse correlation between red wine consumption and incidence of heart disease—a phenomenon known as the French paradox. The highest concentration of *trans*-RSV has been reported in wines prepared from Pinot noir grapes (averaging 5.13 mg/L). White wines contain 1–5% of the RSV content present in most red wines.

Phytoalexins, such as RSV, are produced in especially great amounts in response to injury or fungal infection. Their synthesis is stimulated by exposure to ultraviolet (UV) rays. UV exposure also results in conversion from the trans to cis form of the molecule (see illustration). *Trans*-RSV (which has a molecular weight of 228 daltons) is commercially available and is relatively stable if protected from high pH and light. The difference in absorption maxima for the trans and cis isomers (307 nm and 288 nm, respectively) allows for their separation and detection by high-performance liquid chromatography. More recently, gas chromatography has been used to measure concentrations of *cis*- and *trans*-RSV isomers in plasma and cells. The synthetic potential of *trans*-RSV is highest just before the grapes reach maturity and is low in buds, flowers, and mature fruits. Higher amounts of *trans*-RSV are also observed in healthy areas around necrotic lesions following fungal infection of grape berry skins, and this seems to be an inherent protective mechanism to limit the infection.

Biological Activity and Beneficial Effects

The beneficial effects of extracts prepared from the plant *P. cuspidatum* against a host of disease conditions had been documented decades before the identification of RSV as a potent ingredient of this folk medicine. After this reported observation in the mid-1970s, a lot of research focused on understanding the mechanisms underlying the synthesis of RSV in plants and in studying the effects of environmental conditions on RSV synthesis. Subsequent identification of RSV as a major phenolic constituent of wines and its association with decreased cardiovascular disease risk in moderate wine consumers have resulted in a tremendous increase in interest in elucidating the effects of this polyphenolic compound on human health and disease. Consequently, the diverse biological activities and potential beneficial effects of RSV have come to light over the past 5 years.

Potential cardioprotective effects. Epidemiological studies have linked moderate intake of wine with an appreciable decrease in the risk of coronary artery disease (see **table**), particularly in subjects living in regions of France where the diet is rich in fat.

Antioxidant activity. There is evidence indicating that RSV is a potent inhibitor of the oxidation of polyunsaturated fatty acids found in low-density lipoproteins (LDL). Oxidized lipoproteins provide a permissive environment for atheroma (yellowish plaque in the arterial inner wall) formation and platelet aggregation, thereby fueling the process of atherosclerosis and coronary artery disease.

Regulation of nitric oxide. In addition to its antioxidant activity, RSV may also exert its potential cardioprotective effects via its regulation of nitric oxide (NO) production. Increased levels of NO (a nitrogen species involved in inflammatory responses) cause vascular damage, thereby contributing to the development of atheromatous plaques. Wine phenolics, such as RSV, have been shown to modulate the production of nitric oxide from vascular endothelium.

Inhibition of platelet aggregation. Another observed biological effect of RSV that may support its potential cardioprotective ability is its inhibition of platelet aggregation. Platelet aggregation activates the process of thrombus formation and thus facilitates the process of atherosclerosis.

Neuroprotective effects. Through its inhibitory effect on lipid peroxidation, RSV has also been shown to protect neuronal cells from the toxic effects of oxidized lipoproteins, indicating neuroprotective activity (see table). For example, in the presence of RSV, rat adrenal medulla tumor cells were remarkably protected from the death-inducing effect of ethanol.

Immunomodulatory activity. Among the various salubrious effects of *P. cuspidatum* root extract are its beneficial effects against allergic and inflammatory diseases (see table).

Inhibition of cyclooxygenase and lipooxygenase pathways. Invariably, allergy and inflammation are the result of an increase in activity of leukocytes that release a host of biological response modifiers (chemicals that enhance the immune system's response to disease). The two enzyme systems involved in the synthesis of proinflammatory mediators are the cyclooxygenase (COX) and the lipoxygenase pathways. COX-2 activity, usually undetected in normal tissues, generates proinflammatory substances by the oxygenation of arachidonic acid to prostaglandins D_2 and E_2 (PGD_2 and PGE_2, respectively). In addition, COX-2 and the lipoxygenase pathway catalyze the formation of chemotactic substances that attract leukocytes and induce platelet aggregation and inflammation. Interestingly, RSV has been shown to inhibit both these proinflammatory pathways, thereby blocking the generation of inflammatory mediators. This inhibitory effect on COX and lipoxygenase activities has also been proposed as a possible mechanism for the antitumor activity of RSV. (However, recent studies provide evidence that the antitumor activity may, in part, be due to the ability of RSV to trigger apoptosis, programmed cell death, in tumor cells. Nevertheless, given the role that inflammatory mediators play in the induction and promotion of carcinogenesis, it is plausible that both these observed activities may play a role in cancer chemoprevention.)

Inhibitory effects on macrophage and polymorphonuclear cell activity. The effect of RSV on macrophages and polymorphonuclear cells has also been evaluated. These cells orchestrate the body's response to immunogenic challenges, and biological response modifiers secreted from these cells could contribute to the development of disease states such as allergy and inflammation. One classical model of macrophage activation is via bacterial lipopolysaccharide, an endotoxin that is a component of bacterial gram-negative membranes. Under normal physiological conditions,

Biological effects of resveratrol	
Physiological effects	Mechanism of action
Prevents atherosclerosis and coronary heart disease	Inhibits polyunsaturated fatty acids oxidation Reduces the rate of secretion of cholesterol and triglycerides Modulates NO production from vascular endothelium Inhibits platelet aggregation Inhibits synthesis of inflammatory mediators
Inhibits neuronal cell death	Inhibits membrane lipid peroxidation
Reduces allergic and inflammatory reaction	Induces antiplatelet aggregation effect Inhibits acute and chronic phases of inflammatory process Inhibits iNOS production Inhibits ROS production and proinflammatory mediators and their effects
Inhibits cellular proliferation and growth	Inhibits DNA synthesis Inhibits polyamine synthesis Inhibits intracellular pro-oxidant state Induces cell cycle arrest and increases in p53 and p21$^{Waf1/Cip1}$ (apoptosis-inducing proteins) expression Regulates apoptosis in cancer cells

this activation leads to a moderate increase in the activity of the inducible enzyme nitric oxide synthase (iNOS) within macrophages, resulting in the production of NO, which has bactericidal effects. However, abnormally high concentrations of NO and its derivatives, peroxynitrite or nitrogen dioxide, give rise to inflammation and could contribute to the process of carcinogenesis. Preincubation of cells with RSV results in a dose-dependent inhibition of the induction of the iNOS gene (*iNOS*) and a decrease in the steady-state levels of iNOS messenger ribonucleic acid and the iNOS enzyme. RSV has also been reported to strongly inhibit polymorphonuclear cell-induced proinflammatory signals and cell surface expression of a adhesion molecules.

These results strongly indicate that RSV elicits an inhibitory effect at all physiological phases of the inflammatory response, that is, from the initial recruitment of polymorphonuclear cells to their activation and the subsequent release of inflammatory mediators. Since the inflammatory response is a critical common denominator in the development of many systemic disorders, such as atherosclerosis and carcinogenesis, the strong anti-inflammatory activity of RSV could have tremendous clinical implications.

Cancer preventive activity: regulation of cell growth and proliferation. Since its reported cancer chemopreventive activity in a mouse model of carcinogenesis, there has been a flurry of papers reporting the effects of RSV on critical events that regulate cellular proliferation and growth (see table). In this regard, biochemical pathways involved in cell differentiation, cell transformation, cell cycle regulation, and apoptosis induction have all been demonstrated as potential targets of RSV. The regulatory/inhibitory effect of RSV on these signal transduction pathways has generated tremendous interest in its clinical chemopreventive and chemotherapeutic potential.

Depending upon its concentration, RSV can either stimulate (as shown with breast cancer and pituitary cells) or inhibit cell proliferation. Generally, at the concentrations used in vitro, the effect is predominantly antiproliferative.

Inhibition of DNA synthesis. The mechanism(s) for this growth inhibitory activity of RSV could be due to its ability to block ribonucleotide reductase, a complex enzyme that catalyzes the reduction of ribonucleotides into the corresponding deoxyribonucleotides. Inhibitors of ribonucleotide reductases, such as gemcitabine (2'-difluoro-2'-deoxycytidine), have been in clinical use as chemotherapeutic agents due to their inhibitory effect on deoxyribonucleic acid (DNA) synthesis.

Inhibition of polyamine synthesis. A second probable mechanism for the observed antiproliferative activity of RSV could be its ability to inhibit the enzyme DNA polymerase, or ornithine decarboxylase, a key enzyme involved in polyamine synthesis that is increased in cancer growth.

Interference with cell cycle. A number of studies have also demonstrated that RSV interferes with cancer cell-cycle progression via direct or indirect effects on cell-cycle regulatory pathways.

Inhibition of pro-oxidant state. The antioxidant activity of RSV could be yet another mechanism for growth inhibition, because a slight pro-oxidant intracellular environment, an invariable finding in cancer cells, is a strong stimulus for proliferation. However, recent findings provide evidence that RSV can inhibit or stimulate intracellular production of reactive oxygen species (ROS), depending upon the cell type. These data may help to explain the conflicting results demonstrating pro- and anti-proliferative effects of RSV on mammalian cells.

Evidence from animal models. Although some of the in-vitro biological effects of RSV have not been corroborated in vivo, there is evidence to support the antiproliferative and growth inhibitory activity in animal models of carcinogenesis. It is plausible that the decrease in incidence of tumors in RSV-treated mice could be a function of targeted killing of tumor cells by RSV. In addition, there are reports suggesting that RSV could enhance the effectiveness of treatment regimens that combine chemotherapy agents in vitro. This could indeed have tremendous potential for the clinical management of cancer.

Conclusion

The relatively simple chemical structure of RSV enables it to interact with receptors and enzymes, giving rise to biological effects such as suppression of cell growth, induction of cell differentiation, inhibition of ROS production (antioxidant effect), cell-cycle regulation, inhibition of lipid peroxidation, inhibition of synthesis of proinflammatory mediators (or blocking their effects), regulation of gene expression (via its impacts on transcription factor activity), and modulation of apoptotic signaling (see table). These in-vitro effects have been corroborated in some studies demonstrating the beneficial effects of RSV on the cardiovascular, neurological, and hepatic systems; however, the most exciting in-vivo data relates to its cancer chemopreventive and chemotherapeutic activity. Since it is a natural constituent of wine, fruits, and nuts and has no demonstrated harmful effects on normal cells or tissues, RSV is under preclinical scrutiny for use in the prevention and treatment of disease. The outcome of these studies will elucidate the real clinical potential of this remarkable compound.

For background information *see* ANTIOXIDANT; CANCER (MEDICINE); CELL CYCLE; EICOSANOIDS; IMMUNOLOGIC CYTOTOXICITY; INFLAMMATION; NITRIC OXIDE; PHYTOALEXINS in the McGraw-Hill Encyclopedia of Science & Technology.   Shazib Pervaiz

Bibliography. M. V. Clement et al., Chemopreventive agent resveratrol, a natural product derived from grapes, triggers CD95 signaling-dependent apoptosis in human tumor cells, *Blood*, 92:996–1002, 1998; E. N. Frankel, A. L. Waterhouse, and J. E. Kinsella, Inhibition of human LDL oxidation by resveratrol, *Lancet*, 341:1103–1104, 1993; B. Fuhrman, A. Lavy, and M. Aviram, Consumption of red wine with meals reduces the susceptibility of human plasma and low-density lipoprotein to lipid peroxidation, *Amer. J. Clin. Nutrit.*, 61:549–554, 1995; D. M. Goldberg,

S. E. Hahn, and J. G. Parkes, Beyond alcohol: Beverage consumption and cardiovascular mortality, *Clin. Chim. Acta*, 237:155-187, 1995; J. Gusman, H. Malonne, and G. Atassi, A reappraisal of the potential chemopreventive and chemotherapeutic properties of resveratrol, *Carcinogenesis*, 22:1111-1117, 2001; M. Jang et al., Cancer chemopreventive activity of resveratrol, a natural product derived from grapes, *Science*, 275:218-220, 1997; J. Martinez and J. J. Moreno, Effect of resveratrol, a natural polyphenolic compound, on reactive oxygen species and prostaglandin production, *Biochem. Pharmacol.*, 59:865-870, 2000; S. Pervaiz, Resveratrol: From grapevines to mammalian biology, *FASEB J.*, 17:1975-1985, 2004; S. Pervaiz, Resveratrol—from the bottle to the bedside?, *Leuk. Lymphoma*, 40:491-498, 2001; G. J. Soleas, E. P. Diamandis, and D. M. Goldberg, Resveratrol: A molecule whose time has come? And gone?, *Clin. Biochem.*, 30:91-113, 1997; G. J. Soleas, E. P. Diamandis, and D. M. Goldberg, The world of resveratrol, *Adv. Exp. Med. Biol.*, 492:159-182, 2001.

Ridge-flank hydrothermal circulation

The discovery of high-temperature ("black smoker") hydrothermal vents at sea-floor spreading centers in the late 1970s was a revelation. Fluids at temperatures of 300-400°C (572-752°F) exit these vents and mix with the overlying ocean, creating spectacular mineral deposits and allowing development of unusual ecosystems. Hydrothermal vents have since been found along many sections of the global mid-ocean ridge (MOR) system, a 60,000-km (37,280-mi) chain of undersea volcanoes that circles the Earth like the seams on a baseball. MORs are places where new tectonic plates are created, so there is plenty of magma and heat available in these locations to drive hot, reactive fluids through the crust.

The hydrothermal circulation of seawater through the crust extracts much of the heat associated with formation and cooling of oceanic lithospheric (tectonic) plates, and continues to be thermally significant out to an average sea-floor age of 65 million years. Sea-floor hydrothermal fluxes of fluid, heat, and solutes help to maintain the chemistry of the oceans, change the physical and geochemical state of the plates, transport volatiles into the mantle at subduction zones, influence earthquake processes, and support a vast subsurface biosphere that scientists are just beginning to explore. New research reveals that fluids within the crust can flow much longer distances below the sea floor and at much higher rates than was previously realized.

Ridge-flank circulation. Although it is not widely appreciated among nonspecialists, the dramatic high-temperature vent systems found at many mid-ocean ridge sites represent only a small fraction of the fluid circulation through oceanic crust. Considerably more fluid circulates through the sea floor on ridge flanks, the parts of oceanic plates located away from MORs. The primary driving force for hydrothermal circulation on ridge flanks is created by heat associated with cooling of the lithosphere as it ages. Although ridge-flank hydrothermal circulation generally occurs at lower temperatures than that on MORs, typically 5-70°C (41-158°F), the total fluid and heat fluxes through ridge flanks are at least 10 to 100 times greater than those at MORs. In fact, ridge-flank circulation is so intense that a fluid volume equivalent to that of the entire ocean is recirculated through the ocean crust once every 200,000-500,000 yr, a very short time in a geological sense.

The volcanic oceanic crust (the basaltic rock layers upon which marine sediments are subsequently deposited) constitutes the largest aquifer on Earth. The uppermost crust is composed mainly of porous extrusive rocks: basaltic pillows, flows, and breccia that are erupted on or close to the sea floor and cool relatively quickly. Fluids flow relatively easily through the upper layers of oceanic crust, which are very permeable, and additional pathways are created as the crust is faulted and cracked during formation and spreading. Underlying these extrusive rocks are basaltic dikes and gabbroic intrusions, rocks that have a chemistry similar to that of the surface layers but that cool much more slowly on average because they are more isolated from the cold ocean above. These deeper layers are involved in hydrothermal circulation at MORs, but most circulation on ridge flanks is generally restricted to shallower layers.

Despite the importance of ridge-crest and ridge-flank hydrothermal processes, remarkably little is known about the nature of fluid circulation pathways, the depth extent of circulation, or how hydrogeologic, tectonic, and magmatic processes influence each other and control hydrothermal processes. One difficulty is that these systems are remote, occurring far below the sea floor in deep water. Another challenge is that, like many hydrogeologic systems in Earth's crust, sea-floor hydrothermal circulation is dynamic, operating at numerous length and time scales. Researchers have begun to map out the patterns and rates of fluid circulation, and several recent studies lead to the surprising conclusion that hydrothermal fluids within the sea floor may travel tens of kilometers or more between recharge (entry) and discharge (exit) sites. In fact, such large-scale flow paths seem to be required to explain how ridge-flank hydrothermal circulation can extract significant quantities of lithospheric heat over large parts of the sea floor.

Large-scale transport. Although the upper volcanic ocean crust is highly permeable, the sediments blanketing the crust are much less so. As the sea floor ages, marine sediments accumulate and isolate hydrologically active basement rocks from the overlying ocean. On initial consideration, it would seem that this process should limit the extent of hydrothermal circulation as the crust ages, but thermal observations demonstrate that the oceanic crust hosts vigorous fluid flow throughout its life. One explanation is that the presence of a nearly continuous marine sediment layer forces the creation of

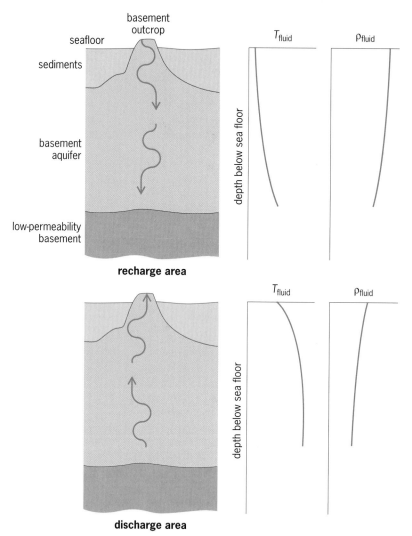

Fig. 1. Thermal and fluid density conditions in areas of sea-floor hydrothermal fluid recharge and discharge. In the recharge area cool fluids penetrate the crustal aquifer, bringing low temperatures and relatively high fluid densities to depth within the upper crust. In the discharge area warm fluids that have reacted with crust rocks and absorbed lithospheric heat exit the sea floor where basement is exposed. Temperatures are relatively high within the discharging water column, and fluid densities are relatively low.

4000 m or 8200–13120 ft) and the range of temperatures characteristic of ridge-flank fluid circulation (~5–70°C or 41–158°F), one may neglect the affects of phase changes, variations in fluid chemistry, and fluid compressibility with pressure. Temperature differences have the greatest impact on fluid density and thus generate the driving forces for large-scale fluid circulation on ridge flanks.

Additional considerations are necessary to understand how long-distance fluid transport occurs in the crust. First, if permeable upper basement rocks are not hydrogeologically isolated from the overlying ocean, either by a nearly continuous blanket of low-permeability sediments or low-permeability basement rocks, then seawater that flows into the crust and is heated slightly will return immediately to the ocean. In fact, shallow circulation of this kind will occur wherever permeable bare rock is exposed, including most areas of young sea floor. These fluids will not spend much time in the crust, will have a chemistry and temperature that is little different from that of bottom seawater, and will not be responsible for thermally or chemically significant crustal fluxes.

Where there is a low-permeability sediment layer above most of the basement aquifer on ridge flanks, a fluid flow system can develop within the crust between basement outcrop locations. Characteristic temperature differences between recharging and discharging water columns on ridge flanks result in maximum pressure differences on the order of only a few tens to hundreds of kilopascals, equivalent to no more than a few atmospheres, even for circulation extending 1000 m (3280 ft) into the crust. These pressure differences are potentially available to drive large-scale fluid circulation, but only if hydrothermal fluids can move easily into and out of the crust. Basement permeabilities must be very high, on the order of 10^{-12} to 10^{-9} m^2 (10^{-11} to 10^{-8} ft^2). These are extremely high permeabilities, equivalent to those found in very productive ground-water aquifers or oil reservoirs.

long-distance fluid flow pathways in underlying basement rocks. Understanding why this occurs requires consideration of the physics of ridge-flank hydrothermal circulation.

Active volcanism at MORs leads to enormous differences in fluid density and pressure associated with phase changes and large variations in heat content and chemistry. In contrast, large-scale fluid flow on ridge flanks is driven mainly by differences in pressure at the base of recharging (cool) and discharging (warm) columns of water (**Fig. 1**). This pressure difference can be approximated by the equation below,

$$\Delta P = g \int_0^{z'} \Delta \rho(z)\, dz$$

where $\Delta \rho(z)$ is the difference in fluid density, and z' is the height of recharging and discharging water columns within the crust. In this general formulation, fluid density can vary with pressure, temperature, and composition. Given the typical depths to the top of basement rocks on ridge flanks (~2500–

Fig. 2. Long-scale fluid transport on the eastern flank of the Juan de Fuca Ridge. (a) Site map showing locations of Baby Bare and Grizzly Bare outcrops in Cascadia Basin, northeast Pacific Ocean. (b) Thermal and Seismic data across Baby Bare outcrop. TWT = two-way traveltime. (c) Thermal and seismic data across Grizzly Bare outcrop. (d) Estimated driving pressure available to move hydrothermal fluid from Grizzly Bare to Baby Bare, based on the equation in the text, as a function of vertical basement permeability. Green curves indicate pressure remaining after vertical fluid flow through recharge and discharge sites, at a range of possible fluid flow rates. Hatched area shows reasonable range of driving pressures, bounded by black lines. Vertical permeability must be greater than 10^{-12} m^2 (10^{-11} ft^2) to allow significant driving forces to remain in the basement. (e) Relations between basement permeability and flow rate between the two outcrops, given a reasonable range of available driving pressures (black lines). Hatched area indicates reasonable range of flow rates and corresponding basement permeabilities that allow large-scale fluid flow, 10^{-12} to 10^{-9} m^2 (10^{-11} to 10^{-8} ft^2). (*Adapted from A. T. Fisher et al., Hydrothermal recharge and discharge across 50 km (31.3 mi) guided by seamounts on a young ridge flank, Nature, 421:618–621, 2003*)

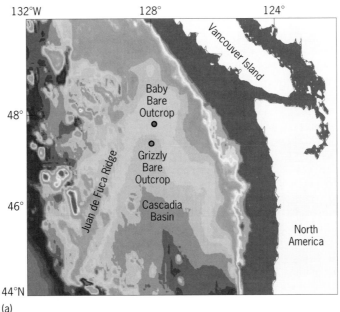

(a)

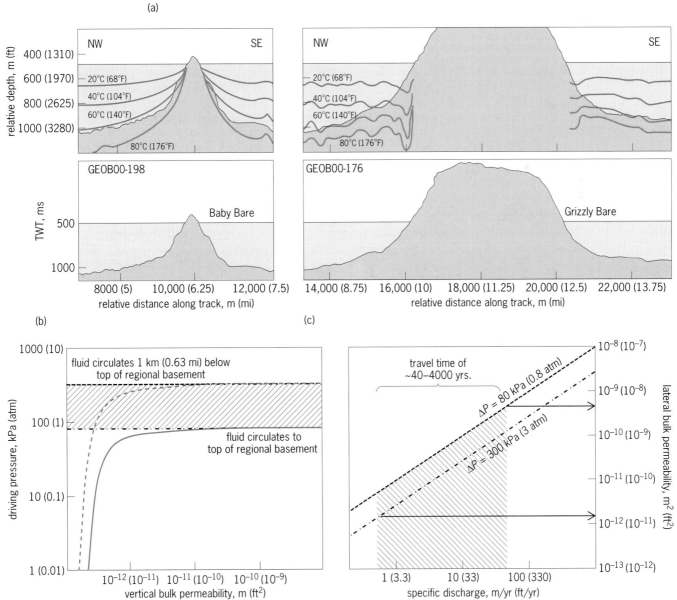

(b)

(c)

(d)

(e)

Two recent examples illustrate the processes and conditions associated with large-scale hydrothermal circulation on ridge flanks.

Minimal basement outcropping. One ridge-flank hydrothermal system that has been studied extensively is on the eastern flank of the Juan de Fuca Ridge in the northeastern Pacific Ocean (**Fig. 2**). Sediment cover in this area of young (<4 million years old) sea floor is almost continuous because of its proximity to the North American continental margin. Basaltic crust is exposed in only a few places on this ridge flank, where seamounts and other basement outcrops penetrate low-permeability sediments (Fig. 2a). The Baby Bare outcrop has been surveyed numerous times by surface ship, submersible, and remotely operated vehicle, and significant quantities of warm crustal fluids are known to discharge from its peak. Although Baby Bare outcrop rises just 65 m (203 ft) above the surrounding sea floor, the basaltic edifice is substantial, rising 600 m (1970 ft) above regional basement (Fig. 2b). Fluids exiting from Baby Bare outcrop are known to be very young, despite being highly altered, and geochemical considerations preclude the possibility that these fluids recharged through nearby sediment or basement rocks. Instead, recharge of fluids that eventually exit through Baby Bare outcrop is thought to occur through Grizzly Bare outcrop, 52 km (32 mi) to the south (Fig. 2c). Thermal surveys of these outcrops show strongly contrasting conditions in basement: temperature contours are swept upward by venting fluids at Baby Bare outcrop and swept downward by recharging fluids at Grizzly Bare outcrop (Fig. 2b and c).

Application of the earlier equation allows calculation of the forces available to drive this large-scale hydrothermal circulation system (Fig. 2d). Given observed rates of fluid discharge at Baby Bare outcrop, the pressure difference at the base of recharging and discharging columns of fluid in the crust must be just 80–300 kPa (0.8–3.0 atm). Given these pressure differences, and the relatively short travel time for fluid flow between the outcrops (~40–4000 years), the bulk permeability of the basement rocks must be on the order of 10^{-12} to 10^{-9} m^2 (10^{-11} to 10^{-8} ft^2; Fig. 2e).

Despite the high crustal permeabilities in this area, hydrothermal circulation has a thermal influence only very close to the basement outcrops. Circulation is relatively inefficient at extracting crust heat on a regional basis because there are few basement outcrops that penetrate the sediment, so total fluid fluxes are limited.

Greater basement outcropping. Areas of basement exposure are much more common on the 18–24-million-year-old Cocos plate approaching the Middle America Trench, offshore of Costa Rica. Recent surveys in this area reveal contrasting thermal conditions within two parts of the plate. One part of the plate was produced at the East Pacific Rise (EPR), a MOR to the west, whereas the other part of the plate was created at the Cocos-Nazca spreading center (CNS) to the south. EPR-generated sea floor is dotted with numerous seamounts and outcrops that penetrate through otherwise regionally extensive sediments, generally 400–500 m (1310–1640 ft) thick. In contrast, CNS-generated sea floor adjacent to the EPR-generated sea floor has no outcrops, and basement is completely isolated from the overlying ocean by thick sediments. Thermal surveys demonstrate that most of the EPR-generated sea floor is anomalously cool, with 70–90% of lithospheric heat being extracted by rapidly flowing hydrothermal fluids. CNS-generated sea floor is anomalously warm, with virtually all heat loss occurring conductively. It appears that the most important factor in explaining the contrast in thermal conditions is the presence of basement outcrops and seamounts, typically separated by 10–50 km (6–31 mi). Thermal studies around these basement features have identified several outcrops that recharge cool bottom sea-water, and other outcrops that vent warmed fluids.

Outlook. Hydrothermal fluids travel long distances through oceanic crust, extracting significant quantities of heat and exchanging solutes with the overlying ocean. Seamounts and other basement outcrops are common features on the sea floor, and help to explain how hydrothermal fluids can move easily between the crust and the ocean. These features allow limited forces to drive crustal fluids rapidly across vast distances. Relatively little work has been done mapping the geology, chemistry, biology, and hydrogeology of seamounts in the deep ocean, but it is known from global (satellite) surveys that there are at least 15,000 of these features distributed around the world—there may actually be as many as 80,000–100,000 of them, considering the limited resolution of satellite data. Additional work is needed to explore these features, model associated hydrologic systems, and determine what properties and processes contribute to fluid circulation in the crust. As surveying and modeling provide improved understanding of deep-sea environments, scientists will develop better ways to map out fluid flow pathways within the enormous ocean crustal aquifer and to assess the importance of large-scale fluid transport within the sea-floor.

For background information *see* HYDROTHERMAL VENT; LITHOSPHERE; MAGMA; MARINE GEOLOGY; MID-OCEANIC RIDGE; ORE AND MINERAL DEPOSITS; SEAMOUNT AND GUYOT in the McGraw-Hill Encyclopedia of Science & Technology.   A. T. Fisher

Bibliography. A. T. Fisher et al., Hydrothermal recharge and discharge across 50 km guided by seamounts on a young ridge flank, *Nature*, 421:618–621, 2003; R. N. Harris, A. T. Fisher, and D. Chapman, Fluid flow through seamounts and implications for global mass fluxes, *Geology*, 8:725–728, 2004; M. Hutnak, The thermal state of 18–24 Ma upper lithosphere subducting below the Nicoya Peninsula, northern Costa Rica margin, in T. Dixon et al. (eds.), *MARGINS Theoretical Institute: SIEZE Volume*, Columbia University Press, New York, in press, 2004.

Satellite climatology

Satellite climatology, or more accurately satellite-based remote sensing of climate, is a subfield of applied climatology. It occupies the overlap between the technical discipline concerned with acquiring information about Earth's environment from a distance (remote sensing), and the study of climate and its variations (climate dynamics). This combination is logical physically because radiative transfer is the basis of both climate and remote sensing.

Satellite climatology began in the 1960s with the deployment of electrooptical sensors into Earth orbit, primarily to improve weather analysis and forecasting, and subsequently to archive radiance (that is, reflectance and emittance) data for periods exceeding the seasonal and interannual variations of cloud cover and the Earth radiation budget. The discipline evolved as new generations of satellite platforms and sensors were deployed, as many more climate variables were acquired remotely, and as societal issues connected with climate variability and change (such as drought, desertification, deforestation, air pollution, and stratospheric ozone depletion) came to the forefront of public debate. Today, satellite climatology is concerned with retrieving information on Earth-surface features important in climate (such as vegetation distribution, soil moisture, land use and land cover, snow cover, polar ice, and sea surface temperatures), and monitoring the atmospheric signatures associated with climate long-distance connections, particularly the El Niño Southern Oscillation (ENSO), for use in seasonal climate prediction.

Measurable atmospheric variables include clouds and their space-time regimes, radiation interactions with clouds or cloud forcing, water-ice content, and microphysics; water vapor and precipitable water; temperature in the free atmosphere; particulates (dust, soot, and volcanic ash) from natural and anthropogenic sources; precipitation occurrence, rate, and type; lightning; and wind speed and direction close to the Earth's surface and in the free atmosphere. In addition, satellites yield integrated (that is, combined Earth-atmosphere) parameters for climate monitoring, notably the planetary albedo (total solar reflectance) and the upwelling infrared (IR) emittance, or outgoing longwave radiation. Measurements of these parameters permit detection of deep convective clouds and heavy precipitation in the tropics, associated with sea-surface temperature changes such as ENSO, and covariations of snow/ice cover with clouds in high latitudes.

Given the extent of the satellite radiance record, evaluation of contemporary climate changes are possible. These include human impacts on the atmosphere (such as its gaseous composition, particulate loading, and clouds) and surface climate resulting from ozone depletion in the stratosphere and troposphere; seasonal biomass burning; overgrazing and desertification; urbanization; deforestation; large-scale irrigated agriculture; and aviation, particularly the generation of "false cirrus" contrail clouds.

Satellite versus conventional data. Differences between the conventional and satellite records of recent climate changes (for example, the magnitude of global warming) derive from the ways in which variables are sensed, "bottom up" (conventional) versus "top down" (remotely sensed); from differences in exactly which variables are being sensed; and from system and nonsystem biases in the satellite observations. Conventional climate data are point observations, whose distribution is highly variable and inhomogeneous between ocean and land, as well as between different latitude zones. Satellite image data are spatially continuous, yet comprise small-area averages (pixels) determined by the sensor's instantaneous spatial resolution. Moreover, a given climate variable measured remotely may be different from that measured conventionally. For example, surface temperature is the shade air temperature measured thermometrically at about 3.5 ft (1 m) above Earth's surface; remotely sensed, it is the "skin temperature" of the Earth's surface obtained by inverting mathematically the thermal radiance received at the satellite. Further, some satellite-observed variables have no corresponding conventional observation (such as integrated cloud liquid water content, outgoing longwave radiation, and "ocean color," or the chlorophyll abundance in marine phytoplankton, which is important in the global carbon budget).

Satellite data contain biases, particularly the timing of polar orbiter satellite overpasses, which may not

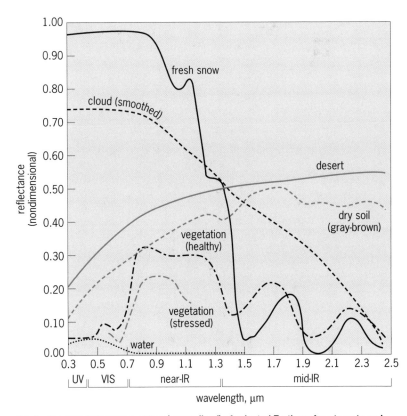

Fig. 1. Spectral reflectance curves (generalized) of selected Earth-surface targets and clouds (all types combined) in the near-ultraviolet through mid-infrared, as compiled from various sources.

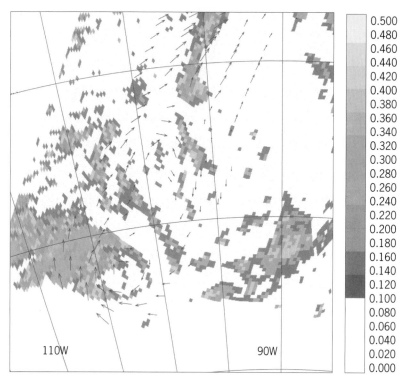

	0.500
	0.480
	0.460
	0.440
	0.420
	0.400
	0.380
	0.360
	0.340
	0.320
	0.300
	0.280
	0.260
	0.240
	0.220
	0.200
	0.180
	0.160
	0.140
	0.120
	0.100
	0.080
	0.060
	0.040
	0.020
	0.000

Fig. 2. Special Sensor Microwave/Imager (SSM/I) 85-GHz retrievals of integrated cloud liquid-water content (kg/m²; see scale at right) and near-coincident *European Remote Sensing Satellite* (*ERS 1*) scatterometer winds (black arrows, where increasing length indicates greater wind speed) for the area just southwest of Drake Passage in early afternoon (UTC) of October 23, 1992. The retrievals indicate mesoscale cyclonic circulations and linear bands of converging atmospheric moisture.

capture the full diurnal range of a climate variable (for example, convective cloudiness and precipitation in the tropics and surface temperature); the increasing path length through the Earth's atmosphere of geostationary satellite observations away from the subsatellite point (for example, when "sounding" water vapor); and changes through time in the satellite overpass characteristics (orbital drift) or calibration of the sensors (sensor drift). Satellite biases must be accommodated when inferring recent climate changes, for providing boundary conditions in numerical models that simulate weather and climate, and when validating model output. Thus, an important theme of satellite climatology involves comparing satellite and conventional observations of a given climate variable with corresponding observations taken from different satellite platforms and sensors (for example, for developing reliable global-scale "Pathfinder" datasets used in climate, ecological, and hydrological modeling).

Satellite remote sensing. Radiation interactions with targets at the Earth's surface and atmosphere form the basis for remotely sensing climate variables (**Fig. 1**). These interactions vary by the wavelength or frequency of radiation considered, and target properties (such as dry versus moist soil, and stressed versus healthy vegetation). Retrieval of target information is optimized for those wavelength bands having the greatest sensitivity. However, the target radiance received at the satellite sensor is influenced by the intervening atmosphere (for example, clear

versus cloudy, small versus large water vapor content, and pollutants). For sensing clouds, particulates, and Earth surface targets, the atmospheric absorption and reemission of radiation for that band should be at a minimum (that is, transmittance is high); when sensing atmospheric gases (for example, water vapor and ozone), thermal IR bands are selected in which absorption by the constituent is at a maximum (that is, transmittance is low). Multispectral sensing in narrow bands permits a large number of climate variables to be determined, such as the midtroposphere water vapor content (around 6.7 micrometers) and clouds distinguished from snow cover (1.6 μm). Moreover, band differencing and band ratioing of radiances yield climatically useful indexes, such as surface skin temperature and the Normalized Difference Vegetation Index. The latter combines the near-IR and visible reflectance values in a composite measure of vegetation activity, involving biomass, leaf-area index, photosynthesis, and phenology, as well as soil moisture and surface wetness.

Nontraditional [that is, microwave (3–300 GHz)] remote sensing comprises passive retrievals (microwave radiometry) and active sensing (radar), which are advantageous in cloudy or nonsunlit areas and over oceans. Although the pixel resolution is coarser than for radar, microwave radiometry permits the acquisition of climate variables such as the extent, concentration and type of sea ice; snow-cover depth and snow-water equivalent; sea surface temperatures; soil moisture; rain rate and near-surface wind speed over ocean areas; atmospheric water vapor; integrated cloud liquid water content; and hail versus rain occurrence. Active microwave imagers (for example, synthetic aperture radar) and nonimagers (radar altimeters and scatterometers) yield high-resolution spatial data on ice characteristics (land and sea), wave heights and relative sea-surface topography, and near-surface wind speed and direction for ocean areas. The last is derived from radiation backscatter changes as the sea surface roughens with increasing wind speed (**Fig. 2**).

Contemporary issues. Many of the research themes in satellite climatology identified as important 15 years ago continue to be emphasized today. These include the need to better retrieve and determine time- and space-scale variations of the Earth radiation budget parameters, clouds and cloud forcing, and precipitation rates (especially their diurnal and subdiurnal variability), as well as to better characterize the important scale linkages, for example, radiation and moisture interactions from cloud particles up to cyclonic circulation systems. These and previously little understood issues in climate science, notably atmospheric particulates and optically thin clouds, are examined using a greater variety of satellite sensors and orbital configurations. Particulates have a direct effect on the Earth radiation budget, primarily through reducing transmission of solar radiation, and an indirect effect when they become incorporated into clouds, enhancing cloud albedo. Because particulate distributions are highly variable spatially

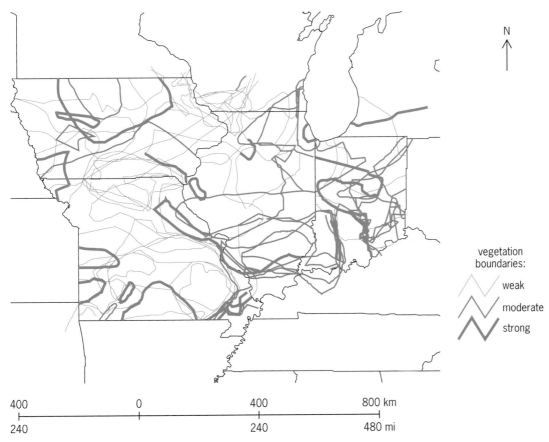

vegetation
boundaries:

weak

moderate

strong

400	0	400	800 km
240		240	480 mi

Fig. 3. Land use and land cover boundaries in the midwestern United States derived from application of GIS to Normalized Difference Vegetation Index images for the late June periods of 1995–2000. High spatial concentrations of lines (for example, in south-central Indiana) indicate permanent boundaries (primarily crop-forest and topography); others are transient (for example, crop-type differences and soil moisture) on these interannual time scales.

and temporally, their timely retrieval is required for more accurate determination of surface skin temperatures. Similarly, optically thin cirrus and contrails influence radiation transfer and target retrieval, yet they are not readily detectable by traditional satellite systems. NASA's Earth Observing System (EOS) comprises a constellation of satellites dedicated to monitoring the Earth's atmosphere, oceans, biosphere, and cryosphere (that is, climate), particularly the Aqua, Terra, TRMM (Tropical Rainfall Measuring Mission), IceSat, and QuickScat satellites. The active and passive sensors on these platforms incorporate advances in hyperspectral remote sensing, which permits more accurate retrieval of a larger number of climate variables [for example, the moderate resolution imaging spectroradiometer (MODIS) on Terra and Aqua]. The promise of space-based active visible light detection and ranging (lidar) for acquiring high-resolution information on vegetation, particulates, and cloud microphysics has been confirmed from aircraft level studies. There is growing emphasis on sophisticated statistical techniques, such as artificial neural networks and wavelet analysis, to extract information from image data. In addition, application of geographic information systems (GIS) helps reveal the spatial relationships among parameters on satellite images (**Fig. 3**). Finally, the explosive growth of the Internet over the last decade has made satellite data products widely available, both in near-real time.

For background information *see* APPLICATIONS SATELLITES; CLIMATOLOGY; EL NIÑO; HEAT BALANCE, TERRESTRIAL ATMOSPHERIC; LIDAR; RADAR METEOROLOGY; SATELLITE METEOROLOGY; TERRESTRIAL RADIATION in the McGraw-Hill Encyclopedia of Science & Technology.   Andrew M. Carleton

Bibliography. A. M. Carleton, *Satellite Remote Sensing in Climatology*, Belhaven Press/CRC Press, 1991; H. Jacobowitz et al., The Advanced Very High Resolution Radiometer Pathfinder Atmosphere (PATMOS) Climate Dataset, *Bull. Amer. Meteorol. Soc.*, 84(6):785–793, 2003; M. Jin, Analysis of land skin temperature using AVHRR observations, *Bull. Amer. Meteorol. Soc.*, 85(4):587–600, 2004; S. Q. Kidder and T. H. Vonder Haar, *Satellite Meteorology: An Introduction*, Academic Press, 1995; National Research Council, *Satellite Observation of the Earth's Environment, Accelerating the Transition of Research to Operations*, National Academies Press, Washington, DC, 2003; W. B. Rossow and E. N. Duenas, The International Satellite Cloud Climatology Project (ISCCP) web site, an online resource for research, *Bull. Amer. Meteorol. Soc.*, 85(2):167–172, 2004.

Segmentation and somitogenesis in vertebrates

Segmentation is a basic characteristic of many animal species ranging from invertebrates to humans. The vertebrate body is built on a metameric organization consisting of a repetition along the antero-posterior (AP) axis of functionally equivalent units, each comprising a vertebra, its associated muscles, peripheral nerves, and blood vessels. Functionally, segmentation is critical to ensure the flexibility of a rodlike structure such as the vertebral column. The segmented distribution of the vertebrae derives from the earlier metameric pattern of the embryonic somites (**Fig. 1**). Somites are epithelial spheres generated one after the other in a rhythmic fashion from the mesenchymal presomitic mesoderm (PSM), and they subsequently differentiate to give rise to the vertebrae and skeletal muscles of the body. Recent evidence from work performed in fish, chick, and mouse embryos indicates that somite formation involves an oscillator—the segmentation clock—whose periodic signal is converted into the periodic array of somite boundaries by a spacing mechanism relying on a threshold of fibroblast growth factor signaling regressing in concert with body axis extension. In humans, mutations in the genes associated to the function of this oscillator result in abnormal segmentation of the vertebral column such as is seen in the spondylocostal dysostosis syndrome.

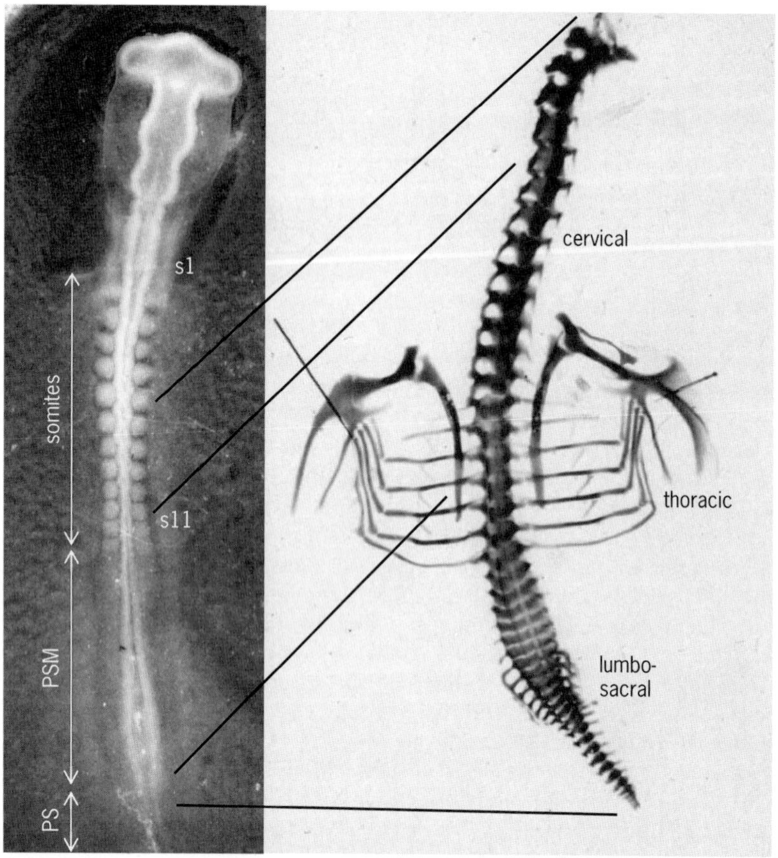

Conservation of the segmented body pattern among very distantly related species provided a strong argument in favor of the idea of the unity of the animal body plan. Because body segmentation is one of the most salient features of the embryo, it was used as a morphological criterion in the pioneering genetic screens performed in the fruit fly by C. Nüsslein-Volhard and E. F. Wieschaus in the late 1970s. These screens led to the identification of the genetic cascade involved in establishing the metameric pattern of the fly embryo. Many of the genes identified in these screens, such as *wingless* or *hedgehog*, proved to be part of major signaling systems which are deregulated in diseases such as cancer. This work, which pioneered the identification of developmental genes using body segmentation as an experimental paradigm, was awarded the Nobel prize in 1995.

In contrast to the fly embryo, in which segments are determined simultaneously, vertebrate segmentation is a sequential process which proceeds synchronously with the posterior extension of the embryo. After the completion of gastrulation, during which the superficial tissues are internalized to form the mesoderm (which gives rise to the muscles and vertebrae) and the endoderm (which gives rise to the lining of the gut), the embryo starts to elongate at its posterior end. This elongation process leads to the sequential formation of embryonic tissues in an anterior-to-posterior sequence. Therefore, the somites that will form the neck vertebrae are formed before those that will form the ribs (Fig. 1). Once they are formed, embryonic structures begin their differentiation. Thus this mode of progressive formation of the body results in the establishment of a gradient of maturation along the antero-posterior axis, the anterior structures being more advanced in their development than the posterior ones. Somite formation follows this differentiation gradient and proceeds rhythmically from head to tail. In the chick embryo, for instance, a new pair of somites is added immediately posterior to the last formed somite pair in 90-min intervals until a fixed number of somites is reached. The somitic series begins posteriorly to the otic vesicle (which forms the ear) and extends to the tip of the tail. The total number of somites produced is highly invariable within a given species and is usually around 50. However, some species such as snakes can produce up to 400 somites. The speed of somite production is also species-specific and can vary depending on the temperature. One pair of somites is produced every 20 min at 25°C in zebrafish or every 90 min at 37°C in the chick embryo, while it takes one day in the salmon at 4°C.

In the past few years, knowledge of the molecular mechanisms underlying vertebrate segmentation

Fig. 1. Vertebrae derive from embryonic somites. The left panel shows a 2-day-old chicken embryo possessing 11 somites. The right panel shows a 10-day-old embryo in which the skeleton is now fully developed. The fate of the somites is indicated by the black lines. The five most anterior somites are incorporated into the skull, which is not shown on the right panel. Half of the cervical and thoracic region is still contained in the presomitic mesoderm (PSM), while the rest of the posterior skeleton is still found in the primitive streak (PS). This developmental sequence illustrates the head-to-tail gradient development characteristic of the vertebrates. Anterior is to the top.

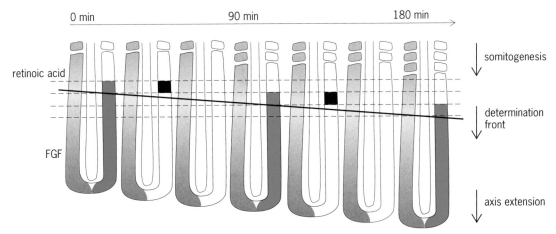

Fig. 2. Clock and wavefront model for somitogenesis. The determination front (wavefront) marks the level at which cells respond to the segmentation clock, and is defined by mutually antagonizing gradient of the diffusible signaling molecules, retinoic acid (in light green) and fibroblast growth factor (FGF) (in gray). When cells reaching the front level receive a periodic signal from the segmentation clock (in dark green), they activate expression of specific genes (*mesp*, in black) in a segment-wide domain. This mechanism translates the oscillating signal of the clock into the spatial periodicity of segments. This mechanism is tightly coupled to the axis extension process. Anterior is to the top.

has increased considerably. Early theoretical models of vertebrate segmentation proposed the existence of a molecular oscillator operating in the presomitic mesoderm to generate a temporal periodicity that could be translated into the spatial periodicity of somite boundaries. Evidence for an oscillator associated with the segmentation process was first recognized in the chick embryo as rhythmic waves of expression in PSM cells of the messenger RNA (mRNA) coding for the transcription factor c-hairy1, a vertebrate homolog of the protein encoded by the fly segmentation gene *hairy*. This oscillator, which was termed the segmentation clock, controls the rhythmic transcription of a group of genes called cyclic genes with a periodicity equal to that of somitogenesis. This periodic expression begins during gastrulation in the somitic precursors and their descendants, and is maintained throughout somitogenesis. The clockwork of the oscillator appears to involve a series of negative feedback loops involving Notch and Wnt signaling, two major pathways at play during development. The role of the segmentation clock in the somitogenesis process is still unclear. One output of the oscillator is the periodic Notch activation in the PSM, which could act as a signal periodically initiating the process of somite boundary specification.

Whereas the segmentation clock is believed to set the rhythm of somitogenesis, it does not specify the positioning of the somite boundaries along the antero-posterior axis. Recent experiments demonstrated that the mechanism controlling the spacing of the future somite boundaries in the forming PSM relies on a traveling threshold of FGF signaling. The *fgf8* mRNA is constantly transcribed in the precursor cell area of the tail bud during axis extension, and its transcription stops when cells enter the PSM. The progressive decay of the *fgf8* mRNA in the PSM results in the formation of an mRNA gradient along the antero-posterior axis. Due to the axis elongation process which results in the addition of new cells

expressing high levels of *fgf8* mRNA selectively in the posterior PSM and to the decay of *fgf8* mRNA in the PSM, the gradient is dynamic and is constantly displaced caudally. This mRNA gradient is translated into a signaling gradient with its maximum located in the most posterior embryonic region. High fibroblast growth factor signaling was shown to maintain the cells in an immature state and prevent their activation of the segmentation program. It is only when cells reach a defined level of the PSM (the determination front) corresponding to a fibroblast growth factor signaling level that they can respond to a periodic signal from the segmentation clock, thus resulting in boundary specification (**Fig. 2**). In this view, the segments are first specified at the level of determination front in response to the segmentation clock. The dynamics of the system is responsible for the caudal movement of the determination front and ensures that the response to the clock is properly spaced along the AP axis.

According to this model, the somite size is defined by the distance traversed by the determination front during one period of the clock. The definition of the position of the determination front was recently shown to be further refined by a gradient of retinoic acid antagonizing the fibroblast growth factor signaling gradient. Retinoic acid signaling limits the rostral extension of the fibroblast growth factor gradient and controls the positioning of the determination front (Fig. 2). Retinoic acid acts as a transcriptional activator for genes involved in segmentation and maturation of the PSM, such as the *mesp* genes. Therefore, an important role of the fibroblast growth factor signaling gradient is not only to maintain PSM cells immature by repressing the activation of PSM maturation genes, but also to position the level at which retinoic acid can begin to act as a transcriptional activator for these genes.

The first genes to be expressed with a strict segmental pattern in the presomitic mesoderm are the transcription factors of the mesp/meso/thylacine family. The *mesp* genes are repressed by high levels of fibroblast growth factor signaling and become activated at the front level in a segment-wide domain in response to retinoic acid and to periodic Notch signaling. Once specified at the determination front level, segments become subdivided genetically into an anterior and a posterior compartment. This subdivision is of critical importance (1) for the segmentation of the peripheral nervous system, since axons and neural crest cells can migrate only in the anterior somite, and (2) for vertebrae formation, since one vertebra forms from the fusion of the posterior part of a somite with the anterior part of the consecutive somite. Finally, in the anteriormost part of the presomitic mesoderm, the posterior somitic boundary is formed and the morphological somite becomes recognizable.

Congenital vertebral malformations in humans represent a major therapeutic challenge, due to the intricate neural and musculoskeletal anatomy of the spine. Identification and early surgical correction in pediatric patients is essential to prevent spinal cord compression, neurological complications, and progressive scoliosis and kyphosis. Understanding the genetic and developmental mechanisms which control somitic/vertebral patterning would be invaluable toward prevention of these birth defects.

For background information *see* ANIMAL MORPHOGENESIS; EMBRYONIC; DIFFERENTIATION; EMBRYONIC INDUCTION; MOLECULAR BIOLOGY in the McGraw-Hill Encyclopedia of Science & Technology.

Olivier Pourquié

Bibliography. J. Cooke and E. C. Zeeman, A clock and wavefront model for control of the number of repeated structures during animal morphogenesis, *J. Theor. Biol.*, 58:455–476, 1976; R. Diez del Corral et al., Opposing FGF and retinoid pathways control ventral neural pattern, neuronal differentiation, and segmentation during body axis extension, *Neuron*, 40:65–79, 2003; J. Dubrulle, M. J. McGrew, and O. Pourquié, FGF signaling controls somite boundary position and regulates segmentation clock control of spatiotemporal Hox gene activation, *Cell*, 106:219–232, 2001; J. Dubrulle and O. Pourquié, fgf8 mRNA decay establishes a gradient that couples axial elongation to patterning in the vertebrate embryo, *Nature*, 427:419–422, 2004; I. Palmeirim et al., Avian hairy gene expression identifies a molecular clock linked to vertebrate segmentation and somitogenesis, *Cell*, 91:639–648, 1997; O. Pourquié, The segmentation clock: Converting embryonic time into spatial pattern, *Science*, 301:328–330, 2003; A. S. Wilkins, *The Evolution of Developmental Pathways*, Sinauer Associates, Sunderland, England, 2001.

Self-cleaning surfaces

Many types of contaminants affect solid surfaces, forcing recurrent cleaning. For example, dust or soot particles or oil aerosols are transported by air until they meet a solid surface on which they will often adhere. Rain may help remove these particles from outdoor surfaces, but it only partially solves the problem. Because solids are imperfect (rough and chemically heterogeneous), the contact angle (measured by depositing a drop and looking at the angle with which it meets its substrate) of water drops can vary in a large interval. The amplitude of this interval is called the contact angle hysteresis. As a consequence, drops can stick to solids, even when the surfaces are inclined, which affects the transparency of the materials (for example, windshields or window panes) or contributes to degrade them (for example, concrete). If the drop is pinned (stuck), the dirt it contains gets concentrated at the contact line (edge), which eventually leads after evaporation to a ring of dirt much more visible than the initial distribution of dust. This is known as the coffee stain effect: as a drop evaporates, a flow takes place, driving matter from the center to the periphery of the drop with higher surface/volume ratio.

There are two ways to achieve a self-cleaning material: using a mechanism that directly degrades the contaminants (which strictly corresponds to the term self-cleaning) or finding a strategy to leave the material dry after a rainstorm. Two kinds of materials, photocatalytic and superhydrophobic solids, respectively, realize such behaviors and as a consequence remain clean longer than usual solids.

Photocatalytic materials. These are solids coated with a mesoporous (2- to 50-nanometer pores) layer of titanium dioxide (TiO_2) or zinc oxide (ZnO). Ultraviolet (UV) irradiation creates pairs of electrons and holes, which survive hours in semiconductors such as titanium dioxide. Because of the strong oxidizing power of the holes and the presence of water close to the surface, hydroxyl radicals are formed which react with organic compounds, transforming them (in the presence of oxygen) into carbon dioxide and water. That is, the organic contaminants are literally burned. Titanium dioxide coatings similarly were found to degrade bacteria, and in Japan antibacterial tiles were produced for sterile environments.

In 2003, several companies marketed photocatalytic glass, which was shown to remain clean longer than untreated glass. One might think that this is due to the degradation of organic compounds, but mineral particles were found to be absent as well. It seems that the major effect is the enhanced hydrophilicity of these materials. Unlike untreated glass, photocatalytic glass remains hydrophilic for several weeks. This allows water films to flow and carry particles, and prevents the formation of contact lines along which dust concentrates as water evaporates. The main effect responsible for the enhanced hydrophilicity seems to be the presence of photoinduced oxygen vacancies at the solid surface which are replaced by dissociated water molecules, making the surface hydrophilic. In addition, these surfaces remain clear if exposed to a dew: water microdrops spread on the surface, preserving the clarity and making these materials applicable for glasses or mirrors. Note, however, the photocatalytic effect takes place only if the material is exposed to UV radiation (that is, mainly outside or possibly inside, if illuminated with fluorescent lamps).

Superhydrophobic solids. These are another class of self-cleaning materials (**Fig. 1**). Chemistry alone cannot provide such a property. The contact angle of water on waxes or fluorinated polymers generally never exceeds 120 to 125°, a value quite far from the maximum one of 180°. However, some natural wax-covered materials (such as lotus, tulip, or gingko biloba leaves, water strider legs, or duck feathers) exhibit contact angles of the order of 160°. This high value was shown to result from the presence of textures at the surface (**Fig. 2**), which enhances hydrophobicity. Inspired by these natural examples, many water-repellent solids have been synthesized during the last 5 years. It is expected that these materials will efficiently repel water drops, which will take with them the dirt deposited on the solids. This mechanism is often referred to as the lotus effect. Following this idea, a paint for the outside of the buildings, called Lotusan, was successfully launched in Germany.

There are two possible reasons why roughness dramatically enhances the hydrophobicity of a solid. First, a rough material has a higher surface area than a smooth one, which increases the hydrophobic power of this material (Wenzel model). Second, air can be trapped in the surface roughness, which also enhances water repellency, since the drop sits partially on air (Cassie model). However, the two scenarios largely differ if one considers the sticking properties of the drops. A Wenzel drop trying to escape a surface is stuck because of the portion of it trapped inside the solid cavities. Such drops are indeed observed to adhere to their substrates in spite of a large contact angle.

A Cassie drop hardly interacts with its substrate, since it floats on it as a fakir does on a bed of nails (Fig. 1). This leads to a nonstick state, characterized by both a large contact angle (as much as 170 to 175°) and a very small contact angle hysteresis (smaller than 5°). This helps explain how fakir drops can be generated; the design of the surface textures must promote air trapping. Different designs were shown to reach this goal, as fractal surfaces (of very large roughness), honeycomb structures (for which it is very difficult to evacuate air as the drop contacts the surface), or tall micropillars (either produced by nanolithography or obtained as a collection of vertical carbon nanotubes). However, it is still not clear which of these designs is optimum.

The dynamics of drops on superhydrophobic surfaces is dramatically different from what is observed for droplets creeping down window panes. Because of a very small contact, they run down much faster (by a factor which can be typically 100 or 1000), and their speed is limited only by the friction due to air (as for a raindrop in free fall). In addition, these drops roll as they move. The combination of rotation and high speed produces a strong deformation of the

Fig. 1. Millimeter-scale drop on a superhydrophobic surface made of vertical carbon nanotubes (coated with a fluoropolymer). Such a surface is rough and hydrophobic, so that the drop sits on a mixture of solid and air, as a fakir on a bed of nails. (*Courtesy of José Bico*)

drop (**Fig. 3**), which tends to be expelled from the surface by centrifugation, leaving the solid without any traces. Another specific property of these "pearl" drops is that they bounce off the surfaces that they hit. The kinetic energy of the impinging drop can be stored during the impact because of its surface deformation, so that the drop behaves as a spring whose stiffness is the liquid surface tension. Superhydrophobic materials thus deserve to be qualified as water-repellent. Rain hitting such a material can be completely scattered.

These fascinating solids raise many questions such as, for a given design, how does the superhydrophobic behavior respond to a reduction of the texture size. Practically, it is very important to reach scales as small as a fraction of a micrometer, for transparency. By the same kind of techniques, it is possible to achieve antidew materials. In nature, superhydrophobic leaves often exhibit two scales of roughness

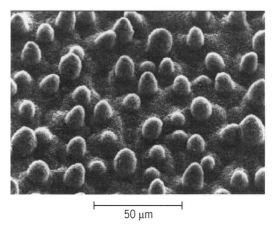

Fig. 2. Scanning electron microscope image of a lotus leaf. The surface is decorated with bumps at the scale of 10 μm, and microfibers at the scale of 1 μm. Water on this surface makes a contact angle of 170° and does not stick. It rolls off and takes with it the dirt present on the surface. This is known as the lotus effect. (*Courtesy of Wilhelm Barthlott*)

Fig. 3. Millimeter-scale water drop moving at a high speed (about 1 m/s) on a superhydrophobic solid. Because of the very high speed, the drop is strongly deformed (whale-shaped). However, this deformation is such that it preserves the dryness of the solid after the drop has passed. (*Courtesy of Denis Richard*)

(typically, bumps at the scale of 10 μm and micro-fibers at the scale of 1 μm), which could correspond to antirain and antidew properties, respectively.

The question of the aging of these substrates also needs to be clarified. Because of the presence of cavities at the surface, these surfaces tend to get (irreversibly) polluted by oil aerosols, which destroys their superhydrophobic properties. And finally, what happens if superhydrophobic solids are fully immersed and moved in water? Does the water slide off them? Except a few controversial papers, this question is still open. However, an efficient slip/slide property would make these materials very attractive for microfluidic applications.

Outlook. Photocatalytic coatings and textured materials have promising characteristics. Recently, it was proposed to mix these properties. L. Jiang and coworkers (Beijing University) succeeded in achieving a solid decorated with zinc oxide micrometric pillars, and showed that such a substrate has the ability to have a reversible wettability. Kept a few days in the dark, this solid is observed to be superhydrophobic; exposed to UV light, it becomes hydrophilic, and indeed superhydrophilic, in particular because of the possibility for the liquid to invade the space between the textures. This example suggests that surfaces properties of solids can be smartly tuned by appropriate textures.

For background information *see* FLUIDS; HOLE STATES IN SOLIDS; INTERFACE OF PHASES; NANO-STRUCTURE; SURFACE TENSION in the McGraw-Hill Encyclopedia of Science & Technology.   David Quéré

Bibliography. R. Blossey, Self-cleaning surfaces—virtual realities, *Nat. Mater.*, 2:301, 2003; A. Fujishima, K. Hashimoto, and T. Watanabe, *TiO$_2$ Photocatalysis: Fundamentals and Applications*, BKC Inc., Tokyo, 1999; A. Nakajima, K. Hashimoto, and T. Watanabe, Recent studies on super-hydrophobic film, *Monatshefte für Chemie*, 132:31–41, 2001; D. Quéré, Non-sticking drops, *Rep. Prog. Phys.*, to be published, 2005.

Serial data transmission

Transmitting data in serial format is the latest trend in high-speed communication applications. Used not only in multicard backplanes (circuit boards), this shift from parallel to serial format is also occurring in devices as common as cell phones and flat-panel displays. What follows is an introduction to serial technology, economic and technology-related trade-offs, and related applications. Before beginning, it is necessary to first know the history for the previous 10 years in backplane communications.

A single-ended parallel bus consists of a series of printed-circuit traces interconnecting one system component with one or more other components. These components could consist of chips on a single circuit board, or they could be separate circuit cards. The number of traces on the bus is called its bus width; thus a 64-bit bus contains 64 traces. In parallel operating mode, only one device can transmit signals over the bus at a time, communicating with one or more receiving devices. In early "high-speed" backplanes, data frequencies were limited to about 1 Mhz. The throughput of a 64-bit bus operating at a 1-Mhz rate is 128 megabits per second, since 2 bits can be transmitted per frequency cycle. To increase data throughput, it is necessary either to increase the data rate or to expand the width of the backplane. Advances in semiconductors have pushed data transmission rates up over the years, but improvements in these single-ended technologies are nearly exhausted. A wider bus introduces problems associated with finding space to place more traces on a circuit board, and with minimizing "skew," that is, keeping the data on the traces aligned in time. Caused by the inherent physical mismatches that exist among the electronic components driving each bit line, skew limits achievable data rates on parallel buses.

Differential signaling—using two traces for each signal—is becoming the transmission technology of choice. Though higher-speed single-ended parallel buses can reach about 66 MHz, differential circuits, specially designed for high-speed transmission, can run at thousands of megahertz. The two most common technologies used for this purpose are called low-voltage differential signaling (LVDS) and common-mode logic (CML). The circuits are simple—they use just a few resistors and transistors—but they must be laid out carefully on an integrated circuit chip and the chip must be fabricated using the proper high-speed semiconductor process.

Exploiting serial differential signaling requires constructing a transceiver that can capture an existing relatively slow rate parallel data stream and pass it along at a higher rate over differential traces. To illustrate, over a single pair of wires, a 1200-MHz differential bus can transmit 20 times the information than can be sent over a single 60-MHz parallel trace. Thus, to transmit the same amount of data in the same time, the differential scheme would need only one-tenth the number of traces required by a parallel system of one-twentieth the speed (2 instead of 20 in this example).

Serial-deserializer characteristics. Semiconductor chips used in pairs to convert parallel bus data to

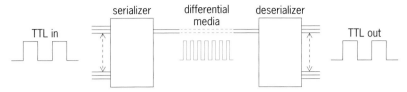

Fig. 1. Data transmission with serializer-deserializer (SerDes) technology. TTL = transistor-transistor logic.

serial form, send it over a high-speed serial differential pair, and then recover the original parallel data are called SerDes (serializer-deserializer) transceivers.

SerDes technology is composed of a dedicated serializer/deserializer interface on each end of the transmission path. Typically, parallel input signals enter the serializer "horizontally" and are then "vertically" aligned such that in one clock period one set of parallel bits (one character or one word) is transmitted (**Fig. 1**). On reception, the serial bits are captured, word-aligned (placed in their proper bit positions on the parallel bus), and placed on the parallel output lines of the transceiver. Clearly, serializer and deserializer internal operating frequencies must be faster than the data rate on the parallel side of the link. At these high data rates, error-free data detection and recovery requires accurate timing and reproducible signal positioning.

In order to meet SerDes timing and signal positioning requirements, two high-performance phase lock loops (PLLs) are required. A PLL has the property that its frequency, or a fraction thereof, can be controlled by an external source and held to close tolerances. One PLL is required for data transmission to coordinate internal operations with the input parallel data rate, to perform the correct parallel-to-serial rate multiplication, and to pace and regulate the serial output data stream. Since each transceiver will have a serial transmission data rate that differs slightly from any other, a SerDes receiver must be able to adapt to the transmitter on the other side of the link. Another PLL is used for this purpose. The design of the SerDes receiver PLL has its built-in frequency tracking capability enhanced by carefully encoding the transmitted data so it will carry both the ("embedded") transmitted signal clock frequency and the serial data stream (**Fig. 2**). Internal operations of the SerDes also use various forms of "elastic buffering" to compensate for these slightly varying transmit and receive data rates.

Device switching. Low-voltage differential signaling (LVDS) cannot drive multiple receivers. To communicate with multiple destinations, a switch interface must be employed. Although these LVDS switch matrixes result in increased design cost and resources, they have the desirable feature of allowing any source to be connected to any receiver. Thus, multiple data exchanges between pairs of SerDes devices can be conducted simultaneously, and data sources do not have to queue up waiting their turn to get transmit access on a parallel bus with its one-transmit-to-many-listen restriction.

Applications. A number of current and forthcoming applications implement the SerDes concept, including the following.

Serial ATA. On personal computers (PCs), serial ATA (Advanced Technology Attachment) hard-disk interfaces replace the wide ribbon cables of parallel interfaces with small, round conductors. Smaller cabling simplifies assembly and reduces internal cooling problems for PCs while substantially increasing the

rate at which data can be transferred between the disk and the computer's central processing unit (CPU).

PCI Express. PCI Express is the latest candidate for a new interconnection scheme for PC peripherals (USB, IEEE 1394, and others). Here the classic parallel bus PCI computer interface will be replaced by a high-speed serial one, simplifying and increasing the speed of data transfer. Since SerDes devices can be switched, it will be possible for several peripherals to communicate directly without having to pass data through the computer's CPU, eliminating a serious speed bottleneck.

Cellular telephony. SerDes devices have found their way into wire rooms located at the bottom of major cell towers. Expanding use of cellular telephones for voice and data communication has required that cellular switches transfer increasing quantities of data from the wireless to the wired telecommunications network. Although the burden can be handled by placing more cell towers at a site, the most economic solution is to widen the data rate though the backplanes handling the call traffic.

Liquid crystal displays. Liquid crystal displays, ranging from flat-panel displays (FPD), to cellular telephone displays, to high-speed digital video signals are operating using SerDes technology. Currently, the highest-speed display interfaces for computer workstations use multiple PCI Express SerDes channels (lanes) to deliver display data from the display card to the display at a combined data rate of up to 32 gigabytes per second.

Portable LCD displays on a hinged frame, such as flip cellular telephones, laptop computers, and video cameras, also benefit from SerDes technology. These hinged frame mounts pass large amounts of data through a cable that must remain physically small and must withstand numerous openings and closings. This cable is kept thin and narrow using SerDes transceivers at each end. Cellular telephones

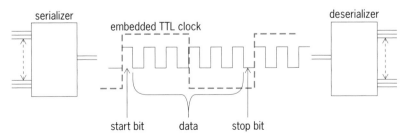

Fig. 2. Encoding of transmitted data in SerDes so that it will carry both an embedded signal clock frequency and the serial data stream.

also benefit from the differential LVDS levels, allowing them to adhere to strict electromagnetic interference regulations from the Federal Communications Commission (FCC).

Some other SerDes technologies that are currently in the initial stages of implementation include InfiniBand, various realizations of 10 Gigabit Ethernet, Fibre Channel, IEEE 1394 FireWire, RapidIO, HyperTransport, and high-speed expansions of SONET, the synchronous optical network standard.

Future. The recent availability in LVDS and CML differential transmission devices has given versatile SerDes technology the tools it needs for future expansion. More and more backplane designs and high-speed communications interfaces will incorporate SerDes technology, and cross-point switches will become more ubiquitous. SerDes LCD applications will continue to grow, particularly when television sets with large liquid crystal displays begin to be produced in significant quantities. The technology is constantly evolving and diversifying and will become an essential component of future electronic devices.

For background information *see* COMPUTER PERIPHERAL DEVICES; COMPUTER SYSTEMS ARCHITECTURE; DATA COMMUNICATIONS in the McGraw-Hill Encyclopedia of Science & Technology.

Edmund H. Suckow

Bibliography. A. Elahi, *Network Communications Technology*, Delmar Learning, 2000; H. W. Johnson, *High-Speed Digital Design: A Handbook of Black Magic*, Prentice Hall PTR, 1993; H. W. Johnson, *High Speed Signal Propagation: Advanced Black Magic*, Prentice Hall PTR, 2003; B. Razavi, *Phase-Locking in High-Performance Systems: From Devices to Architectures*, Wiley-IEEE Computer Society, 2003.

Severe acute respiratory syndrome (SARS)

Severe acute respiratory syndrome was first recognized in November 2002. An international outbreak involving 26 countries, 8098 cases, and 774 deaths subsequently developed, ending in July 2003. Since then, two isolated cases of SARS and one cluster of 11 cases including one death have been identified, resulting from the "escape" of SARS-associated coronavirus (CoV) from research laboratories. In addition, four isolated cases of SARS with no secondary cases were identified in December 2003 and January 2004. Aside from these scenarios, to date there has been no evidence of a large-scale reemergence of SARS. Regardless, much interest and research continues into better understanding SARS in order to be prepared for its possible reemergence and/or the emergence of other similar pathogens.

Etiologic agent. With unprecedented speed after the recognition of SARS, a newly identified coronavirus, SARS-CoV, was pinpointed as the causal agent of the disease. Coronaviruses are enveloped, single-stranded RNA viruses that cause a variety of diseases in animals. Prior to the discovery of SARS-CoV, only two coronaviruses were recognized as causing human disease, coronavirus OC43 and 229E, both associated with the common cold. In general, coronaviruses can be divided into three groups based on their genetic and protein content. However, SARS-CoV does not readily fit into any of these three groups and has been placed into a new fourth group.

While the origins of SARS-CoV are still being investigated, data suggest that SARS-CoV evolved from animal SARS-like viruses that are closely related to SARS-CoV. SARS-like viruses have been identified in a number of different animals, including most frequently Himalayan palm civets found in food markets and eaten as a delicacy in Guangdong Province of the People's Republic of China. Genotypic evidence suggests that cases early in the outbreak were infected with viruses more closely resembling animal SARS-like viruses than SARS-CoV and that SARS-CoV evolved from positive selective pressure acting on animal SARS-like viruses, ultimately culminating in the emergence of the predominant SARS-CoV genotype responsible for the 2002–2003 outbreak.

There is evidence to suggest that animal SARS-like viruses existed and were transmitted to humans, remaining clinically undetected, at least 2 years predating the 2002–2003 outbreak. In addition, it appears that such transmission likely is ongoing today. For example, in Hong Kong 1.8% of frozen sera from 938 adults collected in 2001 were positive for antibodies against animal SARS-like viruses. More recently, at least one of the four isolated cases of SARS that occurred in December 2003 was due to a virus more closely resembling animal SARS-like viruses than SARS-CoV. Present data, while sparse, suggest that human infection due to animal SARS-like viruses is milder and associated with less human-to-human transmission than infection with SARS-CoV.

Epidemiology. Whether or not SARS will reemerge as it did in the 2002–2003 outbreak is up for debate. At the time of this writing, the only known reservoirs of SARS-CoV are laboratories in which live SARS-CoV is being handled. While animal reservoirs of animal SARS-like viruses exist, human or animal reservoirs of SARS-CoV have not been found. Thus the only likely ways for SARS to reemerge are for the circumstances that permitted the selection and purification of SARS-CoV from animal SARS-like viruses to be repeated or for laboratory accidents exposing workers to SARS-CoV to occur. The risk of the former is difficult to predict until more is understood about the circumstances that permitted the selection of SARS-CoV to have occurred in the first case. The risk of the latter will be reduced by continued vigilance and reinforcement of the importance of appropriate biosafety standards to be used in laboratories in which research with SARS-CoV is being conducted. Despite the fact that it is not clear whether or not new outbreaks of SARS will occur, it is important to be prepared for the possibility that they may.

Transmission. SARS-CoV is typically transmitted by direct contact with people with SARS or by mucous membrane exposure to respiratory droplets from infected people. Some evidence suggests that airborne transmission, or transmission via fomites (inanimate contaminated objects) may occur, but only very rarely. Transmission appears not to occur before the onset of symptoms, and is most common from severely ill people in the second week of illness. In the 2003 outbreaks, transmission was almost entirely limited to households, hospitals, and immediate prehospital care.

Pathogenesis. The mechanisms by which SARS-CoV causes disease are still being investigated. Angiotensin-converting enzyme 2, predominantly found in heart and kidney tissues, has been identified as a cellular receptor for SARS-CoV. However, given that SARS-CoV is found in highest numbers in lung and gastrointestinal tissue, other unidentified receptors likely also play a role. Based on clinical, epidemiologic, and virologic data, it appears that SARS-CoV, after entry via the respiratory system, is first associated with viremia (presence of viral particles in the blood), followed by predominant replication in the lung and gastrointestinal tract. However, dissemination to and replication in other organs likely also occurs, given that SARS-CoV ribonucleic acid (RNA) has been found in a wide range of organs at autopsy. Viral replication is thought to be maximal during the second week of symptoms, a time during which SARS-CoV can be most readily detected in tissues and most efficiently transmitted to others. It has been postulated that the immune response to SARS-CoV replication in the lung is primarily responsible for the worsening of respiratory symptoms seen in some patients, typically during the third week of illness. The associated lymphopenia (a reduction of lymphocutes in the circulating blood) observed in many patients suggests that lymphocytes play a role in either fighting the infection or possibly in being infected themselves. Further study is needed to better understand the pathogenesis of and immune response to SARS.

Clinical manifestations. Patients infected with SARS-CoV may be asymptomatic, may have only mild nonspecific symptoms that do not progress, or may develop typical SARS. Patients with typical SARS usually present two to ten days following exposure with nonspecific symptoms including fever, myalgia (muscle pain), headache, malaise, and chills. Three to five days later, these patients develop a nonproductive (dry) cough and dyspnea (difficult or labored breathing) with approximately 25% developing watery diarrhea. Approximately 20% of patients subsequently develop worsening respiratory distress requiring admission to an intensive care unit. Overall 15% of patients require tracheal intubation (insertion of a tube into the trachea to maintain an airway and permit suction of the respiratory tract) and mechanical ventilation support. Approximately 10% of all patients die from progressive respiratory distress or

complications of their hospital admission, typically during the third or fourth week of symptomatic illness.

Typical laboratory abnormalities at the time of presentation include a normal total white blood cell count, lymphopenia, and increased levels of the enzymes lactate dehydrogenase and creatinine kinase (most likely due to associated lung tissue injury). However, such laboratory abnormalities are not uncommon in patients with community-acquired pneumonia due to causes other than SARS-CoV, so that a diagnosis of SARS cannot accurately be made based on clinical or laboratory features alone. Approximately 75% of patients present with infiltrates on a chest radiograph (that is, the appearance of ill-defined opacities on a chest x-ray) but virtually all infected patients will eventually develop chest infiltrates, usually in the second week of illness. Multifocal infiltrates (opacities in more than one lung zone) with ground-glass appearance are more common in SARS than bacterial pneumonia. High-resolution computerized tomography may detect ground-glass opacities not detected by standard radiography. Both laboratory abnormalities and infiltrates usually progress after admission and peak at, or just after, the end of the second week of disease.

Increased age, comorbidity (having a concurrent but unrelated disease), elevated levels of lactate dehydrogenase, and elevated neutrophil (large granular white blood cell) counts at the time of presentation have been associated with an increased risk of death; the case fatality rate in persons over the age of 60 approaches 50%. Infants and children appear to be protected against acquiring the infection and typically have a milder course; associated rhinorrhea (mucous discharge from the nose) is seen in 50% of those who develop symptoms.

Diagnosis. Multiple assays for the diagnosis of SARS have been developed. Reverse transcription-polymerase chain reaction (RT-PCR), which detects SARS-CoV RNA in specimens of blood, stool, and respiratory secretions taken from patients, and serologic tests, which detect SARS-CoV antibodies in the serum produced after infection, have become the mainstay of laboratory diagnosis. Viral culture appears to be less sensitive.

Seroconversion (development of antibodies to a particular antigen) occurs in patients with SARS from 1 to 4 weeks after symptom onset. The majority of patients do not develop either immunoglobulin G (IgG) or IgM antibodies until the third or fourth week of illness.

Treatment. During the 2002–2003 outbreak many antiviral agents, such as ribavirin and lopinavir/ritonavir (which inhibit viral replication), and immunomodulatory durgs, such as interferon, steroids, and intravenous immunoglobulin (which modify the immune response to an infection), were used in varying doses and combinations in different regions of the world. Essentially, all of these treatments have had some reports of associated anecdotal clinical improvement. However, no definitive conclusions regarding the efficacy of any of these treatments can be made.

Many other treatments have shown promise in in-vitro or animal models. These include antiviral agents specifically directed against viral entry into cells and agents directed against expression of viral genes. Further in-vitro and in-vivo studies are needed to better determine the potential role of each of these agents as prophylaxis or treatment for SARS.

Prevention. Many candidate SARS vaccines are currently in the process of being developed, including vaccines based on inactivated virus, recombinant subunits of virus, DNA, and viral vector delivery. Preliminary work based on the immunization of animals has been published showing the ability of many of these vaccines to induce immune responses to SARS-CoV. Much more is expected to be understood about the possible usefulness of these candidate vaccines in the future.

Until effective vaccines are available for human use, the key components to prevention include early identification of cases, rapid implementation of control measures including isolation, contact tracing, and, possibly, quarantine.

Lessons learned. Many lessons learned from the 2002–2003 SARS outbreak can be used toward improving preparedness nationally and internationally for other emerging infectious diseases. The international cooperation of laboratory and clinical working groups during SARS sets a new standard for collaboration. Progress was also made in sharing information across international boundaries that may permit greater transparency and better international collaboration in the management of future outbreaks.

Syndromic surveillance systems, now being piloted in emergency departments and health units in several countries, will likely be useful in the detection of new infectious disease outbreaks. Having a prepared communications system in place to enable real-time communication between health care institutions, front-line physicians, emergency health care workers, public health and other health care workers is also important, as is an overall outbreak public communication plan based on the science of risk communication.

For background information *see* ANIMAL VIRUS; INFECTIOUS DISEASE; RESPIRATORY SYSTEM DISORDERS; VACCINATION; VIRUS in the McGraw-Hill Encyclopedia of Science & Technology.

Susan M. Poutanen; Allison J. McGeer

Bibliography. M. D. Christian et al., Severe acute respiratory syndrome, *Clin. Infect. Dis.*, 38:1420–1427, 2004; J. S. M. Peiris et al., The severe acute respiratory syndrome, *N. Engl. J. Med.*, 349:2431–2441, 2003; S. M. Poutanen and D. E. Low, Severe acute respiratory syndrome: An update, *Curr. Opin. Infect. Dis.*, 17:287–294, 2004; S. M. Poutanen and A. J. McGeer, Transmission and control of SARS, *Curr. Infect. Dis. Rep.*, 6:220–227, 2004; J. T. Wang and S. C. Chang, Severe acute respiratory syndrome, *Curr. Opin. Infect. Dis.*, 17:143–148, 2004.

Shape memory alloys

After undergoing plastic deformation at low temperature, some metallic alloys such as NiTi, CuZnAl, and CuAlNi can restore their original shape upon heating. These alloys are said to "remember" their original shape and are called shape memory alloys. **Figure 1** shows the effect in a NiTi wire. (The remembered shape is the logo of the author's institution, the Technical University Berlin.) In this case, the shape was induced by clamping the wire in that shape at 750 K (890°F) for 5 hours, followed by slow cooling.

Shape memory effect. The memory effect is a result of the strong temperature dependence of the load deformation characteristics of such materials. **Figure 2***a* shows four load deformation (P, D) diagrams for increasing temperatures $(T_1$ through $T_4)$. All the diagrams exhibit hysteresis; that is, on increasing and decreasing the load, the change in shape is not linear. For T_1 and T_2 the diagrams show quasiplastic behavior, and for T_3 and T_4 they show pseudoelastic behavior. Although the load deformation curves are similar to of those of plastic and elastic wires, respectively, in true plasticity the lateral steep lines are absent and yield proceeds along the horizontal lines until fracture occurs; and in true elasticity there is no hysteresis. The memory effect is implied by Fig. 2*a* because the unloaded deformed state D_1, reached from $D = 0$ at low temperature, does not exist at high temperature. At high temperature, the unloaded state has $D = 0$, so the material must return to the undeformed state upon heating.

Typically, the temperature interval T_1-T_4 in Fig. 2*a* is 40 K (72°F), and the center of that interval may be placed at some temperature between 100 and 400 K (−280 and 260°F) by properly choosing the alloy composition and heat treatment. The maximal recoverable deformation is typically 6 to 8%.

Many load deformation isotherms, such as those of Fig. 2*a*, allow us to construct a deformation temperature diagram for a fixed load. Figure 2*b* shows such a diagram for a positive fixed load P_1, as constructed from the load deformation curves of Fig. 2*a*.

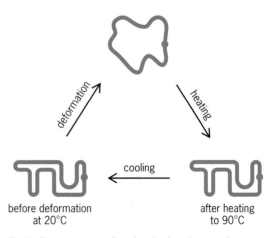

Fig. 1. Shape memory alloy showing how it remembers shape.

Figure 2*b* shows that a shape memory alloy stores two deformations in its memory: a small one for high temperature and a large one for low temperature. If a shape memory wire is subjected to an alternating temperature, it will alternately elongate and shorten. Therefore shape memory wires, springs, or foils may be used to build temperature-controlled switches or actuators. If a suitable temperature gradient is used, one can construct a heat engine. **Figure 3** shows two functioning constructions.

Often, repeated loading and unloading can induce an internal stress field which forces the externally unloaded specimen to alternate between two deformations at alternating temperatures. This phenomenon is called the two-way shape memory effect.

A medical application is illustrated in **Fig. 4***a*. The figure shows a broken jaw bone which must be fixed by a splint. In the application of the splint, it is important that the bone ends are tightly pressed against each other. Much surgical skill is required when a steel splint is used, and even then it is difficult to maintain sufficient compressive force. That is where a shape memory alloy can help. The memory splint is screwed into place in the state of large deformation on the upper border of the hysteresis loop at 36°C (96.8°F). It is then heated to, say, 42°C (107.6°F), whereupon it contracts and pushes the bone ends firmly against each other. The contraction persists even after the splint assumes body temperature again, because now the state of the material lies on the lower border of the hysteresis loop. The relevant states are marked by circles in Fig. 4*a*.

Another application occurs in dentistry when braces are used to readjust the position of teeth. The brace is pseudoelastic, as well as predeformed and prestressed to be on the recovery line upon installation (Fig. 4*b*). The teeth move under the load, and the deformation *D* decreases. As it does, the load *P* is unchanged for some considerable period of contraction. Therefore, a readjustment of the brace is not necessary for a long time.

Phase transformations. The key to understanding the observed phenomena lies in the presence of a phase transition in the lattice. At high temperature, the highly symmetric austenitic phase prevails, whereas at low temperature the lattice is in the less symmetric martensitic phase which is prone to forming twins (mirrored crystal regions). In the unloaded state at low temperature, different twins are ideally present in equal proportions. By uniaxial loading, one of the twins is favored by the direction of the load. That twin will then form at the expense of the others in a process called twinning. At high temperature, no martensite occurs in an unloaded specimen, but the application of a load may force the prevailing austenite into a martensitic twin variant. That happens on the pseudoelastic yield line. On the recovery line the reverse transition occurs.

Modeling. Based on these observations, three mathematical models have been constructed which are capable of simulating the observed phenomena: a thermodynamic model, a kinetic model, and a molec-

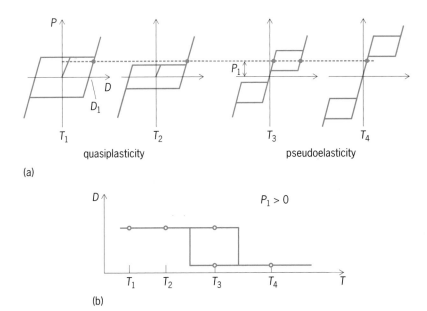

(a)

(b)

Fig. 2. Diagrams of (*a*) load [*P*] deformation [*D*] characteristics for different temperatures, $T_1 < T_2 < T_3 < T_4$; and (*b*) deformation [*D*] temperature [*T*] for a fixed load.

ular dynamic model. Of most interest is the kinetic model, which is capable of simulating shape memory behavior even quantitatively after a few easily measurable parameters have been determined.

This model is a Java simulation which allows one to prescribe load and temperatures as a function of time and to obtain phase fractions and the deformation as a function of time. Alternatively, one can prescribe deformation and obtain the necessary load. The relevant load deformation diagrams can also be created in this manner, and they show all the main features of the diagrams in Fig. 2*a*.

There is also a two-dimensional molecular dynamic simulation of a shape memory alloy in which the austenitic to martensitic phase transition can be viewed along with the twinning phenomenon in an animation. **Figure 5** shows some screen shots taken during the transition. Thermodynamically the high-temperature phase, the highly symmetric austenite,

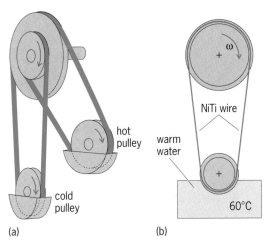

(a)   (b)

Fig. 3. Use of shape memory alloys allows construction of heat engines. (*a*) By A. D. Johnson. (*b*) By F. E. Wang.

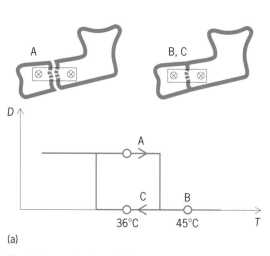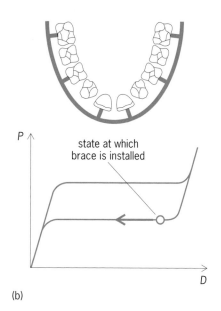

Fig. 4. Two applications of shape memory alloy. (*a*) Deformation (*D*) as a function of temperature (*T*) for a shape memory splint. Installed at condition A, the splint can be heated to B, causing contraction, which remains when cooled to C. (*b*) Use of memory metal in dentistry. The load (*P*) deformation (*D*) characteristics allow the braces to contract (decreasing *D*) without a change in the load (*P*).

is entropically stabilized. That is, although the energy of the austenite is higher than the energy of the martensite, the austenite is still stable because its Gibbs (free) energy is lower, due to higher entropy.

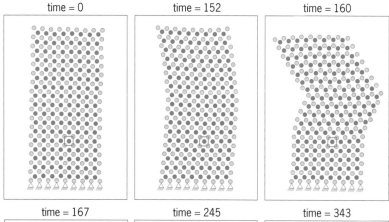

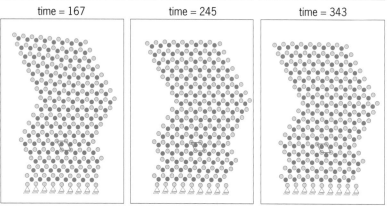

Fig. 5. Results of a two-dimensional molecular dynamics simulation of a temperature-induced austenite to martensite phase transition, showing how a block of atoms of types dark and light color in an austenitic (square) lattice and the phase transition toward the martensitic (rhombic) lattice. The phase transition proceeds from the top to the bottom of the lattice as the temperature decreases, and twinning is evident as not all the rows are oriented in the same direction.

This is a common phenomenon in phase transitions, and it is explicitly confirmed by the molecular dynamic simulation.

For background information *see* ENTROPY; HYSTERESIS; METAL, MECHANICAL PROPERTIES OF; PHASE TRANSITIONS; SHAPE MEMORY ALLOYS; STRESS AND STRAIN; TWINNING (CRYSTALLOGRAPHY) in the McGraw-Hill Encyclopedia of Science & Technology.

Ingo Müller

Bibliography. C. T. Liu et al. (eds.), Shape memory materials and phenomena—Fundamental aspects and applications, *Proceedings of the Materials Research Society Symposium*, vol. 246, Fall Meeting, Boston, 1991; J. Perkins (ed.), *Shape Memory Effect in Alloys*, Plenum Press, New York, 1976; I. Tamura (ed.), *Proceedings of the International on Martensitic Transformation*, Japan Institute of Metals, Nara, 1986.

Sloan Digital Sky Survey

Many decades of observational and theoretical research have converged on a unified picture of the nature of the universe. Instrumental in these advances have been detailed measurements of the expansion rate of the universe by scientists using the Hubble Space Telescope, of the radiation from the cosmic microwave background radiation by the Wilkinson Microwave Anisotropy Probe, and of the distribution of galaxies by several redshift surveys, of which the largest to date have been the Two Degree Field Survey and the Sloan Digital Sky Survey (SDSS).

These last projects probe the universe by measuring the redshifts of galaxies, that is, the shift in wavelength of spectral features from their known rest wavelengths to longer wavelengths. Redshifts are caused by the expansion of the universe. Because of this expansion, the farther away from us a galaxy lies the more rapidly it is receding from us. This relation can be turned around, for once the proportionality between distance and redshift (the Hubble "constant" or parameter) has been measured, it can be used together with a galaxy's redshift to give its distance. Thus observations of the positions on the sky and of the redshifts of a sample of galaxies can be used to measure the three-dimensional distribution of galaxies.

Three-dimensional galaxy distribution. Why is this of interest? Galaxies turn out not to be randomly distributed in space, but their distribution is structured in gravitationally bound clusters, sheets, and filaments. The first three-dimensional maps of significant volumes of the universe were made in the 1980s, and incorporated observations of 1000–2000 galaxies. These maps revealed a frothy, filamentary distribution of galaxies, with the largest feature being a "Great Wall" (in a map made by Margaret Geller, John Huchra, and Valerie de Lapparent) of galaxies more than 600 million light years across and extending to the boundaries of the survey volume, so that it could have been even larger.

Encoded in these maps is the result of the initial mass fluctuations in the early stages of the universe and subsequent evolution due to the expansion of the universe, galaxy formation, and gravity: Material falls into dense regions of space, which thereby accrete more mass as the universe evolves. The exciting results discussed above pointed out the need for surveys of much larger volumes of space, so that a representative sample of the universe could be studied. Current theories of the universe say that, in a statistical sense, it is uniform at all locations and in all directions. The requirement on a redshift survey that can fairly be said to sample the universe is, then, that the volume is large enough to contain many tens of examples of the largest possible structures. It was clear from the early results that to do this would involve measuring redshifts for many hundreds of thousands of galaxies. Galaxies are faint, and accumulating enough light to measure a single redshift can take of order an hour of telescope time. The possibility of conducting surveys large enough to map the local universe in the necessary detail developed with the advent of technology that could measure thousands of redshifts in a single night, and of fast powerful computers to acquire and reduce the data automatically.

Observing strategy. The Sloan Digital Sky Survey is one of several large surveys recently finished or underway. It is unique in several important ways: It is the largest redshift survey, with more than 400,000 galaxy redshifts measured as of June 2004 and plans to measure a million; and it acquires the imaging data from which the positions of the galaxies can be measured during the same time period as it measures the redshifts.

The reason for this strategy is that the Sloan Digital Sky Survey acquires a much better image of the sky than has previously been available for more than very small areas of sky, so it is both an imaging and a redshift survey. This improved imaging is necessary to understand better the intrinsic properties of each observed galaxy (brightness, dominant stellar population, and so forth). The Sloan Digital Sky Survey images the sky through five filters essentially simultaneously, measuring five brightnesses for each object. The filters cover the ultraviolet, green, red, far-red, and near-infrared regions of the optical spectrum. A star that is very hot, and therefore blue, will be much brighter in the ultraviolet filter than in the far-red filter, and vice versa for a cool, red star. Thus objects can be sorted and selected by their colors automatically. The filters are designed to separate quasars from stars by color, for the Sloan Digital Sky Survey is also a redshift survey of 100,000 quasars. Galaxies can be separated from stars and quasars because they appear to be fuzzy rather than pointlike on images (see **illustration**). The data are acquired by electronic imagers called charge-coupled devices (CCDs) similar to those used in digital cameras. The word digital applies both to cameras and the Sloan Survey because the imager converts light to electrons, which can be directly measured and counted as integers by computers.

The normal way in which redshift surveys are done is to image the sky (and before Sloan Digital Sky Survey, all large-area imaging of the sky had been done by photographic plates), select the objects for which redshifts will be measured, and then measure the redshifts. To do this with the Sloan Digital Sky Survey would have been prohibitively time-consuming and expensive. All optical observing needs clear weather, but imaging needs more: pristine clarity that does not change over the night, and a steady atmosphere, so that the images of stars are as close to pointlike as possible. Even the best sites have such conditions only a small fraction of the time. They do, however, have good atmospheric conditions much more frequently. Such conditions are perfectly adequate for obtaining the spectra that give redshifts. Accordingly, the Sloan Digital Sky Survey follows a dual observing strategy: (1) imaging is done in the best weather; (2) objects are selected from the imaging according to carefully defined criteria by powerful computer software; and their spectra are observed during the less-than-perfect weather. In this way the Sloan Digital Sky Survey efficiently uses all the available observing time and can be completed in a reasonable time (5–7 years). The spectra are observed with devices that can obtain 640 spectra simultaneously; 7 or 8 sets, or several thousand spectra, can be observed in a single night. To do this, the Sloan Digital Sky Survey uses a dedicated 2.5-m (98-in.) telescope at the Apache Point Observatory, New Mexico. The telescope is equipped with a multi-charge-coupled-device camera and two fiber-fed spectrographs, and observes every possible clear, dark night year-round except for a short maintenance period in the summer.

Results. The Sloan Digital Sky Survey has obtained several fundamental results. A "Great Wall" of galaxies, 1.4×10^9 light-years long and 80% longer than the "Great Wall" found by Geller, Huchra, and Lapparent,

Distribution of galaxies from the Sloan Digital Sky Survey (SDSS). The image on the left shows the redshift-space distribution for galaxies in the SDSS. The dark cones to the left and right are areas of the universe which cannot be observed because they lie behind the obscuring dust of the Milky Way Galaxy. Each point represents one galaxy. The image on the right is a small piece of sky constructed from the SDSS multicolor imaging data. The extended fuzzy objects are galaxies, and the small pointlike objects are stars in the Milky Way Galaxy. (*Courtesy of M. Tegmark and R. Lupton*)

is the largest structure in the universe observed so far. The distribution of galaxies is in excellent agreement with the current cosmological model, in which only 4% of the universe is normal matter, with the rest being dark matter and dark energy. The signature of the formation of galaxies in lumps of dark matter is seen, and comparison of the distribution of the galaxies with that of the cosmic microwave background confirms the presence of dark energy. Cosmology with the Sloan Digital Sky Survey is only just beginning to yield significant information, and many more results, measuring the universe in great detail and with great accuracy, will be forthcoming in the next few years. The current map of the universe, together with a Sloan Digital Sky Survey image of a cluster of nearby galaxies, is shown in the illustration.

The Sloan Digital Sky Survey quasar work has been enormously successful. Quasars are extremely high luminosity objects, thought to be due to the infall of gas into massive black holes. Because they are so luminous, quasars can be seen to great distances, and we observe distant objects as they were a long time ago because of the light travel time. The most distant quasars are seen when the universe was about 5% of its present age. The distant quasars have allowed the discovery of the Gunn-Peterson effect, predicted in the 1960s, whereby the light from distant objects is absorbed by hydrogen if that hydrogen is atomic. This would prevent our seeing the distant universe unless the intervening hydrogen were

ionized, that is, the electrons separated from the protons, in which case the atoms could no longer absorb light. The detection of the Gunn-Peterson effect therefore locates the epoch at which the dark ages of the universe ended and the hydrogen was ionized by light from newly formed stars.

A survey as powerful as the Sloan Digital Sky Survey produces results in many other areas. The Sloan Digital Sky Survey has been instrumental in discovering that the Milky Way Galaxy is surrounded by streams of stars from the tidal destruction of small satellite galaxies that used to orbit the Galaxy. The Sloan Digital Sky Survey has discovered many gravitational lenses, in which light from distant quasars is bent by the warping of space by lumps of mass (the lenses) between the Earth and the quasar. This effect is predicted by Einstein's theory of general relativity. The Sloan Digital Sky Survey has discovered the most massive lens yet found. All masses will distort the light passing them, and this effect has allowed the Sloan Digital Sky Survey to measure the masses of galaxies directly; the more massive the galaxy the more the light path is bent. The Sloan Digital Sky Survey has discovered the lowest-luminosity galaxy ever found, a satellite of the nearby Andromeda Galaxy. It has discovered large numbers of brown dwarfs, objects intermediate in mass between planets and stars. It has produced the largest sample of small bodies in the solar system with accurate colors, allowing their compositions to be deduced. It can confidently expected that these findings are only the beginning of decades of discovery from the Sloan Digital Sky Survey data.

The data from the Sloan Digital Sky Survey are made freely available to all, with a regular schedule of data releases. As well as being available to professional astronomers in support of their research, they are available for educational purposes and come with a large set of educational tools.

For background information *see* ASTRONOMICAL SPECTROSCOPY; BROWN DWARF; COSMOLOGY; GALAXY, EXTERNAL; GRAVITATIONAL LENS; HUBBLE CONSTANT; QUASAR; REDSHIFT; TELESCOPE; UNIVERSE in the McGraw-Hill Encyclopedia of Science & Technology.   Gillian R. Knapp

Bibliography. B. S. Ryden, *Introduction to Cosmology*, Addison-Wesley, 2003; M. A. Strauss, Reading the blueprints of creation, *Sci. Amer.*, 290(2):42, February 2004; M. Tegmark et al., The three-dimensional power spectrum of galaxies from the Sloan Digital Sky Survey, *Astrophys. J.*, 606:702–740, 2004; D. G. York et al., The Sloan Digital Sky Survey: Technical Summary, *Astron. J.*, 120:1579–1587, 2000.

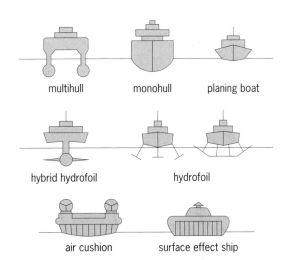

Fig. 1. Alternative hull form geometries.

Small warship design

The world's navies continue to explore a wide variety of ship design types to meet an expanding set of mission requirements. At the same time, severe constraints are being imposed on procurement budgets and operating funds. Application of small, high-speed naval surface combatant ships for combat operations in littoral (shoreline) regions has seen increased emphasis in recent years, including the U.S. Navy's planned acquisition of the littoral combatant ship (LCS). Some key design considerations for such ships will be discussed, and examples of alternative hull forms will be highlighted.

Hull form alternatives. There is a wide range of ship hull form alternatives for small, high-speed naval combatant ships. These can be divided into several categories depending on how they keep a boat afloat: buoyant lift (also called displacement hullforms), powered lift, dynamic lift, and hybrids. Within these categories there are multiple hull form alternatives (**Table 1**).

The alternative hull form geometries are shown in **Fig. 1**. Each hull form has advantages and disadvantages such as the degree of water resistance, powering requirements, seakeeping, ship arrangements (layout), and ship system complexity and flexibility

for installation of weapons and sensors. Monohulls and planing hulls are low-cost options for most naval missions, but are limited to sea states 3–4 due to ship motions and structural loads. Sea state definitions and probability of occurrence are shown in **Table 2**. Catamarans have large deck areas that are conducive to low-density systems and aviation operations, with acceptable powering characteristics at high speeds in sea states up to 4–5. Trimarans, with long, slender center hulls, offer reduction in required power at higher speeds compared to monohulls of similar displacement. A small water plane area twin hull (SWATH) ship exhibits excellent seakeeping at modest speeds, with a penalty of deep draft (depth of penetration into water). A surface effect ship (SES) offers perhaps the best hull form for much higher speeds (50–60 knots or 25–30 m/s), but is limited to sea state 2–3 at top speed, depending on ship size. Air-cushion vehicles (ACVs) can reach speeds of 50–60 knots in sea state 2 or less and provide amphibious assault capability, but are not suitable for open ocean surface combatant missions. Hydrofoils can attain speeds of 50–60 knots in sea state 5–6, but are limited in size due to foil system dimensions and fabrication—a practical limit is on the order of 2500 tons displacement. Hybrids, such as the hydrofoil small waterplane area ship (HYSWAS), can be built at larger sizes and offer a combination of speed and seakeeping approaching that of "pure" hydrofoils, with a disadvantage of deep navigational draft.

The above summary is not exhaustive, as each year sees new hull forms appearing, particularly small

TABLE 1. Hull form categories			
Buoyant lift	Powered lift	Dynamic lift	Hybrids
Monohull	Surface effect ship (SES)	Hydrofoil	FoilCat
Planing hull (slow speed)	Air-cushion vehicle (ACV)	Planing hull (high speed)	Captured air catamaran
Catamaran			—
Trimaran			
Small waterplane area twin hull (SWATH)			Hydrofoil small waterplane area ship (HYSWAS)

Fig. 2. *Visby*, a small monohull warship in the Swedish Navy.

hybrid craft and multihulls of various types. While smaller naval combatant ships with less than 1000 tons displacement and with speeds of 30–40 knots (15–20 m/s) can use existing technology, ships of up to 3000 tons displacement and speeds of 50–60 knots require extrapolation beyond current technology in key areas such as structural load estimation, resistance and powering, and seakeeping and propulsion systems. This requires model testing, new analysis techniques and tools, and develop-

ment of design standards and practices that can be approved by ship classification societies, such as the American Bureau of Shipping (ABS). Technology extrapolations are similar in magnitude for monohulls, trimarans, and SES.

Monohull. While the U.S. shipbuilding industry has experience designing and building monohulls of the size needed for surface combat missions, speeds of these ships are generally below 40 knots (20 m/s). The 72-m (236-ft), 650-metric-ton *Visby* class Swedish Navy corvette is one example of the current state of the art in Europe; it has a top speed of approximately 35 knots (18 m/s) and incorporates stealth technology (**Fig. 2**).

Hydrofoil. Hydrofoils offer the best combination of speed and seakeeping of any hull form alternative. The U.S. Navy operated a squadron of six patrol hydrofoil missile (PHM) class ships in the Caribbean Basin for a period of about 10 years in counterdrug operations (**Fig. 3**). These 240-metric-ton, 36-m (118-ft) ships were capable of speeds in excess of

Fig. 3. Patrol hydrofoil missile (PHM) class ship that was operated by the U.S. Navy.

TABLE 2. Sea state definitions and annual probabilities for open ocean in the Northern Hemisphere				
Sea state	Percentage probability	Significant* wave heights, ft†	Modal wave periods, seconds	Sustained wind speed, knots‡
0–1	0	0.0–0.3	—	0–6
2	5.7	0.3–1.6	3.0–15.0	7–10
3	19.7	1.6–4.1	5.2–15.5	11–16
4	28.3	4.1–8.2	5.9–15.5	17–21
5	19.5	8.2–13.1	7.2–16.5	22–27
6	17.5	13.1–19.7	9.3–16.5	28–47
7	7.6	19.7–29.5	10.0–17.2	48–55
8	1.7	29.5–45.5	13.0–18.4	56–63
>8	0.1	>45.5	20.0	>63

*Significant wave height = average of 1/3 highest waves.
†1 ft = 0.3 m.
‡1 knot = 0.51 m/s.

Fig. 4. HSV-2 *Swift*, a catamaran leased by the U.S. Navy.

45 knots in sea state 5. Recent commercial hydrofoils have employed catamaran hulls, which offer the foil system designer a higher aspect ratio and improved lift/drag ratio. Hybrid concepts, such as HYSWAS, offer improved range at high speed.

Catamaran. The commercial catamaran industry produces high-speed ferries of 2000–4000 tons displacement with speeds up to 40 knots. Therefore, only modest technology investment would be necessary to produce catamaran designs for small, high-speed naval missions. The key design areas are structural loads and structural criteria for unrestricted open-ocean operations. There is high interest in the U.S. Navy, Marine Corps, and Army on analysis and experimentation with high-speed catamarans for logistics and mine warfare missions. The Navy has entered into a lease of the 96-m (315-ft) HSV-2, *Swift* (**Fig. 4**), the Marine Corps continues its lease of the 101-m (331-ft) *Westpac Express*, and the Army is experimenting with its 97-m (318-ft) TSV-1X, *Spearhead*.

Trimaran. The trimaran is of increasing interest for future naval combatant missions. The U.K. and U.S. navies recently completed a cooperative development and trials effort on research vessel *RV Triton* with a full load displacement of about 1300 metric tons and speed of 20 knots (10 m/s; **Fig. 5**). The trimaran offers savings of approximately 20% in required propulsion power at speeds of 30 knots.

Surface effect ship (SES). The SES has about 40 years of developmental and operational experience in the U.S. and overseas, with several hundred built and operated. The current state of the art in naval SES is the Norwegian Navy *Skjold* class of fast reaction craft (**Fig. 6**). These craft are 47 m (155 ft) long, with a full load displacement of 270 tons displacement and speed of 45 knots (23 m/s) in sea state 3.

Weapons systems. Mission flexibility is the primary benefit of small, fast naval surface combatant vessels. Using open systems architectures and modular mission packages is essential for such warships. The Danish Navy's Stanflex 300 program is an excellent example. It allowed the navy to replace 22 ships with 14 Stanflex units. The ships are 54 m (177 ft) in length and 9 m (29.5 ft) in beam, displace 320–485 tons depending on configuration, and are capable of 30 knots (15 m/s). Basic crew size is 19, with a maximum of 29, depending on specific mission systems. Each ship has four modular payload bays and allows the following mission variants: surface attack; antisubmarine warfare (ASW);

Fig. 5. Research vessel *RV Triton*, a trimaran.

Fig. 6. *Skjold*, a surface-effect ship in the Norwegian Navy.

mine countermeasures (MCM); minelayer; patrol/surveillance; and pollution control.

The Danish Navy has put into service over 100 Stanflex mission modules. Each module includes standard interfaces and connections for mounting, local area network, video, degaussing, interior communications, weapons fire control and cooling water. Modules can be changed in 1 hour pierside, using standard cranes and tools. The ship's crew is trained on most module systems, but specialty crews are used for missions such as MCM and ASW. Ship upgrades are accomplished by module upgrades. Industry teams involved with the U.S. Navy's Littoral Combat Ship (LCS) program have looked closely at the Stanflex 300 approach.

Machinery systems. Small (less than 1000 tons displacement) naval vessels reach speeds in the 30–40-knot regime using current state of the art high-speed marine diesel engines and propellers or marine waterjets for propulsion. Higher speeds (50–60 knots) and larger sizes (up to 3000 tons displacement) will require major improvements in machinery technology. Propulsion machinery for such ships must be compact and lightweight, yet provide high power levels with acceptable fuel efficiency. Main engines for small, high-speed naval ships range from gas turbines rated at approximately 10 MW (13,000 hp) to large gas turbines—still in development—rated at up to 43 MW (57,000 hp). Existing gas turbines with ratings up to 32 MW (42,600 hp) are adequate for most applications. Only the largest and fastest designs would require ratings to 43 MW. Gas turbines using more complex thermodynamic cycles and exploiting intercooling and recuperation (ICR) technologies may achieve specific fuel consumption rates approximating those of larger diesel engines. However, ICR gas turbines are heavier and require more internal volume than current engines and may not result in the best design for small, fast naval combatant ships.

Waterjets are the preferred propulsors for high-speed ships above 500 tons. The most powerful waterjets available today, or in development, for commercial high-speed ferries are sufficient for small, high-speed naval vessels. However, the large transom dimensions required for waterjet outflow may degrade speed-power performance and seakeeping. One very large operational unit is a mixed-flow design with an impeller inlet diameter of 2.0 m (6.6 ft) and a rating of 26 MW (34,600 hp). Two other high-speed propulsor alternatives for ships of 500–1000 tons and speeds of 50–60 knots are supercavitating (fully cavitating) propellers and partially submerged (surface-piercing) propellers. However, supercavitating propellers are not compatible with mission profiles that include a range of speeds, and partially submerged propellers have not been developed at sizes needed for high-speed naval vessels.

For background information *see* AIR-CUSHION VEHICLE; HYDROFOIL CRAFT; MARINE ENGINE; NAVAL ARMAMENT; NAVAL SURFACE SHIP in the McGraw-Hill Encyclopedia of Science & Technology.

Mark R. Bebar

Bibliography. J. Allison, Marine waterjet propulsion, Society of Naval Architects and Marine Engineers (SNAME), *Transactions*, vol. 101, 1993; M. Fan and M. Pinchin, Structural design of high speed craft—A comparative study of classification requirements, *Proceedings of FAST '97 Conference*, Sydney, July 1997; J. Jackson et al., Materials considerations for high speed ships, *Proceedings of FAST '99 Conference*, Seattle, September 1999; *Naval Engineers Journal—Special Edition*, American Society of Naval Engineers, February 1985; Lt. (junior grade) J. F. Solomon, U.S. Navy, Lethal in the littoral—A smaller, meaner LCS, *U.S. Naval Institute Proceedings*, pp. 36–39, January 2004.

Smart card

A smart card is a plastic card, the size of a credit card, that contains an embedded silicon computer chip. Its primary purpose is the portable storage and retrieval of data used to authorize various types of electronic transactions. Compared to their predecessors, magnetic stripe cards (for example, credit cards), smart cards can store data relating to more than one institution, facilitate more than one type of use, carry substantially larger volumes of data, process transactions at higher rates, and provide logical and physical security to the card's data. Smart cards work by interacting with a card reader—an interface between the smart card and an institution.

History. Smart cards were developed in the late 1960s to 1970s, when inventors in France, Germany, and Japan formed the idea of cards holding microchips (**Table 1**). Throughout the 1980s, smart card implementation was mainly confined to Europe and Japan. The United States favored wired telecommunication, which supported on-line cardholder/transaction fraud detection, reducing the need to adopt stringent forms of identify verification associated with batch transaction processing. As a result,

TABLE 1. Invention and deployment of the smart card

Year	Event
1968	Jorgen Dethloff and Helmut Grotrupp (Germany) patent the idea of combining computer chips with plastic cards.
1970	Kunitaka Arimura (Japan) receives Japanese patent related to the IC card.
1974	Roland Moreno (France), journalist, receives patent in France for chip cards. Considered father of smart cards, he founded Innovatron, which issued first global licenses.
1976	Honeywell Bull (France) receives licenses as a result of DGT initiative.
1979	Flonic Schlumberger (France) and Honeywell Bull & Philips (Netherlands) receive first Innovatron licenses.
1980	GIE sponsors first smart card trials in three French cities.
1982	First USA trials held in North Dakota and New Jersey.
1996	First USA trials held on university campus.

smart card developments were limited to trials by a few banks, the Department of Defense, and a few universities. Since the September 11, 2001, terrorist attacks, there has been a race to implement stringent identification procedures involving smart cards, primarily in high-liability operating environments.

Applications. Smart cards support a wide range of transactions based on their ability to identify and store information about the cardholder, card issuer, authorized accounts and their activities, authorized merchants, and authorized transactions. The validation of a cardholder's identity may be knowledge-based, where the individual knows something, such as a personal identification number (PIN). It may also be possession-based, where the presence of a "token" (for example, an ID card or a bank card) validates the person's identity. In addition, smart cards can carry biometric information (for example, fingerprints or iris scans), providing greater security for the identification process.

Some common smart card applications include credit transactions, debit transactions, stored-value transactions, information management applications, loyalty/affinity transactions, and multiple applications.

Credit transactions. Smart cards reduce fraud by securing merchant transactions. Smart credit cards give cardholders financial purchasing flexibility in the form of preapproved cash advances and loans (initiated at the point of sale), which are associated with specific electronic "purses" contained on the card.

Debit transactions. Smart cards provide electronic direct-debit at automated teller machines and at specific points of sale. This improves the cardholder's access to specific goods and services in locations where the cardholder's identity and credit history are unknown.

Stored-value transactions. Institutions can create smart cards with a fixed value initially encoded in the card chip's memory. This stored value may be disposable (debit "read-only") or reloaded (debit/credit

"rewrite"). From this sum, purchases can be deducted. These transactions do not require access to the cardholder's personal identity, so no PIN is required. Examples include telephone cards and retail merchant gift cards.

Information management applications. Smart cards can be used for the portable storage and use of cardholder's personal information such as bank credit/debit accounts, insurance information, medical history, emergency contact information, and travel documentation.

Loyalty/affinity transactions. Smart cards can apply vendor incentives (such as points, credits, discounts, or direct delivery of products or services) at the point of sale, which are tied to the purchase of goods or services.

Multiple applications. Smart cards are able to combine two or more applications on a single card (some of which are able to share PIN numbers). At present, this form of combined identity management requires the capabilities of an integrated circuit.

Physical specifications. The size of a smart card is the same as that of a typical magnetic stripe credit, debit, or ID card. The International Standards Organization specification, ISO 7810, established the size of these cards, called ID-1 cards. ISO 7810 specifies that cards be 85.7 mm (3.4 in.) wide, 55.2 mm (2.1 in.) high, and 0.76 mm (0.03 in.) deep (**Table 2**). Smart cards adhere to this standard to promote backward compatibility with the existing transaction-processing infrastructure.

Contact versus contactless smart cards. Contact cards require that the card remain in physical contact with the reader. ISO 7816 specifies the requirements for contact cards. In general, they have a gold plate on one side of the card, covering contact points between the card's chip and the reader (**Fig. 1**). A contact smart card must be physically inserted into the reader and remain in contact with it during the session. The reader verifies that the card is properly situated and that the power supplies of both card and reader are compatible. It then supplies the card with the power necessary to complete the transaction.

The contactless card (ISO 10536, 14443, and 15693) transmits information to and from the card reader via an antenna embedded in the card (Fig. 1). Contactless smart cards rely on electrical coupling for their power. At present, contactless cards tend to be slower, less reliable, and more expensive than

TABLE 2. Smart card standards

Standard	Description
ISO 7810	ID cards—physical characteristics
ISO 7811	Identification cards, magnetic stripe, embossing
ISO 7816	Design and use of contact ICC cards
ISO/IEC 10536	Contactless ICC cards
ISO/IEC 14443	Proximity cards
ISO/IEC 15693	Vicinity cards

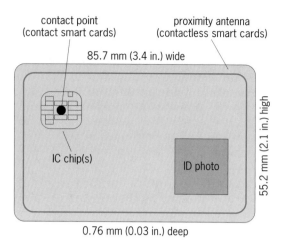

contact point
(contact smart cards)

proximity antenna
(contactless smart cards)

85.7 mm (3.4 in.) wide

55.2 mm (2.1 in.) high

IC chip(s)

ID photo

0.76 mm (0.03 in.) deep

Fig. 1. Smart card showing both the contact point and the proximity antenna.

contact cards. However, this technology is rapidly improving and will eventually replace contact smart cards.

Integrated circuits. The computer chips embedded in smart cards can be either memory chips or microprocessors. At present, memory chips have 1–5 kilobits of memory. Less expensive than microprocessors, they depend on the card reader for processing and security.

Smart cards with microprocessors are "true" smart cards. In many ways, the microprocessor makes them a miniature computer, complete with input/output ports, an operating system, and a hard drive. Microprocessor chips are available with 8-, 16-, and 32-bit architectures. They contain both short- and long-term memory cells, ranging from a few hundred bytes of read-only memory (ROM) to several megabytes of random-access memory (RAM). Usually, the operating system is loaded in the ROM, while the RAM provides a "scratch pad" for processing information.

While smart cards are programmed using a variety of languages, the most popular is Java, which can be

compiled directly on the card. Current innovations center on the development of minidatabases that can be stored on the card. As a result, smart card applications are increasingly important enablers of identify-based transactions.

Transaction. A typical smart card transaction session consists of communication, validation, initiation, transaction, and termination (**Fig. 2**).

Communication. Communication with the reader may be direct or in the vicinity of the reader, depending on the type of card (contact or contactless) and the capability of the reader.

Validation. The card must be validated in order to establish a session. If the card is valid, the system captures an identification number to establish an audit trail and processes the transaction using the authorized "value" that is stored on the chip (for example, removing cash from an electronic purse). This activity is handled by a secure access module (SAM) on a chip housed in the reader. The SAM has the electronic keys necessary to establish communication with the reader's operating system.

Initiation. A reset command is issued to establish communication between the card and the reader, and the clock speed is established to control the session. In the case of a both contact-and-contactless smart card, the reader obtains the required data from the card and initiates the requested transaction.

Transaction. Data are exchanged between the card and the reader through the defined contact (input/output) points. A record of the transaction is stored on both card and reader.

Termination. For contact smart cards, the contacts are set to a stable level, the supply voltage is terminated, and the card is ejected from the terminal. For contactless cards, a termination protocol ends the session and resets the reader.

Security. There are two levels of smart card security: physical and data.

Physical security. This involves protecting the smart card against damage from cardholders, damage from the environment, and physical tampering. Bending a card, scratching its contacts, or breaking chip connections will destroy the card. New manufacturing technologies and protective sleeves are being designed to protect cards.

A range of environmental conditions can damage or destroy smart cards. Examples include electric voltage, frequency of use, humidity, light, temperature, and radiation exposure.

Various methods of tampering with smart cards have been suggested or tried. Smart card manufacturers have protected the layout of the IC chip by hiding its data pathways. Cards that show physical traces of tampering can be identified and their use blocked by readers. Manufacturers also use other technologies including holograms, hidden characters, and fingerprints to make counterfeiting unprofitable for criminals.

Data security. Smart card systems are susceptible to attempts to invade systems to steal or destroy data. This may be done by visually capturing the keystrokes of cardholders in order to learn the PIN

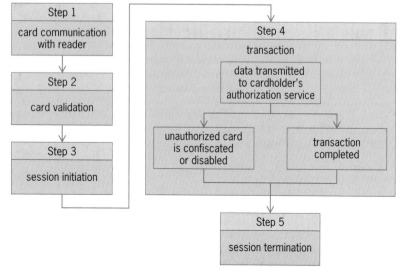

Step 1
card communication with reader

Step 2
card validation

Step 3
session initiation

Step 4
transaction
data transmitted to cardholder's authorization service

unauthorized card is confiscated or disabled

transaction completed

Step 5
session termination

Fig. 2. Typical smart card transaction.

associated with the card. A more sophisticated threat is infiltration of the system, with the goal of finding and stealing valuable data. In the case of contactless smart cards, data that are wirelessly transmitted between cards and readers potentially may be intercepted.

In response to these threats, smart cards use key-based cryptography for data encryption. There are various methods for this based on the logic of the "one-time pad," a randomly generated single-use ciphertext (encrypted text) that is decrypted by both the reader and the card and required to complete a transaction.

In a symmetric-key-encrypted transaction, the card accesses a public-key directory (stored into the reader), called a certificate authority (CA), to encrypt the public key with a "hash" (a document digest of the key). The information used to create this hash includes version number, serial number, signature algorithm, card issuer's name, and so on. Both the card and the CA receive a decrypted copy of the hash. Next, the reader accesses the CA to get the decrypted copy of the hash. This validates and authorizes the identity of both parties, allowing the transaction to take place. The transaction recipient accesses the CA for a final hash, which is encrypted and acts as a response to the card, identifying and authorizing the reader and notifying the card of the close of the transaction.

Asymmetric encryption makes use of both public and private keys to provide additional security for sensitive transactions.

Institutions using smart cards have developed several measures to deal with insecure cards. The most popular among these is "hot listing" of insecure cards. Early on, when only contact smart cards were available, the reader was programmed to retain problematic cards. This proved problematic when readers "ate" damaged but secure cards. Hot listing of contactless cards usually involves notifying the readers of insecure cards and programming the system to refuse transactions.

For background information *see* COMPUTER SECURITY; COMPUTER STORAGE TECHNOLOGY; CRYPTOGRAPHY; INTEGRATED CIRCUITS; MICROPROCESSOR in the McGraw-Hill Encyclopedia of Science & Technology.   Katherine M. Shelfer; Kevin Meadows

Bibliography. *Open Smart Card Infrastructure for Europe V 2*, Vol. 6: *Contactless Technology*, Part 1: *White Paper on Requirements for the Interoperability of Contactless Cards*, eESC TB 8 Contactless Smart Cards, March 2003; K.M.Shelfer et al., Smart cards, *Advances in Computer Science*, vol. 60, 2003.

SMART 1

SMART 1 is the first European mission to the Moon, and at the moment the lonely precursor to a future fleet of international lunar missions. It was launched on September 27, 2003, for a lunar capture on November 15, 2004. *SMART 1* is the first Small Mission for Advanced Research in Technology (SMART) in the science program of the European Space Agency (ESA). The SMART missions are designed to test key technologies in preparation for their later use on major scientific or application missions. *SMART 1* is designed to test such technologies while performing an unprecedented scientific study of the Moon. One of its principal tests is that of solar-electric propulsion, in which a continuous low-thrust engine uses electricity derived from solar panels to produce a beam of charged particles. This beam can be expelled from the spacecraft, pushing it forward, with an efficiency 10 times better than classical chemical propulsion. Such engines are commonly called ion engines.

Technology and science objectives. *SMART 1* not only has programmatic objectives, showing new ways of developing smaller, faster, and more cost-efficient spacecraft, but also demonstrates new technologies and enables scientific experiments to be performed. The miniaturized instruments onboard are designed to test technologies relevant to scientific investigations and also test the spacecraft itself, so that engineers and scientists can prepare space technologies for the future. While cruising from the Earth to the Moon, *SMART 1* has performed technology experiments such as:

(1) demonstrating the operation of an ion engine that can travel into deep space, and constantly monitoring its performance;

(2) communicating with the Earth at high radio frequencies, which will allow larger bandwidths for future planetary communication;

(3) communicating with the Earth using a laser beam shot from Tenerife in the Canary Islands; and

(4) testing a navigation system that, in the future, will allow spacecraft to autonomously navigate through the solar system.

Despite a concerted effort by the Americans during the 1960s and early 1970s, which culminated in six successful crewed landings, and the more recent *Clementine* and *Lunar Prospector* missions, much remains unknown about the Moon. Once at the Moon, *SMART 1* is designed to address key fundamental science questions, including:

(1) what the Moon is made of, and how it formed out of the debris of a massive collision between a Mars-sized object and the Earth, over 4.5×10^9 years ago;

(2) how rocky planets form, accrete, collide, and evolve;

(3) how a small rocky world works compared to the Earth in terms of volcanism, tectonics, and geochemistry;

(4) how the preserved craters of the Moon can be used as a history book of asteroid and comet bombardment of the Earth-Moon system; and

(5) what methods can be used to search for signs of water ice in craters near the Moon's poles.

SMART 1 data could also be used to select sites for future landing and Moon exploration, and serve to prepare the way with the fleet of future lunar missions including *Lunar A* or *Selene* (Japan), *Chang'e* (China), *Chandrayaan 1* (India), and the Lunar Reconnaissance Orbiter and South Pole Aitken Basin Sample Return mission (United States), under the coordination of the International Lunar Exploration Working Group (ILEWG). *See* CHINESE SPACE PROGRAM.

Launch, journey, and lunar mission. *SMART 1* was launched from Kourou, French Guiana, on September 27, 2003. It was an auxiliary passenger, together with main commercial passengers, *INSAT-3E* and *E-BIRD*, onboard an Ariane 5. The *SMART 1* voyage to the Moon was neither quick nor direct. After launch, *SMART 1* went into an elliptical orbit around the Earth, typically used by telecommunications satellites. In this orbit the spacecraft fired its ion engine, gradually expanding its elliptical orbit and spiraling out in the direction of the Moon's orbital plane. Each month, it got closer to the Moon's orbit, between 350,000 and 400,000 km (220,000 and 250,000 mi) from Earth. As *SMART 1* neared the Moon, it began using the gravity of the Moon to nudge it into a position where it could be captured by the Moon's gravitational field.

This complicated and slow journey was necessary because ion engines do not provide the instant power that chemical rockets do. However, because they are more efficient and require little fuel, ion engines are more flexible and allow space probes to reach places where chemical rockets would not be able to go. ESA is investing in this technology so that it can mount missions to Mercury and the Sun. The ion engine used 60 liters (16 gallons) or 84 kg (185 lb) of xenon fuel to go to the Moon, covering a spiral journey of more than 10^8 km (6×10^7 mi) in length.

SMART 1, the first European spacecraft to the Moon. (*ESA, Medialab*)

SMART 1 had its first encounter with the Moon on August 19, 2004, when it was at its apogee (greatest distance from the Earth). *SMART 1* was then about 197,000 km (122,000 mi) from the Moon and about 230,000 km (143,000 mi) from the Earth. Due to the much larger mass of the Earth, *SMART 1* was still completely within the sphere of influence of the Earth, even though it was closer to the Moon. The second lunar resonance approach took place on September 15, 2004, and the last one on October 12, 2004. A pause in the operation of the electric propulsion system was planned between October 14 and November 15, provided the trajectory before capture was verified to be within acceptable limits. Lunar capture was planned for November 15.

The cruise phase from Earth to the Moon took around 14 months, 2 months shorter than initially planned, due to efficient operations and saving of the xenon fuel. After capture by the Moon, the *SMART 1* ion engine is designed to gradually lower the highest altitude. An elliptical orbit is planned over the north and south lunar poles, whose height ranges from approximately 300 to 3000 km (200 to 2000 mi). The total planned mission duration is 2 to 3 years.

Spacecraft design. *SMART 1* is a cubic, three-axis-stabilized spacecraft (see **illustration**). Its solar panels can rotate so that they always keep in an optimum configuration with respect to the Sun. The design is inexpensive and emphasizes miniaturization wherever possible, especially for the payload. The total mass is 367 kg (809 lb), including 19 kg (42 lb) of payload. The dimensions are $1 \times 1 \times 1$ m ($3.3 \times 3.3 \times 3.3$ ft), the size of a washing machine (excluding solar panels). With solar panels deployed, *SMART 1* measures about 14 m (46 ft) across. Around 180 people from industry have worked directly on *SMART 1*, and several hundred more have worked indirectly on products which have been used in the spacecraft. Another 170 engineers and scientists are involved from ESA and other scientific institutes.

Instruments and experiments. *SMART 1* carries several instruments and experiments, with principal investigators from European countries and coinvestigators from Europe, the United States, Japan, India, and Russia.

The Electric Propulsion Diagnostic Package (EPDP) is fed by a selection of sensors, mounted outside the spacecraft. It is designed to monitor the ion engine's effects on the spacecraft. Ion-engine technology can cause surface temperatures to rise and create unwanted electric currents on the spacecraft, so the effects must be carefully watched.

The Spacecraft Potential, Electron, and Dust Experiment (SPEDE) consists of two electrical sensors mounted on the ends of 60-cm (2-ft) booms fixed outside the spacecraft. They, too, will monitor the effects of the solar-electric propulsion unit on the spacecraft. During *SMART 1*'s cruise phase, the experiment is designed to map the plasma-density distribution around the Earth and, when *SMART 1* is in lunar orbit, to study how the solar wind affects the Moon.

The X/Ka-band Telemetry and Telecommand Experiment (KaTE) uses very sensitive receivers onboard the spacecraft, and is designed to test new digital radio communications technology. It is designed to demonstrate, for the first time on a science mission, the performance of a new higher range of communication frequencies in the X-band (8 GHz) and the Ka-band (32–34 GHz). It is also designed to test new data encoding techniques and to validate the corresponding ground-based infrastructure needed to receive these signals.

The Asteroid-Moon Micro-Imager Experiment (AMIE) is a miniature camera, capable of taking color images and storing them in a memory. It can perform some automatic image processing. As well as imaging the Moon, AMIE is designed to support the laser-link experiment and OBAN (discussed below), and to assist with RSIS (discussed below). AMIE's lunar images will be used for educational and science communication as well.

The Radio Science Investigation with *SMART 1* (RSIS) is designed to use KaTE and AMIE to perform a painstaking investigation into the way the Moon wobbles. This would be the first time a spacecraft in orbit has performed such an experiment. It is therefore an essential test for future missions, such as *Bepi-Colombo*, that will investigate Einstein's theories of relativity.

The On-board Autonomous Navigation (OBAN) unit is designed to use AMIE to gather images of celestial objects, such as Earth, the Moon, and asteroids, to work out exactly where *SMART 1* is in space. This is the first step toward a spacecraft that will be able to navigate for itself.

The Infrared Spectrometer (SIR) is designed to perform a detailed analysis of the Moon's surface mineral composition. The results should provide greater insight into the processes of crater and maria formation and the phenomenon of space weathering on the Moon's surface.

An instrument named Demonstration of a Compact Imaging X-ray Spectrometer (D-CIXS) is designed to provide the first global map of the lunar surface's composition. Its observations should allow scientists to confirm theories of the origin of the Moon and of the evolution of lunar terrains and impact basins. This is a test instrument for future x-ray investigations of Mercury, using ESA's *BepiColombo* mission.

The X-ray Solar Monitor (XSM) will monitor the Sun's output of x-rays so that solar storms do not confuse the results from D-CIXS, and it will observe the Sun as an x-ray star during the cruise.

Operations and ground control. *SMART 1* is controlled from the European Space Operations Centre (ESOC) in Darmstadt, Germany. Communication with *SMART 1* takes place, on average, for 8 hours, twice a week, using various ESA network ground stations around the world. The Science and Technology Operations Coordination (STOC) of the mission, including instrument operations and the management of receipt, processing, and distribution of *SMART 1* data, takes place at the European Space Research and Technology Center (ESTEC) at Noordwijk in The Netherlands.

For background information *see* ION PROPULSION; MOON; SPACE PROBE in the McGraw-Hill Encyclopedia of Science & Technology.   Bernard H. Foing

Bibliography.   B. H. Foing et al. (eds.), Astronomy and space science from the Moon, *Adv. Space Res.*, 14(6):1–290, June 1994; B. H. Foing and SMART-1 Science Technology Working Team, The science goals of ESA's *SMART-1* mission to the Moon, *Earth, Moon, and Planets*, 85–86:523–531, 1999; G. D. Racca, B. H. Foing, and SMART-1 Team, *SMART-1* mission description and development status, *Planet. Space Sci.*, 50:1323–1337, 2002.

Solar storms

The dramatic solar storms of October and November 2003 reminded the world of the awesome power of the Sun and its effects on the Earth. The first hint of the impending series of solar storms occurred when active region 10486 erupted on October 23, producing a solar flare and a coronal mass ejection (CME) that only glanced the Earth. As the Sun rotated this active region toward the center of the solar disk on October 28, the region erupted again, producing a giant solar flare with an associated CME that was directed right at the Earth. While the effects of this first flare-CME storm were impacting the Earth's environment, region 10486 erupted again, producing a flare and another CME that also sent its ejecta toward Earth. Region 10486 produced the largest flare ever recorded on November 4, just before it rotated out of view (**Fig. 1**).

Fig. 1. Composite image of the solar flare and coronal mass ejection of November 4, 2003, as observed by the Extreme-ultraviolet Imaging Telescope (EIT) [inner region] and Large Angle Spectroscopic Coronagraph (LASCO) C2 [outer region] instruments on the *Solar and Heliospheric Observatory (SOHO)* spacecraft. (*EIT/SOHO and LASCO/ SOHO Consortia, ESA, and NASA*)

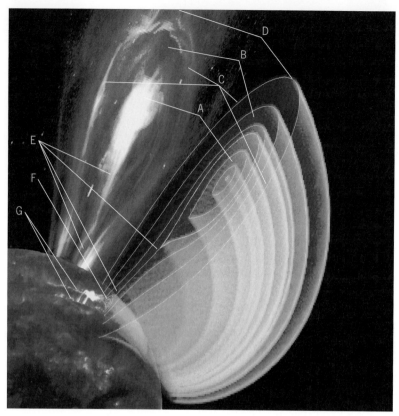

Fig. 2. Composite image of a two-ribbon flare and a cross section of the coronal mass ejection (CME) of February 5, 1999, as observed by the *SOHO* spacecraft (the CME image is offset from the flare image for clarity). The diagram depicts the inferred magnetic field structure according to the theoretical model of J. Lin and T. G. Forbes. The coronal mass ejection comprises (A) the driver gas, (B) the dark void, (C) the outward-moving bright bubble, and (D) the shock front. The (E) current sheet, (F) arcade of magnetic field loops, and (G) their foot points, which form the two-ribbon flare, also are shown. (*After A. Panasyuk, Harvard-Smithsonian Center for Astrophysics, and LASCO/SOHO Consortium*)

The energy that drives both solar flares and CMEs is believed to come from the interplay between the electrically charged gas in the solar atmosphere and the Sun's local magnetic field at the site of the event. Magnetic energy is converted to that of particle acceleration and heat. A flare or a CME can occur without the other, or they can occur together. When the energy available to a CME is large enough, there is essentially always an associated solar flare.

In this article, solar flares, coronal mass ejections, and their effects in space, in the Earth's atmosphere, and on its surface are described. A summary of the present and potential capability to predict such events and their effects also is provided.

Solar flares. Solar flares occur in the neighborhood of sunspots at solar altitudes near the base of the outer atmospheric layer called the corona. The probability of a solar flare occurring depends on the magnetic complexity of the underlying sunspot group. The most distinguishing feature of solar flares is bright emission from long ribbons on the solar disk. The ribbons lie at the foot points of a system of magnetic loops that form an arcade. Solar flares produce sudden bursts of energetic particles and high-energy radiation that persist for minutes to hours and

travel at or near the speed of light. Production of x-radiation and gamma radiation implies that electrons and ions are accelerated to extremely high energies near the flare site. Protons are accelerated to nearly the speed of light, and high-energy neutrons are produced by the interaction of the protons with alpha particles (helium nuclei) in the flaring corona.

A typical model of a solar flare consists of a preflare magnetic configuration that is then stressed by motions in the underlying photosphere. The energy of the stressed field steadily increases until an instability occurs that leads to an opening of magnetic field loops that become layers of opposite-polarity magnetic sheets. These sheets collide and annihilate one another, releasing their magnetic energy in a process called magnetic reconnection (**Fig. 2**).

Coronal mass ejections. CMEs can occur with or without a solar flare and with or without the eruption of a solar prominence (a bright structure that can be seen above the limb of the Sun during a natural solar eclipse), but most CMEs involve the latter. Prior to the CME onset, a coronal streamer is typically seen with foot points surrounding a prominence. In visible coronagraph images, CMEs often appear as a brightening at the coronal base and an outward-moving bright bubble surrounding a dark region called the void (Fig. 2). The leading edge of the CME bubble accelerates away from the Sun at speeds ranging from 100 to 1000 km/s (60 to 600 mi/s) or even faster. The fine structure of the CME often appears to have a twisted shape. Ultraviolet observations show bright emissions from relatively cool gas originating in ejected prominences. Doppler velocities sometimes reveal helical structures and rapid rotation of the ejecta. The shapes of the structures mimic the geometry of the magnetic field that is carried along with the CME. The orientation of the magnetic field (that is, how the polarity varies along the helical structure) is probably determined by conditions near the Sun. The strength of the field can be increased by compression in interplanetary space. CMEs arrive near the Earth (**Fig. 3**) at speeds ranging from 100 to 2500 km/s (60 to 1500 mi/s). The orientation, strength, and speed of the CME's magnetic field are important in determining its impact on the Earth's magnetosphere.

The largest flares and the most energetic CMEs appear to come from active regions with a great amount of energy stored in stressed magnetic structures. CMEs associated with the largest flares differ from most CMEs in several respects: (1) more rapid disruption of the pre-CME streamer, (2) higher Doppler shifts, indicating large transverse velocities of the streamer material, and (3) higher-temperature plasma, evidenced by ultraviolet emissions from 17-, 18-, and 20-times ionized iron ions, indicating temperatures above 6×10^6 K. The iron ion emissions provide evidence of a high-temperature current sheet that is the site of magnetic reconnection processes that heat a high-temperature bubble, and also heat magnetically confined loops of gas at the coronal base (Fig. 2).

For these large flare-CME events, the CME speed eventually results in a shock wave as it moves outward from the Sun. Shock-heated oxygen ions have been observed behind the leading edge of the CME. Particles (such as protons) may be accelerated to 10^6–10^9-electronvolt energies by CME shock waves. As a CME-driven shock wave propagates away from the Sun, particles accelerated at the shock wave can escape upstream and downstream into the interplanetary medium and propagate along the interplanetary magnetic field. Depending on the speed of the CME, it can take up to 4 days to reach the vicinity of the Earth, during which time it continues to produce very high energy particles.

Hazards to astronauts and spacecraft. The greatest hazard to astronauts and their equipment from solar storms is believed to be energetic particles produced by CMEs and solar flares. In Earth orbit, the magnetosphere partially shields the astronauts and their spacecraft. Outside the magnetosphere and on the lunar and Martian surfaces, the hazard is greater.

The 10^8-electronvolt protons pose a biological radiation hazard to astronauts that can persist for several days. These particles also can penetrate the skins of spacecraft, damaging electronic components and corrupting computer memories. High-energy electrons also penetrate spacecraft skins and build up static charge on insulators to the point where a high-voltage discharge is generated that can damage components. The x-rays and gamma rays produced in the flaring corona also pose a radiation hazard to astronauts. During October 2003, astronauts in the *International Space Station* took protective cover for several 20-min periods in the Russian-built Zvezda service module. Two Japanese satellites were damaged. Other satellites were put into safe configurations, and sensitive solar panels were reoriented to minimize damage.

Geomagnetic storms. The outermost extension of a CME is a fast shock wave (Fig. 2). It is followed by interplanetary material that is swept up, compressed, and heated by the shock. Behind this is the CME driver gas that originated at the Sun. It is composed of electrically charged atomic ions, protons, and electrons, and carries with it a magnetic field. In the case of highly "geo-effective" CMEs, the field has a high intensity and it moves forward with high velocity (1000–2000 km/s or 600–1200 mi/s). The magnetic configuration can be complex with a helical geometry and large out-of-the-ecliptic components (Fig. 3). If one were to cross the magnetic field along the Sun-Earth line, the field would change from north to south or south to north.

As the CME approaches the Earth, the magnetosphere, which is formed by the Earth's magnetic field and the solar wind, deflects the CME magnetic field and the charged particles, and they flow along the boundary of the magnetosphere. If the rapidly moving CME field is directed southward (that is, opposite to the Earth's magnetic field), the fields can annihilate one another (reconnect) causing large changes in the Earth's field in a short time and over a large

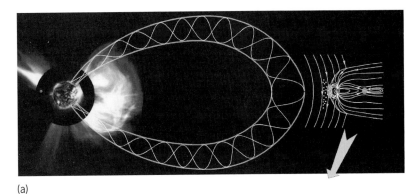

(a)

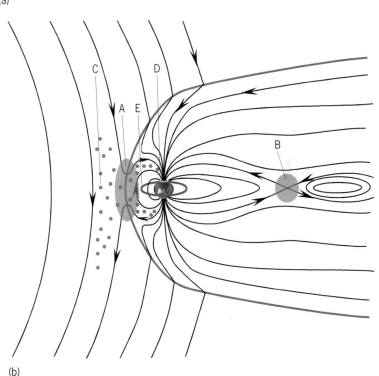

(b)

Fig. 3. Diagram of (*a*) the magnetic configuration of a coronal mass ejection as it approaches the Earth, and (*b*) the interaction of this magnetic field with the Earth's magnetosphere. Arrows indicate the direction of the magnetic fields. Magnetic reconnection sites (A) at the sunward boundary of the magnetosphere and (B) at the magnetotail, (C) accelerated charged electrons and ions, (D) the auroral ovals, and (E) the ring current are depicted. (*After A. Panasyuk, Harvard-Smithsonian Center for Astrophysics*)

area. The rapidly fluctuating changes to the Earth's magnetic field as the complex CME field moves forward is known as a geomagnetic storm. The reconnection can occur at the sunward boundary of the magnetosphere and also in the magnetotail, contributing to the production of aurorae. This system of changing magnetic fields and charged-particle currents acts like an electric generator, creating voltages as high as 10^5 V that, in turn, accelerate the charged particles and energize the ionosphere. Charged particles are accelerated toward the night side of the magnetosphere and subjected to forces due to the field's curvature and spatial variation that set up a ring current around the equator. This current decreases the intensity of the Earth's magnetic field.

Aurora. The most spectacular products of geomagnetic storms are the visual displays of aurorae in the Northern and Southern hemispheres. They are produced primarily at altitudes of 90 to 250 km (55 to 155 mi) in regions called the auroral ovals that encompass the geomagnetic poles. The boundary of the auroral oval expands to lower latitudes as the intensity of the geomagnetic storm increases.

Impressive aurorae are produced by magnetic reconnection between the CME magnetic field and the leading edge of the Earth's magnetosphere, where charged particles are accelerated toward the auroral ovals. The CME-magnetosphere generator also accelerates charged particles and magnetic field to the magnetotail, where more magnetic reconnection occurs. Electrons are accelerated back toward the polar field lines, following them to lower altitudes. Electron collisions with molecular oxygen and nitrogen, and with atomic oxygen, produce excited states of those systems that radiatively decay producing emissions of blue, magenta, and yellow-green, respectively. Atomic oxygen also produces a red emission at altitudes above 200 km (120 mi). Aurorae may also result from ring-current ions and electrons being slowed down by collisions with lower-energy ions and then trapped in orbits. These particles may eventually be released and accelerated at reconnection sites, ultimately colliding with atoms and molecules. The production of shimmering curtains and other patterns of light is under active investigation.

Air, ground, and communication hazards. Geomagnetically induced currents can flow through good conductors like pipelines and electrical power transmission lines. Pipeline currents cause increased corrosion and malfunction of meters, while surges in power grids can overload transformers. High-frequency communication systems use the ionosphere to reflect radio signals over long distances. Because the ionosphere is altered during geomagnetic storms, these signals can be distorted or completely blocked. Examples are ground-to-air and ship-to-shore communications. Radio communication through the ionosphere with spacecraft also is affected.

Biological effects are minimal for persons on the ground in the midlatitudes. The magnetosphere shields the Earth's surface from protons and electrons. However, aircraft flying high-altitude polar routes are subject to a flux of protons because the magnetic shielding is weak near the poles. Radiation levels for high-altitude aircraft at lower latitudes are also of concern.

Hazard predictions. The *Solar and Heliospheric Observatory* (*SOHO*) spacecraft has enabled the detection and characterization of CMEs. The *RHESSI* and *TRACE* satellites are gaining new insights into the production of large flares. Such remote sensing combined with models of relativistic proton acceleration and transport can be used to predict the levels and production sites of the high-energy particle hazards. The Ultraviolet Coronagraph Spectrometer (UVCS) instrument aboard *SOHO* has detected and characterized the current sheets and CME shocks that are believed to produce the proton hazard. Such input in the initial phase of an event is needed for theoretical models aimed at predicting the following 1–2 days of proton production, including its geometrical extent. However, more work is needed to take this capability to a practical level. There has been some success in extrapolating the characteristics of CMEs near the Sun to the magnetosphere and predicting the intensity of geomagnetic storms.

Predictions of high-energy particles and radiation produced at the flare site are more difficult. Remote sensing information, high-energy particles, and radiation travel at or near the speed of light, and so knowledge of the hazard's existence arrives at about the same time as the hazard. The understanding of how the details of the magnetic complexity of an active region relates to its affinity for producing flares is improving, but it is still not possible to predict with high confidence when an active region will flare.

For background information *see* AURORA; GEOMAGNETIC VARIATIONS; MAGNETOSPHERE; RADIOWAVE PROPAGATION; SOLAR CORONA; SOLAR WIND; SPACE BIOLOGY; SUN in the McGraw-Hill Encyclopedia of Science & Technology.　　　John Kohl

Bibliography. S.-I. Akasofu, Aurora, in S. Suess and B. T. Tsurutani (eds.), *From the Sun: Auroras, Magnetic Storms, Solar Flares and Cosmic Rays*, American Geophysical Union, Washington, DC, 1998; D. Alexander and L. W. Acton, The active sun, in J. A. M. Bleeker et al. (eds.), *The Century of Space Science*, Kluwer Academic Publishers, Dordrecht, 2001; L. Golub and J. M. Pasachoff, *Nearest Star: The Exciting Science of Our Sun*, Harvard University Press, Cambridge, 2001; K. R. Lang, *The Sun from Space*, New York, Springer, 2000; J. Lin, W. Soon, and S. L. Baliunas, Theories of solar eruptions: A review, *New Astron. Rev.*, 47:53–84, 2003; B. T. Tsurutani and W. D. Gonzalez, Magnetic storms, in S. Suess and B. T. Tsurutani (eds.), *From the Sun: Auroras, Magnetic Storms, Solar Flares and Cosmic Rays*, American Geophysical Union, Washington, DC, 1998.

Space flight

While world aviation in 2003 celebrated "the next century of flight," observing Orville and Wilbur Wright's pioneering first flight in a motor-driven machine 100 years earlier, space flight, with the loss of the space shuttle *Columbia* and the crew, suffered a tragedy that rivaled the loss of the *Challenger* in 1986. This loss shocked the world and overshadowed a number of highlights both in human space missions and in automated space exploration and commercial utilization.

While the United States space budget stayed level, international space activities continued prior-year trends of reduced public spending and modest launch services. In general, launch activities in 2003

TABLE 1. Some significant space events in 2003

Designation	Date	Country	Event
ICESat	January 12	United States	Successful launch of NASA's Ice, Cloud, and Land Elevation Satellite into a 370-mi (600-km) retrograde near-polar orbit for Earth features observation.
CHIPSat	January 12	United States	Successful launch of NASA's Cosmic Hot Interstellar Plasma Spectrometer Satellite for all-sky spectroscopy into a 370-mi (600-km) retrograde near-polar orbit.
STS-107 (Columbia)	January 16	United States	28th and last flight of Shuttle *Columbia*. Lost on Feb. 1 during reentry with its seven crew members, Husband, McCool, Brown, Chawla, Anderson, Clark, and Ramon.
SORCE	January 25	United States	Successful launch of NASA's Solar Radiation and Climate Experiment Satellite for radiation measurements into a 400-mi (645-km), 40° orbit, by a Pegasus XL.
Progress M-47/10P	February 2	Russia	Crewless logistics cargo/resupply mission to the *International Space Station* on a Soyuz-U rocket.
Soyuz TMA-2/ISS-6S	April 26	Russia	Launch of the first *ISS* crewr otation flight on a Soyuz, bringing the Expedition 6 caretaker crew of Yuri Malenchenko and Edward Lu. Returned on *TMA-2* in October.
GALEX	April 28	United States	Successful launch of NASA's Galaxy Evolution Explorer telescope platform for ultraviolet observation of galaxies, into a 431 x 435 mi (694 x 700 km), 28.99° orbit, by a Pegasus XL.
Hayabusa (Muses-C)	May 9	Japan	Asteroid sample return mission, launched on an M5 solid-propellant rocket toward the asteroid 25143 Itokawa/1998 SF36.
Mars Express	June 2	(Europe)	Successful launch of first ESA mission to Mars, on a Soyuz/Fregat. Lander *Beagle 2* was lost on entry into Mars atmosphere, but *Mars Express* entered nominal orbit on December 25.
Progress M1-10/11P	June 8	Russia	Crewless logistics cargo/resupply mission to the *International Space Station,* on a Soyuz-U rocket.
MER-A/Spirit	June 10	United States	Successful launch of NASA's Mars Exploration Rover-A, named *Spirit*, toward the Red Planet, on a Delta 2 rocket. The six-wheeled rover had a mass of ~400 lb (180 kg).
MER-B/Opportunity	July 7	United States	Successful launch of NASA's Mars Exploration Rover-B, named *Opportunity*, toward the Red Planet, on a Delta 2 rocket. The six-wheeled rover had a mass of ~400 lb (180 kg).
SST (SIRTF)	August 25	United States	Successful launch of NASA's *Spitzer Space Telescope* (formerly *Space Infrared Telescope Facility*) into an Earth-trailing orbit around the Sun (first time).
Progress M-48/12P	August 28	Russia	Crewless logistics cargo/resupply mission to the *International Space Station,* a Soyuz-U rocket.
Shenzhou 5	October 15	P.R. of China	Successful launch of first Chinese astronaut, "taikonaut" Lt. Col. Yang Liwei, in "Divine Vessel 5," with safe recovery after circling Earth 14 times in 21 h 23 min.
Soyuz TMA-3/ISS-7S	October 18	Russia	Launch of the second *ISS* crew rotation flight on a Soyuz, bringing the Expedition 7 crew of Michael Foale and Alex Kaleri, plus 10-day visitor Pedro Duque from ESA.
H-2A	November 29	Japan	Launch of sixth H-2A heavy lifter; failed shortly after liftoff, with two reconnaissance satellites.

remained more or less on par with 2002, without showing much promise of any sizable rebound in 2004. A total of 60 successful launches worldwide carried 86 payloads, compared to 61 flights in 2002. There also were three launch failures (down from four in 2002), including *STS 107/Columbia*.

The U.S. National Aeronautics and Space Administration (NASA) again reached a number of milestones, with the successful launches of planetary probes such as the two Mars Exploration Rovers MER-A *Spirit* and MER-B *Opportunity*; science missions such as *SORCE* (Solar Radiation and Climate Experiment), *GALEX* (Galaxy Evolution Explorer), and the infrared space telescope *SIRTF* (Space Infrared Telescope Facility); and interplanetary milestones such as the two *Voyager* missions, the *Cassini-Huygens* mission to Saturn and Titan, and the spacecraft *Galileo* before its final plunge into Jupiter on September 21, 2003.

In 2003, the commercial space market declined again after its recovery in 2002. Out of the 60 successful launches worldwide, about 20 were commercial launches, compared to 28 in 2002. In the civil science satellite area, worldwide launches totaled 14, up 4 from the preceding year.

Russia's space program showed continued dependable participation in the buildup of the *International Space Station*. This partnership became particularly important after the shuttle stand-down caused by the loss of *Columbia*. Europe's space activities in 2003 dropped below the previous year's; the last Ariane 4 was followed by three successful missions of the Ariane 5 heavy-lift launch vehicle, which brought the number of successes of this vehicle to 13.

During 2003 a total of three crewed flights from the two major space-faring nations (down from seven flights in 2002) carried 12 humans into space (down from 40), but 7 of them were lost with the shuttle *Columbia*. An additional crewed flight was conducted by the People's Republic of China, carrying the first Chinese into space on the *Shenzhou 5* spaceship. This brought the total number of people launched into space since 1958 (counting repeaters) to 971, including 100 women; not counting repeaters, 435 people have been launched, including 38 female. Some significant space events in 2003 are listed in **Table 1**, and the launches and attempts are enumerated by country in **Table 2**.

TABLE 2. Successful launches in 2003 (Earth-orbit and beyond)

Country	Number of launches (and attempts)
United States (NASA/DOD/commercial)	25 (26)
Russia	21 (21)
People's Republic of China	6 (7)
Europe (ESA/Arianespace)	4 (4)
Japan	2 (3)
India	2 (2)
TOTAL	**60** (63)

International Space Station

Goals of the *International Space Station* (*ISS*) are to establish and maintain a permanent habitable residence and laboratory for science and research. The *ISS* is already providing a unique platform for making observations of the Earth's surface and atmosphere, the Sun, and other astronomical objects.

The *ISS* is the largest and most complex international scientific project in history. The station, projected for completion by 2010, will have a mass of about 1,040,000 lb (470 metric tons). It will measure 356 ft (109 m) across and 290 ft (88 m) long, with almost an acre of solar panels to provide up to 110 kilowatts of power to six state-of-the-art laboratories. The *ISS* draws upon the scientific and technological resources of 16 nations: the United States, Canada, Japan, Russia, 11 nations of the European Space Agency (ESA), and Brazil.

Operations and assembly. A continuing partnership issue of 2003 was the debate over assured crew return capability after the Russian obligation to supply Soyuz lifeboats expires in April 2006. In NASA's space transportation planning, a United States crew rescue capability (other than via space shuttle) will be available only in 2010. Of much greater significance to the continuation of *ISS* assembly and operation was Russia's shouldering the burden of providing crew rotation and consumables resupply flights to the station after the loss of space shuttle *Columbia* on February 1 brought shuttle operations to a standstill.

In 2002, following the recommendations of an independent advisory panel of research scientists called Remap (Research Maximization and Prioritization), NASA established the formal position of a Science Officer for one crew member aboard the *ISS*, responsible for expanding scientific endeavors on the station. After Flight Engineer Peggy Whitson of Expedition 5 became NASA's first Science Officer, Donald Pettit of Expedition 6, Ed Lu of Expedition 7, and Michael Foale of Expedition 8 were the other Science Officers in 2003.

After the initial milestones for the *ISS* program following the beginning of orbital assembly in 1998, buildup and early operations of the permanently crewed station had continued through 2001 at a rapid pace. In April 2002, the first of several truss elements, S0 (S-Zero), was attached on top of the United States laboratory module *Destiny*, becoming the centerpiece of the 356-ft-long (109-m) truss for carrying the solar cell arrays of the station. In June 2002, the Expedition 4 crew was rotated with the new station crew of Expedition 5 and delivered cargo including the Mobile Service System [to provide mobility for the Space Station Remote Manipulator System (SSRMS)] and the Italian-built Multi-Purpose Logistics Module (MPLM) *Leonardo* for cargo and equipment transport. The second truss segment, S1, arrived in October 2002 and was attached to S0 on the starboard side. Its counterpart on port, P1, followed in November. The same shuttle mission brought the Expedition 6 crew of U.S. Commander Kenneth Bowersox, Russian Flight Engineer Nikolay Budarin, and U.S. Flight Engineer/Science Officer Pettit, and returned the Expedition 5 crew to Earth.

Early in 2003, further progress in *ISS* assembly was brought to a halt by the stand-down of the space shuttles after the *Columbia* loss. As an immediate consequence of the unavoidable reduction in resupply missions to the station, which now could be supported only by Russian crewless automated Progress cargo ships, station crew size was reduced from three to a two-person "caretaker" crew per expedition (also known as increment). Operations procedures had to be revised accordingly, and such vital areas as onboard systems maintenance and spares provision had to be replanned carefully to continue crewed occupancy and a viable science research program onboard despite the sudden constriction in logistics.

After appropriate training for the new emergency situation, Expedition 7 was launched to the *ISS* in April with Russian Commander Yuri Malenchenko and U.S. Flight Engineer Lu on a Soyuz TMA spacecraft, while the three members of Expedition 6 returned on the previous Soyuz that had served as a contingency crew return vehicle (CRV). The replacement crew, Expedition 8, came 6 months later in a fresh Soyuz TMA, consisting of U.S. Commander Foale and Russian Flight Engineer Alexander Kaleri, to continue station operations into 2004. By the end of 2003, 38 carriers had been launched to the *ISS*: 16 shuttles, 2 heavy Protons (carrying the modules FGB/*Zarya* and SM/*Zvezda*), and 20 Soyuz rockets (12 crewless Progress cargo ships, the DC-1 docking module, and 7 crewed Soyuz spaceships).

Progress M-47. Designated ISS-10P, the first of three crewless cargo ships to the *ISS* in 2003 lifted off on a Soyuz-U rocket at the Baikonur Cosmodrome in Kazakhstan on February 2, one day after the *Columbia* accident. Like all Progress transports, it carried about 2 tons of resupply for the station, including maneuver propellants, water, food, science payloads, equipment, and spares.

Soyuz TMA-2. *Soyuz TMA-2* (#212), *ISS* Mission 6S (April 26–October 28), was the first crew rotation flight by a Soyuz. It carried Expedition 7, the two-man caretaker crew, to stretch out *ISS* consumables during the current shuttle stand-down. *TMA-2* docked to the *ISS* on April 28, replacing the previous CRV, *Soyuz TMA-1/5S*. The Expedition 6 crew members returned to Earth, experiencing an unexpected switch of their onboard computer to the backup reentry mode of pure ballistic descent, which almost doubled their peak deceleration ($\sim 9\,g$), and missed the primary landing site by an undershoot of $\sim$300 mi (480 km), landing on May 3. Recovery forces reached the landing site after a delay of 4.5 h, finding the crew in good health outside the capsule, which had fallen on its side. Expedition 6 had spent 162 days in space (160 days onboard the *ISS*).

Progress M1-10. *ISS-11P* was the next crewless cargo ship, launched in Baikonur on a Soyuz-U on June 8 and arriving at the station with fresh supplies on June 10.

Progress M-48. *ISS-12P*, the third automated logistics transport in 2003, lifted off on its Soyuz-U on August 28, docking at the *ISS* 2 days later.

Soyuz TMA-3. *Soyuz TMA-3* (#213), *ISS* mission 7S (October 18–April 29, 2004), was the second *ISS* crew rotation flight by a Soyuz. Its crew comprised Expedition 8 (the second caretaker crew) and visiting European Space Agency (ESA) crew member Pedro Duque from Spain. On October 28 the previous CRV, *Soyuz TMA-2/6S*, undocked from the of the FGB/*Zarya* module nadir port, where it had stayed for 182 days, and landed safely in Kazakhstan on October 29 local time with Malenchenko, Lu, and Duque. Expedition 7's total mission elapsed time from launch to landing was 184 days 22 h 47 min.

United States Space Activities

Launch activities in the United States in 2003 showed an increase from the relatively low level of the previous year. There were 26 NASA, DOD, and commercial launches, with one failure, *Columbia* (2002: 18 out of 18 attempts).

Space shuttle. Because of the loss of orbiter *Columbia* on the first shuttle mission in 2003 (STS-107), operations with the reusable shuttle vehicles of the U.S. Space Transportation System (STS) came to a halt for the remainder of the year (and also for 2004). Resupply and crew rotation flights to the *ISS* were taken over by Russian Soyuz and Progress vehicles.

Columbia, on its twenty-eighth flight, lifted off on January 16, 2003, on a research mission staffed by researchers working 32 payloads with 59 separate investigations. With Commander Rick D. Husband, Pilot William C. McCool, and Mission Specialists David M. Brown, Kalpana Chawla, Michael P. Anderson, Laurel B. Clark, and Ilan Ramon, *Columbia* circled Earth for nearly 16 days in an orbit of 173 mi (277 km) altitude. Both in the shuttle middeck and in the *Spacehab* Research Double Module (RDM), on its first flight in the cargo bay, the international crew, with the first Israeli astronaut (Ramon) and two women (Chawla and Clark), worked 24 h a day in two alternating shifts on a mixed complement of competitively selected and commercially sponsored research in the space, life, and physical sciences. In addition to the RDM, payloads in the shuttle cargo bay included the FREESTAR (Fast Reaction Experiments Enabling Science, Technology, Applications, and Research) with six payloads, the MSTRS (Miniature Satellite Threat Reporting System), and the STARNAV (Star Navigation).

After completing a successful mission, *Columbia* returned to Earth on February 1 but was lost with its crew during reentry. The Columbia Accident Investigation Board (CAIB) was chaired by Admiral (ret.) Harold W. Gehman, Jr. It later concluded that, unbeknown to crew and ground, one of the left wing's leading-edge reinforced carbon-carbon (RCC) elements had been punctured during ascent to orbit by a "suitcase-sized" chunk of foam insulation, blown off the external tank (ET) by the supersonic air-stream, hitting the leading edge and rendering the wing unable to withstand reentry heating longer than about 8 min after entry interface. Further shuttle operations and subsequent intensive return to flight (RTF) efforts by NASA and its contractors were halted for the duration of the CAIB investigation. *See* SPACE SHUTTLE.

Advanced transportation systems activities. NASA's new 5-year Space Launch Initiative (SLI) project, announced in 2001, continued in 2003, with 22 contracts awarded to industry in 2001 for developing the technologies that would be used to build an operational reusable spacelaunch vehicle (RLV) before 2015. (These plans were later changed, when in 2004 President Bush announced NASA's new long-range vision for space exploration.)

Space sciences and astronomy. In 2003, the United States launched five civil science spacecraft, two more than in the previous year: *CHIPSat*, *ICESat*, *SORCE*, *GALEX*, and *SIRTF*.

CHIPSat. CHIPSat (Cosmic Hot Interstellar Plasma Spectrometer) satellite is a University Class Explorer (UNEX) mission funded by NASA, designed for all-sky spectroscopy of the diffuse background at wavelengths λ from 9 to 26 nm with a peak resolution of $\lambda/150$ (about 0.5 eV). *CHIPSat* was launched on January 12 on a Delta 2 rocket into a $94°$ inclination, 370-mi (600-km) circular orbit, along with the environmental satellite *ICESat*. Its observations are helping scientists determine the electron temperature, ionization conditions, and cooling mechanisms of the million-degree plasma believed to fill the local interstellar bubble.

SORCE. SORCE (Solar Radiation and Climate Experiment) is a NASA-sponsored satellite to provide state-of-the-art measurements of incoming x-ray, ultraviolet, visible, near-infrared, and total solar radiation. The spacecraft was launched on January 25 on a Pegasus XL into a 400-mi (645-km), $40°$ orbit of the Earth, carrying four instruments, including the Total Irradiance Monitor (TIM), Solar/Stellar Irradiance Comparison Experiment (SOLSTICE), Spectral Irradiance Monitor (SIM), and Extreme Ultraviolet Photometer System (XPD).

GALEX. GALEX (Galaxy Evolution Explorer) is an orbiting space telescope for observing tens of millions of star-forming galaxies in ultraviolet light across 10^{10} years of cosmic history. Additionally, *GALEX* probes the causes of star formation during a period when most of the stars and elements we see today had their origins. *GALEX* was launched on April 28 by a Pegasus XL rocket into a nearly circular Earth orbit. Its telescope has a basic design similar to the *Hubble Space Telescope* (*HST*), but while *HST* captures the sky in exquisite detail in a narrow field of view the *GALEX* telescope is tailored to view hundreds of galaxies in each observation. Thus, it requires a large field of view, rather than high resolution, in order to efficiently perform the mission's surveys.

SIRTF. NASA's *Space Infrared Telescope Facility* (*SIRTF*), renamed the *Spitzer Space Telescope* (*SST*) in December, was launched on August 24 aboard a

Delta 2 into an Earth-trailing orbit—the first of its kind—around the Sun. The *SST* is the fourth and final element in NASA's family of Great Observatories and represents an important scientific and technical bridge to NASA's Astronomical Search for Origins program. The observatory carries a 33-in. (85-cm) cryogenic telescope and three cryogenically cooled science instruments capable of performing imaging and spectroscopy in the 3.6–160 μm range. Its supply of liquid helium for radiative-cryogenic cooling was estimated postlaunch to last for about 5.8 years, assuming optimized operation. *See* SPITZER SPACE TELESCOPE.

RHESSI. RHESSI (Reuven Ramaty High Energy Solar Spectroscopic Imager), launched on February 5, 2002, in 2003 continued its operation in Earth orbit, providing advanced images and spectra to explore the basic physics of particle acceleration and explosive energy release in solar flares.

Hubble Space Telescope. Thirteen years after it was placed in orbit, the *Hubble Space Telescope* (*HST*) continued to probe far beyond the solar system, producing imagery and data useful across a range of astronomical disciplines. In 2003, astronomers employed the *HST*'s new Advanced Camera for Surveys (ACS) to obtain the clearest view yet of the dust disk around a young, 5-million-year-old star, the birthplace of planets. The ACS used a distant galaxy cluster called Abell 1689 as a giant "lens" to bend and magnify the light of galaxies located far behind it, thus extending the *HST*'s range deeper into the universe. Scientists found faint objects that may have started to shine at the end of the "dark ages" of the universe, about 13 billion years ago. Other momentous accomplishments of the *HST* in 2003 were the identification of the oldest known planet in the Milky Way Galaxy (about 13 billion years old, more than twice the Earth's 4.5 billion years); targeting data for Europe's 2004 *Rosetta* mission to Comet 67P/Churyumov/Gerasimenko; the biggest, brightest, and hottest megastar birth ever seen, with a million blue-white newborn stars in the "Lynx arc"; and a firestorm of star birth in the colorful nebula NGC 604.

In 2003, development started on the *HST*'s successor, the *James Webb Space Telescope* (*JWST*). The giant new cosmic telescope (11,880 lb or 5400 kg) is planned to launch in 2011 on a European Ariane 5 toward the second Lagrangian point (L2), 930,000 mi (1.5 million kilometers) beyond Earth's orbit on the Sun-Earth line, where effects of their light on its optics are minimized and gravitational pull is relatively well balanced.

Chandra Observatory. Launched on shuttle mission *STS 93* on July 23, 1999, the massive (12,930 lb or 5870 kg) *Chandra X-ray Observatory* uses a high-resolution camera, high-resolution mirrors, and a charge-coupled-detector (CCD) imaging spectrometer to observe x-rays of some of the most violent phenomena in the universe which cannot be seen by the *Hubble*'s visual-range telescope. After NASA had formally extended the operational mission of

Chandra from 5 years to 10 years in September 2001, in 2003 *Chandra*'s most popular image was the Crab Nebula with its remarkable pulsar and dazzling tornado of high-energy particles and magnetic fields. *Chandra* discovered two supermassive black holes orbiting each other in the nucleus of the galaxy NGC 6240, the first definitive identification of a binary supermassive black hole system. Seven mysterious sources of x-rays are candidates for the most distant supermassive black holes ever observed. *Chandra* provided the best image yet of x-rays produced by fluorescent radiation from oxygen atoms in the sparse upper atmosphere of Mars; it also stunned observers with images of the spiral nebula M83 with its ethereal beauty of neutron stars and black holes around a blazing, starburst heart of multimillion-degree gas.

Galileo. In 2003, *Galileo* continued to return unprecedented data on Jupiter and its satellites before ending its 14-year mission on September 21, when the spacecraft passed into Jupiter's shadow, then disintegrated in the planet's dense atmosphere (to ensure that there was no chance the long-lived spacecraft could hit and possibly contaminate the moon Europa). *Galileo*'s prime mission ended 7 years ago, after 2 years of orbiting Jupiter, and NASA extended the mission three times to continue taking advantage of the probe's unique capabilities. During its mission, the spacecraft used 2040 lb (925 kg) or 246 gallons (931 liters) of propellant, returned more than 30 gigabytes of data, and transmitted about 14,000 pictures. It covered a travel distance of about 2.9 billion miles (4,631,778,000 km) from launch to impact.

Cassini. NASA's 6-ton (5.4-metric-ton) spacecraft *Cassini* continued its epic 6.7-year, 2-billion-mile (3.2-billion-km) journey to the planet Saturn. During 2003, the spacecraft remained in excellent health, with 1 year to go before becoming the first Earth envoy to enter orbit around the ringed planet Saturn, on June 30, 2004. In July 2004, after a close flyby (1243 mi or 2000 km) of the farthest of Saturn's moons, Phoebe, and orbit insertion, *Cassini* began a 4-year tour of the planet, its moons, rings, and complex magnetic environment, during which the spacecraft is scheduled to complete 74 orbits of the planet, 44 close flybys of the moon Titan, and numerous flybys of Saturn's other icy moons. On December 25, 2004, it is scheduled to release a European-built *Huygens* probe for descent through the thick atmosphere of the moon Titan on January 14, 2005. In 2003, using a sensitive new imaging instrument on the spacecraft, the probe discovered a large and surprisingly dense gas cloud at Jupiter, sharing an orbit with the planet's icy moon Europa. The tool, the Magnetospheric Imaging Instrument, is one of 12 science instruments on the main spacecraft and one of 6 designed to investigate the environments around Saturn and its moons. Another highlight of *Cassini*'s voyage in 2003 was the discovery of a dark cloud swirling around Jupiter's north pole, which rivals the Great Dark Spot in size. Images taken from a distance of 69 million miles (111 million kilometers), or from about three-fourths of the distance between Earth and the Sun, showed enhanced details in the rings and atmosphere as well as five of Saturn's icy moons.

WMAP. NASA's *Wilkinson Microwave Anisotropy Probe* (formerly called the *Microwave Anisotropy Mission*, or *MAP*) was launched on June 30, 2001, on a Delta 2. Now located in an orbit around the second Lagrangian libration point L2, its differential radiometers measure the temperature fluctuations of the cosmic microwave background radiation (CMBR), the light left over from the big bang, with unprecedented accuracy. Since start of WMAP operations, scientists have produced the first version of a full sky map of the faint anisotropy or variations in the cosmic microwave background radiation temperature. One surprise revealed in the data is that the first generation of stars to shine in the universe ignited only 200 million years after the big bang, over 13 billion years ago, much earlier than scientists had expected.

Genesis. The solar probe *Genesis* was launched on August 8, 2001, on a Delta 2 rocket into a perfect orbit about the first Earth-Sun Lagrangian libration point L1 about 930,000 mi (1.5 million kilometers) from Earth and 92.3 million miles (148.5 million kilometers) from the Sun on November 16, 2001. After the unconventional Lissajous orbit insertion, *Genesis* began the first of five "halo" loops around L1, lasting about 30 months. Collection of samples of solar wind material started on October 21, 2001. On December 10, 2002, its orbit around L1 was fine-tuned with the seventh of 15 planned station-keeping maneuvers during the lifetime of the mission. Throughout 2003, *Genesis* continued its mission of collecting solar wind material. In April 2004, the sample collectors were deactivated and stowed, and the spacecraft returned to Earth.

ACE and Wind. The *Advanced Composition Explorer* (*ACE*), launched on August 25, 1997, is positioned in a halo orbit around L1, where gravitational forces are in equilibrium. *ACE* in 2003 continued to observe, determine, and compare the isotopic and elemental composition of several distinct samples of matter, including the solar corona, the interplanetary medium, the local interstellar medium, and galactic matter. *Wind*, launched on November 1, 1994, as part of the International Solar-Terrestrial Project (ISTP), was first placed in a sunward, multiple double-lunar swingby orbit with a maximum apogee of 350 earth radii, followed by a halo orbit at the L1 point. The spacecraft carries an array of scientific instruments for measuring the charged particles and electric and magnetic fields that characterize the interplanetary medium (or solar wind)—a plasma environment. Nearly continuous plasma measurements made by *Wind* near Earth are being used to investigate the disturbances and changes in the solar wind that drive important geomagnetic phenomena in the near-Earth geospace (such as aurorae and magnetic storms), as detected by other satellites and ground-based instruments.

Stardust. NASA's comet probe *Stardust* was launched on February 3, 1999, to begin its mission to intercept a comet and return closeup imagery to Earth. Also, for the first time, comet dust and interstellar dust particles were to be collected during a close encounter with Comet P/Wild-2 in 2004 and returned to Earth for analysis. *Stardust*'s trajectory made three loops around the Sun before closest approach to the comet in January 2004. A second orbit of the Sun was completed in mid-2003, and the comet P/Wild 2 encounter took place on January 2, 2004. The sample collector was deployed in late December 2003 and was retracted, stowed, and sealed in the vault of the sample reentry capsule after the Wild flyby. Images of the comet nucleus were also obtained. On January 15, 2006, a capsule is scheduled to separate from the main craft and return to Earth.

Ulysses. The joint European/NASA solar polar mission *Ulysses*, launched in 1990, continues to study the Sun's polar regions. All spacecraft systems and the nine sets of scientific instruments remain in excellent health. *Ulysses* arrived over the Sun's south polar regions for the second time in November 2000, followed by a rapid transit from maximum southern to maximum northern helio-latitudes completed in October 2001. The spacecraft then headed away from the Sun toward aphelion at the end of June 2004, after passing through its critical eighth conjunction on August 30, 2003 (where Earth, Sun, and spacecraft are aligned, with the Sun in the middle).

Pioneer 10. Launched from Cape Kennedy, in Florida, in 1972, the 570-lb (258-kg) *Pioneer 10* had become the Earth's longest-lived interplanetary explorer, as it continued on its epic voyage. By the end of 2002 at a distance of 7.52 billion miles (12.1 billion kilometers) from Earth and 81.86 astronomical units (AU) from the Sun, *Pioneer 10* was passing through the transitional region between the farthest traces of the Sun's atmosphere, the heliosphere, and free intergalactic space. Signals transmitted by the spacecraft needed 11 h 12 min to reach Earth. What now appears to have been the space probe's last signal was received by NASA's Deep Space Network on January 23, 2003, a very weak signal without telemetry, indicating that *Pioneer 10*'s radioisotope power source had decayed.

Voyager. The *Voyager* mission, now in its twenty-seventh year, continues its quest to push the bounds of space exploration. On November 5, 2003, *Voyager 1* reached 90 AU from the Sun (about 8.4 billion miles or 13.5 billion kilometers). Now the most distant human-made object in the universe, *Voyager 1* is the only spacecraft to have made measurements in the solar wind from such a great distance from the source of the dynamic solar environment. *Voyager 2*, which reached a distance from the Sun of 6.6 billion miles (10.6 billion kilometers or 70 AU) in July 2003, also continues the groundbreaking journey with the current mission to study the region in space where the Sun's influence ends and the dark recesses of interstellar space begin.

Mars exploration. After the stunning failures of two Mars probes in 1999, NASA's Mars exploration program rebounded in 2001–2003. After NASA in 2002 had narrowed the list of possible landing sites for the next two *Mars Exploration Rover* (*MER*) missions, their successful launches made headlines in 2003.

MER-A. The first *Mars Exploration Rover* was launched on June 10 on a Delta 2 Heavy rocket, weighing about 400 lb (180 kg) and carrying a six-wheeled rover vehicle, Spirit. The explorer landed on Mars on January 3, 2004 (Eastern Standard Time), touching down almost exactly at its intended landing site in Gusev Crater in excellent condition.

MER-B. NASA's second Mars explorer, twin to *MER-A*, followed on July 7, also on a Delta 2 Heavy. *Opportunity*, the lander, touched down on January 25, 2004, right on target on Meridiani Planum, halfway around the planet from the Gusev Crater site of its twin, also in excellent condition.

Mars Odyssey. The *Mars Odyssey* probe reached Mars in 2001. Entering a highly elliptical orbit around the poles of the Red Planet, it began to reduce its ellipticity to a circular orbit at 250 mi (400 km) by the end of January 2002. The orbiter is circling Mars for at least 3 years, conducting a detailed mineralogical analysis of the planet's surface from space and measuring the radiation environment. During 2003, its instruments collected a huge volume of data and transmitted detailed observations to Earth highlighting water ice distribution and infrared images of the Martian surface. Its instruments are giving scientists an unprecedented look at the processes that continue to change the planet's surface, once thought to be a dead dust bowl. The mission has as its primary science goals to gather data to help determine whether the environment of Mars was ever conducive to life, to characterize the climate and geology of the planet, and to study potential radiation hazards to possible future astronaut missions. The orbiter also acts as a communications relay for other missions at Mars over a period of 5 years.

Mars Global Surveyor (MGS). *MGS* completed its primary mission at the end of January 2001 and entered an extended mission. The spacecraft has returned more data about Mars than all other missions combined. Through 2003, imagery and transmissions continued. On May 8, *MGS* succeeded in capturing six other celestial bodies in a single photographic frame: taking advantage of an alignment in the orbits of Earth and Jupiter, *MGS* delivered a picture that included the two planets, plus the Moon and three of Jupiter's Galilean satellites—Callisto, Ganymede, and Europa.

Earth science. In 2003, NASA launched the *ICESat*.

ICESat. ICESat (Ice, Cloud, and land Elevation Satellite) is the latest Earth Observing System (EOS) spacecraft and the benchmark mission for measuring ice sheet mass balance, cloud and aerosol heights, as well as land topography and vegetation characteristics. Launched on January 12 on a Delta 2 Expendable Launch Vehicle (ELV) into a near-polar orbit, the spacecraft carries only one instrument, the Geoscience Laser Altimeter System (GLAS).

Aqua. *Aqua* was launched by NASA in 2002. Formerly named *EOS PM* (signifying its afternoon equatorial crossing time), *Aqua* is part of the NASA-centered international Earth Observing System. Since May 2002, the 3858-lb (1750-kg) satellite, carrying six instruments weighing 2385 lb (1082 kg) designed to collect information on water-related activities worldwide, has been circling Earth in a polar, Sun-synchronous orbit of 438 mi (705 km) altitude. During its 6-year mission, *Aqua* is observing changes in ocean circulation, and studies how clouds and surface water processes affect climate. *Aqua* joined *Terra*, launched in 1999, and was followed by *Aura* in 2004.

POES-M (NOAA-M). The operational weather satellite *POES-M* (Polar-orbiting Operational Environmental Satellites M) was launched from Vandenberg Air Force Base, in California, on a commercial Titan 2 rocket on June 24, 2002. The satellite, later renamed *NOAA-M*, is part of the POES program, a cooperative effort between NASA and the National Oceanic and Atmospheric Administration (NOAA), the United Kingdom, and France. It joined the *GOES-M* launched in July 2001. Both satellites, operated by NOAA, provide global coverage of numerous atmospheric and surface parameters for weather forecasting and meteorological research.

GRACE. Launched on March 17, 2002, on a Russian Rockot carrier, the twin satellites *GRACE* (Gravity Recovery and Climate Experiment), named "Tom" and "Jerry," are mapping in detail the Earth's gravity fields by taking accurate measurements of the distance between the two satellites, using the Global Positioning System (GPS) and a microwave ranging system. The project is a joint partnership between NASA and the German DLR (Deutsches Zentrum für Luft- und Raumfahrt).

Department of Defense space activities. United States military space organizations continued their efforts to make space a routine part of military operations across all service lines. One focus is to shift the advanced technology base toward space in order to continue building a new foundation for more integrated air and space operations in the twenty-first century. Space is becoming more dominant in military reconnaissance, communications, warning, navigation, missile defense, and weather-related areas. The use of space systems within military operations reached a distinct mark in 2002 for the war on terrorism and operations in Afghanistan, and in Iraq in 2003. The increased use of satellites for communications, observations, and—through the Global Positioning System—navigation and high-precision weapons targeting was of decisive importance for the military command structure.

In 2003, there were 11 military space launches. These included two Titan 4/Centaur vehicles from Cape Canaveral, Florida, with the sixth Milstar FLT satellite, to complete the ring of communications satellites around the Earth and provide ultrasecure, jam-resistant transmission for troops and government leaders virtually anywhere on the planet, and an NRO signal intelligence (sigint) satellite. Eleven other satellites, for communications, navigation, sigint, weather/environment, and technology development, were lifted into Earth orbit by three Delta 2's, the second and third new heavy lifter Delta 4, one Atlas 2 AS, one Russian-powered Atlas 3B, and the last two Titan 2G SLV converted ICBMs.

Commercial space activities. In 2003, commercial space activities in the United States exhibited a sluggish increase over prior years, after the 2001–2002 slump in the communications space market. In addition to the financial crisis, some difficulties remained due to the export restrictions imposed on sensitive technologies in United States industry. In general, commercial ventures continue to play a relatively minor role in United State space activities.

Of the 26 launch attempts by the United States in 2003 (versus 18 in 2002), 8 were commercial missions (NASA 7; military 11). In the launch services area, Boeing sold seven Delta 2 vehicles, while ILS/Lockheed Martin flew one Atlas 2AS and two Atlas 3B (with Russian engines). Both companies also had successful launches of their next-generation evolved expendable launch vehicle (EELV) rockets, for example Lockheed Martin with the second and third Atlas 5s (comsats *Hellas Sat 2* and *Rainbow 1*), and Boeing with the third and fourth Delta 4 heavy launchers (comsats *Eutelsat W5* and *DSCS III B-6*). Orbital Science Corp. had four successful flights of the Pegasus XL airplane-launched rocket, carrying NASA's *SORCE*, *ICESat*, and *GALEX*, Canada's *Scisat 1* aeronomy satellite, and the commercial *Orbview 3* for imaging; while the partnership of Boeing, RSC-Energia (Russia), NPO Yushnoye (Ukraine), and Kvaerner Group (Norway) successfully launched three Russian Zenit 3SL rockets, carrying United Arab Emirates (UAE) *Thuraya 2*, the *Echostar 9/Telstar 13*, and the *Galaxy 13/Horizons 1* comsats, from the *Odyssey* sea launch platform floating at the Equator (first launch 1999).

On May 20, the Scaled Composites company, currently the only commercial enterprise independently engaged in advanced development of human space flight capability, made the first flight of its spaceship, the piloted mother ship/airplane *White Knight* carrying the *SpaceShipOne* rocket-powered glider to nearly 50,000 ft (15,000 m), both remaining joined. The first gliding flight of the latter followed on August 7, after separation from the carrier aircraft at 47,000 ft (14,300 m). Both *White Knight* and *Space-ShipOne* landed smoothly under test pilot control at the Mojave, CA, test range.

Russian Space Activities

Russia showed relatively unchanged activity in space operations from 2002 to 2003. Its total of 21 successful launches (out of 21 attempts) was three less than in 2002: four Soyuz-U (one crewed), four Soyuz-FG (one crewed), five Protons, two Rockots (first launch 1994), three Zenit-3SL (sea launch; counted above under United States activities), two Molniya, three Kosmos-3M, and one Strela (launcher

test with dummy payload). The upgraded Soyuz-FG rocket's new fuel injection system provides a 5% increase in thrust over the Soyuz-U, enhancing its lift capability by 440 lb (200 kg) and enabling it to carry the new Soyuz-TMA spacecraft. Soyuz-TMA was flown for the first time in 2002, as *ISS* Mission 5S. It was followed in 2003 by Soyuz *TMA-2* (6S) and *TMA-3* (7S) [see International Space Station, above].

The Russian space program's major push to enter into the world's commercial arena leveled off in 2003. First launched in July 1965, the Proton heavy lifter, originally intended as a ballistic missile (UR500), by the end of 2003 had flown 223 times since 1980, with 14 failures (reliability 0.937). Its launch rate in recent years has been as high as 13 per year. Of the five Protons launched in 2003 (2002: 9), three were for commercial customers (*AMC-9/SES*, *Yamal-201* and *-202*, *Ekspress AM-21*), and the others for the state, including military. Between 1985 and 2003, 171 Proton and 396 Soyuz rockets were launched, with 10 failures of the Proton and 10 of the Soyuz, giving a combined reliability index of 0.965. Until a launch failure on October 15, 2002, the Soyuz rocket had flown 74 consecutive successful missions, including 12 with human crews onboard; subsequently, another eight successful flights were added, including two carrying five humans.

European Space Activities

Europe's efforts to reinvigorate its faltering space activities, after the long decline since the mid-1990s, in 2003 continued their modest pace of 2002. Work was still underway by the European Union (EU) on a new European space strategy for European Space Agency (ESA) to achieve an autonomous Europe in space, under Europe's new constitution that makes space and defense a European Union responsibility.

However, 2003 did not bring the much-needed breakthrough of Europe's commercial space industry in its attempts at recovery, given particular emphasis by the last flight of Arianespace's Ariane 4 workhorse on February 15, carrying *Intelsat 907*. Although Arianespace was able to close its 2003 accounts with a very small net profit after 3 years of losses, it still labored under the impact of the failure of the upgraded Ariane 5 EC-A in December 2002 with two high-value comsats, shortly after liftoff. Thus, in 2003, the heavy-lift Ariane 5G (generic) was launched only three times (down from four in 2002), bringing its total to 17. Its new EC version, designed to lift 10 tons to geostationary transfer orbit, enough for two big communications satellites at once, uses a new cryogenic upper stage, an improved Vulcain 2 main stage engine, and solid boosters loaded with 10% more propellant. After the EC-A failure, European industry quickly developed an Ariane 5 recovery plan, and on June 20 Arianespace concluded a preliminary contract with European aerospace conglomerate EADS Space Transportation for a batch of 30 Ariane 5 launchers (signed in May 2004). Also in 2003, the French and Russian governments reached an agreement that would allow a Soyuz launch pad to be installed at the European spaceport in Kourou, French Guiana, in return for Russian cooperation on future launcher technologies.

In 2003, the most significant space undertaking for the 15 European countries engaged in space continued to be the development of the *Galileo* navigation and global positioning system. Starting in 2008, this will enable Europe to be independent of the U.S. Global Positioning System. In its final configuration, *Galileo* will consist of a constellation of 30 small satellites placed in medium orbit (15,000 mi or 24,000 km above Earth). It will be independent of but compatible with GPS. *See* GPS MODERNIZATION.

The *ISS* remains ESA's biggest ongoing program in the human spaceflight area. European *ISS* share (totaling 8.6%) remains unchanged due to an agreement signed by previous governments of the participating nations. France has a relatively large and active national space program, including bilateral (outside ESA) activities with the United States and Russia. The Italian Space Agency, ASI, participates in the *ISS* program through ESA, but also had entered a protocol with NASA for the delivery of three multipurpose logistics modules (MPLM) for the *ISS*. Two MPLMs have already been delivered, *Leonardo* and *Raffaello*, and both flew in 2001. The third MPLM is *Donatello*, and Italy has also developed a second *ISS* Node, which was delivered to NASA in June 2003. The faltering interest of Germany's government in this field continued in 2003. Germany is the second major ESA contributor after France, but it has essentially no national space program remaining. In the space science area, there were only two new European launches in 2003, the lunar probe *Smart 1* and the partially successful *Mars Express* orbiter/lander mission.

Envisat. In 2003, ESA's operational environmental satellite *Envisat* continued its observations after its launch on March 1, 2002, on the eleventh Ariane 5. The 18,100-lb (8200-kg) satellite circles Earth in a polar orbit at 500 mi (800 km) altitude, completing a revolution of Earth every 100 min. Because of its polar Sun-synchronous orbit, it flies over and examines the same region of the Earth every 35 days under identical conditions of lighting. The 82-ft-long (25-m) and 33-ft-wide (10-m) satellite is equipped with 10 advanced instruments, including an Advanced Synthetic Aperture Radar (ASAR), a Medium Resolution Imaging Spectrometer (MERIS), an Advanced Along Track Scanning Radiometer (AATSR), a Radio Altimeter (RA-2), a Global Ozone Monitoring by Occultation of Stars (GOMOS) instrument, a Michelson Interferometer for Passive Atmosphere Sounding (MIPAS), and a Scanning Imaging Absorption Spectrometer for Atmospheric Cartography (SCIAMACHY).

Spot 5. Launched on May 4, 2002, by an Ariane 4, the fifth imaging satellite of the commercial Spot Image Company in 2003 continued operations in its polar Sun-synchronous orbit of 505 mi (813 km) altitude.

Integral. ESA's *INTEGRAL* (International Gamma-Ray Astrophysics Laboratory), a cooperative project with Russia and the United States, continued successful operations in 2003. Launched on October 17, 2002, the sensitive gamma-ray observatory provides new insights into objects such as black holes, neutron stars, active galactic nuclei, and supernovae. *INTEGRAL* in 2003 resolved the long-standing question as to the nature of the diffuse glow of soft gamma rays seen from the central region of our Galaxy. Its observations have shown that most of the emission is produced by individual point sources.

XMM-Newton. Europe's *XMM* (X-ray Multi Mirror)-*Newton* observatory, launched on December 10, 1999, on an Ariane 5, is the largest European science research satellite ever built. The telescope has a length of nearly 11 m (36 ft), with a mass of almost 8800 lb (4 metric tons). After the observatory's numerous significant discoveries, in November 2003 an extension of the *XMM-Newton* mission up to March 31, 2008, was unanimously approved by the Science Program Committee, with funding secured up to March 2006 and a provisional budget for an additional 2 years.

SMART 1. *SMART 1* (*Small Missions for Advanced Research in Technology 1*) is Europe's first lunar spacecraft. The 816-lb (370-kg) spacecraft was launched on September 27 with two commercial communications satellites (*Insat 3E*, *e-Bird*) on an Ariane 5G. Built by Swedish Space Corp., it is intended to demonstrate new technologies for future missions, in this case the use of solar-electric propulsion as the primary power source for its ion engine, fueled by xenon gas. *SMART 1* is scheduled to arrive at its goal in April 2005 to begin its science program, executed with spectrometers for x-rays and near infrared as well as a camera for color imaging. *See* SMART 1.

Mars Express. *Mars Express* was launched on June 2, 2003, from the Baikonur launch site by a Russian Soyuz/Fregat rocket. After a 6-month journey, it arrived at Mars in December. Six days before arrival, *Mars Express* ejected the *Beagle 2* lander, which was to have made its own way to the correct landing site on the surface but was lost, failing to make contact with orbiting spacecraft and Earth-based radio telescopes. The *Mars Express* orbiter successfully entered Martian orbit on December 25, first maneuvering into a highly elliptical capture orbit, from which it moved into its operational near-polar orbit in January 2004. *Mars Express* is remotely exploring the planet with a sophisticated instrument package comprising the High Resolution Stereo Camera (HRSC), Energetic Neutral Atoms Analyzer (ASPERA), Planetary Fourier Spectrometer (PFS), Visible and Infrared Mineralogical Mapping Spectrometer (OMEGA), Sub-Surface Sounding Radar Altimeter (MARSIS), Mars Radio Science Experiment (MaRS), and Ultraviolet and Infrared Atmospheric Spectrometer (SPICAM).

Asian Space Activities

China, India, and Japan have space programs capable of launch and satellite development and operations.

China. With a total of 7 launches in 2003, China moved into third place of space-faring nations, after the United States and Russia. In fact, the People's Republic's space program made worldwide headlines in 2003 with its successful orbital launch of the first Chinese "taikonaut," 38-year-old Lieutenant Colonel Yang Liwei. His spacecraft, *Shenzou 5* ("Divine Vessel 5"), was launched on October 15, followed by a reentry and landing 21 h 23 min later about 600 mi (1000 km) east of the launch area of the Jiuquan Satellite Launch Center launch site in the Gobi Desert. *See* CHINESE SPACE PROGRAM.

The flight of the first Chinese into space had been preceded by several years of developments, including four crewless test flights and the construction of major new facilities. In past years, China had been training 14 potential taikonauts, some of them at Russia's Gagarin Cosmonaut Training Center (GCTC).

The launch vehicle of the *Shenzhou* spaceships is the new human-rated Long March 2F rocket. China's Long March (Chang Zheng, or CZ) series of launch vehicles consists of 12 different versions, which by the end of 2003 had made 75 flights, sending 85 payloads (satellites and spacecraft) into space, with 90% success rate. China has three modern (but land-locked, thus azimuth-restricted) launch facilities: at Jiuquan (Base 20, also known as Shuang Cheng-Tzu/East Wind) for low Earth orbit (LEO) missions, Taiyuan (Base 25) for Sun-synchronous missions, and Xichang (Base 27) for geostationary missions.

Five other major launches in 2003 served to demonstrate China's growing space maturity. On May 24, a CZ-3A launched the *Beidou* navigation satellite, followed on October 21 by the launch of the Chinese/Brazilian Earth resources imaging satellite *CBERS 2*, along with the small *Chuangxin 1* communications research satellite on a CZ-4B; on November 3 by the military/civilian imaging recoverable spacecraft *FSW 18* on a CZ-2D; on November 14 by the comsat *Zhongxing 20* on a CZ-3A; and on December 29 by *Tan Ce 1*, the first of two Chinese/European DoubleStar magnetospheric spacecraft (ESA designation: *DSP-E*), on a CZ-2C.

India. India continued its development programs for satellites and launch vehicles through the Indian Space Research Organization (ISRO, created in 1969), part of the Department of Space (DOS). In 2003, the country successfully conducted two launches. One was the second developmental test flight of the Delta 2 class GSLV (Geosynchronous Satellite Launch Vehicle) designated GSAT-D2, carrying the 2000-kg (4400-lb)-class experimental communication satellite *GSat 2*. The other was a four-stage PSLV-C5, the eighth Polar Space Launch Vehicle (PSLV), with the *IRS-P6/ResourceSat 1* for multispectral remote sensing of the Earth. The launch took place from India's Sriharikota Space Center, renamed Satish Dhawan Space Centre-SHAR in 2002. Also in 2003, India offered to make the PSLV available for launching two small Brazilian technology satellites that were originally to be orbited by Brazil's own VLS 3 rocket, destroyed on August 22. India has also formally expressed its desire to participate in the European Union's *Galileo* satellite navigation system.

Japan. Starting on October 1, 2003, Japan's three space organizations were consolidated into one new space agency, the Japan Aerospace Exploration Agency (JAXA), with a workforce of about 1800.

In past years, the National Space Development Agency (NASDA) developed the launchers N1, N2, H1, and H2. The H2-A vehicle, an upgraded and more cost-effective version of the H-2, had its maiden flight in 2001. In 2002, the H-2A executed three missions, all successful, launching eight satellites, including one for communications, two for remote sensing, one for microgravity research, and several for technology development. In 2003, however, after the successful launch of two reconnaissance satellites (*IGS-1A/Optical 1* and *IGS-1B/Radar 1*) on the fifth H-2A, on March 28, the November 29 launch of the sixth H-2A failed when ground controllers were forced to send a destruct command after one of its two solid-rocket boosters failed to separate from the first stage. Lost with the heavy lifter were *Optical 2* and *Radar 2*, the second set of the orbiting reconnaissance system.

In its longer-range view, JAXA is studying versions of a "new generation" launch vehicle, essentially a heavier-lift version of the H-2A with 10–20% greater lift capacity than its predecessor, which would put it into the Delta 4 class.

One area of great promise for Japan continues to be the *ISS* Program, in which the country is participating with a sizable 12.6% share. Its contributions to the *ISS* are the 15-ton pressurized Japanese Experiment Module JEM called *Kibo*, along with its ancillary remote manipulator arm and unpressurized porchlike exposed facility for external payloads, and the H-2 transfer vehicle (HTV), which will carry about 6 metric tons (6.6 short tons) of provisions to the *ISS* once or twice a year, launched on an H-2A. In 2003, the Mitsubishi-built JEM left Yokohama harbor by ship, arriving at NASA's Kennedy Space Center in Florida on May 30. *Kibo* is to be launched to the *ISS* on the space shuttle.

Hayabusa (Muses-C). Japan's only other successful launch in 2003 was the first asteroid sample return mission aboard an ISAS solid-propellant M-5 rocket that lifted off at the Kagoshima Space Center in southern Japan on May 9, placing the spacecraft *Hayabusa (Muses-C)* on a trajectory toward the asteroid 25143 Itokawa/1998 SF36. The deep-space probe is scheduled to arrive at the asteroid in mid-2005. The launch was the first for the M-5 since the launch failure of the *ASTRO-E* astronomy spacecraft in February 2000.

Nozomi. In 2003, efforts to put the *Nozomi* ("Hope") spacecraft into Martian orbit were abandoned, when an attempt to fire thrusters to orient the craft for a Mars orbit insertion burn failed on December 9. Smaller thrusters were successfully fired, and *Nozomi* flew past Mars at a distance of 600 mi (1000 km) on December 14, going into a heliocentric orbit with a period of roughly 2 years.

For background information *see* COMET; COMMUNICATIONS SATELLITE; COSMIC BACKGROUND RADIATION; INFRARED ASTRONOMY; JUPITER; MARS; METEOROLOGICAL SATELLITES; MILITARY SATELLITES; REMOTE SENSING; SATELLITE ASTRONOMY; SATELLITE NAVIGATION SYSTEMS; SATURN; SOLAR WIND; SPACE FLIGHT; SPACE PROBE; SPACE STATION; SPACE TECHNOLOGY; SPACE TELESCOPE, HUBBLE; SUN; X-RAY ASTRONOMY in the McGraw-Hill Encyclopedia of Science & Technology.   Jesco von Puttkamer

Bibliography. *AIAA Aerospace America*, November 2003; *Aerospace Daily*; *Aviation Week & Space Technology*; ESA Press Releases; NASA Public Affairs Office News Releases; *Space News*.

Space shuttle

On February 1, 2003, the space shuttle *Columbia* disintegrated over the southern United States during reentry. The Columbia Accident Investigation Board (CAIB) was formed and issued their report in August 2003. The report was organized into three sections: (1) "The Accident," which summarized the shuttle program, the *Columbia*'s final flight, and the accident's physical cause; (2) "Why the Accident Occurred," which compared *Challenger* to *Columbia* with emphasis on organizational reasons for the accident; and (3) "A Look Ahead," which looked at near-, mid-, and long-term implications of the accident for United State space flight. In each section, the report presented many findings and recommendations.

The accident. The shuttle is attached to its external fuel tank via a bipod support structure located at the top of the shuttle just forward of the crew compartment. A piece of foam insulation on the left bipod separated 81.7 seconds after launch and struck reinforced carbon-carbon (RCC) panel 8 on the forward part of the left wing. This panel is one of many that are part of the shuttle's reentry Thermal Protection System. Upon reentry, superheated air penetrated the damaged panel and started to melt the shuttle's aluminum structure. The air simultaneously increased aerodynamic forces on the shuttle, causing loss of control and eventual breakup of the shuttle.

External tank and foam. NASA does not fully understand the mechanisms that caused the foam loss. There was no indication that the age of the tank or the total prelaunch exposure to the elements contributed to the loss of foam. There was no evidence of any negligence during the application of the foam that could lead to its loss. The Board also found shedding from the left bipod during launch that NASA was not aware of. There were 7 known instances of such foam shedding out of 72 launch or external tank separation events for which photographic imagery was available. Foam loss occurred on more than 80% of the 79 missions for which imagery was available. Thirty percent of all missions did not have imagery that could help determine if there was foam loss.

RCC panels. The original design of the RCC panels required essentially zero resistance to impacts. NASA used spare RCC panels to test impact resistance. During these tests, pieces of foam were fired at velocities comparable to those that occurred during the launch and at various angles to the RCC panels. The results of the tests showed that the panels were "remarkably tough and have impact capabilities that far exceed the minimal impact resistance specified." However, RCC components are weakened by mass loss caused by oxidation, and get weaker with repeated use. The Board found that the visual inspection techniques currently employed are not sufficient to assess the structural integrity of the RCC panels. Only two RCC panels—flown for 15 and 19 missions, respectively—were destructively tested to determine actual loss of strength due to oxidation.

Imagery and analyses. Photographic imagery of the *Columbia* during launch indicated that foam strikes were the likely cause of the failure. The cameras used were not high-speed or high-resolution and made the engineering analysis of the impact more difficult. However, the photos were able to show the location of the failure and its effect on the RCC tiles when the foam struck the shuttle at 775 ft (236 m) per second.

While the shuttle was on orbit during the second day of flight, the Air Force made radar and optical observations. When analyzed after the accident, these images showed a small object in orbit with *Columbia*. The radar cross section (how much radar energy the object scatters) and the ballistic coefficient (which indicates how fast the orbit will decay) of the object were measured and found to match that expected of an RCC panel fragment.

Debris analysis was another source of information. Once the debris was recovered, the pieces were identified and assembled on a hangar floor to show their original conformation. The total amount of left wing debris was significantly smaller than that of the right wing. Individual pieces were analyzed for indications of superheating, molten deposits, and erosion. This analysis was key to understanding the accident, as it established the point of superheated air intrusion as RCC panel 8.

Finally, the CAIB performed a fault tree analysis to determine if there were any other possible causes of the accident. They looked at everything from solid rocket booster bolt catchers to willful damage, and found no significant problems.

Why the accident occurred. The CAIB found that the cause of the *Columbia* accident was more than its physical mechanisms.

Historical context. The CAIB looked at the *Challenger* disaster in 1986 and compared it with the *Columbia* breakup. Regarding the *Columbia* accident, it found that "The causal roots of the accident can also be traced, in part, to the turbulent post-Cold War policy environment in which NASA functioned during most of the years between the destruction of Challenger and the loss of Columbia." With no Cold War, there was no political reason for NASA's Human Space Flight Program. The U.S.–Soviet space competition was over. Thus, NASA could not obtain budget increases in the 1990s. During the 1990s, the budget and workforce decreased by 40%.

The shuttle was considered to be a mature system, and therefore was transitioned into an operational program. This means that the focus was on operating the shuttle, and not understanding the mission-by-mission problems inherent in a developmental spacecraft.

In the 1990s, the shuttle replacement date was moved from 2006 to 2012 and then to 2015 or later. This caused confusion and ambivalence as to whether or not to invest money for upgrading the shuttle for safety, operational support, and infrastructure.

Decision making at NASA. This is the most important part of the CAIB report. There are more findings in this area than any other, many critical of NASA.

Foam loss was considered a serious problem when the shuttle was first designed. The engineers were worried about any potential damage to the Thermal Protection System, which is inherently fragile. A thumbnail can easily damage the tile with slight pressure. The baseline design requirement therefore stated that there would be no shedding of ice or other debris during prelaunch or flight. However, NASA did not follow this design requirement. *Columbia* was the first shuttle to fly, and it sustained damage from debris loss. More than 300 tiles had to be replaced after that mission. Foam loss was discussed after the *Challenger* accident. However, foam continued to be shed during subsequent missions. NASA engineers and managers began to regard the foam shedding as inevitable.

In fact, NASA did not follow its own rules on foam shedding. Although NASA worked on this continuously, the debris shedding requirement was not met on any mission. Further, the shedding went from a serious safety concern to an "in-family" or "accepted risk" concern. Foam shedding on two missions, STS-52 and STS-62, was not noticed until the CAIB directed NASA to examine External Tank separation photos more closely. Finally, despite the shedding of the foam, NASA managers did not strengthen the shuttle tile against these impacts. Moreover, there were scheduling problems. NASA Headquarters wanted to meet the Space Station Node 2 launch date of February 19, 2004. This pressure may have influenced how NASA managers handled the foam strike on STS-112 and *Columbia*'s foam strike. There was no schedule flexibility to fix unforeseen problems.

In-flight decision making. This section of the CAIB report contains actual email messages and log reports that occurred during the mission. On flight day 2 the Intercenter Photo Working Group noticed the large foam strike and began the process of requesting imagery from the Department of Defense (DOD). There were eight times where the DOD could have helped them, but NASA failed to ask for support due to various reasons. NASA managers were asked by the CAIB to look at a repair on orbit or a rescue using the shuttle *Atlantis*. The repair was considered

to be high-risk. The rescue, however, was concluded to be feasible: *Atlantis* ground processing would be accelerated, the astronauts on *Columbia* would reduce activities to conserve air, and *Atlantis* could have docked with *Columbia*.

The CAIB report summarized NASA's internal problems as: "Management decisions made during Columbia's final flight reflect missed opportunities, blocked or ineffective communications channels, flawed analysis, and ineffective leadership." Managers at all levels seemed uninterested in the foam-strike problem and its implications. Because managers failed to use the wide range of expertise and opinions available, they failed to meet their own safety-of-flight criteria. Some Space Shuttle Program managers failed to fulfill the implicit contract to do whatever is possible to ensure the safety of the crew. In fact, their actions unknowingly imposed barriers, keeping at bay both engineering concerns and dissenting views. This helped create "blind spots," so the managers could not see the danger the foam strike posed.

Looking ahead. After the *Challenger* accident, the Rogers Commission recommended large changes in NASA's safety organization to make it an independent oversight organization. The CAIB found that this never happened. One finding sums it up well. "The Associate Administrator for Safety and Mission Assurance is not responsible for safety and mission assurance execution, as intended by the Rogers Commission."

Although the CAIB did recommend that NASA initiate an aggressive program to eliminate all debris-shedding at its source (with emphasis on the region where the bipod struts attach to the external tank), the Board saw that fixing the foam loss, although important, is not the major item that needs remedy before the shuttle flies again. It found that the causes of the institutional failure responsible for the *Challenger* had not been fixed, and that if these systemic flaws are not resolved, another accident is likely. Therefore, their recommendations for change were "not only for fixing the Shuttle's technical system, but also for fixing each part of the organizational system that produced Columbia's failure."

For background information *see* SPACE SHUTTLE in the McGraw-Hill Encyclopedia of Science & Technology.   Jeffrey C. Mitchell

Bibliography. *Columbia Accident Investigation Board Report*, 6 vols., Government Printing Office, 2003.

Spherical microphone arrays

The most common microphone in use today is omnidirectional, meaning that it responds to signals from all directions with equal sensitivity. An omnidirectional microphone is appropriate for cellular and telephone handset use where the microphone is relatively close to the desired sound source. It is also used for specific close microphone recording of

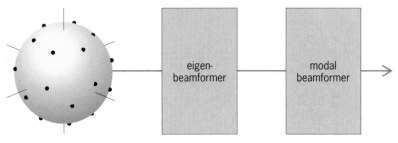

Fig. 1. Spherical microphone array system.

individual instruments and to record the background reverberation in sound recording. However, its use in applications where the microphone is distant from a desired speech source typically results in a recording that is too reverberant and sensitive to background room noise. A common solution for the distant microphone problem is to use a directional microphone. A directional microphone spatially filters the sound field to emphasize the desired sound signal over signals propagating from different directions. Common commercial directional microphones use a single microphone element that has openings to the sound field on both sides of the microphone membrane. Since a microphone membrane moves proportionally to the net force (the acoustic pressure difference) on the microphone, this simple mechanical design can exhibit a directional response to input acoustic waves. However, the amount of directional gain is limited, and the preferred listening direction depends on the physical orientation of the microphone.

A more general way to obtain a directional response is to use multiple microphones and combine them in a way that results in a more selective sensitivity to sounds propagating from any preferred direction. Microphone arrays consist of an arrangement of two or more microphones and an arithmetic processor (called a beamformer) that linearly combines the microphone signals. A combination of the

multiple microphone signals allows one to pick up the desired sound signals depending on their direction of propagation and also to electronically steer the preferred direction of the microphone. An advantage of using adjustable beamforming arrays over conventional fixed-directional microphones, like a shotgun microphone, is their high flexibility due to the added degrees of freedom offered by the multitude of elements and the associated beamformer. The directional pattern of a microphone array is adjustable over a wide range by modifying the linear combinations of the beamformer, which typically are implemented in software. Therefore, no mechanical alteration of the system is needed to change the beam pattern or the steering direction.

These basic characteristics are also true for a spherical microphone array. In addition, a spherical array has several advantages over many other geometries: Due to the inherent symmetry of the sphere, the beam pattern can be steered to any direction in space without changing the shape of the pattern. Spherical arrays also allow one to easily realize an array with full spatial control of the beam-pattern shape independent of steering direction.

A computationally efficient implementation of a spherical microphone array is one where the beamformer is split into two parts. In the first stage, called the eigenbeamformer, the incoming sound field is divided by special beamformers that have the highly desired property that their responses are all power-normalized, and that the spatial integral of the product of any two incommensurate beam patterns is zero. These beamformers are said to decompose the recorded signal into spatially orthonormal components. A second beam-shaping stage, called the modal beamformer, combines these components to form the output signal. Such an implementation allows for a simple, efficient, and flexible design (**Fig. 1**). The microphones, which can be numbered 1 through S, are arranged on a spherical surface. These microphone signals are fed to the decomposition stage of the beamformer that transforms the S microphone signals into N spatially orthonormal beam components. These components can then be uniquely combined to form an output beam or multiple simultaneous output beams.

Eigenbeamformer. When using a spherical coordinate system, with angles θ and ϕ, a sound field is best expressed as a series of spherical harmonics, $Y_n^m (\theta, \phi)$. Spherical harmonics are mathematical functions of order n ($n = 0, 1, 2, 3, \ldots$) and degree m ($m = -n, -n + 1, \ldots, -1, 0, 1, \ldots, n - 1, n$) that depend on the angles θ and ϕ and that possess the highly desirable property of being mutually orthonormal. [This means that the spherical integral of the square of the amplitude of a spherical harmonic (equal to the product of the spherical harmonic and its complex conjugate) is 1, and the spherical integral of the product of a spherical harmonic and the complex conjugate of a different spherical harmonic is zero.] **Figure 2** shows spherical harmonics for orders 0, 1, and 2 and degree 0. The job of the eigenbeamformer is to extract these orthonormal

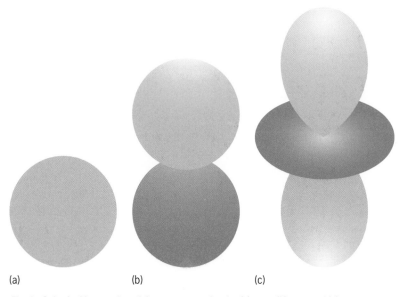

(a)   (b)   (c)

Fig. 2. Spherical harmonics of degree zero and order (*a*) zero, (*b*) one, and (*c*) two.

components from the sound field. Using the general integral definition of orthogonal functions (that the spatial integral of the product of incommensurate orthogonal beampatterns is zero), it is straightforward to see that the eigenbeamformer is itself a set of beamformers. To extract a certain mode, the eigenbeamformer has to realize this mode as a beam pattern by using that mode's spatial response as the eigenbeamformer weighting function. At the output of the eigenbeamformer, all spatial information of the original sound field is maintained up to the degree and order of the sampled spherical harmonic. For each order n, there are $2n + 1$ modes. This means that the overall number of modes for a maximum order of N is $(N + 1)^2$. It can also be shown that the number of microphones must be equal or greater than the number of modes.

Modal beamformer. The modal beamformer uses the N output beams of the eigenbeamformer, which are called eigenbeams, as input signals. Due to the orthonormal property of the eigenbeam signals, it is very simple to extract some directional information about the sound field at this stage. For example, by applying the weights $\sqrt{1}$, $\sqrt{3}$, and $\sqrt{5}$ to the first three modes of degree zero and then adding these signals, the modal beamformer generates the pattern depicted in **Fig. 3**. This pattern is sensitive to sound coming from the top direction, while sound impinging from all other directions is attenuated. The shape of the beam pattern can be dynamically changed by varying the beamformer weights. The beam pattern can be easily steered to any angle. This is accomplished by applying scalar weights that are derived from trigonometric functions to the individual eigenbeams. Spatial resolution depends on the highest-order mode that is extracted by the eigenbeamformer.

Microphone arrangement. The microphones are located on the surface of a rigid sphere in a specific arrangement that fulfills what is termed discrete orthogonality. A discrete sampling of a sound field needs to preserve the orthonormality that is required for the beamformer design to operate as desired. There are specific spatial arrangements that fulfill the requirement of discrete orthonormality. Since the design specifically limits the order of the spherical harmonic, there are arrangements with finite distributions that can be found to fulfill the discrete orthonormality up to a maximum order. The theory for the set of sample points that meet the orthonormality is complex. For an array that operates up to fourth-order spherical harmonics, one arrangement that fulfills the discrete orthonormality requirement comprises 32 pressure microphones positioned at the center of the faces of a truncated icosahedron.

Applications. The spherical array is useful for a variety of applications. Directional sound pickup and spatial sound field analysis are the most obvious application. Other possible applications are 3-D audio and spatial post editing. These applications are the subject of ongoing research.

Adaptable directional sound pickup. For a fixed setup, conventional directional microphones may be sufficient.

Fig. 3. Second-order hypercardioid pattern.

But in case of a changing acoustic scene, for example, a moving talker, a dynamically steerable array that tracks multiple desired sources can be advantageous since it allows a dynamic adaptation to the environment. Adaptive arrays can also self-optimize their spatial responses to minimize noise signals propagating in different directions to the desired sources. Also, it might be desirable to be able to dynamically change the beam-pattern shape of the microphone to accommodate audio scenes that vary in time and position.

Spatial sound-field analysis. In applications of room acoustics, measurements of many spatial sound-field properties are carried out, such as direct-to-reverberant sound energy or lateral versus vertical sound energy. A spherical array allows these measures to be determined with a single recording of all eigenbeams. An analysis can be done offline by applying the appropriate beam pattern. The array also makes it possible to steer a narrow beam through all directions to visualize the output energy as a function of direction.

Spatial post editing. Two-channel stereo recording techniques using directional microphones have been accepted for many decades. The rapid growth of DVD and home theater systems capable of 5-, 6-, and 7-channel audio playback has spawned a renewed interest in spatially accurate multichannel playback. Unfortunately, standard recording techniques for spatial multichannel playback have not kept up with this new playback capability. Spherical microphone arrays make it possible to record the spatial extent of the sound field, giving sound engineers the ability to tailor the precise auditory impression that they would like to convey.

Spatially accurate audio. An ability to record audio in a form that preserves accurate spatial information about a sound field would be an advance over current stereo audio recording techniques. Also, allowing multiple listeners to have the ability to adjust their own spatial audio perspective or orientation in a sound field would give a new level of sound playback control. For example, the ability to modify the playback of spatial audio might be important where architectural constraints preclude using the standard positioning geometry for surround-audio playback. Flexible control of a surround sound field would allow the listener to handle arbitrary playback system geometries. Also, having a general way of recording a high-resolution version of the spatial sound field would allow for compatibility of future audio playback systems that have more loudspeakers placed around and above as well as below the listener.

Higher-order ambisonics is an example. Ambisonics is a name given to a specific surround-sound recording technique. It is an extension of a mid-side (pressure and pressure-difference combination) stereo recording technique first described in a 1931 patent by Alan Dower Blumlein. Ambisonics is based on the recording of the acoustic pressure and the three orthogonal pressure-difference signals. By combining these signals through simple multiplications, additions, and subtractions, multiple output signals can be formed for surround-sound playback.

For background information *see* DIRECTIVITY; MICROPHONE; SOUND RECORDING; SPHERICAL HARMONICS in the McGraw-Hill Encyclopedia of Science & Technology.   Gary W. Elko; Jens Meyer

Bibliography. John Eargle, *The Microphone Book*, 2d ed., Focal Press, 2004; J. Meyer and G. Elko, Spherical microphone arrays for 3D sound recording, Chap. 2 in Y. Huang and J. Benesty (eds.), *Audio Signal Processing for Next-Generation Multimedia Communication Systems*, Kluwer Academic, Boston, 2004; E. G. Williams, *Fourier Acoustics*, Academic Press, San Diego, 1999.

Spitzer Space Telescope

The *Spitzer Telescope* takes advantage of dramatic advances in infrared detectors that have occurred over the last 20 years; it utilizes modern detector arrays in space, where they are limited only by the faint glow of the zodiacal dust cloud. Ground-based infrared telescopes can operate only at the wavelengths where the atmosphere is transparent, lying between 1 and 25 micrometers. Even within these windows, the thermal emission of the atmosphere is more than a million times greater than the dilute emission of the zodiacal cloud; there is additional foreground thermal emission from the telescope itself. High-sensitivity detectors are blinded by these bright foreground signals. Operating in space eliminates the atmospheric absorption and emission; also, a telescope in the vacuum of space can be cooled sufficiently to virtually eliminate its emission.

Spitzer was known for most of its life as *SIRTF* (*Space Infrared Telescope Facility*). *SIRTF* was launched on August 25, 2003, and was renamed after Lyman Spitzer, Jr., the first prominent astronomer to advocate putting telescopes into space.

Development. The *Infrared Astronomy Satellite* (*IRAS*) and the *Infrared Space Observatory* (*ISO*) were important predecessors of the *Spitzer Telescope*. *IRAS*, launched in 1983, had four sets of infrared detectors, operating at wavelengths of 12, 25,

60, and 100 μm. The detectors and a 60-cm (24-in.) telescope were mounted inside a liquid-helium dewar vessel that maintained them at a temperature of ~2 K (−456°F) for the 10-month mission, allowing a very sensitive all-sky survey. The European Space Agency built *ISO* to follow up on the *IRAS* survey. *ISO* too had a 60-cm telescope mounted in a liquid helium dewar. It carried four instruments designed for detailed study of individual objects through imaging, photometry, and spectroscopy. *ISO* was launched in 1995 and had a 28-month mission that returned a wealth of data over the 3–200-μm spectral range.

The *Spitzer Telescope* was also conceived to follow up on *IRAS*. Although its science team was selected in 1984, the mission was repeatedly delayed. Throughout this period, the mission was revived repeatedly with new technical concepts. Finally, in 1996 the telescope got started officially toward construction under the direction of the Jet Propulsion Laboratory. By that time, the National Aeronautics and Space Administration had downsized all its science missions under the guidelines of its "faster-better-cheaper" philosophy. Spitzer had been descoped by about a factor of 6 in cost from the original plans.

Design. The approved concept introduces a number of innovations developed in response to the pressure to reduce cost. It uses an Earth-trailing orbit to get far from the thermal radiation of the Earth (**Fig. 1**). The instruments are cooled inside a dewar containing liquid helium, as with *IRAS* and *ISO*. However, the telescope is mounted on the outside of this dewar and was launched warm. The very cold

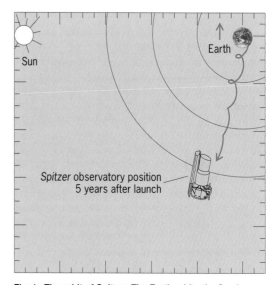

Fig. 1. The orbit of *Spitzer*. The Earth orbits the Sun in a counterclockwise direction in this figure, and the frame of reference rotates with the Earth-Sun line to show how *Spitzer* trails behind. Viewed in a solar frame of reference, the *Spitzer* orbit is a normal Keplerian ellipse. In the Earth-centered frame here, the observatory makes a *loop* or *kink* every year when it partially overtakes the Earth. The observatory will fall farther and farther behind during its 5-year mission. The circles centered on the Earth are 0.2, 0.4, and 0.6 astronomical unit in radius, respectively.

environment away from the Earth allows the outer shell of the satellite to cool passively, greatly reducing the heat load on the telescope. The thermal loads on the helium dewar are also very small. This concept of a telescope launched warm that cools by radiating into space provides the technical foundation for future, larger infrared telescopes such as the *James Webb Space Telescope (JWST)*.

The 85-cm (33.5-in.) telescope and 360-liter (95-gallon) dewar are held off the spacecraft by a thermally insulating truss, and a thermal shield blocks radiation from the spacecraft (**Fig. 2**). The solar panel is cantilevered off the spacecraft, and the telescope is protected from its heat by another thermal shield. The dewar protected the liquid helium from thermal loads while the observatory was on the ground. The instruments are mounted on top of the helium vessel, and a vacuum-tight aperture door sealed out the air when on the ground. Upon reaching orbit, the aperture door was opened to allow light from the telescope to reach the instruments. The outer shell cooled to 34 K (−398°F) during the

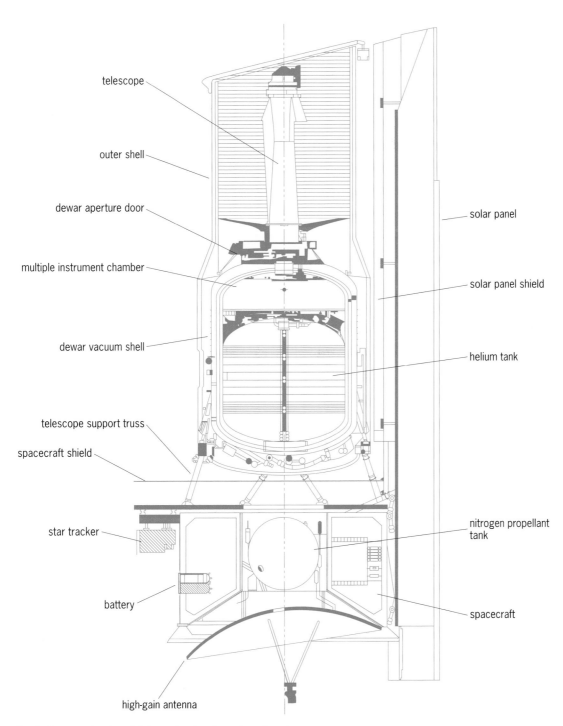

telescope

outer shell

dewar aperture door

multiple instrument chamber

dewar vacuum shell

telescope support truss

spacecraft shield

star tracker

battery

high-gain antenna

solar panel

solar panel shield

helium tank

nitrogen propellant tank

spacecraft

Fig. 2. Cutaway drawing of *Spitzer*, showing the major subsystems.

first month of the mission. Venting the cold helium vapor from the dewar through heat exchangers cools the telescope further, to about 6 K ($-449°$F). The extremely efficient thermal design is expected to yield a 5-year lifetime at these temperatures.

Spitzer has three instruments. The Infrared Array Camera (IRAC) images in bands at wavelengths of 3.6, 4.5, 5.8, and 8 μm. The Infrared Spectrograph (IRS) provides spectra from 5 to 40 μm at low resolution (1–2%) and from 10 to 38 μm at moderate resolution (about 0.15%). The Multiband Imaging Photometer for *Spitzer* (MIPS) provides imaging at 24, 70, and 160 μm. The instruments use advanced infrared detector arrays that operate at or near the fundamental natural background limit in space. The IRAC arrays are 256×256 pixels in format, using indium antimonide (InSb) photodiodes for the two shorter bands and silicon:arsenic (Si:As) impurity band conduction (IBC) devices for the longer ones. IRS uses 128×128 arrays of Si:As and silicon:antimony (Si:Sb) IBC devices. MIPS uses a similar Si:As array at 24 μm and germanium:gallium (Ge:Ga) arrays for the two longer bands. The high performance of these detector arrays allows *Spitzer* to deliver its anticipated breakthrough in science capability, despite its history of descopes. *ISO* had about 50 times more detectors than *IRAS*, and *Spitzer* has about 100 times more than *ISO*. The sensitivity per detector has also improved by an order of magnitude with each new mission.

Objectives. *Spitzer* is operated as a general-access observatory from the Jet Propulsion Laboratory and a science center at the California Institute of Technology. Seventy-five percent of the observing time will be assigned by competitive application from the astronomical community. It is therefore not possible to predict the entire science program. However, the initial objectives include investigations of star formation, planetary systems, and the assembly and evolution of galaxies.

Star formation. Young stars form in dense interstellar molecular clouds, carrying a heavy burden of interstellar dust. Initially, forming stars are cold, tens of kelvins. Even after gravitational contraction has warmed them to more typical stellar temperatures, they are hidden by the surrounding cocoons of interstellar gas and dust and can be seen only in the infrared. *Spitzer* will map many nearby star-forming regions to identify very new stars that are too cold and faint to be known from other types of observation, documenting the very first stages of star formation. It will find luminous massive stars and the outflows of gas that occur as accretion stops on these objects. It will also locate very-low-mass substellar brown dwarfs, revealing the processes by which stellar-mass clumps in molecular clouds can fragment into much smaller pieces before collapsing.

Figure 3*a* shows a star-forming region about 14,000 light-years away that has a classic structure. Within the core of a molecular cloud, clumps of gas have collapsed into massive stars. Winds from these stars have cleared away the remaining gas, punching a hole into the cloud. From the *Spitzer* images, more than 300 newly formed stars have been identified in the central hole and surrounding molecular cloud. The remains of the cloud are heated by the young stars and glow at wavelengths of 4.5, 5.8, and 8 μm in this false-color coded image.

Planetary systems. Embryo planets are believed to form during the first few million years of the life of their star, from a protoplanetary disk made of dusty and gaseous material with too much angular momentum to fall into the star itself. This stage in the planet formation process stops as the disk is dissipated by evaporation of grains of material within it, by accretion of material into the star or the embryo planets, and by ejection of material from the system. Thereafter, planets grow by accretion of material from the embryos that is released when they collide with each other. A by-product of these collisions is circumstellar disks of dust and debris. The dust grains in the circumstellar disk are warmed by the star and glow in the infrared, where *Spitzer* can detect them. Initial *Spitzer* measurements of debris disks show a broad variety of behavior, indicating that some of them are dominated by recent and dramatic collisions. The disk signals damp down after about 150 million years. These observations are reminiscent of theories for the formation of the planets in the solar system, and in particular of the extreme bombardment of the Earth over the first few hundred million years of its existence.

Figure 3*b* and *c* shows Spitzer images of Fomalhaut, a bright star about 25 light-years away and about 200 million years old. The image of the star itself has been removed from the figure panels. Figure 3*b*, at a wavelength of 70 μm, shows the dusty debris from a shattering collision of asteroid-sized embryo planets. Their remnants are orbiting the star in a distorted ring, seen edge-on; the distortion may arise from the gravitational action of a massive planet. At a wavelength of 24 μm (Fig. 3*c*), the ring is filled in with warmer dust that is probably falling into the star.

Assembly and evolution of galaxies. In nearby galaxies, the shortest-wavelength *Spitzer* bands reveal the distribution of the stars that dominate the visible mass, improving understanding of the structure of the galaxies. The longer-wavelength bands glow in the infrared emission of various forms of interstellar dust, from tiny polyaromatic hydrocarbon powder heated to hundreds of degrees from the absorption of a single ultraviolet photon, to large grains at about 20 K ($-424°$F).

Figure 3*d* shows how Spitzer can see into the nearest large elliptical radio galaxy, Centaurus A, about 10 million light-years away. A striking trapezoidal structure glows in the 8-μm band, which traces the distribution of polyaromatic hydrocarbon powder. A computer model indicates that this beautiful structure remains from a spiral galaxy that has been engulfed by the larger elliptical galaxy, a process that may have been responsible for triggering the nuclear activity and radio emission.

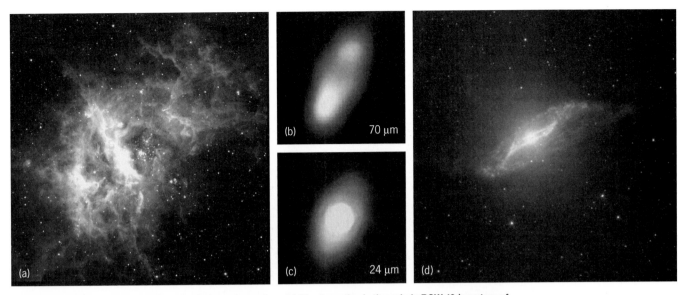

Fig. 3. Illustrative science results from the *Spitzer* observatory. (*a*) Star formation in the nebula RCW 49 (*courtesy of NASA/JPL-Caltech; E. Churchwell, University of Wisconsin*). (*b*) Fomalhaut circumstellar disk at wavelengths of 70 μm and (*c*) 24 μm (*courtesy of NASA/JPL-Caltech; K. Stapelfeldt, JPL*). (*d*) Dusty, elliptical galaxy Centaurus A (*courtesy of NASA/JPL-Caltech; J. Keene, Spitzer Science Center & Caltech*).

In the distant universe, *Spitzer* is locating a population of very luminous, dusty galaxies at redshifts of $z = 1$-4. These galaxies are identified because of their very red colors in the IRAC bands (resulting from the combination of high redshift and absorption by interstellar dust) or their detection in the MIPS bands, along with information on the redshift from spectra, or inferred from the IRAC colors. They represent a stage when smaller galaxies were merging at a high rate to form what have become large elliptical galaxies and spiral galaxy bulges. *Spitzer* is also detecting galaxies and quasars at even higher redshifts and will contribute to knowledge of the most distant objects known in the universe.

For background information *see* BROWN DWARF; GALAXY, EXTERNAL; INFRARED ASTRONOMY; MOLECULAR CLOUD; PLANET; PROTOSTAR; QUASAR; SOLAR SYSTEM; STELLAR EVOLUTION in the McGraw-Hill Encyclopedia of Science & Technology.   George Rieke

Bibliography. J. C. Mather (ed.), *Optical, Infrared, and Millimeter Space Telescopes*, Proc. SPIE, vol. 5487, 2004; *Spitzer Space Telescope* Mission (special issue), *Astrophys. J. Suppl. Ser.*,154(1):1–474, September 2004.

Stereoscopic displays

Stereoscopic displays allow observers to perceive depth effects and enable them to visualize information with ever-increasing complexity in three-dimensional (3D) space. The proliferation of 3D displays has been driven mainly by the tremendous potential of virtual reality and augmented reality (VR/AR) for a wide spectrum of application areas such as scientific visualization, engineering design, training and education, and entertainment. Rapid developments in 3D graphics capabilities on personal computers have further expedited the popularity of stereoscopic display techniques.

Principle of 3D viewing. Human eyes rely on many visual cues, both monocular and binocular, to perceive and interpret depth in the real world. Monocular depth cues are observed with only one eye; common examples include perspective, occlusion, texture gradients, distribution of light and shadows, and motion parallax. Binocular depth perception is based on displacements (that is, binocular disparity) between the projections of a scene object onto the left and right retinas due to eye separation. The binocular disparity is processed by the brain, giving the impression of relief in an effect known as stereopsis. Stereoscopic displays enable depth sensation by exploiting binocular disparity.

Taxonomy of 3D displays. The majority of the existing 3D displays recreate stereoscopic depth sensation by presenting the eyes with 2D image pairs of the same scene generated from two slightly different viewpoints. The key to such displays is a mechanism to present the left and right images to the corresponding eyes without crosstalk. A taxonomy of 3D display techniques is shown in **Fig. 1**, and the displays are classified as either eye-aided or autostereoscopic displays.

The eye-aided displays require a user to wear special goggles that enable proper separation of the stereo images. Such displays can be further categorized into head-attached and spatial displays. Head-attached displays mostly provide separate image elements for each eye and thus are referred to as non-shuttering displays. Spatial displays usually present a stereo pair on the same screen surfaces; thus a multiplexing technique is required to make each image of the stereo pair visible only to one eye. These are known as shuttered displays.

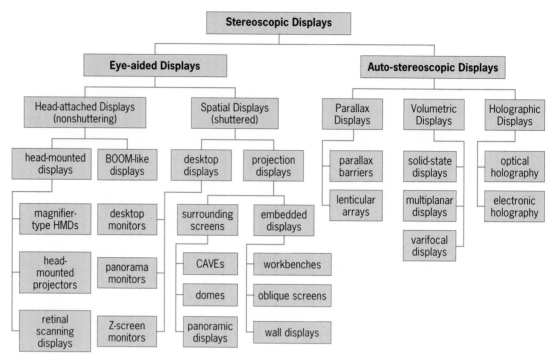

Fig. 1. A taxonomy of stereoscopic displays.

Auto-stereoscopic displays have image-separation techniques integrated into the display units and do not require a user to wear goggles. Such approaches can be divided into parallax displays, volumetric displays, and holographic displays. Parallax displays present stereo pairs simultaneously and deliver multiple views directly to the correct eyes by direction-multiplexed mechanisms. Volumetric displays directly illuminate spatial points within a display volume by filling or sweeping out a volumetric image space. Holographic displays reconstruct light information emitted by a 3D object from interference fringes generated through a holographic recording process.

Head-attached displays. A typical head-attached display consists, for each eye, of an image source, an optical system, a housing unit by which the image source and the optics are attached to the user, and a tracking system to couple a dynamic viewpoint to the user's head and eye motions. These devices are typically worn on the head and are often referred to as head-mounted displays (HMDs). HMD designs may be further classified as immersive or see-through: The former refers to designs that block the direct real-world view; the latter refers to designs that allow superposition of synthetic images onto the real world. Alternatively, some head-attached displays are floor- or ceiling-mounted; an observer uses these devices by holding a handle rather than directly wearing heavy displays. One such example is a BOOM-like (binocular omni-orientation monitor) system, which offers stereoscopic capability on a counterbalanced, motion-tracking floor-support structure for practically weightless viewing.

The inherent portability of HMDs finds its application in wearable computing and outdoor visualiza-tion, which in turn demand brighter displays with improved portability. The image sources in HMDs mostly rely upon the advancement of microdisplays. The emergence of novel microdisplays, such as LCOS (liquid crystal on silicon), OLEDs (organic light-emitting diodes), and TMOS (time multiplex optical shutter), offer potentially brighter imaging at higher resolution. Alternatively, the retinal scanning display replaces the microdisplay image source with a modulated, low-power laser beam and associated scanning systems to write individual pixels directly onto the retina of the human eye.

The basic forms of HMD optical design are eyepiece and objective-eyepiece combination magnifiers. One challenge in HMD design is the trade-off between display resolution and field of view. Several approaches—for instance, physical or optical tiling of multiple displays and fovea-contingent high-resolution inset schemes—have been researched to pursue high-quality designs. Another recent advancement is the replacement of eyepiece optics with projection optics accompanied by retro-reflective screens, leading to the miniaturization of the HMD optics as well as a wider field of view than traditional HMD designs. Overall advances in HMD optical design capitalize on more readily available emerging technologies such as plastic lenses, aspheric surfaces, and diffractive optical elements.

Spatial displays. A stereo image pair is generally presented on the same screen surface for spatial displays. The major display elements are installed within the environment and thus are not physically coupled with the user's head movement. Accompanied by appropriate shuttering techniques, stereo images are either displayed sequentially at a doubled frame rate (field-sequential schemes) or concurrently at a regular frame rate (field-concurrent schemes). When a field-sequential scheme is used, to avoid image flickering and ghosting effects, the image source must have a high refresh rate and each image must decay completely before the next field is displayed.

The major shuttering techniques include color filters, polarization filters, and liquid crystal display (LCD) shutter glasses. In color-multiplexed displays, the left- and right-eye images are displayed as monochromatic pairs, typically in near-complementary colors (for example, red-green or blue-red), and observers wear corresponding color-filtering glasses. In polarization-multiplexed displays, stereo images are polarized in perpendicular directions before they are projected onto a screen surface, and observers wear polarizer glasses with polarization directions in harmony with that of the polarized images. With LCD shutter glasses, the left and right images are actively synchronized with the on-off status of the shutters. The color and polarization filters are known as passive shuttering and can be used in both field-sequential and field-concurrent modes, while the LCD shutters are known as active shuttering and can be used only in sequential mode.

Desktop displays. Using desktop monitors, often with field-sequential shutters, is the traditional "fish tank" VR approach. Multiple monitors can be tiled together to create a panoramic display system. Mirrors or beam splitters can be employed with desktop display configurations to overlay 3D graphics with a physical workspace or merge views from multiple monitors. The desktop stereo systems are nonimmersive, and their applications are limited to near-field operations and personal usage.

Projection displays. The mainstream spatial displays extensively employ video projectors to cast stereo images onto single or multiple, planar or curved screen surfaces. Metallic screen surfaces that do not depolarize incoming light are required for systems using polarization filters, as every organic material would reverse or depolarize the polarization direction and consequently would fail to separate stereo pairs. Since the introduction of CAVE Automated Virtual Environments, many types of projection displays have been developed, including highly immersive surrounding displays and various embedded display systems. Surrounding displays, such as CAVEs, CUBEs, domes, and panoramic displays, are featured with multiple planar or single curved screen surfaces to encapsulate multiple users in an immersive virtual environment (VE). Embedded display systems, such as workbench and wall displays, integrate a single or a small number of screens to create a semi-immersive VE that is embedded within the surrounding real world.

Auto-stereoscopic displays. Unlike eye-aided displays, auto-stereoscopic displays send stereo pairs directly to the correct eyes, light spatial points within a display volume, or reconstruct light information emitted from a 3D object.

Parallax displays. For a parallax display, a 2D base display is overlaid with an array of elements that direct the emitted light from a screen pixel to only the correct eye. LCD pixels typically have high positional accuracy and stability; thus LCD panels are usually the primary choice for base displays. The pixels of a base display are divided into two or multiple groups, one group per viewpoint. The array of light-directing elements generates a set of viewing windows through which stereo images are observed by the corresponding eyes. Typical examples of light-directing elements include the parallax barrier and the lenticular lens array. Parallax barrier is the simplest approach to light directing, and its principle is illustrated in **Fig. 2***a*. The left and right images are interlaced in columns on the base display, and the parallax barrier is positioned so that the left and right image pixels are blocked except in the region of the left and right viewing windows. Lenticular sheet displays, in Fig. 2*b*, apply an array of optical elements such as cylindrical lenses that are arranged vertically relative to a 2D base display. The cylindrical lenses direct the diffuse light from a pixel so that it can be seen only in a limited angle; thus they allow different pixels to be directed to a limited number of defined viewing windows.

Volumetric displays. Instead of presenting two separate 2D images, volumetric displays directly illuminate spatial points within a display volume by filling or sweeping out a volumetric image space, and they appear to create a transparent volume in physical space. Many of the volumetric display technologies impose minimal restriction on the viewing angle, and thus an observer can move around and view 3D content from practically arbitrary orientation. Examples of volumetric displays include (1) solid-state devices that display voxel data within a transparent substrate by generating light points with an external source (for example, by using two intersecting infrared laser beams with different wavelength to excite the electrons to a higher energy level and thus emit visible light); (2) multiplanar volumetric displays, which build up a 3D volume from a time-multiplexed series of 2D images via a swiftly moving or spinning display element; and (3) varifocal mirror displays, which apply flexible mirrors to sweep an image of a CRT screen through different depth planes of an image volume.

Holographic displays. Taking a fundamentally different approach, holographic displays reconstruct light emitted by a 3D object from interference fringes generated through a holographic recording process. The interference fringes, when appropriately illuminated, function as a complex diffractive grating that reconstructs both the direction and intensity of light reflected off the original object. However, an optical hologram cannot be produced in real time and thus is not appropriate for dynamic displaying. An electronic holographic display computes 3D holographic images from a 3D scene description, and can potentially lead to real-time electronic holography, known as holovideo. It involves two main processes:

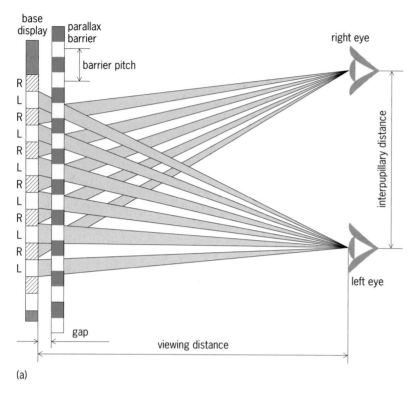

(a)

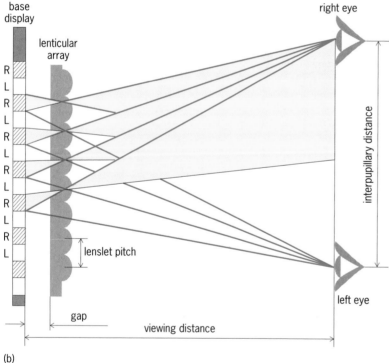

(b)

Fig. 2. Parallax displays. (a) Principle of parallax barrier displays. (b) Principle of lenticular array displays.

have led to the computation of fairly complex scene contents at interactive rates.

Conclusion. Stereoscopic displays are intriguing subjects of research owing to their tremendous potential for applications. However, none of the existing technologies yet match the powerful capabilities of the human visual system in any aspect. For instance, the majority of the existing stereoscopic displays decouple the physiological actions of accommodation and convergence (by which the eyes adjust to changes in distance to the object), few of them offer resolvability comparable to foveal visual acuity, and few are capable of presenting natural occlusion cues cohesively and correctly. Nevertheless, it is exciting to anticipate potential uses of emerging display technologies. Stereoscopic display systems are reshaping how we explore science, conduct business, and advance technology, ultimately improving the ways we live and work.

For background information *see* ELECTRONIC DISPLAY; EYE (VERTERATE); HOLOGRAPHY; STEREOSCOPY; VIRTUAL REALITY; VISION in the McGraw-Hill Encyclopedia of Science & Technology.   Hong Hua

Bibliography. B. Blundell and A. Schwarz, *Volumetric Three Dimensional Display Systems*, Wiley, New York, 2000; C. Cruz-Neira, D. J. Sandin, and T. A. DeFanti, Surround-screen projection-based virtual reality: The design and implementation of the CAVE, *Computer Graphics (Proc. SIGGRAPH 1993)*, pp. 135–142, 1993; M. Halle, Autostereoscopic displays and computer graphics, *Computer Graphics (Proc. SIGGRAPH '97)*, 31(2):58–62, 1997; J. E. Melzer and K. Moffitt (eds.), *Head Mounted Displays: Designing for the User*, McGraw-Hill, New York, 1997.

Submarine hydrodynamics

Submarines have been around for about 400 years. Their stealth plays an important role in a modern naval force. However, the lack of a significant civilian or industrial requirement for this mode of transportation has restricted the scope of its development and left the technology largely shrouded in the military. Compared with airplane aerodynamics, for example, submarine hydrodynamics receives little attention in the open scientific literature.

From a fluid dynamics perspective, submarines are more like dirigibles than airplanes. Both submarines and dirigibles maintain a balance between static buoyancy and weight forces for stability, whereas airplanes rely entirely on dynamic forces to overcome their weight and maintain stability.

Hydrostatic stability in a submerged vehicle is achieved by keeping its center of buoyancy (CB; the volume centroid of an envelope defined by the outer surface of the vehicle) above its center of gravity (CG; the center of the mass contained within the buoyancy envelope). The pressure of the water on the outside of the envelope generates a buoyancy force that always acts vertically upward through the CB, and is generally constant. The weight within the envelope

fringe computation, in which the 3D description is converted into digital holographic fringes, and optical modulation, in which light is modulated by the fringes and output as 3D images. The grand challenge lies in the enormous amount of computation required by holography. Various experimental methods, such as horizontal-parallax-only, holographic bandwidth compression, and faster digital hardware, always acts vertically downward through the CG. The vehicle is neutrally buoyant when the buoyancy and weight are equal. When the CB and CG are not vertically aligned, a static moment results which acts to realign the centers. Hydrodynamic forces aside, the equilibrium is attained when the CG hangs directly below the CB. Operators cannot change the submarine's buoyancy because they cannot change the shape or size of its exterior, but they can change the boat's weight. Water and other onboard fluids are routinely pumped between trim tanks to move the CG around to keep the boat level. Onboard compressed air is used to evacuate water (weight) from the ballast tanks to allow the buoyancy, which has not changed, to surface the vessel.

All submarine hydrodynamics issues are superimposed on a boat's inherent hydrostatic stability. This results in added complexity because the hydrostatic forces are constant, while the hydrodynamic forces vary as the square of the speed of the boat. Therefore, handling characteristics are speed-dependent when hydrostatic forces are involved. This occurs during rolling and pitching motions when the vertical orientation of the CG-CB line changes.

In the early days of submarine development, submerged speeds were low and hydrodynamics was a performance rather than a safety issue. The development of high-speed, nuclear-powered boats has brought safety to the forefront and led to improved tools and analysis methods to deal with hydrodynamic stability, predictability, and recoverability. Modern nuclear boats have submerged speeds around 30 knots (15 m/s), lengths from 70 to 170 m (230 to 560 ft), and displacements from 3000 to 20,000 tons. Modern diesel-electric submarines are also faster, requiring that more attention be paid to the impact of hydrodynamics on safety. Modern diesel boats have submerged speeds around 20 knots (10 m/s), lengths from 50 to 70 m (160 to 230 ft), and displacements from 500 to 3000 tons.

To investigate submarine safety and determine optimal operational limits, designers and operators use computers to simulate maneuvering submarines. Realistic simulations require accurate hydrodynamic modeling.

Hydrodynamics issues. Unlike airplanes or dirigibles, submarines operate in a layer of water only about three hull lengths thick, so depth control is very important. The "sail" on a submarine (otherwise known as its conning tower or fairwater) introduces a hydrodynamic asymmetry that makes depth control an issue during simple horizontal plane turns. **Figure 1** shows what happens using a simplified, uniform drift angle. (Real turns require a rotation of the submarine about a vertical axis, and this makes the drift angle vary along the length.) The submarine is experiencing an angle of drift β caused by a turn to starboard. This generates a side force on the sail, just as an angle of attack generates lift on an airplane wing. Associated with this force is the horseshoe vortex shown trailing from the sail. The hull-bound leg of the vortex causes circulation around the

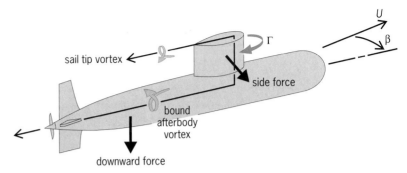

Fig. 1. A submarine at an angle of drift β generates circulation Γ that results in a side force on the sail and a downward force on the hull. U = submarine velocity.

hull, which interacts with the crossflow generating a downward force on the afterbody. This phenomenon is known as the Magnus effect. It does not matter whether the turn is to port or starboard; a downward

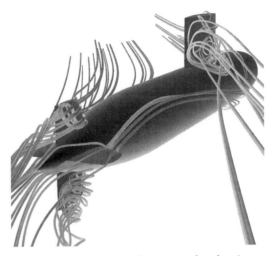

Fig. 2. CFD prediction of the flow over a submarine at a uniform angle of drift of 30°. (*Courtesy of ANSYS Canada Ltd.*)

Fig. 3. Flow visualization of the leeside of the rudder at a high angle of drift. (*With thanks to the Institute for Aerospace Research, Ottawa*)

force on the afterbody always results, and the helmsman must compensate with appropriate depth control.

To complicate matters further, whenever the sail experiences side force, the boat rolls until the hydrodynamic moment on the sail is balanced by the hydrostatic righting moment from the vertical misalignment of the CG and CB. Submarines do not have roll control to compensate for this, as do airplanes.

Predicting forces on maneuvering submarines. Submarine hydrodynamicists have three basic methods for predicting the forces used in design and simulation.

Analytical/semiempirical methods. These are the easiest, most general, fastest, and cheapest tools to use. There is an extensive body of aerodynamics literature, including slender-body and thin-airfoil theory, which is directly applicable to submarines. These flow prediction methods need to be augmented with semiempirical corrections that account for viscous effects, especially for predicting the dominant hull forces. These methods are often the only affordable means of analyzing the hydrodynamics of industrial underwater vehicles, usually crewless robots.

Experimental methods. This is the most reliable tool if properly used, but also the most time-consuming and expensive. Large models and test facilities are required to ensure realism. Captive-model testing provides a direct measure of the forces and therefore the best results, for the cases that can be tested. Captive-model facilities include wind tunnels and towing tanks (translational motion) and rotating-arm facilities (yaw and pitch rotations). Free-swimming models are also used. These have the advantage of 6-degree-of-freedom motion but do not provide a direct measure of the hydrodynamic forces.

Computational fluid dynamics (CFD). This is an increasingly practical tool as computational power increases and becomes more affordable. There are many CFD methods, but at present the most practical for predicting the general viscous flow around a vehicle is the Reynolds-averaged Navier-Stokes (RANS) solver. The advantages are low cost, relative to experiments, and the fact that rotation is modeled as easily as translation. In addition, a general, unsteady, 6-degree-of-freedom simulation capability is limited only by the available computational power, a limitation that is continually receding. Disadvantages of the method are the necessity for semiempirical approximations to model turbulence, and the dependency of the method on a complicated spatial grid that enables a numerical solution of the complex governing equations. As a result, engineers still require some degree of experimental validation for new or complex flow situations.

The prudent hydrodynamicist maintains a capability in all three of the above methods. The methods complement each other; together they are more than the sum of what they can do individually.

Validating CFD simulations with experiments. It is always desirable to validate a CFD calculation against experimental data when possible.

Figure 2 shows a CFD prediction (CFX-TASCflow) of the flow around a generic submarine shape at a high drift angle, where complex vortices flowing from the sail and tailplanes (fins) are clearly visible. Flow visualization from a wind-tunnel test is shown in **Fig. 3**, where pigmented oil (painted on the model tail prior to running the tunnel) shows the direction of the local surface flow. This provides a good qualitative indication that the CFD prediction of tailplane stall is realistic. However, some aspects of the flow,

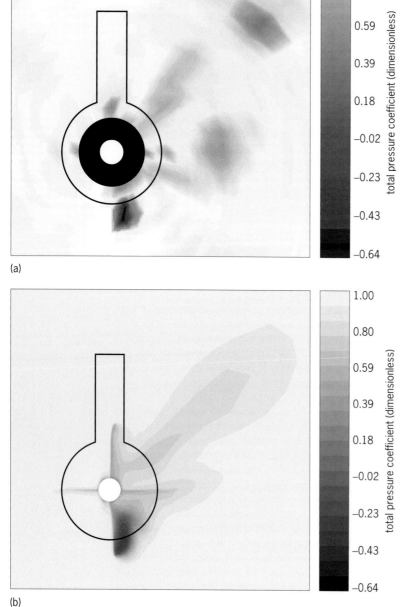

(a)

(b)

Fig. 4. Comparison of (*a*) experimental measurements and (*b*) CFD predictions of the trailing vortex field on a submarine model at an angle of drift $\beta = 20°$. **(*Part a with thanks to the Institute for Aerospace Research, Ottawa; part b courtesy of ANSYS Canada Ltd.*)**

such as the vortex trailing from the sail in Fig. 2, are not well predicted.

Figure 4 shows experimentally measured and predicted total pressure (a measure of energy) in a flow similar to Fig. 2. The view is looking forward along the hull axis at a transverse plane, just aft of the tailplanes. The upstream hull and sail profiles are drawn in to provide scale. The small white disk in the middle of the hull profile is the local hull diameter in this transverse plane. Energy maps such as these are a good wayk to locate vorticity, which characteristically exhibits high energy loss.

The prominent features in Fig. 4a are the vortex trailing from the sail tip, seen at 1 to 2 o'clock relative to the hull, and the hull afterbody vortex, which has separated from the hull and is at 3 o'clock, slightly inboard of the sail-tip vortex. It is necessary to model this latter vortex correctly to predict the downward force on the afterbody of a turning submarine (Fig. 1). [The feature at 6 o'clock results from a stalled lower rudder. The black annulus around the hull is a region where measurements were not made.]

Compared to experimental results, CFD does not model the vorticity in the outer flow well at all (Fig. 4b), nor does it predict the forces for this high-flow-incidence case very well. The main reasons are the poor grid resolution in the outer regions of the flow and the turbulence model, which is known to influence separation.

Promise of CFD. Despite its shortcomings, CFD does have promise as a tool for augmenting current knowledge. When a CFD calculation has been well validated, it becomes an economic tool for optimization, that is, for assessing the effect of minor changes in the flow and geometry on some relevant characteristic. Also, CFD predictions readily provide flow details which are expensive or impossible to extract experimentally.

Unsteady flows and rotation are also hard to model experimentally, but are straightforward to implement on CFD. Confidence in the CFD predictions is enhanced by a uniform flow validation at similar incidence angles.

Scale effects are another area where CFD may provide solutions that experiment cannot. Model testing is necessarily done at less than at full scale, resulting in viscous effects that are too large. CFD should be able to reliably correct for this, provided the necessary extra computational power and memory (finer grids are needed) are available.

Putting CFD to use. Two current projects are worth mentioning.

Experiments were conducted several years ago to measure the effectiveness of submarine sternplanes, to improve our ability to economically estimate this characteristic. However, the steady-state force measurements contain errors generated by the structure supporting the model in the wind tunnel. CFD can be reliably used to correct these errors, a good example of how different methods can be used to augment each other.

A more ambitious task is the development of a 6-degree-of-freedom unsteady CFD submarine simulation capability. This involves coupling the CFD solver to the differential equations governing the motion of the submarine, and simultaneously solving the RANS and motion equations. This work supports an ongoing effort to understand a roll instability that occurs in many submarines during buoyant ascent.

There are many other new and improved CFD technologies being used and researched. While RANS is the most practical one now, in 10 years that may no longer be true. Indeed, RANS is likely just a stepping-stone on the road to reliable CFD predictions. Submarine hydrodynamicists must stay abreast of this developing technology to help ensure that submarines and their crews are as safe as possible.

For background information *see* BUOYANCY; CENTER OF GRAVITY; COMPUTATIONAL FLUID DYNAMICS; HYDRODYNAMICS; HYDROSTATICS; NAVIER-STOKES EQUATIONS; SUBMARINE; TOWING TANK; VORTICITY; WIND TUNNEL in the McGraw-Hill Encyclopedia of Science & Technology.   George D. Watt

Bibliography. M. Mackay, H. J. Bohlmann, and G. D. Watt, Modelling submarine tailplane efficiency, *NATO RTO SCI-120 Meeting Proceedings, Berlin*, Pap. 2, May 2002 [SCI = Systems Concepts and Integration panel]; G. D. Watt and H. J. Bohlmann, Submarine rising stability: Quasi-steady theory and unsteady effects, *25th Symposium on* Naval Hydrodynamics, St. John's, August 2004; G. D. Watt and K. J. Knill, A preliminary validation of the RANS Code TASCflow against model scale submarine hydrodynamic data, CFD99, *7th Annual Conference of the CFD Society of Canada*, May 1999.

Superconducting motors

Electric motors consume 64% of the electric power generated in the United State, with about half of this amount used by large motors (greater than 100 hp or 75 kW power). The efficiency of the average motor is 92%, and that of large motors about 96%. Conventional electric motors have reached their optimum in terms of power density and efficiency, and only the application of new materials can lead to further improvements. Superconductors offer the ability to carry very high direct electric current density without losses, and rotating apparatus (motors and generators) represent some of the most promising applications for superconductivity.

Superconducting motors first appeared around 1965, more than 50 years after the discovery of superconductivity. These motors had to operate at 4.2 K ($-452°$F) and thus were cooled with liquid helium. For reasons of technology and cost, these low-temperature superconductor motors were considered only for applications requiring greater than 1 MW power. In 1986, the discovery of high-temperature superconductivity (HTS) in copper-oxide ceramics allowed consideration of superconducting motors with operating temperatures in the

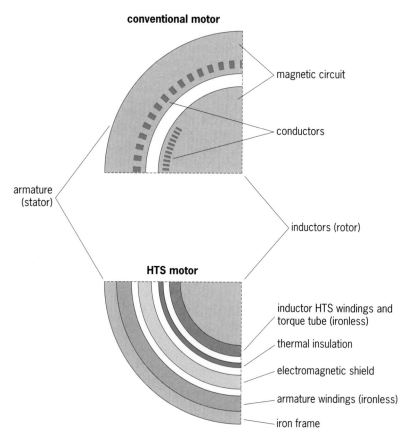

conventional motor

magnetic circuit

conductors

armature
(stator)

inductors (rotor)

HTS motor

inductor HTS windings and
torque tube (ironless)

thermal insulation

electromagnetic shield

armature windings (ironless)

iron frame

Fig. 1. Comparison between a conventional synchronous motor and an HTS synchronous motor.

perconducting motor is virtually an "air core" where the stator is constructed without the iron teeth of a conventional motor. This type of construction eliminates core saturation and iron losses.

Figure 1 shows a comparison between the construction of a conventional motor and an HTS motor. In the HTS motor, the rotor consists of field coils made with HTS wires fed with direct current (dc). The dc inductor windings typically operate at a temperature of 25–40 K (−415 to −388°F), with a dc flux density of up to 4 tesla. Although high-temperature superconducting wires can carry direct current with almost no losses, transient loads on the motor can lead to variations of magnetic field on the coils. HTS wires are sensitive to such field variations, and the resulting losses can lead to a decrease in current density and a loss of the superconducting state. In order to prevent this, an electromagnetic shield is placed in the air gap, usually as part of the rotor assembly. This shield can be a bulk metallic hollowed cylinder or a squirrel cage.

Superconducting wires currently available are too sensitive to flux variation to be used in the armature. Flux variation results in ac losses raising the temperature in the superconductor and thereby limiting the operating current. Therefore, conventional copper wires are used in the armature.

The synchronous motor configuration represents the most feasible topology for superconducting motors; however, many unconventional HTS motor configurations based on the unique properties of superconductors have been developed in laboratories worldwide.

Cooling method. Superconductors need to operate at a temperature around one-third of their critical temperature (the temperature at which they lose superconductivity). For most HTS conductors the current density that can be carried without loss increases when the temperature decreases. High-temperature superconductors therefore show very high

range of 30–80 K (−406 to −316°F). Although HTS motors would still require a cryogenic cooling system, the temperature could be easily achieved using cryocoolers or cheap cryogenic fluids such as liquid nitrogen.

Superconducting motors based on high-temperature superconductors can reduce the losses of equivalent conventional motors by half, resulting in an efficiency exceeding 98%, and can achieve a size less than half that of a conventional motor with one-third the weight. They also have significantly higher efficiency over a wider range of operating speed and are inherently quieter than conventional motors. Moreover, HTS motors exhibit outstanding dynamic performance and stability due to their small synchronous reactance (15 times smaller than conventional motors).

Principle of operation. Motors consist of two main parts, a rotor and a stator. The interaction between the magnetic field provided by the rotor and the alternating currents (ac) flowing in the windings result in electromagnetic torque. Conventional motors use copper windings and an iron core to increase the magnitude of the air gap flux density created by copper windings. Iron has a nonlinear magnetic behavior and saturates at a flux density of 2 tesla, thus limiting the electromagnetic torque. In HTS motors, the copper rotor windings can be replaced with superconducting windings with a resulting current density 10 times greater without any resistive losses. The su-

High-temperature superconducting motor projects		
Maker	Power and speed	Type
Kyoto University (Korea)	100 hp (75 kW), 1800 rpm	Synchronous, cooled by solid nitrogen
Reliance Electric–AMSC	200 hp (150 kW), 1800 rpm	Synchronous, cooled by gas helium
Siemens	535 hp (400 kW), 1500 rpm	Synchronous, cooled by liquid neon
Rockwell-AMSC	1000 hp (750 kW), 1800 rpm	Synchronous, cooled with liquid helium
General Atomics	5330 hp (4 MW)	Homopolar, cooled by gas helium
AMSC	5000 hp (3.75 MW), 1800 rpm	Synchronous, cooled by gas helium
AMSC-ALSTOM	6700 hp (5 MW), 230 rpm	Synchronous, cooled by gas helium
GE (Generator)	100 MVA	Synchronous

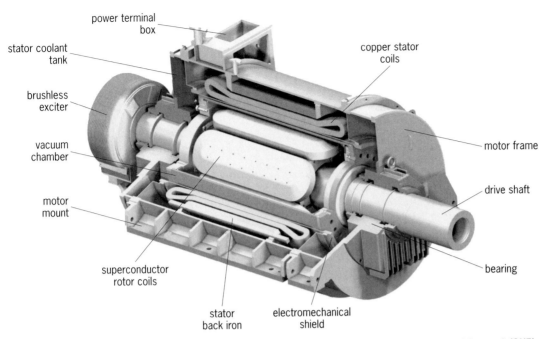

Fig. 2. Cutaway view of 5-MW HTS motor built by American Superconductor (AMSC) for the Office of Naval Research (ONR). It is the most powerful HTS motor ever built.

performance in the range 30–50 K (-406 to $-370\,°$F), which enables the use of cryogenic gases such as neon, helium, and hydrogen. These temperatures can be achieved with a cryocooler that provides a cold head at the desired temperature. The cooling system usually consists of a closed loop of gas and requires only electric power and chilled water to run.

Fig. 3. AMSC/ONR 5-MW HTS motor.

Design challenges. Current research on HTS motors deals with optimization in terms of power density. The challenge is to develop HTS motors that are as compact as possible even for low power levels. HTS motors need to be designed as a whole system, including the electromechanical conversion part and the cooling system. HTS power device performances depend on the superconductor properties. Because the ac losses in currently available HTS conductors are too high, only the rotor windings of the HTS motors are superconducting. The development of ac wires that could be used in the armature will enable the design of very high power density, fully superconducting motors. In order to achieve a high level of compactness, cryocoolers also need to be optimized.

HTS motor development. Many research programs exist around the world. For large motors, these include 5-MW and 36.5-MW ship propulsion motors; a 100-MVA HTS superconducting generator rotor; and a 4-MW generator. These actual machines use HTS wires made of BiSrCaCuO on a silver matrix.

Most HTS motors are synchronous motors. An HTS homopolar motor rated at 4 MW is under development for the U.S. Navy. The **table** presents some of the recent and existing HTS motors projects.

Figures 2 and **3** show an actual 5-MW HTS motor designed and built by American Superconductor (AMSC) and ALSTOM for the U.S. Navy's Office of Naval Research. This 230-rpm HTS ship propulsion motor is a model for larger motors to be used in ship propulsion. The motor has completed factory acceptance tests and is being prepared for extensive testing at the Center for Advanced Power Systems at Florida State University, Tallahassee. The tests will include full power test at 5 MW and dynamic long-term testing emulating ship propeller action to determine how an HTS motor performs under realistic service conditions onboard a ship.

AMSC is presently designing a 36.5-MW (50,000-hp) HTS motor, also for the U.S. Navy, as a full-scale propulsion motor for use on surface ships.

Potential applications. Applications that could benefit from HTS motor development include those where high efficiency, compactness, and outstanding dynamic performances are required. Prime examples would be large, continuously operated pumps and fans for utilities and industrial systems, and propulsion motors for ships, either as in-hull propulsion motors or as pod-mounted motors. The latter application has the potential to reduce ship weight and improve fuel consumption.

High-temperature superconducting motors represent one of the most promising applications of superconductors. The next revolution in superconducting power devices will come with the availability of long lengths of coated conductors (second-generation conductors) usable at 77 K ($-321°$F) and made of YBCO ($YBa_2Cu_3O_{7-x}$).

For background information *see* CRYOGENICS; ELECTRIC ROTATING MACHINERY; MOTOR; SUPERCONDUCTING DEVICES; SUPERCONDUCTIVITY in the McGraw-Hill Encyclopedia of Science & Technology.   Philippe J. Masson; Steinar J. Dale

Bibliography. D. Aized et al., Status of the 1000 hp HTS motor development, *IEEE Trans. ASC*, vol. 9, no. 2, June 1999; J. R. Bumby, *Superconducting Rotating Electrical Machines*, Monographs in Electrical and Electronic Engineering, Oxford Science, 1983; R. Jha, *Superconductor Technology, Applications to Microwave, Electro-optics, Electrical Machines and Propulsion Systems*, Wiley-Interscience, New York, 1998; S. S. Kalsi, *Development Status of Superconducting Rotating Machines*, IEEE, 0-7803-7322-7/02, 2002; W. Nick et al., 380 kW synchronous machine with HTS rotor windings—Development at Siemens and first test results, *Physica C*, vol. 372–376, pp. 1506–1512, 2002; B. Seeber, *Handbook of Applied Superconductivity*, Institute of Physics Publishing, Bristol and Philadelphia, 1998; P. Tixador, Superconducting electrical motors, *Int. J. Refrig.*, 22:150–157, 1999.

Superheavy elements

The nuclear shell model is one of the fundamental theoretical models used to understand the internal structure of the nucleus, which is defined by the arrangement of neutrons and protons within the nucleus. This model, with proper inclusion of spin-orbit coupling, pairing, and deformation parameters, successfully describes many aspects of the vast majority of nuclei that are currently known, such as the spins and parities of ground states, expected configurations of excited states, and the enhanced stability or "magicity" of certain combinations of neutrons or protons. The nuclear shell model currently explains that those nuclides with proton number Z or neutron number N equal to 2, 8, 20, 28, 50, or 82, or neutrons with $N = 126$, are more stable than nuclei with even just one more or less nucleon. This is analogous to the enhanced chemical stability exhibited by the noble gas elements, and the periodicity exhibited by electrons filling various atomic orbitals. In the nuclear shell model, the neutrons and protons in the nucleus also fill orbitals.

Furthermore, this model can be used to predict the properties of unknown elements and nuclides. A key core extrapolation of the shell model is the prediction of the next doubly magic nucleus. In doubly magic nuclei, both neutrons and protons completely fill orbitals, like ^{208}Pb with $Z = 82$ and $N = 126$, the heaviest known doubly magic nuclide. Indeed, since the mid-1960s, the existence of a region of spherical, long-lived nuclei with extra stability near $Z = 114$ and $N = 184$ has been predicted. Current predictions are fuzzier on the next proton magic number, with $Z = 114$, 120, or even 126, predicted due to the sparse, relatively evenly spaced states in this region of nuclear levels. This region has been called variously the island of stability or magic island, and is indicated for one set of model parameters as the dark blue region near $Z = 114$ and $N = 184$ in **Fig. 1**.

Search motivations. This prediction of long-lived nuclides on the island of stability initiated intensive searches for these elements in many different kinds of experiments. The chance to critically test such a fundamental nuclear physics theory as the shell model, together with the prospects of discovering new superheavy elements with unmeasured nuclear properties and with unknown but certainly fascinating chemistry, has captured the imagination of many scientists working in this field. For the chemists, adding elements to the periodic table and investigating the novel chemical properties of these elements as a result of relativistic effects or involvement of the $5g$ orbitals in the valence electrons are compelling reasons for attempting to produce superheavy elements. (The relativistic effects arise from electrons in the innermost atomic orbitals having velocities that are a significant fraction of the speed of light, resulting in an increase in their mass and a change in the orbital location. This change affects the position of the valence orbitals and hence the chemical properties exhibited by the element.) For physicists, the testing of basic nuclear structure models and the investigation of new long-lived and heavy isotopes offer the prospects of producing and studying the heaviest matter synthesizable. When americium was produced in 1945, the practical application of its use in smoke detectors was unimaginable. Thus scientists are particularly attracted to the investigation of heavy elements by prospects of discovering a new element with new chemical properties that provides a novel technology for humankind.

Production of new elements. A variety of experimental techniques have been used to make new chemical elements, including heavy-ion transfer reactions, cold- or hot-fusion evaporation reactions, neutron captures, and light-ion charged particle-induced reactions. In a heavy-ion transfer reaction, typically, a heavy nucleus hits a target nucleus and transfers one or more (generally a few) nucleons (protons or neutrons). A light-ion charged particle-induced reaction involves a light nucleus (like a deuteron, proton, or ^{4}He nucleus) hitting a target nucleus and transferring a particle with charge such as a proton or alpha particle. Neutron capture reactions involve a neutron hitting a target nucleus and being absorbed into that nucleus. All of these techniques involve transmuting one kind of nucleus into another. Two new elements, einsteinium (Es) and fermium (Fm), with $Z = 99$ and 100, respectively, were discovered in the debris of an atmospheric nuclear explosion. This entailed rapid capture of many neutrons by uranium, followed subsequently by up to about 20 beta decays of the extremely neutron-rich uranium isotope and its daughters to more stable elements, a process similar to r-process nucleosynthesis in some types of stars. Each of these techniques has advantages and disadvantages making them suitable for studying nuclei in certain regions.

Cold- and hot-fusion reactions. Modern "nuclear alchemists" working on the synthesis of superheavy elements "transmute" one element into another using particle accelerators and smashing a beam of one element into a target of another element to produce the desired element. The types of nuclear reactions that have been successfully used to produce new elements in the last decade or so are cold-fusion reactions and hot-fusion reactions.

Cold-fusion reactions use more symmetric beam and target nuclei, produce a compound nucleus with generally lower excitation energy that typically requires evaporation of one or no neutrons, and generate less neutron-rich isotopes of an element. They have higher survival probabilities with respect to fission but have lower fusion probabilities. An example of this type of reaction is ^{70}Zn + ^{208}Pb → ^{277}112 + $1n$ with a cross-section of about 1 picobarn (10^{-36} cm^2). Because the 112 isotope ultimately decays by alpha emission to known nuclei, namely, isotopes of elements 102 (nobelium) and 104 (rutherfordium), identification of this element is straightforward.

Hot-fusion reactions use more asymmetric beam and target nuclei, produce a compound nucleus with generally higher excitation energy that typically requires evaporation of three to five neutrons, and generate more neutron-rich isotopes of an element. They have lower survival probabilities with respect to fission but have higher fusion probabilities. An example of this type of reaction is ^{48}Ca + ^{244}Pu → ^{288}114 + $4n$, also with a cross section of about 1 pb. Because of the neutron richness of this isotope of element 114, it never subsequently decays to any known isotope, and thus its identification is more problematic.

Both of these types of reactions utilize doubly magic nuclei, as either target or projectile to attempt to increase the stability of the compound nucleus. Cold-fusion reactions have been successful in producing elements 104–112, and hot-fusion reactions have recently provided evidence for elements 113–116 and 118 (Fig. 1).

Separation and detection techniques. Often the produced isotope decays rapidly and is produced inefficiently, so sophisticated separation and detection techniques are required for studying the reaction products. Typical experiments involve the use of post-target separators to remove the unused beam particles and unwanted reaction products from other types of reactions occurring simultaneously with the fusion-evaporation reaction of interest. Position-sensitive solid-state semiconductor detectors measure the energy and decay mode of the product of interest, called the evaporation residue.

Several types of separators have been used, including velocity filters such as SHIP at the Gesellschaft für Schwerionenforschung (GSI) in Darmstadt, Germany, and VASSILISSA at the Joint Institute for Nuclear Research (JINR) in Dubna, Russia, and gas-filled separators such as the Berkeley Gas-filled Separator (BGS) at Lawrence Berkeley National Laboratory

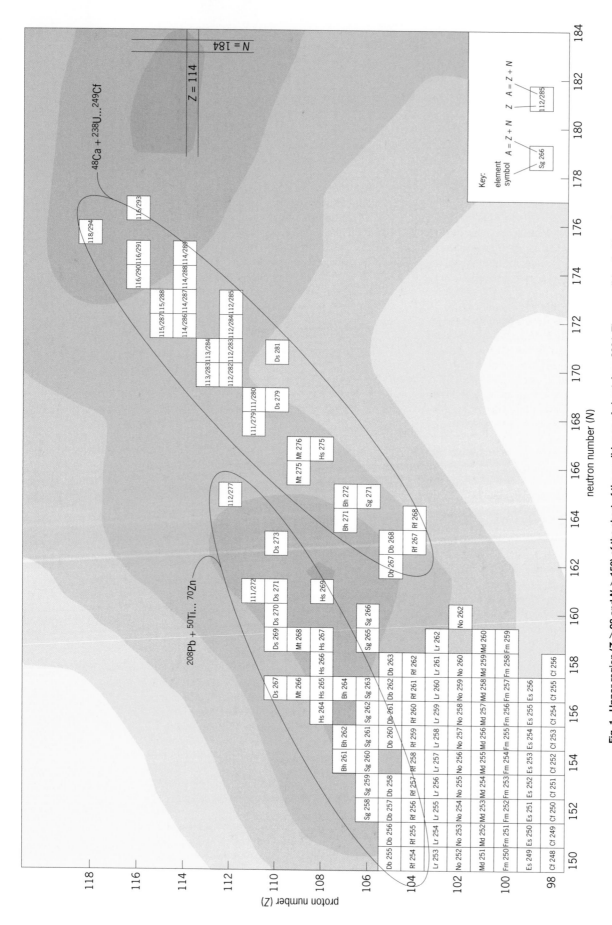

Fig. 1. Upper region ($Z > 98$ and $N > 150$) of the chart of the nuclides as of about June 2004. Those nuclides that have been synthesized by cold-fusion reactions (less neutron-rich) are indicated in the circle toward the upper left in the figure, and those that have been synthesized by hot-fusion reactions (more neutron-rich) are shown in the circle toward the upper right. The background color shades indicate strengths of the shell correction terms for a particular nuclear model, with the darker shades indicating higher (more stable) values. (*JINR, Dubna, Russia*)

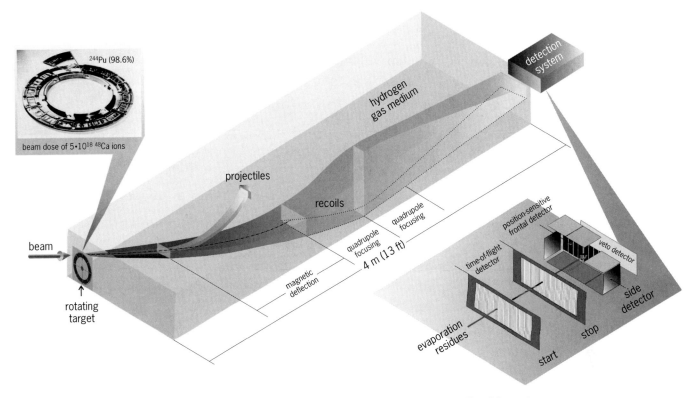

Fig. 2. Example of a gas-filled separator used for producing superheavy elements: the Dubna Gas-Filled Recoil Separator (DGFRS) at the Joint Institute for Nuclear Research. (*JINR, Dubna, Russia*)

(LBNL) and the Dubna Gas-Filled Recoil Separator (DGFRS) at JINR (**Fig. 2**). Velocity filters are electromagnetic separators designed for in-flight separation of unretarded complete fusion reaction products, namely those products in which a projectile nucleus and a target nucleus fuse together and do not scatter in the target or off beam-line components. The separator's magnets and electrostatic deflectors can be tuned to permit passage of the desired reaction products. Gas-filled separators fill the gaps between dipole and quadrupole magnets with a dilute gas such as hydrogen or helium that efficiently exchanges electrons with the evaporation residue via collisions, which do not alter the evaporation residue's path or energy much, until a unique and well-defined charge state is achieved. Thus, the recoiling evaporation residue rapidly achieves a well-defined charge-to-mass ratio and the separator magnets can be tuned to permit passage of these evaporation residues but reject many other types of fragments.

Depending on the purpose of the experiment, evaporation residues are typically implanted into a silicon detector that is position-sensitive in two dimensions. Evaporation residues hitting the detector are identified with multiwire proportional counters that detect the passage of ions just before implantation occurs. This distinguishes signals that arise from the implantation of evaporation residues from those that arise from their alpha decays. Typical detectors used for such purposes include a focal plane detector (the one the implantation occurs in) and detectors surrounding the sides to catch alpha particles that can escape from the focal plane detector because the evaporation residues are implanted at shallow depths. Fissions are distinguished by high-energy deposition in the focal plane detector and side detectors. (The two fission fragments recoil nearly 180° apart, so that one fragment is detected by the focal plane detector and one is detected by one of the side detectors.) Typical alpha-particle energies are 9–11 MeV and fission energies are 170–230 MeV for the heaviest nuclei. Position correlations make it possible to determine genetically related events in a decay chain occurring within the detector, and the times between such events can be measured to obtain lifetime or half-life information about the isotopes observed. Rapid identification and characterization of interesting coincidence events involving evaporation residues and alphaparticles sometimes allows the beam to be interrupted for a short time so that subsequent alpha decays or fissions occur in the detectors under lower background conditions.

An example of the setup for a typical superheavy experiment is shown in Fig. 2.

Current status. Elements up to $Z = 112$ have been confirmed. Once confirmed, element names are then suggested by the scientists first synthesizing them. Scientists at GSI have been credited with discovering elements 107–112. A name for element 112 has not been proposed yet, but scientists at GSI have proposed roentgenium (chemical symbol Rg) for element 111. Elements with $Z = 113$–116 and 118 were synthesized by a team from Dubna and Lawrence Livermore National Laboratory but have yet to be confirmed. Another isotope of element 113 produced in the cold-fusion reaction $^{70}Zn + {}^{209}Bi$ was recently reported by a Japanese group at RIKEN, but is also unconfirmed.

For background information *see* MAGIC NUMBERS; NUCLEAR REACTION; NUCLEAR STRUCTURE; PARTICLE ACCELERATOR; PARTICLE DETECTOR; TRANSURANIUM ELEMENTS in the McGraw-Hill Encyclopedia of Science & Technology.   Mark A. Stoyer

Bibliography. D. C. Hoffman, A. Ghiorso, and G. T. Seaborg, *The Transuranium People: The inside Story*, ICP, London, 2000; Yu. Ts. Oganessian, V. K. Utyonkov, and K. J. Moody, Voyage to the Superheavy Island, *Sci. Amer.*, 282(1):45–49, January 2000; M. Schädel (ed.), *The Chemistry of Superheavy Elements*, Kluwer Academic, Dordrecht, 2003.

Surface plasmon resonance

Surface plasmon resonance (SPR) spectroscopy is a real-time, optical technique that has gained widespread use in surface science and biomolecular interaction analysis since the early 1990s. In SPR experiments, molecules of interest are immobilized on a solid substrate, and their interaction with ligands is monitored to elicit information such as the binding affinity. The SPR data—reflectivity spectra or sensorgrams—can be used to determine bulk concentration and rate constants, as well as to extract parameters on binding energy, recognition sites, interaction mechanisms, and stoichiometry. The key advantage of SPR spectroscopy over conventional fluorescence or radiological assays in biomolecular interaction analysis is that no labeling is required. Although fluorescent labels allow high detection sensitivity, they have the drawback of potentially altering the native state of the studied species, particularly in proteins and peptides.

Theory. Surface plasmons (SPs) are electron density fluctuations propagating in a thin metal layer confined to a metal-dielectric interface. The electric field accompanying the oscillating charges decays exponentially from the metal surface, typically about 200 nanometers for red light sources, making SPR exceptionally surface-sensitive. Surface plasmons can be excited by parallel-polarized light if the wavevector of the incident light matches the wavevector of the propagating plasmons. **Figure 1** shows the most commonly used Kretschmann configuration for surface plasmon excitation. A thin metal layer of a noble metal (for example, 50 nm of gold or silver) is deposited onto a glass substrate that is index-matched to a prism. The wavevector of the incoming light is scanned by rotating the prism to obtain a spectrum of wavevectors versus reflected light intensity. Where resonance occurs, the light energy transferred into surface plasmons manifests itself as a sharp dip in the reflected light intensity. The resulting shape of the obtained spectra is dependent on the refractive index of the biointeractive molecules, the thickness of the molecular layers, and the refractive index of the medium. Changes in the refractive index of the material covering the metal surface result in a shift of the angular position of the maximum energy transfer. Using empirically determined conversion factors, one can express the material change (by ligand adsorption or desorption) in terms of mass per surface area. A commonly used measuring unit is the resonance unit (RU), which expresses an angular position change of the reflectivity minimum by $0.0001°$, corresponding to 1 picogram per square millimeter of organic material.

Two arrangements of SPR spectroscopy are commonly encountered. One tracks the position of the reflectance minimum θ_{min} over time, while the other monitors the reflected intensity at a fixed angle (**Fig. 2**). The second approach is faster and thus frequently used in kinetic analysis and SPR imaging (also known as SPR microscopy). SPR imaging is currently revolutionizing biomolecular interaction analysis. The ability to perform comparative studies at high throughput under identical conditions allows for immediate analysis of hundreds of samples.

Quantification. A major advantage of SPR is the ability to obtain real-time binding information. While fluorescence techniques typically require measurements over a range of concentrations to determine the affinity of a ligand, SPR can reveal the binding constant (k_a) and dissociation constant

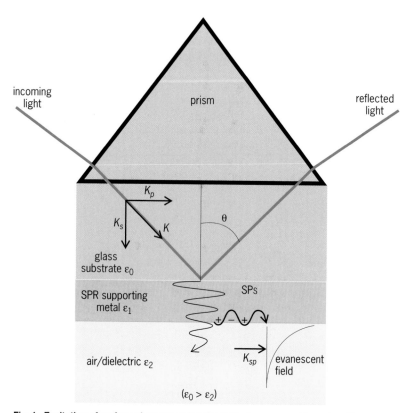

Fig. 1. Excitation of surface plasmons using the Kretschmann coupling prism. The wavevector of the incoming light beam K_p has to match the wavevector of the propagating surface plasmons K_{sp}. This can be obtained only when ε_0 is larger than μ_2.

(k_d) directly. Analysis of the reflectivity spectra is performed by comparing the experimental data with theoretically predicted values of an optical multilayer stack. By using a nonlinear least-square fit, thin-film parameters, such as refractive index and thickness of each individual layer, can be obtained. These parameters are most precise when multiple spectra at different wavelengths or in different media have been collected and analyzed together.

Kinetic analysis in SPR typically employs mathematical modeling using parameters associated with analyte concentration, valency of the interaction, and mass transfer. The differential equations are then numerically solved by computers to obtain the best fit. Consider the case in which species A in solution binds with species B on the substrate to form the complex AB (**Fig. 3**). If the amount of species A is known, monitoring the amount of complex AB formed over time is sufficient to extract the association constant and the dissociation constant, based on the equation below.

$$d[AB]/dt = k_a[A][B] - k_d[AB]$$

The sensorgram is routinely corrected with data from reference measurements to eliminate contribution from nonspecific binding and changes in the background refractive index. Great progress has been made in the interpretation of sensorgrams, using elaborate models to account for complicating effects such as mass transport limitations, heterogeneous species, intermediate binding steps, and multivalent binding sites. Selecting the appropriate model is important for obtaining meaningful data. Global fitting of several sensorgrams, collected at different concentrations, provides more accurate values than using individual fits.

Recent technological developments. Several recent innovations promise to substantially improve the analytical power of SPR-based instruments. Employing a bicell photodiode detector simplifies the instrument and vastly increases the temporal resolution without compromising the sensitivity. The bicell position is adjusted so that the photovoltage from each cell is equal before the measurement, and therefore a shift in the minimum angle results in an increase of the photovoltage in one cell and a decrease in the other. The SPR reflectivity minimum is detected by using a signal amplifier to track the changes, affording sub-microsecond temporal resolution.

On thin metal films, coupling fluorescence spectroscopy with SPR can improve the detection sensitivity by several orders of magnitude. A new technique known as surface plasmon coupled emission, which focuses the anisotropic fluorophore emission into a very small angular range, allows for much higher collection efficiencies. Since the radiative coupling is limited to a narrow distance range, background fluorescence is significantly minimized.

Another promising technique is coupled waveguiding surface plasmon resonance spectroscopy,

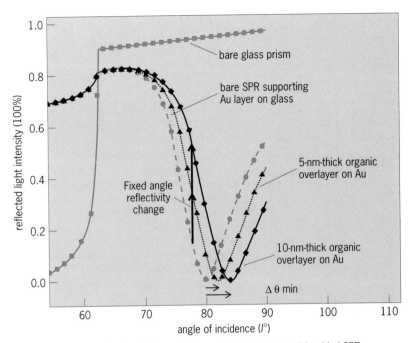

Fig. 2. Comparison of reflectivity spectra for a bare glass prism, with added SPR supporting metal, with organic thin-film overlayers covering the metal. The organic overlayer shifts the angle θ minimum, where the minimum amount of light is reflected, to higher angles with increasing thickness.

which utilizes waveguiding substrates consisting of an SPR supporting metal layer with a silicon oxide overlayer. The silicon oxide layer acts as an optical amplifier and a phase shifter, making it possible to excite plasmons with both parallel- and perpendicular-polarized light. The two modes of polarization result in distinct reflectivity spectra that have much narrower half-widths and thus are intrinsically more

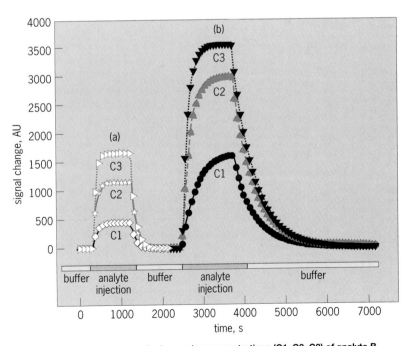

Fig. 3. Simulated sensorgrams for increasing concentrations (C1, C2, C3) of analyte B that forms an ideal complex AB with the immobilized species A. (a) Curves show ligand binding with a dissociation rate three times higher than (b) curves under otherwise identical conditions. AU = arbitrary units.

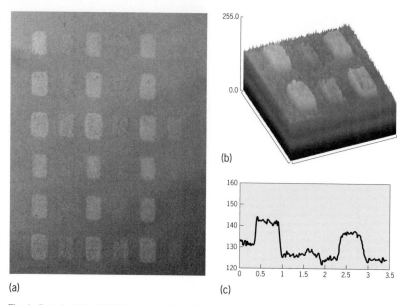

(b)

(c)

(a)

Fig. 4. Protein chip. (*a*) SPR image of the chip with alternating cholera toxin (CT) and bovine serum albumin (BSA) columns after incubation with CT antibody. (*b*) 3D rendition of the protein chip. (*c*) Intensity profile of the protein dots.

sensitive than conventional SPR on metal films. The method allows the study of optically anisotropic thin films, which possess a different refractive index in the plane of immobilization than in the perpendicular plane. One prominent application of coupled waveguiding SPR spectroscopy is the study of light-activated conformational change in the highly optically anisotropic receptor rhodopsin.

Applications. Because of the advantages in high sensitivity and label-free quantification, SPR spectroscopy has become a preferred method for the study of a wide range of biological interactions. Recently, the technique has been used for studying interactions that were difficult or impossible to follow with other techniques. A good example is in nucleic acid research, where a number of SPR studies with DNA are helping to elucidate interaction mechanisms involving mutation and repair. Biosensing applications including DNA single-base mismatch discrimination, detection of polymerase chain reaction (PCR) products, and discovery of transcription initiators have been reported.

The study of proteins and proteomics by SPR has profound implications for academic and industry research. This is evident in SPR's increasing use for protein expression and protein interaction studies with DNA, RNA, carbohydrates, receptors, and drugs. The label-free nature of SPR measurements and the ability to determine binding affinity directly translate into less sample preparation and fewer experiments, which leads to increased productivity in labor-intensive endeavors such as protein interaction mapping and ligand screening. An example is the use of SPR to "fish" for protein receptor ligands of potential pharmaceutical value.

SPR has also found applications in many other areas, such as biomimetic membranes. Self-organized membranes, created by assembly of a lipid bilayer

on a solid surface, allow for reconstitution of natural receptors in their native conformation for optimal binding efficiency with minimal nonspecific surface interactions. This technique has matured over the past 10 years into a simple and effective strategy in biomimetics. SPR is being used to advance membrane technology by obtaining information on the assembly, stability, and interactions of these structures on planar surfaces. It has also been used as a valuable tool to evaluate properties of membrane proteins, such as ion channels, on supported membranes.

Outlook. SPR imaging is an emerging method that has the potential to observe biomolecular interactions on a large scale or in a high-throughput fashion. It is particularly attractive as an optical scanning device for microarray detection because it can quantify the changes due to binding without the need of labeling (**Fig. 4**). At present, it has been demonstrated for applications such as detection of DNA/RNA hybridization and characterization of peptide-antibody interactions. Introduction of commercial SPR imaging instruments will engender proficient analysis of protein microarrays, with a technical impact on the burgeoning field of proteomics similar to that of fluorescence arrays on genomics.

In addition, combined techniques, such as SPR–mass spectrometry, capillary electrophoresis–SPR, and electrochemistry–SPR, hold great promise for biological applications. Although the theoretical underpinnings of SPR can be quite complex for nonspecialists, instrument makers and software programmers have made it convenient to use SPR technology successfully for an increasing range of applications. As an example in its application to disease research, SPR has been applied to cancer, arthritis, anthrax, cystic fibrosis, scrapie, multiple sclerosis, and human immunodeficiency virus (HIV). Thus, one can only begin to imagine its future possibilities.

For background information *see* BIOELECTRONICS; BIOSENSOR; LASER SPECTROSCOPY; LIGAND; OPTICAL PRISM; PLASMON; SIGNAL TRANSDUCTION; SPECTROSCOPY; SURFACE AND INTERFACIAL CHEMISTRY; TRANSDUCER in the McGraw-Hill Encyclopedia of Science & Technology.

Thomas Wilkop; K. Scott Phillips; Quan Jason Cheng

Bibliography. E. Gizeli and C. R. Lowe, *Biomolecular Sensors*, pp. 87–120, 241–268, Taylor and Francis, London, 2002; J. M. Mcdonnell, Surface plasmon resonance: Towards an understanding of mechanisms of biological molecular recognition, *Curr. Opin. Chem. Biol.*, 5:572–577, 2001; Z. Salamon, H. A. Macleod, and G. Tollin, Coupled plasmon-waveguide resonators: A new spectroscopic tool for probing proteolipid film structure and properties, *Biophys. J.*, 73(5):2791–2797, 1997; E. A. Smith and R. M. Corn, Surface plasmon resonance imaging as a tool to monitor biomolecular interactions in an array based format, *Appl. Spectros.*, 57:320A, 2003; N. J. Tao et al., High resolution surface plasmon resonance spectroscopy, *Rev. Sci. Instrum.*, 70(12):4656–4660, 1999.

Systematics in forensic science

For over a decade, molecular biology methods have been used to resolve identity issues in violent crimes, acts of terrorism, missing persons cases, mass disasters, and paternity testing. One very sensitive method is the direct sequencing of mitochondrial DNA (mtDNA). A mitochondrion contains 2–10 copies of mtDNA, and there can be up to 1000 mitochondria per somatic cell. Because of the large number of copies of mtDNA, tissues such as hair, bones, and teeth can be typed more readily than with nuclear-based DNA markers. mtDNA sequencing is the primary genetic analysis used to identify individuals from bones recovered from American war casualties. Victim remains from the World Trade Center tragedy have had their mtDNA sequenced. In addition, mtDNA analysis has been used in evolutionary and anthropological studies, including analyses of ancient DNA.

Due to the low fidelity of mtDNA polymerase (when replicating the mtDNA genome) and the apparent lack of mtDNA repair mechanisms, these sequences generally have a higher rate of mutation compared with nuclear DNA. Most of the sequence variation among individuals is found within two specific segments of the genome termed hypervariable region 1 (HV1) and hypervariable region 2 (HV2). The small size of each of these regions allows for amplification by the PCR (polymerase chain reaction) method, and hence HV1 and HV2 are routinely typed for identity testing purposes.

Unlike nuclear DNA, mtDNA is inherited from the maternal side only. This characteristic often is helpful in forensic cases, such as analysis of the remains of a missing person, where known maternal relatives can provide reference samples for direct comparison to the questioned mtDNA type (for example, the bones of a child). mtDNA can also be used to compare an evidence sample from a crime scene to a suspect. When the DNA profile—that is, the specific nucleotide sequence—from the evidence and the DNA profile from the suspect are different, the interpretation is that the evidence sample could not have arisen from the suspect. If the evidence and suspect have the same DNA profiles, the suspect cannot be excluded as a potential source of the evidence. (The suspect cannot be definitively pinpointed as the source because all maternal relatives also carry the same mtDNA type, and in a small fraction of cases unrelated individuals may share the same mtDNA sequence; the most common mtDNA sequences are shared by 6% of humans.)

Population databases. In a forensic case, the weight of evidence is primarily based on the number of times that a profile is observed in a reference population database, which may shed light on the ethnicity or other attributes of the source. Typically, the population databases used in forensics are composed of unrelated samples that represent major population groups (Asian, African, European, Hispanic, and Native American). Such databases are sometimes criticized as being not large enough to represent the true degree of variation present in a population, or as being inaccurate in their representation of the members of a population. Phylogenetic systematics may be used to assess the validity of the databases for forensic applications.

Phylogenetic analyses. Determining the phylogenetic (evolutionary) relationships among human mtDNA lineages provides a great deal of detailed information about the structure of human genetic variation. The mtDNA type of an individual is described as a haplotype. A haplogroup refers to a cluster of haplotypes that share common variable characters that define them as having a shared ancestry. Phylogenetic methods have been used to identify the major human haplogroups and the important differences, called variable single-nucleotide polymorphisms (SNPs), that define these groups.

There are a number of alternate methods for reconstructing phylogenetic trees of the sequence data. These methods are well characterized by the systematics community, and there are numerous references that provide detailed descriptions of them and the assumptions that each requires. Because it is important to determine the specific characters that define human lineages, a character-based approach is emphasized. A character refers to a specific position in the DNA sequence and the variable nucleotide that is present there. Sometimes these characters are referred to as SNPs. The sites of SNPs can be assessed by a variety of molecular biology techniques, such as PCR, mass spectrometry, and hybridization techniques, which differ in their ability to assess sites quickly (ranging from 10 to 1000 sites). When the number of SNPs that can be assessed effectively is limited by the technology, time, or cost, systematic methods are helpful in determining the relative value of one SNP over another. Parsimony (ancestry-based) methods have been used extensively to examine and characterize human mtDNA variation, and thus a large body of published data is available for comparison. In addition, the assumption of shared variation due to shared ancestry is preferred over alternative clustering strategies (such as clustering by overall sequence similarity). When using parsimony methods, the recently derived (evolved) characters are used to identify groups. An outgroup is selected to determine the primitive versus derived state of the character. For the human tree, either an early human lineage (generally African L haplogroup) or an Old World primate is used as an outgroup, both to root the tree and to determine the polarity of character change on each branch of the tree.

Forensic relevance. Phylogenetic analyses of human mtDNA variation were carried out on the Scientific Working Group on DNA Analysis Methods (SWGDAM) reference mtDNA population data sets—a nationally funded database supported by the Department of Justice and used in all Federal Bureau of Investigations (FBI) cases as well as many state and local crime labs—using the computer programs Winclada and Nona. One goal was to compare the results of these phylogenetic analyses with other published studies from comparable human lineages. If the same major lineages and subbranches are supported and the same relevant SNPs are observed in other published data sets, this would support the relevance of the SWGDAM reference data set as being representative of human variation for a particular group.

Some of the general results of phylogenetic analysis include a list of all derived characters (gene position and nucleotide observed) that subdivide the targeted population (**Fig. 1**); a topological tree structure that shows all major and minor lineages (**Fig. 2**); and the characters that diagnose each of the particular groups (see **table**). If each lineage is named and counted as a separate haplogroup, it would provide frequency distributions of all subclusters (haplogroups) observed. Additional information acquired through phylogenetic analysis includes the relative amount that each character changes across the phylo-

genetic tree. Rapidly varying characters may change so fast that they are uniquely derived in every instance of their occurrence. Alternatively, a particular character may occur only once on the tree and diagnose a haplogroup. Using this information, SNPs can be ranked as to their importance in defining the structure of variation for a particular genetic data set.

Previous analyses of European genetic variation outlined the major branches of the tree of West Eurasian mtDNA haplogroups, as well as the SNPs that define these haplogroups. Comparisons of the SWGDAM European data sets with other European analyses show concordance with the basic phylogenetic structure and defining sites. Approximately 99% of the known European mtDNA variation can be categorized into 10 major haplogroup lineages H, I, J, K, M, T, U, V, W, and X. For the European SWGDAM data set, 229 parsimony-informative SNPs (that is, characters having the same nucleotide found in two or more individuals in the data set) were observed, of which 72 SNPs defined clades (groups)

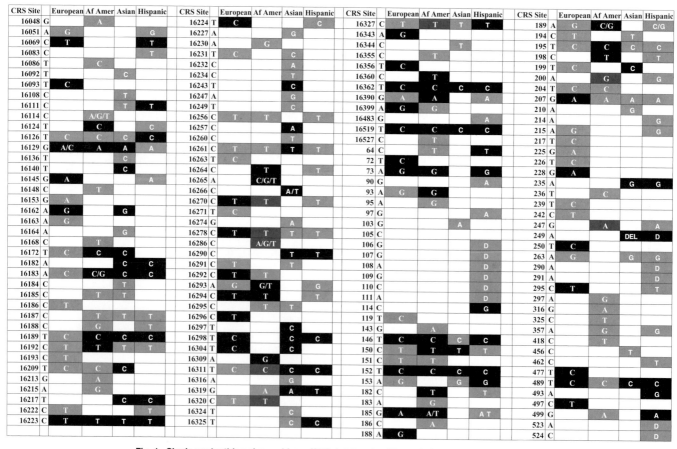

Fig. 1. Single nucleotide polymorphisms (SNPs) determined from phylogenetic analysis of the mtDNA control region sequences of the African American, Asian, European, and Hispanic SWGDAM data sets. The numbers refer to the revised Cambridge Reference Sequence (rCRS) nomenclature system for sites. White bars refer to the SNP being either absent or observed in less than 1% of individuals of the data set. Medium-gray and black bars are SNPs that defined groups by at least 1% or more of the individuals in the data set. Black bars refer to the most informative SNPs based on phylogenetic analysis and a close examination of the evolution of the character data on a tree; this entailed removal of some of the SNPs that defined the same groups. Character states were listed both for the rCRS and for the more common variable sites (medium gray and black bars). Nucleotides that were observed as defining characters were listed. Listing more than one character referred to multiple state changes at a site.

Diagnostic characters of major haplogroups in the SWGDAM data sets

Haplogroup	Poly 1	Poly 2	Poly 3	Poly 4	Poly 5	Poly 6	Poly 7	Poly 8	Poly 9	Poly 10	Poly 11	Poly 12	Poly 13	Poly 14	Poly 15
A	16223T	16290T	16319A	16362C	235G										
B	16189C	16217C	499A												
B4a	16183C	16189C	16217C	16261T											
B4b	16136C	16183C	16189C	16217C	499A										
B5a	16140C	16189C	16266A	210G											
B5b	16140C	16183C	16189C	16243C											
C	16223T	16298C	16327T	249D	290D	291D	489C								
D	16223T	16362C	489C												
D4a	16129A	16223T	16362C	152C											
D4b	16223T	16319A	16362C												
D5a	16182C	16183C	16189C	16223T	16266T	16362C	150T								
F1a	16129A	16172C	16304C	249D											
F1b	16183C	16189C	16304C	249D											
F1c	16111T	16129A	16304C	152C	249D										
F2a	16291T	16304C	249D												
G2a	16223T	16227G	16278T	16362C											
H	73A														
I	16223T	199C	204C	250C											
J	16069T	16126C	295T												
K	16224C	16311C													
L1a	16129A	16148T	16172C	16187T	16188G	16189C	16223T	16230G	16311C	16320T	93G	185A	189G	236C	247A
L1b	16126C	16187T	16189C	16223T	16264T	16270T	16278T	16311C	152C	182T	185T	195C	247A	357G	
L1c	16129A	16189C	16223T	16278T	16294T	16311C	16360T	151T	152C	182T	186A	189C	247A	316A	
L2a	16223T	16278T	16294T	16390A	146C	152C	195C								
L2b	16114A	16129A	16213A	16223T	16278T	16390A	150T	152C	182T	195C	198T	204C			
L2c	16223T	16278T	16390A	93G	146C	150T	152C	182T	195C	198T	325T				
L3b	16223T	16278T	16362C												
L3d	16223T	152C													
L3e1	16223T	16327T	189G	200G											
L3e1a	16185T	16223T	16311C	16327T	189G	200G									
L3e2	16223T	16320T	195C												
L3e2a	16223T	16320T	195C	198T											
L3e2b	16172C	16189C	16223T	16320T	195C										
L3e3	16223T	16265T	195C												
L3e4	16051G	16223T	16264T												
L3f	16209C	16223T	16311C	189G	200G										
L3f1	16209C	16223T	16292T	16311C	189G	200G									
M10	16223T	16311C	16519C												
M7a1	16209C	16223T	16324C												
M7b1	16129A	16192T	16223T	16297C	150T	199C									
M7b2	16129A	16189C	16223T	16297C	16298C	150T	199C								
M7c	16223T	16295T	146C	199C											
M8a	16184T	16223T	16298C	16319A											
M9	16223T	16234T	16316G	16362C											
N9a	16223T	16257A	16261T	150T											
R9a	16298C	16355T	16362C	207A	249D										
T	16126C	16294T													
U5	16270T														
V	16298C	72C													
W	16223T	189G	195C	204C	207A										
X	16189C	16223T	16278T	195C											
Y	16126C	16231C	16266T	146C											
Z	16185T	16223T	16260T	16298C	152C	249D									

of 10 or more individuals. After the removal of some of the redundant sites from closely associated SNPs, a minimum set of 32 SNPs were defined, and this largely overlaps with that of previous descriptions of the structurally informative SNPs for Europeans.

The phylogenetic analyses also show that the SWGDAM African American data set contains variation consistent with that described in other African populations. Sixteen of the 18 haplogroups previously observed in African populations were identified in the SWGDAM reference data set and include L1a, L1b, L1c, L1e, L2a, L2b, L2c, L3*, L3b, L3d, L3e1*, L3e1a, L3e2a, L3e2b, L3e3, and L3e4. Approximately 12% of the haplogroups observed within African Americans were common in European or Asian populations. There were 217 parsimony-informative SNPs observed in two or more individuals within the African American SWGDAM data set, and 71 SNPs defined clades of 10 or more individuals. A minimum set of 34 SNPs could partition most of the African American haplogroups (see table). The most informative SNPs for defining major haplogroups are consistent with previous studies. Similar work is ongoing to characterize Asian, Native American, and Hispanic SWGDAM data sets, and these analyses are shaping the selection of SNPs to be used in design of analytical techniques. Moreover, these data lend additional support that the forensic data sets are useful for applying weight to an observed

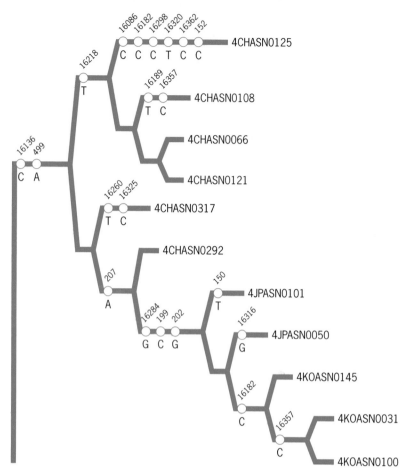

Fig. 2. Representative portion of the phylogenetic tree for Asian individuals in the SWGDAM data set (B4b haplogroup). The numbers refer to the revised Cambridge Reference Sequence (rCRS) nomenclature system for site positions, and states are listed below the branches.

mtDNA match in a criminal case or human identification.

Microbial applications. While human genetic variation has been used widely in a criminal context, there has been development of a new application of molecular biology in response to terrorism and the growing threat of biological agents. The goal is to determine to which lineage a particular microbial forensic sample belongs and to determine the origin of the sample. Many microbial species have been characterized genetically. A number of government agencies are actively surveying the genetic diversity of these select agents and their close relatives. Phylogenetic analysis of these genetic data sets can provide similar information both for identifying the presence of an agent and for determining the major lineages found among related organisms. The standard methods and protocols that have served the forensic community well for collecting sequences, databasing the information, and introducing the genetic evidence into United States courts will likely be followed for all major forensic biological samples. In this way, forensic examiners will follow a well-established path to add more genetic databases to their arsenal of forensic tools.

For background information *see* FORENSIC BIOLOGY; HUMAN GENETICS; MOLECULAR BIOLOGY; PHYLOGENY; POLYMORPHISM in the McGraw-Hill Encyclopedia of Science & Technology.   Marc Allard; Bruce Budowle; Mark R. Wilson

Bibliography.   B. Hall, *Phylogenetic Trees Made Easy: A How-To Manual for Molecular Biologists*, Sinauer Assoc. Inc., Sunderland, MA, 2001; D. M. Hillis et al., *Molecular Systematics*, 2d ed., Sinauer Assoc. Inc., Sunderland, MA, 1996; I. J. Kitching et al., *Cladistics: The Theory and Practice of Parsimony*, 2d ed., Systematics Association, Oxford University Press, New York, 1998; K. Monson et al., The mtDNA Population Database: An integrated software and database resource for forensic comparison, *Forensic Sci. Commun.*, 4(2), 2002; M. Wilson and M. W. Allard, Phylogenetic and mitochondrial DNA analysis in the forensic sciences, *Forensic Sci. Rev.*, 16:37–62, 2004.

Systems architecting

Most organizations depend on the effective integration of a number of systems in order to quickly adapt to changes of all types, such as changing technology, changing customers and customer needs, and changing business partners. To accomplish this, an organization's systems must form an enterprise architecture that is in effect a system of systems (SOS) or systems family. The perspectives of all the stakeholders, from the chief executive to the technology developer to various implementation contractors, must be considered in developing the architecture. The systems that constitute the system family must be evolvable and adaptive in order to enhance the ability of the organization to cope with emerging needs.

Christopher Alexander has developed much of architectural relevance by considering architecture as a synthesis of form or patterns that serves as a formal plan to guide construction of a system. This plan, or architecture, is obtained by using an appropriate framework of principles, standards, models, tools, and methods that enable subsequent development and deployment of a system. The architecture must be detailed and it must be organized such that it supports knowledge about the structural properties of the system. Thus, the system's architecture provides a medium for communication of information about the system and its environment. This supports the development of action plans for technically directing the system, managing its evolution and integration with legacy (preexisting) systems and organizational functions, and building the operational system. Architecting refers to the process of bringing about an architecture, just as engineering initially referred to the process of bringing about an engine.

An architecting framework is crucial for determination of a systems architecture. Architectural frameworks are high-level guidelines for creating an instantiated architecture that is effective and efficient. An architectural development process is needed to obtain an instantiated architecture from an architectural framework. In this way, an architectural framework is a necessary step in engineering a system as well as its subsequent integration, adaptation, and evolution over time.

Architectural perspectives. A standard dictionary definition of architecture is that it is the style and method of design and construction. A common definition found in the literature is that the architecture of a system is the structure of the components of a system, their interrelationships, and the principles that govern their evolution over time. A common theme in most definitions of architecture is that it relates to the structural properties of a system, and potentially to this structure as the system evolves over time. These structural properties may be stated in terms of system components, interrelationships, and the principles or guidelines that are established about the use of the system over time. The precise structural properties of the system which need specification, and the best ways to represent this structure, depend on what is of interest to the users of the architecture.

There may be many uses and users of a system's architecture. Thus, a variety of perspectives must be considered in order to develop the most appropriate single composite architecture for the system. Each of these architectural perspectives, or views, is intended to show the system from a different level of abstraction, each appropriate to different stakeholders to the systems architecture. There are three obvious major perspectives: system users, systems engineers and managers, and system implementers. The functional, physical, and implementation architectures are associated with these perspectives. They depict, respectively: the functions that the enterprise itself and enterprise system users would like to have achieved; the physical structure that the systems engineers and managers envision to accomplish these functions; and the detailed implementation or operational factors that the builders of the system deal with. There are, of course, a number of other stakeholder perspectives that are also of importance.

Each of these groups has different needs that an appropriate architecture must address. For the user of the system, the objective is to enhance the effectiveness and efficiency in using the system. To this end, the users will be concerned with the various interfaces between them and the system. They will wish to know that the system has been designed for human interaction, and will expect the system architect to know the basic functional features that will be required for this. The systems engineers and systems managers need physical constructs that will enable them to direct technical efforts to bring about a reliable system. The implementers, or builders, of the various parts of the system will need to have sufficient implementation details (such as design specifications) that enable their construction efforts.

Eberhardt Rechtin suggests that architecture is the road map of, or a top-down description of, the structure of the system and is the tool used by the developer to interface with the customer. The architecture of the system must take into account all external constraints on the system including socio-political, environmental, perceptual, and other real-world factors. It represents the structure of the system and expands as the system emerges from conceptual to functional to physical to implementation architecture. One purpose of an architecture is to manage and control the effects of system complexity by limiting interfaces and avoiding a design that may lead to interface defects or discontinuities. The architecture of a system is created and expanded upon by the process of decomposition. The decomposition of a system includes the identification, characterization, and specification of functional and physical interfaces that must be represented by the architecture. There are operational, or implementation, interfaces at the level of the specific components used to construct the system. Most importantly, there are many interfaces with humans and organizations who may own, set policy over, utilize, engineer, manage, and maintain systems.

Joint Technical Architecture. There are a number of related architectural frameworks in use today. An overview of one of these is given here. The Department of Defense Joint Technical Architecture (JTA), Version 6, dated October 6, 2003, is truly an enterprise integration effort for information technology and information systems. Its objective is to provide a framework for the interoperability of current and future information systems. The purpose of the JTA is to enable and improve the ability of Department of Defense (DOD) systems to support joint and combined operations by integrating information systems. According to the JTA documentation, the JTA provides the foundation for interoperability, mandates standards and guidelines for information system development and acquisition, communicates to industry the DOD's intent to use open systems, and reflects the direction of industry's standards-based product development.

The JTA mandates a minimum set of standards and guidelines for the acquisition of DOD systems that produce, use, or exchange information. These standards and guidelines must be addressed during any activity involved in the management, development, or acquisition of new or upgraded systems in the DOD. It is very closely related to the DOD Command, Control, Communications, Computers, Intelligence, Surveillance, and Reconnaissance (C4ISR) architectural framework effort, which has been active for several years, and which is now called the Department of Defense Architectural Framework (DODAF).

In these efforts, three primary architectural views are espoused. These are based on varying levels of concern regarding information system integration. The highest level of concern is the operational level, with a focus on a means to integrate information for the conducting of military operations. The second level of concern deals with the integration at the system level and the exchange of information between those applicable systems. The third level of concern pertains to the interconnection of specific system components. The following outlines these three architectural views as described in the JTA and DODAF documents.

1. The Operational Architecture (OA) is a description of the activities, operational elements, and information flows required to support military operations. This includes the detailed definition of the parameters and specific requirements of information exchange. This architectural level centers on tactical and strategic information, from the point it is obtained to its utilization in operations.

2. The System Architecture (SA) is a description of system components and their interconnections required for supporting war-fighting functions. This includes multiple system links and interoperations to the level of physical connections, locations, key node identifications, circuits, networks, and platforms. This level of architecture is clearly technical in nature and is the focus of the majority of the JTA document.

3. The Technical Architecture (TA) is set of technical rules defining the interactions and interdependencies of system parts and elements to ensure that compatible systems satisfy a specified set of requirements. This architectural level provides greater detail regarding information types, content, and the nature and timing of information movement. This could also include processing and repository requirements of information.

The JTA and the DODAF serve as both an integration architecture and as an integration architecture framework, especially from the point of view of total force integration. The scope is extremely broad, from the information systems required by strategic and tactical operational organizations to the specific standards to be employed for the development of those information systems.

Framework types. Systems engineering is a multiphase process. Each of these phases can be viewed at a number of levels: "system of systems" or "family of systems" (SOS or FOS), system, subsystem, component, and part. These are generally defined in a functional block diagram structure for the system-based constructs being engineered. At each of these levels, the various phases of the systems engineering process need to be enabled through identification of appropriate work efforts; a work breakdown structure (WBS) is one appropriate way to display this information. We may identify a two-dimensional matrix framework representation of the phases and levels, such as shown in **Fig. 1**. In this figure, the darker shades of activity cells are intended to represent greater intensity of effort and activity. While there is activity at the system and family of systems (and enterprise level, not shown) across each of these phases, activity at the component and part levels are to be found primarily during the development phase of effort. When we recall that this framework needs to apply to each of the three major architectural perspectives (functional, physical, and implementation) and that the family of systems may be composed of a large number of systems, the complexity of the effort to engineer these systems becomes apparent. This strongly recommends modeling and simulation as a part of the architectural effort.

There is no process inherently associated with an architectural framework that enables the engineering of an appropriate architecture. Instead of thinking of the columns of the activity framework in Fig. 1 in terms of the phases of the systems engineering life cycle, we might think of the questions that need be asked during the process of engineering a system. These include questions that concern structure: what, where, when, and who. They also include questions that concern function and purpose: how and why. The use of such questions to define the abstractions associated with the various perspectives is characteristic of a framework called the Zachman framework for architecting.

Since architecting is a phase within the systems acquisition or production process life cycle, we need to associate the steps of formulation, analysis, and assessment and interpretation with a framework. Generally, these steps are employed in an iterative manner through use of a process such as the multistroke SOS level decomposition process **(Fig. 2)**. There is a strong presence of integration in the architecting formulation step, in terms of determining what needs to exist in the ultimate enterprise system in regard to technological products, processes, people, and facilities.

For background information *see* SYSTEMS ARCHITECTURE; SYSTEMS ENGINEERING; SYSTEMS INTEGRATION in the McGraw-Hill Encyclopedia of Science & Technology.   Andrew P. Sage

Bibliography. C. Alexander, S. Ishikawa, and M. Silverstein, *A Pattern Language: Towns, Buildings, Construction*, Oxford University Press, 1997; B. Boar, *Constructing Blueprints for Enterprise IT Architectures*, Wiley, 1999; M. W. Maier and E. Rechtin, *The Art of Systems Architecting*, 2d ed.,

Level or phase	Definition	Development	Deployment
Family			
System			
Subsystem			
Component			
Part			

Fig. 1. Systems engineering phases across the system family elements. Shading indicates level of activity.

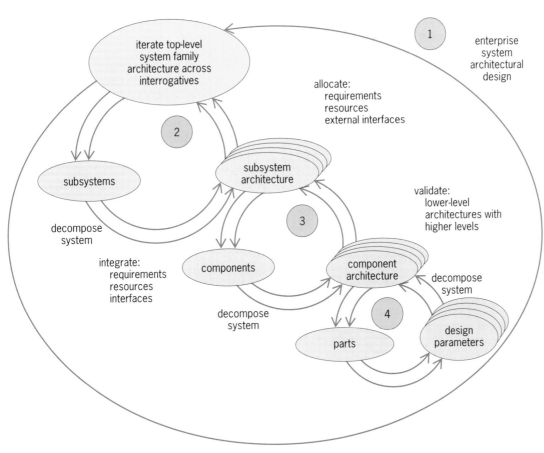

Fig. 2. Decomposition to identify the architecture for a system of systems.

CRC Press, 2000; E. Rechtin, *Systems Architecting: Creating and Building Complex Systems*, Prentice Hall, 1991.

Tetraquarks

Quarks are fundamental particles that respond to the strong force. The strong force is one of the four fundamental forces, and dominates in strength over the other three forces—electromagnetic, weak, and gravity—at very short distances (of the order of 10^{-15} m, the size of an atomic nucleus). Quarks do not appear in isolation but are confined in clusters known as hadrons. The nucleus of the hydrogen atom, the proton, is the most familiar example of such a hadron. The proton requires three quarks to combine to account for its overall electric charge and other properties. The family of such triplets is known as baryons.

Another way that quarks combine is when a single quark is attracted to an antiquark—the antimatter analog of a quark. Such states are known as mesons, of which the lightest example is the pion.

There are six types (or flavors) of quark: the up (u) and down (d) quarks, which make up the proton, neutron, and pion, and are of very small mass; and the strange (s), charmed (c), bottom (b), and top (t) quarks, with successively larger masses. The strong force cannot transform a quark into a different quark (although it can create a quark-antiquark pair of the same flavor); such a transformation requires the weak force. This property of the strong force can be expressed by associating with each quark a quantum number (similar in nature to electric charge) that is conserved (cannot be changed) by the strong force. The quantum numbers associated with the strange and charmed quarks are called strangeness and charm, respectively.

For over 50 years, hundreds of hadrons have been discovered which have fit into this scheme as either quark triplets or quark-antiquark pairs. There have been no hadrons confirmed that definitively require a more complicated explanation, although there have been a handful that have hinted that the big picture may be more involved.

Discoveries of multiquark hadrons. In 2003–2004, this interpretation changed with the discovery of four different hadrons that are not easily described within this historical picture. One of these is a baryon known as the theta, whose positive electric charge and positive amount of strangeness cannot be simultaneously accounted for without having at least four quarks and an antiquark clustered together. Because of this structure, this state is believed to be the first example of what has become known as a pentaquark. This evidence that nature builds more complex structures in the baryons supports suggestions that certain mesons also might be better understood if they are made from two quarks and two antiquarks rather than just a single one of each. Such states, made from a total of four quarks or antiquarks, are known as tetraquarks. *See* PENTAQUARKS.

Three varieties of such tetraquarks have been identified; one has hidden charm (that is, it has a charmed quark, but this quark is paired with a charmed antiquark, so that the total charm of the hadron is zero); one has both charm and strangeness; and the final ones contain neither of these attributes. The discoveries in 2003–2004 fit into the first two classes and consist of three mesons. These mesons will be referred to by their masses (energies at rest), which are 3872, 2460, and 2317 megaelectronvolts (MeV).

X(3872) state. There is a large family of mesons with hidden charm, the meson being made from a single charmed quark and a charmed antiquark, forming a family generically known as charmonium. The first example, known as the J/psi (ψ) meson, was discovered in 1974 and was like a strongly interacting analog of the positronium atom. Here the constituents can spin and orbit in various energy states, giving rise to a tower of levels. Several of these are metastable (long-lived) and as such give rise to sharp lines in the spectra of gamma rays formed when one state loses energy by radiating light and converting to a lower energy state—analogous to what is familiar in atomic spectroscopy, although at energy scales millions of times larger.

These metastable states of charmonium have masses (energies at rest) that span the range from about 3000 to about 3700 MeV. Charmonium states whose masses are greater than this had been expected to decay rapidly into a pair of charmed D mesons, where a charmed quark or antiquark combines with a lightweight quark of the up or down variety. Such states are seen and indeed are not metastable. However, recently a further metastable state has been found with a mass of 3872 MeV. Now the debate is what it is made of.

It is possible that this is a charmonium state whose total spin (J) and behavior under mirror reflection or parity (P) and matter-antimatter inversion (C) give a correlation in $J(PC)$ such as $2(--)$ that is forbidden to decay rapidly into the two lightest charmed D mesons. This would then explain its metastability. However, there is another interpretation inspired by a numerical coincidence. The mass of this meson agrees to better than one part in 2500 with that of a D meson and the next-lightest charmed meson, known as D^*. More specifically, this outcome occurs if they are electrically neutral: the D^0, made from a charmed quark and an antiquark of the up variety, and analogously for its D^* analog. Thus, it is suspected that the meson in question is actually a combined state of D^0 and D^{*0}, or in quark language a tetraquark made of a charmed quark, an up quark, and their respective antiquarks. If this is the case, then the spin, parity, and matter inversion values, $J(PC)$, are predicted to be $1(++)$ rather than $2(--)$. Hence, it should be pos-

sible with the accumulation of data in coming years to tell which interpretation is correct. If the meson is a tetraquark, $J(PC)$ should turn out to be $1(++)$.

D_s(2317) and D_s(2460) states. The charmed varieties of quark or antiquark can also combine with their lighter cousins of the strange variety. The ensuing mesons are known as c-s-bar ($c\bar{s}$) states, denoted D_s, and have both strangeness and charm. Whereas the charmonium states are strongly bound analogs of atomic positronium atom, these $c\bar{s}$ states, consisting of a heavy (charmed) and a light (strange) constituent, are strongly interacting analogs of the hydrogen atom.

Several examples of such $c\bar{s}$ excited states are known, and their measured masses have tended to agree with the theoretical predictions to better than 1 or 2%. However, in 2003 two examples were discovered, with masses of 2317 and 2460 MeV, which were up to 10% lower than had been predicted, even while some closely related states had been observed with masses as expected.

The masses of both of these states are similar to the combined masses of a charmed meson and a strange meson. As in the case of the 3872-MeV meson, these two mesons with charm and strangeness appear to be more closely related to tetraquarks (two quarks and two antiquarks) than to simply a conventional $c\bar{s}$ description. The lighter of the pair has no spin and is known as a scalar. Its mass of 2317 MeV is less than 2% below that of a charmed D meson (mass 1865 MeV) and a strange meson, K (mass 495 MeV). These two mesons at rest would naturally form a spinless scalar, in accord with the 2317-MeV state. The second state is heavier, at 2460 MeV, and appears to be an analogous combination of a strange K meson and a heavier charmed meson, the D^*. In this case the 2460-MeV state would have total spin of 1 and positive parity; a meson with these properties is known as an axial meson. This is very much like the 3872-MeV meson, which appears to be an axial meson made of D and D^*; the 2460-MeV axial state is made of K and D^*. The 2460-MeV state mass is within 2% of that of a K and D^*, and 2317 MeV is within 2% of that of a K and D. In summary, each of the 3872-, 2460-, and 2317-Mev states has the characteristics of being a tetraquark.

Confirming that these states have these values of spin should be done soon, but confirming that they are indeed tetraquarks requires more extensive studies. It is interesting, therefore, to see whether there are tetraquark analogs involving just the light up, down, and strange flavors of quark and antiquark. It was suggested long ago that this should be so, and the evidence now appears to support this.

Light-quark states. There is a set of mesons whose masses are comparable to or less than a proton and which have no overall spin and have positive parity (are scalar). In the simple quark model, such states could be made as a quark and an antiquark orbiting around one another. However, this orbital motion causes the state to have rather large mass, typically 50% or more greater than that of a proton, and does not easily explain the existence of the $f_0(980)$ and $a_0(980)$ mesons, the 980 being their masses in MeV, which are just a few MeV below that of a strange K and $\bar{K}$ meson at rest. This similarity in masses has long caused them to be suspected as bound states or "molecules" of a K and $\bar{K}$. Furthermore, R. L. Jaffe pointed out in 1977 that one could make relatively light scalar mesons for two quarks and two antiquarks without need for any orbital motion. In the quantum chromodynamic theory of the forces between quarks, there is an analog of a magnetic attraction that naturally causes such scalar tetraquark configurations to have masses below 1 GeV (10^9 electronvolts), in accord with the data. It is now generally agreed that these mesons are composed of two quarks and two antiquarks; presently the debate centers on whether these are tightly compacted into a tetraquark cluster or whether they are more loosely linked, and described as a K-$\bar{K}$ molecule.

In any event, the multiquark nature of these scalar mesons fits in rather naturally with the tetraquark picture of the heavier states, $D_s(2317)$, $D_s(2460)$, and $X(3872)$. Thus, there begins to appear evidence that quarks and antiquarks can cluster together in more complex ways than the simplest "one quark and one antiquark" and "three quarks" known hitherto. The spectroscopy of such states will reveal richer information on the nature of the strong forces that ultimately build the nuclei of atoms and are the seeds of matter as we know it.

For background information *see* BARYON; ELEMENTARY PARTICLE; FUNDAMENTAL INTERACTIONS; MESON; PARITY (QUANTUM MECHANICS); QUANTUM CHROMODYNAMICS; QUARKS; STRANGE PARTICLES; SYMMETRY LAWS (PHYSICS) in the McGraw-Hill Encyclopedia of Science & Technology.　　Frank Close

Bibliography. F. Close, *Particle Physics: A Very Short Introduction*, Oxford University Press, 2004; F. Close, Strange days, *Nature*, 424:376–377, 2003.

Thermoacoustic mixture separation

The separation of gas mixtures by thermoacoustic waves is due to a complex cycle of thermal diffusion and viscous motion. This recently discovered process has been described theoretically and demonstrated in experiments which show that the method is well suited for isotope enrichment. The basic phenomena are well understood, and technology development is underway.

Microscopic process. The fundamental processes responsible for thermoacoustic mixture separation are thermal diffusion and viscous motion, in perpendicular directions (**Fig. 1**). A single-frequency sound wave propagating in a gas mixture in a tube produces sinusoidal oscillations of pressure and velocity. These oscillations, and their thermal and viscous interactions with the wall of the tube, produce a complex time sequence of events leading to net separation of the mixture.

To understand this sequence, it is convenient to subdivide the period τ of the oscillations into 8 steps. At time $t = \tau/8$, the pressure is high, and the time-dependent part of the temperature has a steep gradient near the wall (that is, in this region it changes rapidly with distance from the wall), due to the pressure-induced adiabatic temperature rise in the gas far from the wall and the large heat capacity of the solid wall itself. During this time, thermal diffusion drives the heavy molecules down the temperature gradient toward the wall and the light molecules up the temperature gradient away from the wall. Following this step, at $2\tau/8$, the gas near the wall is enriched in heavy molecules and that far from the wall is enriched in light molecules. Next, at $3\tau/8$, the gas moves down in the center of the tube but with viscosity impeding its motion near the wall of the tube. Hence, the downward motion carries primarily light-enriched gas, because the heavy-enriched gas closer to the wall is relatively immobilized by viscosity. At $5\tau/8$, the pressure is low and the temperature gradient and thermal diffusion are reversed relative to $\tau/8$, so that at $6\tau/8$ the gas near the wall is enriched in light molecules and that far from the wall is enriched in heavy molecules. Finally, at $7\tau/8$, motion upward carries predominantly heavy-enriched gas while light-enriched gas is immobilized in the boundary layer. The overall effect of these steps, illustrated by the difference between the molecule positions at $t = 0$ and $t = 8\tau/8$, is that the heavy molecules move up and the light molecules move down.

The distance δ over which the temperature gradient and thermal diffusion occur depends on the acoustic frequency and the properties of the gas, but is typically less than 1 mm (0.04 in.). For most gas mixtures, the viscous boundary layer thickness is comparable to the thermal boundary layer thickness. The distance δ is much smaller than the acoustic wavelength λ, which is typically of the order of 1–10 m (3–30 ft).

History and development. Remarkably, this simple acoustic phenomenon remained undiscovered until 1998, despite the fact that most physical acoustics research has used a gas mixture (that is, air). The discovery occurred accidentally, in the course of an experiment to study mechanical vibration of a thermoacoustic engine, in which a helium-xenon mixture was used to provide both high density and high thermal conductivity.

Theoretical understanding of the phenomenon and experimental confirmation of the theory followed rapidly. The first experiments used a helium-argon mixture in a 1-m-long (3-ft) tube, but at an exceptionally low frequency so that this tube length was much shorter than the acoustic wavelength. Subsequent experiments in a 2.5-m (8-ft) tube were at higher frequencies so that the acoustic wavelength was shorter than the tube length. In the long tube, two gas mixtures were tested. A mixture of half helium and half argon was separated to create 30% helium and 70% argon at one end of the tube and

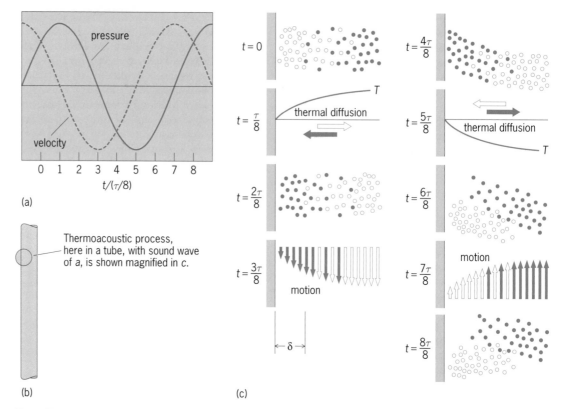

Fig. 1. Thermoacoustic mixture separation. (*a*) Sinusoidal oscillations of the pressure and velocity of the sound wave as functions of time *t*, with period τ. (*b*) Separation tube, with a small portion near the wall identified. (*c*) Time history of phenomena in the gas mixture near the tube wall. The viscous and thermal boundary layers have thicknesses of order δ. Light molecules are indicated by open circles and heavy molecules by filled circles. Variation of temperature *T* with distance from the wall is plotted at $t = \tau/8$ and $t = 5\tau/8$.

70% helium and 30% argon at the other end. Neon gas, an isotopic mixture of 9% neon-22 (^{22}Ne) and 91% neon-20 (^{20}Ne), was separated to create 1% isotope-fraction differences from end to end.

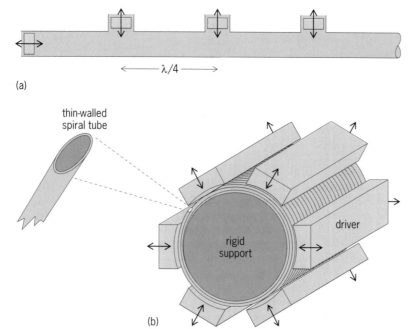

Fig. 2. Practical configurations for long separation tubes. (*a*) Straight tube with a side-branch driver every quarter wavelength, $\lambda/4$. (*b*) Extremely long, spiral-wrapped tube with only six drivers.

Fundamental limitations. The purities can be made as high as desired by extending the length of the tube to many wavelengths of the sound wave: Each wavelength of tube can produce a concentration difference of the order of $n_H n_L \Delta m/m_{\text{avg}}$, where n_H and $n_L = 1 - n_H$ are the local average fractions of the heavy and light molecules, respectively, Δm is the mass difference between the heavy and light molecules, and m_{avg} is their average mass.

The rate at which the molecules are separated increases in proportion to the square of the amplitude of the acoustic wave, up to amplitudes at which the oscillating motion becomes turbulent. A pressure amplitude approximately 5% of the average gas pressure is best. The separation rate does not increase in proportion to the cross-sectional area of the tube, because the useful phenomena occur only close to the tube wall. Tube diameters around 10δ are best. Thus, high separation rates demand many tubes in parallel or large tubes subdivided by internal structures, perhaps with honeycomb geometry.

The energy efficiency of any mixture separation process is defined as the rate at which the Gibbs free energy of the mixture is increased divided by the power consumed by the process. In the case of thermoacoustic mixture separation, the power consumed is acoustic power, and the energy efficiency is limited because the separation depends on two intrinsically irreversible processes: viscous flow and thermal diffusion. Under reasonable circumstances, these processes limit the efficiency to about $10^{-2} n_H n_L (\Delta m/m_{avg})^2$.

Prospects for practical use. Practical needs for mixture separation are extremely diverse. Large-scale applications include the separation of air into oxygen, nitrogen, and argon; the separation of crude oil into an array of products; and the enrichment of the fissile uranium isotope for nuclear power. Small-scale needs include isotope production for medical diagnostics.

Some mixtures can be separated by almost perfectly efficient methods. For example, if a semipermeable membrane exists, one component of a mixture can be purified by doing only the work of compression against its partial pressure. Distillation of mixtures in which the boiling points of the components differ significantly is also very efficient. However, many mixtures cannot be separated by such efficient methods. Most isotopes and isomers are among the difficult mixtures that must be separated by inherently irreversible processes.

In energy efficiency, thermoacoustic mixture separation falls far short of distillation, so it is unlikely that thermoacoustics will ever be useful for large-scale air separation or petroleum refining. However, its efficiency is comparable to that of other widely used irreversible methods (for example, gaseous diffusion for isotope enrichment), and the apparent convenience of an ambient-temperature, ambient-pressure separation method seems attractive for small-scale applications, irrespective of its energy efficiency. Hence, there is motivation to develop practical engineering of this new method, especially for isotope enrichment.

To obtain reasonably high purities from isotopic mixtures, a practical way to use tubes of great length must be developed. Creating the sound wave in the tube is the principal challenge, because acoustic attenuation in a long, narrow tube is large enough that a sound wave injected at one end cannot travel far. The long-tube separator that was used to separate a helium-argon mixture and in the isotopic separation of neon gas had side branches (tees) every quarter wavelength, at which drivers (bellows-sealed pistons, moved by loudspeakers) injected acoustic power to maintain a nearly constant acoustic amplitude along the tube (**Fig. 2a**). As such a tube is lengthened, it can be folded for compactness, at each tee. However, if hundreds of wavelengths are required, then the number of parts to assemble and the number of joints to make leak-tight could become prohibitive. One alternative now under consideration is a long, seamless tube that is spirally wrapped around a rigid support (Fig. 2b). The tube has an elliptical cross section, and its wall is thin enough that it is flexible, so squeezing it changes its cross-sectional area. The operating frequency is chosen so that one wavelength of sound equals one turn of the spiral. Six drivers spaced $60°$ apart around the spiral, and driven with time phases also $60°$ apart, will create a peristaltic wave in the tube's cross-sectional area that

will maintain a traveling wave in the gas throughout the entire length of the tube.

For background information *see* ADIABATIC PROCESS; DIFFUSION; DISTILLATION; FREE ENERGY; ISOTOPE SEPARATION; SOUND; THERMOACOUSTICS; VISCOSITY; WAVE (PHYSICS); WAVE MOTION in the McGraw-Hill Encyclopedia of Science and Technology.   Gregory W. Swift

Bibliography. D. A. Geller and G. W. Swift, Saturation of thermoacoustic mixture separation, *J. Acous. Soc. Amer.*, 111:1675–1684, 2002; D. A. Geller and G. W. Swift, Thermoacoustic enrichment of the isotopes of neon, *J. Acous. Soc. Amer.*, 115:2059–2070, 2004; D. A. Geller and G. W. Swift, Thermodynamic efficiency of thermoacoustic mixture separation, *J. Acous. Soc. Amer.*, 112:504–510, 2002; P. S. Spoor and G. W. Swift, Thermoacoustic separation of a He-Ar mixture, *Phys. Rev. Lett.*, 85:1646–1649, 2000; G. W. Swift and P. S. Spoor, Thermal diffusion and mixture separation in the acoustic boundary layer, *J. Acous. Soc. Amer.*, 106:1794–1800, 1999, 107:2299 (E), 2000, 109:1261(E), 2001.

Transits of Venus

A transit of Venus, when Venus crosses the face of the Sun as seen from Earth, is one of the most unusual events in astronomy. Until 2004, only five transits had been seen in history: in 1639, 1771, 1779, 1874, and 1882. Nobody alive had seen one. But on June 8, 2004, a transit of Venus was widely visible and was seen by millions.

Venus is only 1/30 the apparent diameter of the Sun in the sky, so it covers only 0.1% of the Sun's disk. Still, the dark silhouette was very striking to observers. It was barely visible to the unaided eye, but it showed clearly in telescopes of all sizes and with telephoto lenses (**Fig. 1**).

Analysis. Finding the distance to the Sun by analyzing transits of Venus was for hundreds of years

Fig. 1. June 8, 2004, transit of Venus across the face of the Sun. (*From J. M. Pasachoff et al., Committee for Research and Exploration of the National Geographic Society*)

the most important problem in astronomy. Analysis by Edmond Halley in 1715 showed that they were a way of finding the size and scale of the solar system. Nicholas Copernicus had shown in 1543 that only Mercury and Venus had orbits around the Sun interior to Earth's, so that only those two planets could show transits to Earth's inhabitants. Johannes Kepler in 1609 had shown that the planets orbit the Sun in ellipses with the Sun at one focus and that the speed in their orbits depends in a certain way on their distance from the Sun. Most importantly for this purpose, Kepler had shown in 1619 that the square of the period of a planet's orbit is proportional to the cube of the size of the planet's orbit (technically, to the cube of the semimajor axis of the ellipse). Since the periods with respect to the Earth can be observed directly, and they can be transformed into periods with respect to the stars using mathematics worked out by Copernicus, Kepler's methods gave proportionalities for the sizes of the planets' orbits. Mars's orbit, for example, has a size of about 1.5 astronomical units, and Venus's has a size of about 0.7 astronomical unit, where an astronomical unit is essentially the size of Earth's orbit.

With proportionalities known, if any one distance in the solar system could be found, then all the other distances immediately would be known. With those distances, the actual sizes of the planets and the Sun could be found by transforming the apparent angles they subtended in the sky. Halley showed how timing the duration of a transit of Venus from different latitudes on Earth could, using the concept of parallax, reveal the distance from the Earth to Venus. The effect is much like the one you see when you look at your thumb alternately through one eye and then through the other. If you hold your thumb up close to your face, it appears to jump widely from side to side between views, while if you extend your arm, your thumb jumps much less. Similarly, the difference in the projection of Venus's path against the Sun depends on how far away Venus is.

Historical transits. Kepler, using his Rudolphine Tables of 1627 (which were the first tables of planetary positions to make use of his laws of planetary motion, and were dedicated to his late patron, Rudolph II), had predicted transits of Mercury and of Venus in 1631. The transit of Mercury was observed from Europe, confirming the tables' accuracy, but the transit of Venus had been predicted to be on the other side of the Earth and was not seen. After Kepler's death, Jeremiah Horrocks, aged about 20, in Hoole near Liverpool in England, realized from Kepler's tables that there would be another transit of Venus in 1639. Only Horrocks and a friend of his near Manchester, William Crabtree, who had been advised by Horrocks to watch, saw this event.

Transits of Venus occur in pairs separated by 8 years, and then with intervals of 105.5 or 121.5 years before the first transit of the next pair. These intervals arise because of the inclination of Venus's orbit with respect to the Earth and the commensurability of the two orbits, given that the Sun, Venus, and the Earth have to be in syzygy (three objects in a line) for a transit to occur.

Transit of 1761. When the next transit occurred, in 1761, dozens of expeditions were sent all over the globe to observe it. Halley's method, to get the desired accuracy, required timing of the contacts to about a second. (First contact is when Venus's silhouette first touches the Sun, the beginning of ingress; second contact is when it entirely enters the Sun; third contact is when it internally touches the egress side of the Sun; and fourth contact is when it entirely leaves the Sun.) But during the observations, when Venus should have entirely entered the Sun at second contact, a dark band or ligature seemed to link Venus's silhouette with the dark background sky outside the Sun. This linking darkness is known as the black drop.

This black-drop effect lasted 30 seconds to a minute, severely limiting the accuracy of the timing and therefore of the determination of the distance to Venus and, still further, of the size and scale of the solar system. A bright rim around Venus before it entered and after it left the Sun's disk was identified as Venus's atmosphere. The atmosphere then and since was often erroneously cited as the cause of the black-drop effect.

Transit of 1769. Many expeditions again went all over the world for the transit of 1769. The British Admiralty gave a ship, *Endeavour*, to the lieutenant James Cook (making him Captain Cook) to travel to Tahiti to observe the transit. Cook and the astronomer onboard, Charles Green, observed in clear skies but reported that their timing disagreed because of the black-drop effect. Cook's subsequent investigations of New Zealand and Australia can be considered spinoffs of this astronomical research.

Transits of 1874 and 1882. These transits were widely studied with the then new methods of photography. The black-drop effect again prevented more accurate measures, and eventually the distance from the Earth to the Sun and the size of the solar system were known from parallaxes of asteroids and by other means. Now, indeed, radar has been bounced off Venus and even off the solar corona, so the distances are known quite well. Still, observing the 2004 transit of Venus had historic interest.

Explanation of black-drop effect. In 1999, NASA's *Transition Region and Coronal Explorer* (*TRACE*) observed a transit of Mercury. Data reduction by Jay Pasachoff and Glenn Schneider revealed a slight black-drop effect, even though the observations were obtained from outside the Earth's atmosphere and were of a planet that essentially has no atmosphere. The observation thus proved that an atmosphere was not necessary to have a black-drop effect. Further, the detailed data reduction showed that the black-drop effect for Mercury was a combination of the inherent blurring that results from the finite size of the telescope and the solar limb-darkening, the lessening of the Sun's intensity near its edge. The limb-darkening effect is extreme very

Fig. 2. Venus's silhouette on the disk and an arc of Venus's atmosphere outside the Sun's limb at ingress on June 8, 2004, imaged with NASA's *TRACE* spacecraft. (*From J. M. Pasachoff and G. Schneider, Committee for Research and Exploration of the National Geographic Society, NASA, Lockheed Martin Solar and Astrophysics Laboratory, Smithsonian Astrophysical Observatory*)

close to the limb (edge), in just the region of the black drop.

Transit of 2004. The June 8, 2004, transit was widely anticipated. The European Southern Observatory, the National Aeronautics and Space Administration (NASA), and other organizations provided educational efforts, enlisting school students and others. The transit was entirely visible from almost all of Europe, Africa, and Asia. Only the end was visible from the United States, and only the beginning was visible from Australia. On the day of the transit, clear skies prevailed over most of the observing regions. In Hoole, where Horrocks had observed the 1639 event, skies were sufficiently clear to allow participants in an International Astronomical Union colloquium on the transit of Venus to see the event.

Thousands of photographs—both on film and with digital cameras—are posted on the Web. Children's drawings are posted on the European Southern Observatory's sites along with other observations and with the value of the astronomical unit derived from the many students' observations reported to them.

Pasachoff and Schneider obtained observations with the *TRACE* spacecraft, in collaboration with Karel Schrijver, Leon Golub, Ed DeLuca, and others (**Fig. 2**).

Application to exoplanet transits. Transits of Venus have in recent years gained a new importance that replaces the historical importance for the solar system. Since 1995, over 135 planets have been found orbiting other Sun-like stars. These exoplanets have largely been found by their gravitational effects on their stars, but a handful have been detected in transit across their stars. New efforts to use the transit method, and even planned spacecraft to exploit the

method, guarantee that hundreds of such exoplanet transits will be visible. Careful observation and analysis of the 2004 transit are providing basic data that may help explain details of the exoplanet transits. In particular, understanding how the atmosphere of Venus affects the total light of the Sun + Venus may help in the discovery of atmospheres around exoplanets and in the analysis of the temperature structure of the parent stars. Pasachoff, Schneider, and Richard Willson used the *ACRIM* satellite in Earth orbit to detect the 0.1% drop in the solar intensity, as well as the solar limb-darkening, as Venus entered and proceeded across the face of the Sun.

Future transits. The June 5–6, 2012, transit of Venus will be in progress at sunset in the continental United States and most of Canada, and it will be entirely visible from Hawaii and Alaska. The following transits of Venus will not occur until 2117 and 2125.

For background information *see* CELESTIAL MECHANICS; KEPLER'S LAWS; MERCURY; ORBITAL MOTION; PLANET; SYZYGY; TRANSIT (ASTRONOMY); VENUS in the McGraw-Hill Encyclopedia of Science & Technology.   Jay M. Pasachoff

Bibliography. L. Golub and J. M. Pasachoff, *Nearest Star: The Surprising Science of Our Sun*, Harvard University Press, 2002; E. Maor, *Venus in Transit*, Princeton University Press, 2000; J. M. Pasachoff, *A Field Guide to the Stars and Planets*, Houghton Mifflin, 2000; D. Sellers, *The Transit of Venus: The Quest To Find the True Distance of the Sun*, Maga Velda, 2001; W. Sheehan and J. Westfall, *The Transits of Venus*, Prometheus, 2004.

Transportation system security

Recent events worldwide have radically changed how security is viewed in the transportation environment. Previously, most transportation security operations were concerned with vandalism and theft. Now they focus on preventing terrorist attacks and mitigating the potential consequences of such attacks. A broad range of conventional and innovative technologies are being applied to address such threats and improve transportation security.

Transportation security challenges. Applying technology solutions is particularly challenging in the United States due to the unique characteristics of the nation's transportation systems.

Most transportation systems are open and easily accessible to the public. Millions of miles of air, rail, highway, and shipping networks crisscross the country. These systems are intermodal and are increasingly interdependent. They are focused on efficiency and competitiveness, and are owned, operated, and used by local, state, and federal government agencies, as well as the private sector. Each of these factors adds complexities that affect the deployment of security technologies and make many solutions unsuitable.

In addition, transportation systems are the backbone of the economy, which further complicates the insertion of security measures. People depend on transportation systems for their livelihood. According to the *U. S. Census 2000 Briefs*, 120 million people use highways to commute to work. In the New York City metropolitan area, over 7 million people take public transportation daily. On the commercial side, industry depends on transporting goods in a timely manner. In 2002, Booz Allen Hamilton brought together key business and government leaders to address the potential implications of a simulated attack against the transportation system and supply chains. This port security "war game" simulated a radiological dirty bomb discovered in a cargo container just unloaded from a ship. The economic loss of closing ports in response to the attack for less than two weeks was over $58 billion. These statistics show that solutions for preventing or responding to a terrorist threat or event can have serious economic consequences.

Security system solutions. Because of the characteristics of the transportation infrastructure, the best technology solutions must balance security with efficiency, choose between short-term and long-term solutions, account for operational and maintenance needs, consider privacy implications and ethical issues, and deal with the human component of the system. That is, they must use technology wisely in a systems approach to effectively address these challenges.

The National Academies' Transportation Research Board advocates the concept of a layered security system with multiple and integrated backup features. Transit agencies are applying this concept for securing facilities, for example, by installing closed-circuit television (CCTV) systems, alarmed doors, and motion detectors for intrusion detection. Some camera systems incorporate image analysis software to automatically identify suspicious behavior and warn transit police. The use of magnetic cards for entry adds another layer to facility security. If an intruder enters without a card, an alarm sounds in the security control center (and in some cases activates a CCTV monitor) so that transit police can determine the nature of the intrusion and react appropriately.

Technologies for enhancing security. Security technology systems can be applied in different ways and for a variety of purposes in transportation, including prevention, detection, response, and recovery. For example, pager-sized radiation detectors can help in identifying potential hidden radiological sources. Both vapor and nonvapor technologies can identify explosives and are particularly useful for baggage screening. Various transportation systems have taken Department of Defense chemical agent detector technologies and deployed them in public transportation environments to facilitate response by quickly identifying the agent and tracking its dispersion.

Transportation facilities and containers can be protected with a variety of intrusion-detection and access-control technologies. Some mix conventional motion detectors, heat-imaging and volumetric monitoring systems, and "smart" CCTVs. In addition, biometric access-control technologies that match fingerprints and retina scans, and electronic seals for monitoring access to cargo containers are being tested and used.

Wireless and traditional communication systems link remote detection systems to other systems to transmit information in real time for immediate analysis and response. Robust communication systems that can operate during power outages and multiple system failures can greatly improve communication for first responders during an emergency.

Reinforcing ("hardening") a tunnel with special materials to resist or mitigate the effects of an explosive device can help transportation systems restore normal services more rapidly after an attack. Decontamination technologies, such as air curtains and fume burners, can be used to help restore a transportation system and meet long-term recovery and public health expectations.

Tracking devices, such as those based on the Global Positioning System (GPS) technologies, can be linked to "geofencing" and exclusion zone software that identifies vehicles deviating from an approved route or entering an inappropriate area. These can be coupled with technologies for remotely shutting off the engine of a suspect vehicle.

New techniques for cybersecurity are becoming even more critical as people depend more and more on information systems in transportation. U.S. Customs has deployed information management systems, such as the Automated Manifest System and its Automated Targeting System, that sift through massive amounts of container manifest data and patterns to identify high-risk cargoes or anomalies for further inspection.

Figures 1, **2**, and **3** demonstrate selected technologies in various transportation environments (freight, transit, and response), and illustrate how disparate technologies can be linked together into a secure transportation system.

Security technology systems. There are numerous success stories of technologies being incorporated into transportation security systems. The first three examples below describe complete security systems that are at various stages of deployment. The other examples describe specific technologies, including some that are currently being tested and refined, that are components of a complete security system.

Operation Safe Commerce (OSC). This is a good example of a systems approach to security. OSC is a federal program managed through the Transportation Security Administration with the aim of deploying numerous technology suites along the entire supply chain, from manufacturing to distribution, and then testing how well these technologies perform their functions. The OSC is a public-private partnership that was created to enhance supply chain security, while facilitating efficient movement of goods. OSC

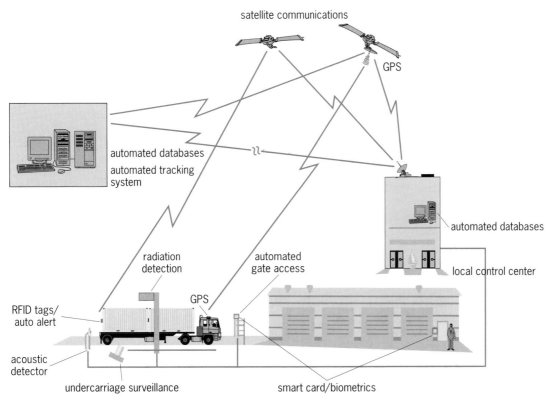

Fig. 1. Selected freight security technologies.

is testing technologies and associated procedures such as electronic seals to better prevent tampering with cargo containers, radio-frequency identification tags, assisted GPS (which combines GPS systems with wireless networks), integrated intrusion detection systems to secure warehouses and terminals, and automated tracking and detection systems for supporting smart and secure trade lanes. The ports of Tacoma/Seattle and Los Angeles/Long Beach and the Port Authority of New York and New Jersey are participating in the OSC program.

Transportation Worker Identification Credential (TWIC). This federal program will give a common credential to all transportation workers in all modes across the United States. It uses smart card technology that will have biometric identifiers to positively identify transportation workers who need access to secured areas within transportation systems.

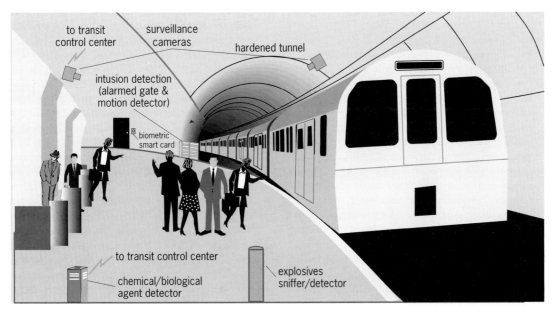

Fig. 2. Selected transit security technologies.

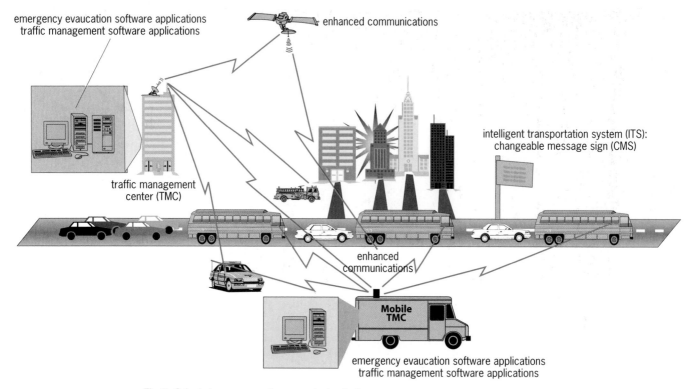

emergency evacuation software applications
traffic management software applications

enhanced communications

intelligent transportation system (ITS):
changeable message sign (CMS)

traffic management
center (TMC)

enhanced
communications

Mobile
TMC

emergency evacuation software applications
traffic management software applications

Fig. 3. Selected response and recovery technologies.

Program for Response Options and Technological Enhancements for Chemical/Biological Terrorism (PROTECT). This is a joint effort of the Department of Energy, Department of Justice, Federal Transit Administration, and Argonne National Laboratories. The system uses chemical agent detectors, CCTV cameras, digital video, real-time networking, contaminant dispersion modeling and simulation, and the Chemical Biological Emergency Management Information System database to collect and share critical information about an attack within minutes, and to allow rail system operators and emergency responders to take organized and effective response actions. This significantly reduces response time in the event of a chemical attack, thus saving lives. A prototype system is currently operational at the Washington Metropolitan Area Transit Authority subway system in Washington, DC.

Airline passenger screening system. The Federal Aviation Administration has been working on new technology to further improve airline passenger screening. Through a partnership with the Department of Energy's Pacific Northwest National Laboratory, a system is being developed to rapidly identify hidden weapons, explosives, and other contraband, including plastic, ceramic, and other nonmetallic objects. The technology uses millimeter waves (microwaves) to generate holographic images. This is an improvement over current systems such as metal detectors which cannot identify nonmetallic objects or explosives, and over x-ray systems which may present a health risk to the public.

Vehicle and Cargo Inspection System (VACIS®). This is a new technology being used by the U.S. Customs Service at locations throughout the United States for noninvasive imaging of sea containers, trucks, and other vehicles that may contain contraband, improperly manifested cargo, or other threats. It uses a proprietary gamma-ray imaging technique, requiring a very low radiation dose rate. This has advantages over older techniques for large-object inspection and can be operated without a special protective building or enclosure. The system helps identify and segregate suspect cargo for further inspection.

Decontamination system. The Soldier's Biological Chemical Command is investigating and testing battlefield technologies for application in civilian transportation systems to better decontaminate facilities and infrastructure after a chemical, biological, or radiological attack without adversely affecting the environment. These solutions will have direct applications for transportation systems such as subways. Some of these technologies are being investigated for how to better decontaminate sensitive equipment.

SAFECOM. This federal project is developing standard architectures and platforms that allow incompatible radio systems to communicate and fuse together voice and data systems through a wireless system. It uses various technologies such as software-defined radio, voice over IP (internet protocol) systems, and highly configurable communication systems. Demonstration projects are underway in over a dozen locations.

Outlook. Although security technologies are rapidly being improved and deployed, we must address related challenges to assure that we can continue to use technology wisely. Newer technologies are leaping over legacy systems and presenting us with the challenge of developing systems that can be easily upgraded. This may be as simple as assuring that control software is designed with the capability to incorporate information from yet-to-be-developed sensor systems. Other challenges may be far greater; for example, designing modular systems that will be compatible with future nanotechnologies. Continued improvements are needed in general areas such as system cost reduction, miniaturization, wireless communication systems, data fusion techniques, image analysis methodologies, materials, and predictive algorithms.

The next few years should show an increased standardization of security technologies, giving the industry the added benefit of more affordable off-the-shelf solutions rather than our current costly customized applications. We will be required to continue to think in nontraditional ways and be prepared to develop and deploy technologies to protect against future threats that have not been identified.

For background information *see* AIRPORT ENGINEERING; CLOSED-CIRCUIT TELEVISION; COMPUTER SECURITY; HARBORS AND PORTS; PARTICLE DETECTOR; RAILROAD ENGINEERING; SATELLITE NAVIGATION SYSTEMS; SUBWAY ENGINEERING; TRANSPORTATION ENGINEERING; VESSEL TRAFFIC SERVICE in the McGraw-Hill Encyclopedia of Science & Technology.

Joyce Wenger; Nicholas Bahr

Bibliography. *Cybersecurity of Freight Information Systems, a Scoping Study*, Transportation Research Board of the National Academies, Committee on Freight Transportation Information Systems Security of the Computer Sciences and Telecommunications Board, Spec. Rep. 274, 2003; *Deterrence, Protection, and Preparation, the New Transportation Security Imperative*, Transportation Research Board of the National Academies, Panel on Transportation, Committee on Science and Technology for Countering Terrorism, Spec. Rep. 270, 2002; War game scenario shows economic impact of terror, *Wall Street Journal*, Dec. 4, 2002.

Two-photon emission

Two-photon radiation experiments with metastable atomic hydrogen provide a unique method of studying the Einstein-Podolsky-Rosen (EPR) paradox concerning the incompleteness of quantum mechanics. In 1935, the authors of this paradox argued that, in certain hypothetical experiments which they constructed, there was a correlation between the results of measurements at spatially separated points that could not be understood by quantum mechanics, and that quantum mechanics could therefore not be considered a complete description of physical reality. In the experiments with metastable hydrogen, one detects, in coincidence, two photons emitted from a common source. The coherent photons are emitted together in a single atomic transition. This contrasts with other experiments to study the EPR

paradox, such as those of A. Aspect and colleagues, where the two photons arise from two separate transitions and are emitted in cascade, delayed with respect to each other by the lifetime of the intermediate state. Experimental sensitivity can be further enhanced by a three-polarizer arrangement, which provides the largest difference observed so far (more than 40%) between the predictions of quantum mechanics and classical local realistic theories.

Quantum theory. The roots of the quantum theory of radiation go back to the work of Max Planck (1900) and Albert Einstein (1905). The familiar transitions observed in atomic and molecular spectra arise from the interaction of an atomic electron with the radiation field in its lowest order. This interaction is responsible for one-photon electric and magnetic dipole transitions as well as higher-order multipole transitions. Multiphoton transitions, however, represent second-order radiation processes. Pioneering papers on the quantum theory of multiphoton processes were published by Maria Göppert-Mayer in 1929 and 1931. These papers led to an interest in the two-photon emission process in the simplest atom, hydrogen. *See* MULTIPHOTON IONIZATION.

In 1940, Gregory Breit and Edward Teller calculated the rate of two-photon electric dipole decay of the metastable $2^2S_{1/2}$ state in the hydrogen atom using the quantum theory of multiphoton processes introduced by Göppert-Mayer. **Figure 1** is an energy-level diagram that shows the ground state and first excited states of the hydrogen atom. The $2^2P_{1/2}$ state lies lower in energy than the $2^2S_{1/2}$ state due to the Lamb shift and, in principle, the $2^2S_{1/2}$ state can decay

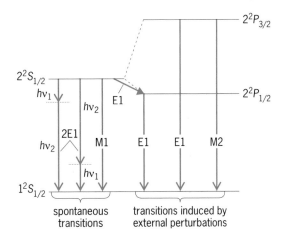

Fig. 1. Energy-level diagram showing the ground state ($1^2S_{1/2}$) and first excited states of the hydrogen atom. Spontaneous and electric-field-induced transitions out of the metastable $2^2S_{1/2}$ state are represented by the vertical and diagonal arrows, respectively. The arrows are labeled by the decay modes involved. All transitions are single-photon, except for the two-photon transition labeled 2E1. The two photons have energies $h\nu_1$ and $h\nu_2$, where ν_1 and ν_2 are the corresponding light frequencies and h is Planck's constant. The single-photon decay modes are labeled E1 (electric dipole), M1 (magnetic dipole), and M2 (magnetic quadrupole). (*After A. J. Duncan, M. O. Scully, and H. Kleinpoppen, Polarization and coherence analysis of the optical two-photon radiation from the metastable $2^2S_{1/2}$ state of atomic hydrogen, Adv. Atom. Mol. Opt. Phys., 45:99–147, 2001*)

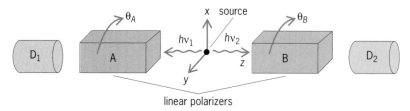

Fig. 2. Coordinate system with reference to the emission and polarization correlations of the two-photon emission ($h\nu_1$ and $h\nu_2$) of metastable atomic hydrogen, H(2S), detected in the $+z$ and $-z$ directions by the detectors D_1 and D_2. The transmission axes of the two polarizers, A and B, are at angles θ_A and θ_B with respect to the x axis. (*After W. Perrie, Stirling two-photon apparatus, Ph.D. thesis, Stirling University, 1985*)

via a one-photon electric dipole (E1) transition to the $2^2P_{1/2}$ state. However, due to the very small energy separation of the two states, the E1 transition probability is negligibly small. The lifetime of the $2^2S_{1/2}$ state would be about 20 years if it decayed solely via this transition. A relativistic treatment of the problem using the Dirac theory indicates that one-photon magnetic dipole (M1) transitions between the $2^2S_{1/2}$ and the $1^2S_{1/2}$ state (ground state) are allowed. The lifetime of the $2^2S_{1/2}$ state against such M1 decays is calculated to be about 2 days. The fact that the $2^2S_{1/2}$ state is shorter-lived is the result of its decaying via a two-photon electric dipole process to the $1^2S_{1/2}$ state. Breit and Teller estimated the mean lifetime against this type of decay to be about one-eighth of a second, which is comparable to the measured value. (Two-photon emission is not restricted to atomic spectroscopy. It has been observed in many areas of physics, including nuclear and particle physics, the laser cooling of atoms, solid-state physics, and biophysics.)

The two photons emitted in opposite directions in the decay of the metastable $2^2S_{1/2}$ state in the hydrogen atom have correlated properties since they have a common source, and can be used to carry out fundamental experiments of the type envisioned in the EPR paper. For example, the polarization of the two emitted photons can be studied using a pair of linear polarization analyzers placed in the oppositely directed paths of the photons (**Fig. 2**). The outputs of the detectors that are placed behind the polarizers are measured in coincidence in order to search for correlations between the emitted photons. The analysis of the coincidence data proves the coherent nature of the two-photon process in atomic hydrogen. By placing a multiwave plate in the path of one of the photons in the correlated pair, the spatial and temporal coherence of a single photon can be measured. These quantities are found to be extremely small, roughly 400 nm and 10^{-15} s, respectively. An even more sensitive EPR-type test of the predictions of quantum mechanics and local realistic theories has been made by using three linear polarizers, one in the path of one photon and two in the path of the oppositely directed photon.

The quantum electrodynamical theory of the simultaneous emission of two photons is based on the electron's linear momentum and the magnetic vector potential. The two photons do not carry off

any angular momentum in the transition between the $2^2S_{1/2}$ and $1^1S_{1/2}$ states. Therefore, the two photons must have equal helicities. This means that the two circularly polarized photons are either both right-handed (so that the two-photon state is $[R_1> [R_2>)$, both left-handed ($[L_1> [L_2>$), or in a superposition of these states. Parity conservation, however, requires the superposition state with the plus sign, given by Eq. (1).

$$[\Psi>_+ = 2^{-1/2}([R_1> [R_2> + [L_1> [L_2>)   (1)$$

Alternatively, the state vector describing the two photons can be written in terms of a linear polarization representation, as in Eq. (2), where $[x>$ and $[y>$

$$[\Psi>_+ = 2^{-1/2}([x_1> [x_2> + [y_1> [y_2>)   (2)$$

correspond to the linear polarization components of the photons being parallel to the x and y directions, respectively, in the coordinate system of Fig. 2.

Coincidence experiments. In a typical coincidence experiment, either the circular or the linear polarization correlations are measured for the two photons that move in opposite directions along the z axis (Fig. 2). The measurement forces the state vector of the pair to "collapse" into the circularly polarized components $[R_1> [R_2>$ or $[L_1> [L_2>$ or alternatively into the linearly polarized components $[x_1> [x_2>$ or $[y_1> [y_2>$. Each result would occur with a probability of 1/2. The collapse of the state vector implies that a measurement of the polarization of a photon traveling in the $+z$ direction determines the result of a subsequent measurement of the polarization of the photon moving in the $-z$ direction, irrespective of the distances between the two measuring devices. This clearly violates the principle of locality, which is well known from classical and relativistic physics. According to this principle, the value of a physical quantity

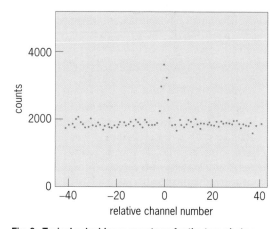

Fig. 3. Typical coincidence spectrum for the two-photon emission from the decay of the metastable $2^2S_{1/2}$ state in atomic deuterium [D(2S)]. The time-correlated spectrum shown is obtained after subtracting a background contribution arising from the quenching of the D(2S) atoms in the beam by use of an external electric field. The time delay per channel number is 0.8 ns. (*After W. Perrie, Polarization correlation of the two photons emitted by metastable atomic deuterium: A test of Bell's inequality, Phys. Rev. Lett., 54:1790–1793, 1985*)

measured at point A cannot depend on the choice of measurement made at another point B unless the physical quantities measured at A and B are correlated. This discussion leads to the Bohm-Aharanov version of the EPR paradox.

Measurements have been performed on the pair of correlated photons emitted in the two-photon decay of the $2^1S_{1/2}$ of hydrogen. In these experiments, the heavier isotope of hydrogen, deuterium, was used for technical reasons. A beam of deuterium atoms in the metastable $2^2S_{1/2}$ state was produced in a charge-exchange reaction as positively charged deuterium ions from a radio-frequency ion source were passed through a cesium vapor cell. The reaction, $D^+ + Cs \rightarrow D(2S) + Cs^+$ has a large cross section for capture into the excited state of D at a D^+ energy of about 1 keV. The metastable D(2S) atoms in the beam were quenched by passing them through an external electric field. The electric field in the quench region Stark mixes the metastable $2^2S_{1/2}$ state with the much shorter-lived $2^2P_{1/2}$ state. This mixed state decays by a one-photon electric dipole process producing the well-known Lyman-alpha line. The quench signal is proportional to the number of D(2S) atoms in the beam, and it was used to normalize the two-photon coincidence signal arising from the decay of unquenched D(2S) atoms. The two-photon coincidence radiation was detected at right angles to the D(2S) beam. **Figure 3** shows a coincidence spectrum. The shape of the peak is symmetric, which is to be expected in the case of the simultaneous emission of two photons.

Results. The first experimental study of the two-photon decay of the metastable $2^2S_{1/2}$ state of hydrogen was performed by D. O'Connell and colleagues in 1975. In this experiment the angle α between the directions of the two photons was varied. The data, based on three values of the angle α, approached the theoretical prediction of $1 + \cos^2 \alpha$ and clearly disagreed with an isotropic angular correlation in the detection plane. In subsequent experiments that placed linear polarization analyzers in the paths of the two oppositely directed photons ($\alpha = \pi$), the coincidence rate was measured as a function of the angle θ between the transmission axes of the two linear polarizers. The detected coincidences demonstrated that the correlation between the photon polarizations was in essential agreement with the predictions of quantum mechanics but not those of local realistic theories. This result agreed with other types of polarization correlation experiments such as ones based on a pair of photons emitted in two cascading transitions, laser interferences, positronium decay, and radiation processes in nuclear and particle physics.

The EPR arguments concerned the problematic question of the completeness of quantum mechanics. The controversy between local realistic theories and quantum mechanics was highlighted in the debates between Einstein and Niels Bohr in 1935. In a local theory, a measurement of a physical quantity at one point in space-time is not influenced by

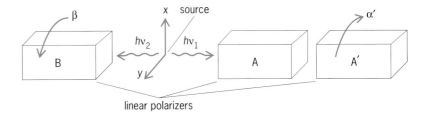

(a)

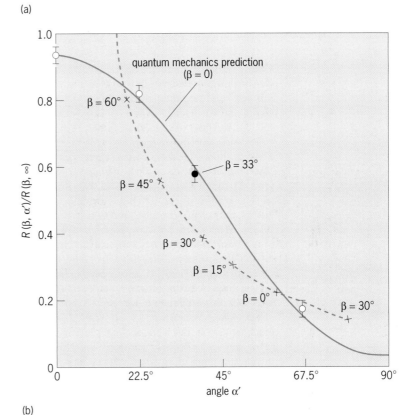

(b)

Fig. 4. Three-polarizer experiment. (*a*) Schematic arrangement. The orientation of polarizer A is fixed with its transmission axis parallel to the *x* axis. The transmission axes of polarizers B and A′ are rotated through angles β and α', respectively, with respect to the *x* axis. (*b*) The ratio $R(\beta,\alpha')/R(\beta,\infty)$ is plotted as a function of the angle α' for $\beta = -30°$, $0°$, $15°$, $45°$, and $60°$. The open circles and black circle represent the data for $\beta = 0°$ and $33°$, respectively. The solid curve represents the quantum-mechanical prediction for $\beta = 0°$, and the broken line shows the upper limit for the ratio set by the local realistic model of A. Garuccio and F. Selleri for various values of the angle β. (*After A. J. Duncan, M. O. Scully, and H. Kleinpoppen, Polarization and coherence analysis of the optical two-photon radiation from the metastable $2^2S_{1/2}$ state of atomic hydrogen, Adv. Atom. Mol. Opt. Phys., 45:99–147, 2001*)

a measurement made at another spatially separated point. In a realistic theory, objects exist independent of a measurement or observation. This problem lay dormant for some time. In 1964, John Bell showed that local realistic theories and quantum mechanics predicted different results for the outcome of experiments designed to measure the correlation of physical properties in a two-particle system. This opened the door for experimental tests of the different predictions. In a series of experiments that culminated in the work of Aspect and colleagues (1982), it was shown that the Bell inequality was violated. The measured correlations were found to be stronger than predicted by local realistic theories but were in good agreement with those predicted by quantum mechanics.

In a standard arrangement for studying correlations between the polarization states of two photons, linear polarizers are placed in the paths of the photons as they move in opposite directions from a common source (Fig. 2). In 1984, A. Garuccio and F. Selleri proposed an extension of this arrangement. They suggested placing a second linear polarizer in one of the arms. This novel experiment makes it possible to test more sensitively local and realistic theories in a hitherto unexplored manner. An experimental scheme and the relevant geometries for the directions of the three polarizers are shown in **Fig. 4a**. In the experiment the plane of polarizer A was held fixed while that of polarizer B was rotated through an angle β in a clockwise sense and polarizer A$'$ through an angle α' in a counterclockwise direction. The ratio of coincidence rates, $R(\beta,\alpha')/R(\beta, \infty)$, was measured (Fig. 4b); here $R(\beta,\alpha')$ is the rate for two photons detected with all three polarizers in place, and $R(\beta, \infty)$ is the rate with polarizer A$'$ removed. Figure 4b shows that the prediction of local realistic theories for this coincidence rate ratio is 0.413, while the measured value is seen to be 0.585 ± 0.029, which disagrees with the prediction of the local realistic theory of Garuccio and Selleri by about six standard deviations or a difference of more than 40% relative to the prediction of quantum mechanics. The result also violates Bell's inequality. This relative difference between the predictions of quantum mechanics and local realistic theory is much larger that that obtained in experiments such as the one by Aspect and colleagues, involving a pair of photons emitted in cascade transitions. (In this case the difference was only a few percent.)

Prospects. The three-polarizer technique could be promising for future two-photon studies using hydrogenlike ions with nuclear charge Z greater than 1. New theoretical and experimental investigations of coherence effects of two-photon radiation might be required in order to perform and interpret such experiments. Interesting coherence properties of two-photon emission have, indeed, been recently reported. These include a coherence time (on the femtosecond time scale) and a coherence length (of wave packets on the nanometer scale).

For background information *see* ATOMIC STRUCTURE AND SPECTRA; COHERENCE; ENERGY LEVEL (QUANTUM MECHANICS); HELICITY (QUANTUM MECHANICS); HIDDEN VARIABLES; NONRELATIVISTIC QUANTUM THEORY; QUANTUM ELECTRODYNAMICS; QUANTUM MECHANICS in the McGraw-Hill Encyclopedia of Science & Technology.　　Hans Kleinpoppen

Bibliography. U. Becker and A. Crowe, *Complete Scattering Experiments*, Kluwer Academic/Plenum, 2001; A. J. Duncan et al., Two-photon polarization Fourier spectroscopy of metastable atomic hydrogen, *J. Phys. B*, 30:1347–1359, 1997; A. J. Duncan, M. O. Scully, and H. Kleinpoppen, Polarization and coherence analysis of the optical two-photon radiation from the metastable $2^2S_{1/2}$ state of atomic hydrogen, *Adv. Atom. Mol. Opt. Phys.*, 45:99–147, 2001; F. Selleri, *Quantum Mechanics Versus Local Realism*, Kluwer Academic/Plenum, 1988; G. Weihs et al., Violation of Bell's inequality under strict Einstein locality conditions, *Phys. Rev. Lett.*, 81:5039–5043, 1998.

Ultrasonic aerosol concentration and positioning

The detection, identification, and separation of aerosols have potential applications in a wide variety of fields ranging from environmental science to homeland security. Information about airborne particle sizes, particle distribution, and particle composition is important, and sometimes even critical, to researchers studying the environmental impact of industrial smog, or to military specialists and emergency responders trying to detect chemical or biological agents in the air. To aid in the detection and classification of aerosols, many researchers have turned to acoustics to help concentrate and position aerosol particles in flowing airstreams. This article describes a new class of acoustic concentration and particle manipulation devices that has been developed for use in aerosol sampling applications.

Concentration and positioning devices. Scientists at Los Alamos National Laboratory have recently created a new class of acoustic devices that can concentrate and position aerosols quickly and inexpensively using sound waves created inside a vibrating tube. These solid-state devices require very little power (about 0.1 W) to operate and have the potential to greatly increase the efficacy of current-generation aerosol detection devices (such as particle sizers, optical classifiers, and particle fluorescence monitors) by significantly increasing the number of aerosol particles that enter the detector. They also have the ability to precisely position aerosol particles in a flowing airstream for detectors that require knowledge of the aerosol particle location for proper detection and classification.

It has long been known that aerosol particles experience a force in a high-intensity acoustic standing wave that transports them to the vicinity of either a pressure node or an antinode. This effect was first observed in 1874 by August Kundt, who noticed striations of dust particles in organ pipes. Since then, scientists have implemented acoustics to position macrosocopic samples (usually on the order of a millimeter or larger) in applications where the sample cannot contact the wall of a containment vessel. Many studies in containerless processing have been conducted to observe, for example, the growth of freely forming crystals and measurements of the physical properties of substances in both terrestrial and low-gravity environments.

Recent global events have stimulated government interest in better detection and classification of aerosols. The greatest interest is in the detection and identification of aerosols in the human respirable size range of 1–10 micrometers. In response to this interest, a set of novel acoustic concentration devices in

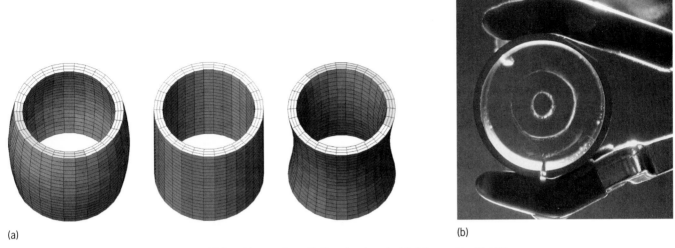

(a)   (b)

Fig. 1. Cylindrical-tube acoustic concentrator. (*a*) Breathing-mode oscillation of a piezoelectric tube used to drive the acoustic concentration field within the inner cavity of the tube. (*b*) Aerosol acoustically concentrated into three rings inside a piezoelectric tube. The tube is driven at 67 kHz, and the rings correspond approximately to the pressure nodes of the standing-wave field in the cylindrical cavity.

the form of hollow cylinders (tubes) has been developed. Constructed from a commercially available piezoelectric material, the devices work by changing their shape by alternately expanding and contracting the tube wall (vibrating) when stimulated by an alternating voltage. At a specific vibration frequency—governed by the physical dimensions and properties of the tube—the shape changes are radially uniform and it appears as if the tube is breathing; for this reason, this type of mode is designated a breathing mode (**Fig. 1**). The breathing mode can be a very energy-efficient mode to drive large-amplitude vibrations, known as resonances, in a structure.

The vibration of the tube's interior wall generates sound waves within the tube's cavity. Amplification of the pressure waves inside the air cavity results from wave interference and resonance. At certain well-defined excitation frequencies that depend on the physical properties of the air in the tube (such as temperature), the pressure vibrations within the tube's cavity produce standing waves. A resonance is defined as a standing wave where many waves interact simultaneously to amplify the maximum amplitude of the pressure field. There are two different resonances involved: (1) the resonance of the tube itself and (2) the resonance in the air cavity inside

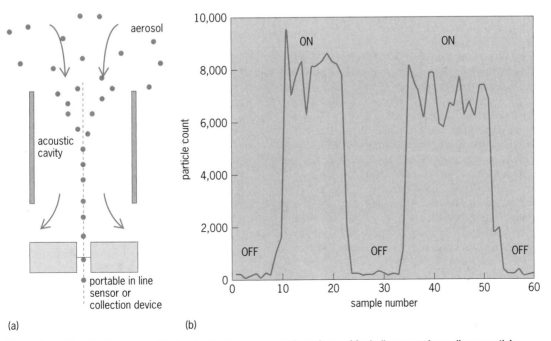

(a)   (b)

Fig. 2. Acoustic collection and positioning. (*a*) Design of an acoustic cavity used for in-line aerosol sampling or particle collection. Aerosol particles are concentrated along the central axis of the device and are siphoned into a detector or collected for later analysis. (*b*) Particle counts taken from the central axis in this device when the acoustic field is ON or OFF.

the tube. By carefully making the two resonance frequencies the same (by appropriately selecting the tube's size and other properties), a significantly high-intensity standing-wave pressure field can be generated inside the tube. This matching of the two resonance frequencies provides the high efficiency of this class of devices.

Within the cavity, aerosol particles experience a time-averaged force that directs them to the vicinity of a pressure node or antinode (depending on the properties of the aerosol particles) within the resident sound field. The force experienced by the particle is a balance between acoustic radiation pressure and viscous forces (and other higher-order effects) that act on the particle. These forces depend on particle size, particle density and compressibility, excitation frequency, and the sound pressure level within the cavity, and these properties determine the attraction of the particles to either a pressure node or antinode.

Aerosol concentration. Aerosol concentration is a means by which aerosol particles in a flowing airstream are bunched together for extraction or sampling purposes. For example, in detecting dangerous chemical or biological toxins in the air, a quantity of just one harmful particle per liter of air is considered the lower detection limit by officials in homeland security applications. From a volumetric standpoint, a 2-μm particle has a volume 10^{14} times smaller than a liter of air. It would be advantageous to selectively siphon off only the particles from an airstream and store them in a small region of space for analysis while letting the large volumes of clean air pass through. That is the principle behind aerosol concentrators in general, as well as the principle behind acoustic concentration.

In an acoustic concentrator, aerosol particles are forced to a localized position within the acoustic field for particle extraction. The simplest form of the acoustic concentrator is driven in an axisymmetric mode (Fig 1). In this mode, pressure nodes within the tube cavity are concentric rings that are centered about the axis of the cylinder. Aerosol particles are driven to the vicinity of these positions for later extraction. The extraction can be done by drawing off a small amount of the flow stream at these localized positions. The particles may then be transported to an in-line detector, trapped in a filter, or impacted into a small volume of liquid for later analysis. Concentration factors of greater than 40 have been observed with the current line of acoustic concentrators (**Fig. 2**).

As mentioned earlier, matching the breathing-mode structural resonance of the tube to the resonance frequency of the air-filled cavity gives the tube concentrator high power efficiency. For example, concentration of particles in the 10-μm size range with input electrical power on the order of 100 mW in a device 20 mm (0.8 in.) in diameter has been demonstrated. Such low power and small size will be especially useful for compact personal monitoring devices that will benefit from battery-powered aerosol concentration devices.

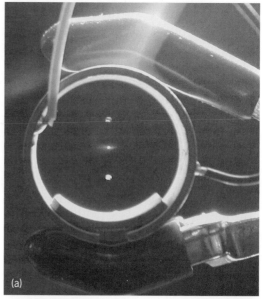

Fig. 3. Noncylindrical cavity positioners. (*a*) A small insert breaking the circular symmetry forces the aerosol to agglomerate at three distinct positions within the cavity. (*b*) Cavity modification is made to create four focused aerosol streams flowing through the acoustic cavity.

Aerosol positioning. Acoustic aerosol positioning is very similar to acoustic concentration in that the aerosol stream is forced to a given position within the acoustic chamber. In acoustic positioning, the particles are positioned for in-line analysis as opposed to collection (Fig. 2). Particle-laden air enters the acoustic device, and the particles are forced to a specified location for analysis. Many optical techniques (such as fluorescence spectroscopy) use probe beams with high optical power densities at the position of the particle. To accommodate the large optical powers, the light beams are focused to very small focal dimensions typically on the order of 500 μm. The acoustic aerosol positioner is used to locate particles within the focal area of the beam.

Very localized concentration regions can be generated by breaking the symmetry of the circular cross section of the tube. Aerosol particles can then be forced into several parallel streams for parallel processing in detection and identification platforms. The number of aerosol streams can be adjusted by altering the cavity design. Designs with as few as one and as many as six streams have been demonstrated (**Fig. 3**).

Scientific and technical impact. Acoustic concentration and positioning has a wide variety of applications in aerosol monitoring. With the global resolve to better detect chemical and biological attacks, these acoustic techniques will prove useful for homeland security detection and identification platforms. There are also applications unrelated to detecting toxins in the air. Given its ability to concentrate particulates, an acoustic concentrator can be used as a filterless filter; that is, it can be used to separate particulate matter from exhaust streams without the use of filter media (such as Hepa filters). This results in a filtering mechanism with a near-zero pressure drop across the device. There are many other applications, including the possibility of filtering bacteria or allergens from ventilation systems. The use of acoustic concentration outside the laboratory setting is still in its infancy and will see a wide array of applications in the future.

For background information *see* ACOUSTIC LEVITATION; ACOUSTIC RADIATION PRESSURE; AEROSOL; PARTICULATES; PIEZOELECTRICITY; RESONANCE (ACOUSTICS AND MECHANICS); ULTRASONICS; VIBRATION in the McGraw-HILL Encyclopedia of Science & Technology.   Gregory Kaduchak;
Christopher Kwiatkowski; Dipen Sinha

Bibliography. A. A. Doinikov, Acoustic radiation force on a spherical particle in a viscous heat-conducting fluid, II. Force on a rigid sphere, *J. Acous. Soc. Amer.*, 101:722–730, 1997; G. Kaduchak, D. N. Sinha, and D. C. Lizon, Novel cylindrical, air-coupled acoustic levitation/concentration devices, *Rev. Sci. Instrum.*, 73:1332–1336, 2002; E. H. Trinh, Compact acoustic levitation device for studies in fluid dynamics and material science in the laboratory and microgravity, *Rev. Sci. Instrum.*, 56:2059–2065, 1985; R. Tuckermann et al., Trapping of heavy gases in stationary ultrasonic fields, *Chem. Phys. Lett.*, 363:349–354, 2002.

Visual illusions and perception

To the casual observer, visual illusions may be mere curiosities, amusing tricks for the eye concocted by clever artists. But to the serious student of perception, visual illusions reveal important secrets.

Role in vision research. Vision seems effortless. We simply open our eyes and see the world as it is. This apparent ease is itself an illusion. Tens of billions of neurons and trillions of synapses labor together in the simple act of opening our eyes and looking. Visual illusions, far from being mere parlor tricks, allow researchers to peer inside this complexity and discover what all these neurons are doing. Most of us assume that our vision is much like a camera, simply taking snapshots or movies, and providing an objective report of what is around us. Visual illusions, and the research that builds on them, reveal that vision

does not report objective truth but weaves elaborate interpretations based on sophisticated detective work. Certainly the eye itself is a miniature camera, focusing an image on the retina at the back of the eye. But this retina is composed of hundreds of millions of neurons performing sophisticated operations, and the signals they send to the brain engage a network of billions of neurons. This is where vision really takes place and where the sophisticated detective work proceeds. The role of visual illusions in vision research is to reveal the clues and trains of reasoning that spawn our vision of the world.

Visual construction. Why should vision proceed by collecting clues and assembling sophisticated chains of reasoning? The reason is simple. Each eye has about 120 million photoreceptors, cells specialized to catch particles of light called photons. The eye cannot record shapes, depths, objects, colors, edges, motions, and all the other features of our visual world. It records only the number of photons caught by each of the 120 million photoreceptors and how these numbers vary with time. Thus, we must construct objects, depths, and colors from the clues hidden in a haystack of 120 million numbers. This becomes apparent when trying to build a computer vision system that sees objects and depth. The input to the computer comes from an attached video camera, and this camera provides a list of numbers that say, in effect, "It's this bright at this point in the image, and it's that bright at that point in the image . . ." for each of the millions of points of the image. The job is to find objects and three-dimensional shapes. It becomes clear that the only way to find them is to construct them, which requires many megabytes of software in computer vision systems and billions of neurons and trillions of synapses in the human brain.

It is because we must construct our visual world from such sparse clues that visual illusions are

Fig. 1. Trading towers illusion. We switch rapidly between seeing four tall towers and four shorter towers. This rapid switching occurs because both interpretations violate one of our rules of visual construction. (*Courtesy of D. D. Hoffman*)

possible and, indeed, inevitable. The clues available do not, in general, dictate a single construction that is logically guaranteed to be correct. When our visual system decides to interpret the available clues in a certain way, it necessarily makes assumptions that may prove to be false. If they do prove false, we have an illusion. This illusion becomes an important clue to the assumptions that, in this case, led the visual system astray. Thus, researchers study visual illusions, which are particularly insightful guides to the invisible machinations of the visual system.

Probabilistic reasoning. A central insight to emerge from the study of illusions is that our visual constructions rely on sophisticated probabilistic reasoning.

Trading towers illusion. As an example, consider the trading towers illusion in **Fig. 1.** We see this as a three-dimensional shape with four tall rectangular towers poking up. However, the image sometimes switches, revealing four shorter towers poking up. It shifts rapidly between these interpretations, leaving the visual system a bit unsettled.

There are a couple of points to observe about this illusion. First, we see it as a three-dimensional shape, even though it is a flat figure printed on the page. The visual system falsely fabricates a three-dimensional shape, despite our knowledge that it should be seen as flat. The reason is that the image that forms on the retina of the eye has only two dimensions. Therefore, the visual system must fabricate all the depth that we ever see. Once the visual system has evidence that it deems to warrant a three-dimensional interpretation, it fabricates that interpretation and does not usually listen to other parts of the brain that may want to contradict it.

Second, we see the various lines in the figure as straight lines in space in our three-dimensional interpretation. The edges of the towers, for instance, look perfectly straight. The visual system has a rule that says, in effect, "Always interpret a straight line in an image as a straight line in three-dimensional space." The reason for this rule is probabilistic. Suppose we see what looks like a straight line. Even though the curve in three-dimensional space that we are viewing looks straight, it is conceivable that in fact it is wiggly. However, to be visible, the wiggles would have to be precisely directed along our line of sight. In this case, if we moved the head slightly to get a different view of the curve, suddenly the wiggles

would become visible. Only from just the right viewpoint would the wiggles disappear and the curve look perfectly straight. It is highly unlikely, reasons the visual system (unconsciously), that we happen to be looking from just the right viewpoint to make the wiggles disappear. Therefore, it is highly likely that there are, in fact, no wiggles. And this leads to the rule that straight lines in an image must be interpreted as straight lines in three-dimensional space.

The rules of construction in vision are usually not inviolable. Instead, different rules interact, sometimes overriding each other in the process of coming to an interpretation. The trading towers illusion also demonstrates this point. At the central point of the figure, the lines constituting the three-dimensional interpretation appear to bend sharply, in contradiction to the rule that straight lines in an image should be seen as straight lines in three-dimensional space. Other rules of construction prompt the visual system to violate this rule at this one point. But the violation is unsettling and leads to the constant switching of interpretations we experience when viewing this figure.

Generic views principle. In the trading towers illusion, probabilistic considerations dictate the rule of visual construction. Using another form of probabilistic reasoning, one can also show that if two lines in an image terminate at precisely the same point to form a corner, the visual system must interpret the two lines as also meeting to form a corner in three dimensions. For if the lines did not meet in three dimensions but just happened to look like they meet from our current viewpoint, then any slight change in viewpoint would make them no longer look like they meet. This type of probabilistic reasoning is known as the "generic views principle" among vision researchers, and it has proved a powerful method for understanding visual processes and the illusions they can engender.

This rule makes sense from an evolutionary perspective. Since the probability is nearly zero that a three-dimensional curve will look straight in our retinal image, in our daily experience we are almost always right if we interpret each straight line in our retinal image as a straight line in three-dimensional space. As we move about and see each straight line from different points of view, we receive confirmation that our interpretation is correct. This pattern of experience, over many generations, will tend to adapt us phylogenetically to interpret the world in accordance with this rule. Over the life of an individual, this pattern of experience will tend to fine-tune the individual's specific neural responses to agree with this rule.

Neon worm illusion. Similar probabilistic considerations, using the generic views principle, explain why we see a glowing green worm on the right in **Fig. 2** but not on the left. If we look carefully on the right, we realize that the glowing green seen between the lines is entirely our construction, and not literally printed on the page. This glowing green disappears if the endpoints of the black and green lines

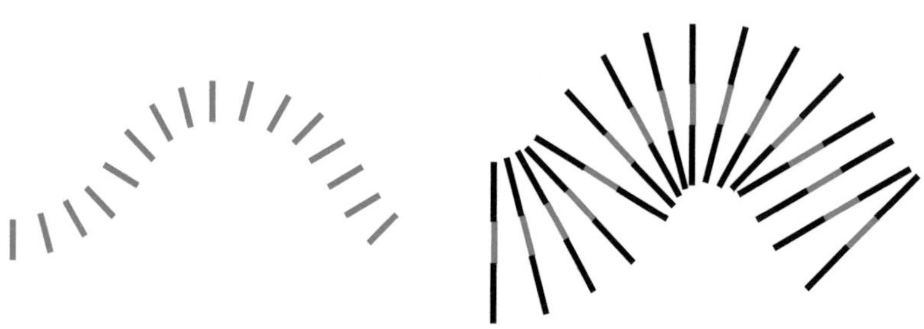

Fig. 2. Neon worm illusion. Our visual system constructs a pale-green surface between the green lines of the right figure but not between the green lines of the left figure. In the right figure, precise alignment of the black-and-green lines, end to end, is required for us to see the illusory green surface. (*Courtesy of D. D. Hoffman*)

are not precisely aligned. The visual system reasons (unconsciously) that the remarkable alignments of black-to-green and green-to-black transitions across the lines on the right of Fig. 2 are not simply an accident of viewpoint but must be due to some systematic process. The best explanation it can find for that systematic process is that there is a green transparent surface sitting in front of black lines, and so it constructs that visual interpretation. Physiologically this is, in one sense, nothing unusual. Recall that the retina has only a discrete array of photoreceptors. Since surfaces are continuous, the visual system must always construct the continuous surfaces we see from discrete information provided by the retina. The neon worm just makes more visible and undeniable this constructive process that is always at work.

Conclusion. By applying probabilistic reasoning principles to the study of visual illusions such as the trading towers and neon worm, vision researchers are discovering the clues and chains of reasoning used by the visual system to create our visual worlds.

For background information *see* EYE (VERTEBRATE); NEUROBIOLOGY; PERCEPTION; PHOTORECEPTION; VISION in the McGraw-Hill Encyclopedia of Science & Technology.   Donald D. Hoffman

Bibliography. D. Hoffman, *Visual Intelligence*, W. W. Norton, New York, 1998; M. Livingstone, *Vision and Art*, Harry Abrams, New York, 2002; S. Palmer, *Vision Science*, MIT Press, Cambridge, MA, 1999; J. Rothenstein, *The Playful Eye*, Chronicle Books, San Francisco, 2000; A. Seckel, *The Great Book of Optical Illusions*, Firefly Books, Westport, CT, 2002.

Vitamin D

The term vitamin D refers to a family of related fat-soluble, sterollike compounds that are essential for normal calcium and phosphorus deposition in bones and teeth. There are two major forms, vitamin D_2 and vitamin D_3. Vitamin D_3 (cholecalciferol) and its provitamin, 7-dehydrocholesterol are derivatives of animal sterols; their chemical structure resembles cholesterol, bile acids, and the steroid hormones (**Fig. 1**). Vitamin D_2 (ergocalciferol) is formed by the ultraviolet irradiation of its precursor ergosterol. Ergosterol is found in yeast, fungi, and plants. However, the synthesis of ergocalciferol from ergosterol hardly takes place in nature. Moreover, it has been demonstrated that vitamin D_2 has much lower bioavailability than vitamin D_3.

Metabolism and cellular actions. Sunlight is the major provider of vitamin D for humans. In the skin, the ultraviolet B (UVB) spectrum of sunlight (290–315 nm) induces the synthesis of vitamin D_3 (via an intermediate product, previtamin D_3) from its provitamin 7-dehydrocholesterol (Fig. 1). Food is a secondary source of vitamin D, but only a few foods such as eel, herring, and salmon provide sufficient levels (15–30 μg per 100 g edible portion). Conse-

quently, the dietary contribution of vitamin D usually constitutes only 10–20% of the human vitamin D supply. Cutaneously synthesized or orally ingested vitamin D is biologically inactive. Conversion to its active form begins in the liver, where it is hydroxylated via the enzyme 25-hydroxylase to produce 25-hydroxyvitamin D [25(OH)D], the major circulating form of vitamin D. In the kidney, 25(OH)D undergoes a second hydroxylation, catalyzed by the enzyme 1α-hydroxylase to produce the final active form of vitamin D, calcitriol [1,25-dihydroxyvitamin D] (**Fig. 2**). Plasma calcitriol levels are regulated homeostatically; therefore, if adequate plasma calcitriol levels are achieved, 25(OH)D is converted by a renal 24-hydroxylase into 24,25-dihydroxyvitamin D [24,25(OH)$_2$D] (Fig. 2). Calcitriol is also produced by local 1α-hydroxylases in various extrarenal tissues. (Moreover, vitamin D receptors exist in more than 30 different tissues.)

Calcitriol is an important regulator of systemic calcium and phosphorus metabolism, especially of serum calcium homeostasis. In this context, calcitriol is responsible for active calcium transport across the duodenal mucosa, for calcium resorption (release) from bone, and for active calcium reabsorption in the kidney. Moreover, calcitriol is known to regulate intracellular calcium metabolism of various tissues and cellular secretion of cytokines (chemical messengers).

Assessment of vitamin D nutritional status. Since 25(OH)D is the major circulating form of vitamin D and the precursor of the active vitamin D metabolite, calcitriol, the serum concentration of 25(OH)D is considered to be the hallmark for determining vitamin D nutritional status (that is, deficiency, insufficiency, hypovitaminosis, sufficiency, and toxicity). Low serum levels of 25(OH)D can occur as a

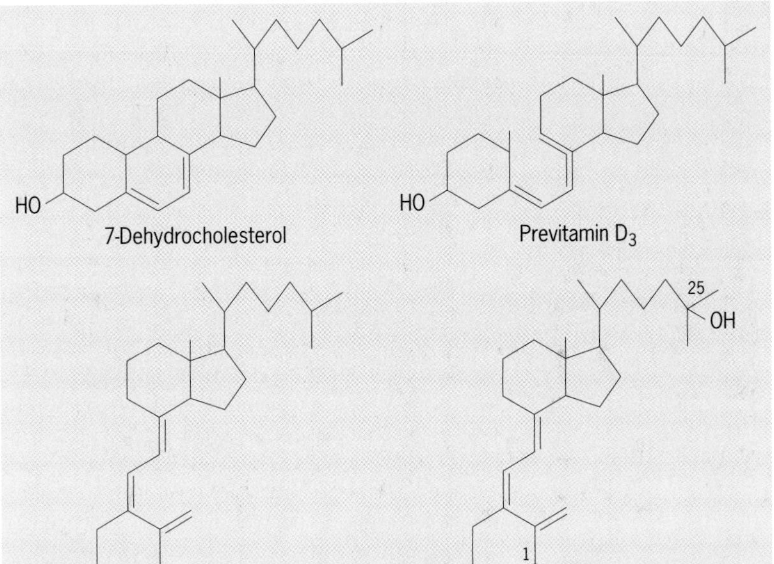

Fig. 1. Vitamin D precursors and metabolites.

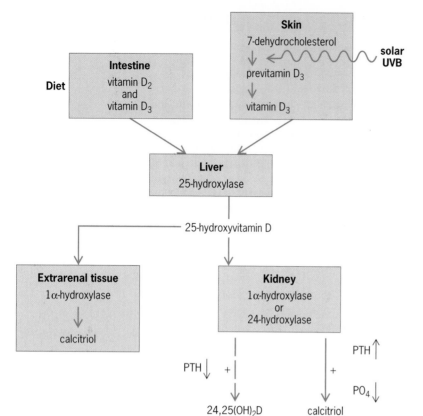

Fig. 2. Major metabolic pathways of vitamin D. + indicates stimulation; PTH, parathyroid hormone; 24,25(OH)₂ D, 24,25-dihydroxyvitamin D; PO₄, phosphate; up arrow, increase; down arrow, decrease.

consequence of inadequate UVB light–mediated skin synthesis and/or inadequate dietary vitamin D intake. Serum levels of 25(OH)D below 25 nmol/L (10 ng/ml) indicate vitamin D deficiency, which is associated with severe secondary hyperparathyroidism (characterized by excessive secretion of parathyroid hormone in response to reduction in blood calcium) and calcium malabsorption. Levels between 25 and 50 nmol/L indicate vitamin D insufficiency, resulting in mild hyperparathyroidism and low intestinal calcium absorption rates. Levels of 25(OH)D between 50 and 80–100 nmol/L reflect hypovitaminosis D, a condition in which body stores of vitamin D are low and parathyroid hormone levels can be slightly elevated. When levels of vitamin D are sufficient [serum concentrations of 25(OH)D between 80–100 and 200 nmol/L], no disturbances of vitamin D–dependent functions occur. Serum levels of 25(OH)D above 200–300 nmol/L indicate vitamin D intoxication, which causes intestinal calcium hyperabsorption and increased net bone resorption, leading to hypercalcemia (a condition characterized by excessive amounts of calcium in the blood).

Usually, the serum calcitriol level is not a reliable measure of vitamin D nutritional status, since in a vitamin D–deficient person the increase of parathyroid hormone, which increases renal production of calcitriol (but not extrarenal calcitriol synthesis), leads to normal or even elevated circulating calcitriol levels (Fig. 2). In specific diseases such as renal insuffi-

ciency, however, a low serum calcitriol level is indicative of reduced renal 1α-hydroxylase activity. Renal insufficiency is usually associated with hyperparathyroidism, despite normal 25(OH)D levels.

Prevalence of vitamin D deficiency and insufficiency. Inadequate vitamin D supply is a worldwide problem that affects all age groups. The primary cause is insufficient skin synthesis of vitamin D. Exposure to the solar UVB spectrum is negligible from November until February at 40°N latitude, and from October until March between 50° and 60°N latitude. Moreover, skin synthesis of vitamin D is possible only between the hours of 10 a.m. and 6 p.m., with maximal capacity of vitamin D synthesis occurring between 12 noon and 2 p.m. Since infants are usually kept out of the sun, they have to be supplemented with vitamin D. However, even in North America and Europe, an apparent resurgence of vitamin D deficiency during infancy is observed, since an increasing number of parents are not aware of or neglect preventive measures. Healthy subjects living between 40° and 60°N latitude, for instance in parts of North America and Europe, have seasonal fluctuations in circulating 25(OH)D levels, with insufficient 25(OH)D concentrations during wintertime.

Even in countries that have high levels of sun exposure, such as India, urbanization in association with low levels of outdoor activity during the daytime can lead to vitamin D deficiency. Moreover, women who for religious and cultural reasons dress in garments covering the whole body, including the hands and face, are at high risk for vitamin D deficiency. A high prevalence of inadequate vitamin D supply is also observed in dark-skinned people living in the Northern Hemisphere [50% or more of this population has insufficient plasma levels of 25(OH)D]. In elderly subjects, the capacity of the skin to produce vitamin D is markedly lower, and participation in outdoor activities is often reduced compared with younger people. Therefore, elderly subjects often have mean 25(OH)D levels in the insufficiency range throughout the year. In institutionalized elderly subjects, 25(OH)D levels are often in the deficiency range. Undetectable 25(OH)D levels have been found in more than 95% of those 100 years of age or older.

Musculo-skeletal diseases. Severe vitamin D deficiency results in an undermineralization of the growing skeleton and in demineralization of the adult skeleton, leading to rickets and osteomalacia, respectively. This is due to an imbalance of serum calcium levels. Moreover, there is now general agreement that an insufficient vitamin D status contributes to osteoporotic fractures in elderly people. Vitamin D deficiency and insufficiency also leads to myopathy (muscle abnormalities) and subclinical myopathy (characterized by symptoms such as reduced muscle strength and reduced muscle tone), respectively, which are both frequently found in elderly people. The antifracture effect of vitamin D is due to an improvement in bone mineralization and better neuromuscular coordination (due to the restoration of calcium balance, which is necessary for nervous system function), which decreases the risk of falling.

Other diseases. There is increasing evidence that vitamin D deficiency and insufficiency is also involved in the pathogenesis of various autoimmune diseases, including type 1 diabetes, rheumatoid arthritis, and multiple sclerosis. Moreover, vitamin D insufficiency is linked to an increased risk of specific types of cancer such as colon, prostate, and breast cancer. Vitamin D is also able to lower blood pressure, probably by downregulating the renin-angiotensin system. Most of the above-mentioned diseases are frequently observed in people living at higher latitudes and in highly urbanized countries.

Hereditary forms of vitamin D deficiency. Two hereditary forms of rickets with impaired vitamin D action are known: vitamin D–dependent rickets type I (VDDR I) and VDDR II. VDDR I results from several inactivating mutations in the 1α-hydroxylase gene. VDDR II is caused by mutations in the gene of the vitamin D receptor, which prevent calcitriol action in target cells. In all these cases, absent calcitriol action leads to severe secondary hyperparathyroidism, calcium malabsorption, hypocalcemia, bone diseases, myopathy, and probably to an upregulation (increased activity) of proinflammatory cytokines and a suppression of anti-inflammatory cytokines. Patients with VDDR I and II often experience growth retardation. Patients with VDDR II may also have alopecia (hair loss). VDDR I leads to very low serum calcitriol levels, while calcitriol levels are markedly elevated in patients with VDDR II.

Prevention of vitamin D deficiency and insufficiency. In infants, rickets can be effectively prevented by a daily vitamin D supplement of 10 μg (400 IU). In elderly people, children, and adults, exposure of the hands, arms, face, or back to sunlight two to three times per week for a short period (roughly equivalent to 25% of the time that it would take to turn the skin slightly pink) seems to be adequate to satisfy one's vitamin D requirement. In the absence of solar UVB exposure, the currently recommended daily intake of 5–15 μg (200–600 IU) of vitamin D is, however, obviously far too low to achieve adequate circulating 25(OH)D levels in adults. Data from several recent studies suggest that in the absence of solar UVB exposure, adequate daily intake of vitamin D would have to be 50–100 μg (2000–4000 IU).

Toxicity. Generally, vitamin D excess is the result of very high oral vitamin D intake. Intoxications are the consequence of an unregulated intestinal vitamin D uptake in association with uncontrolled 25-hydroxylation in the liver, leading to high circulating 25(OH)D levels. There are several known case reports of vitamin D intoxication due to very high therapeutic doses of vitamin D, overfortification of foods with vitamin D, and over-the-counter supplements with excessive vitamin D content. There are no reports of vitamin D intoxication in healthy adults

after intensive sunlight exposure. In the skin, previtamin D_3 and vitamin D_3 are effectively converted to a multitude of vitamin D–inactive photoproducts. Thus, the skin can never generate quantities of vitamin D excessive enough to cause vitamin D intoxication. Daily oral doses of 25 μg (1000 IU) of vitamin D are safe in infants, and daily doses of 100 μg (4000 IU) do not cause toxicity in adults.

For background information *see* BONE; CALCIUM METABOLISM; SKELETAL SYSTEM DISORDERS; VITAMIN; VITAMIN D in the McGraw-Hill Encyclopedia of Science & Technology.   Armin Zittermann

Bibliography. R. Vieth, Vitamin D supplementation, 25-hydroxyvitamin D concentrations, and safety, *Amer. J. Clin. Nutr.*, 69:842–856,1999; M. F. Holick, Vitamin D: The underappreciated D-lightful hormone that is important for skeletal and cellular health, *Curr. Opin. Endocrinol. Diabetes*, 9:87–98, 2002; A. Zittermann, Vitamin D in preventive medicine—Are we ignoring the evidence?, *Brit. J. Nutr.*, 89:552–572, 2003.

Voice over IP

Voice over the Internet Protocol, or VoIP, is the next phase in the advancement of the public telephone system. VoIP has been around for more than 10 years, but only recently has the technology advanced from being used by home computer operators to being used as a replacement for the public switched telephone network (PSTN) used by most telephone service providers. This article describes how VoIP operates and examines some of the critical issues being solved to turn it into mainstream technology.

Comparison of VoIP and PSTN. The present public switched telephone network transfers voice by converting speech received from the telephone into 64-kbps (kilobits per second) digital data and transporting it in a timeslot (limited time interval) that is periodically inserted in a higher-capacity signal, a procedure referred to as time-division multiplexing (TDM). Two timeslots, one for each direction, are allocated for each phone call. These timeslots are set up by the signaling function of the public switched telephone network and kept in use for the duration of the call. The timeslots are switched in and out of use based on the calls taking place at any point in time and are referred to as being circuit-switched.

Although the public switched telephone network can carry nonvoice data, it does so through the use of a modem. The modem fits into the analog bandwidth of a standard telephone, allowing it to fit into the same two 64-kbps timeslots required by voice. Again these timeslots are reserved for use during the length of the call and required even when an end user does not have data to send or receive, a procedure which is not very efficient for data service, which is typically asymmetrical and bursty.

By constrast, voice over IP uses IP packets to carry speech over a data network. Only those packets that contain speech need to be transported, thereby

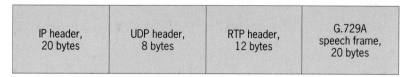

Fig. 1. Internet Protocol (IP) packet with voice over IP speech frame and headers. UDP = User Datagram Protocol. RTP = real-time protocol.

allowing voice over IP to improve bandwidth efficiency by transporting packets only in the direction of the call participant who is listening. To further improve efficiency, voice over IP uses speech-compression algorithms to reduce speech from 64 kbps to 2.4–8 kbps. This helps to offset the overhead required by the packet headers (**Fig. 1**). These packets are transmitted every 10 milliseconds, resulting in a bandwidth of 48 kbps during speech.

Unlike the public switched telephone network using time-division multiplexing, the IP network does not allocate specific timeslots for a particular user, although newer technology does make it possible to guarantee bandwidth over the network. This allows the IP network to take advantage of silent periods of a normal call by not sending any packets, further increasing its efficiency.

While the public switched telephone network was developed for voice and backfitted for data, IP was developed for data and is being backfitted for voice. The ubiquity of IP allows the convergence of new services dependent on voice and data. Voice over IP is being expanded to support applications such as video phone, video conferencing, and whiteboard conferencing (teleconferencing involving the use of whiteboards, which are areas on display screens on which multiple users can write or draw).

VoIP-PSTN interoperation. Voice over IP can take place over any portion of the transmission path of a particular call, even over the public switched telephone network. In its original instantiation, voice over IP used the public switched telephone network by compressing speech, placing it into IP packets, and transporting it from one computer to another using a modem signal. Today service providers are using packet technology in addition to the public switched telephone network for transmitting voice, using gateways to provide the interface between circuit-switched and packet-switched technology.

The most critical development has been in technology that allows the public switched telephone network to communicate with the IP network, transforming an IP call to a public switched telephone network call, and vice versa. **Figure 2** illustrates one example of a network that combines voice over IP and the public switched telephone network. In this example, a standard analog phone user, caller A, calls an IP phone user, caller B. Once the number is dialed, the switch processing the number uses the Signaling System 7 (SS7) protocol to send this request to the signaling gateway, which converts the SS7 messages into a Session Initiation Protocol (SIP) message. The call proceeds to caller B, who answers the call on either an SIP telephone or a personal computer running SIP client software. Once the media gateway controller detects that caller B picked up the telephone, it enables the packet voice gateway. Caller B's telephony equipment (either a voice over IP phone or a personal computer) negotiates capabilities between itself and the media gateway in the form of security and speech coding algorithms.

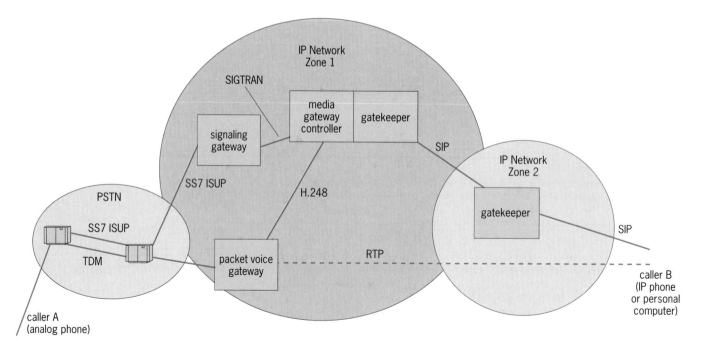

Fig. 2. Example of network operation between voice over IP and the public switched telephone network (PSTN). SS7 = Signaling System 7. ISUP = ISDN User Part. TDM = time-division multiplexing. SIGTRAN = Signaling Transport. RTP = real-time protocol. SIP = Session Initiation Protocol.

At the completion of these negotiations, the connection between the media gateway and caller B is completed and voice is carried between callers A and B. The packet voice gateway or media gateway converts time-division multiplexing used on the public switched telephone network to and from voice being carried in real-time protocol (RTP) packets in the Internet network.

This example happens to use SIP for the voice over Internet Protocol within the Internet network, but either the Media Gateway Control Protocol (MGCP) or H.323 might also be used. The Media Gateway Control Protocol attempts to look very similar to the public switched telephone network by adding services and control information to a central network, allowing the endpoints to be dumb. Both H.323 and SIP use a distributed architecture with intelligent endpoints and call control elements such as signaling gateways and gatekeepers, which help make sure that a call is routed to the appropriate destination. The distributed model provides for increased flexibility of applications, while the centralized approach provides an easier management of call control and provisioning (programming or instructing equipment to operate in a desired way).

Other common examples for voice over IP network configurations are end-to-end, business or enterprise services, and trunking. End-to-end, which was the initial application for voice over IP, is typically associated with low-quality calls, as the calls often go over the public switched telephone network with the end users being bandwidth-constrained. Business or enterprise applications are associated with advanced services, and their operation over high-speed local-area networks provides plenty of bandwidth. Trunking is the most recent application for voice over IP and allows service providers to transport efficiently both voice and data over their networks.

Security. Most people think of the public switched telephone network as being secure (which it is, relatively speaking), since the network resources are dedicated for the duration of a call. Packet networks or the Internet are much less secure. They are more often compared to party lines, where a piece of equipment is told to listen only for packets intended for it, but can operate in a promiscuous mode where all packets on the network can be received. This is undesirable for end users, but even more problematic is the ease with which hackers have been able to attack IP networks.

Internet Protocol networks have been broken into throughout the world, and in numerous cases caused equipment to be taken off line or rendered temporarily useless. In order to build reliability into a voice over IP network, service providers must be able to increase its security level to the level of the public switched telephone network. Numerous tools and protocols have been developed for helping to secure IP networks such as firewalls, intrusion detection and prevention elements, virus scanners, and virtual private network (VPN) appliances. IP Security (IPSec),

Secure Sockets Layer (SSL), and voice-over-IP-specific protocols are being developed, but new advances in voice over Internet Protocols and applications constantly pose new challenges to IP networks that must repeatedly upgrade their security measures to prevent network outages.

Quality of service. The public switched telephone network based on time-division multiplexing has fixed delays, while IP networks have variable delays based on the path that the data take through the network and the degree of congestion of the network. These issues can disrupt the voice by causing packets to arrive out of order or be dropped. Voice over IP software is designed to make adjustments for out-of-order packets, and speech algorithms help to adjust for lost packets, but they cannot overcome these impairments by themselves.

When delays become too great, the normal flow of a conversation is disrupted. For example, when one party finishes talking there will be an inordinate amount of time between the end of speech and a response from the far end. To address this issue, protocols [such as Multiprotocol Label Switching (MPLS) and differentiated services] are being implemented to allow for the transport of delay-sensitive data to improve performance. These protocols use information contained within packet headers to help prioritize packets so that those carrying speech (or other delay-sensitive traffic) can be routed prior to other traffic. In the case of MPLS, more overhead is added to the packet header.

Companies are also starting to increase the available bandwidth for Internet traffic and limiting the amount of traffic that can go across a particular link to improve real-time performance. When few people are trying to access or use any link, the delay is small enough not to be noticeable; but when more data go across the network, delays increase to the point where call quality is unacceptable. Home users are improving performance by using cable modems and digital subscriber lines (DSL) to increase the available bandwidth for voice over IP.

Cost savings. The first reason for using voice over IP was the cost savings that resulted from allowing home users to bypass the charges associated with long-distance calls. Voice over IP allows service providers to bypass access fees charged for using the final mile to the home since it is presently charged as data. Ultimately, though, these fees may change and voice over IP may become as regulated as the present public switched telephone network, thereby reducing much of today's cost advantage.

For background information *see* DATA COMMUNICATIONS; INTERNET; PACKET SWITCHING; TELECONFERENCING; TELEPHONE SERVICE; VIDEOTELEPHONY in the McGraw-Hill Encyclopedia of Science & Technology.   Daniel Heer

Bibliography. K. Camp, *IP Telephone Demystified*, McGraw-Hill, 2002; D. Collins, *Carrier Grade Voice over IP*, 2d ed., McGraw-Hill, 2002; D. Minoli and E. Minoli, *Delivering Voice over IP Networks*, 2d ed., Wiley, 2002.

West Nile fever (encephalitis)

West Nile fever is a mosquito-borne viral infection. The causative agent, West Nile virus (WNV), was first isolated in the West Nile area of Uganda, Africa, in the early 1930s. It was later recognized as a cause of serious meningoencephalitis (inflammation of the brain and meninges) in an outbreak in Israel. This virus found its way to the shores of North America in 1999, possibly through international travel, the importation of infected birds or mosquitoes, or the migration of infected birds. Serious infection with WNV can result in viral encephalitis, a sometimes fatal inflammation of the brain. WNV infects horses and several species of domestic and wild birds and can cause death in these species. In 2002, there were 4156 reported cases of human WNV infection in the United States, with the virus reaching 44 states.

Epidemiology of arboviruses. WNV is an arbovirus— that is, a virus that is transmitted by an arthropod vector. Some other arboviruses include yellow fever virus, St. Louis encephalitis virus, and dengue fever virus. These viruses are carried by mosquito vectors from host to host. Arboviruses must be able to (1) infect vertebrates, such as a human host, as well as invertebrates; (2) cause viremia (presence of viral particles in the blood) for enough time to allow the invertebrate host to acquire the virus; and (3) productively infect the invertebrate's salivary gland to allow infection of other vertebrate hosts. Surprisingly, humans are usually "dead-end hosts," which means that they cannot spread the virus back to the arthropod vector because they do not maintain a persistent viremia. This is true of WNV; the reservoirs (susceptible hosts that allow the reinfection of other arthopods) for this virus are birds, not humans. Therefore, there is no transmission of WNV from human to mosquito.

The arboviruses are usually restricted to a specific arthropod vector. Even in tropical areas, where mosquitoes are numerous and variable, the spread of an arbovirus is restricted to one genus of mosquitoes. For example, WNV is spread by the *Culex* mosquito, while dengue virus is spread by the *Aedes* mosquito. Arboviruses that cause encephalitis in the United States include WNV, western equine encephalitis, eastern equine encephalitis, and St. Louis encephalitis. The mosquito vectors for these viruses include the genera *Culex*, *Culiseta*, and *Aedes*.

Transmission. WNV is a member of the family of flaviviruses, a group of enveloped viruses that contain ribonucleic acid (RNA) as their nucleic acid. Flaviviruses (*see* **table**) are typically associated with mild systemic disease but can cause severe infections, including encephalitis. Mosquitoes acquire the flaviviruses through feeding on the blood of a viremic vertebrate host. The virus then travels from the midgut of the mosquito to the salivary glands, where it reproduces itself in great numbers. The salivary glands then release virus into the saliva, where it is transferred to the next victim during a blood meal. On biting a host, the mosquito regurgitates the virus-containing saliva into the victim's bloodstream. The virus travels through the host's plasma to tissue sites for which it has a tropism, or preference.

Pathogenesis. The disease that results is determined by the virus' tissue tropism, the virus concentration, and the immune response of the host to the infection. The initial symptoms include fever, chills, headache, and backaches caused by the resulting viremia. Most viral infections do not progress beyond this stage; however, a secondary viremia may occur, which can result in infection of organs such as the brain, liver, and skin. This can lead to fatality in the infected host.

Detection. Detection methods for WNV include surveillance and reporting of human, bird, and horse cases as well as laboratory diagnostic tests.

Surveillance. Surveillance for human cases by local, state, and national agencies is important in the detection of WNV. However, since the number of clinically identified cases of WNV infection in humans is small, additional means of surveillance are central to the control of this disease. Therefore, surveillance of cases in horses and birds is important. Using these animals as sentinels, or alarm animals, can alert health officials to the probable occurrence of human disease in the area. Public health officials have been able to use bird mortality, particularly in crows, to track the movement of WNV. In 2000, the Centers for Disease Control (CDC) developed an electronic-based surveillance and reporting system to track WNV activity in humans, horses, birds, and mosquitoes. This system, known as the ArboNet surveillance system, was updated in 2003 to allow streamlined reporting to the CDC of WNV activity by state public health departments.

Laboratory diagnosis. The current tests for human serologic diagnosis of WNV are the CDC-defined immunoglobulin M (IgM) and immunoglobulin G (IgG) enzyme-linked immunosorbent assays (ELISA). These tests are the most sensitive screening tests available at this time. (Currently, commercial kits for diagnosis of human WNV infection are in development but are not yet available.) An ELISA uses antigens, or "pieces," of the virus to detect the presence of immunoglobulins or antibodies that may be present in a patient's serum (the liquid component of blood). When these antibodies are present, a positive ELISA test will result. However, because the ELISA can cross-react with other flaviviruses (such as St. Louis encephalitis virus), it is considered a screening test only and must be confirmed by a virus neutralization test showing that WNV is indeed present in the patient. The procedures for animal serology follow those used for humans.

In addition to serologic testing, the virus can be isolated in the laboratory using susceptible mammalian or mosquito cell lines. Virus isolation from clinical specimens can be performed using cerebrospinal fluid from clinically ill humans; human brain tissue from biopsy or postmortem; postmortem brain tissue from horses; and kidney, brain, and heart tissue from birds. Confirmation of virus isolate identity is

Flaviviruses				
Disease name	Vector	Host	Distribution	Disease manifestations
Dengue	*Aedes*	Humans, monkeys	Worldwide (especially tropics)	Mild systemic disease, including break-bone fever, dengue hemorrhagic fever, dengue shock syndrome
Yellow fever	*Aedes*	Humans, monkeys	Africa, South America	Hepatitis, hemorrhagic fever
Japanese encephalitis	*Culex*	Pigs, birds	Asia	Encephalitis
West Nile encephalitis	*Culex*	Birds	Africa, Europe, Central Asia, North America	Fever, encephalitis, hepatitis
St. Louis encephalitis	*Culex*	Birds	North America	Encephalitis

SOURCE: Adapted from P. R. Murray et al., *Medical Microbiology,* 4th ed., 2002.

by indirect immunofluorescence assay (IFA), which uses virus-specific monoclonal antibodies to detect the presence of the virus in the cells. Nucleic acid detection methods, including the reverse-transcriptase polymerase chain reaction (RT-PCR) to make multiple deoxyribonucleic acid (DNA) copies of the viral RNA, may also be used on these specimens. Finally, virus neutralization assays may be used to differentiate viruses in cases of human disease found in areas where more than one virus is known to be present in the mosquito population.

Control. Control of arbovirus disease is best accomplished through a comprehensive mosquito management program that includes mosquito surveillance and reduction. These programs must be established at the local level, and surveillance must be sensitive enough to detect transmission of WNV in wild animal populations, as this is associated with increased risk of disease in humans or domestic animals.

Mosquito surveillance. Effective mosquito control must include a sustained surveillance program to detect vector species and to identify and map their immature habitats by season. Larval mosquito surveillance is performed by sampling a wide range of aquatic habitats for the presence of pest and vector species. Adult mosquito surveillance is used to monitor species presence and relative abundance of adult mosquitoes in an area. Since WNV is carried by a specific genus, *Culex*, detection of this genus in an area indicates the possible presence of WNV as well. Traps for mosquitoes use carbon dioxide as bait (because mosquitoes are attracted to the carbon dioxide that humans exhale; thus, the traps simulate a living human for the mosquito to feed upon). WNV can then be detected in individual mosquitoes using the methods listed above.

Mosquito reduction. Source reduction of mosquitoes is the elimination of mosquito larval habitats. This is the most effective method for controlling mosquitoes, and can be as simple as proper disposal of used tires and cleaning of rain gutters, bird baths, and unused swimming pools. Mosquito breeding areas can range from items as small as a bottle cap to the foundation of a demolished building. Sanitation is a major part of integrated vector management programs and is

central to controlling WNV infection in human and animal populations.

Prevention. Education of the general public should be the first goal of health agencies, as key changes in human behavior are essential to the prevention of WNV infection. However, programs that include improved access to repellents and window screening materials and target high-risk groups will be more successful in preventing disease.

Personal, household, and community prevention. Individuals can use DEET (*N,N*-diethyl-*m*-toluamide or *N,N*-diethly-3-methylbenamide) based repellents on skin and clothing as their first defense against WNV infection, in addition to avoiding the outdoors during prime mosquito-biting hours (dusk to dawn). Households may be protected by eliminating mosquito breeding sites and repairing or installing window screens. Communities can protect against WNV disease through active reporting of dead birds and organized mosquito control measures.

Targeted prevention. Targeted prevention includes seeking out those groups that are most at risk for severe disease with WNV, including those over age 50 and those who are immunocompromised. Individuals who work outside extensively are also at greater risk for WNV infection and should be targeted with disease-preventing information.

Vaccine. No vaccine is available to prevent WNV disease in humans. One research group has successfully used recombinant DNA vaccines to induce protective immune responses in mice and horses; however, this success has yet to be transferred to humans.

For background information *see* ANIMAL VIRUS; ARBOVIRAL ENCEPHALITIDES; INFECTIOUS DISEASE; MOSQUITO; VIRUS in the McGraw-Hill Encyclopedia of Science & Technology.   Marcia M. Pierce

Bibliography. B. S. Davis et al., West Nile virus recombinant DNA vaccine protects mouse and horse from virus challenge and expresses in vitro a noninfectious recombinant antigen that can be used in enzyme-linked immunosorbent assays, *J. Virol.,* 75(9):4040–4047, 2001; A. K. Malan et al., Evaluations of commercial West Nile Virus immunoglobulin G (IgG) and IgM enzyme immunoassays show the value of continuous validation, *J. Clin. Micro.,* 42(2):727–733, 2004; P. R. Murray et al., *Medical*

Microbiology, 4th ed., Mosby, St. Louis, 2002; E. Nester et al., *Microbiology: A Human Perspective*, 4th ed., McGraw-Hill, New York, 2004; H. E. Prince and W. R. Hogrefe, Detection of West Nile virus (WNV)–specific immunoglobulin M in a reference laboratory setting during the 2002 WNV season in the United States, *Clin. Diag. Lab. Immunol.*, 10(5):764–768, 2003; B. Shrestha, D. Gottlieb, and M. S. Diamond, Infection and injury of neurons by West Nile encephalitis virus, *J. Virol.*, 77(24):13203–13213, 2003; S. J. Wong et al., Detection of human anti-flavivirus antibodies with a West Nile Virus recombinant antigen microsphere immunoassay, *J. Clin. Micro.*, 42(1):65–72, 2004.

Wide-band-gap III-nitride semiconductors

The band gap, or more precisely energy band gap, denotes the difference in energy between the highest valence electron energy states (valence band maximum) and the lowest conduction energy states (conduction band minimum). In optoelectronic devices, the band gap dictates the wavelength of light that a semiconductor absorbs and emits. The size of band gap also determines the ultimate robustness of electronic devices under high ambient temperature or excessive power loads. Semiconductors, including II–VI compounds, such as ZnSe, ZnS, and ZnO, have at times shared the label of wide-band-gap semiconductors by providing important proof-of-concept demonstrations of blue-green lasers and light-emitting diodes (LEDs) in the early 1990s. In a relatively short time, however, the nitride compounds with three of the column IIIA elements (AlN, GaN, and InN) have emerged as exceptionally versatile semiconductors, with functionalities unattainable from traditional Si and GaAs technologies. Compared to GaAs with an energy gap corresponding to infrared emission (890 nanometers), the AlGaInN family covers the entire visible spectrum from infrared (InN) to blue (InGaN), and extending into deep ultraviolet (AlGaN), creating opportunities in display, illumination, high-density optical storage, and biological and medical photonics (**Fig 1**). Performance of traditional Si devices is known to deteriorate at device temperatures above 100°C (212°F) or when the applied voltage exceeds certain critical electrical fields. Transistors and diodes made from wide-band-gap GaN are expected to create opportunities beyond low-power digital electronics, including applications in automobiles, aircraft, utility distribution, and wireless communications.

Optoelectronic applications. Semiconductor light-emitting devices are used in optical communication, information display, lighting/illumination, and information storage. Propagation of coded light waves or pulses in the infrared and red spectrum through dielectric fibers is the backbone of modern communication and is exclusively supported by GaAs- and InP-based technologies. The availability of wide-band-gap III-nitride semiconductors has already sig-

nificantly impacted the other three categories of photonic applications. The emergence of compact ultraviolet (UV) nitride emitters is expected to enable new and portable biochemical and biological analytic instrumentation technologies.

Display technology. Display devices rely on line-of-sight transmission of photons for visual conveyance of information. Practical examples vary greatly in complexity from simple indicator lights on instrument panels (on/off), traffic lights (red/yellow/green), and backlighting for liquid crystal displays (LCDs), to sophisticated two-dimensional panels such as computer/television screens and outdoor bulletin boards/signs. Since the band-to-band electronic transitions in semiconductors tend to yield monochromatic emission, red, green, and blue (RGB) sources needed to be developed. Red LEDs from the InGaP family reached respectable performance in late 1980s. And the arrival of high-brightness blue and green LEDs from AlGaInN materials have provided the other two sources for RGB displays.

Solid-state lighting. According to the U.S. Department of Energy, general illumination accounts for more than 20% of the national energy consumption, so research for an efficient lighting strategy bears significant implications for the overall energy infrastructure. The benchmark parameter used in lighting is luminous efficiency with units of lumen/watt (lm/W). The lighting industry is currently dominated by traditional technologies of incandescent and fluorescent lamps. An incandescent lamp has a very low efficiency in light conversion (~15 lm/W, or a power conversion efficiency of 5%). Fluorescent lamps, based on a combined plasma discharge and wavelength conversion through phosphor coating, represent a much more efficient process (70 lm/W). GaN-based blue and green LEDs have exhibited monochromatic efficiency of more than 50 lm/W in the laboratory. These light sources will ease the demand on power plants. Additionally, the III-nitride LEDs have a standard lifetime exceeding 100,000 h, representing a hundredfold increase in lifetime, compared to conventional lamps' lifetime of about 1000 h. The use of LEDs for illumination would also allow for flexible configurations (for example, sheet or wallpaper-like lamps), directional projection, and color tunability.

High-density optical storage. The first CD-ROMs used an 850-nm infrared laser for writing/reading 650 Mb of data. Today's rewritable DVDs for video are powered by a red diode laser with disk information capacity of approximately 5 Gb. In the past 3 years, the emergence of the violet/blue InGaN laser (near 410 nm) as a technologically mature device has led to the emergence of the third generation of optical disk storage. The new technology is now entering the marketplace, with standard-size, single-sided 120/130-mm-diameter (5-in.) optical disks having 25 Gb of storage. This striking advance follows in part from fundamentals set by optical diffraction limits, namely that the area density of optical

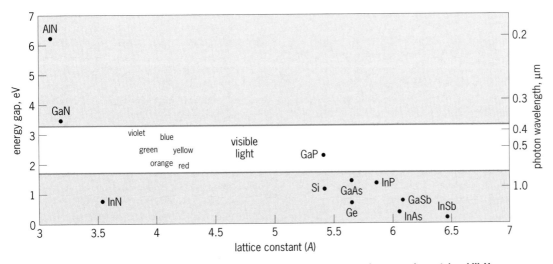

Fig. 1. Energy gap and corresponding photon wavelength versus lattice parameter for some elemental and III–V semiconductors.

information scales as the inverse square of the wavelength. In parallel, important advances have been made in choosing sophisticated optical engineering tools for reaching the limits allowed by lightwave theory. Very large numerical aperture optics have been developed in the recording head for focusing the 410-nm laser for writing/reading, in conjunction with innovations of the recording media and rapid data transfer. Currently, two consumer electronics industry consortia, each with its own disk format scheme, are struggling to define a unified standard. The two formats are known as Blu-ray and HD-DVD, respectively, enabling up to several hours of high-definition digital video and 20–30 h of standard television broadcast recording depending on the formatting details. Future DVD products are likely to include units that house a multiwavelength semiconductor laser and are thus "backward"-compatible, that is, able to record and play back Blu-ray Disc (or HD-DVD), DVD, and CD.

High-power, high-temperature, and high-frequency electronics. Failure of power electronic devices ultimately occurs due to an avalanchelike multiplication of the current. For these devices, the breakdown threshold, and hence strength under load, is proportional to the energy gap. In addition, a high operating temperature leads to thermal excitation of intrinsic carriers across the band gap, which offsets the intentional carrier profiles and causes deviation of device characteristics. Compared with Si and GaAs, a two- to threefold increase of band gap in employing GaN corresponds to a fivefold increase in the voltage-bearing capability and an increase of hundreds of degrees in temperature ceiling. Among GaN electronic devices that have been studied, the AlGaN/GaN heterostructure field-effect transistors (HFET) have received the most attention due to a spontaneous formation of a conducting channel with high carrier concentration and mobility. Microwave (1–10 GHz) amplifiers with superior power-handling ability (10 W/mm) have been demonstrated and are

finding applications in high-capacity wireless links among base stations and local/metro area networks. Taking advantage of the ability of GaN to withstand very high breakdown fields (up to 10 times that of silicon), researchers have built GaN-based Schottky rectifiers for primary and subsidiary power distribution systems.

Substrates. Ideally, semiconductor substrates should serve the dual purpose of providing electrical contacts to devices and offering paths for heat dissipation. Bulk substrates are normally synthesized near the melting temperature of the material to facilitate controlled crystallization from liquid into solid phases. GaN possesses both a high melting point (>2500°C or 4500°F) and concurrently a very high vapor pressure (~45,000 atm or 4500 kPa), making the synthesis of large-area substrates a daunting task. Most research, development, and manufacture

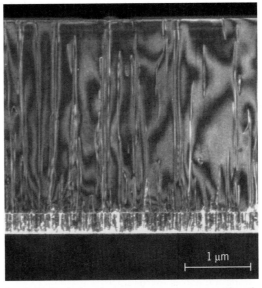

Fig. 2. Transmission electron microscope cross section of a III-nitride LED on sapphire, where the vertical lines are dislocations.

of III-nitride devices relies on sapphire (Al_2O_3) substrates, a crystal that produces necessary symmetry yet with substantial mismatches in both atomic spacing and crystallographic arrangement, leading to structures with a high density of dislocations (**Fig. 2**). It is now established that these dislocations create deleterious effects on the performance and reliability of III-nitride devices. SiC substrates exhibit excellent thermal and electrical conductivity, and represent an improved matching to III-nitride, though not close enough for significant reduction in defects. The full potential of III-nitride wide-band-gap semiconductors, including applications in general lighting, cannot be realized without the mass production of single-crystalline III-nitride substrates. Recent progress in epitaxial GaN substrates and bulk AlN substrates offers promising routes to III-nitride substrates. At this time, their impact remains unclear given their limited availability and a prohibitive cost.

Epitaxial growth. A typical epitaxial process involves the injection and transport of vapor-phase reactant species into a controlled heat zone where the substrate is located. The reactants undergo chemical reactions in the vapor phase and on the substrate's surface upon heating. Under optimum conditions, surface kinetic processes, such as adsorption, diffusion, desorption, and incorporation, work together to facilitate an ordered deposition of atoms in a way similar to the construction of LEGO blocks with perfect atomistic registry. So far, the linkages among growth processes, material properties, and device performance are treated primarily in an empirical manner. Important issues include the nucleation and growth of AlGaInN, growth and design of highly efficient GaInN active regions, compensation and activation of *p*-type magnesium acceptors, and control of mismatched strain in building heterostructures.

Device development. While device progress with nitride semiconductor blue/violet lasers, blue/green LEDs, and high-power microwave transistors has been much faster than forecast a decade ago, many exciting challenges and opportunities remain. Among these are the development of advanced light-emitter geometries for high-brightness directional LEDs by concepts such as vertical cavity and photonic crystal device structures and optimization of III-nitride UV emitters. Demonstration diode UV (360 nm) lasers have been introduced in continuous-wave operation, while exploratory LEDs have been reported in the sub-300-nm regime, albeit with a low efficiency. At UV wavelengths, the ever-higher spatial resolution provides attractive prospects for nitride light emitters as sources in advanced microscopy, ultrahigh density optical storage including holography, and specialized lithography. *See* PHOTONIC CRYSTAL DEVICES.

The major challenge in the push to UV for nitride devices is in part fundamental, since truly wide-gap semiconductors prefer to be electrical insulators. AlGaN is the material used now, where the relative Al/Ga fraction is used to control the electronic and optical energy landscape across the multilayer device structure. High-efficiency electrical injection of electrons and holes, their radiative recombination in the active device medium, and control of crystalline defects form an intertwined set of problems, where material synthesis techniques, device science, and device fabrication approaches require considerable synergy in order to achieve the next breakthroughs.

White light strategy. The emission spectrum of LEDs typically is monochromatic, in contrast to typical white light sources that have either a broad spectrum or multiple color components. Several strategies have been proposed to achieve a white light spectrum using LEDs. One uses blue LEDs with a semiabsorbing yellow phosphor as a two-color approximation to the white light spectrum; this represents the least complicated strategy. Another uses a UV LED with multiple phosphors of different colors to achieve a broad spectrum with increased flexibility. These two strategies share the advantage of single-chip implementation for white LEDs, yet suffer from reduced efficiency due to downconversion of photon energy. For the ultimate flexibility and efficiency, though at the expense of complex fabrication process, three or more LEDs (red, green, and blue) could be integrated.

Outlook. Just as transistor-based radios and gadgets quietly marked the end of vacuum-tube-based electronics and the birth of semiconductor microelectronics, solid-state lighting is poised to replace current lighting technology. Already, there has been a steady and pervasive penetration of high-power white LEDs in special lighting sectors, including automobile headlights, landscape/contour lighting, and commercial lighting. The two driving factors for market acceptance are efficacy and affordability. GaN LEDs compete well in the former category, yet are about 100 times more expensive than tube-based light lamps, a common phenomenon at the inception of a new technology. It is expected that rapid technological progress will be made under fierce international competition among industries, academia, and governments.

For background information *see* BAND THEORY OF SOLIDS; COMPACT DISK; ILLUMINATION; LASER; LIGHT-EMITTING DIODE; LUMINOUS EFFICIENCY; MICROWAVE SOLID-STATE DEVICES; POWER INTEGRATED CIRCUITS; SEMICONDUCTOR; SEMICONDUCTOR HETEROSTRUCTURES in the McGraw-Hill Encyclopedia of Science & Technology.

Jung Han; Arto V. Nurmikko

Bibliography. A. Bergh et al., The promise and challenge of solid-state lighting, *Phys. Today*, pp. 42–47, December 2001; L. F. Eastman and U. K. Mishra, The toughest transistor yet, *IEEE Spectrum*, pp. 28–33, May 2002; N. Johnson, A. Nurmikko, and S. DenBaars, Blue diode lasers, *Phys. Today*, pp. 31–37, October 2000; S. Nakamura, G. Fasol, and S. J. Pearton, *The Blue Laser Diode*, 2d ed., Springer, 2000; F. Ren and J. C. Zolper, *Wide Energy Bandgap Electronic Devices*, World Scientific, 2003; G. Zorpette, Let there be light, *IEEE Spectrum*, pp. 70–74, September 2002.

Wireless sensors

The number of sensors of all types installed in vehicles and industrial plants worldwide for operational safety and process control is in the billions and increasing exponentially. Historically, sensors have been linked by cables to electronic modules, which provide excitation, amplification, signal conditioning, and display. There are severe problems, however, when sensing parameters such as strain, pressure, torque, and temperature with sensors mounted on rotating components or operating in hazardous or extreme environments. For these applications, wireless sensing is advancing rapidly.

At present, wireless sensors for measuring strain and pressure are finding use in automotive and potentially aeronautical applications. The real and present safety-related issue of tire pressure sensing in automobiles has stimulated the first large-scale application of wireless sensing. The effective realization of wireless sensing on a global scale, however, will depend on corporate or even intergovernmental agreement on common standards.

Sensor systems. Wireless sensors may be active (require an independent energy source or power supply) or passive (derive their power from the interrogation signal).

Active devices may be battery-powered, or powered from local energy harvesting (for example, a flexible piezoelectric device generating charge through physical deformation, which is stored in a capacitor or miniature rechargeable battery), or powered from an input carrier wave [for example, via inductive coupling between adjacent coils at about 125 kHz, followed by demodulation, rectification (RF to DC conversion), and storage]. The sensor signal, representing a quantity such as strain, pressure, or temperature, modulates a stable radio-frequency (RF) carrier wave which, in response to an interrogation signal (sent by the receiver antenna), is transmitted via a local antenna to a receiving antenna and associated electronics, where it is demodulated and conditioned for display, control, or data logging.

In passive devices, strain-sensitive sensors (piezoelectric devices) can be directly excited by an incident RF interrogation wave and transmit a low-level RF (backscattered) signal via a local antenna to a receiving antenna and associated electronics, to demodulate and condition it for display, control, or data logging.

Permitted frequencies. Currently, there is little standardization in frequencies for wireless sensors, except at the highest (UHF) frequency (2.45 GHz),

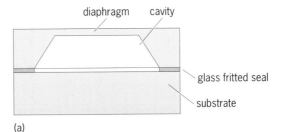

(a)

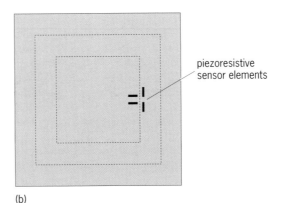

(b)

Fig. 1. Bulk-micromachined piezoresistive pressure sensor, fabricated from a single-crystal silicon die. (*a*) Cross section. Potassium hydroxide (KOH) anisotropic etching is used to create a diaphragm above a cavity. The cavity is closed with a glass fritted seal to the silicon substrate base. (*b*) Top view. The piezoresistive sensor elements in a region of high diaphragm tensile strain are connected to a Wheatstone bridge network.

where Bluetooth resides (see **table**). Care should be taken in interpreting the table, since permitted power levels vary significantly and with equipment and signal type; for example, Japan limits radiated power levels more severely than other countries. In the United States, where Federal Communications Commission (FCC) Part 15 Rules govern the operation of radio-frequency devices, it is possible to operate sensors outside the industrial-scientific-medical (ISM) bands if the power levels are sufficiently low.

Sensor types. While the ubiquitous foil strain gauge is still relevant for certain applications needing wireless data transfer, such as for shaft torsional strain for torque measurement, the need for reduced cost and size to meet automotive needs has led to the development of silicon-based piezoresistive, capacitive, and piezoelectric (single-crystal quartz) devices. These can be manufactured by the high-volume microelectronic fabrication technique known as photolithography, thereby reducing cost by yielding more devices per wafer as line widths decrease and component densities increase. Devices of this kind can be classed as micro-electro-mechanical systems (MEMS).

Figure 1 shows the configuration of a typical bulk-micromachined silicon pressure sensor in which a die is deep-etched to create a framed diaphragm and then bonded to a base element, creating a hermetically sealed cavity. A piezoresistive pressure transducer is formed by surface deposition on the diaphragm.

Nominal frequencies of the industrial, scientific, and medical (ISM) bands.					
	315 MHz	433 MHz	868 MHz	915 MHz	2.45 GHz
U.S.	—	—	—	√	√
Europe	—	√	√	—	√
Japan	√	—	—	—	√

Figure 2 shows a smaller surface-micromachined silicon device in which a variation in capacitance is created by pressure acting on the diaphragm, modifying its separation from the surface below. In addition to lowering cost, this design offers improved temperature performance and lower power consumption.

Figure 3 shows a passive, single-port surface-acoustic-wave (SAW) resonator. An incident RF signal excites the central interdigital transducer, generating an outwardly propagating mechanical acoustic wave. The wave is reflected by gratings to set up a standing wave whose resonant frequency, which is a function of strain, is influenced by both temperature and pressure. When the interrogating signal is interrupted, a return RF signal is regenerated by the piezoelectric effect, and transmitted. In one passive tire pressure monitoring system (TPMS) sensor configuration, a group of three SAW resonators generates three resonant frequencies and two frequency differences from which pressure and temperature may be uniquely determined.

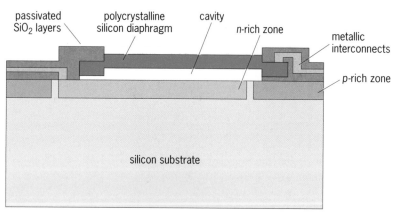

Fig. 2. Cross section of surface-micromachined capacitive pressure sensor. The cavity is selectively etched after deposition of the diaphragm. The device is circular in plan.

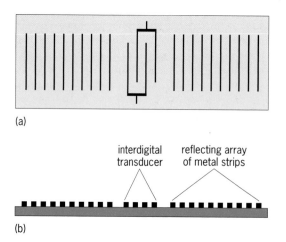

Fig. 3. Single-port surface-acoustic-wave (SAW) resonator on a single-crystal quartz substrate. (*a*) Top view. (*b*) Cross section. Reflecting array of approximately 200 metal strips numbers approximately 200, with spacing of approximately 2 μm between successive strips. Aluminum metallization layer is approximately 0.1 μm thick.

System configurations. In active tire pressure monitoring systems, each wheel carries a sensor module comprising a pressure sensor (Fig. 1 or 2), a temperature sensor, a battery, a transceiver, a microcontroller, and an electronically erasable programmable read-only memory (EEPROM) to provide storage for sensor calibration constants and unique identification. Much development effort has been expended on lithium ion batteries and their associated management systems in order to ensure the 10-year operational life expected by automobile manufacturers. The high-level digitally encoded RF output signals of milliwatt power are typically transmitted from all four tires to chassis-mounted receivers, where after demodulation they may be sent by data buses to a central display.

In a passive tire pressure monitoring system, each wheel carries only a sensor, which may be mounted in the wheel drop center, on the valve, or in/on the tire. The sensor is excited by an RF signal (at a power level of about 1 mW) from a wheel arch antenna. The low-power return RF signals (about 1 nW) have a range of up to 0.6 m (2 ft), which limits their causing RF interference, and are received by the same wheel arch antennas from where coaxial or twisted pair cables connect to a central electronic control unit for demodulation, conditioning, and display.

Packaging. Sensors and batteries for automotive applications must perform to their operational specification over a wide range of environmental conditions, including temperatures from $-40°C$ to $125°C$ ($-40°F$ to $257°F$) and centrifugal accelerations up to 2500 g (24,500 m/s^2), resulting from a nominal maximum speed of 250 km/h (155 mi/h) permitted on the German Autobahn with a 16-in. wheel. MEMS devices are sensitive to contamination, so the packaging must remain hermetic for the sensor lifetime. The stresses due to differential thermal expansion and mechanical loading must be safely accommodated, while fatigue resistance and minimal drift due to component aging must be demonstrated.

Automotive applications. Currently 55 million new vehicles (including 15 million in the United States) are sold annually. The demand for tire pressure monitoring systems was accelerated by a spate of sport utility vehicle (SUV) rollovers in the United States, following sudden tire deflations, which resulted in a number of fatalities. The U.S. TREAD Act led the National Highway Traffic Safety Administration (NHTSA) to mandate that from November 2003 warning of tire underinflation should be progressively introduced for new cars, SUVs, light trucks, and buses of gross weight up to 10,000 lb (4540 kg). Initially two threshold options were indicated: (1) up to four tires at 25% below recommended (placard) pressure, or (2) one tire at 30% under placard pressure.

The second option effectively permitted indirect tire pressure monitoring systems based on antilock braking systems (ABS) which monitor wheel rotation speeds. Since an underinflated wheel rotates faster, it is possible to detect it by comparison with correctly inflated wheels. However, the inherent inaccuracies of the ABS method and its inability to detect more than one wheel simultaneously deflating led to a U.S. Court of Appeals judgment in 2003 excluding indirect methods from compliance with the TREAD Act. This decision effectively mandated direct tire pressure monitoring; and since it is very difficult to make wired connections to a rotating wheel, wireless sensor systems were needed. In September 2004 NHTSA issued a revised proposal requiring a four-tire TPMS capable of detecting a tire more than 25% underinflated and warning the driver. The new standard will now apply progressively from September 2005.

First-generation direct tire pressure monitoring systems installed in passenger cars have overwhelmingly been active battery-powered systems, utilizing piezoresistive MEMS sensors. For the next generation, suppliers and buyers are debating active versus passive sensors. The argument for passive sensors is based partly on concerns about obtaining a 10-year life and the safe environmental disposal of lithium batteries. Currently, 1.2 billion tires are sold annually, and the number equipped with tire pressure monitoring systems will rise rapidly, especially in the United States because of legislation, and in Europe and Japan where perceived safety and quality are strong car-buying motives.

Low sensor bulk and weight (and counterweight for wheel balance) is a serious concern for some types of vechicles, such as motorcycles and motor sport vehicles, where passive devices are favored. Passive devices also lower the cost of sensors (in the wheel), which is important in the replacement market. However, overall system cost will tend to dominate automobile industry purchasing decisions for the supply of new vehicles, and here the jury is still out.

Another key automotive requirement is measuring torque for electrical power-assisted steering (EPAS), powertrain, and driveline. Torque can be sensed using lightweight passive wireless sensors to measure the principal tensile and compressive strains acting $45°$ to the shaft axis. The availability of torque as a control parameter will enable further improvements in fuel consumption, handling, and ride comfort to be made.

Aeronautical applications. At present, wireless sensors for this sector are at the applied research stage. Embedded passive resonant pressure sensors, utilizing capacitance change across an evacuated cavity, plus antennas have been proposed for hot (greater than $600°C$ or $1100°F$) gas turbine conditions and can also be envisaged for shaft torque monitoring. Similarly, single-port SAW resonators plus antennas can address strain and temperature monitoring in airfoil sections, with operating temperatures in excess of $300°C$ ($600°F$). The benefits of direct parameter measurement, as opposed to calculated or inferred values, will be seen in improved fuel consumption arising from both improved combustion control and the reduced mass of optimized rotating components.

For background information *see* MICRO-ELECTRO-MECHANICAL SYSTEMS (MEMS); MICROSENSOR; PIEZO-ELECTRICITY; PRESSURE TRANSDUCER; SURFACE-ACOUSTIC-WAVE DEVICES; TRANSDUCER in the McGraw-Hill Encyclopedia of Science & Technology.

Ray Lohr

Bibliography. J. Beckley et al., Non-contact torque sensors based on SAW resonators, *IEEE Frequency Control Symposium*, pp. 202–213, New Orleans, May 29–31, 2002; T. Costlow, Sensors proliferate, *Auto. Eng. Int.*, 11(9):38–44, September 2003; V. Kalinin, Passive wireless strain and temperature sensors based on SAW devices, *IEEE Radio & Wireless Conference*, Atlanta, Sept. 19–22, 2004; F. Schmidt and G. Scholl, Wireless SAW identification and sensor systems, *Int. J. High Speed Electr. Sys.*, 10(4):1143–1191, 2000; R. Verma, B. P. Gogoi, and D. Mladenovic, *MEMS Pressure and Acceleration Sensors for Automotive Applications*, SAE SP-1782, 2003-01-0204, pp. 51–58, 2003.

Xenoturbella (invertebrate systematics)

Xenoturbella bocki is a simple marine worm whose natural habitat is in soft mud bottoms, at a depth of around 60 m (200 ft), off the coasts of Sweden and Norway. This worm has been the subject of controversy for almost a century.

Despite numerous attempts at assigning *Xenoturbella* a position in the evolutionary tree of life (that is, its phylogenetic placement), this animal has remained enigmatic due to its extreme morphological simplicity. However, recent analyses using deoxyribonucleic acid (DNA) have found *Xenoturbella* to be a primitive member of the deuterostomes, animals whose mouth develops from a secondary opening in the embryo, such as the vertebrates.

Morphology. *Xenoturbella* is a ciliated worm up to 3 cm (1.2 in.) long with a mouth on the ventral side (the underside) opening into a gastral cavity (gut), but without an anus; simply described, it is no more than a hollow bag (**Fig. 1**). It lacks defined excretory structures, body cavities, and reproductive organs. It has a gravity-sensing organ called a statocyst and two sensory grooves (Fig. 1) with a dense underlying nerve plexus. *Xenoturbella* has neither a brain nor a condensed nerve cord; its nervous system simply consists of a nerve net under the epidermis. The gastral cavity is surrounded by a gastrodermis, and between the gastrodermis and the outer epidermis there are some muscular and soft cellular tissues. Externally, the worm is completely covered with cilia, which it uses to glide along the muddy sea floor. The reproduction and embryology of *Xenoturbella* are still a mystery.

Previous phylogenetic placement. *Xenoturbella* was discovered in 1915 by the Swedish biologist S. Bock, but it was not described until 30 years later by E. Westblad. Due to its simplicity of form, Westblad initially classified the worm as a primitive turbellarian, a class of ciliated nonparasitic flatworms, and named it accordingly (*Xenoturbella* = strange

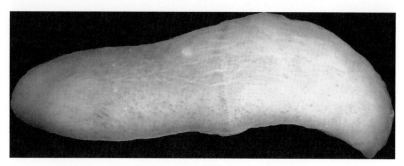

Fig. 1. Photograph of a live specimen of *Xenoturbella* (1–2 cm), showing the sensory circumferential groove around the middle of the animal.

flatworm). Later authors thought *Xenoturbella* to be the most primitive bilaterally symmetrical animal, placing it at the base of the Bilateria, a group which includes most of the animals except the cnidarians (sea anemones, jellyfish, and corals), which are radially symmetrical, and the poriferans (sponges), which lack body symmetry. *Xenoturbella* has also been thought to be related to the acoelomorph flatworms, due to their similar simple body plan and cilia. However, acoelomorph flatworms are now considered to be basal bilaterians. A final hypothesis based on *Xenoturbella* morphology has supported a relationship with the hemichordates (acorn worms) and echinoderms (starfishes and sea urchins), based on ultrastructural similarities of the epidermis, the presence of a nerve net, and a resemblance between the statocysts of *Xenoturbella* and those of a group of echinoderms, the sea cucumbers.

In 1997 a molecular study based on the analysis of ribosomal DNA and eggs found inside *Xenoturbella* claimed to have resolved its phylogenetic position. The authors of the study found DNA sequences belonging to bivalve mollusks of the genus *Nucula*, and concluded that *Xenoturbella* was a very unusual mollusk that had lost all specialized molluscan morphology. Eggs and embryos resembling those of mollusks were also found within the gastrodermis of *Xenoturbella*. These embryos were assumed to have developed from gametes found inside specimens of *Xenoturbella*. However, this conclusion seemed particularly surprising, as these embryos would have to lose all molluscan features, such as a brain, reproductive organs, and an anus, during their development into adult *Xenoturbella*. The fact that nuculid bivalves were abundant in the same habitat as *Xenoturbella* suggested, instead, that the mollusk eggs and embryos were ingested as food by *Xenoturbella*.

Current phylogenetic placement. Since DNA is the blueprint of all organisms, it can be used to study evolutionary relationships. All organisms share many genes, which are assumed to be homologous, that is, inherited from a common ancestor. Many of these genes are involved in basic cellular functions, such as protein synthesis and cell respiration. By comparing homologous genes from different organisms (using a mathematical formula to take into account the regular pattern of evolutionary change), it is possible to find out how closely related organisms are. To put it

simply, the more similar the DNA sequences of two animals, the more closely related they are.

Molecular analysis. A molecular analysis was recently carried out using new data from the ribosomal genes of *Xenoturbella*. The ribosomal genes code for ribosomal RNA (rRNA), which (along with ribosomal proteins) forms the structure of ribosomes, the protein-synthesizing organelles of living cells. These genes have frequently been used to reconstruct phylogenetic relationships, because they contain conserved and variable regions, which are useful in determining evolutionary relationships between species. Ribosomal gene sequences are known for thousands of animals, allowing the reconstruction of detailed phylogenetic trees for the whole animal kingdom. The mitochondrial cytochrome oxidase genes were also sequenced in this analysis. Cytochrome oxidase, an oxidizing enzyme important in cell respiration, is found in mitochondria, the power houses of cells. The 16,000 nucleotide base pairs of the mitochondrial genome mainly contain genes involved in protein synthesis and respiration.

The ribosomal and mitochondrial genes from *Xenoturbella* were amplified using the polymerase chain reaction (PCR), and two different types of sequences were consistently obtained from the reactions. One type was identical to that of bivalve mollusks of the genus *Nucula*, while the second type was an entirely different sequence. When the animal was dissected to minimize contamination from the gut contents, this new sequence was the one predominantly found in the PCR, indicating that it derives from *Xenoturbella* itself. Comparing this sequence to ribosomal DNA sequences from other animals using molecular phylogenetic tools revealed a relationship to the echinoderms and the hemichordates (**Fig. 2**).

To put this result in context, current molecular phylogenies support three main bilaterian divisions or superclades: the deuterostomes, the ecdysozoans, and the lophotrochozoans. The deuterostomes contain the chordates (the phylum to which vertebrates and humans belong), the hemichordates (acorn worms), and the echinoderms (starfishes). The ecdysozoans include animals with a molting cuticle (ecdysis), such as the arthropods, nematodes, priapulids, onychophorans, and tardigrades. The lophotrochozoans contain animals that all seem to have either a trochophore larva or a ciliated feeding structure called a lophophore, such as the annelids, the mollusks, sipunculans, bryozoans, entoprocts, and phoronids. The new position of *Xenoturbella* within the deuterostomes implies that the group containing *Xenoturbella* (the phylum Xenoturbellida) is now a fourth deuterostome phylum placed at the base of the echinoderms and hemichordates (collectively called the Ambulacraria) [Fig. 2]. This result also concurs with the morphological similarity observed between the epidermis, nerve net, and statocyst of *Xenoturbella* and certain ambulacrarians.

Mitochondrial gene order. The circular genome of mitochondria codes for 13 proteins, 2 rRNAs, and 22

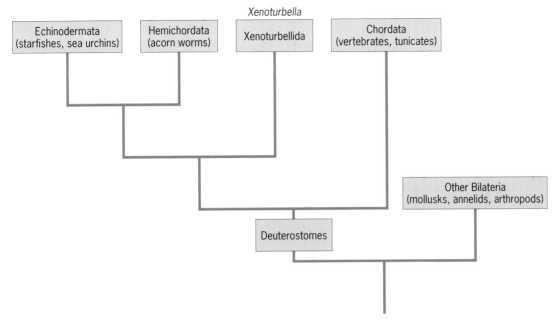

Fig. 2. Phylogenetic tree based on the 18s ribosomal DNA sequence, showing the position of *Xenoturbella* at the base of the hemichordates and echinoderms.

transfer RNAs (small RNA molecules responsible for binding amino acids and transferring them to the ribosomes during protein synthesis). The order and arrangement of these genes can be a powerful tool for elucidating phylogenetic relationships. The particular order of mitochondrial genes in *Xenoturbella* is seen in only certain deuterostomes (cephalochordates, vertebrates, and hemichordates) and has not been found in any other metazoan studied to date. This is further evidence that *Xenoturbella* is indeed a deuterostome. The fact that this gene arrangement is absent in echinoderms indicates that *Xenoturbella* did not evolve from within the echinoderms. Further comparisons of the genetic code in *Xenoturbella* also suggest that *Xenoturbella* cannot be a hemichordate either and is indeed a member of a separate phylum (Xenoturbellida).

Elucidating deuterostome evolution. It is surprising to find that such a morphologically simple animal belongs within the deuterostomes. The simple body plan of *Xenoturbella* might provide insights into what the earliest common ancestor of the deuterostomes looked like. However, ambulacrarians and chordates share anatomical features that are absent in *Xenoturbella* (such as body cavities, a complete gut with separate mouth and anus, a centralized nervous system, organized reproductive organs, and gill slits), implying that some aspects of *Xenoturbella* may be secondarily simplified. Due to the great diversity of body plans among the deuterostome phyla, defining their interrelationships has always been

problematic, and studies based on paleontological, anatomical, and embryological data have generated contrasting hypotheses. Further studies of *Xenoturbella*, particularly of its embryogenesis, might well be expected to reveal the presence of various deuterostome characteristics, such as radial cleavage of cells in the embryo and the possession of a ciliated larva.

Because of the pivotal phylogenetic position of *Xenoturbella* as the closest sister group to the hemichordates and echinoderms and a group which is evolutionarily close to the chordates, future studies of the genetics, morphology, and embryology of *Xenoturbella* have the potential to provide great insight into the evolution of the deuterostomes.

For background information *see* ANIMAL EVOLUTION; ANIMAL SYSTEMATICS; ECHINODERMATA; GENE AMPLIFICATION; HEMICHORDATA; PHYLOGENY in the McGraw-Hill Encyclopedia of Science & Technology.
Sarah J. Bourlat

Bibliography. S. J. Bourlat et al., *Xenoturbella* is a deuterostome that eats molluscs, *Nature*, 424:925–928, 2003; H. Gee, You aren't what you eat, *Nature*, 424:885–886, 2003; M. Noren and U. Jondelius, *Xenoturbella*'s molluscan relatives, *Nature*, 390:21–32, 1997; O. Israelsson, . . . and molluscan embryogenesis [*sic*], *Nature*, 390:32, 1997; C. Nielsen, *Animal Evolution: Interrelationships of the Living Phyla*, Oxford University Press, 2001; M. Noren and U. Jondelius, *Xenoturbella*'s molluscan relatives. . . , *Nature*, 390:21–32, 1997.

Contributors

Contributors

The affiliation of each Yearbook contributor is given, followed by the title of his or her article. An article title with the notation "coauthored" indicates that two or more authors jointly prepared an article or section.

A

Adams, Dr. William, W. III. *Professor, University of Colorado, Boulder, Department of Ecology and Evolutionary Biology.* PHOTOPROTECTION IN PLANTS.

Aizenberg, Dr. Joanna. *Bell Laboratories, Lucent Technologies.* NATURAL OPTICAL FIBERS—coauthored.

Albertsson, Prof. Ann-Christine. *Department of Fibre and Polymer Technology, Royal Institute of Technology (KTH), Stockholm, Sweden.* POLYMER RECYCLING AND DEGRADATION—coauthored.

Allard, Dr. Marc. *Department of Biological Science, George Washington University, Washington, DC.* SYSTEMATICS IN FORENSIC SCIENCE—coauthored.

Antón, Dr. Susan. *Department of Anthropology, New York University, New York.* HOMO ERECTUS.

B

Bahr, Nicholas. *Booz Allen Hamilton, McLean, Virgina.* TRANSPORTATION SYSTEM SECURITY—coauthored.

Baird, Prof. Henry S. *Professor of Computer Science and Engineering, Lehigh University, Bethlehem, Pennsylvania.* HUMAN/MACHINE DIFFERENTIATION—coauthored.

Barney Smith, Dr. Elisa H. *Electrical and Computer Engineering Department, Boise State University, Boise, Idaho.* DOCUMENT SCANNING.

Basso, Dr. Thomas. *National Renewable Energy Laboratory, Golden, Colorado.* DISTRIBUTED GENERATION (ELECTRIC POWER SYSTEMS).

Bebar, Mark R. *Professional Advisor, Computer Sciences Corporation/Advanced Marine Center, Washington, DC.* SMALL WARSHIP DESIGN.

Bercovici, Dr. David. *Yale University, New Haven, Connecticut.* MANTLE TRANSITION-ZONE WATER FILTER—coauthored.

Bock, Prof. Jane H. *University of Colorado, Boulder.* FORENSIC BOTANY—coauthored.

Boppart, Dr. Stephen A. *Beckman Institute, University of Illinois at Urbana-Champaign.* OPTICAL COHERENCE TOMOGRAPHY—coauthored.

Borrmann, Prof. Stephan. *Director, Department of Cloud Physics and Chemistry, Max Planck Institute for Chemistry, Mainz, Germany.* HIGH-ALTITUDE ATMOSPHERIC OBSERVATION.

Bourlat, Dr. Sarah J. *University Museum of Zoology, Cambridge, United Kingdom.* XENOTURBELLA (INVERTEBRATE SYSTEMATICS).

Branham, Dr. Marc. *Assistant Professor, Department of Entomology and Nematology, University of Florida, Gainesville.* FIREFLY COMMUNICATION.

Budowle, Bruce. *Laboratory Division, FBI Academy, Quantico, Virginia.* SYSTEMATICS IN FORENSIC SCIENCE—coauthored.

C

Carleton, Dr. Andrew M. *Pennsylvania State University, University Park.* SATELLITE CLIMATOLOGY.

Chang, Dr. Gary C. *Washington State University, Pullman.* BIOLOGICAL PEST CONTROL.

Chapman, Prof. Michael S. *School of Physics, Georgia Institute of Technology, Atlanta, Georgia.* NEUTRAL-ATOM STORAGE RING—coauthored.

Cheng, Dr. Quan Jason. *Assistant Professor of Chemistry, University of California, Riverside.* SURFACE PLASMON RESONANCE—coauthored.

Close, Prof. Frank. *Theoretical Physics, Oxford University, United Kingdom.* TETRAQUARKS.

Cohen, Dr. Zoë. *Department of Blood Transfusion Medicine Research, Saint Michael's Hospital, Toronto, Ontario, Canada.* AUTOIMMUNITY TO PLATELETS—coauthored.

Corda, Dr. Daniela. *Consorzio Mario Negri Sud, Italy.* PHOSPHOINOSITIDES—coauthored.

Creagh, Dr. Linda T. *Business Development Director, Spectra, Inc., Denton, Texas.* DISPLAY MANUFACTURING BY INKJET PRINTING.

D

Dale, Dr. Steinar J. *Center for Advanced Power Systems, Florida State University, Tallahassee.* SUPERCONDUCTING MOTORS—coauthored.

Daniel, Sheila Ellen. *Senior Environmental Consultant, AMEC Earth & Environmental, Ontario, Canada.* ENVIRONMENTAL MANAGEMENT (MINING).

De Matteis, Dr. Maria Antonietta. *Consorzio Mario Negri Sud, Italy.* PHOSPHOINOSITIDES—coauthored.

DeLaurier, Dr. James. *Institute for Aerospace Studies, University of Toronto, Ontario, Canada.* LOW-SPEED AIRCRAFT.

Desaire, Dr. Heather. *Department of Chemistry, University of Kansas, Lawrence.* MASS SPECTROMETRY (CARBOHYDRATE ANALYSIS).

Dörner, Prof. Reinhard. *Institut für Kernphysik, Johann Wolfgang Goethe-Universität, Frankfurt am Main, Germany.* MULTIPLE IONIZATION (STRONG FIELDS).

Dosoretz, Carlos G. *Israel Institute of Technology, Haifa, Israel.* LIGNIN-DEGRADING FUNGI—coauthored.

Dumas, Dr. Paul. *QED Technologies, Rochester, New York.* MAGNETORHEOLOGICAL FINISHING (OPTICS).

Duncan, Dr. Don. *Wikoff Color Corporation, Fort Mill, South Carolina.* RADIATION CURABLE INKS AND COATINGS.

Dyakonov, Dr. Vladimir. *Department of Physics, Energy and Semiconductor Research Laboratory, University of Oldenburg, Germany.* POLYMER SOLAR CELLS—coauthored.

E

Eisch, Dr. Amelia J. *Assistant Professor of Psychiatry, University of Texas Southwestern Medical Center at Dallas.* NEUROTROPHIC FACTORS.

Elko, Dr. Gary. *Summit, New Jersey.* SPHERICAL MICROPHONE ARRAYS—coauthored.

Elliott, Dr. David K. *Department of Geology, Northern Arizona University, Flagstaff.* ORIGIN OF VERTEBRATES.

Erickson, Dr. Gregory M. *Department of Biological Science, Florida State University, Tallahassee.* DINOSAUR GROWTH.

F

Faghri, Prof. Ardeshir. *Associate Chairman, Director, Delaware Center for Transportation, Newark, Delaware.* INTELLIGENT TRANSPORTATION SYSTEMS.

Finley, Dr. Russell L., Jr. *Center for Molecular Medicine and Genetics, Wayne State University School of Medicine, Detroit, Michigan.* PROTEIN NETWORKS—coauthored.

Fisher, Dr. Andrew T. *Professor, Earth Sciences Department, Institute for Geophysics and Planetary Physics, University of California, Santa Cruz.* RIDGE-FLANK HYDROTHERMAL CIRCULATION.

Foing, Dr. Bernard. *Chief Scientist, ESA Science Programme, Noordwijk, The Netherlands.* SMART 1.

Fox, Dr. Christopher J. *Graduate Program Director, James Madison University, Department of Computer Science, Harrisonburg, Virginia.* EXTREME PROGRAMMING AND AGILE METHODS.

Franzen, Iver C. *Naval Architect, Annapolis, Maryland.* MODERN TRADITIONAL SAILING SHIPS.

G

Gad-el-Hak, Dr. Mohamed. *Department of Mechanical Engineering, Virginia Commonwealth University, Richmond.* ADAPTIVE WINGS.

Gibb, Dr. Robbin. *Department of Psychology and Neuroscience, University of Lethbridge, Alberta, Canada.* EXPERIENCE AND THE DEVELOPING BRAIN—coauthored.

Ginley, Dr. David. *National Renewable Energy Laboratory, Golden, Colorado.* COMBINATORIAL MATERIALS SCIENCE—coauthored.

Godfrey, Dr. Laurie. *University of Massachusetts, Anthropology Department, Amherst.* MADAGASCAN PRIMATES.

Gomez, Dr. Alessandro. *Mason Laboratory, Yale University, New Haven, Connecticut.* MICROCOMBUSTION.

Gonzalez, Juan E. *Associate Professor of Molecular Genetics and Microbiology, University of Texas at Dallas, Richardson.* QUORUM SENSING IN BACTERIA.

Grandmason, Dr. Terra. *The Nature Conservancy, Seattle, Washington.* KILLER WHALES.

Griffith, Dr. Andrew J. *Division of Intramural Research, National Institutes of Health, Rockville, Maryland.* HEARING DISORDERS.

Gröning, Mikael. *Department of Fibre and Polymer Technology, Royal Institute of Technology (KTH), Stockholm, Sweden.* POLYMER RECYCLING AND DEGRADATION—coauthored.

H

Hadar, Dr. Yatzhak. *Faculty of Agriculture, Hebrew University of Jerusalem, Rehovot, Israel.* LIGNIN-DEGRADING FUNGI—coauthored.

Hadjichristidis, Prof. Nikos. *University of Athens, Department of Chemistry, Director of Industrial Laboratory and Head of Polymers Group, Athens, Greece.* MACROMOLECULAR ENGINEERING—coauthored.

Hall, Prof. Lawerence. *Department of Computer Science and Engineering, University of South Florida, Tampa.* COMPUTATIONAL INTELLIGENCE.

Han, Dr. Jung. *Yale University, New Haven, Connecticut.* WIDE-BAND-GAP III-NITRIDE SEMICONDUCTORS—coauthored.

Harley, Dr. John P. *Department of Biological Sciences, Eastern Kentucky University, Richmond.* BIOTERRORISM.

Harris, Scott. *Managing Partner, Harris Wiltshire and Grannis LLP, Washington, DC.* RADIO-FREQUENCY SPECTRUM MANAGEMENT—coauthored.

Hartman, Dr. Joe. *Electrical and Computer Engineering Department, Boise State University, Boise, Idaho.* ENVIRONMENTAL SENSORS.

Heer, Daniel. *Lucent-Bell Laboratories, Westford, Massachusetts.* VOICE OVER IP.

Hicks, Prof. Kenneth. *Department of Physics and Astronomy, Ohio University, Athens.* PENTAQUARKS.

Hoffman, Dr. Donald D. *Department of Cognitive Science, University of California, Irvine.* VISUAL ILLUSIONS AND PERCEPTION.

Holbrey, Dr. John D. *Department of Chemistry, University of Alabama, Tuscaloosa.* IONIC LIQUIDS—coauthored.

Hua, Dr. Hong. *Assistant Professor of Optical Sciences, University of Arizona, Tucson.* STEREOSCOPIC DISPLAYS.

Hughes, Prof. Terrence J. *Climate Change Institute, University of Maine, Bryand Global Sciences Center, Orono.* GLACIOLOGY.

I

Iatrou, Dr. Hermis. *Assistant Professor, University of Athens, Department of Chemistry, Athens, Greece.* MACROMOLECULAR ENGINEERING—coauthored.

J

Jeffrey, Dr. Jonathan. *International School of Amsterdam, The Netherlands.* EVOLUTIONARY DEVELOPMENTAL BIOLOGY (VERTEBRATE).

K

Kaduchak, Dr. Gregory. *Los Alamos National Laboratory, Electronic and Electrochemical Materials and Devices Group, Los Alamos, New Mexico.* ULTRASONIC AEROSOL CONCENTRATION AND POSITIONING—coauthored.

Karabinis, Dr. Peter. *Vice President and Chief Technical Officer, Mobile Satellite Ventures LP, Reston, Virginia.* MOBILE SATELLITE SERVICES.

Karato, Dr. Shun-ichiro. *Yale University, New Haven, Connecticut.* MANTLE TRANSITION-ZONE WATER FILTER—coauthored.

Kasunic, Dr. K. J. *Ball Aerospace and Technologies Corp., Boulder, Colorado.* PHOTONIC CRYSTAL DEVICES.

Keh, Dr. Charlene C. K. *Department of Chemistry, Tulane University, New Orleans, Louisiana.* ATOM ECONOMY—coauthored.

Ketcham, Dr. Richard A. *Research Scientist, University of Texas at Austin, Department of Geological Sciences.* CT SCANNING (VERTEBRATE PALEONTOLOGY).

Kleinpoppen, Prof. Hans. *Fritz-Haber-Institut der Max-Planck-Gesellschaft, Berlin, Germany.* TWO-PHOTON EMISSION.

Knapp, Dr. Gillian. *Department of Astrophysical Sciences, Princeton University, Princeton, New Jersey.* SLOAN DIGITAL SKY SURVEY.

Ko, Dr. Frank K. *Fibrous Materials Research Laboratory, Department of Materials Science and Engineering, Drexel University, Philadelphia, Pennsylvania.* ELECTROSPINNING.

Kobayashi, Dr. George S. *Professor Emeritus, Department of Internal Medicine, Washington University School of Medicine, St. Louis, Missouri.* CLINICAL YEAST IDENTIFICATION.

Kohl, Dr. John L. *Senior Astrophysicist, Harvard-Smithsonian Center for Astrophysics, Cambridge, Massachusetts.* SOLAR STORMS.

Kolb, Dr. Bryan. *Department of Psychology and Neuroscience, University of Lethbridge, Alberta, Canada.* EXPERIENCE AND THE DEVELOPING BRAIN—coauthored.

Krolick, Dr. Cyril F. *Syntek Technologies, Inc., Arlington, Virginia.* INTEGRATED ELECTRIC SHIP POWER SYSTEMS.

Kwiatkowski, Christopher. *Los Alamos National Laboratory, Los Alamos, New Mexico.* ULTRASONIC AEROSOL CONCENTRATION AND POSITIONING—coauthored.

L

Ladson, Damon. *Technology Policy Advisor, Harris, Wiltshire and Grannis LLP, Washington, DC.* RADIO-FREQUENCY SPECTRUM MANAGEMENT—coauthored.

Lee, Prof. Ka Yee Christina. *Associate Professor, Chemistry, University of Chicago, Illinois.* CELL MEMBRANE SEALING.

Leighton, Dr. Chris. *Assistant Professor, University of Minnesota, Department of Chemical Engineering and Materials Science, Minneapolis.* EXCHANGE BIAS—coauthored.

Li, Prof. Chao-Jun. *Department of Chemistry, McGill University, Montreal, Quebec, Canada.* ATOM ECONOMY—coauthored.

Lin, Prof. Yuh-Lang. *Department of Marine, Earth, and Atmospheric Sciences, North Carolina State University, Raleigh.* OROGRAPHIC PRECIPITATION (METEOROLOGY).

Liou, Prof. Kou-Nan. *Department of Atmospheric Sciences, University of California, Los Angeles.* CIRRUS CLOUDS AND CLIMATE.

Lohr, Dr. Raymond. *Transense Technologies plc, Upper Heyford, Bicester, Oxfordshire, United Kingdom.* WIRELESS SENSORS.

Lopresti, Dr. Daniel P. *Associate Professor of Computer Science and Engineering, Lehigh University, Bethlehem, Pennsylvania.* HUMAN/MACHINE DIFFERENTIATION—coauthored.

Lund, Michael S. *University of Minnesota, Department of Chemical Engineering and Materials Science, Minneapolis.* EXCHANGE BIAS—coauthored.

M

Ma, Dr. Minghong. *Assistant Professor, Department of Neuroscience, University of Pennsylvania, Philadelphia.* OLFACTORY SYSTEM CODING.

Marvier, Dr. Michelle. *Biology and Environmental Studies, Santa Clara University, Santa Clara, California.* PHARMACEUTICAL CROPS.

Maskell, Prof. Duncan I. *Centre for Veterinary Science, Department of Veterinary Medicine, University of Cambridge, United Kingdom.* COMPARATIVE BACTERIAL GENOME SEQUENCING.

Masson, Dr. Phillipe. *Center for Advanced Power Systems, Florida State University, Tallahassee.* SUPERCONDUCTING MOTORS—coauthored.

McGeer, Dr. Allison. *Director of Infection Control, Mount Sinai Hospital; Microbiologist and Infectious Disease Consultant, Toronto Medical Laboratories and Mount Sinai Hospital; Professor, Department of Laboratory Medicine and Pathobiology, University of Toronto, Ontario, Canada.* SEVERE ACUTE RESPIRATORY SYNDROME (SARS)—coauthored.

McGinnis, Dr. Daniel. *EAWAG, Limnological Research Center, Kastanienbaum, Switzerland.* LAKE HYDRODYNAMICS—coauthored.

Meadows, Kevin. *Drexel University, Philadelphia, Pennsylvania.* SMART CARD—coauthored.

Mehl, Dr. James B. *Orcas, Washington.* ACOUSTIC VISCOMETER.

Meyer, Dr. Jens. *New York, New York.* SPHERICAL MICROPHONE ARRAYS—coauthored.

Mitani, Dr. John C. *Department of Anthropology, University of Michigan, Ann Arbor.* CHIMPANZEE BEHAVIOR.

Mitchell, Dr. Jeffrey. *SPACEHAB, Inc., Webster, Texas.* SPACE SHUTTLE.

Molander, Dr. Gary. *Department of Chemistry, University of Pennsylvania, Philadelphia.* CATALYTIC HYDROAMINATION—coauthored.

Moncrief, Bill. *Hood-Patterson & Dewar, Inc. Decatur, Georgia.* HARMONICS IN ELECTRIC POWER SYSTEMS.

Müller, Prof. Dr. Dr. h. c. Ingo. *Technical University of Berlin, Germany.* SHAPE-MEMORY ALLOYS.

N

Neu, Dr. Wayne L. *Associate Professor and Assistant Department Head, Department of Aerospace and Ocean Engineering, Virginia Tech, Blacksburg.* HUMAN-POWERED SUBMARINES.

Norris, David O. *University of Colorado, Boulder.* FORENSIC BOTANY—coauthored.

Nurmikko, Arto V. *Professor of Engineering, Brown University, Providence, Rhode Island.* WIDE-BAND-GAP III-NITRIDE SEMICONDUCTORS—coauthored.

O

Oberg, Dr. James. *Soaring Hawk Productions, Dickinson, Texas.* CHINESE SPACE PROGRAM.

O'Hagan, Dr. Derek T. *Senior Director, Chiron Vaccines, Emeryville, California.* HIV VACCINES.

Oldenburg, Dr. Amy L. *Beckman Institute, University of Illinois at Urbana-Champaign.* OPTICAL COHERENCE TOMOGRAPHY—coauthored.

P

Parisi, Prof. Dr. Jürgen. *Department of Physics, Energy and Semiconductor Research Laboratory, University of Oldenburg, Germany.* POLYMER SOLAR CELLS—coauthored.

Parrish, Jodi R. *Center for Molecular Medicine and Genetics, Wayne State University School of Medicine, Detroit, Michigan.* PROTEIN NETWORKS—coauthored.

Pasachoff, Prof. Jay M. *Director, Hopkins Observatory, Williams College, Willamstown, Massachusetts.* TRANSITS OF VENUS.

Patáková, Dr. Petra. *Department of Fermentation Chemistry and Bioengineering, Institute of Chemical Technology, Prague, Czech Republic.* RED YEAST RICE.

Pearl, Dr. Thomas P. *Department of Physics, North Carolina State University, Raleigh.* ATOMIC-SCALE SURFACE IMAGING.

Peccei, Dr. Jocelyn Scott. *Los Angeles, California.* MENOPAUSE.

Peirce, Dr. J. Jeffrey. *Department of Civil and Environmental Engineering, Duke University, Durham, North Carolina.* HURRICANE-RELATED POLLUTION.

Perkins, Dr. John. *National Renewable Energy Laboratory, Golden, Colorado.* COMBINATORIAL MATERIALS SCIENCE—coauthored.

Pervaiz, Dr. Shazib. *Graduate School of Integrative Sciences and Engineering, National University Medical Institutes, National University of Singapore.* RESVERATROL.

Phillips, Dr. Colin. *Associate Professor, Department of Linguistics, University of Maryland, College Park.* LANGUAGE AND THE BRAIN—coauthored.

Phillips, K. Scott. *Department of Chemistry, University of California, Riverside.* SURFACE PLASMON RESONANCE—coauthored.

Piccirillo, Dr. Ciriaco A. *Professor of Immunology and Associate Member of the Center for the Study of Host Resistance, McGill University, Montreal, Quebec, Canada.* HELPER AND REGULATORY T CELLS.

Pierce, Dr. Marcia. *Eastern Kentucky University, Richmond.* WEST NILE FEVER (ENCEPHALITIS).

Pitsikalis, Dr. Marinos. *University of Athens, Greece.* MACROMOLECULAR ENGINEERING—coauthored.

Pourquié, Dr. Olivier. *Stowers Institute for Medical Research, Kansas City, Missouri.* SEGMENTATION AND SOMITOGENESIS IN VERTEBRATES.

Poutanen, Dr. Susan M. *Microbiologist and Infectious Disease Consultant, Toronto Medical Laboratories and Mount Sinai Hospital; Assistant Professor, Department of Laboratory Medicine and Pathobiology, University of Toronto, Ontario, Canada.* SEVERE ACUTE RESPIRATORY SYNDROME (SARS)—coauthored.

Prakash, Dr. C. S. *Tuskegee University, College of Agriculture, Tuskegee, Alabama.* GENETICALLY MODIFIED CROPS.

Pratt, Dr. Brian R. *Department of Geological Sciences, University of Saskatchewan, Saskatoon, Canada.* NEOPROTEROZIC PREDATOR-PREY DYNAMICS.

Prothero, Prof. Donald R. *Department of Geology, Occidental College, Los Angeles, California.* MASS EXTINCTIONS.

Q

Quéré, Dr. David. *Laboratoire de Physique de la Matière Condensée, Collège de France, Paris.* SELF-CLEANING SURFACES.

R

Rauchwerk, Dr. Michael. *Director, Bell Labs Advanced Technologies, Holmdel, New Jersey.* ADVANCED WIRELESS TECHNOLOGY.

Reid, Dr. Gregor. *Lawson Health Research Institute, Ontario, Canada.* PROBIOTICS.

Reike, Prof. Marcia. *University of Arizona, Department of Astronomy, Steward Observatory, Tucson, Arizona.* SPITZER SPACE TELESCOPE.

Rensink, Dr. Ronald A. *Department of Psychology, University of British Columbia, Canada.* CHANGE BLINDNESS (PSYCHOLOGY).

Riedel, Dr. Ingo. *Department of Physics, Energy and Semiconductor Research Laboratory, University of Oldenburg, Germany.* POLYMER SOLAR CELLS—coauthored.

Rogers, Prof. Robin D. *Department of Chemistry, University of Alabama, Tuscaloosa.* IONIC LIQUIDS—coauthored.

Romero, Jan Antoinette C. *Department of Chemistry, University of Pennsylvania, Philadelphia.* CATALYTIC HYDROAMINATION—coauthored.

Rozhdestvensky, Prof. Kirill. *FIMarEST Distinguished Scientist of the Russian Federation, Saint Petersburg State Marine Technical University, Saint Petersburg, Russia.* FLAPPING-WING PROPULSION.

Rubin, Prof. Charles M. *Department of Geological Sciences, Central Washington University, Ellensburg.* DENALI EARTHQUAKE.

Ruscic, Dr. Branko. *Chemistry Division, Argonne National Laboratory, Argonne, Illinois.* ACTIVE THERMOCHEMICAL TABLES.

Ryzhov, Prof. Vladimir. *Saint Petersburg State Marine Technical University, Saint Petersburg, Russia.* FLAPPING-WING PROPULSION.

S

Sage, Dr. Andrew P. *Founding Dean Emeritus and First American Bank Professor, University Professor, School of Information Technology and Engineering, George Mason University, Fairfax, Virginia.* SYSTEMS ARCHITECTING.

Sakai, Kuniyoshi L. *University of Tokyo.* LANGUAGE AND THE BRAIN—coauthored.

Sauer, Jacob A. *School of Physics, Georgia Institute of Technology, Atlanta, Georgia.* NEUTRAL-ATOM STORAGE RING—coauthored.

Schliwa, Prof. Dr. Manfred. *Adolf-Butenandt-Institut, Zellbiologie, Universität München, Germany.* MOLECULAR MOTORS.

Scott, Logan. *Breckenridge, Colorado.* LOCATION-BASED SECURITY.

Serhan, Dr. Charles N. *Professor, Brigham and Women's Hospital/Harvard Medical School, Boston, Massachusetts.* LIPIDOMICS.

Shelfer, Dr. Katherine M. *Drexel University, Philadelphia, Pennsylvania.* SMART CARD—coauthored.

Sheridan, Dr. Thomas B. *Professor Emeritus, Massachusetts Institute of Technology, Cambridge, Massachusetts.* HUMANS AND AUTOMATION.

Siegert, Dr. Martin J. *Bristol Glaciology Centre, School of Geographical Sciences, University of Bristol, United Kingdom.* RADIO-ECHO SOUNDING.

Sinha, Dipen. *Los Alamos National Laboratory, Los Alamos, New Mexico.* ULTRASONIC AEROSOL CONCENTRATION AND POSITIONING—coauthored.

Slansky, Dr. Jill E. *Assistant Professor, University of Colorado Health Sciences Center, Denver.* ADOPTIVE TUMOR IMMUNOTHERAPY.

Smith, Dr. John E. *Department of Bioscience, University of Strathclyde, United Kingdom.* MEDICINAL MUSHROOMS.

Soja, Dr. Constance M. *Professor of Geology, Colgate University, Hamilton, New York.* FOSSIL MICROBIAL REEFS.

Solie, Stacey. *Seattle, Washington.* MISSISSIPPI RIVER DEGRADATION.

Spear, Dr. Scott K. *Department of Chemistry, University of Alabama, Tuscaloosa.* IONIC LIQUIDS—coauthored.

Stansell, Tom. *Stansell Consulting, Rancho Palos Verdes, California.* GPS MODERNIZATION.

Starkey, Alison. *Canadian Blood Services, Toronto, Ontario, Canada.* AUTOIMMUNITY TO PLATELETS—coauthored.

Stoyer, Dr. Mark. *Lawrence Livermore National Laboratory, Livermore, California.* SUPERHEAVY ELEMENTS.

Straube, Dr. John F. *Department of Civil Engineering, University of Waterloo, Ontario, Canada.* MOISTURE-RESISTANT HOUSING.

Suckow, Edward. *Fairchild Semiconductor, South Portland, Maine.* SERIAL DATA TRANSMISSION.

Sweet, Dr. Kathleen M. *Associate Professor of Aviation Technology, Purdue University, West Lafayette, Indiana.* AVIATION SECURITY.

Swift, Dr. Gregory W. *Fellow, Los Alamos National Laboratory, Los Alamos, New Mexico.* THERMOACOUSTIC MIXTURE SEPARATION.

T

Taylor, Matthew. *National Renewable Energy Laboratory, Golden, Colorado.* COMBINATORIAL MATERIALS SCIENCE—coauthored.

Teplin, Dr. Charles. *National Renewable Energy Laboratory, Golden, Colorado.* COMBINATORIAL MATERIALS SCIENCE—coauthored.

Tsuboi, Dr. Seiji. *Institute for Frontier Research on Earth Evolution, Japan Marine Science and Technology Center, Yokohama, Japan.* EARTH SIMULATOR.

Tucker, Dr. Robert. *Assistant Department Head, Department of Computer Science, James Madison University, Harrisonburg, Virginia.* EXTENSIBLE MARKUP LANGUAGE (XML) DATABASES.

Turner, Megan B. *Department of Chemistry, University of Alabama, Tuscaloosa.* IONIC LIQUIDS—coauthored.

U

Udd, Dr. John E. *Principal Scientist, Mining, Natural Resources Canada.* ARCTIC MINING TECHNOLOGY.

V

Valentine, Dr. James W. *Department of Integrative Biology, University of California, Berkeley.* EVOLUTIONARY DEVELOPMENTAL BIOLOGY (INVERTEBRATE).

Vallero, Dr. Daniel. *Adjunct Associate Professor, Duke University, Department of Civil and Environmental Engineering, Durham, North Carolina.* PERSISTENT, BIOACCUMULATIVE, AND TOXIC POLLUTANTS.

van Hest, Dr. Maikel. *National Renewable Energy Laboratory, Golden, Colorado.* COMBINATORIAL MATERIALS SCIENCE—coauthored.

Vicent, Prof. José Luis. *Departamento de Física de Materiales, Facultad de Ciencias Físicas, Universidad Complutense de Madrid, Spain.* COLLECTIVE FLUX PINNING.

von Puttkamer, Dr. Jesco. *Office of Space Flight, NASA Headquarters, Washington, DC.* SPACE FLIGHT.

W

Wang, Prof. Pao K. *Department of Atmospheric and Oceanic Studies, University of Wisconsin-Madison.* ATMOSPHERIC MODELING, ISENTROPIC.

Watt, Dr. George D. *Defence R&D Canada—Atlantic, Dartmouth, Nova Scotia, Canada.* SUBMARINE HYDRODYNAMICS.

Weissman, Dr. Kira J. *Department of Biochemistry, University of Cambridge, United Kingdom.* DIRECTED EVOLUTION.

Wenger, Joyce. *Booz Allen Hamilton, McLean, Virginia.* TRANSPORTATION SYSTEM SECURITY—coauthored.

Whishaw, Dr. Ian. *Department of Psychology and Neuroscience, Canadian Centre for Behavioral Neuroscience, Lethbridge, Alberta, Canada.* EXPERIENCE AND THE DEVELOPING BRAIN—coauthored.

Wilkop, Dr. Thomas. *Department of Chemistry, University of California, Riverside.* SURFACE PLASMON RESONANCE—coauthored.

Williams, Dr. Arthur. *Faculty of Construction, Computing and Technology, Nottingham Trent University, Nottingham, United Kingdom.* MICRO HYDROPOWER.

Wilson, Dr. Mark R. *Laboratory Division, FBI, Washington, DC.* SYSTEMATICS IN FORENSIC SCIENCE—coauthored.

Wolf, Dr. Don P. *Senior Scientist, Division of Reproductive Sciences, Oregon National Primate Center, Beaverton, Oregon.* CLONING.

Wüest, Prof. Alfred. *EAWAG, Limnological Research Center, Kastanienbaum, Switzerland.* LAKE HYDRODYNAMICS—coauthored.

Y

Yablon, Dr. Andrew D. *OFS Laboratories, Murray Hill, New Jersey.* NATURAL OPTICAL FIBERS—coauthored.

Ye, Dr. Nong. *Director, Information and Systems Assurance Laboratory, Professor of Industrial Engineering, Affiliated Professor of Computer Science & Engineering, Arizona State University, Tempe.* NETWORK SECURITY AND QUALITY OF SERVICE.

Z

Zhang, Dr. Jianzhi. *Assistant Professor, Department of Ecology and Evolutionary Biology, University of Michigan, Ann Arbor.* BRAIN SIZE (GENETICS).

Zittermann, Dr. Armin. *Associate Professor, Department of Cardio-Thoracic Surgery, Heart Center NRW, Bad Oeynhausen, Germany.* VITAMIN D.

Index

Asterisks indicate page references to article titles.

G

503   Magistrates court,

    Law.

503   Magistrates court,

    Law.